Advanced Digital Design with the Verilog HDL

PRENTICE HALL XILINX DESIGN SERIES

Value is determined during simulation by the driver of the net; e.g., a primitive or a continuous assignment. (Example: **wire** Y = A + B.)

Registers: Store information and retain value until reassigned.

Value is determined by an assignment made by a procedural statement.

Value is retained until a new assignment is made; e.g., **reg**, **integer**, **real**, **realtime**, **time**.

Example:

```
always @ (posedge clock)
  if (reset) q_out <=0;
    else q_out <= data_in;
```

Procedural Statements

Describe logic abstractly; statements execute sequentially to assign value to variables.

```
if (expression_is_true) statement_1; else
statement_2;
case (case_expression)
case_item: statement;
…
default: statement;
endcase
for (conditions ) statement;
repeat constant_expression statement;
while (expression_is_true) statement;
forever statement;
fork statements join // execute in parallel
```

Assignments

Continuous: Continuously assigns the value of an expression to a net.

Procedural (Blocked): Uses the = operator; executes statements sequentially; a statement cannot execute until the preceding statement completes execution. Value is assigned immediately.

Procedural (Nonblocking): Uses the <= operator; executes statements concurrently, independent of the order in which they are listed. Values are assigned concurrently.

Procedural (Continuous):

assign … **deassign** overrides procedural assignments to a net.

force … **release** overrides all other assignments to a net or a register.

Operators

{ }, {{ }}	concatenation
+ – * /	arithmetic
%	modulus
> >= < <=	relational
!	logical negation
&&	logical and
||	logical or
==	logical equality
!=	logical inequality
===	case equality
!==	case inequality
~	bitwise negation
&	bitwise and
|	bitwise or
^	bitwise exclusive-or
^~ or ~^	bitwise equivalence
&	reduction and
~&	reduction nand
!	or
~|	reduction nor
^	reduction exclusive-or
~^ or ^~	reduction xnor
≪	left shift
≫	right shift
?:	conditional
or	Event or

Specify Block

Example: Module Path Delays

```
specify
// specparam declarations (min: typ: max)
specparam t_r = 3;4:5, t_f = 4:5:6;
(A, B) *> Y) = (t_r, t_f);      // full
(Bus_1 => Bus_1) = (t_r, t_f); // parallel
if (state == S0) (a, b *> y) = 2; // state dep
(posedge clk => (y -: d_in)) = (3. 4); // edge
endspecify
```

Example: Timing Checks

```
specify
specparam t_setup = 3:4:5, t_hold = 4:5:6;

$setup (data, posedge clock, t_setup);
$hold (posedge clock, data, t_hold);
endspecify
```

Memory

Declares an array of words.

Example: Memory declaration and readout

```
module memory_read_display();
  reg [31: 0] mem_array [1: 1024];
  integer k;
```

Advanced Digital Design with the Verilog HDL

Michael D. Ciletti

Department of Electrical and Computer Engineering
University of Colorado at Colorado Springs

Pearson Education, Inc.
Upper Saddle River, New Jersey 07458

Library of Congress Cataloging-in-Publication Data

Ciletti, Michael D.
   Advanced digital design with Verilog HDL / Michael Ciletti.-- 1st ed.
     p. cm. -- (Prentice Hall Xilinx design series)
   Includes bibliographical references and index.
   ISBN 0-13-089161-4
    1. Digital electronics. 2. Logic circuits--Computer-aided design. 3. Verilog (Computer hardware description language) I. Title. II. Series.

TK7868.D5 .C48 2002
621.39'5--dc21   2002074816

Vice President and Editorial Director, ECS: *Marcia J. Horton*
Publisher: *Tom Robbins*
Editorial Assistant: *Jody McDonnell*
Vice President and Director of Production and Manufacturing, ESM: *David W. Riccardi*
Executive Managing Editor: *Vince O'Brien*
Managing Editor: *David A. George*
Production Editor: *Kevin Bradley*
Director of Creative Services: *Paul Belfanti*
Creative Director: *Carole Anson*
Art Director: *Jayne Conte*
Cover Designer: *Bruce Kenselaar*
Art Editor: *Greg Dulles*
Manufacturing Manager: *Trudy Pisciotti*
Manufacturing Buyer: *Lynda Castillo*
Marketing Manager: *Holly Stark*

©2003 by Pearson Education, Inc.
Pearson Education, Inc.
Upper Saddle River, NJ 07458

The author and publisher of this book have used their best efforts in preparing this book. These efforts include the development, research, and testing of the theories and programs to determine their effectiveness. The author and publisher make no warranty of any kind, expressed or implied, with regard to these programs or the documentation contained in this book. The author and publisher shall not be liable in any event for incidental or consequential damages in connection with, or arising out of, the furnishing, performance, or use of these programs.

Silos and Simucad are registered trademarks of Simucad, Inc., 32970 Alvarado-Niles Road, Union City, CA 94587.
Verilog is a registered trademark of Cadence Design Systems, Inc., 2655 Seely Avenue, San Jose, CA 95134.

Printed in the United States of America
10 9 8 7 6 5 4 3 2 1

ISBN 0-13-089161-4

Pearson Education Ltd., *London*
Pearson Education Australia Pty. Ltd., *Sydney*
Pearson Education Singapore, Pte. Ltd.
Pearson Education North Asia Ltd., *Hong Kong*
Pearson Education Canada, Inc., *Toronto*
Pearson Educación de Mexico, S.A. de C.V.
Pearson Education—Japan, *Tokyo*
Pearson Education Malaysia, Pte. Ltd.
Pearson Education, Inc., *Upper Saddle River, New Jersey*

Contents

10 Architectures for Arithmetic Processors   651

Preface

Simplify, Clarify, and Verify

Behavioral modeling with a hardware description language (HDL) is the key to modern design of application-specific integrated circuits (ASICs). Today, most designers use an HDL-based design method to create a high-level, language-based, abstract description of a circuit, synthesize a hardware realization in a selected technology, and verify its functionality and timing.

Students preparing to contribute to a productive design team must know how to use an HDL at key stages of the design flow. Thus, there is a need for a course that goes beyond the basic principles and methods learned in a first course in digital design. This book is written for such a course.

Many books discussing HDLs are now available, but most are oriented toward robust explanations of language syntax, and are not well-suited for classroom use. Our focus is on design methodology enabled by an HDL.

Our goal in this book is to build on a student's background from a first course in logic design by (1) reviewing basic principles of combinational and sequential logic, (2) introducing the use of HDLs in design, (3) emphasizing descriptive styles that will allow the reader to quickly design working circuits suitable for ASICs and/or field-programmable gate array (FPGA) implementation, and (4) providing in-depth design examples using modern design tools. Readers will be encouraged to simplify, clarify, and verify their designs.

The widely used Verilog hardware description language (IEEE Standard 1364) serves as a common framework supporting the design activities treated in this book, **but our focus is on developing, verifying, and synthesizing designs of digital circuits, not on the Verilog language**. Most students taking a second course in digital design will be familiar with at least one programming language and will be able to draw on that background in reading this textbook. We cover only the core and most widely used features of Verilog. In order to emphasize *using* the language in a synthesis-oriented design environment, we have purposely placed many details, features, and explanations of syntax in the Appendices for reference on an "as-needed" basis.

Most entry-level courses in digital design introduce state machines, state-transition graphs, and algorithmic-state machine (ASM) charts. We make heavy use of ASM charts and demonstrate their utility in developing behavioral models of sequential machines. The important problem of designing a finite-state machine to control a complex datapath in a digital machine is treated in-depth with ASMD charts (i.e., ASM charts annotated to display the register operations of the controlled datapath). The design of a reduced intruction-set computer central processing unit (RISC CPU) and other important hardware units are given as examples. Our companion website includes the RISC machine's source code and an assembler that can be used to develop programs for applications. The machine also serves as a starting point for developing a more robust instruction set and architectural variants.

The Verilog language is introduced in an integrated, but selective manner, only as needed to support design examples. The text has a large set of examples illustrating how to address the key steps in a very large scale integrated (VLSI) circuit design methodology using the Verilog HDL. Examples are complete, and include source code that has been verified with the Silos-III simulator to be correct. Source code for all of the examples will be available (with important test suites) at our website.

The Intended Audience

This book is for students in an advanced course in digital design, and for professional engineers interested in learning Verilog by example, in the context of its use in the design flow of modern integrated circuits. The level of presentation is appropriate for seniors and first-year graduate students in electrical engineering, computer engineering, and computer science, as well as for professional engineers who have had an introductory course in logic design. The book presumes a basic background in Boolean algebra and its use in logic circuit design and a familiarity with finite-state machines. Building on this foundation, the book addresses the design of several important circuits used in computer systems, digital signal processing, image processing, data transfer across clock domains, built-in self-test (BIST), and other applications. The book covers the key design problems of modeling, architectural tradeoffs, functional verification, timing analysis, test generation, fault simulation, design for testability, logic synthesis, and postsynthesis verification.

Special Features of the Book

- Begins with a brief review of basic principles in combinational and sequential logic
- Focuses on modern digital design methodology
- Illustrates and promotes a synthesis-ready style of register transfer level (RTL) and algorithmic modeling with Verilog
- Demonstrates the utility of ASM charts for behavioral modeling
- In-depth treatment of algorithms and architectures for digital machines (e.g., an image processor, digital filters and circular buffers)
- In-depth treatment of synthesis for cell-based ASICs and FPGAs
- A practical treatment of timing analysis, fault simulation, testing, and design for testability, with examples
- Comprehensive treatment of behavioral modeling
- Comprehensive design examples, including a RISC machine and datapath controller
- Numerous graphical illustrations
- Provides several problems with a wide range of difficulty after each chapter
- Contains a worked example with JTAG and BIST for testing

- Contains over 250 fully verified examples
- An indexed list of all models developed in the examples
- A set of Xilinx FPGA-based laboratory-ready exercises linked to the book (e.g., arithmetic and logic unit [ALU], a programmable lock, a key pad scanner with a FIFO, a serial communications link with error correction, an SRAM controller, and first in, first out [FIFO] memory)
- Contains an up-to-date chapter on programmable logic device (PLDs) and FPGAs
- Contains a packaged CD-ROM with the popular Silos-III Verilog design environment and simulator and the Xilinx integrated synthesis environment (ISE) synthesis tool for FPGAs
- Contains an Appendix with full formal syntax of the Verilog HDL
- Covers major features of Verilog 2001, with examples
- Supported by an ongoing website containing:

1. Source files of models developed in the examples
2. Source files of testbenches for simulating examples
3. An Instructor's Classroom Kit containing transparency files for a course based on the subject matter
4. Solutions to selected problems
5. Jump-start tutorials helping students get immediate results with the Silos-III simulation environment, the Xilinx FPGA synthesis tool, the Synopsys synthesis tools, and the Synopsys Prime Time static timing analyzer
6. ASIC standard-cell library with synthesis and timing database
7. Answers to frequently asked questions (FAQs)
8. Clever examples submitted by readers
9. Revisions

Sequences for Course Presentation

The material in the text begins with a review of combinational and sequential logic design, but then progresses in the order dictated by the design flow for an ASIC or an FPGA. Chapters 1 to 6 treat design topics through synthesis, and should be covered in order, but Chapters 7 to 10 can be covered in any order. The homework exercises are challenging, and the laboratory-ready Xilinx-based exercises are suitable for a companion laboratory or for end-of-semester projects. Chapter 10 presents several architectures for arithmetic operations, affording a diversity of coverage. Chapter 11 treats postsynthesis design validation, timing analysis, fault simulation, and design for testability. The coverage of these topics can be omitted, depending on the level and focus of the course. Tools supporting Verilog 2001 are emerging, so an appendix discusses and illustrates the important new features of the language.

Chapter Descriptions

Chapter 1 briefly discusses the role of HDLs in design flows for cell-based ASICs and FPGAs. Chapters 2 and 3 review mainstream topics that would be covered in a first course in digital design, using classical methods (i.e. Karnaugh maps). This material will refresh the reader's background, and the examples will be used later to introduce HDL-based methods of design. Chapters 4 and 5 introduce modeling of combinational and sequential logic with the Verilog HDL, and place emphasis on coding styles that are used in behavioral modeling. Chapter 6 addresses cell-based synthesis of ASICs, and introduces synthesis of combinational and sequential logic. Here we pursue two main objectives: (1) present synthesis-friendly coding styles, and (2) form a foundation that will enable the reader to anticipate the results of synthesis, especially when synthesizing sequential machines. Many sequential machines are partitioned into a datapath and a controller. Chapter 7 covers examples that illustrate how to design a controller for a datapath. The designs of a simple RISC CPU and a UART[1] serve as platforms for the subject matter. Chapter 8 covers PLDs, complex PLDs (CPLDs), ROMs, and static random-access memories (SRAMs), then expands the synthesis target to include FPGAs. Verilog has been used extensively to design computers and signal processors. Chapter 9 treats the modeling and synthesis of computational units and algorithms found in computer architectures, digital filters, and other processors. Chapter 10 develops and refines algorithms and architectures for the arithmetic units of digital machines. In Chapter 11 we use the Verilog HDL in conjunction with fault simulators and timing analyzers to revisit a selection of previously designed machines and consider performance/timing issues and testability, to complete the treatment of design flow tasks that rely heavily on designer intervention. Chapter 11 models the test access port (TAP) controller defined by the IEEE 1149.1 standard (commonly known as the JTAG standard), and presents an example of its use. Another elaborate example covers built-in self test (BIST).

Acknowledgments

The author is grateful for the support of colleagues and students who expanded his vision of Verilog and contributed to this textbook. The reviewers of the original manuscript provided encouragement, critical judgment, and many helpful suggestions. Stu Sutherland helped the author gain a deeper appreciation for the issue of race conditions that can creep into the models of a digital system. These insights led to the disciplined style of adhering to nonblocking assignments for modeling edge-sensitive behavior and blocked assignments for modeling level-sensitive behavior. I owe a debt of gratitude to Dr. Jim Tracy and Dr. Rodger Ziemer, who supported my efforts to develop courses in VLSI circuit design; to Bill Fuchs, who introduced me to the Silos-III Verilog simulator from Simucad, Inc., and placed a user-friendly design environment in the hands of our students. Kirk Sprague and Scott Kukel were helpful in developing a Hamming encoder to work with the UART. Cris Hagan's thesis led to the models presented in Chapter 9 for decimators and other functional units found in digital signal processors. Rex Anderson proofread several chapters and scrubbed down my work. Terry Hansen and Lisa Horton provided the inspiration for the coffee vending machine example, and developed the assembler that supports the RISC CPU. Dr. Greg Sajdak developed material relating chip defects to test coverage and process yield. Dr. Bruce Harmon provided material for a FIR filter example. My editors, Tom Robbins and Eric Frank, have been a delight to work with. They supported the concept, encouraged my work and guided this book through the production process. My deep thanks to all of you.

---

[1]Universal asynchronous receiver and transmitter (UART), a circuit used in data transmission between systems.

Dedication

This book is dedicated to the memory of Sr. Laurencia Rihn, RSM, and Fr. Jerry Wilson, CSC. My life has been shaped by their faith, encouragement, and love. To my wife, Jerilynn, and our children, Monica, Lucy, Rebecca, Christine, and Michael and their spouses, Mike McCormick, David Steigerwald, Peter Van Dusen, and Michelle Puhr Ciletti, and our grandchildren, Michael, Katherine, Brigid, David, Jackson, Samantha, Peter, Anthony, and Matthew—thank you for the journey and the love we've shared.

CHAPTER 1   Introduction to Digital Design Methodology

Classical design methods relied on schematics and manual methods to design a circuit, but today computer-based languages are widely used to design circuits of enormous size and complexity. There are several reasons for this shift in practice. No team of engineers can correctly design and manage, by manual methods, the details of state-of-the-art integrated circuits (ICs) containing several million gates, but using hardware description languages (HDLs) designers easily manage the complexity of large designs. Even small designs rely on language-based descriptions, because designers have to quickly produce correct designs targeted for an ever-shrinking window of opportunity in the marketplace.

Language-based designs are portable and independent of technology, allowing design teams to modify and re-use designs to keep pace with improvements in technology. As physical dimensions of devices shrink, denser circuits with better performance can be synthesized from an original HDL-based model.

HDLs are a convenient medium for integrating intellectual property (IP) from a variety of sources with a proprietary design. By relying on a common design language, models can be integrated for testing and synthesized separately or together, with a net reduction in time for the design cycle. Some simulators also support mixed descriptions based on multiple languages.

The most significant gain that results from the use of an HDL is that a working circuit can be synthesized automatically from a language-based description, bypassing the laborious steps that characterize manual design methods (e.g., logic minimization with Karnaugh maps).

HDL-based synthesis is now the dominant design paradigm used by industry. Today, designers build a software prototype/model of the design, verify its functionality, and then use a synthesis tool to automatically optimize the circuit and create a netlist in a physical technology.

HDLs and synthesis tools focus an engineer's attention on functionality rather than on individual transistors or gates; they synthesize a circuit that will realize the desired functionality, and satisfy area and/or performance constraints. Moreover, alternative architectures can be generated from a single HDL model and evaluated quickly to perform design tradeoffs. Functional models are also referred to as behavioral models.

HDLs serve as a platform for several tools: design entry, design verification, test generation, fault analysis and simulation, timing analysis and/or verification, synthesis, and automatic generation of schematics. This breadth of use improves the efficiency of the design flow by eliminating translations of design descriptions as the design moves through the tool chain.

Two languages enjoy widespread industry support: Verilog™ [1] and VHDL [2]. Both languages are IEEE (Institute of Electrical and Electronics Engineers) standards; both are supported by synthesis tools for ASICs (application-specific integrated circuits) and FPGAs (field-programmable gate arrays). Languages for analog circuit design, such as Spice [3], play an important role in verifying critical timing paths of a circuit, but these languages impose a prohibitive computational burden on large designs, cannot support abstract styles of design, and become impractical when used on a large scale. Hybrid languages (e.g., Verilog-A) [4] are used in designing mixed-signal circuits, which have both digital and analog circuitry. System-level design languages, such as SystemC [5] and Superlog™ [6], are now emerging to support a higher level of design abstraction than can be supported by Verilog or VHDL.

1.1   Design Methodology—An Introduction

ASICs and FPGAs are designed systematically to maximize the likelihood that a design will be correct and will be fabricated without fatal flaws. Designers follow a "design flow" like that shown in Figure 1-1, which specifies a sequence of major steps that will be taken to design, verify, synthesize, and test a digital circuit. ASIC design flows involve several activities, from specification and design entry, to place-and-route and timing closure of the circuit in silicon. Timing closure is attained when all of the signal paths in the design satisfy the timing constraints imposed by the interface circuitry, the circuit's sequential elements, and the system clock. Although the design flow appears to be linear, in practice it is not. Various steps might be revisited as design errors are discovered, requirements change, or performance and design constraints are violated. For example, if a circuit fails to meet timing constraints, a new placement and routing step will have to be taken, perhaps including redesign of critical paths.

Design flows for standard-cell-based ASICs are more complex than those for FPGAs because the architecture of an ASIC is not fixed. Consequently, the performance that can be realized from a design depends on the physical placement and routing of the cells on the die, as well as the underlying device properties. Interconnect delays play a significant role in determining performance in submicron designs below 0.18 μm, in which prelayout estimates of path delays do not guarantee timing closure of the routed design.

The following sections will clarify the design flow described in Figure 1-1.

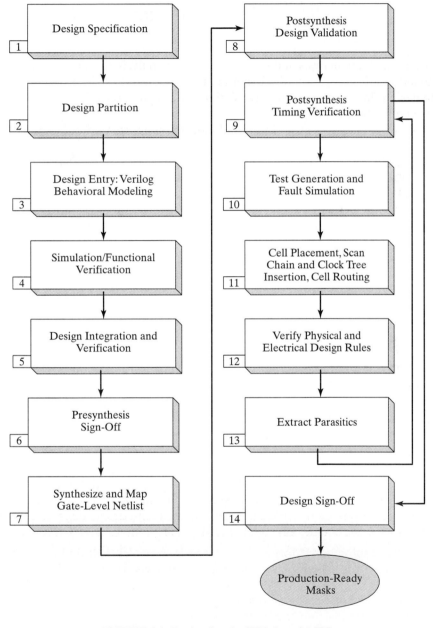

FIGURE 1-1 Design flow for HDL-based ASICs.

1.1.1 Design Specification

The design flow begins with a written specification for the design. The specification document can be a very elaborate statement of functionality, timing, silicon area, power consumption, testability, fault coverage, and other criteria that govern the design. At a minimum, the specification describes the functional characteristics that are to be implemented in a design. Typically, state transition graphs, timing charts, and algorithmic-state machine (ASM) charts are used to describe sequential machines, but interpretation of the specification can be problematic, because the HDL-based model might actually implement an unintended interpretation of the specification. The emerging high-level languages, like SystemC [5], and Superlog [6] hold the promise that the language itself provides an executable specification of the design, which can then be translated and synthesized into a circuit.

1.1.2 Design Partition

In today's methodologies for designing ASICs and FPGAs, large circuits are partitioned to form an *architecture*—a configuration of interacting functional units, such that each is described by a behavioral model of its functionality. The process by which a complex design is progressively partitioned into smaller and simpler functional units is called *top-down design* or *hierarchical design*. HDLs support top-down design with mixed levels of abstraction by providing a common framework for partitioning, synthesizing, and verifying large, complex systems. Parts of large designs can be linked together for verification of overall functionality and performance. The partitioned architecture consists of functional units that are simpler than the whole, and each can be described by an HDL-based model. The aggregate description is often too large to synthesize directly, but each functional unit of the partition can be synthesized in a reasonable amount of time.

1.1.3 Design Entry

Design entry means composing a language-based description of the design and storing it in an electronic format in a computer. Modern designs are described by hardware description languages, like Verilog, because it takes significantly less time to write a Verilog behavioral description and synthesize a gate-level realization of a large circuit than it does to develop the gate-level realization by other means, such as bottom-up manual entry. This saves time that can be put to better use in other parts of the design cycle. The ease of writing, changing, or substituting Verilog descriptions encourages architectural exploration. Moreover, a synthesis tool itself will find alternative realizations of the same functionality and generate reports describing the attributes of the design.

Synthesis tools create an optimal internal representation of a circuit before mapping the description into the target technology. The internal database at this stage is generic, which allows it to be mapped into a variety of technologies. For example, the technology mapping engine of a synthesis tool will use the internal format to migrate a design from an FPGA technology to an ASIC standard cell library, without having to reoptimize the generic description.

HDL-based designs are easier to debug than schematics. A behavioral description encapsulating complex functionality hides underlying gate-level detail, so there is less information to cope with in trying to isolate problems in the functionality of the design. Furthermore, if the behavioral description is functionally correct, it is a gold standard for subsequent gate-level realizations.

HDL-based designs incorporate documentation within the design by using descriptive names, by including comments to clarify intent, and by explicitly specifying architectural relationships, thereby reducing the volume of documentation that must be kept in other archives. Simulation of a language-based model explicitly specifies the functionality of the design. Since the language is a standard, documentation of a design can be decoupled from a particular vendor's tools.

Behavioral modeling is the predominant descriptive style used by industry, enabling the design of massive chips. *Behavioral modeling describes the functionality of a design* by specifying what the designed circuit will do, not how to build it in hardware. It specifies the input–output model of a logic circuit and suppresses details about physical, gate-level implementation.

Behavioral modeling encourages designers to (1) rapidly create a behavioral prototype of a design (without binding it to hardware details), (2) verify its functionality, and then (3) use a synthesis tool to optimize and map the design into a selected physical technology. If the model has been written in a synthesis-ready style, the synthesis tool will remove redundant logic, perform tradeoffs between alternative architectures and/or multilevel equivalent circuits, and ultimately achieve a design that is compatible with area or timing constraints. By focusing the designer's attention on the functionality that is to be implemented rather than on individual logic gates and their interconnections, behavioral modeling provides the freedom to explore alternatives to a design before committing it to production.

Aside from its importance in synthesis, behavioral modeling provides flexibility to a design project by allowing parts of the design to be modeled at different levels of abstraction. The Verilog language accommodates mixed levels of abstraction so that portions of the design that are implemented at the gate level (i.e., structurally) can be integrated and simulated concurrently with other parts of the design that are represented by behavioral descriptions.

1.1.4   Simulation and Functional Verification

The functionality of a design is verified (Step 4 in Figure 1-1) either by simulation or by formal methods [7]. Our discussion will focus on simulation that is reasonable for the size of circuits we can present here. The design flow iterates back to Step 3 until the functionality of the design has been verified. The verification process is threefold; it includes (1) development of a test plan, (2) development of a testbench, and (3) execution of the test.

1.1.4.1   Test Plan Development   A carefully documented *test plan* is developed to specify what functional features are to be tested and how they are to be tested. For example, the test plan might specify that the instruction set of an arithmetic and logic unit (ALU) will be verified by an exhaustive simulation of its behavior, for a specific set of input data. Test plans for sequential machines must be more elaborate to ensure a high level of confidence in the design, because they may have a large number of states. A test plan identifies the stimulus generators, response monitors, and the gold standard response against which the model will be tested.

1.1.4.2   Testbench Development   The *testbench* is a Verilog module in which the unit under test (UUT) has been instantiated, together with pattern generators that are to be applied to the inputs of the model during simulation. Graphical displays and/or response monitors are part of the testbench. The testbench is documented to identify the goals and sequential activity that will be observed during simulation (e.g., "Testing the opcodes"). If a design is formed as an architecture of multiple modules, each must be verified separately, beginning with the lowest level of the design hierarchy, then the integrated design must be tested to verify that the modules interact correctly. In this case, the test plan must describe the functional features of each module and the process by which they will be tested, but the plan must also specify how the aggregate is to be tested.

1.1.4.3   Test Execution and Model Verification   The testbench is exercised according to the test plan and the response is verified against the original specification for the design, e.g. does the response match that of the prescribed ALU? This step is intended to reveal errors in the design, confirm the syntax of the description, verify style conventions, and eliminate barriers to synthesis. Verification of a model requires a systematic, thorough demonstration of its behavior. *There is no point in proceeding further into the design flow until the model has been verified.*

1.1.5   Design Integration and Verification

After each of the functional subunits of a partitioned design have been verified to have correct functionality, the architecture must be integrated and verified to have the correct functionality. This requires development of a separate testbench whose stimulus generators exercise the input–output functionality of the top-level module, monitor port and bus activity across module boundaries, and observe state activity in any embedded state machines. *This step in the design flow is crucial* and must be executed thoroughly to ensure that the design that is being signed off for synthesis is correct.

1.1.6   Presynthesis Sign-Off

A demonstration of full functionality is to be provided by the testbench, and any discrepancies between the functionality of the Verilog behavioral model and the design specification must be resolved. *Sign-off* occurs after all known functional errors have been eliminated.

1.1.7   Gate-Level Synthesis and Technology Mapping

After all syntax and functional errors have been eliminated from the design and sign-off has occurred, a synthesis tool is used to create an optimal Boolean description and compose it in an available technology. In general, a synthesis tool removes redundant logic and seeks to reduce the area of the logic needed to implement the functionality and satisfy performance (speed) specifications. This step produces a netlist of standard cells or a database that will configure a target FPGA.

1.1.8 Postsynthesis Design Validation

Design validation compares the response of the synthesized gate-level description to the response of the behavioral model. This can be done by a testbench that instantiates both models, and drives them with a common stimulus, as shown in Figure 1-2. The responses can be monitored by software and/or by visual/graphical means to see whether they have identical functionality. For synchronous designs, the match must hold at the boundaries of the machine's cycle—intermediate activity is of no consequence. If the functionality of the behavioral description and the synthesized realization do not match, painstaking work must be done to understand and resolve the discrepancy. Postsynthesis design validation can reveal software race conditions in the behavioral model that cause events to occur in a different clock cycle than expected.[1] We will discuss how good modeling techniques can prevent this outcome.

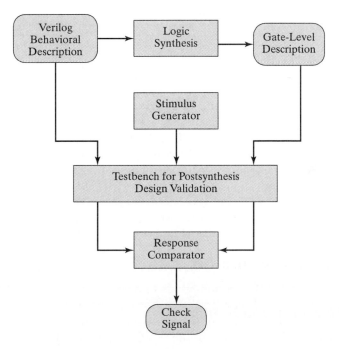

FIGURE 1-2 Postsynthesis design validation.

---

[1]Postsynthesis validation in an ASIC design flow is followed by a step for postlayout timing verification.

1.1.9   Postsynthesis Timing Verification

Although the synthesis process is intended to produce a circuit that meets timing specifications, the circuit's timing margins must be checked to verify that speeds are adequate on critical paths (Step 9). This step is repeated after Step 13, because synthesis tools do not accurately anticipate the effect of the capacitive delays induced by interconnect metalization in the layout. Ultimately, these delays must be extracted from the properties of the materials and the geometric details of the fabrication masks. The extracted delays are used by a static timing analyzer to verify that the longest paths do not violate timing constraints. The circuit might have to be resynthesized or re-placed and rerouted to meet specifications. Resynthesis might require (1) transistor resizing, (2) architectural modifications/substitutions, and (3) device substitution (more speed at the cost of more area).

1.1.10   Test Generation and Fault Simulation

After fabrication, integrated circuits must be tested to verify that they are free of defects and operate correctly. Contaminants in the clean-room environment can cause defects in the circuit and render it useless. In this step of the design flow a set of test vectors is applied to the circuit and the response of the circuit is measured. Testing considers process-induced faults, not design errors. Design errors should be detected before presynthesis sign-off. Testing is daunting, for an ASIC chip might have millions of transistors, but only a few hundred package pins that can be used to probe the internal circuits. The designer might have to embed additional, special circuits that will enable a tester to use only a few external pins to test the entire internal circuitry of the ASIC, either alone or on a printed circuit board.

   The patterns that are used to verify a behavioral model can be used to test the fabricated part that results from synthesis, but they might not be robust enough to detect a sufficiently high level of manufacturing defects. Combinational logic can be tested for faults exhaustively, but sequential machines present special challenges, as we will see in Chapter 11. Fault simulation questions whether the chips that come off the fabrication line can, in fact, be tested to verify that they operate correctly. Fault simulation is conducted to determine whether a set of test vectors will detect a set of faults. The results of fault simulation guide the use of software tools for generating additional test patterns. To eliminate the possibility that a part could be produced but not tested, test patterns are generated before the device is fabricated, to allow for possible changes in the design, such as a scan path.[2]

1.1.11   Placement and Routing

The placement and routing step of the ASIC design flow arranges the cells on the die and connects their signal paths. In cell-based technology the individual cells are integrated to form a global mask that will be used to pattern the silicon wafer with gates.

---

[2] Scan paths are formed by replacing ordinary flip-flops with specially designed flip-flops that can be connected together in test mode to form a shift register. Test patterns can be scanned into the design, and applied to the internal circuitry. The response of the circuit can be captured in the scan chain and shifted out for analysis.

This step also might involve inserting a clock tree into the layout, to provide a skew-free distribution of the clock signal to the sequential elements of the design. If a scan path is to be used, it will be inserted in this step too.

1.1.12 Physical and Electrical Design Rule Checks

The physical layout of a design must be checked to verify that constraints on material widths, overlaps, and separations are satisfied. Electrical rules are checked to verify that fanout constraints are met and that signal integrity is not compromised by electrical crosstalk and power-grid drop. Noise levels are also checked to determine whether electrical transients are problematic. Power dissipation is modeled and analyzed in this step to verify that the heat generated by the chip will not damage the circuitry.

1.1.13 Parasitic Extraction

Parasitic capacitance induced by the layout is extracted by a software tool and then used to produce a more accurate verification of the electrical characteristics and timing performance of the design (Step 13). The results of the extraction step are used to update the loading models that are used in timing calculations. Then the timing constraints are checked again to confirm that the design, as laid out, will function at the specified clock speed.

1.1.14 Design Sign-Off

Final sign-off occurs after all of the design constraints have been satisfied and timing closure has been achieved. The mask set is ready for fabrication. The description consists of the geometric data (usually in GDS-II format) that will determine the photomasking steps of the fabrication process. At this point significant resources have been expended to ensure that the fabricated chip will meet the specifications for its functionality and performance.

1.2 IC Technology Options

Figure 1-3 shows various options for creating the physical realization of a digital circuit in silicon, ranging from programmable logic devices (PLDs) to full-custom ICs. Fixed-architecture programmable logic devices serve the low end of the market (i.e., low volume and low performance requirements). They are relatively cheap commodity parts, targeted for low-volume designs.

The physical database of a design might be implemented as (1) a full-custom layout of high-performance circuitry, (2) a configuration of standard cells, or (3) gate arrays (field- or mask-programmable), depending on whether the anticipated market for the ASIC offsets the cost of designing it, and the required profit. Full-custom ICs occupy the high end of the cost–performance domain, where sufficient volume or a customer with corporate objectives and sufficient resources warrant the development time and investment required to produce fully custom designs having minimal area and maximum speed. FPGAs have a fixed, but electrically programmable architecture

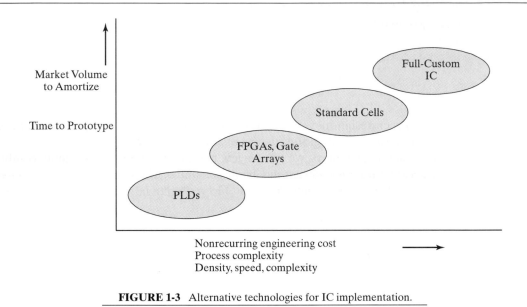

FIGURE 1-3 Alternative technologies for IC implementation.

for implementing modest-sized designs. The tools supporting this technology allow a designer to write and synthesize a Verilog description into a working physical part on a prototype board in a matter of minutes. Consequently, design revisions can be made at very low cost. Board layout can proceed concurrently with the development of the part because the footprint and pin configuration of an FPGA are known. Low-volume proto-typing sets the stage for the migration of a design to mask-programmable and stan-dard-cell-based parts.

In mask-programmable gate-array technology a wafer is populated with an array of individual transistors that can be interconnected to create logic gates that imple-ment a desired functionality. The wafers are prefabricated and later personalized with metal interconnect for a customer. All but the metalization masks are common to all wafers, so the time and cost required to complete masks is greatly reduced, and the other nonrecurring engineering (NRE) costs are amortized over the entire customer base of a silicon foundry.

Standard cell technology predesigns and characterizes individual logic gates to the mask level and assembles them in a shared library. A place-and-route tool places the cells in channels on the wafer, interconnects them, and integrates their masks to create the functionality for a specific application. The mask set for a customer is specific to the logic being implemented and can cost over $500K for large circuits, but the NRE costs associated with designing and characterizing the cell library are amortized over the entire customer base. In high-volume applications, the unit cost of the parts can be relatively cheap compared to the unit cost of PLDs and FPGAs.

1.3   Overview

The following chapters will cover most of the steps in the design flow presented in Figure 1-1, but not cell placement and routing, design-rule checking, or parasitic extraction. These steps are conducted by separate tools, which operate on the physical mask database rather than on an HDL model of the design, and they presume that a functionally correct design has been synthesized successfully. The steps we cover are the mainstream designer-driven steps in the overall ASIC flow.

In the remaining chapters, we will review manual methods for designing combinational and sequential logic design in Chapters 2 and 3. Then we will treat combinational logic design (Chapter 4) and sequential logic design (Chapter 5) using Verilog, and by example, contrast manual and HDL-based methods. This chapter also introduces the use of ASM charts and algorithmic state machine and datapath (ASMD) charts, which prove to be very useful in writing behavioral models of sequential machines. Chapter 6 covers synthesis of combinational and sequential logic with Verilog models. This chapter equips the designer with the background to compose synthesis-friendly designs and to avoid common pitfalls that can thwart a design. Chapter 7 continues with a treatment of datapath controllers, including a RISC CPU and a UART. Chapter 8 introduces PLDs, CPLDs, RAMS and ROMS, and FPGAs. The problems at the end of this chapter specify designs that can be implemented on a widely available prototyping board. Chapter 9 covers algorithms and architectures for digital processors, and Chapter 10 treats architectures for arithmetic operations. Chapter 11 treats the postsynthesis issues of timing verification, test generation, and fault simulation, including JTAG and BIST.

Three things matter in learning design with an HDL: examples, examples, and examples. We present several examples, with increasing difficulty, and make available their Verilog descriptions. Several challenging problems are included at the end of each chapter that require design with Verilog. We urge the reader to embrace the mantra: simplify, clarify, verify.

REFERENCES

1. *IEEE Standard Hardware Description Language Based on the Verilog Hardware Description Language*, Language Reference Manual (LRM), IEEE Std.1364–1995. Piscataway, NJ: Institute of Electrical and Electronic Engineers, 1996.
2. *IEEE Standard VHDL Language Reference Manual* (LRM), IEEE Std, 1076–1987. Piscataway, NJ: Institute of Electrical and Electronic Engineers, 1988.
3. Negel LW. *SPICE2: A Computer Program to Simulate Semiconductor Circuits*, Memo ERL-M520, Department of Electrical Engineering and Computer Science, University of California at Berkeley, May 9, 1975.
4. Fitzpatrick D, Miller I. *Analog Behavioral Modeling with the Verilog-A Language*, Boston: Kluwer, 1998.
5. SystemC Draft Specification, Mountain View, CA: Synopsys, 1999.

6. Rich, D., Fitzpatrick, T., "Advanced Verification Using the Superlog Language," Proc. Int. HDL Conference, San Jose, March 2002.
7. Chang H, et al. *Surviving the SOC Revolution*, Boston: Kluwer, 1999.

CHAPTER 2

Review of Combinational Logic Design

This chapter will review manual methods for designing combinational logic. In Chapter 6 we will see how these steps can be automated with modern design tools.

2.1 Combinational Logic and Boolean Algebra

Combinational logic forms its outputs as Boolean functions of its input variables on an instantaneous basis. That is, at any time t the outputs y_1, y_2, and y_3 in Figure 2-1 depend on only the values of a, b, c, and d at time t. The outputs of combinational logic at any time t are a function of only the inputs at time t. The outputs of other circuits may depend on the history of the inputs up to time t, and they are called sequential circuits. Sequential circuits require memory elements in hardware.

The variables in a logic circuit are binary—they may have a value of 0 or 1. Hardware implementations of logic circuits use either positive logic, in which a high voltage level, say 5 volts, corresponds to a logical value of 1, and a low voltage, say 0, corresponds to a logical 0. In negative logic, a low electrical level corresponds to a logical 1, and a high electrical level corresponds to a 0.

Some common logic gates are shown in Figure 2-2, together with the Boolean equation that determines the value of the output of the gate as a function of its inputs, and Table 2-1 lists common symbols for hardware-based Boolean logic operations.[1]

2.1.1 ASIC Library Cells

Logic gates are implemented physically by a transistor-level circuit. For example, in CMOS (complementary metal-oxide semiconductor) technology, a logic inverter consists

---

[1]*Note*: The schematic symbol for the three-state buffer uses the symbol z to indicate the high impedance condition of the device.

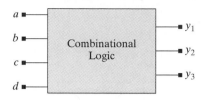

FIGURE 2-1 Block diagram symbol for combinational logic having four inputs and three outputs.

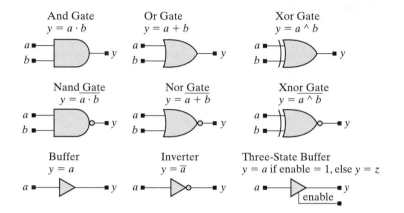

FIGURE 2-2 Schematic symbols and Boolean relationships for some common logic gates.

TABLE 2-1 Common Boolean logic symbols and operations.

Symbol	Logic Operation
+	Logic "or"
·	Logical "and"
⊕	Exclusive "or"
∧	Exclusive "or"
'	Logical negation
−	Logical negation (overbar)

of a series connection of *p*-channel and *n*-channel MOS transistors having a common drain that serves as the output, and a common gate that serves as the input. When the input is low, the *p*-channel device conducts and the *n*-channel device is an open circuit. In this mode the output capacitor charges to V_{dd}. When the input is high, the *n*-channel device conducts, and the *p*-channel device is an open circuit. This discharges the output node capacitor to ground. Figure 2-3 shows (b) the pull-up and (c) pull-down paths for current in the inverter in (a).

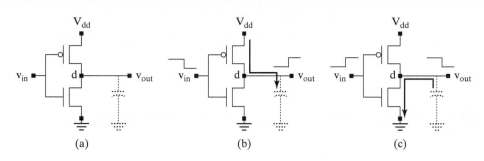

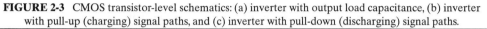

FIGURE 2-3 CMOS transistor-level schematics: (a) inverter with output load capacitance, (b) inverter with pull-up (charging) signal paths, and (c) inverter with pull-down (discharging) signal paths.

Other logic gates can be implemented using the same basic principles of pull-up and pull-down logic. Figure 2-4 shows the transistor-level schematic for a three-input NAND gate. If one or more of the inputs is low, the output node Y is pulled up to V_{dd}; all inputs must be high to pull the output to ground.

Very-large-scale integrated (VLSI) circuits that implement logic gates are fabricated by a series of processing steps in which photomasks are used to selectively dope a silicon wafer to form and connect transistors. Figure 2-5 shows a composite view of the basic masks used in an elementary process to fabricate a CMOS inverter by implanting semiconductor dopants and depositing metal and polycrystalline silicon. The masking steps are performed in a well-defined sequence, beginning with the implantation of a dopant to form the n-well, a region that is heavily doped with an n-type material (e.g., arsenic). A p-channel transistor is formed in the n-well by implanting a p-type (e.g., boron) source and drain regions, and an n-channel transistor is formed in the host silicon substrate. Polycrystalline silicon is deposited to form the gates of the transistors, and metal is deposited to form interconnections in and between devices. The actual processes involve many more steps and can have several more layers of metal than these simple structures. Figure 2-6 shows a cross section of a simple ASIC cell for an inverter, revealing the doped regions.

Circuits that implement basic and moderately complex Boolean functions are characterized for their functional, electrical, and timing properties, and packaged in

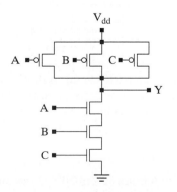

FIGURE 2-4 Transistor-level schematic for three-input CMOS Nand gate.

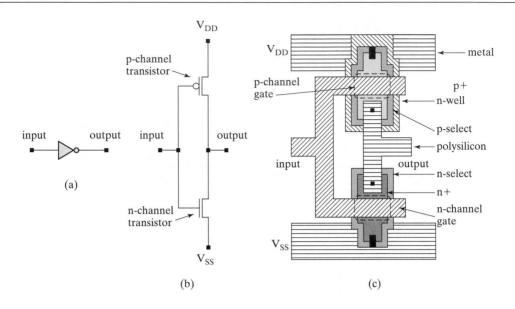

FIGURE 2-5 Views of a CMOS inverter: (a) circuit-symbol view, (b) transistor-schematic view, and (c) simplified composite fabrication mask view.

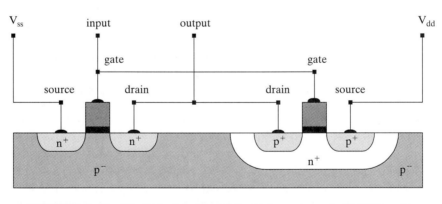

FIGURE 2-6 Simplified side view showing the doping regions of a CMOS inverter.

standard-cell libraries for repeated use in multiple designs. Such libraries commonly contain basic logic gates, flip-flops, latches, muxes, and adders. Synthesis tools build complex integrated circuits by mapping the end result of logic synthesis onto the parts of a cell library to implement the specified functionality with acceptable performance.

2.1.2   Boolean Algebra

The operations of logic circuits are described by Boolean algebra. A Boolean algebra consists of a set of values **B** = $\{0, 1\}$ and the operators "+" and "·". The operator "+" is

TABLE 2-2   Laws of Boolean algebra.

Laws of Boolean Algebra	SOP Form	POS Form
Combinations with 0, 1	$a + 0 = a$	$a \cdot 1 = a$
	$a + 1 = 1$	$a \cdot 0 = 0$
Commutative	$a + b = b + a$	$ab = ba$
Associative	$(a + b) + c = a + (b + c)$ $= a + b + c$	$(ab)c = a(bc) = abc$
Distributive	$a(b + c) = ab + ac$	$a + bc = (a + b)(a + c)$
Idempote	$a + a = a$	$a \cdot a = a$
Involution	$(a')' = a$	
Complementarity	$a + a' = 1$	$a \cdot a' = 0$

called the sum operator, the "OR" operator, or the disjunction operator. The operator " $\cdot$ " is called the product operator, the "AND" operator, or the conjunction operator. The operators in a Boolean algebra have commutative and distributive properties such that for two Boolean variables A and B having values in $\mathbf{B}$, $a + b = b + a$, and $a \cdot b = b \cdot a$. The operators "+" and " $\cdot$ " have identity elements 0 and 1, respectively, such that for any Boolean variable a, $a + 0 = a$, and $a \cdot 1 = a$. Each Boolean variable a has a complement, denoted by a', such that $a + a' = 1$, and $a \cdot a' = 0$. Table 2-2 summarizes the laws of Boolean algebra for sum-of-products (SOP) and product-of-sums (POS) Boolean expressions (more on this later). For simplicity, we have omitted showing the " $\cdot$ " operator and will do so freely in the remaining examples.

A multidimensional space spanned by a set of n Boolean variables is denoted by $\mathbf{B}^n$. A point in $\mathbf{B}^n$ is called a *vertex* and is represented by an n-dimensional vector of binary valued elements, for example, (100). A binary variable can be associated with the dimensions of a Boolean space, and a point is identified with the values of the variables. A Boolean variable is represented symbolically by a literal, such as a. A literal is an instance (e.g., a) of a variable or its complement (e.g., a'). Boolean expressions are formed by strings of literals and Boolean operators. A product of literals, such as $ab'c$ is a *cube*. A cube is associated with a set of vertices, and a cube is said to "contain" one or more vertices. Figure 2-7 illustrates how each point in $\mathbf{B}^3$ can be represented (a) by a vector of binary values (e.g., its coordinates), and (b) by a cube of literals.

A *completely specified* m-dimensional Boolean function with n inputs is a mapping from $\mathbf{B}^n$ into $\mathbf{B}^m$, denoted by f: $\mathbf{B}^n \rightarrow \mathbf{B}^m$. An *incompletely specified* function is defined over a subset of $\mathbf{B}^n$, and is considered to have a value of *don't-care* at points outside the domain of definition: f: $\mathbf{B}^n \rightarrow \{0, 1, *\}$, where * denotes don't-care.

The *On_Set* of a Boolean function consists of the vertices at which the function is asserted (logically true), that is, *On-Set* $= \{x\colon x \in B^n \text{ and } f(x) = 1\}$. The *Off_Set* is the set of vertices at which the function is de-asserted (logically false): *Off_Set* $= \{x\colon x \in B^n \text{ and } f(x) = 0\}$. The *Don't-Care-Set* is the set of vertices at which no significance is attached to the value of the function, so *Don't-Care-Set* $= \{x\colon x \in B^n \text{ and } f(x) = *\}$. The Don't-Care-Set set accommodates input patterns that never occur or outputs that will not be observed.

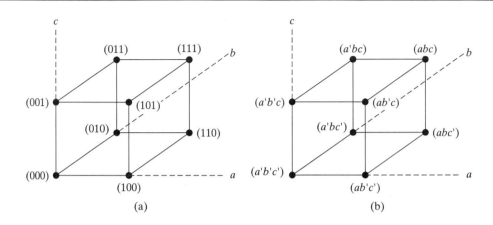

FIGURE 2-7   Points in a Boolean space: (a) represented by vectors of binary variables,
and (b) represented symbolically.

2.1.3   DeMorgan's Laws

DeMorgan's laws allow us to transform a circuit from an SOP form to a POS form, and
vice versa. The first form of the law specifies the complement of a sum of terms:

$$(a + b + c + \dots)' = a' \cdot b' \cdot c' \cdot \dots$$

For two variables the relationship specifies that:

$$(a + b)' = a' \cdot b'$$

The Venn diagrams in Figure 2-8 illustrate the operations of DeMorgan's laws for two
variables.

The second form of DeMorgan's laws specifies the complement of the product of
terms:

$$(a \cdot b \cdot c \dots)' = a' + b' + c' + \dots$$

For two variables, the law states that:

$$(a \cdot b)' = a' + b'$$

These relationships are illustrated by the Venn diagrams in Figure 2-9.

2.2   Theorems for Boolean Algebraic Minimization

Important theorems that are used to minimize Boolean algebraic expressions to produce
efficient circuit realizations are shown in Figure 2-10, in POS and SOP form. Logical adjacency and the consensus theorem are illustrated by the Venn diagrams in Figure 2-11. The
consensus term, bc, is redundant because it is covered by the union of ab and $a'c$. Laws
that apply specifically to the exclusive-or operation are shown in Figure 2-12.

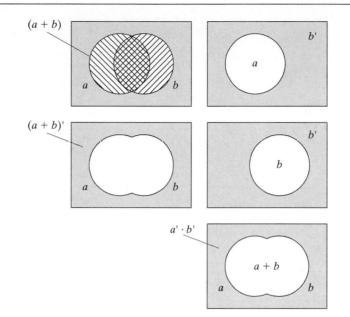

FIGURE 2-8 Venn diagrams illustrating DeMorgan's law: $(a + b)' = a' \cdot b'$.

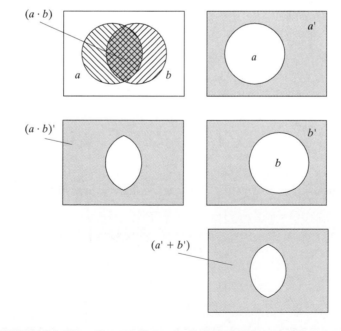

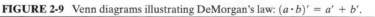

FIGURE 2-9 Venn diagrams illustrating DeMorgan's law: $(a \cdot b)' = a' + b'$.

Theorem	SOP Form	POS Form
Logical Adjacency	$ab + ab' = a$	$(a + b)(a + b') = a$
Absorption or:	$a + ab = a$ $ab' + b = a + b$ $a + a'b = a + b$	$a(a + b) = a$ $(a + b')b = ab$ $(a' + b)a = ab$
Multiplication and Factoring	$(a + b)(a' + c) = ac + a'b$	$ab + a'c = (a + c)(a' + b)$
Consensus	$ab + bc + a'c = ab + a'c$	$(a + b)(b + c)(a' + c) =$ $(a + b)(a' + c)$

FIGURE 2-10   Theorems for minimizing Boolean expressions.

Given a Boolean function $f(x_1, x_2, \ldots, x_n)$, its cofactor with respect to variable x_i is

$$f_{xi} = f(x_1, x_2, \ldots, x_{i-1}, 1, x_{i+1}, \ldots, x_n),$$

and its cofactor with respect to $x_{i'}$ is

$$f_{xi'} = f(x_1, x_2, \ldots, x_{i-1}, 0, x_{i+1}, \ldots, x_n).$$

A binary-valued Boolean function can be represented by its cofactors as the following so-called Shannon expansion:

$$f(x_1, x_2, \ldots, x_{i-1}, x_i, x_{i+1}, \ldots, x_n) = x_i \cdot f_{xi} + x_{i'} \cdot f_{xi'} = (x_i + f_{xi'}) \cdot (x_{i'} + f_{xi})$$

for all $i = 1, 2, \ldots, n$. The Boolean difference of a Boolean function f is given by

$$\partial f / \partial x_i = f_{xi} \oplus f_{xi'}$$

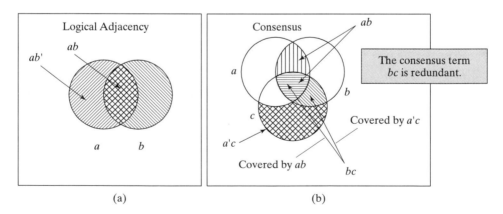

FIGURE 2-11   Venn diagrams: (a) Logical adjacency, and (b) Consensus.

Exclusive-Or Laws	
Combinations with 0, 1	$a \oplus 0 = a$
	$a \oplus 1 = a'$
	$a \oplus a = 0$
	$a \oplus a' = 1$
Commutative	$a \oplus b = b \oplus a$
Associative	$(a \oplus b) \oplus c = a \oplus (b \oplus c) = a \oplus b \oplus c$
Distributive	$a(b \oplus c) = ab \oplus ac$
Complement	$(a \oplus b)' = a \oplus b' = a' \oplus b = ab + a'b'$

FIGURE 2-12 Boolean relationships for the exclusive-or operation.

The Boolean difference of f with respect to x_i determines whether f is sensitive to a change in input variable x_i. This property has utility in algebraic methods to determine whether a test detects a fault in a circuit [1]. Also, a binary tree mapping of a Boolean function can be generated by recursively applying Shannon's expansion [2].

2.3 Representation of Combinational Logic

We will consider three common representations of combinational logic: (a) structural (i.e., gate-level) schematics, (b) truth tables, and (c) Boolean equations. An additional representation, a binary decision diagram (BDD) is a graphical representation of a Boolean function and contains the information needed to implement it [2, 3]. BDDs are used primarily within EDA software tools because they can be more efficient and easier to manipulate than truth tables. They can also be helpful in finding hazard covers [4]. We will not make use of BDDs here, but will rely on truth tables instead.

Example 2.1

The truth table for the combinational logic of a half adder is shown in Figure 2-13. The adder forms a sum and carry out bit from two data bits (without a carry in bit).

The Boolean equations describing the half adder can be derived from the truth table and written in SOP form:

$$sum = a'b + ab' = a \oplus b$$

$$c\_out = a \cdot b$$

End of Example 2.1

Inputs		Outputs	
a	*b*	*c_out*	*sum*
0	0	0	0
0	1	0	1
1	0	0	1
1	1	1	0

(a)

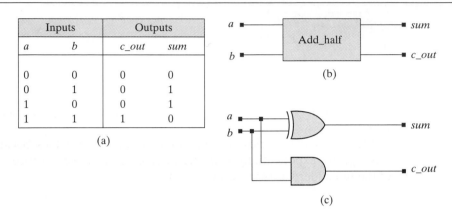

(b)

(c)

FIGURE 2-13 Half adder: (a) truth table, (b) block diagram symbol, and (c) schematic.

Example 2.2

A full adder forms a sum and carry out bit from two data bits and a carry in bit. The truth table for the combinational logic of a full adder is shown in Figure 2-14.

The Boolean equations describing *sum* and *c_out* bits are given by:

$$sum \ = a' \cdot b' \cdot c\_in + a' \cdot b \cdot c\_in' + a \cdot b' \cdot c\_in' + a \cdot b \cdot c\_in$$

$$c\_out = a' \cdot b \cdot c\_in + a \cdot b' \cdot c\_in + a \cdot b \cdot c\_in' + a \cdot b \cdot c\_in$$

Inputs			Outputs	
a	*b*	*c_in*	*c_out*	*sum*
0	0	0	0	0
0	0	1	0	1
0	1	0	0	1
0	1	1	1	0
1	0	0	0	1
1	0	1	1	0
1	1	0	1	0
1	1	1	1	1

(a)

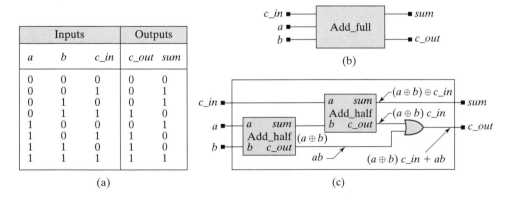

(b)

(c)

FIGURE 2-14 Full adder: (a) truth table, (b) block diagram symbol, and (c) schematic for a full adder composed of half adders and glue logic.

These can be rearranged as:

$$sum = a \oplus b \oplus c\_in$$

$$c\_out = (a \oplus b) \cdot c\_in + a \cdot b$$

The Venn diagram in Figure 2-15 shows the assertions of *sum* and *c_out* as a function of *a*, *b*, and *c_in*.

Truth tables become unwieldy for functions of several variables because the number of rows grows exponentially with the number of variables. Notice that the truth table of the full adder has twice as many rows as the table for the half adder.

End of Example 2.2

2.3.1 Sum of Products Representation

A cube is formed as the product of literals in which a literal appears in either uncomplemented or complemented form. For example, *ab'cd* is a cube but *ab'cbd* is not. A cube need not contain every literal. A *Boolean expression* is a set of cubes, and is typically expressed in *sum-of-products (SOP)* form as the "OR" of product terms (cubes), rather than in set notation.

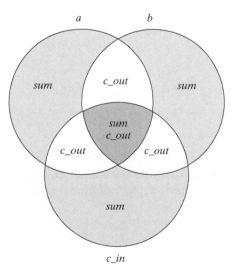

FIGURE 2-15 Venn diagram representation of the truth table for a full adder.

Example 2.3

The following expression is in sum-of-products form: $abc' + bd$.

End of Example 2.3

Each term of a Boolean expression in SOP form is called an *implicant* of the function. A *minterm* is a cube in which every variable appears. The variable will be in either true (uncomplemented) or complemented form (but not both). Thus, a minterm corresponds to a single point (vertex) in $\mathbf{B}^n$. A cube that is not a minterm represents two or more points in $\mathbf{B}^n$. The minterms of a Boolean function correspond to the rows of the truth table at which the function has a value of 1.

Example 2.4

The cube $ab'cd$ is a minterm in $\mathbf{B}^4$. The cube abc is not a minterm. It represents the pair of vertices defined by $abcd + abcd'$.

End of Example 2.4

A Boolean expression in SOP form is said to be canonical if every cube in the expression has a unique representation in which all of the literals are in complemented or uncomplemented form.

Example 2.5

The expression $abcd + a'bcd$ is a canonical SOP.

End of Example 2.5

A *canonic* (standard) SOP function is also called a standard sum-of-products (SSOP). In decimal notation, a minterm is denoted by m_i, and the pattern of 1s and 0s in the binary equivalent of the decimal number indicates the true and complemented literals. For example, $m_7 = a'bcd$.

In $\mathbf{B}^n$ there is a one-to-one correspondence between a minterm and a vertex of an n-dimensional cube, as shown in Figure 2-16. The minterm $m_3 = a'bc$ corresponds to the vertex with coordinates 011.

A *Boolean function* is a set of minterms (vertices) at which the function is asserted. A Boolean function in SOP form is expressed as a sum of minterms.

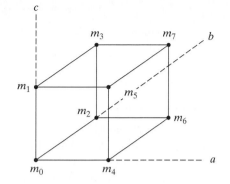

FIGURE 2-16   Correspondence between minterms and vertices in $\mathbf{B}^3$.

Example 2.6

The sum and carry bits of the full adder can be expressed as a sum of minterms, ordered as $\{a, b, c\_in\}$ in $\mathbf{B}^3$:

$$sum = m_1 + m_2 + m_4 + m_7 = \Sigma m(1, 2, 4, 7)$$

$$c\_out = m_3 + m_5 + m_6 + m_7 = \Sigma m(3, 5, 6, 7)$$

End of Example 2.6

Example 2.7

The set of shaded vertices in Figure 2-17 define $f = m_1 + m_2 = m_3 = a'b'c + a'bc' + a'bc$.

End of Example 2.7

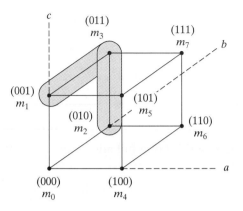

FIGURE 2-17 Set of minterms for $f = a'b'c + a'bc' + a'bc$.

2.3.2 Product-of-Sums Representation

A Boolean function can also be expressed in a POS form in which the expression is written as a product of Boolean factors, each of which is a sum of literals.

Example 2.8

The POS representation of the c_out bit in a full adder circuit is formed by expressing the 0s of the truth table in SOP form (see Figure 2-14):

$$c\_out' = a'b'c\_in' + a'b'c\_in + a'bc\_in' + ab'c\_in'$$

Then the expression for c_out' is complemented, giving

$$c\_out = (a'b'c\_in' + a'b'c\_in + a'bc\_in' + ab'c\_in')'$$

DeMorgan's Laws can be applied to c_out to give the POS expression shown below:

$$c\_out = (a'b'c\_in')' \cdot (a'b'c\_in)' \cdot (a'bc\_in')' \cdot (ab'c\_in')'$$

$$c\_out = (a + b + c\_in) \cdot (a + b + c\_in') \cdot (a + b' + c\_in) \cdot (a' + b + c\_in)$$

End of Example 2.8

A Boolean expression in POS form is said to be *canonical* (i.e., a unique representation for a given function) if each factor has all of the literals in complemented or uncomplemented form, but not both.

A *maxterm* is an OR-ed sum of literals in which each variable appears exactly once in true or complemented form (e.g., $a + b + c\ in$ is a maxterm in the POS expression for c_out). A canonical POS expansion consists of a product of the maxterms of the truth table of a function. The decimal notation of a maxterm is based on the rows of the truth table at which the function is zero (i.e., where f' is asserted). The variables are complemented when forming the POS expression.

Example 2.9

The SOP form of the $c\ out'$ bit of a full adder was given in the previous example. The decimal notation for c_out is given by the product of the maxterms that corresponds to the cubes of $c\ out'$ as

$$c\_out' = a'b'c\_in' + a'b'c\_in + a'bc\_in' + ab'c\_in'$$

$$c\_out = M_0 \cdot M_1 \cdot M_2 \cdot M_4 = \Pi M(0, 1, 2, 4)$$

$$c\_out = (a + b + c\_in) \cdot (a + b + c\_in') \cdot (a + b' + c\_in) \cdot (a' + b + c\_in)$$

A canonical SOP expression can be a very efficient representation of a Boolean function because there might be very few terms at which the function is asserted. Alternatively, f' expressed as a POS expression might be very efficient because there are only a few terms at which the function is de-asserted.

End of Example 2.9

2.4   Simplification of Boolean Expressions

An SOP expression can be implemented in hardware as a two-level AND-OR logic circuit. Although a Boolean expression can always be expressed in a canonical form, with every cube containing every literal (in complemented or uncomplemented form), such descriptions are usually inefficient and waste hardware. In practice, minimization is important because the cost of hardware implementing a Boolean expression is related to the number of terms in the expression and to the number of literals in a term, that is, in a cube in an SOP expression.

A Boolean expression in SOP form is said to be *minimal* if it contains a minimal number of product terms and literals (i.e., a given term cannot be replaced by another that has fewer literals). A minimum SOP form corresponds to a two-level logic circuit having the fewest gates and the fewest number of gate inputs.

There are four common approaches to simplifying a Boolean expression. The first is a manual graphical method that is guided by Karnaugh maps displaying logical adjacencies of the function. Manual methods are feasible only for functions that have no more than six inputs [5]. The Quine–McCluskey minimization algorithm relies on logical adjacency and the same principles as logic minimization with Karnaugh maps [6–9]. It can be applied manually on small functions, but can also be implemented as a computer program that is effective for larger circuits. A third method, Boolean minimization [5], is also manual, and relies on clever application of the theorems describing relationships between Boolean variables to find simpler, equivalent expressions. The method is not straightforward, can be difficult, and requires experience. As a fourth alternative, the theorems that are used in Boolean minimization are now embedded in modern synthesis tools and programs such as Espresso-II [10] and mis-II (multilevel interactive synthesis) [4], in which they are used to perform logic minimization and form efficient realizations of two-level and multilevel logic circuits.

Logic minimization searches for efficient representations of Boolean functions. In a Boolean expression, a cube that is contained in another cube is said to be redundant—a cube is *redundant* if its set of vertices is properly contained in the set of vertices of another cube of the function. A Boolean expression is *nonredundant (irredundant)* if no cube contains another cube.

Example 2.10

The following Boolean expression is redundant because the vertex set of *ab* is a subset of the vertex set of *a*:

$$f(a, b) = a + ab$$

The redundant cube can be removed to give an equivalent, but more efficient representation:

$$f(a, b) = a$$

End of Example 2.10

Example 2.11

The Boolean function given by $f(c, d) = c'd' + cd$ is irredundant.

End of Example 2.11

The cubes of an irredundant expression do not share a common vertex, that is, their corresponding sets of vertices are pairwise disjoint. Boolean minimization is difficult because the minimum SOP form and minimum POS forms of a Boolean expression are not unique. Boolean minimization/simplification exploits logical adjacency by (1) repeatedly combining cubes that differ in only one literal, and (2) eliminating redundant implicants.

Example 2.12

With the objective of illustrating Boolean minimization of the *function* $f(a, b, c) = abc + a'bc + abc' + a'b'c + ab'c' + a'b'c'$, we start by combining cubes that differ by only one literal, as shown in Figure 2-18, which shows the Boolean expression for f and illustrates the adjacent vertices graphically. A pair of adjacent shaded vertices can be combined into a single cube that covers both of them.

An equivalent, minimal expression is formed by removing the cubes $ab + a'b'$ and adding the cube ac' to the expression, as shown in Figure 2-19:

$$f(a, b, c) = ac' + a'c + bc + b'c'$$

Note that $f(a, b, c) = ac' + a'c + a'b'c' + abc$ is also an equivalent expression, but not minimal. Figure 2-20 shows how another equivalent and minimal expression can be obtained by applying logical adjacency to different terms of the original SOP expression.

$$f(a,b,c) = \overbrace{abc + a'bc}^{ab}\ \underbrace{+ abc'}_{bc} + \overbrace{a'b'c + ab'c'}^{a'b'} + a'b'c'_{b'c'}$$

$$f(a, b, c) = bc + ab + a'b' + b'c'$$

Each term (cube) of a Boolean expression in SOP form is called an implicant of the function. An implicant covers a vertex if the vertex is included in the set of vertices at which the implicant is asserted. An implicant may cover more than one vertex of the function. The fewer the number of literals in a cube, the larger the set of vertices covered by the cube. So the hardware implementation is minimized if a cube has as few literals

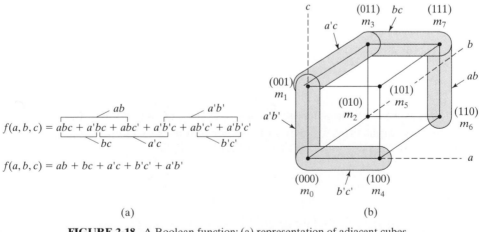

(a)   (b)

FIGURE 2-18 A Boolean function: (a) representation of adjacent cubes, and (b) a graphical representation of adjacent vertices.

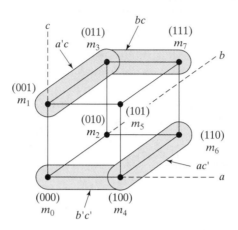

FIGURE 2-19 A second, equivalent, minimal expression for the function in Figure 2-18.

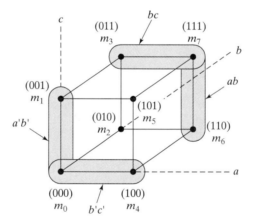

FIGURE 2-20 A third, equivalent minimal expression for the function in Figure 2-18.

as possible. The function $f = abc + abc'$ shown in Figure 2-21 has two vertices that can be combined by logical adjacency into one implicant, ab. A representation in terms of just one implicant is preferred to another that uses two implicants because the hardware of the former realization will be simpler and cheaper.

End of Example 2.12

The vertices of the On-Set of a Boolean function provide a complete, but inefficient, description of the function because each minterm includes every literal that is used by the function. We can exploit logical adjacency to reduce the size of the terms (cubes)

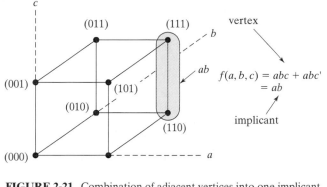

FIGURE 2-21 Combination of adjacent vertices into one implicant.

that are used in the SOP form of the function. Terms that are logically adjacent can always be combined into a single term that has fewer literals. But be aware that merely applying logical adjacency to the cubes of a function does not necessarily minimize the function. It is essential that every vertex be covered by a cube, that is, each must belong to the vertex set of a cube; but how many such cubes are necessary to completely cover the function, and which cover is the most efficient? We will now formalize these concepts.

A *prime implicant* of the On-Set of a Boolean function is an implicant whose assertion does not imply assertion of any other implicant of the function. A prime implicant is a cube whose vertices are not properly contained in the set of vertices of some other cube of the function.

Example 2.13

Consider the function $f(a, b, c, d) = a'b'cd + a'bcd + ab'cd + abcd + a'b'c'd'$ in SOP form:

$$f(a, b, c, d) = a'b'cd + a'bcd + ab'cd + abcd + a'b'c'd'$$
$$= a'cd + acd + a'b'c'd'$$
$$= cd + a'b'c'd'$$

Note that $a'cd$ and acd both imply cd, so they are not prime implicants. The term $a'b'c'd$ is a prime implicant.

End of Example 2.13

A prime implicant cannot be combined with another implicant to eliminate a literal or to be eliminated from the expression by absorption. An implicant that implies another implicant is said to be covered by it; the set of the vertices of the covered implicant is a subset of the vertices of the implicant that covers it. The covering implicant, having fewer literals, has more vertices. The set of prime implicants of a Boolean expression is unique.

Example 2.14

The expression, vertices, and prime implicants of a Boolean function in $\mathbf{B}^3$ are shown in Figure 2-22.

End of Example 2.14

A prime implicant that is not covered by any *set* of other implicants is an *essential prime implicant*. An essential prime implicant must be retained in a cover of the function.

Example 2.15

The vertices and implicants of $f(a, b, c) = a'bc + abc + ab'c' + abc'$ are shown in Figure 2-23. The set of prime implicants of f is $\{ac', ab, bc\}$. The set of essential prime implicants, an SOP expression for f, and a minimal SOP expression are also listed below.

Essential Prime Implicants: $\{ac', bc\}$

SOP Expression $f(a, b, c) = ac' + ab + bc$

Minimal SOP Expression: $f(a, b, c) = ac' + bc$

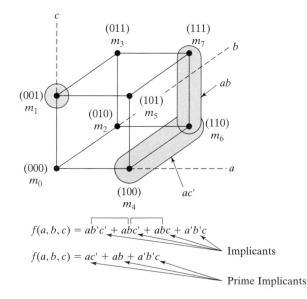

FIGURE 2-22   The vertices and prime implicants of $f = ab'c' + abc' + abc + a'b'c$.

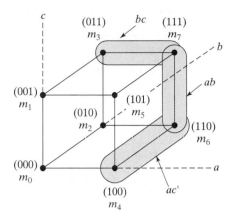

FIGURE 2-23 Vertices and implicants of $f(a, b, c) = a'bc + abc + ab'c' + abc'$.

End of Example 2.15

The process for minimizing a Boolean expression follows these steps: (1) find the set of all prime implicants, (2) find a minimal subset of implicants that covers all of the prime implicants (including the essential prime implicants). The *minimal cover* of a Boolean expression is a subset of prime implicants that covers all of its prime implicants.

Example 2.16

Consider the function given by:

$$f(a, b, c, d) = a'b'cd + a'bcd + ab'cd + abcd + a'b'c'd'$$

By combining adjacent terms we get

$$f(a, b, c, d) = a'cd + acd + a'b'c'd'$$
$$f(a, b, c, d) = cd + a'b'c'd'$$

The set of prime implicants is $\{cd, a'b'c'd'\}$, and the minimal cover is $\{cd, a'b'c'd'\}$.

End of Example 2.16

Boolean minimization combines terms that are logically adjacent, that is, that differ in only one literal.

Example 2.17

Figure 2-24 illustrates how logically adjacent terms in the expression for the carry-out bit of the full adder can be combined to obtain a minimal expression.

End of Example 2.17

Boolean minimization also exploits logical adjacency of complementary expressions.

Example 2.18

Consider the expression $(c + db)(a + e') + c'(d' + b')(a + e')$ and note that $(c + db)' = c'(d' + b')$.

Then

$$(c + db)(a + e') + c'(d' + b')(a + e') = (c + db)(a + e') + (c + db)'(a + e')$$
$$= a + e'.$$

End of Example 2.18

The absorption $(a + ab = a)$ and consensus $(ab + bc + a'c = ab + ac')$ properties of Boolean algebra can be used to eliminate redundant terms in an expression.

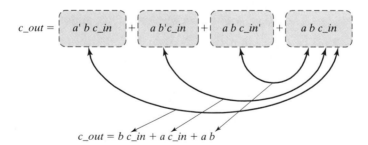

FIGURE 2-24 Boolean minimization of *c_out* in a full adder by combinations of logically adjacent terms.

Example 2.19

The consensus property will be used to reduce the expression: $f = e'fg' + fgh + e'fh$. We rearrange the terms in f to associate with the terms in the consensus law, leading to the result $f = e'fg' + fgh$, as shown in Figure 2-25 below:

End of Example 2.19

The absorption property can be used repeatedly to eliminate literals in the expression.

Example 2.20

Consider $f = efgh' + e'f'g'h' + e'f$ and rearrange it to get:

$$f = efgh' + e'f'g'h' + e'f = efgh' + e'(f + f'g'h')$$

$$= efgh' + e'(f + g'h')$$

$$= f(egh' + e') + e'g'h'$$

$$= f(gh' + e') + e'g'h'$$

$$= fgh' + e'f + e'g'h'$$

End of Example 2.20

Sometimes it is helpful to introduce redundant terms into an expression to support absorption and logical adjacency. A Boolean expression in SOP form is preserved under the following operations: (1) adding the product of a literal and its complement (e.g., the term: aa'), (2) adding the consensus term (e.g., adding bc to $ab + a'c$), (3) adding to a literal its product with any other literal (e.g., adding ab to a to form $ab + a$).

Consensus term: $fe'h$

FIGURE 2-25 The consensus property and simplification of a Boolean expression.

A Boolean expression in POS form is preserved under the operations: (1) multiplying the expression by a factor consisting of the sum of any literal and its complement [e.g., multiplying by the factor $(a + a')$], (2) introducing the product consensus factor [e.g., introduce the factor $(b + c)$ in $(a + b)(a' + c)$], multiplying a literal by its sum with any other literal [e.g., $a (a + b)$]. Expanding with the consensus term is helpful when it can absorb other terms or eliminate a literal.

Example 2.21

The consensus theorem $(ab + bc + a'c = ab + a'c)$ is usually used to eliminate a redundant term (bc) that is covered by two other terms in an expression. But it can be used to add a redundant term, thereby leading to simplification of a larger expression. As an example of how to add a redundant term, consider the expression $f = bcd + bce + ab + a'c$ and note that the terms $ab + a'c$ are the result of eliminating the consensus term from $ab + bc + a'c$. So, adding the consensus term (bc) back into f gives the expression $f = b'c + bcd + bce + ab + bc + a'c$. Now it is possible to absorb terms with bc and then delete bc, leaving $f = ab + c$.

End of Example 2.21

2.4.1   Simplification with Exclusive-Or

The properties that were listed in Figure 2-12 for the exclusive-or can be used to simplify expressions.

Example 2.22

The expression for the *sum* output of a full adder is $sum = a' \cdot b' \cdot c\_in + a' \cdot b \cdot c\_in' + a \cdot b' \cdot c\_in' + a \cdot b \cdot c\_in$. This expression simplifies to $sum = (a \oplus b) \oplus c\_in = a \oplus b \oplus c\_in$, which requires a pair of two-input exclusive-or gates in hardware.

End of Example 2.22

2.4.2   Karnaugh Maps (SOP Form)

Karnaugh maps (K-maps) provide a graphical/visual representation of a Boolean function of up to five or six variables. The map of an expression reveals logical adjacencies and opportunities for eliminating a literal from two or more cubes. The columns and rows of the map are arranged so that they are logically adjacent over the space of the input variables of the function. Each vertex (point) in the Boolean domain of the function is represented by a square in the map. Each cell of the map has an entry to indicate where the vertex is in the *On-Set* (1), *Off-Set* (0), or the *Don't-Care-Set* (x). K-maps facilitate finding the largest possible cubes that cover all 1s without redundancy. Their application requires manual effort.

The K-map of a function of four variables shows all 16 possible vertices. Also, observe the ordering of the rows and columns, and that the topmost and bottommost rows are logically adjacent, and the leftmost and rightmost columns are logically adjacent. Logically adjacent cells that contain a 1 can be combined. A rectangular cluster of cells that are logically adjacent can be combined. Any don't-cares of the function can be used to form prime implicants and create additional possibilities for reduction.

Example 2.23

The K-map in Figure 2-26 shows how its corresponding Boolean function can be reduced by logical adjacency to give $f = bd + b'c'$. The SOP expression corresponding to the minterms at the four corners of the map can be simplified successively as shown below:

$$a'b'c'd' + a'b'cd' + a'b'c'd' + ab'c'd' = a'b'd' + b'c'd'$$

The resulting expression can also be reduced to

$$ab'cd' + ab'c'd' + ab'cd' + a'b'cd' = ab'd' + b'cd'$$

The four terms on the left of the above expression can be combined to give:

$$a'b'd' + b'c'd' + ab'd' + b'cd' = b'd'$$

The shaded inner quad of minterms in Figure 2-26 can also be reduced:

$$a'bc'd + a'bcd + abc'd + abcd = bc'd + bcd = bd$$

and so

$$f = b'd' + bd = (b \oplus d)'$$

Note that each corner minterm implies $b'd'$, and that $b'd'$ does not imply another implicant. Therefore, it is a prime implicant. It is also an essential prime implicant. Similarly, bd is an essential prime implicant.

End of Example 2.23

To form a minimal realization from a Karnaugh map, (1) identify all of the essential prime implicants, using don't-cares as needed, and (2) use the prime implicants to form a cover of the remaining 1s in the map (ignoring don't-cares). In general, the covering set of prime implicants is not unique.

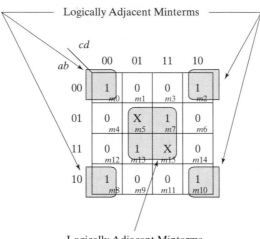

Logically Adjacent Minterms

Logically Adjacent Minterms

FIGURE 2-26 K-map of $f = a'b'c'd' + a'b'cd + ab'cd' + ab'c'd' + abcd + a'bcd$.

Example 2.24

Minimal covers will be found for the Boolean function whose K-map is shown in Figure 2-27.

    The prime implicants of f are listed below, and those that are essential are identified.

Prime Implicants: $m3$, $m2$, $m7$, $m6 \rightarrow a'c$ (essential)

         $m2$, $m10 \rightarrow b'cd'$ (essential)

         $m7$, $m15 \rightarrow bcd$

         $m13$, $m15 \rightarrow abd$

         $m12$, $m13 \rightarrow abc'$ (essential)

    The minimal covers of f are formed by including the essential prime implicants with implicants that cover the remaining vertices.

Minimal Covers: (1) $a'c$, $b'cd'$, bcd, abc'

                (2) $a'c$, $b'cd'$, abd, abc'

End of Example 2.24

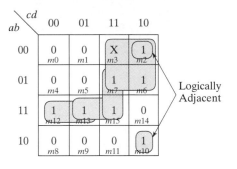

FIGURE 2-27 K-map for $f = abc'd' + abc'd + abcd + a'bcd + a'b'cd + a'b'cd' + a'bcd' + ab'cd'$.

The following steps will form a minimal cover: (1) select an uncovered minterm, (2) identify all adjacent cells containing a 1 or an x, and (3) a single term (not necessarily a minterm) that covers the minterm and all of its neighboring cells having a 1 or an x is an essential prime implicant. Add the term to the set of essential prime implicants. Step 1 is repeated until all of the essential prime implicants have been selected. After Step 1 is complete, find a minimal set of prime implicants that cover the other 1s in the map (do not cover cells containing x). These steps may produce more than one possible minimal cover. Select the cover that has the fewest literals.

2.4.3   Karnaugh Maps (POS Form)

The minimal product of sums form of a Boolean expression is formed by finding a minimal cover of the 0s in the Karnaugh map, then applying DeMorgan's Theorem to the result.

Example 2.25

The 0-cells of the K-map shown in Figure 2-28 can be combined through logical adjacency):

$$m0, m1, m4, m5: a'b'c'd' + a'b'c'd + a'bc'd' + a'bc'd \rightarrow a'c'$$

$$m0, m1, m8, m9: a'b'c'd' + a'b'c'd + ab'c'd' + ab'c'd \rightarrow b'c'$$

$$m9, m11 \qquad : ab'c'd + ab'cd \rightarrow ab'd$$

$$m14 \qquad\quad : abcd'$$

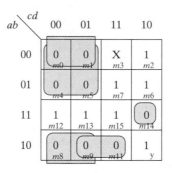

FIGURE 2-28 K-map for a minimal POS expression.

The expression for the 0-cells becomes:

$$f'(a, b, c, d) = a'c' + b'c' + ab'd + abcd'$$

After DeMorgan's law is applied to f', the minimal POS expression for f becomes:

$$f(a, b, c, d) = (a + c)(b + c)(a' + b + d')(a' + b' + c' + d)$$

End of Example 2.25

2.4.4 Karnaugh Maps and Don't-Cares

Don't-cares represent situations in which an input cannot occur or the output does not matter. The general rule is that don't-cares can be used when covering them leads to an improved representation.

Example 2.26

A binary coded decimal (BCD) word is a 4-bit word whose values correspond to the digits $0, \ldots, 9$. The BCD code, also known as a 8421 code, uses only the first 10 patterns, beginning with 0000_2 and ending with 1001_2. The code for each decimal digit N is obtained by adding 1 to the code of the preceding digit, $N - 1$. Suppose a function f is asserted when the BCD representation of a 4-variable input is 0, 3, 6 or 9 [5]. The K-map in Figure 2-29(a) does not make use of don't-cares (denoted by x).

The function obtained without exploiting don't-cares has 16 literals:

$$f(a, b, c, d) = a'b'c'd' + a'b'cd + a'bcd' + ab'c'd$$

If the don't-cares are included, f has the SOP form $f(a, b, c, d) = a'b'c'd'$ $+ a'b'cd + abc'd' + abc'd + abcd + abcd' + ab'c'd + ab'cd$, which has 32 literals. The K-map in Figure 2-29(b) shows how f can be reduced to obtain an SOP form that has 12 literals:

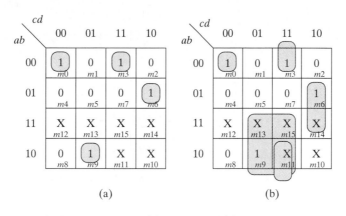

FIGURE 2-29   K-maps (a) without, and (b) with don't-cares.

$$f(a, b, c, d) = a'b'c'd' + b'cd + bcd' + ad$$

which can be reduced to:

$$f(a, b, c, d) = a'b'c'd' + b'cd + ab + ac'd$$

End of Example 2.26

2.4.5   Extended Karnaugh Maps

A 4-variable Karnaugh map can be extended by entering variables to indicate that the represented function is asserted if the variable is asserted. No entry indicates that the function is not asserted if the variable is asserted. The process for finding a minimal representation of a Boolean function using an extended K-map is (1) to find the minimal cover with the extension variables de-asserted, then (2) for each variable, to separately find the minimal sum with all 1s changed to x in the map, and all other variables set to 0, and form the product of the minimal sum and the extension variable. Form the sum obtained by combining (1) with the sum of the results of (2). The result is a minimal representation if the extension variables can be assigned independently.

Example 2.27

The K-map in Figure 2-30(a) indicates where function F asserts independently of variables f and e, and where it asserts with them. In Figure 2-30(b) we consider logical adjacencies with f and e both set to 0, then in Figure 2-30(c) we show f asserted with all 1s

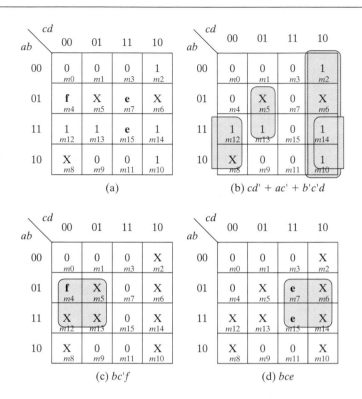

(a)

(b) $cd' + ac' + b'c'd$

(c) $bc'f$

(d) bce

FIGURE 2-30 Extended Karnaugh maps: (a) F asserts independently of f and e, and with them, (b) logical adjacencies with don't-cares and with e and f both 0, (c) f asserted, and x replacing all 1s of the original map, and (d) with e asserted, f de-asserted, and all 1s of the original map set to 0.

of the original map replaced by an x, and in Figure 2-30(d) we show variable e asserted with f de-asserted, and all 1s of the original map set to x. The sum of the cubes obtained from these steps gives $F = cd' + ac' + b'c'd + bc'f + bce$. *Note*: fe is contained in e and f, so the simultaneous assertion of f and e has also been considered.

End of Example 2.27

2.5   Glitches and Hazards

The output of a combinational circuit may make a transition even though the logical values applied at its inputs do not imply a change. These unwanted switching transients are called "glitches." Glitches are a consequence of the circuit structure, the delays of an actual implementation, and the application of patterns that cause the glitch to occur. A circuit in which a glitch may occur under the application of appropriate inputs signals is

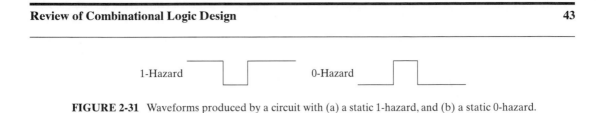

FIGURE 2-31 Waveforms produced by a circuit with (a) a static 1-hazard, and (b) a static 0-hazard.

said to have a "hazard." If a circuit has a hazard it could exhibit a glitch under certain conditions. There are two types of hazards, *static* and *dynamic*. The term *static* refers to a circuit in which the output should not change under the application of certain inputs, but does.

A circuit has a *static 1-hazard* if an output has an initial value of 1, and an input pattern that does not imply an output transition causes the output to change to 0 and then return to 1. A circuit has a *static 0-hazard* if an output has an initial value of 0, and an input pattern that does not imply an output transition causes the output to change to 1 and then return to 0. The waveforms that result from these hazards are shown in Figure 2-31. Whether a static hazard occurs or not depends on the application of the appropriate input pattern.

Static hazards are caused by differential propagation delays on reconvergent fanout paths. The signal applied at C in Figure 2-32 reconverges at the OR gate whose output is F. Thus, the signal propagates along two different paths to reach the output. The logic that forms the signals that arrive at the inputs of the gate implies that the two inputs are complementary. However, if the propagation delays along the signal paths are different, the output will have a hazard. Hazards might not be significant in a synchronous sequential circuit if the clock period can be extended. A hazard is problematic if the signal serves as the input to an asynchronous subsystem (e.g., a counter or a reset circuit).

The steps that form a "minimal" realization of a circuit do not imply that it will be hazard-free. If hazards are problematic, then more work needs to be done to detect and remove them. Static hazards can be eliminated by introducing redundant cubes in the cover of the output expression (the added cubes are called a "hazard cover"). Note that the treatment of hazards assumes that the output glitch is caused by the transition of a single bit of an input signal. Methods that eliminate hazards in two-level and multilevel circuits apply only if this condition is satisfied.

Consider the circuit shown in Figure 2-32, where $F = AC + BC'$. If the initial inputs to the circuit are $A = 1$, $B = 1$, $C = 1$, the output is $F = 1$. Next, if the inputs are $A = 1$, $B = 1$, $C = 0$, the output should still be $F = 1$. In a physical realization of the circuit (i.e., nonzero propagation delays), the delay of the path to $F1$ will be greater than the delay of the path to $F0$, because the signal travels through an additional gate, causing a change in C to reach $F1$ later than it reaches $F0$ (the path with greater delay is said to be longer), that is, AC de-asserts before BC' asserts. Consequently, when C changes from 1 to 0, the output undergoes a single, momentary transition to 0 and then returns to 1. The presence of a static hazard is apparent in the simulated waveforms in Figure 2-33 and in the Karnaugh map of the output signal shown in Figure 2-34.[2]

---

[2]We will consider simulation of digital logic in Chapter 4. The simulator used to obtain the results in Figure 2.33 is bundled on the CD-ROM that accompanies this book.

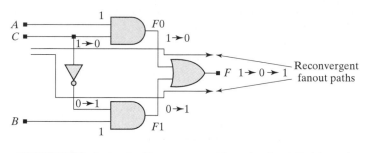

FIGURE 2-32   A circuit with reconvergent fanout and a static 1-hazard.

The Karnaugh map in Figure 2-34 reveals how a change of input C from 1 to 0 in the circuit of Figure 2-32 causes the cube AC to de-assert and the cube BC' to assert. However, AC de-asserts before BC' can assert. In the circuit of Figure 2-32, for example, the hazard occurs because the cube AC is initially asserted, while BC' is not. The hazard can be removed by adding a redundant cube, AB, to cover the adjacent 1s of the adjacent prime implicants associated with the hazard. The redundant cube is referred to as a "hazard cover." It eliminates the dependency of the output on the input C (the boundary between the cubes is now covered). *The cover of a hazard introduces redundant logic and requires additional hardware.*

Example 2.28

The hazard-free cover of the circuit in Figure 2-32 is given by: $F = AC + BC' + AB$. The circuit-level realization of the covered function in Figure 2-35 has an additional AND gate.

End of Example 2.28

2.5.1   Elimination of Static Hazards (SOP Form)

When a signal changes value at the input of a circuit, a static 1-hazard could occur at the output of the circuit if three conditions are satisfied: (1) an output remains asserted (i.e., its value is 1 before and after the input signal changes value), (2) the cube that is asserted in the SOP expression of the output by the initial value of the signal is different from the cube that is asserted by the final value of the signal, and (3) the cubes that are asserted by the initial and final values of the signal are not covered by the same prime implicant. If the output cubes asserted by the initial and final values of the input

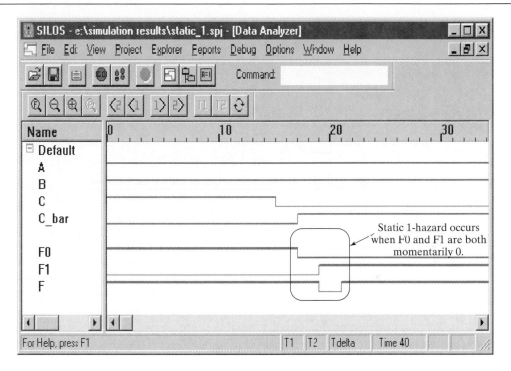

FIGURE 2-33 Results of simulating the circuit shown in Figure 2.32, which has reconvergent fanout and a static 1-hazard.

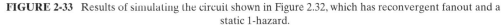

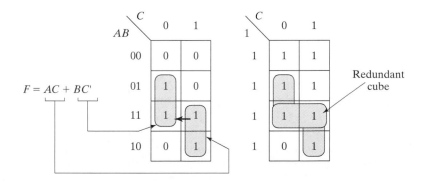

FIGURE 2-34 Karnaugh map of the logic for the circuit shown in Figure 2.32, which has reconvergent fanout and a static 1-hazard. *Note*: The arrow indicates a transition that causes the hazard to occur.

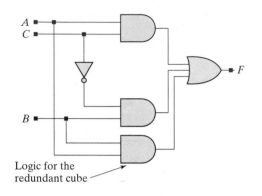

FIGURE 2-35   Circuit modified to remove a static 1-hazard.

signal are covered by the same prime implicant, a glitch cannot occur when the input signal changes value. Whether a static 1-hazard actually occurs depends on the accumulated delays along the signal propagation path from the inputs to the output.

If a static 1-hazard can be caused by changing the value of a single input signal, the cell that is asserted by the initial value of the input signal must be logically adjacent to the cell that is asserted by the final value of the input signal, because only one input signal is allowed to change. Consequently, the addition of a redundant cube covering both cells will cover the boundary between them, and cover the hazard. So, to eliminate a static 1-hazard caused by changing the value of a single input signal, form an SOP cover that covers every pair of adjacent 1s that reside in adjacent cubes. This guarantees that every single-bit input change is covered by a prime implicant. The set of such prime implicants is a hazard-free cover for a two-level (And-Or) realization of the circuit, but a better alternative might be found, and should be sought.

Example 2.29

The expression $f = \Sigma\, m(0, 1, 4, 5, 6, 7, 14, 15) = a'c' + bc$, whose K-map is shown in Figure 2-36, has a static 1-hazard, because f is asserted by the cubes $a'c'$ and bc, and with $a = 0$, $b = 1$, and $d = 1$, a glitch can occur as c changes from 1 to 0 or vice versa. Note that adjacent cells of adjacent cubes are asserted, depending on whether $c = 0$ or $c = 1$. The hazard can be removed by adding either the cube $a'bd$ or $a'b$ to the expression. Both are redundant prime implicants, but $a'b$ is chosen to given a minimal resulting expression: $f = a'c' + bc + a'b$. Also, observe that the redundant cube that is added to the SOP expression to cover the hazard does not depend on the input signal whose change causes the hazard.

There are two approaches to eliminating a static 0-hazard. The first detects where the transition of a single input causes a transition across the boundary between adjacent prime implicants and adds redundant prime implicants to f' as needed. The second method (1) eliminates the static 1-hazards of f, (2) considers whether the implicants of the 0s of the expression that is free of static 1-hazards also cover all adjacent 0s of the

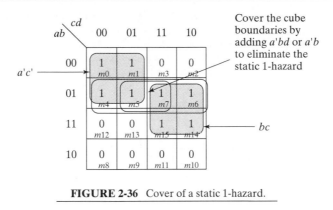

FIGURE 2-36 Cover of a static 1-hazard.

original function, then (3) adds redundant prime implicant factors to the complement of the static 1-hazard-free expression in POS form, as needed.

End of Example 2.29

Example 2.30

The Karnaugh map shown in Figure 2-37 for the function $f = a'c' + bc$ (see Example 2.29) has a static 0-hazard, because f is de-asserted in cubes ac' and $b'c$, and with $a = 1$, $b = 0$, and $d = 1$, switching c from 1 to 0, or vice versa, crosses a boundary between adjacent 0s in adjacent de-assertion cubes of f. By following the first method for eliminating a static 0-hazard, we consider the 0s of the K-map in Figure 2-37 and apply DeMorgan's law to obtain $f = (a' + c)(b + c')$. We cover the hazard by adding ab' to f', and including the redundant prime implicant product factor $(a' + b)$ in the POS form of f. The factor $ab'd$ would also cover the hazard, but it is not minimal. The resulting, equivalent, hazard-free POS expression is $f = (a' + c)(b + c')(a' + b)$.

   Also observe that, in this example, the POS expression that eliminates the static 0-hazard is equivalent to the expression that eliminates the static 1-hazard, for

$$f = (a' + c)(b + c')(a' + b)$$
$$= a'ba' + a'bb + a'c'a' + a'c'b + cba' + cbb + cc'a' + cc'b$$

$$= a'b + a'c' + bc$$

End of Example 2.30

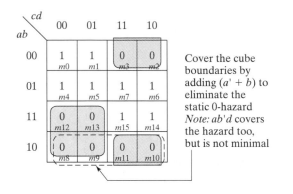

FIGURE 2-37   Cover of the 0s in the K-map of a static 1-hazard in $f = a'c' + bc$.

Example 2.31

The alternative method for eliminating a static 0-hazard from the expression given in Example 2.29 begins with the static 1-hazard-free function: $f = a'c' + bc + a'b$. Now consider the K-map for $f' = (a + c)(b' + c'(a + b'))$ and examine it for coverage of the 0s in the K-map of the original function.

$$f' = ab'a + ab'b' + ac'a + ac'b + cb'a + cb'b' + cc'a + cc'b'$$

$$= ab' + ac' + b'c$$

The K-maps of complement of the static 1-hazard-free function, and the original function are shown in Figure 2-38.

All the adjacent 0s of the K-map for $f = a'c' + bc$ are covered by 0s of the complement of the static 1-hazard-free function, $f' = ab' + ac' + b'c$, and no boundary at which a transition could occur between adjacent 0-cubes is uncovered. Therefore there is no static 0-hazard.

In this example there are no static 0-hazards or static 1-hazards in the expression with the added redundant cube. In general, eliminating the static 1-hazards might not eliminate the static 0-hazards.

End of Example 2.31

2.5.2   Summary: Elimination of Static Hazards in Two-Level Circuits

The methodology for eliminating static 1-hazards in a two-level circuit is (1) to cover with prime implicants all 1s in adjacent cells of adjacent cubes of the K-map of the SOP form of the function and (2) to add redundant prime implicants as needed to complete

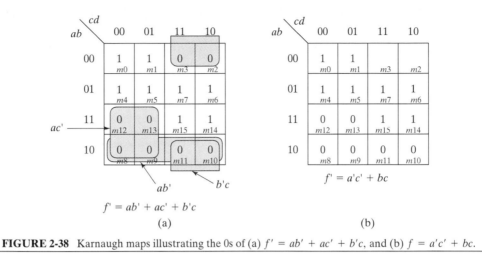

FIGURE 2-38 Karnaugh maps illustrating the 0s of (a) $f' = ab' + ac' + b'c$, and (b) $f = a'c' + bc$.

the cover of the function. To cover a static 0-hazard, we cover all adjacent 0s in the POS form of the static 1-hazard-free function, adding prime implicants in the POS form of the static 1-hazard function as needed to cover any uncovered adjacencies.

2.5.3   Static Hazards in Multilevel Circuits

Like two-level circuits, multilevel circuits are subject to static hazards, but the outputs of multilevel circuits are not written in SOP or POS forms, which have two levels of logic. In a multilevel circuit there may be multiple paths from an input to an output of the circuit, with each path having a different propagation delay. When a circuit has propagation delays, a Boolean variable and its complement might not change value at exactly the same time. For example, a transition in the variable will precede a transition in its complement, a', by the propagation delay of the inverter whose input is a and whose output is a'. Thus, the Boolean cube aa' has a transient interval over which its value is not 0, and the circuit could have a static 0-hazard. Similarly, the factor $(b + b')$ might have a transient during which the value of the factor is 0, rather than 1.

The Boolean expression for an output of a multilevel circuit can always be flattened into a two-level form by multiplying its product factors. To eliminate a static hazard in a multilevel circuit, we begin by flattening the multilevel description of the output expression into an SOP form, f_{tof}, called the "transient output function" [4], taking care not to eliminate either the product or the sum of a literal with its complement. Each input variable and its complement are treated as independent variables in f_{tof}. For example, we do not cancel aa' from an SOP form and do not cancel a factor like $a + a'$ in a POS form. Preserving factors such as aa' and $(a + a')$ exposes the possible transients in which the indicated variables are not the complement of each other. The presence of a product of a variable and its complement reveals a static 0-hazard in that input; a sum of a variable and its complement indicates a static 1-hazard in that input. After forming the transient output function, check for static 1-hazards in the two-level expression (terms such as aa'

can be ignored in this check), and add redundant prime implicants to cover adjacent 1s in the K-map. Then check to see whether the 0s of the static 1-hazard-free function cover the 0s of the original function. Introduce redundant terms as needed to create a hazard-free cover (terms such as aa' reveal the variable that causes a static 0-hazard).

Example 2.32

Consider the possibility of a static 1-hazard in the multilevel function

$$f = bcd + (a + b)(b' + d') = bcd + ab' + ad' + bb' + bd'$$

with the transient output function:

$$f_{tof} = bcd + ab' + ad' + bd'$$

Note that f_{tof} does not include the cube bb'—it implies a static 0-hazard and has no influence on a possible 1-hazard. The K-map of f_{tof} without the bb' term is shown in Figure 2-39.

The transient output function has three static 1-hazards, that is, three cube boundaries across which the transition of a single input might cause a hazard, depending on the propagation delays of the circuit. The three possibilities are listed below, where $(1111) \leftrightarrow (1011)$ denotes a transition of $(abcd)$ between initial and final values.

$$(a, b, c, d) = (1111) \leftrightarrow (1011)$$

$$(a, b, c, d) = (1111) \leftrightarrow (1110)$$

$$(a, b, c, d) = (0111) \leftrightarrow (0110)$$

By adding two additional cubes, bc and ac, to f we cover the hazards and form the static 1-hazard-free expression, f_{1HF}:

$$f_{1HF} = bcd + ab' + ad' + bd' + bc + ac$$

The final, minimal, form is obtained by removing the redundant cube bcd:

$$f_{1HF} = ab' + ad' + bd' + bc + ac$$

Next, we illustrate the removal of a static 0-hazard in a multilevel circuit.

End of Example 2.32

Example 2.33

The multilevel function $f = bcd + (a + b)(b' + d')$ has its complement given by:

$$f' = [bcd + (a + b)(b' + d')]' = [bcd]'[(a + b)(b' + d')]'.$$

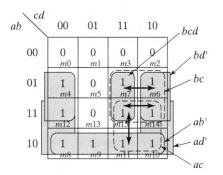

FIGURE 2-39 Karnaugh map of the transient output function $f_{tof} = bcd + ab' + ad' + bd'$.

By applying DeMorgan's law to the above expression we obtain

$$f' = (b' + c' + d')(a'b' + bd)$$
$$= a'b' + a'b'c' + bc'd + a'b'd'$$
$$= a'b' + bc'd$$

and

$$f = (a'b' + bc'd)' = (a + b)(b' + c + d')$$

The cubes of the expression for f' indicate where f will be 0. Now consider the 0s in the K-map of f, as shown in Figure 2-40.

The boundary between the cubes of the (logically and physically) adjacent 0s in the map indicates that a static 0-hazard exists when the inputs make the transitions $(a, b, c, d) = (0101) \leftrightarrow (0001)$. We add to f' a redundant cube, $a'c'd$, to cover the hazard and form the complement of a static 0-hazard-free function, f'_{0HF}. The augmented expression becomes

$$f'_{0HF} = a'b' + bc'd + a'c'd$$

and so f_{0HF} has the POS form

$$f_{0HF} = (a + b)(b' + c + d')(a + c + d')$$

The final expression for f_{0HF} is free of static 0-hazards. Also, using the results obtained in Example 2.32, observe that

$$f_{0HF} = ab' + ad' + bd' + bc + ac = f_{1HF}$$

We conclude that f_{0HF} and f_{1HF} are both free of static 0-hazards and static 1-hazards.

End of Example 2.33

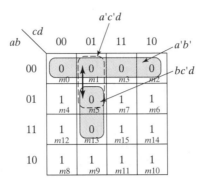

FIGURE 2-40   0s of the K-map of $f = (a'b' + bc'd)'$.

2.5.4   Summary: Elimination of Static Hazards in Multilevel Circuits

Static hazards can be eliminated in multilevel circuits by (1) forming f_{tof}, the transient output function, by collapsing the multilevel logic into an SOP form (while ignoring complement relationships, e.g., aa'), (2) covering every group of adjacent 1s of f_{tof} in the K-map to form f_1, a function free of static 1-hazards, (3) applying DeMorgan's Law to f_1 and simplifying with Boolean relationships (treating each variable and its complement as independent variables), and (4) forming f_0 in SOP form by covering any groups of adjacent 0s. If no term of the resulting expression contains the product of a variable and its complement, the expression will be free of static 1-hazards and static 0-hazards.

2.5.5   Dynamic Hazards

A circuit has a dynamic hazard if an input transition is supposed to cause a single transition in an output, but causes it two or more transitions before it reaches its expected value. Typical waveforms of a dynamic hazard are shown in Figure 2-41. Such hazards are a consequence of multiple static hazards caused by multiple reconvergent paths in a multilevel circuit. They are not easy to eliminate, but if a circuit is free of all static hazards it will be free of dynamic hazards. Consequently, a method for eliminating dynamic hazards is to (1) transform the circuit into a two-level form, and then (2) detect and eliminate all static hazards.

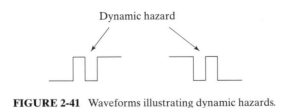

FIGURE 2-41   Waveforms illustrating dynamic hazards.

Example 2.34

The circuit in Figure 2-42 has two nodes where input *C* reconverges. The output *F_static* has a static hazard, and *F_dynamic* (the location of the second reconvergence) has a dynamic hazard. The simulation results in Figure 2-43 display the effect of the hazards.

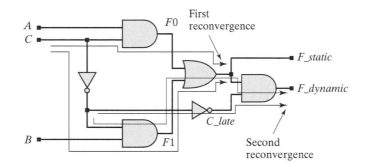

FIGURE 2-42 A circuit having two nodes of reconvergence, a static hazard, and a dynamic hazard.

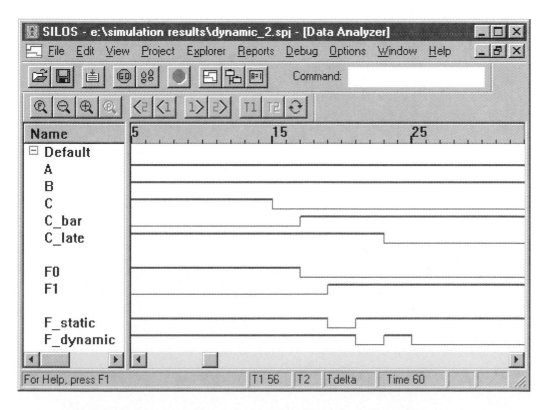

FIGURE 2-43 Simulation results showing the effects of static and dynamic hazards.

The Karnaugh map for *F_static* is shown in Figure 2-44, where it is apparent that the redundant prime implicant, *AB*, covers the boundary between cubes *BC'* and *AC*. The redundant cube eliminates the static 1-hazard and assures that *F_dynamic* will not depend on the arrival of the effect of the transition in *C*. The additional logic for the redundant cube is shown in Figure 2-44. The hazard-free circuit and its simulated waveforms are shown in Figures. 2-45 and 2-46, respectively.

End of Example 2.34

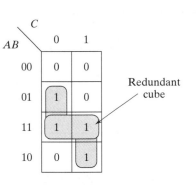

FIGURE 2-44 Karnaugh map of *F_static* in Figure 2-42.

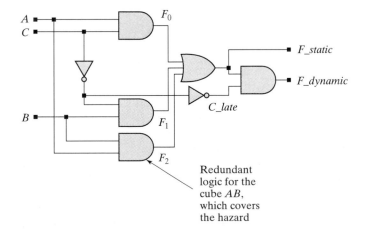

FIGURE 2-45 Hazard-free equivalent of the circuit in Figure 2-42. Redundant logic forming F_2 has been added to the original circuit.

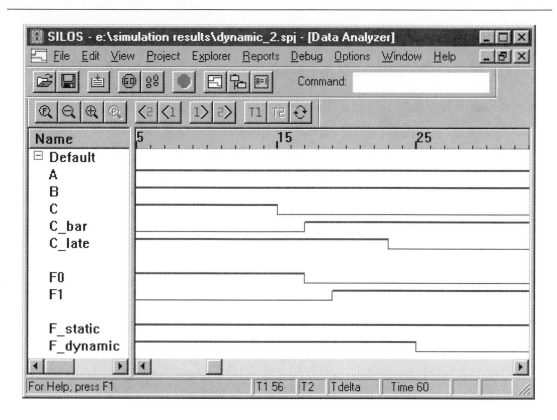

FIGURE 2-46 Simulation results for the hazard-free circuit in Figure 2-45.

2.6 Building Blocks for Logic Design

Combinational logic encompasses a wide range of functionality and circuit structures, but certain structures and circuits are commonly used in many applications, and it is worthwhile to gain familiarity with them.

2.6.1 NAND–NOR Structures

In complementary metal-oxide semiconductor (CMOS) technology, AND gates and OR gates are not implemented as efficiently as NAND gates and NOR gates. An SOP form or a POS form can always be converted to a NAND logic structure or a NOR logic structure. The NAND gate and NOR gate are universal logic gates—any Boolean function can be realized from only NAND gates or only NOR gates. DeMorgan's laws provide equivalent structures for NAND and NOR gates, shown in Figure 2-47.

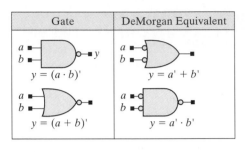

FIGURE 2-47   Equivalent circuits that result from DeMorgan's laws.

Networks realizing SOP expressions[3] can be transformed by DeMorgan's laws into a circuit that uses only NAND gates and inverters by (1) replacing AND gates by NAND gates in the original AND–OR structure, (2) placing inversion bubbles at the inputs of the OR gates, (3) inserting inverters where needed to match bubbles at the inputs of the OR gates, and (4) substituting a NAND gate for a NOR gate that has inversion bubbles at its inputs. A circuit in POS form[4] can be transformed into an equivalent circuit that uses only NOR gates and inverters by (1) replacing OR gates by NOR gates, (2) placing inversion bubbles at the inputs of the AND gates, (3) inserting inverters where needed to match bubbles at the inputs of the AND gates, and (4) substituting a NOR gate for a NAND gate that has inversion bubbles at its inputs.

Example 2.35

The function $Y = G + EF + AB'D + CD$ has the two-level circuit realization shown in Figure 2-48(a), which can be transformed into the circuit in Figure 2-48(b) by placing bubbles at the inputs to the OR gate that forms Y and an inverter at input G to match the bubble at the input to the OR gate driven by G. Then DeMorgan's law is applied to replace the OR gate with input bubbles by the NAND gate shown in Figure 2-48(c).

To verify that that circuit of (c) is equivalent to that of (a), we note that

$$Y = [(G')(EF)'(AB'D)'(CD)']'$$
$$= (G')' + [(EF)']' + [(AB'D)']' + [(CD)']'$$
$$= G + EF + AB'D + CD$$

End of Example 2.35

<hr style="width:20%">

[3]The output of the network must be the output of an OR gate.
[4]The output of the network must be the output of an AND gate.

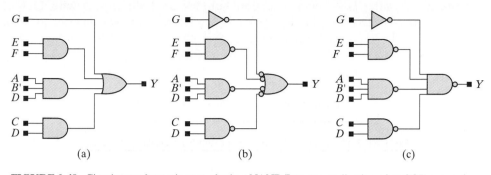

FIGURE 2-48 Circuit transformations to obtain a NAND/Inverter realization of an SOP expression.

Example 2.36

Now consider the POS expression $Y = D(B + C)(A + E + F')(A + G)$. The circuit in Figure 2-49(b) is formed by replacing OR gates with NOR gates, adding inversion bubbles to the input of the AND gate, and adding an inverter to D to match its input bubble. Then the NAND gate with inversion bubbles at its inputs is replaced by an equivalent NOR gate to form the circuit in Figure 2-49(c).

A check reveals that the transformed circuit is equivalent to the original circuit:

$$Y = [D' + (B + C)' + (A + E + F')' + (A + G)']'$$
$$= (D')'[(B + C)']'[(A + E + F')']'[(A + G)']'$$
$$= D(B + C)(A + E + F')(A + G)$$

A circuit whose structure does not consist of alternating AND gates and OR gates can still be transformed into an equivalent structure that uses only NAND gates and inverters or a structure that uses only NOR gates and inverters. To transform such a circuit into a NAND structure: (1) replace all AND gates by NAND gates (Figure 2-50a), (2) place inversion bubbles at the inputs of all OR gates (Figure 2-50b), and (3)

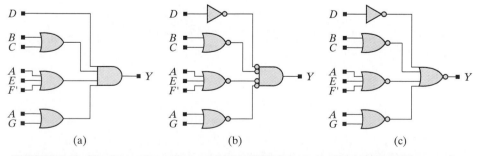

FIGURE 2-49 Circuit transformations to obtain a NOR/Inverter realization of a POS expression.

replace OR gates that have inversion bubbles on their inputs with DeMorgan-equivalent NAND gates (Figure 2-50c). If, after these changes have been made, the output of a NAND gate drives the input of another NAND gate, place an inverter at the input of the driven NAND gate (Figure 2-50d); if the output of an OR gate having bubbles at its inputs drives another OR gate that has inversion bubbles at its inputs, place an inverter on the path connecting them (see Figure 2-50e). Then replace OR gates with inversion bubbles at their inputs by equivalent NAND gates. These steps ensure that inversions caused by the replacement of gates will be matched, and that the final circuit will be equivalent to the original circuit.

Alternatively, to transform the circuit into a NOR structure: (1) replace OR gates by NOR gates (Figure 2-51a), (2) place inversion bubbles at the inputs of any AND gates (Figure 2-51b), and (3) replace AND gates with bubble inputs by DeMorgan-equivalent

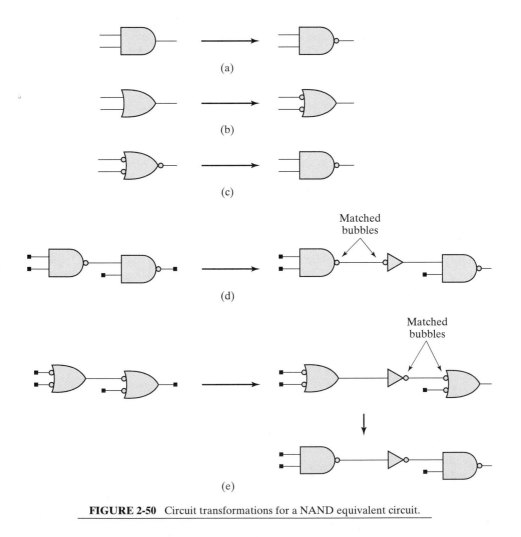

FIGURE 2-50   Circuit transformations for a NAND equivalent circuit.

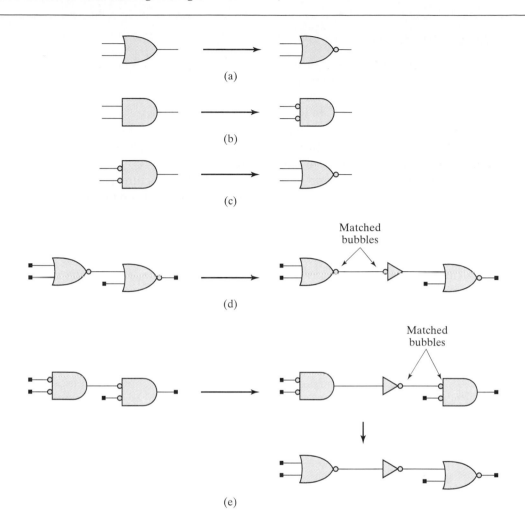

FIGURE 2-51 Circuit transformations for a NOR equivalent circuit.

NOR gates (Figure 2-51c). If, after these changes have been made, the output of a NOR gate drives the input of another NOR gate, place an inverter at the input of the driven NOR gate (Figure 2-51d); if the output of an AND gate with bubbles at its inputs drives another AND gate that has inversion bubbles at its inputs, place an inverter on the path connecting them (see Figure 2-51e). Then replace the AND gate that has inversion bubbles at its inputs with an equivalent NOR gate.

These rules ensure that inversion bubbles will be matched and guarantee that the transformed (NAND or NOR) circuit is equivalent to the original circuit.

End of Example 2.36

2.6.2   Multiplexers

Multiplexer circuits are used to steer data through functional units of computers and other digital systems. For example, a multiplexer can be used to steer the contents of a particular storage register to the inputs of an arithmetic and logic unit (ALU), and to steer the output of the ALU to the same or a different register. A gate-level schematic of a two-channel multiplexer is shown in Figure 2-52. When *sel* = 0 the data at input *a* passes through the circuit (with some propagation delay) to *y_out*; likewise, if *sel* = 1 the data at input *b* goes to *y_out*. The Boolean expression describing the function of the circuit is given by: $y\ out = sel' \cdot a + sel \cdot b$.

In general, a multiplexer has *n* datapath input channels and a single output channel. An *m*-bit address determines which input channel is connected to the output channel. The input channel selected by the multiplexer shown symbolically in Figure 2-53 is governed by: $Data\_Out = Data\_In\,[Address[k]]$, where *k* is an index into the address space.

Multiplexers can also be used to implement combinational logic. The values of a Boolean function can be assigned to the input lines and decoded by the select lines. This implementation might be inefficient because the mux must fully decode the truth table of all of the input bits.

Example 2.37

The truth table in Figure 2-54 describes a 4-bit majority function, which asserts its output if a majority of its inputs are asserted. The schematic shows how to implement the function with a 16-input mux, using its four select lines to decode the possible bit patterns of the inputs to the function

---

End of Example 2.37

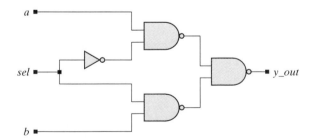

FIGURE 2-52   Gate-level schematic for a two-channel multiplexer circuit.

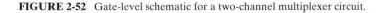

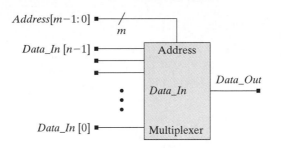

FIGURE 2-53 Schematic symbol for an *n*-channel multiplexer with an *m*-bit channel selector address.

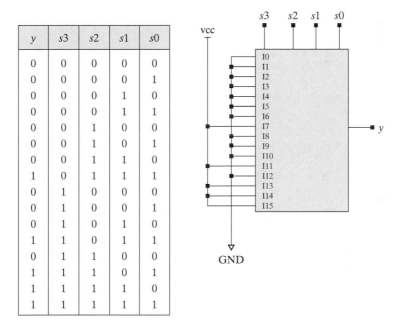

y	s3	s2	s1	s0
0	0	0	0	0
0	0	0	0	1
0	0	0	1	0
0	0	0	1	1
0	0	1	0	0
0	0	1	0	1
0	0	1	1	0
1	0	1	1	1
0	1	0	0	0
0	1	0	0	1
0	1	0	1	0
1	1	0	1	1
0	1	1	0	0
1	1	1	0	1
1	1	1	1	0
1	1	1	1	1

FIGURE 2-54 Truth table and circuit for a 16-input mux implementation of a 4-bit majority function.

2.6.3 Demultiplexers

A demultiplexer circuit implements the reverse functionality of a multiplexer. It has a single input datapath, *n* output datapaths, and an input *m*-bit address that determines which of the *n* outputs is connected to the input. The output channel selected by the demultiplexer shown in Figure 2-55 is determined by $Data\_Out\,[n - 1:0] = Data\_In$ [Address[k]], where k is an index into the address space.

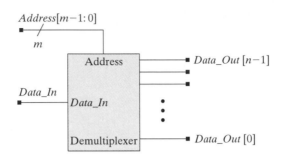

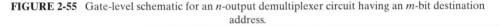

FIGURE 2-55 Gate-level schematic for an n-output demultiplexer circuit having an m-bit destination address.

2.6.4 Encoders

Multiplexer and demultiplexer circuits dynamically establish connectivity between datapaths in a system. A data pattern that passes through a multiplexer or a demultiplexer is not altered by the circuit. On the other hand, an encoder circuit acts to transform an input data word into a different output data word. Input data words are typically wide in comparison to the encoded output word, so encoders serve to reduce the size of a datapath in a system. An encoder assigns a unique bit pattern to each input line. Usually, a device whose output code is smaller than its input code is referred to as an encoder. If the size of the output word is larger than the size of the input word, the circuit is referred to as a decoder. One typical application of an encoder is in a client–server polling circuit whose output code indicates which of n clients requesting service from a server is to be granted service.

An encoder has n inputs and m outputs, with $n = 2^m$. An encoder could transform up to 2^m different input words into unique output codes, treating the remaining input patterns as don't-care conditions, but ordinarily only one of the inputs is asserted at a time, and a unique output bit pattern (code) is assigned to each of the n inputs. The asserted output is determined by the index of the asserted bit of the n-bit binary input word. Block diagram symbols for encoders are shown in Figure 2-56.

Example 2.38

A 5:3 encoder having a 5-bit input word is to generate a 3-bit output code indicating the number of bits that are asserted in the input word. The input words and encoded output bit patterns are shown in Figure 2-57. Boolean logic equations can be derived for each bit of the output word.

End of Example 2.38

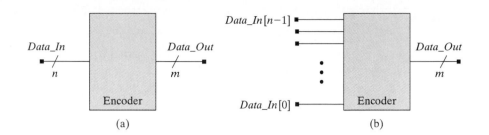

FIGURE 2-56 Schematic symbols for an encoder: (a) an encoder with an n-bit input bus, and (b) an encoder with individual bit-line inputs.

Input	Output	Input	Output
00000	000	10000	001
00001	001	10001	010
00010	001	10010	010
00011	010	10011	011
00100	001	10100	010
00101	010	10101	011
00110	010	10110	011
00111	011	10111	100
01000	100	11000	010
01001	010	11001	011
01010	010	11010	011
01011	011	11011	100
01100	010	11100	011
01101	011	11101	100
01110	011	11110	100
01111	100	11111	101

FIGURE 2-57 Input–output words for a 5:3 encoder that indicates the number of asserted bits in the input word.

2.6.5   Priority Encoder

A priority encoder allows multiple input bits to be asserted simultaneously and uses a priority rule to form an output bit pattern. A priority encoder in a client–server system would identify the client that has the highest priority among multiple clients requesting service.

Example 2.39

The input–output patterns for an eight-client priority encoder are shown in Figure 2-58. (*Note*: x denotes a don't-care condition.) The client corresponding to the leftmost 1-bit of the input word has highest priority. This is a combinational scheme; a sequential machine could impose some rule providing all clients with some level of service.

Input Word	Output Word
1 x x x x x x x	0 0 0
0 1 x x x x x x	0 0 1
0 0 1 x x x x x	0 1 0
0 0 0 1 x x x x	0 1 1
0 0 0 0 1 x x x	1 0 0
0 0 0 0 0 1 x x	1 0 1
0 0 0 0 0 0 1 x	1 1 0
0 0 0 0 0 0 0 1	1 1 1

FIGURE 2-58 Input–output words for an 8:3 priority encoder.

End of Example 2.39

2.6.6   Decoder

A binary decoder interprets an input pattern of bits and forms a unique output word in which only 1 bit is asserted. Decoders are commonly used to extract the opcode from an instruction in a digital computer; row and column address decoders are used to locate a word in memory from its address. Figure 2-59 shows block diagram symbols for

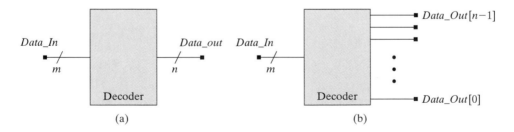

FIGURE 2-59 Block diagram symbols for decoders: (a) decoder with input/output busses, and (b) decoder with an expanded output bus.

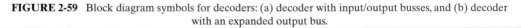

a decoder. A binary decoder has m inputs and n outputs, with $n = 2^m$. There are many different possible mappings between input words and output words. An encoder can be built from combinational logic that forms the input–output mapping. (Sequential encoders and decoders are widely used in communication and video transmission circuits.)

Example 2.40

The input–output patterns for an eight-client decoder are shown in Figure 2-60. The arrangement of bits in the output words identifies the client that will be served. This decoder does not resolve contention between multiple clients, and it assumes that only one client at a time will request service.

A binary decoder generates all of the minterms of its inputs. All of the output lines are available to form as many functions of the same inputs as are needed by an application. Binary decoders can be used for small implementations with multiple outputs, but the number of outputs precludes their use in applications with a large number of inputs.

Input Word	Output Word
0 0 0	1 0 0 0 0 0 0 0
0 0 1	0 1 0 0 0 0 0 0
0 1 0	0 0 1 0 0 0 0 0
0 1 1	0 0 0 1 0 0 0 0
1 0 0	0 0 0 0 1 0 0 0
1 0 1	0 0 0 0 0 1 0 0
1 1 0	0 0 0 0 0 0 1 0
1 1 1	0 0 0 0 0 0 0 1

FIGURE 2-60 The input–output patterns for an eight-client decoder.

End of Example 2.40

Example 2.41

A decoder can be used to implement multiple Boolean functions of the same inputs. The truth table in Figure 2-61 describes $f1$, a majority function, along with some other function, $f2$. Additional logic is used with the decoder to combine its outputs to form the two functions.

End of Example 2.41

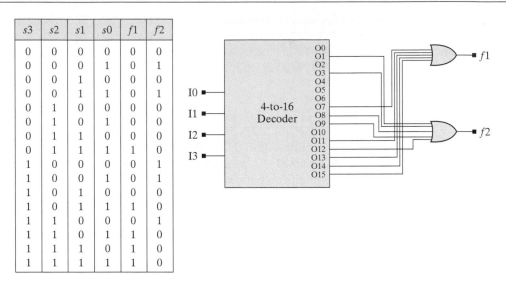

s3	s2	s1	s0	f1	f2
0	0	0	0	0	0
0	0	0	1	0	1
0	0	1	0	0	0
0	0	1	1	0	1
0	1	0	0	0	0
0	1	0	1	0	0
0	1	1	0	0	0
0	1	1	1	1	0
1	0	0	0	0	1
1	0	0	1	0	1
1	0	1	0	0	0
1	0	1	1	1	0
1	1	0	0	0	1
1	1	0	1	1	0
1	1	1	0	1	0
1	1	1	1	1	0

FIGURE 2-61   A truth table for two functions implemented by a single 4-to-16 decoder.

2.6.7   Priority Decoder

A priority decoder can be used in applications in which multiple input codes might imply contention.

Example 2.42

The input code for a client–server system that is to serve eight clients will contain a 1 in any bit position that corresponds to a request from a client for service. The server must determine which of multiple clients is to be served. One simple rule assigns a unique priority to each client. The input–output codes in Figure 2-62 assign the highest priority to the client associated with the leftmost bit of the input code. The table accounts for all possible input patterns. A sequential decoder circuit could base service on other considerations, such as whether a client has been blocked from service by higher-priority clients for too long.

End of Example 2.42

Input Word	Output Word
1 x x x x x x x	1 0 0 0 0 0 0 0
0 1 x x x x x x	0 1 0 0 0 0 0 0
0 0 1 x x x x x	0 0 1 0 0 0 0 0
0 0 0 1 x x x x	0 0 0 1 0 0 0 0
0 0 0 0 1 x x x	0 0 0 0 1 0 0 0
0 0 0 0 0 1 x x	0 0 0 0 0 1 0 0
0 0 0 0 0 0 1 x	0 0 0 0 0 0 1 0
0 0 0 0 0 0 0 1	0 0 0 0 0 0 0 1

FIGURE 2-62 The input–output patterns for an eight-client priority decoder.

REFERENCES

1. Breuer MA, Friedman AD. *Diagnosis & Design of Reliable Digital Systems*. Rockville, MD: Computer Science Press, 1976.
2. Fabricius ED. *Introduction to VLSI Design*. New York: McGraw-Hill, 1990.
3. Bryant RE. "Graph-Based Algorithms for Boolean Function Manipulation," *IEEE Transactions on Computers, C-35*, 677–691, 1986.
4. Tinder RF. *Engineering Digital Design*, 2nd ed. San Diego: Academic Press, 2000.
5. Katz RH. *Contemporary Logic Design*. Redwood City, CA: Benjamin Cummings, 1994.
6. McCluskey EJ. "Minimization of Boolean Functions," *Bell Systems Technical Journal, 35*, 1417–1444, 1956.
7. McCluskey EJ. *Introduction to the Theory of Switching Circuits*. New York: McGraw-Hill, 1965.
8. McCluskey EJ. *Logic Design Principles*, Upper Saddle River, NJ: Prentice-Hall, 1986.
9. Wakerly JF. *Digital Design Principles and Practices*, 3rd ed. Upper Saddle River, NJ: Prentice-Hall, 2000.
10. Brayton RK. et al. *Logic Minimization Algorithms for VLSI Synthesis*. Boston: Kluwer, 1984.

PROBLEMS

1. Find the canonical SOP form of the following Boolean function: $F(a, b, c) = \Sigma\,(1, 3, 5, 7)$.
2. Find the canonical POS form of the following Boolean function: $F(a, b, c, d) = \Pi(0, 1, 2, 3, 4, 5, 12)$.
3. Express the function $F = a'b + c$ as a sum of minterms.
4. Express the function $F = a'bcd' + a'bcd + a'b'c'd' + a'b'c'd$ as a sum of minterms.
5. Express the function $G = (a'bcd' + a'bcd + a'b'c'd' + a'b'c'd)'$ as a sum of minterms.

6. Find a NAND circuit realization of the function $f = ac' + bcd + a'd$.
7. Find a NOR circuit realization of the function $f = (b + c + d)(a' + b + c)(a' + d)$.
8. Find the complement of the following expressions:
   a. $ab' + a'b$
   b. $b + (cd' + e)a'$
   c. $(a' + b + c)(b' + c')(a + c)$

9. Simplify the following Boolean functions to a minimum number of literals:
   a. $F = a + a'b$
   b. $F = a(a' + b)$
   c. $F = ac + bc' + ab$

10. Using Karnaugh maps, simplify the following Boolean functions:
    a. $F(a, b, c) = \Sigma (0, 2, 4, 5, 6)$
    b. $F(a, b, c) = \Sigma (2, 3, 4, 5)$
    c. $F = bc' + ac' + a'bc + ab$
    d. $F = \Sigma (0, 1, 2, 4, 5, 6, 8, 9, 12, 13, 14)$
    e. $F = a'b'c' + b'cd' + a'bcd' + ab'c'$

11. Find a NAND gate realization of the following Boolean function: $F(a, b, c) = \Sigma (0, 6)$

12. Using the K-map diagram below:

    a. Draw a K-map for $f = \Sigma m(0, 4, 6, 8, 9, 11, 12, 14, 15)$

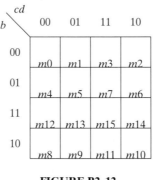

FIGURE P2–12

b. Identify the prime implicants of f.
c. Identify the essential prime implicants of f.
d. Find all minimal expressions of f and identify those that use only essential prime implicants.

13. Design a two-level circuit that implements a 4-bit majority function, that is, the output is a 1 if three or more inputs are asserted.
14. Example 2.37 showed how to implement a 4-bit majority function with a 16-input mux. Show that it is also possible to implement this function with a 8-input mux.

CHAPTER 3

Fundamentals of Sequential Logic Design

Computers and other digital systems that have memory or that execute a sequence of operations under the direction of stored information are referred to as "sequential machines," and their circuitry is modeled by sequential logic. Sequential machines do not behave like combinational logic because the outputs of a sequential machine depend on the history of the applied inputs as well as on their present value.

The history of the inputs applied to a sequential machine is represented by the state of the machine and requires hardware elements that store information; that is, it requires memory to store the state of the machine as an encoded binary word. For example, a machine whose output is the running count of the number of 1s encountered by a receiver of a serial bit stream must have storage elements to hold the value of the count. Today's electronic systems rely on transistor circuits to store information. Transistors are small, are easy to fabricate, operate reliably, and they have two states, on or off, which can be used to develop voltages representing logical 0 and logical 1.

Sequential machines can be deterministic or probabilistic, and synchronous or asynchronous. We will consider only synchronous, deterministic machines. A common clock acts as a synchronizing signal for the operations of a synchronous sequential machine. This establishes fixed, predictable, intervals for propagating signals through the circuit, leading to a more reliable design and a simpler design methodology. Today's synthesis tools support only synchronous circuits.

3.1 Storage Elements

Storage elements store information in a binary format—that is, as a pattern of 0s and 1s. For example, the opcode for addition in a simple microprocessor might be the pattern 0010. The circuits that store information can be level-sensitive, edge-sensitive, or a combination of both. Level-sensitive storage elements are commonly referred to as latches, and edge-sensitive storage elements are referred to as flip-flops. The outputs of a level-sensitive sequential circuit are immediately affected by a change in the value of one or more inputs, as long as an enabling signal is asserted. The outputs of an edge-sensitive circuit are sensitive to the values of the inputs, but may change value only when a synchronizing signal makes either a rising or falling edge transition. Storage elements may be clocked or unclocked; that is, they may operate in either a synchronous or an asynchronous mode.

3.1.1 Latches

The circuits in Figure 3-1 implement basic S-R (set–reset) latches. Their feedback structure of cross-coupled (a) NOR gates or (b) NAND gates enables the output (Q) of the circuits to have two stable states, 0 and 1, depending on the value of the set (S) and reset (R) inputs. Once the input conditions establish an output value it will remain until it is changed by new input conditions. The truth tables shown with the circuits describe the new (next) state that results from a given state when an input pattern is applied while the latch is in a known state (only one input is allowed to change at a time). In practice, we avoid applying 11 to a NOR latch because the outputs of the latch will not be logical complements of each other and because (in a physical circuit) a race

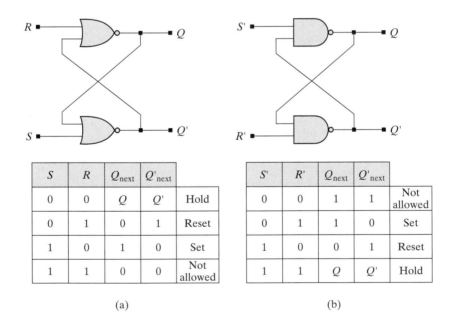

S	R	Q_{next}	Q'_{next}	
0	0	Q	Q'	Hold
0	1	0	1	Reset
1	0	1	0	Set
1	1	0	0	Not allowed

S'	R'	Q_{next}	Q'_{next}	
0	0	1	1	Not allowed
0	1	1	0	Set
1	0	0	1	Reset
1	1	Q	Q'	Hold

(a)   (b)

FIGURE 3-1 Feedback circuit structures implementing latches: (a) cross-coupled NOR gates, and (b) cross-coupled NAND gates.

condition occurs if the inputs are changed from 11 to 00. This makes the output unpredictable.[1] Likewise, we avoid applying 00 to a NAND latch because the outputs of the latch will not be logical complements of each other and because (in a physical circuit) a race condition occurs if the inputs are changed from 00 to 11.

3.1.2 Transparent Latches

Latches are level-sensitive storage elements; the action of data storage is dependent on the level (value) of the input clock (or enable) signal. The output of a transparent latch changes in response to the data input only while the latch is enabled; that is, changes at the input are visible at the output. A transparent latch is also called a D-latch, or a data latch.

A transparent latch results from a minor change to a basic unclocked S-R latch. The latch circuit in Figure 3-2 has additional NAND gates and uses a clock signal to gate the inputs; that is, *Enable* determines whether S' and R' will have an effect on the circuit. When *Enable* is de-asserted the circuit is not affected by the values of S' and R'. An S-R latch with gated inputs is also called a "clocked latch" and a "gated latch." The modified circuit in Figure 3-3(a) retains the signal *Enable* but passes complementary values of *Data* to the S' and R' inputs of the latch. This ensures that an unstable condition will not occur (00 will not be applied to the S-R stage) and that the value of *Q_out* will follow the value of *Data* while *Enable* is asserted. When *Enable* is de-asserted the value of *Q_out* becomes fixed by the feedback loop at its current value and is said to be latched. It remains latched until *Enable* is asserted again. The waveforms in Figure 3-3(b) illustrate the latching behavior of the circuit.

3.2 Flip-Flops

Flip-flops are edge-sensitive storage elements; the action of data storage is synchronized to either the rising or falling edge of a signal, which is commonly referred to as a clock signal. The value of data that is stored depends on the data that is present at the data input(s) when the clock makes a transition at its active (rising or falling) edge; at all other times the value and transitions of the data are ignored. There are various kinds of flip-flops,[2] depending on the action of additional input signals that control the storage of data, such as a reset signal [1–4].

3.2.1 D-Type Flip-Flop

A D-type flip-flop is the simplest type; at each active edge of the clock it stores the value that is present at its D input, independently of the present stored value. A block diagram symbol and truth table for a D-type flip-flop are shown in Figures 3-4(a) and Figures 3-4(b), respectively. The truth table includes an entry for the present state of the flip-flop (Q) and the state (Q_{next}) that will result at the next active edge of the clock signal (*clk*) for a given value of the data input (D). The waveforms shown in

---

[1]The race condition also leads to an indeterminate result in simulation.
[2]Only fundamental-mode flip-flops will be considered here—those in which only one input may change at a time.

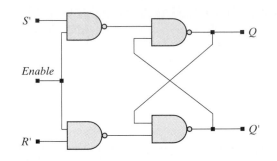

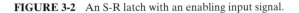

FIGURE 3-2   An S-R latch with an enabling input signal.

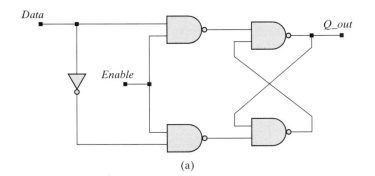

(a)

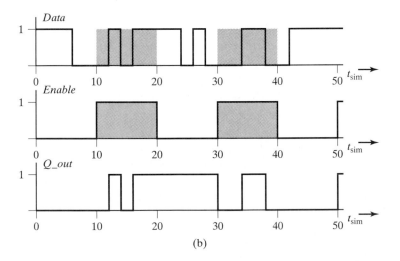

(b)

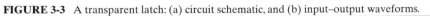

FIGURE 3-3   A transparent latch: (a) circuit schematic, and (b) input–output waveforms.

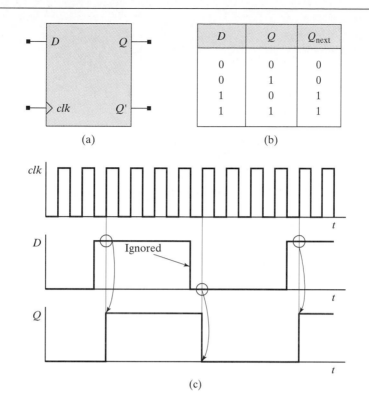

	(a)			(b)	

D	Q	Q_{next}
0	0	0
0	1	0
1	0	1
1	1	1

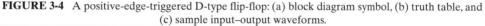

FIGURE 3-4 A positive-edge-triggered D-type flip-flop: (a) block diagram symbol, (b) truth table, and (c) sample input–output waveforms.

Figure 3-4(c) illustrate how data present at D is stored, for this example, on the rising edge of *clk*, and how transitions in D are ignored over the interval between the active edges of *clk*. However, D must be stable for a sufficiently long time prior to the active edge of *clk*; otherwise, the device may not operate properly. The Boolean logic describing a D-type flip-flop obeys the following so-called characteristic equation [2], $Q_{\text{next}} = D$. A D-type flip-flop may also have other (level-sensitive) inputs, such as *set* and *reset* signals, to override the synchronous behavior and initialize the output.

3.2.2 Master–Slave Flip-Flop

A D-type flip-flop can be implemented by a master–slave configuration of two data latches, as shown in Figure 3-5. The transparent latch of the master stage samples the input during the half-cycle beginning at the inactive edge of the clock; the sampled value will be propagated to the output of the latch of the slave stage at the next active edge, during the so-called slave cycle of the circuit. The output of the master stage must settle before the enabling edge of the slave stage. The master stage is enabled on the inactive edge of the clock, and the slave stage is enabled on the active edge. Setup and

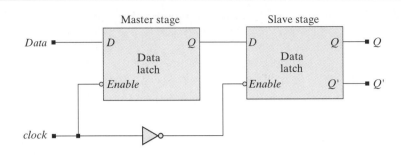

FIGURE 3-5 A master–slave implementation of a negative-edge-triggered D-type flip-flop.

hold conditions apply at the active edge of the clock (see Chapter 10). These specify conditions for stability of the data relative to the clock in order to ensure proper operation of the device.

In complementary metal-oxide semiconductor (CMOS) technology [5, 6] a D-type flip-flop is commonly implemented with transmission gates. D-type flip-flops are popular because they have fewer input signal paths and circuits using them are simpler to design. A transmission gate is formed by a parallel connection of an *n*-channel transistor and *p*-channel transistor, shown in Figure 3-6 with a circuit symbol for a transmission gate. Transmission gates have symmetric noise margins in either direction of transmission, and support bidirectional signal transmission.

The transmission gates and "glue logic" shown in Figure 3-7 form a master–slave circuit that has the functionality of a positive-edge-triggered D-type flip-flop with an additional signal, *Clear_bar*, that forces output *Q* to be de-asserted when *Clear_bar* is 0. The master stage is active while *clock* is low, and the slave stage is active while *clock* is high. While *clock* is low the master stage charges to a value determined by *Data*; when *clock* goes high, the output of the master stage is passed through to the slave stage. The waveforms in Figure 3-7(b) show that *Q* gets the value of *Data* at the rising edges of *clock*.

Figure 3-8 shows the signal paths (a) during the master cycle and (b) during the slave cycle. The output node of the master stage, *w2*, is charged by the input during the master cycle (with *clock* low) and sustained by the feedback loop during the slave

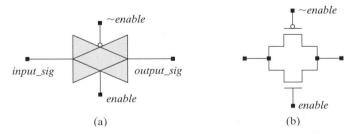

FIGURE 3-6 CMOS transmission gate: (a) circuit symbol, and (b) transistor-level schematic.

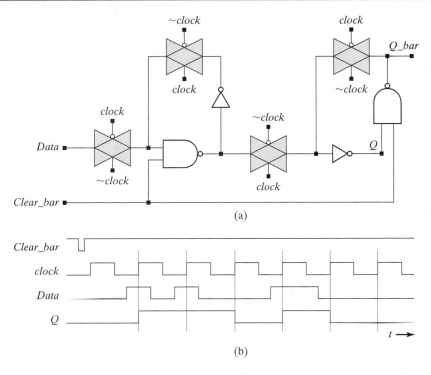

FIGURE 3-7 CMOS master–slave circuit of a D-type flip-flop: (a) circuit schematic, and (b) sample waveforms.

cycle, that is, while *clock* is high. The output of the slave stage is sustained by its feedback loop while the master stage is charging. At the active edge of the flip-flop, the output of the master stage is sustained by its feedback loop, and it charges the output of the slave stage during the slave cycle (clock is high).

3.2.3 J-K Flip-Flops

J-K Flip-flops are edge-sensitive storage elements; data storage is synchronized to an edge of a clock. The value of the data stored is conditional, depending on the data that is present at the J and K inputs when the clock makes a transition at its active edge. The characteristic equation describing the next state of the flip-flop is: $Q_{next} = JQ' + K'Q$. A J-K flip-flop can be implemented by a D-type flip-flop combined with input logic that forms the data input as $D = JQ' + K'Q$. A block diagram symbol, truth table, and sample waveforms of a J-K flip-flop are shown in Figure 3-9.

3.2.4 T Flip-Flop

T flip-flops are edge-sensitive storage elements. If the T (toggle) input is asserted the output will be complemented at the active edge of the clock. Otherwise, the output remains unchanged. A T flip-flop can be efficient in implementing a counter. The characteristic equation of a T flip-flop is given by: $Q_{next} = QT' + Q'T = Q \oplus T$. This type

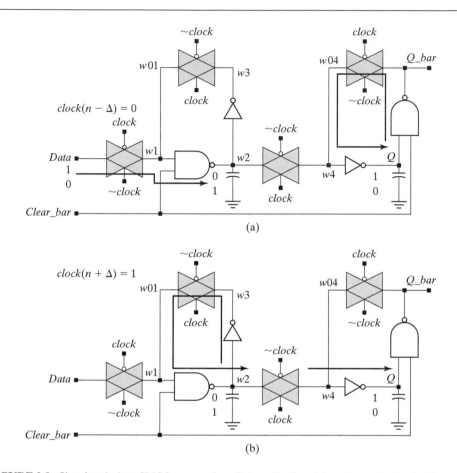

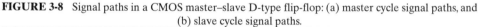

FIGURE 3-8 Signal paths in a CMOS master–slave D-type flip-flop: (a) master cycle signal paths, and
(b) slave cycle signal paths.

of flip-flop can be implemented by connecting the *T*-input to the *J* and *K* inputs of a J-K flip-flop. Figure 3-10 shows the schematic symbol, truth table, and sample waveforms for a T flip-flop. Note that the frequency of toggles in *Q* are one-half those of *clk*.

3.3   Busses and Three-State Devices

Busses are multiwire signal paths that connect multiple functional units in a system. They are the highways for information flow. For example, the motherboard of a personal computer has an address bus that carries the source and destinations of data to be retrieved from or stored in memory, and a data bus that carries data being exchanged between functional units, registers, and memory. Other busses might provide dedicated high-speed datapaths that serve a graphics engine or a special arithmetic processor. By sharing the physical resource of a bus, the overall physical resources and

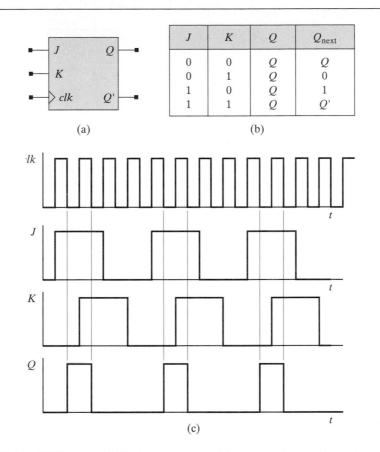

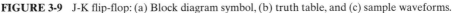

FIGURE 3-9 J-K flip-flop: (a) Block diagram symbol, (b) truth table, and (c) sample waveforms.

board space supporting the architecture of a system can be reduced, compared to a circuit with dedicated signal paths. The tradeoff is that access to the bus must be managed to avoid conflicts. Bus management makes use of hardware and software.

At the hardware level, three-state devices provide a dynamic interface between a bus and a circuit, acting as signal paths when enabled, but are otherwise an open circuit. Multiple drivers can be connected to a common bus, each with its own set of three-state buffers or inverters that interface to the bus. The output of a three-state device is a function of its data input while the controlling input is asserted. Otherwise, the output is said to be in the high-impedance state, or it is disconnected from the circuit. Figure 3-11 shows logic symbols and truth tables for various three-state circuit elements that can buffer or invert an input signal ("Hi-Z" represents a high-impedance state.)

Three-state devices are commonly used to isolate subcircuits from a bus, as shown in Figure 3-12. When *send_data* is high the content of the register is placed on the external bus, *data_to_or_from_bus*; when *rcv_data* is high, data from the external bus is passed into the circuit via *inbound_data*.

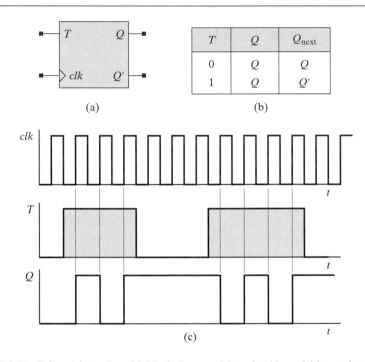

(a)

T	Q	Q_{next}
0	Q	Q
1	Q	Q'

(b)

(c)

FIGURE 3-10 T (toggle) flip-flop: (a) Block diagram, (b) truth table, and (c) sample waveforms.

Busses may operate in a synchronous or an asynchronous manner. At the software level, handshaking protocols are used to establish and support coherent transmission of data. Busses include an arbitration scheme that resolves contention issues between multiple requesters for bus service.

Example 3.1

The registers in Figure 3-13 are connected by a 4-bit bidirectional data bus. Each register can send data to any other register. The signal waveforms shown in Figure 3-14

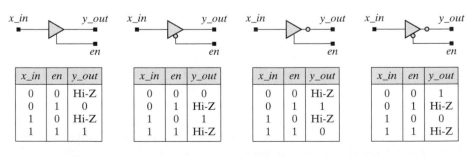

x_in	en	y_out
0	0	Hi-Z
0	1	0
1	0	Hi-Z
1	1	1

x_in	en	y_out
0	0	0
0	1	Hi-Z
1	0	1
1	1	Hi-Z

x_in	en	y_out
0	0	Hi-Z
0	1	1
1	0	Hi-Z
1	1	0

x_in	en	y_out
0	0	1
0	1	Hi-Z
1	0	0
1	1	Hi-Z

FIGURE 3-11 Circuit symbols and truth tables for three-state devices.

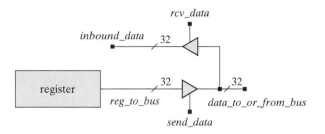

FIGURE 3-12 Bus isolation with three-state devices.

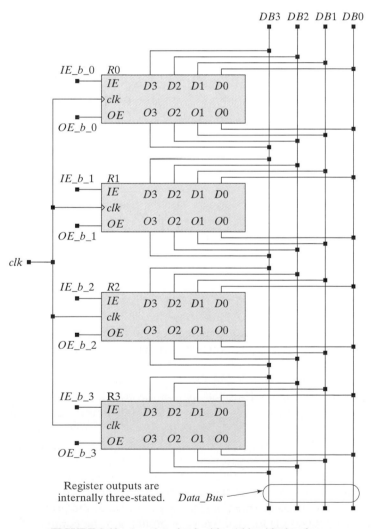

Register outputs are
internally three-stated.   *Data_Bus*

FIGURE 3-13 A register bank with a 4-bit-wide data bus.

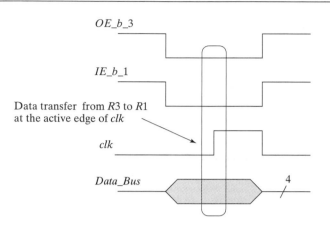

OE_b_3

IE_b_1

Data transfer from *R3* to *R1*
at the active edge of *clk*

clk

Data_Bus

4

FIGURE 3-14   Bus isolation and data transfer with three-state devices.

would establish datapaths connecting the output of register *R3* to the inputs of register *R1* via active-low three-state buffers that are built into the register circuit. To connect the output of *R3* to the input of *R1*, both *OE_3_b* and *IE_b_1* must be low. The other registers are not affected by the bus activity.

End of Example 3.1

3.4   Design of Sequential Machines

Unlike combinational logic, whose output is an immediate function of only its present inputs, sequential logic depends on the history of its inputs. This dependency is expressed by the concept of "state." The future behavior of a sequential machine is completely characterized by its input and its present state. At any time, the state of a system is the minimal information that, together with the inputs to the system, is sufficient to determine the future behavior of the system. For example, knowing the number of 1s that will appear at the input of a machine that counts 1s in a serial bit stream is not enough information to determine the count at any time in the future. The present count must also be known. Thus, the state of the counter is its present count.

Sequential machines are widely used in applications that require prescribed sequential activity. For example, the outputs of a sequential machine control the synchronous datapath and register operations of a computer. All sequential machines have the general feedback structure shown in Figure 3-15, in which the next state of the machine is formed from the present state and the present input. Combinational logic forms the next state (NS) from the primary inputs and the stored value of the present state (PS). A state register (memory) holds the value of the present state (PS), and the value of the next state is formed from the inputs and the content of the state register. In this structure the state transitions are asynchronous.

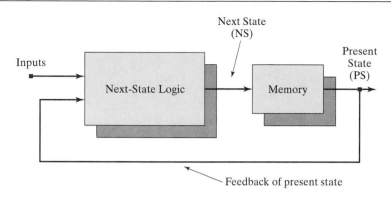

FIGURE 3-15   Block diagram of a sequential machine.

The state transitions of an asynchronous sequential machine are unpredictable. Most application-specific integrated circuits (ASICs) are designed for fast synchronous operation because race conditions are very problematic for asynchronous machines, and they get worse as the physical dimensions of devices and signal paths shrink. Synchronous machines overcome race issues by having a clock period that is sufficient to stabilize the signals in the circuit. In an edge-triggered clocking scheme, the clock isolates a storage register's inputs from its output, thereby allowing feedback without race conditions. In fact, synchronous machines are widely used because timing issues are reduced to (1) ensuring that setup and hold timing constraints[3] are satisfied at flip-flops (for a given system clock), (2) ensuring that clock skew[4] induced by the physical distribution of the clock signal to the storage elements does not compromise the synchronicity of the design, and (3) providing synchronizers at the asynchronous inputs to the system [2].

The state transitions of an edge-triggered flip-flop-based synchronous machine are synchronized by the active edge (i.e., rising or falling) of a common clock. State changes give rise to changes in the outputs of the combinational logic that determines the next state and the outputs of the machine. Clock waveforms may be symmetric or asymmetric. Figure 3-16 illustrates features of an asymmetric clock waveform, that is, the length of the interval in which the clock is low is not equal to the length of the interval in which the clock is high. Register transfers are all made at either the rising or the falling edge of the clock, and input data are synchronized to change between the active edges.

The period of the clock must be long enough to allow all transients activated by a transition of the clock to settle at the outputs of the next-state combinational logic before the next active edge occurs. This establishes a lower bound on the cycle time (period) of the clock of a sequential machine. The inputs to the state register's flip-flops

---

[3]Setup constraints require the data to be stable in an interval before the active edge of the clock; hold constraints require the data to be stable in an interval after the active edge.
[4]"Clock skew" refers to the condition that the active edge of the clock does not occur at exactly the same time at every flip-flop.

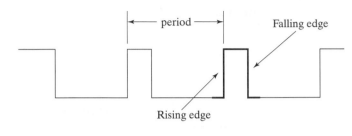

FIGURE 3-16   Waveform of an asymmetric clock signal.

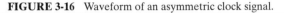

must remain stable for a sufficient interval before and after the active edge of the clock. The constraint imposed before the clock establishes an upper bound on the longest path through the circuit, which constrains the latest allowed arrival of data. The constraint imposed after the clock established a lower bound on the shortest path through the combinational logic that is driving the storage device, by constraining the earliest time at which data from the previous cycle could be overwritten. Together, these constraints ensure that valid data are stored. Otherwise, timing violations may occur at the inputs to the flip-flops and cause a condition of metastability, with the result that invalid data are stored.[5]

The set of states of a sequential machine is always finite, and the number of possible states is determined by the number of bits that represent the state. A machine whose state is encoded as an n-bit binary word can have up to 2^n states. We will use the term *finite-state machine* to refer to a clocked sequential machine that has one of the two structures shown in Figure 3-17. Synchronous (i.e., clocked) finite-state machines (FSMs) have widespread application in digital systems, for example, as datapath controllers in computational units and processors. Synchronous FSMs are characterized by a finite number of states and by clock-driven state transitions.

There are two fundamental types of finite-state machines: Mealy and Moore. The next state and the outputs of a Mealy machine depend on the present state and the inputs; the next state of a Moore machine depends on the present state and the inputs, but the output depends on only the present state. In both machines the next state and outputs are formed by combinational logic.

3.5   State-Transition Graphs

Finite-state machines can be described and designed systematically with the aid of timing diagrams [2], state tables, state graphs [3], and ASM charts [1]. Timing diagrams can be used to specify relationships between assertions and transitions of signals in a system and at its interface to its environment. For example, the write cycle of a static random access memory can be specified by a timing chart that indicates when the address

---

[5]We will consider metastability in Chapter 5.

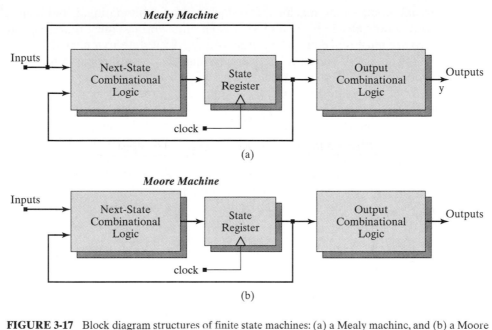

FIGURE 3-17 Block diagram structures of finite state machines: (a) a Mealy machine, and (b) a Moore machine.

of a memory cell must be asserted prior to assertion of a write-enable signal.[6] In a synthesis-oriented design methodology, timing specifications are incorporated as constraints on the circuit that must be realized by the design tools. Our attention here will focus on state tables, state-transition graphs, and ASM charts. Chapter 11 will consider timing analysis.

State tables, or state-transition tables, display in tabular format the next state and output of a state machine for each combination of present state and input. A state-transition graph (STG), or diagram, of an FSM is a directed graph in which the labeled nodes, or vertices, correspond to the machine's states, and the directed edges, or arcs, represent possible transitions under the application of an indicated input signal when the system is in the state from which the arc originates. The vertices of the STG of a Mealy machine are labeled with the states. The edges of the graph are labeled with (1) the input that causes a transition to the indicated next state and (2) the output that is asserted in the present state for that input. The graph for a Moore-type machine is similar, but its outputs are indicated in each state vertex, instead of on the arcs.

Given an STG for a synchronous machine, the design task is to determine the next-state and output logic. If the state of the machine is represented by a binary word, its value can be stored in flip-flops. At each active edge of the clock, the inputs to the state-holding flip-flops become the state for the next cycle of the clock. The design of the machine specifies the logic that forms the inputs to the flip-flops from the state and

---

[6]See Chapter 8.

the external inputs to the machine. This logic will be combinational, and it should be minimized, if possible. To be a valid STG, each of its vertices must represent a unique state, each arc must represent a transition from a given state to a next state under the action of the indicated input, and each arc leaving a node must correspond to a unique input. In general, the Boolean conditions associated with the inputs on the set of arcs leaving a node must sum to 1[7] (i.e., the graph must account for all possible transitions from a node), and each branching condition associated with assertions of the input variables in a given state must correspond to one and only one arc, i.e., the machine may exit a node on only one arc [4]. (See Roth [3] for guidelines for construction of STGs). The state transitions represented by the STG of a synchronous machine are understood to occur at the active edges of a clock signal, based on the values of the state and inputs that are present immediately before the clock.

We will now present two examples of designing a state machine by manual methods and STGs. These examples will be revisited in Chapter 6, in which we will describe the machines using the Verilog hardware description language, synthesize them to obtain their physical implementation, and validate the design by comparing and matching simulation results obtained before and after synthesis.[8]

3.6   Design Example: BCD to Excess-3 Code Converter

In this example, a serially transmitted binary coded decimal (BCD word), B_{in}, is to be converted into an Excess-3 encoded serial bit stream, B_{out}. An Excess-3 code word is obtained by adding 3_{10} to the decimal value of the BCD word and taking the binary equivalent of the result. Table 3-1 shows the decimal digits, their 4-bit BCD code words, and their Excess-3 encoded counterparts. An Excess-3 code is self-complementing [2, 4, 7], that is, the 9s complement[9] of an Excess-3 encoded word is obtained in hardware by complementing the bits of the word (i.e., by taking the 1s complement of the word). For example, the code for 6_{10} is 1001_2; its bitwise complement is 0110_2, which is the code for 3_{10}. This feature of the Excess-3 code makes it possible to easily implement a diminished radix[10] complement scheme for subtracting numbers that are encoded in a BCD form. This is similar to subtraction of signed binary words by adding the 2s complement of the minuend to the subtrahend. The 2s complement is formed by adding 1 to the 1s (diminished radix) complement of the minuend. Thus, the 10s complement of 6_{10} can be obtained by bitwise complementing 1001_2, the Excess-3 code of 6_{10}, and adding 1 to the result: $0110_2 + 0001_2 = 0111_2$, which decodes to 4_{10}.

A BCD to Excess-3 code converter for a serial bit stream can be implemented as a Mealy FSM. Figure 3-18 shows a serial bit stream, B_{in}, entering the converter and the corresponding serial stream of Excess-3 encoded bits, B_{out}, leaving the machine. Note

---

[7]The chart can be simplified by showing only the transitions that leave a state, by omitting arcs that begin and end at the same state, and by omitting return arcs that are activated by a reset signal.
[8]See sections 6.6.1 to 6.6.3.
[9]The 9s complement of a binary number a is the binary value a' such that $a + a' = 9$.
[10]The radix 9 is the diminished radix for a base 10 (decimal) system.

TABLE 3-1   BCD and Excess-3 code words.

Decimal Digit	8-4-2-1 Code (BCD)	Excess-3 Code
0	0000	0011
1	0001	0100
2	0010	0101
3	0011	0110
4	0100	0111
5	0101	1000
6	0110	1001
7	0111	1010
8	1000	1011
9	1001	1100

that the bits of B_{in} are transmitted in sequence, with the least significant bit (LSB) first. Consequently, care must be taken to interpret the waveforms of B_{in} and B_{out} correctly. The order of the bits in the waveforms is shown progressing from right to left (with increasing t), with the LSB at the left and the most significant bit (MSB) at the right. The pattern of bits in the waveform must be reversed, as shown in Figure 3-18, to form the binary values of the transmitted and received words.

The STG[11] of a serial code converter that implements the code in Table 3-1 is shown in Figure 3-19(a), with an asynchronous reset signal that transfers the machine to state S_0 whenever it is asserted, independently of the clock. The machine's action commences in S_0 with the first clock edge after reset and continues indefinitely, repeating the addition of 0011_2 to successive 4-bit slices of the input stream. The LSB of the word is the first bit in the sequence of input samples, and the first bit generated for the output word. The state table in Figure 3-19(b) summarizes the same information as the state transition graph, but in tabular format. The notation "-/-" indicates an unspecified or impossible condition.

Systematic design of a D flip-flop realization of an FSM consists of the following steps: (1) construct an STG for the machine, (2) eliminate equivalent states, (3) select a state code (e.g., a binary code), (4) encode the state table, (5) develop Boolean equations describing the inputs of the D flip-flops that hold the state bits, and (6) using K-maps, optimize the Boolean equations. In general, the number of flip-flops used to represent the state of the machine must be sufficient to accommodate a binary representation of the number of states—that is, a machine that has 12 states requires at least

---

[11]The STG of a completely specified machine with n inputs must have 2^n arcs leaving each node, and the number of its states must be a power of 2. Otherwise, some bit patterns will be unused in the hardware implementation.

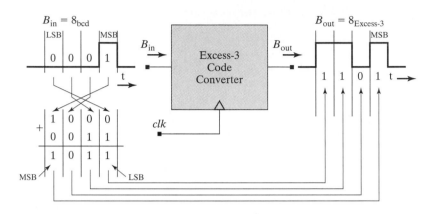

FIGURE 3-18   Input–output bit streams in a BCD to Excess-3 serial code converter.

four flip-flops. For a given set of flop-flops it is then necessary to assign a unique binary code to each state. This problem is difficult because the number of possible codes grows exponentially with the number of available flip-flops. The choice ultimately matters, because it can have an impact on the complexity of the logic required to implement the machine. We will consider this topic in more detail in Chapter 6. The codes for state assignment in our example are shown in Figure 3-20, where a simple (sequential) 3-bit binary code has been used to encode the seven states of the machine. The encoded next state and output table are also shown.

Our next step will be to develop Karnaugh maps for each bit of the encoded state and output, as functions of the present state bits and the input (B_{in}). These maps are

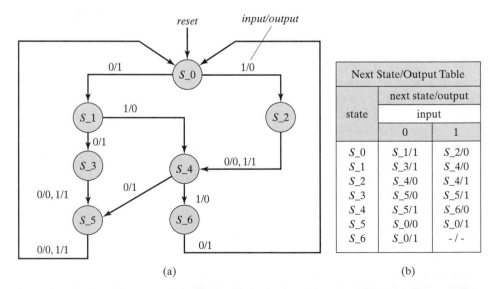

FIGURE 3-19   BCD to Excess-3 serial code converter implemented as a Mealy-type FSM: (a) state transition graph, and (b) the machine's state table.

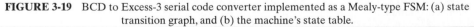

State Assignment			Encoded Next State / Output Table				
			State	Next State		Output	
$q_2 q_1 q_0$	State		$q_2 q_1 q_0$	$q_2^+ q_1^+ q_0^+$			
				Input		Input	
				0	1	0	1
000	S_0	S_0	000	001	101	1	0
001	S_1	S_1	001	111	011	1	0
010	S_6	S_2	101	011	011	0	1
011	S_4	S_3	111	110	110	0	1
100		S_4	011	110	010	1	0
101	S_2	S_5	110	000	000	0	1
110	S_5	S_6	010	000	—	1	—
111	S_3		100	—	—	—	—

(a) (b)

FIGURE 3-20 BCD to Excess-3 code converter implemented as a Mealy-type FSM: (a) state assignment, and (b) encoded next state and output table.

shown in Figure 3-21 with their corresponding Boolean equations. The unspecified entries in the table are treated as don't-care conditions. Each equation has been minimized individually, although this does not necessarily produce the optimal realization (speed vs. area) of the logic. We will consider optimization of a set of Boolean equations in our discussion of logic synthesis in Chapter 6.

The Boolean equations for q_2^+ and B_{out} can be converted into the following NAND gate structure, where, for clarity, we use the symbol " $\cdot$ " to indicate the Boolean AND operator:

$$q_2^+ = q_1'q_0'B_{in} + q_2'q_0B_{in}' + q_2q_1q_0$$

$$\overline{q_2^+} = \overline{q_1'q_0'B_{in} + q_2'q_0B_{in}' + q_2q_1q_0}$$

$$\overline{q_2^+} = \overline{q_1'q_0'B_{in}} \cdot \overline{q_2'q_0B_{in}'} \cdot \overline{q_2q_1q_0}$$

$$q_2^+ = \overline{\overline{q_1'q_0'B_{in}} \cdot \overline{q_2'q_0B_{in}'} \cdot \overline{q_2q_1q_0}}$$

and

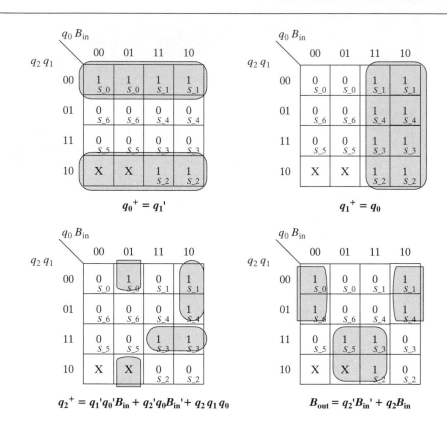

$$q_0^+ = q_1'$$

$$q_1^+ = q_0$$

$$q_2^+ = q_1'q_0'B_{in} + q_2'q_0B_{in}' + q_2q_1q_0$$

$$B_{out} = q_2'B_{in}' + q_2B_{in}$$

FIGURE 3-21   Karnaugh maps for the encoded state bits and output bit (B_{out}) of a BCD to Excess-3 code converter implemented as a Mealy-type FSM with input bit (B_{in}).

$$B_{out} = q_2'B_{in}' + q_2B_{in}$$

$$B_{out}' = \overline{q_2'B_{in}' + q_2B_{in}}$$

$$B_{out}' = (\overline{q_2'B_{in}'}) \cdot (\overline{q_2B_{in}})$$

$$B_{out} = \overline{(\overline{q_2'B_{in}'}) \cdot (\overline{q_2B_{in}})}$$

The schematic of the code converter is shown in Figure 3-22, with three positive edge-triggered flip-flops storing the state bits. The simulation results in Figure 3-23 illustrate the input–output waveforms and the state transitions of the machine. The annotation of the displayed waveforms shows the bit stream of the encoded word produced by the converter for $B_{in} = 0100_2$, where the LSB is asserted first, and the most significant bit is last in the time sequence. Since the output of a Mealy machine depends on the input as well as the state, the transitions of B_{in} affect the waveform of B_{out}. We have aligned the transitions of B_{in} to occur on the inactive edge of the clock. This is a recommended practice, which ensures that the data are stable before the active edge of the clock. Since the input of a Mealy machine can cause the output of the machine to change its value, the valid output of a Mealy machine is taken to be the

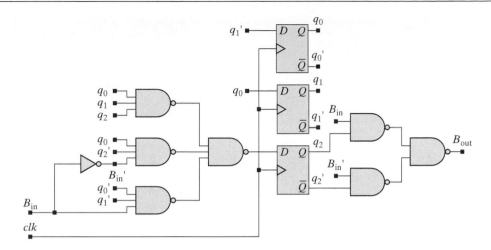

FIGURE 3-22   Circuit for a Mealy-type FSM implementing a BCD to Excess-3 code converter.

value of the output immediately before the active edge of the clock. The value of B_{out} immediately before an active edge of the clock depends on the value of B_{in} immediately before the clock, and it is the valid output of this machine.[12] Thus, the input bit stream 0100_2 generates the output bit stream 0111_2. The waveforms of B_{in} and B_{out} in Figure 3-23 are annotated with bubbles to mark corresponding values of the BCD and Excess-3 encoded bits.

3.7   Serial-Line Code Converter for Data Transmission

Line codes are used in data transmission or storage systems to reduce the effects of noise in serial communications channels and/or to reduce the width of channel data-paths [2]. For example, in data transmission in which bits of a word are encoded and transmitted synchronously over a channel, the receiver of the data must be able to op-erate synchronously with the sending unit, identify the boundaries between words (frames), and distinguish the transmitted bits from each other. One scheme for data re-covery after transmission requires three signals: a clock to define the boundaries of the data bits, a synchronizing signal to define word boundaries, and a data stream. Other implementations using fewer signal channels are possible. For example, a phone system or a disk read/write head will have a single channel for the data and use a coding scheme to enable clock recovery and synchronization. Code converters transform the data stream into a format that has been encoded to enable the receiver to recover the data. Four common serial encoding schemes are described below. A phase lock loop (PLL [3]) can recover the clock from the line data (i.e., synchronize itself to the clock of the data) if there are no long series of 1s or 0s in a data stream with an

---

[12]A physical circuit implementing the machine would have to perform the machine's addition fast enough to have the result ready for the active edge of the clock. Performance is an issue for synthesis and will be ad-dressed in Chapter 10.

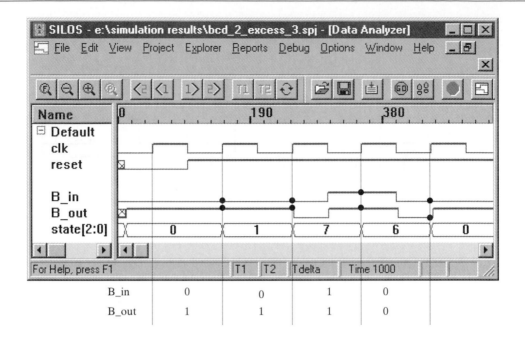

FIGURE 3-23   Simulation results for a BCD to Excess-3 code converter implemented as a Mealy-type FSM, with annotation marking corresponding input and output values.[13]

non–return-to-zero (NRZ) format; the clock can be recovered from a data stream with a (non–return-to-zero invert-on-ones) (NRZI) or return-to-zero (RZ) code format if the data has no long string of 0s. Manchester encoders are attractive because they can recover the clock independently of the pattern of the data, but they require higher bandwidth.

*   **NRZ Code**: The signal waveform of the line value formed by an NRZ code generator duplicates the bit pattern of the input signal, as shown in Figure 3-24. The output waveform makes no transition between two identical successive bits. A Moore machine implementing an NRZ code samples the data at the active edge of *clock_1*; the transitions of the data are synchronized to the inactive edge of *clock_1*.
*   **NRZI Code**: If the input to an NRZI code converter is 0 the sequential output of the converter remains at it previous value. If the input is a 1, the output is the complement of the previous input. So the output remains constant as long as the input remains at 0, and the output toggles if the input is held at 1, as shown in Figure 3-24. The asserted value is held for the entire bit time.

---

[13]Note: *B_in* and *B_out* in the simulator output represent B_{in} and B_{out}, respectively.

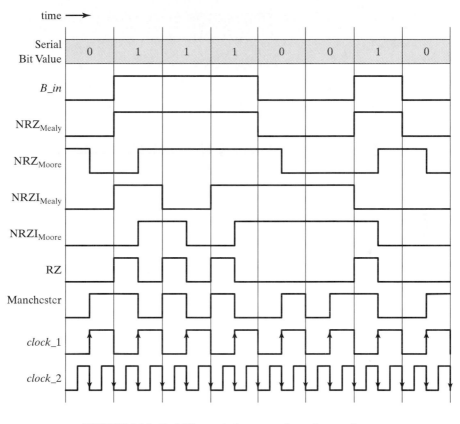

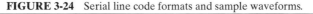

FIGURE 3-24   Serial line code formats and sample waveforms.

- **RZ Code**: A 0 in the input bit stream of RZ code generator is transmitted as a 0 for the entire bit time. A 1 in the bit stream is transmitted as a 1 for the first half of the bit time, and 0 for the remaining bit time (typically).

- **Manchester Code:** A 0 in the bit stream of a Manchester code generator is transmitted as a 0 for the lead half of the bit time and as a 1 for the remaining half. A 1 in the bit stream is transmitted as a 1 for the leading half of the bit time and as a 0 for the remaining bit time.

Figure 3-24 shows the encoded waveforms that are produced by these encoding schemes for a sample of serial bit stream values.[14] The clock frequency (*clock_2*) of Mealy-type machines implementing NRZI, RZ, and Manchester encoders must operate at twice the frequency of the bit stream generator (*clock_1*) in order to assert the line value for a full bit time without latency.

---

[14]The actual waveform produced by the encoder might have latency, depending on the type of state machine that implements the encoder. Latency will shift the transitions of the waveform of the output to occur later than the transitions of the waveform of the input.

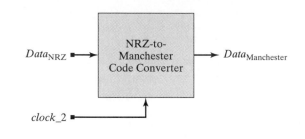

FIGURE 3-25   Input-output datapaths for a NRZ Manchester code converter.

3.7.1   A Mealy-Type FSM for Serial Line-Code Conversion

A serial line-code converter can be implemented by an FSM in which the inbound bit stream controls the actions of the machine to produce an encoded outbound bit stream. As an example, a Mealy-type FSM will be designed to convert a data stream in NRZ format, $Data_{NRZ}$, to a data stream in Manchester code format, $Data_{Manchester}$, as represented by the block diagram in Figure 3-25.

     The state-transition graph and state table for the machine are shown in Figure 3-26, and the state assignment and encoded state table are shown in Figure 3-27. The table shows the state label and code for each state, the codes corresponding to the bits of the next state for each possible value of B_{in}, and the output that will be asserted in the indicated state for each possible value of B_{in}. An asserted bit in the input will be asserted for two clocks of the output. Thus, the arcs leaving S_1 with an input of 1 and from S_2 with an input of 0 are not shown because those inputs sequences cannot occur. The corresponding entries in the next-state table are marked as don't-cares.

     The Karnaugh maps and Boolean equations for the line converter are derived from the state-transition graph, and are shown in Figure 3-28. The circuit schematic shown in Figure 3-29 implements the Boolean equations of the machine and stores its state in two negative edge-triggered D-type flip-flops. The simulation results in Figure 3-30 show the input (B_{in}) and output (B_{out}) waveforms for a sample of data and also show the internal operation of the state machine that accomplishes the code conversion. Note that because the machine's output is generated as a Mealy-type output, there is

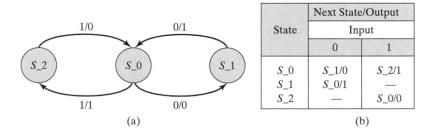

			Next State/Output	
State			Input	
			0	1
S_0			S_1/0	S_2/1
S_1			S_0/1	—
S_2			—	S_0/0

(a)   (b)

FIGURE 3-26   Mealy-type NRZ-to-Manchester encoder: (a) the state transition graph, and (b) the next-state table.

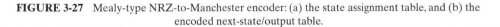

	q_0	
q_1	0	1
0	S_0	S_1
1	S_2	

(a)

State		Next State		Output	
$q_1 q_0$		$q_1^+ q_0^+$			
		Input		Input	
		0	1	0	1
S_0	00	01	10	0	1
S_1	01	00	00	1	—
S_2	10	00	00	—	0

(b)

FIGURE 3-27 Mealy-type NRZ-to-Manchester encoder: (a) the state assignment table, and (b) the encoded next-state/output table.

no latency between the waveform of the inbound bit stream and that of the output; that is the bit times of the input and the output coincide.

3.7.2 A Moore-Type FSM for Serial Line-Code Conversion

The output of a Mealy machine is subject to glitches in the input bit stream. If this cannot be tolerated, a Moore-type machine should be used. A simplified state-transition graph (without impossible arcs) and a state table for a Moore machine NRZ-to-Manchester encoder are shown in Figure 3-31, and the state assignment and encoded state table are shown in Figure 3-32. Note that the data transitions are synchronized by the negative-edge transitions of *clock_1*, and the state transitions of the encoder are synchronized by the negative-edge transitions of *clock_2*.

The Karnaugh maps and Boolean equations for the converter are derived from the state-transition graph, and are shown in Figure 3-33, with don't-care conditions denoted by " − " (for impossible patterns). The circuit schematic shown in Figure 3-34 implements the Boolean equations of the machine with negative-edge triggered flip-flops. The

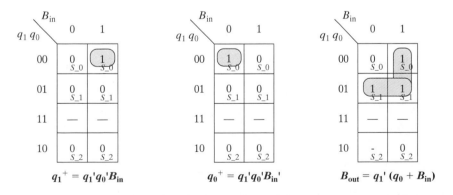

$$q_1^+ = q_1'q_0'B_{in}$$

$$q_0^+ = q_1'q_0'B_{in}'$$

$$B_{out} = q_1' (q_0 + B_{in})$$

FIGURE 3-28 The Karnaugh maps and Boolean equations for the encoded state bits and output bit (B_{out}) of an NRZ-to-Manchester encoder with input bit B_{in}.

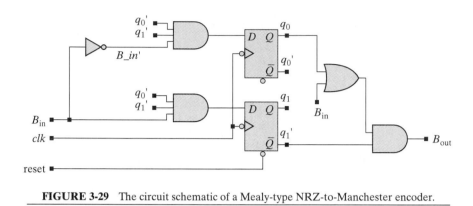

FIGURE 3-29   The circuit schematic of a Mealy-type NRZ-to-Manchester encoder.

simulation results in Figure 3-35 show the input (B_{in}) and output (B_{out}) waveforms for a sample of data and also show the internal operation of the state machine that accomplishes the code conversion. Note that the Manchester encoder must run at twice the frequency of the incoming data and that the output bit stream of the Moore machine lags the input bit stream by one-half the input cycle time. The transitions of the input data stream are made at the falling edges of *clock_1*, and the state machine samples the

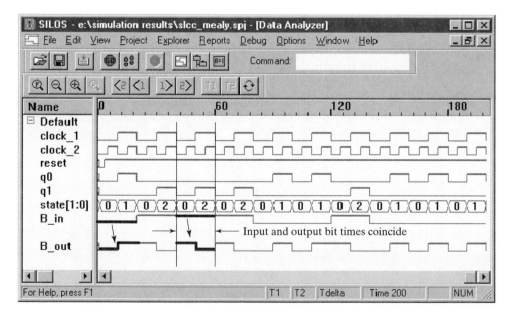

FIGURE 3-30   Simulation results for the Mealy-type NRZ-to-Manchester encoder.[15]

[15]In general, it is not advisable to switch an input at the same time that a sequential machine makes a state transition, but in this circuit the samples that are made when the input is making a transition occur in states (i.e., *S_1* and *S_2*) in which the input is treated as a don't-care. So the input will not affect the transition from *S_1* or *S_2* to *S_0*. The scheme here effectively samples B_{in} in the middle of its bit time, and eliminates a cycle of latency in the output of the converter.

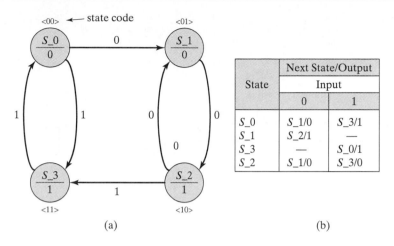

FIGURE 3-31   Moore machine NRZ-to-Manchester encoder: (a) the state-transition graph, and (b) the next-state/output table.

input at the rising edges of *clock_1*—that is, in the middle of the bit time of the input, which corresponds to every other falling edge of *clock_2*. The transitions of the state of the machine coincide with the falling edge of *clock_2*. The latency between the output and the input is a consequence of the fact that the outputs of a Moore machine are dependent on only the state. Thus, a change in the input must first cause a transition to a state before the encoder can assert an output that corresponds to the sampled/detected input.

3.8   State Reduction and Equivalent States

Two states of a sequential machine are equivalent ($\equiv$) if they have the same output sequence for all possible input sequences. Such states of the machine cannot be distinguished from each other on the basis of observed outputs. Equivalent states can be

(a)

q_1	q_0 0	1
0	S_0	S_1
1	S_2	S_3

(b)

State		Next State		Output
$q_1 q_0$		$q_1^+ q_0^+$		
		Input		
		0	1	
S_0	00	01	11	0
S_1	01	10	—	0
S_3	11	—	00	1
S_2	10	01	11	1

FIGURE 3-32   Moore-type NRZ-to-Manchester encoder: (a) the state-assignment table, and (b) encoded next-state/output table.

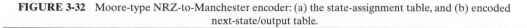

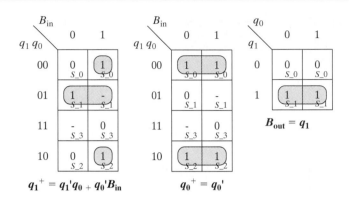

$$q_1^+ = q_1'q_0 + q_0'B_{in} \qquad q_0^+ = q_0' \qquad B_{out} = q_1$$

FIGURE 3-33 The Karnaugh maps and Boolean equations for the encoded state bits and output bit (B_{out}) for a NRZ-to-Manchester encoder implemented as a Moore machine with input bit B_{in}.

combined without changing the input–output behavior of the machine. Identifying and combining equivalent states usually simplifies the state table and the state-transition graph of the machine and leads to a reduction in hardware (because the equivalent states do not have to be decoded) without compromising functionality [8]. In general, for every finite-state machine there is a unique equivalent machine that is minimal.

Example 3.4

The state machine whose next-state table is shown in Figure 3-36 has two states that are equivalent: $S\_4 \equiv S\_5$. Both S_4 and S_5 have the same next states and outputs under the actions of the inputs. If the machine is in S_4 and an input sequence is applied, the outputs will be the same as if the machine was in S_5 and the same input sequence

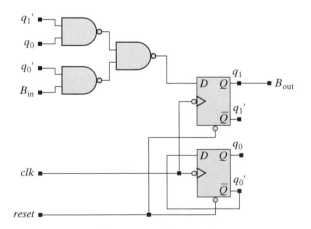

FIGURE 3-34 The circuit schematic of a Moore machine NRZ-to-Manchester encoder.

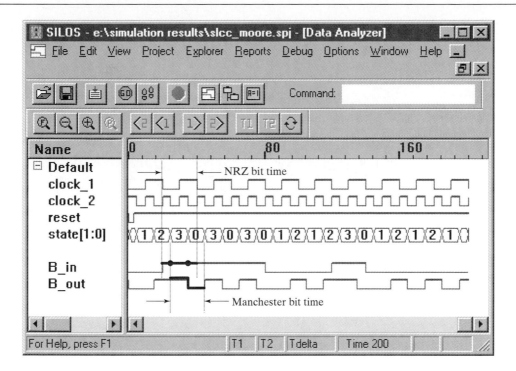

FIGURE 3-35 Simulation results for the Moore machine NRZ-to-Manchester encoder.

is applied. The machine's state-transition graph, given in Figure 3-37(a), shows how S_4 and S_5 map into the same next state, and that they also have the same outputs for all assertions of the input.

Two states of a sequential machine are equivalent if their associated rows in the machine's state table are identical. The state-transition graph of a machine can be

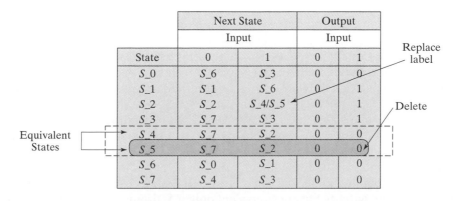

	Next State		Output	
	Input		Input	
State	0	1	0	1
S_0	S_6	S_3	0	0
S_1	S_1	S_6	0	1
S_2	S_2	S_4/S_5	0	1
S_3	S_7	S_3	0	1
S_4	S_7	S_2	0	0
S_5	S_7	S_2	0	0
S_6	S_0	S_1	0	0
S_7	S_4	S_3	0	0

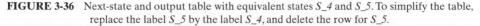

FIGURE 3-36 Next-state and output table with equivalent states S_4 and S_5. To simplify the table, replace the label S_5 by the label S_4, and delete the row for S_5.

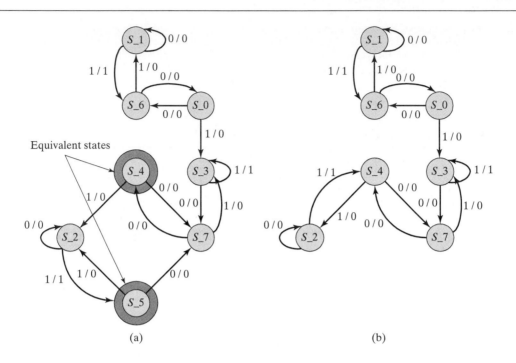

FIGURE 3-37   State-transition graph with equivalent states: (a) graph with S_4 and S_5, and (b) state-transition graph after removal of S_5 and redirection of arcs.

pruned by removing all but one of the states that are equivalent and redirecting the affected arcs to the remaining equivalent state. Likewise, the next-state table can be simplified by eliminating one of the rows and replacing the labels of the deleted state by the label of its equivalent state. But be careful—do not conclude that two states are not equivalent if their state-table rows are different. The condition of matching rows of the next-state table is a sufficient condition for the associated states to be equivalent, but is not a *necessary* condition to ensure their equivalence. Merely comparing the rows of a state table is not a foolproof way to identify equivalent states. Other states that are equivalent might not be detected by such an approach.

A more general procedure for eliminating equivalent states relies on a recursive definition of equivalence: Two states are equivalent if they have the same output for each input value, and the states they reach for the same input value are equivalent. The procedure (1) forms a triangular array (see Figure 3-38) showing the possible pairwise combinations of distinct states, and (2) considers the conditions for the pair of states to be equivalent (we already know that S_4 and S_5 of the original table are equivalent, so we don't consider S_5). For example, S_0 and S_4 can be equivalent only if the next states they reach are equivalent and if they have the same output under the action of each possible input value. The table entries in Figure 3-36 indicate that S_0 and S_4 have the same outputs, but are equivalent only if S_6 and S_7 are equivalent, and if S_2 and S_3 are equivalent. Both conditions must be satisfied—we do not know in advance whether S_2 and S_3 are equivalent and whether S_6 and S_7 are equivalent.

	S_0	S_1	S_2	S_3	S_4	S_6
S_1						
S_2		S_6 - S_4				
S_3		S_1 - S_7 S_6 - S_3	S_2 - S_7 S_4 - S_3			
S_4	S_6 - S_7 S_3 - S_2					
S_6	S_3 - S_1				S_7 - S_0 S_2 - S_1	
S_7	S_6 - S_4				S_2 - S_3	S_0 - S_4 S_1 - S_3

FIGURE 3-38   Triangular array showing possible pairwise combinations of states.

Within a given cell corresponding to a row and column of the table, list the pairs of states that are reached from the states whose labels correspond to the row and column. For example, if the machine is in S_1, its next states is S_1 or S_6 if the input is 0 or 1, respectively. Likewise, if the machine is in S_3 it reaches S_7 and S_3 under the same applied input values. Thus, for S_1 and S_3 to be equivalent, it is necessary and sufficient that S_1 and S_7 be equivalent states, and that S_6 and S_3 be equivalent states, provided of course, that the outputs are the same. The cells of Figure 3-38 are annotated with the pairs of states that must be equivalent in order for the states corresponding to the row and column to be *equivalent*. If a pair of states has a different output for some input, the pair is eliminated and the cell is shaded to signify that the states cannot be equivalent. For example, S_1 and S_4 cannot be equivalent because their output assertions do not match. Next, any cells containing the labels of a state pair corresponding to a shaded state pair are lined out, as shown in Figure 3-39. For example, the cell containing the pairs (S_1, S_7 and S_3, S_6) is lined out because S_1 and S_7 cannot be equivalent. When this process is completed, the remaining labeled cells identify equivalent states. The results in Figure 3-39 indicate that $S\_4 \equiv S\_5$, $S\_0 \equiv S\_7$, $S\_2 \equiv S\_2$, and $S\_4 \equiv S\_6$. This leads to the simplified state-transition graph shown in Figure 3-40, having only four states, instead of eight.

End of Example 3.4

REFERENCES

1. Katz RH. *Contemporary Logic Design*. Redwood City, CA: Benjamin Cummings, 1994.
2. Wakerly JF. *Digital Design Principles and Practices*, 3rd ed. Upper Saddle River, NJ: Prentice-Hall, 2000.

	S_0	S_1	S_2	S_3	S_4	S_6
S_1						
S_2		S_6 - S_4				
S_3		S_1 - S_7 / S_6 - S_3	S_2 - S_7 / S_4 - S_3			
S_4	S_6 - S_7 / S_3 - S_2					
S_6	S_3 - S_1				S_7 - S_0 / S_2 - S_1	
S_7	S_6 - S_4				S_2 - S_3	S_0 - S_4 / S_1 - S_3

FIGURE 3-39   Triangular array showing possible pairwise combinations of states.

3. Roth CH Jr. *Fundamentals of Logic Design*, 4th ed., St. Paul: West, 1992.
4. Tinder RF. *Engineering Digital Design*, 2nd ed. San Diego: Academic Press, 2000.
5. Weste N, Eshraghian K. *Principles of CMOS VLSI Design*. Reading, MA: Addison-Wesley, 1993.
6. Smith MJS. *Application-Specific Integrated Circuits*. Reading, MA: Addison-Wesley, 1997.
7. Breeding KJ. *Digital Design Fundamentals*. Upper Saddle River, NJ: Prentice-Hall, 1989.
8. Hachtel GD, Somenzi F. *Logic Synthesis and Verification Algorithms*. Boston: Kluwer, 1996.

PROBLEMS

1. Using D-type flip-flops, design a synchronous Moore finite-state machine that monitors two inputs, A and B, and asserts a scalar output if the number of 1s observed on the inputs is a multiple of 4.

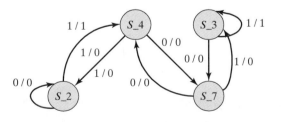

FIGURE 3-40   Completely pruned state-transition graph.

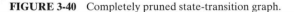

2. Using D-type flip-flops, design a Moore machine whose output indicates the even parity of its serial bit stream of inputs since reset.

3. Design a 3-bit counter that will count in the cyclic pattern of even numbers: $0_{10}\, 2_{10}\, 4_{10}\, 6_{10}$.

4. Draw the state-transition graph of a Mealy machine that samples a serial bit stream and asserts an output if the last three samples are 1. Assume that the samples are made at the inactive edges of the clock. Design the machine with D-type flip-flops.

5. Draw the state-transition graph of a Moore machine that samples a serial bit stream and asserts an output if the last three samples are 1. Assume that the samples are made at the inactive edges of the clock. Design the machine with D-type flip-flops.

6. Examine the Mealy machine described by the K-maps in Figure 3-28 and identify possible glitches that could occur when the input makes a transition.

7. Draw the state-transition graph of a Moore-type state machine that receives a serial bit stream and generates a serial output in which the output (1) is asserted for one clock cycle if the input pattern 0111 is recognized, (2) remains de-asserted until the pattern 0111 is recognized again, and (3) re-asserts on detecting the second occurrence of 0111, and so forth, continuously.

8. Draw the state-transition graph of a machine that behaves like the machine in Problem 7, but terminates its activity after it has detected six occurrences of 0111, until a reset condition is asserted.

9. Draw the state-transition graph of a Moore machine that converts a NRZ bit stream into a NRZI bit stream.

10. Draw the state-transition graph of an NRZI line encoder.

CHAPTER 4   Introduction to Logic Design with Verilog

Designers of application-specific integrated circuits (ASICs) use hardware description languages (HDLs) at key steps in the design flow of a circuit. An HDL model is developed; synthesized into a physical circuit; and verified for functionality, timing, and fault coverage. HDLs have some similarity to ordinary, general-purpose languages like C, but they have additional features for modeling and simulating the functionality of combinational and sequential circuits. Two languages are in widespread use: Verilog and VHDL. Both are Institute of Electrical and Electronic Engineers (IEEE) standards, and both have enthusiastic users. Rather than compare, contrast, and even argue the relative merits of the languages, our focus will be on digital design with the Verilog HDL.

Designers using an HDL (1) write a text-based description (model) of a circuit, (2) compile the description to verify its syntax, and (3) simulate the model to verify the functionality of the design. Verification by simulation requires that designers write a testbench containing descriptions of stimulus waveforms that are applied to exercise the functionality of the circuit. Simulators display waveforms by exercising the circuit's behavior, and some can analyze the waveforms to detect functional errors.

Verilog gives designers several alternative ways to describe a circuit. Some designs can be entered easily as a structural description in which logic gates are connected to other gates, just as they would be on a schematic. But other styles of design can be more abstract, such as an algorithm that describes a lowpass finite impulse response digital filter.

The Verilog HDL serves as a vehicle for designing, verifying, and synthesizing a circuit [1]. It is also a medium for exchanging designs between designers. Intellectual property encapsulated as a Verilog description can be exported and embedded in other designs.

Today's designs are so large that they must be designed with a top-down methodology that systematically partitions a complex design into smaller functional units that can be designed and verified individually and more easily, before reintegrating them and reverifying the aggregate. Verilog lets designers create a design hierarchy of functional units. In today's global design environment, the partitioned design can be distributed across multiple teams located in the same lab or anywhere on the World Wide Web.

Chapter 2 summarized three common descriptions of combinational logic: schematic/gates, truth tables, and Boolean equations. Verilog includes constructs for each of these, and abstract models too. This chapter will present Verilog constructs corresponding to gate-level and truth-table descriptions of combinational logic.

4.1   Structural Models of Combinational Logic

A Verilog model of a circuit encapsulates a description of its functionality as a structural or behavioral view of its input–output (I/O) relationship. A structural view could be a netlist of gates or a high-level architectural partition of the circuit into major functional blocks, such as an arithmetic and logic unit (ALU). A behavioral view could be a simple Boolean equation model, a register transfer level (RTL) model, or an algorithm. We will begin our introduction to HDLs by considering how Verilog supports structural design. We will introduce basic concepts here before moving to the world of abstract models.

Structural design is similar to creating a1 schematic. Figure 4-1 shows a gate-level schematic of a half adder circuit, along with its Verilog description. A schematic consists of icons (symbols) of logic gates, lines representing wires that connect gates, and labels of relevant signal names at I/O pins and internal nodes. Similarly, an *HDL structural model* consists of a list of *declarations*, or statements, that specify the inputs and outputs of the unit and list any gate primitives (e.g., **xor, and**) that are interconnected to implement the desired functionality. The primitives that are listed in a declaration within a module are said to be instantiated in the design.

4.1.1   Verilog Primitives and Design Encapsulation

Verilog includes a set of 26 predefined functional models of common combinational logic gates, called primitives. *Primitives* are the most basic functional objects that can be used to compose a design. Their functionality is built into the language by means of

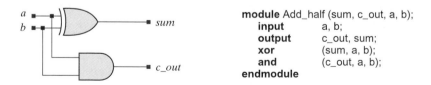

FIGURE 4-1 Schematic and Verilog description of a half adder.

TABLE 4-1   Verilog primitives for modeling
combinational logic gates.

n-Input	*n*-Output, 3-state
and	buf
nand	not
or	bufif0
nor	bufif1
xor	notif0
xnor	notif0

internal truth tables that define the relationship between the scalar (i.e., a single bit) output(s) of each primitive and its scalar input(s). Table 4-1 shows an abbreviated list of the predefined primitives and their reserved keywords.[1] Their names suggest their functionality (the ***not*** primitive corresponds to an inverter). The primitives in Table 4-1 are called *n*-input primitives because the same model keyword (e.g., ***nand***) automatically accommodates any number of scalar inputs, rather than only a pair of inputs. The *n*-output primitives (e.g., the buffer primitive, ***buf***) have a single input but can have multiple scalar outputs (to model a gate that has fanout to more than one location). The primitives ***bufif0*** and ***bufif1*** are three-state buffers; ***notif0*** and ***notif1*** are three-state inverters.[2]

MODELING TIP

The output port of a primitive must be first in the list of ports. The instance name of a primitive is optional.

Each primitive has ports (terminals) that connect to its environment. Figure 4-2 shows a 3-input ***nand*** primitive and a Verilog statement that would imply the use of this primitive in a circuit in which it is connected to input signals *a*, *b*, and *c* and has output signal *y*. The list of ports is placed immediately to the right of the primitive name as a comma-separated list enclosed by parentheses and terminated by a semicolon (;). The output port(s) of a primitive must be first in the list of ports, followed by the primitive's input port(s). A primitive is instantiated within a module by a statement declaring its keyword name, followed by an optional instance name and by a parentheses-enclosed list of its terminals.[3] All identifiers (names) in Verilog have a scope (i.e., domain of definition) that is local to the module, function, task, or named block in which they are declared. They have meaning within that scope and cannot be referenced directly from outside it.[4]

---

[1]Keywords have a predefined meaning and may not be used for any other purpose.
[2]The complete set of primitives is described in Appendix A.
[3]See Appendix E for a formal description of the language syntax.
[4]Verilog does have a mechanism for hierarchical dereferencing of an identifier in a design.

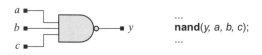

FIGURE 4-2 A three-input ***nand*** gate and an example of its instantiation as a Verilog primitive.

A simulator uses built-in truth tables to form the outputs of primitives during simulation. The values of the inputs attached to a primitive are determined by the circuitry external to the primitive, just as the output of a combinational logic gate is determined by the values of its inputs. Figure 4-3 shows a list of instantiated primitives that are connected by wires[5] to have the functionality of a five-input and-or-invert (AOI) circuit.

A list of primitives describes the functionality of a design, but it has a fixed set of signal names. By itself, the list would require that the model be used only in an environment that has the same signal names. The list does not have a name either, but it would be helpful if it had a descriptive name by which we could identify its functionality. To circumvent these limitations, the functionality of a design and its interface to the environment are encapsulated in a Verilog ***module***. A ***module*** is declared by writing text that describes (1) its functionality, (2) its ports to/from the environment, and (3) its timing and other attributes of the design, such as the physical silicon area, that would be needed to implement it in on a chip.

All Verilog modules have the text format shown in Figure 4-4. The keywords ***module*** and ***endmodule*** encapsulate the text that describes the module having type-name *my_design*. A module's type-name is user-defined and distinguishes the module from others. The ports of the module are listed beside *my_design* in a comma-separated list enclosed by parentheses. Unlike primitives, modules do not have a restriction on the relative ordering of I/O ports in the list. The inputs and outputs can be listed in any order.

The declarations placed within a module determine whether the module will be structural, behavioral, or a combination of these. All the examples in this chapter will be structural models.

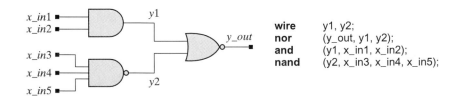

FIGURE 4-3 A list of declarations of wire-connected primitives having the functionality of a five-input AOI gate.

---

[5]We will see that the data-type ***wire*** in Verilog is used to establish connectivity in a design, just as a physical wire establishes connectivity between gates.

```
module my_design (module_ports);
... // Declarations of ports go here
... // Functional details go here
endmodule
```

FIGURE 4-4   The format of a Verilog module, with *module* ... *endmodule* keyword encapsulation.

4.1.2   Verilog Structural Models

A structural model of a logic circuit is declared and encapsulated as a named Verilog module, consisting of (1) a module name accompanied by its ports, (2) a list of operational modes of the ports (e.g., input), (3) an optional list of internal wires and/or other variables used by the model, and (4) a list of interconnected primitives and/or other modules, just as one would place and connect their physical counterparts on a PC board or on a schematic. The *primary* inputs and outputs of a physical circuit connect to its environment, and are the named ports of the model. In operation and in simulation, signals applied at the primary inputs interact with the internal gates to produce the signals at the primary outputs. A designer could apply signal generators to the inputs of the actual circuit and observe the inputs and outputs on an oscilloscope or logic analyzer. A declared module can be referenced (instantiated) within the declaration of some other module to create more elaborate and complex structural models.

A complete Verilog structural model of a five-input AOI circuit is listed in Figure 4-5 to illustrate some Verilog terminology. (*All keywords in the text are shown in boldface italic type.*) The keywords *module* and *endmodule* enclose (encapsulate) the description.[6] The text between these keywords declares (1) the interface between the model and its environment (by declaring the port list and the mode of each port), (2) the wires that are used to connect the logic gates that model the circuit (by declaring $y1$ and $y2$ to have type *wire*), and (3) the logic gate primitives and the configurations of the port signals that connect them together to form the circuit (by listing the gates and their ports).

4.1.3   Module Ports

The ports of a module define its interface to the environment in which it is used. The *mode* of a port determines the direction that information (signal values) may flow through the port. A port's mode may be unidirectional (*input*, *output*) or bidirectional (*inout*). Signals in the environment of a module are made available through its *input* ports; signals generated within a module are made available to the environment through its *output* ports. Bidirectional (**inout**) ports accommodate a flow of information in

---

[6]Appendix B contains a list of Verilog keywords.

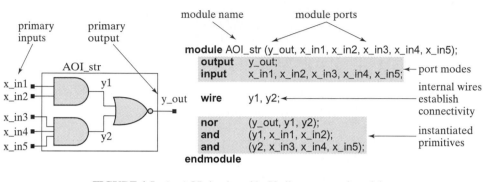

FIGURE 4-5   An AOI circuit and its Verilog structural model.

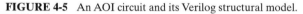

either direction. The mode of a module port is declared explicitly, and is not determined by the order in which a port appears in the port list (but recall that outputs are the left-most entries in the port list of a primitive).

The environment of a module interacts with its ports, but does not have access to the internal description of the module's functionality. Those details are hidden from the surrounding circuitry. The listing of declarations within a Verilog module tells a simulator how to form the output of the circuit from the values of its inputs. We will discuss simulation in more detail below.

4.1.4   Some Language Rules

Verilog is a case-sensitive language, so it matters whether you refer to a signal as *C_out_bar* or *C_OUT_BAR*. Verilog treats these as different names. An identifier (name) in Verilog is composed of a case-sensitive, space-free sequence of upper and lower case letters from the alphabet, the digits (0, 1, ..., 9), the underscore (_), and the $ symbol. The name of a variable may not begin with a digit or $, and may be up to 1024 characters long.[7] White space may be used freely, except in an identifier.

Usually, each line of text in a Verilog description must terminate with a semicolon (;). An exception is the terminating ***endmodule*** keyword. Comments may be imbedded in the source text in two ways. A pair of slashes, //, forms a comment from the text that follows it on the same line; the symbol-pair /* initiates a multiline comment, and must be matched by the symbol-pair */ to terminate the scope of the comment. Multiline comments may not be nested.

4.1.5   Top-Down Design and Nested Modules

Complex systems are designed by systematically and repeatedly partitioning them into simpler functional units whose design can be managed and executed successfully. A high-level partition and organization of the design is sometimes referred to as an

---

[7]The names of predefined system tasks begin with $.
[8]Do not attempt recursive declarations.

architecture. The individual functional units that result from the partition are easier to design and simpler to test than larger, equivalent aggregates. The divide-and-conquer strategy of top-down design makes possible the design of circuits with several million gates. Top-down design is used in the most modern and sophisticated design methodologies that integrate entire systems on a chip (SoC) [2]. Nested modules are the Verilog mechanism supporting top-down design. The instantiation of a module within the declaration of a different[8] module automatically creates a partition of the design.

MODELING TIP

Use nested module instantiations to create a top-down design hierarchy.

Example 4.1

A binary full adder circuit can be formed by combining two half adders and an OR gate as shown in the schematic in Figure 4-6(a). The Verilog hierarchical model of the partitioned design (Figure 4-6b) contains two instances of the module *Add_half.* The module names have been appended with *_0_delay* to indicate that the internal models have not accounted for propagation delays.

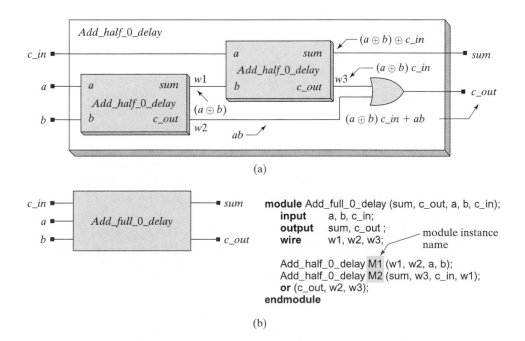

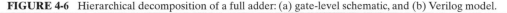

FIGURE 4-6 Hierarchical decomposition of a full adder: (a) gate-level schematic, and (b) Verilog model.

Modules may be nested within other modules, but not in a recursive manner. When a module is referenced by another module (i.e., when a module is listed inside the declaration of another module), a structural hierarchy is formed of the nesting/nested design objects. The hierarchy establishes a partition and represents relationships between the referencing and the referenced modules. The referencing module is called a *parent* module; the referenced module is called a *child* module. The module in which a child module is instantiated is a parent module. The two instances of *Add_half_0_delay* within *Add_full_0_delay* are child modules of *Add_full_0_delay*. Primitives are basic design objects. Although modules may have other modules and primitives nested within them, nothing can be instantiated (nested) within a primitive.

A module hierarchy can have arbitrary depth. It is limited only by the capacity of the host computer's memory. Each instance of an instantiated module must be accompanied by a module instance name that is unique within its parent module. An instantiated primitive may be given a name, but that is not required.

End of Example 4.1

MODELING TIP

The ports of a module may be listed in any order.
An instantiated module must have an instance name.

Example 4.2

A 16-bit ripple-carry adder can be formed by cascading four 4-bit ripple-carry adders in a chain in which the carry generated by a unit is passed to the carry input port of its neighbor, beginning with the least significant bit. Each 4-bit adder is declared as a cascade of full adders. Figure 4-7 shows the partition and the port signals that are associated with each unit of *Add_rca_16_0_delay*, an idealized model that ignores the propagation delay of the gates. Zero-delay models are very useful for verifying functionality; and by substituting other models at the lower levels of the design hierarchy we can account for nonzero propagation delay. For example, we could add unit propagation delay to the primitives in the models of the full and half adders to form a 16-bit adder with unit gate delays. Unit delay models are useful because their simulation reveals the time sequence of signal changes, which can be helpful in debugging a model.

The complete text for *Add_rca_16_0_delay* is given below.

```
module Add_rca_16_0_delay (sum, c_out, a, b, c_in);
    output [15:0]       sum;
    output              c_out;
    input [15:0]        a, b;
    input               c_in;
    wire                c_in4, c_in8, c_in12, c_out;

    Add_rca_4 M1    (sum[3:0],     c_in4,    a[3:0],    b[3:0],    c_in);
    Add_rca_4 M2    (sum[7:4],     c_in8,    a[7:4],    b[7:4],    c_in4);
    Add_rca_4 M3    (sum[11:8],    c_in12,   a[11:8],   b[11:8],   c_in8);
    Add_rca_4 M4    (sum[15:12],   c_out,    a[15:12],  b[15:12],  c_in12);
endmodule

module Add_rca_4 (sum, c_out, a, b, c_in);
    output [3: 0]       sum;
    output              c_out;
    input [3: 0]        a, b;
    input               c_in;
    wire                c_in2, c_in3, c_in4;

    Add_full M1     (sum[0],    c_in2,    a[0], b[0], c_in);
    Add_full M2     (sum[1],    c_in3,    a[1], b[1], c_in2);
    Add_full M3     (sum[2],    c_in4,    a[2], b[2], c_in3);
    Add_full M4     (sum[3],    c_out,    a[3], b[3], c_in4);
endmodule

module Add_full_0_delay(sum, c_out, a, b, c_in);
    output              sum, c_out;
    input               a, b, c_in;
    wire                w1, w2, w3;
    Add_half_0_delay    M1 (w1, w2, a, b);
    Add_half_0_delay    M2 (sum, w3, w2, c_in);
    or                  M3 (c_out, w2, w3);
endmodule

module Add_half_0_delay (sum, c_out, a, b);
    output              sum, c_out;
    input               a, b;
    xor                 M1 (sum, a, b);
    and                 M2 (c_out, a, b);
endmodule
```

4.1.6   Design Hierarchy and Source-Code Organization

The hierarchical model for *Add_rca_16_0_delay* illustrates how Verilog supports top-down structured design by nesting modules within modules. Figure 4-8 shows the hierarchy for *Add_rca_16* (a model without the *0_delay* notation). The top-level functional unit is encapsulated in *Add_rca_16*, and it contains instantiations of other functional units of lesser complexity, and so on. The lowest level of the hierarchy consists of

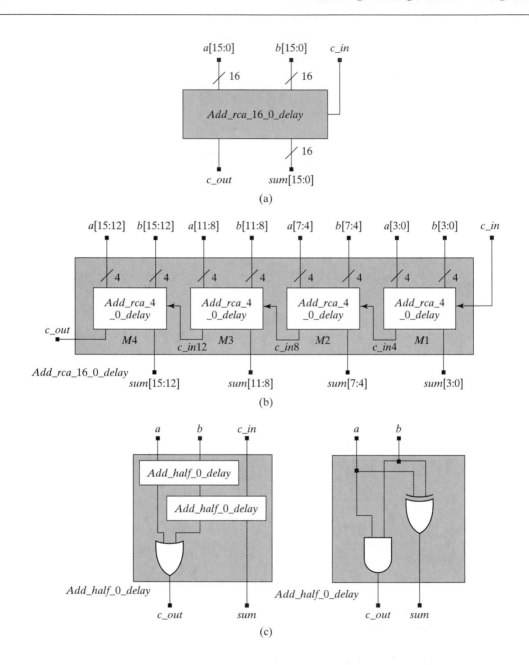

FIGURE 4-7 Hierarchical decomposition of a 16-bit, 0-delay, ripple-carry adder into a chain of four 4-bit-slice adders, each formed by a chain of full adders: (a) top-level schematic symbol, (b) decomposition into four 4-bit adders, and (c) full and half adders.

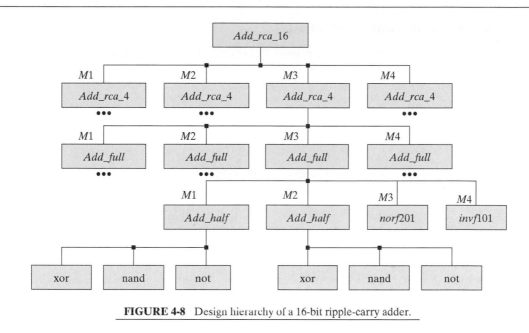

FIGURE 4-8 Design hierarchy of a 16-bit ripple-carry adder.

primitives and/or modules that have no underlying hierarchical detail. All of the modules that compose a design must be placed in one or more text files that, when compiled together, completely describe the functionality of the top-level module. It does not matter how the modules are distributed across multiple source code files, as long as their individual descriptions reside in a single file. A simulator will compile designated source files and extract the modules that it needs in order to integrate a complete description of any design hierarchy that is implied by the port structures and references to nested/instantiated modules.

4.1.7 Vectors in Verilog

A vector in Verilog is denoted by square brackets, enclosing a contiguous range of bits, e.g., *sum[3:0]* represents four bits from *sum*, which was declared in *Add_rca_16_0_delay* as a 16-bit signal. *The language specifies that, for the purpose of calculating the decimal equivalent value of a vector, the leftmost index in the bit range is the most significant bit, and the rightmost is the least significant bit.* An expression can be the index of a part-select. If the index of a part-select is out of bounds the value **x** is returned by a reference to the variable.[9] For example, if an 8-bit word *vect_word[7:0]* has a stored value of decimal 4, then *vect_word[2]* has a value of 1; *vect_word[3:0]* has a value of 4; and *vect_word[5:1]* has a value of 2.

---

[9]In Verilog's logic system, the symbol **x** represents an ambiguous (unknown) value, not a don't-care.

4.1.8   Structural Connectivity

Wires in Verilog establish connectivity between design objects. They connect primitives to other primitives and/or modules, and connect modules to other modules and/or primitives. By themselves, wires have no logic. The variable type *wire* is a member of a family of nets, all of which establish connectivity in a design.[10]

MODELING TIP

Use nets to establish structural connectivity.

The logic value of a ***wire*** (net) is determined dynamically during simulation by what is connected to the wire. If a *wire* is attached to the output of a primitive (module), it is said to be driven by the primitive (module), and the primitive (module) is said to be its driver. For example, in Figure 4-5 *y_out* is driven by a ***nor*** gate (primitive). The logic of the gate and the values of its inputs determine *y_out*. In that example we explicitly declared *y*1 and *y*2 to have type ***wire***, but did not have to do so. *Any identifier that is referenced without having a type declaration is by default of type **wire**.*[11] Consequently, the input and output ports have default type ***wire*** too, unless we specifically declare them to have a different type (e.g., we will see that a variable may have type ***reg***).

MODELING TIP

An undeclared identifier is treated by default as a ***wire***.

The ports of an instantiated module must be connected in a manner that is consistent with the declaration of the module, but the names of the connecting signals need not be the same. In Example 4.2 the formal name of the first port of *Add_half_0_delay* (i.e., the name given in the declaration of *Add_half_0_delay*) is *Sum*, but in instance *M*1 the actual name of the port is *w*1. The actual ports were associated with the formal ports by their position in the port list. This mechanism works well in models that have only a few ports, but when the list of ports is large it is easier and safer to associate ports by their names using the following convention in the port list: *.formal_name(actual_name)*. This connects *actual_name* to *formal_name*, regardless of the position of this entry in the list. The *formal_name* is the name given in the declaration of the instantiated module, and *actual_name* is the name used in the instantiation of the module.

---

[10]The Verilog family of predefined nets is described in Appendix C.
[11]The default net type can be changed by a compiler directive.

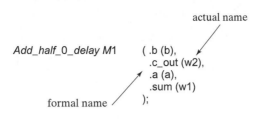

FIGURE 4-9   Formal and actual names for port association by name in module *Add_full.*

Example 4.3

The first (*M*1) instantiation of *Add_half_0_delay* in *Add_full_0_delay* can be written using port name association as shown in Figure 4-9.

End of Example 4.3

Our next example of a structural model will be used as a point of comparison with other examples to illustrate alternative styles of design with Verilog.

Example 4.4

A 2-bit comparator compares two 2-bit binary words, *A* and *B*, and asserts outputs indicating whether the decimal equivalent of word *A* is less than, greater than, or equal to that of word *B*. The functionality of the comparator is described by the set of Boolean equations below, where *A*1 and *A*0 are the bits of *A*, and *B*1 and *B*0 are the bits of *B*.

$A\_lt\_B = A1'\ B1 + A1'\ A0'\ B0 + A0'\ B1\ B0$

$A\_gt\_B = A1\ B1' + A0\ B1'\ B0' + A1\ A0\ B0'$

$A\_eq\_B = A1'\ A0'\ B1'\ B0' + A1'\ A0\ B1'\ B0 + A1\ A0\ B1\ B0 + A1\ A0'\ B1\ B0'$

The Karnaugh map methods of Chapter 2 can be used to eliminate redundant logic from these equations and produce the generic gate-level description of the comparator shown in Figure 4-10. This gate-level, combinational logic implementation of the comparator can be modeled by a structural interconnection of Verilog primitives. Their aggregate behavior is that of the comparator circuit.

A structural Verilog description corresponding directly to the schematic is shown below. Notice that two of the instantiations of the ***and*** gate have three inputs, and the others have two inputs. This feature of the built-in primitives allows the same primitive

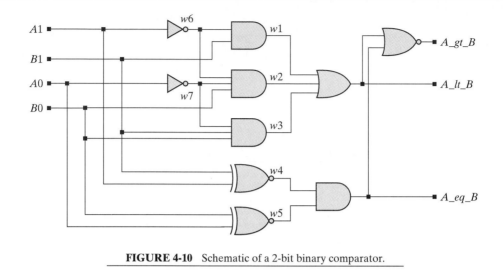

FIGURE 4-10   Schematic of a 2-bit binary comparator.

to be used freely, for whatever number of inputs are required by the context of its usage.

```
module compare_2_str (A_gt_B, A_lt_B, A_eq_B, A0, A1, B0, B1);
    output    A_gt_B, A_lt_B, A_eq_B;
    input     A0, A1, B0, B1;

    nor       (A_gt_B, A_lt_B, A_eq_B);
    or        (A_lt_B, w1, w2, w3);
    and       (A_eq_B, w4, w5);
    and       (w1, w6, B1);
    and       (w2, w6, w7, B0);
    and       (w3, w7, B0, B1);
    not       (w6, A1);
    not       (w7, A0);
    xnor      (w4, A1, B1);
    xnor      (w5, A0, B0);
endmodule
```

We will see in Chapter 6 that a synthesis tool can automatically optimize a gate-level description, remove redundant logic, and draw the resulting schematic. Next, we will use the 2-bit comparator as a building block to form a structural model of a 4-bit comparator.

End of Example 4.4

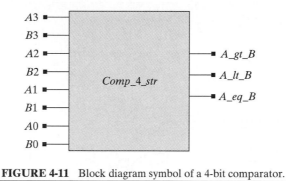

FIGURE 4-11   Block diagram symbol of a 4-bit comparator.

Example 4.5

A 4-bit comparator has the block diagram symbol shown in Figure 4-11. The comparator compares 4-bit binary words and asserts outputs indicating their relative size. It would be cumbersome to write the Boolean equations for the outputs, so we will connect the outputs of two 2-bit comparators with additional logic to generate the appropriate outputs that result from comparing 4-bit words. The logic for connecting the 2-bit comparators is based on the observation that a strict inequality in the higher order bit-pair determines the relative magnitudes of the 4-bit words; on the other hand, if the higher-order bit-pairs are equal, the lower-order bit-pairs determine the output. The hierarchical structure shown in Figure 4-12 implements the 4-bit comparator, and the simulation results in Figure 4-13 display the assertions of the outputs for some values of the bits of the datapaths, with $A\_bus = \{A3, A2, A1, A0\}$ and $B\_bus = \{B3, B2, B1, B0\}$ formed in the testbench for the purpose of simulation.

The source code for module *Comp_4_str* is given below. It contains two instantiations of the module *Comp_2_str* that was declared in Example 4.4.

```
module Comp_4_str (A_gt_B, A_lt_B, A_eq_B, A3, A2, A1, A0, B3, B2, B1, B0);
   output   A_gt_B, A_lt_B, A_eq_B;
   input    A3, A2, A1, A0, B3, B2, B1, B0;
   wire     w1, w0;

   Comp_2_str M1 (A_gt_B_M1, A_lt_B_M1, A_eq_B_M1, A3, A2, B3, B2);
   Comp_2_str M0 (A_gt_B_M0, A_lt_B_M0, A_eq_B_M0, A1, A0, B1, B0);

   or       (A_gt_B, A_gt_B_M1, w1);
   and      (w1, A_eq_B_M1, A_gt_B_M0);
   and      (A_eq_B, A_eq_B_M1, A_eq_B_M0);
   or       (A_lt_B, A_lt_B_M1, w0);
   and      (w0, A_eq_B_M1, A_lt_B_M0);
endmodule
```

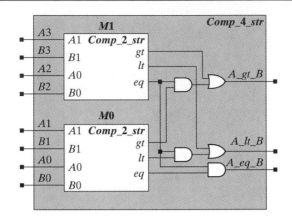

FIGURE 4-12   Hierarchical structure of a 4-bit binary comparator composed of 2-bit comparators and glue logic.

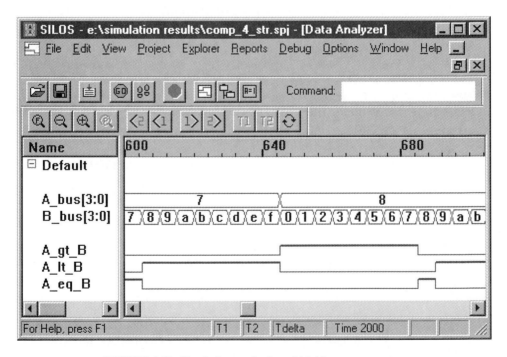

FIGURE 4-13   Simulation results for a 4-bit binary comparator.

End of Example 4.5

4.2 Logic System, Design Verification, and Test Methodology

Language-based models of a circuit must be verified to assure that their functionality conforms to the specification for the design. Two methods of verification are used: logic simulation and formal verification. Logic simulation applies stimulus patterns to a circuit and monitors its simulated behavior to determine whether it is correct. Formal verification uses elaborate mathematical proofs to verify a circuit's functionality without having to apply stimulus patterns. Although the use of formal methods is increasing, due to the difficulty of fully simulating large circuits, logic simulation is still widely used. We will consider only logic simulation.

4.2.1 Four-Value Logic and Signal Resolution in Verilog

Verilog uses a four-valued logic system having the symbols: 0, 1, x, and z. The language's abstract modeling constructs and the truth tables of its built-in primitives are defined for all four values of a primitive's inputs.[12] A simulator creates input waveforms in this four-value logic system and generates the internal and output signals for a circuit.

In Verilog's four-value logic system the values 0 and 1 correspond, respectively, to assertion (True) or de-assertion (False) of a signal. A signal in an actual circuit will have only these values, but simulators can accommodate additional logic values. The value x represents a condition of ambiguity in which the simulator cannot determine whether the value of the signal is 0 or 1. This happens, for example, when a net is driven

---

MODELING TIP

The logic value x denotes an unknown (ambiguous) value.
The logic value z denotes a high impedance.

---

by two primitives that have opposing output values. The primitive gates that are built into Verilog are able to model automatically this kind of contention between signals. (Verilog also has models for open collector and emitter follower logic in which signal contention is resolved by the technology itself to form either wired-and or wired-or structures, respectively [1].)

---

[12]Appendix A describes Verilog's built-in primitives and their truth tables in Verilog's four-value logic system.

The logic value *z* denotes a three-state condition in which a wire is disconnected from its driver. Figure 4-14 shows the waveforms that a simulator would produce in simulating an ***and*** primitive that is driven by signals that range over all of the possible input values. The waveforms in Figure 4-15 demonstrate how a Verilog simulator resolves multiple drivers on a net. Note that the three-state primitives[13] produce a value of *z* when they are not enabled, and that the value of *x* is produced when a ***wire*** is driven by opposing values. In practice, great care is taken to ensure that a bus does not have contending multiple drivers active at the same time.

4.2.2   Test Methodology

A large circuit must be tested and verified systematically to ensure that all of its logic has been exercised and found to be functionally correct. A haphazard test methodology makes debugging very difficult, creates a false sense of security that the model is correct, and risks enormous loss should a product fail as a result of an untested area of logic. In practice, design teams write an elaborate test plan that specifies the features that will be tested and the process by which they will be tested.

Example 4.6

The partition of a 16-bit ripple-carry adder into four bit-slices of 4 bits each was shown in Figure 4-7. Each of the four-bit units was built as a chain of full adders, and each full adder was composed of a simple hierarchy of half-adders and glue logic. An attempt to verify the 16-bit adder by applying 16-bit stimulus patterns to its datapaths and an additional bit for its carry-in bit requires checking the results of applying 2^{33} input patterns! It would be foolhardy to attempt this if a simpler strategy can be found. Applying 2^{33} patterns consumes a lot of processing time, and any error that might be detected would be difficult

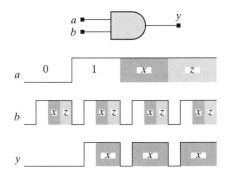

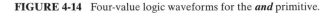

FIGURE 4-14   Four-value logic waveforms for the ***and*** primitive.

---

[13]See Appendix A.

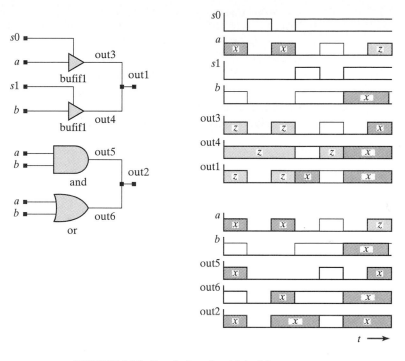

FIGURE 4-15   Resolution of multiple drivers on a net.

to associate with the underlying circuit. A more clever strategy is to verify that the half adder and fulladder each work correctly. Then, the 4-bit slice unit can be verified exhaustively by applying only 2^9 patterns! Once these simpler design units have been verified, the 16-bit adder can be tested to work correctly for a carefully chosen set of patterns that check the connectivity between the four units. This reduces the number of patterns and focuses the debugging effort on a much smaller portion of the overall circuitry. (See the problems section at the end of the chapter.)

We saw that modeling begins with a complex functional unit and partitions it in a top-down fashion to enable the design of simpler units. Systematic verification proceeds in the opposite direction, beginning with the simpler units and moving to the more complex units above them in the design hierarchy. A basic methodology for verifying the functionality of a digital circuit consists of building a testbench that applies stimulus patterns to the circuit and displays the waveforms that result. The user, or software, can verify that the response is correct. The testbench is a separate Verilog module that has the basic organization shown in Figure 4-16. It resides at the top of a new design hierarchy that contains a stimulus generator, a response monitor, and an instantiation of the unit under test (UUT). The stimulus generator uses Verilog statements to define the patterns that are to be applied to the circuit. During simulation the response

Design_Unit_Test_Bench (DUTB)

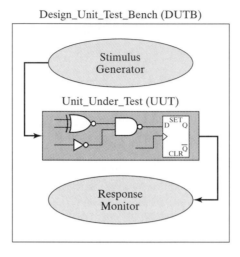

FIGURE 4-16   Organization of a testbench for verifying a unit under test.

monitor selectively gathers data on signals within the design and displays them in a text or graphical format. Testbenches can be very complex, containing a variety of pattern generators and additional software to perform analysis on the gathered data to detect and report functional errors.

A simulator performs three essential tasks: it (1) checks the source code, (2) reports any violations of the language's syntax,[14] and (3) simulates the behavior of the circuit under the application of input signals that are defined in the testbench. Syntax errors must be eliminated before a simulation can run, but be aware that absence of syntax errors does not imply that the functionality of the model is correct.

End of Example 4.6

Example 4.7

The module below, *t_Add_half*, has the basic structure of a testbench for verifying *Add_half_0_delay* in a simulator having a graphical user interface.[15] Notice that it contains an instantiation of the UUT, *Add_half_0_delay*. The waveforms that are to be applied to the UUT are not generated by hardware. Instead, they are generated abstractly by a single-pass Verilog behavior, declared by the keyword ***initial*** and

---

[14]Appendix E presents the formal syntax of the Verilog language.
[15]See "Selected System Tasks and Functions" at the companion web sites for built-in system tasks, such as ***$monitor***, which can be used to display simulation results in a text format.

accompanied by statements enclosed by the block statement keyword pair **begin** ... **end**. In this simple example the user serves as the response monitor by comparing the output waveforms to their expected values. Constructs used in a testbench are introduced in *t_Add_half*, and will be explained in the next section.

```verilog
module t_Add_half();
    wire        sum, c_out;
    reg         a, b;

    Add_half_0_delay M1 (sum, c_out, a, b);            //UUT

    initial begin                                      // Time Out
    #100 $finish;
    end

    initial begin                                      // Stimulus patterns
    #10 a = 0; b = 0;
    #10 b = 1;
    #10 a = 1;
    #10 b = 0;
    end
endmodule
```

End of Example 4.7

4.2.3   Signal Generators for Testbenches

A Verilog behavior is a group of statements that execute during simulation to assign value to simulation variables as though they were driven by hardware. The keyword *initial* declares a single-pass behavior that begins executing when the simulator is activated, at $t_{sim} = 0$ (we use t_{sim} to denote the timebase of the simulator). The statements that are associated with the behavior are listed within the **begin** ... **end** block keywords, and are called procedural statements. When each procedural statement (e.g., $b = 1$;) in the testbench *t_Add_half* in Example 4.7 executes, it assigns a value to a variable. Using the procedural assignment operator, =, such statements are called "procedural assignments."

MODELING TIP

Use procedural assignments to describe stimulus patterns in a testbench.

The time at which a procedural assignment statement in a **begin** ... **end** block executes depends on its order in the list of statements, and on the delay time preceding the statement (e.g., #10). The statements execute sequentially from top to bottom and from left to right across lines of text that may contain multiple statements. In this example each line is preceded by a time delay (e.g., 10 simulator time units) that is prescribed with a delay control operator (#) and a delay value, for example, #10. A delay control operator preceding a procedural assignment statement suspends its execution and, consequently, suspends the execution of the subsequent statements in the behavior until the specified time interval has elapsed during simulation. A single-pass behavior expires when the last statement has executed, but (in general) the simulation does not necessarily terminate, because other behaviors might still be active.

The general structure of a testbench consists of one or more behaviors that generate waveforms at the inputs to the UUT, monitor the simulation data, and control the overall sequence of activity. Note that the inputs to the UUT in *t_Add_half_0* are assigned value abstractly by a single-pass behavior and are declared by the keyword **reg**, which indicates that the variables (signals) *a* and *b* are getting their values from execution of procedural statements (just like in an ordinary procedural language, such as C, but as waveforms that evolve under the control of the simulator). Since hardware is not driving the value of an abstractly generated input, the type declaration **reg** ensures that the value of the variable will exist from the moment it is assigned by a procedural statement until execution of a later procedural statement changes it. The outputs of the UUT are declared as wires. Think of them as providing the ability to observe the output ports of the UUT.

The waveforms that result from simulating *t_Add_half_0_delay* are shown in Figure 4-17. In Verilog, all nets and all variables that have type **reg** are given the value *x* when the simulation begins, and they hold that value until they are assigned a different value. The cross-hatched waveform fill pattern denotes the value *x*. In this example, the inputs are assigned value at time 10, when the first assignment is made. The indicated

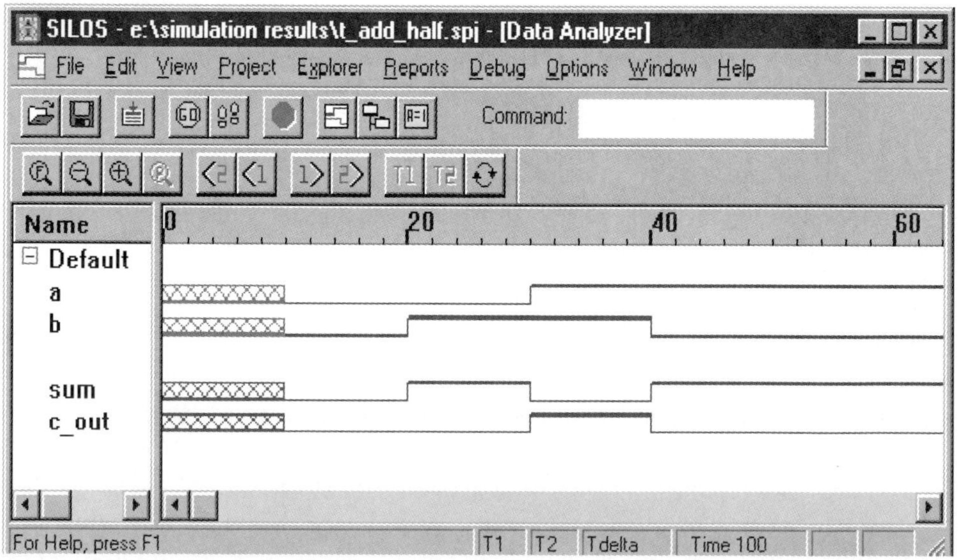

FIGURE 4-17   Waveforms produced by a simulation of *Add_half_0_Delay*, a 0-delay binary half adder.

simulation time steps are dimensionless; a timescale directive can be used to associate physical units with numerical values.[16] Note the correspondence between the waveforms specified by the stimulus generator in the testbench and the simulated waveforms applied to the circuit. Execution of each statement is delayed by 10 time steps, so the final statement in the stimulus generator executes at time step 40 (i.e., in this example the delays accumulate). The testbench includes a separate behavior for a stopwatch that terminates the simulation after 100 time steps. The stopwatch executes a built-in Verilog system task, *$finish*, which surrenders control to the operating system of the host machine when time step 100 is reached. A stopwatch is optional and not needed here, but, in general, it will prevent endless execution if the testbench otherwise fails to terminate the simulation.

MODELING TIP

A Verilog simulator assigns an *initial* value of **x** to all nets and to variables that have type **reg**.

4.2.4 Event-Driven Simulation

A change in the value of a signal (variable) during simulation is referred to as an "event." The one-pass behavior in *t_Add_half_0* programmed the inputs to *Add_half_0_delay* to change at prescribed times, independently of the signals in the circuit. However, the events of the outputs of the UUT, *sum* and *c_out*, depend on the occurrence of events at the inputs to the device, just as the signals in a physical circuit change in response to the signals at the input. By knowing (1) the schedule of events at the inputs and (2) the structure of a circuit, we can create a schedule for the events at the outputs [3].

The simulators used for logic simulation are said to be event-driven because their computational activity is driven by the propagation of events in a circuit. Event-driven simulators are inactive during the interval between events. When an event occurs at the input to the UUT, the simulator schedules updating events for the internal signals and outputs of the UUT. Then the simulator rests until the next triggering event occurs at the input. All of the gates and abstract behaviors are active concurrently [4], and it is the simulator's job to detect events and schedule any new events that result from their occurrence.

4.2.5 Testbench Template

Testbenches are an important tool in the design flow of an ASIC. Much effort is expended to develop a thorough testbench because it is an insurance policy against failure. A test plan should be developed *before* the testbench is written. The plan, at a minimum, should specify what features will be tested and how they will be tested in the testbench. Testbenches are customized to the UUT, but the general structure given below can be used to develop testbenches for the problems at the ends of the chapters.

---

[16]See Verilog's compiler directives at the companion web site.

```
module t_DUTB_name ();          // substitute the name of the UUT
    reg ...;                    // Declaration of register variables for primary
                                // inputs of the UUT
    wire ...;                   // Declaration of primary outputs of the UUT
    parameter                      time_out = // Provide a value

UUT_name M1_instance_name (UUT ports go here);
    initial $monitor ( );       // Specification of signals to be monitored and
                                // displayed as text

    initial #time_out $stop;    // Stopwatch to assure termination of simulation
    initial                     // Develop one or more behaviors for pattern
                                // generation and/or
                                // error detection

    begin

                                // Behavioral statements generating waveforms
                                // to the input ports, and comments documenting
                                // the test. Use the full repertoire of behavioral
                                // constructs for loops and conditionals.
    end
endmodule
```

4.2.6   Sized Numbers

The values assigned to the stimulus waveforms in the testbench in Example 4.7 are sized numbers. *Sized numbers* specify the number of bits that are to be stored for a value. For example, $8'ha$ denotes an 8-bit stored value corresponding to the hexadecimal number a; the binary value in memory will be 0000_1010.[17] Unsized numbers are stored as integers having length determined by the host simulator (usually 32 bits). Four formats are available: binary (b), decimal (d), octal (o), and hexadecimal (h). The format specifier is not case-sensitive, and by default a number is interpreted to be a decimal value.

4.3   Propagation Delay

Physical logic gates have a propagation delay between the time that an input changes and the time that the output responds to the change. The primitives in Verilog have a default delay of 0, meaning that the output responds to the input immediately, but a nonzero delay can also be associated with a primitive. Timing verification ultimately depends on realistic values of the propagation delays in a circuit, but simulation is commonly done with 0 delay to verify the functionality of a model quickly. Simulation with a unit delay is often done, too, because it exposes the time sequence of signal activity, which can be masked by 0-delay simulation.

---

[17]In Verilog, the underscore may be inserted in the representation of a number to make it more readable.

> ## MODELING TIP
>
> All primitives and nets have a default propagation delay of 0.

Example 4.8

The primitives within *Add_full* and *Add_half* are shown below with annotation that assigns to them a unit delay. The delay notation #1 has been inserted before the instance name of each instantiated primitive (# denotes a delay control operator). The effect of the delay is apparent in the simulated transitions of *sum* and *c_out* shown in Figure 4-18. Notice that the 0-delay simulation results do not reveal whether *c_out* is formed before or after *sum*. Both, in fact, appear to change as soon as the inputs change. The unit-delay model reveals the time-ordering of the signal activity.

```
module Add_full(sum, c_out, a, b, c_in);
    output      sum, c_out;
    input       a, b, c_in;
    wire        w1, w2, w3;

    Add_half    M1 (w1, w2, a, b);
    Add_half    M2 (sum, w3, w1, c_in);
    or          #1 M3 (c_out, w2, w3);
endmodule

module Add_half (sum, c_out, a, b);
    output      sum, c_out;
    input       a, b;
    xor         #1 M1 (sum, a, b);
    and         #1 M2 (c_out, a, b);
endmodule
```

ASICs are fabricated by assembling onto a common silicon die the logic cells from a standard-cell library. The library cells are predesigned and precharacterized so that their Verilog model includes accurate timing information, and synthesis tools use this information to optimize the performance (speed) of a design. Our focus in this book will be on synthesizing gate-level structures from technology-independent[18] behavioral models of circuits, using either standard cells or field-programmable gate arrays (FPGAs). The timing characteristics of the latter are embedded within the synthesis tools for FPGAs; the timing characteristics of the former are embedded within the

---

[18]Technology-independent behavioral models describe only the functionality of a circuit, not its propagation delays.

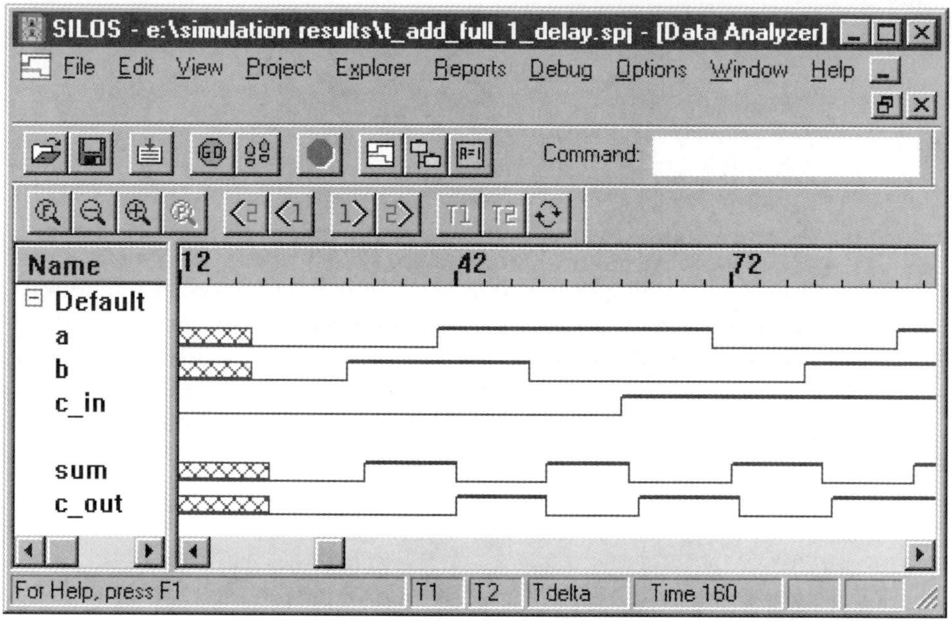

FIGURE 4-18   Results of unit-delay simulation of a 1-bit full adder.

model of the cell and are used by a synthesis tool to analyze the timing of a circuit in conjunction with selecting parts from a cell library to realize specified logic. Circuit designers do not attempt to create accurate gate-level timing models of a circuit by manual methods. Instead, they rely on a synthesis tool to implement a design that will satisfy timing constraints. We will address this topic in Chapter 10.[19]

End of Example 4.8

Example 4.9

The models of *Add_half_ASIC* and *Add_full_ASIC* shown below use parts (*norf201*, *invf101*, *xorf201*, and *nanf201*) from a standard-cell library.[20] Complementary metal-oxide semiconductor (CMOS) *and* gates are commonly implemented by combining a *nand* gate with an inverter; likewise, an *or* gate is implemented by a *nor* gate and an inverter. The Verilog timescale directive 'timescale 1ns / 1ps at the first line of the source

---

[19]See "Additional Features of Verilog" at the companion web site for additional information about modeling propagation delay.
[20]*norf201* is a two-input nor gate standard cell; *invf101* is an inverter, *xorf201* is a two-input excusive-or gate standard cell; and *nanf201* is a two-input nand standard cell.

file directs the simulator to interpret the numerical time variables as having units of nanoseconds, with a resolution of picoseconds.[21]

```
'timescale 1 ns / 1 ps
module Add_full_ASIC
    (sum, c_out, a, b, c_in);
    output      sum, c_out;
    input       a, b, c_in;
    wire        w1, w2, w3;
    wire        c_out_bar;

    Add_half_ASIC   M1 (w1, w2, a, b);
    Add_half_ASIC   M2 (sum, w3, w1, c_in);

    norf201         M3 (c_out_bar, w2, w3);
    invf101         M4 (c_out, c_out_bar);
endmodule
```

```
module Add_half_ASIC (sum, c_out, a, b);
    output      sum, c_out;
    input       a, b;
    wire        c_out_bar;
    xorf201     M1 (sum, a, b);
    nanf201     M2 (c_out_bar, a, b);
    invf101     M3 (c_out, c_out_bar);
endmodule
```

The models for *norf201*, *invf101*, *xorf201*, and *nanf201* include propagation delays based on characterization of the physical standard cells. The effect of the realistic propagation delays is apparent in the waveforms produced by simulating *Add_full_ASIC*, as shown in Figure 4-19.

End of Example 4.9

4.3.1 Inertial Delay

Logic transitions in a digital circuit correspond to transitions in voltage levels caused by the accumulation or dissipation of charge at a physical node/net. The physical behavior of a signal transition is said to have inertia, because every conducting path has some capacitance, as well as resistance, and charge cannot accumulate or dissipate instantly. Hardware description languages must be able to model these effects.

The propagation delay of the primitive gates in Verilog obeys an inertial delay model. This model accounts for the fact that charge must accumulate in the physical circuit before a voltage level can be established corresponding to a 0 or a 1. If an input signal is applied to a gate and then removed before sufficient charge has accumulated the output signal will not reach a voltage level corresponding to a transition. For example, if all the inputs to a **nand** gate are at value 1 for a long time before one of them is changed momentarily to 0, the output will not change to 1 unless the input is held to 0 for a long enough time. The amount of time that the input pulse must be constant in order for the gate to make a transition is the inertial delay of the gate.

---

[21]See "Compiler Directives" at the companion web site for a description of time scales.

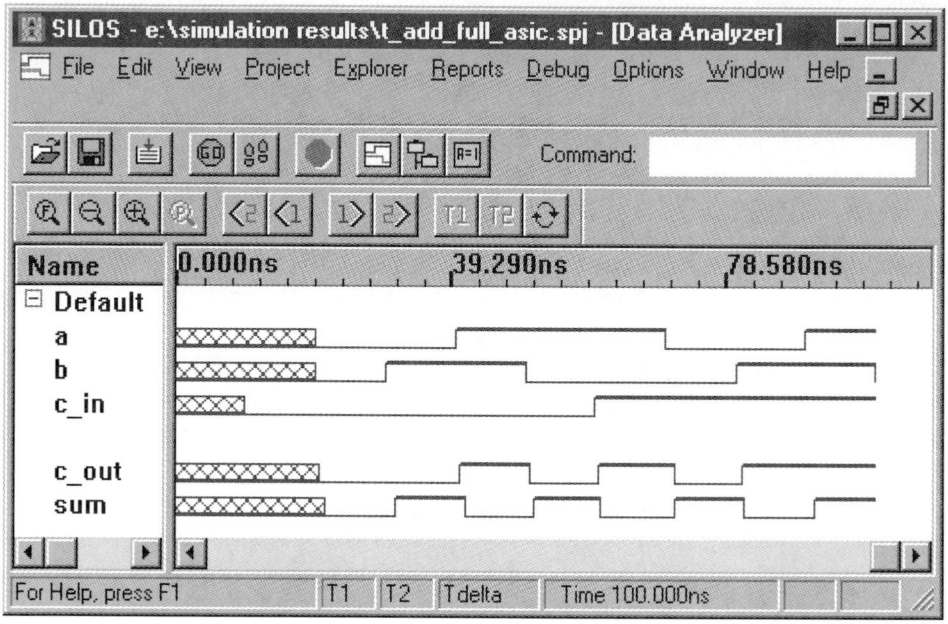

FIGURE 4-19   Results of simulating a 1-bit full adder implemented with ASIC cells having technology-dependent propagation delays.

Verilog uses the propagation delay of a gate as the minimum width of an input pulse that could affect the output; that is the value of propagation delay is also used as the value of the inertial delay. The width of a pulse must be at least as long as the propagation delay of the gate. The Verilog simulation engine detects whether the duration of an input has been too short, and then deschedules previously scheduled outputs triggered by the leading edge of a pulse. Inertial delay has the effect of suppressing input pulses whose duration is shorter than the propagation delay of the gate.

Example 4.10

The input to the inverter in Figure 4-20 changes at $t_{sim} = 3$. Because the inverter has a propagation delay of 2, the effect of this change is to cause the output to be scheduled to change at $t_{sim} = 5$. However, for pulsewidth $\Delta = 1$, the input changes back to the initial value at $t_{sim} = 4$. The simulator cannot anticipate this activity. The effect of the two successive changes is to create a narrow pulse at the input to the inverter. Because the pulsewidth is less than the propagation delay of the inverter, the simulator deschedules the previously scheduled output event corresponding to the leading edge of

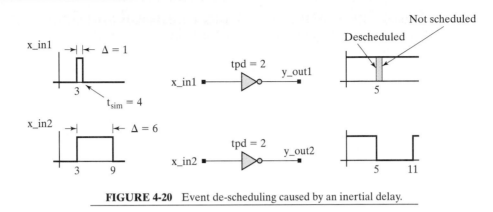

FIGURE 4-20 Event de-scheduling caused by an inertial delay.

the narrow input pulse, and does not schedule the event corresponding to the trailing edge of the pulse. Descheduling is required because the simulator cannot anticipate the falling edge and must wait until it occurs. On the other hand, the pulse with $\Delta = 6$ persists sufficiently long for the output to be affected.

In physical circuits, the propagation delay of a logic gate is affected by its internal structure and by the circuit that it drives. The internal delay is referred to as the intrinsic delay of the gate. In a circuit, the driven gates and the metal interconnect of their fanin nets create additional capacitive loads on the output of the driving gate and affect its timing characteristics. The slew rate of the input signal, which represents the slope of a signal's transition between logic values, can also have an affect on the transitions of the output signal. Accurate standard-cell models account for all these effects.

End of Example 4.10

4.3.2 Transport Delay

The time-of-flight of a signal traveling a wire of a circuit is modeled as a transport delay. With this model narrow pulses are not suppressed, and all transitions at the driving end of a wire appear at the receiving end after a finite time delay. In most ASICs, the physical distances are so small that the time-of-flight on wires can be ignored, because at the speed of light the signal takes only .033 ns to travel a centimeter. However, Verilog can assign delay to individual wires in a circuit to model transport delay effects in circuits where it cannot be neglected, such as in a multichip hardware module or on a printed circuit board. Wire delays are declared with the declaration of a wire. For example, *wire #2 A_long_wire* declares that *A_long_wire* has a transport delay of two time steps.

4.4   Truth Table Models of Combinational and Sequential Logic With Verilog

Verilog supports truth-table models of combinational and sequential logic. Although Verilog's built-in primitives correspond to simple combinational logic gates, and do not include sequential parts, the language has a mechanism for building user-defined primitives (UDPs), which use truth tables to describe sequential behavior and/or more complex combinational complex logic. UDPs are widely used in ASIC cell libraries because they simulate faster and require less storage than modules.

Example 4.11

The text below declares *AOI_UDP*, a truth table version of the five-input AOI circuit that was introduced in Section 4.1.2.

```
primitive AOI_UDP (y, x_in1, x_in2, x_in3, x_in4, x_in5);
  output y;
  input   x_in1, x_in2, x_in3, x_in4, x_in5;
  table
// x1 x2 x3 x4 x5 : y
  0 0 0 0 0 : 1;
  0 0 0 0 1 : 1;
  0 0 0 1 0 : 1;
  0 0 0 1 1 : 1;
  0 0 1 0 0 : 1;
  0 0 1 0 1 : 1;
  0 0 1 1 0 : 1;
  0 0 1 1 1 : 0;
  0 1 0 0 0 : 1;

  0 1 0 0 1 : 1;
  0 1 0 1 0 : 1;
  0 1 0 1 1 : 1;
  0 1 1 0 0 : 1;
  0 1 1 0 1 : 1;
  0 1 1 1 0 : 1;
  0 1 1 1 1 : 0;

  1 0 0 0 0 : 1;
  1 0 0 0 1 : 1;
  1 0 0 1 0 : 1;
  1 0 0 1 1 : 1;
  1 0 1 0 0 : 1;
  1 0 1 0 1 : 1;
```

```
              1 0 1 1 0 : 1;
              1 0 1 1 1 : 0;

              1 1 0 0 0 : 0;
              1 1 0 0 1 : 0;
              1 1 0 1 0 : 0;
              1 1 0 1 1 : 0;
              1 1 1 0 0 : 0;
              1 1 1 0 1 : 0;
              1 1 1 1 0 : 0;
              1 1 1 1 1 : 0;
          endtable
        endprimitive
```

UDPs are declared in a source file in the same way that a module is declared, but with the encapsulating keyword pair ***primitive*** ... ***endprimitive***. They can be instantiated just like built-in primitives, with or without propagation delay. A UDP has only a single, scalar (single-bit), output port; also, the input ports of a UDP must be scalars.

End of Example 4.11

MODELING TIP

The output of a user-defined primitive must be a scalar.

The truth table for a UDP consists of a section of columns, one for each input, followed by a colon and a final column that specifies the output. The order of the input columns must conform to the order in which the input ports are listed in the declaration.[22] A simulator references the table whenever one of its inputs changes, and proceeds from the top toward the bottom of the table searching for a match. The search terminates at the first match.

Example 4.12

The Verilog UDP *mux_prim*, shown in Figure 4-21 describes a two-input multiplexer and includes comments citing some basic rules for UDP models.

A simulator will automatically assign the (default) value *x* to the output of a UDP if its inputs have values that do not match a row of the table. An input value of *z*

---

[22]See "Rules for UDPs" at the companion web site for additional features and rules of use for UDPs.

```
primitive mux_prim (mux_out, select, a, b);
  output mux_out;
  input  select, a, b;
  table
```

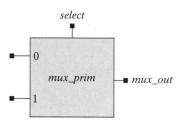

```
// select      a       b       :  mux_out
```

select	a	b	:	mux_out	
0	0	0	:	0 ;	// Order of table columns 5 port order of inputs
0	0	1	:	0 ;	// One output, multiple inputs, no inout
0	0	x	:	0 ;	// Only 0, 1, x on input and output
0	1	0	:	1 ;	// A z input in simulation is treated as x
0	1	1	:	1 ;	// by the simulator
0	1	x	:	1 ;	// Last column is the output

```
// select      a       b       :  mux_out
```

select	a	b	:	mux_out
1	0	0	:	0 ;
1	1	0	:	0 ;
1	x	0	:	0 ;
1	0	1	:	1 ;
1	1	1	:	1 ;
1	x	1	:	1 ;

x	0	0	:	0 ;	// Reduces pessimism
x	1	1	:	1 ;	

```
  endtable          // Note: Combinations not explicitly specified will drive 'x'
  endprimitive      // under simulation.
```

FIGURE 4-21   UDP for a two-input multiplexer.

is treated as *x* by the simulator. The last two rows of the table describing the behavior of the multiplexer in Figure 4-21 reduce the pessimism that might result during simulation. If both data inputs have the same value the output has that value, regardless of the value of the *select* input. When the value of *select* is *x* (ambiguous), the output is 0 if both inputs are 0, and 1 if both inputs are 1. If the UDP table overlooks this additional detail, a simulator will propagate a value of *x* to the output when *select* is *x*. It is generally desirable to reduce the situations under which a primitive propagates a value of *x* because ambiguity may reduce the amount of useful information that can be derived from a simulation.

The entries for the inputs in a truth table can be reduced by using a shorthand notation. The **?** symbol allows an input to take on any of the three values, 0, 1, and *x*. This allows one table row to effectively replace three rows.

End of Example 4.12

Example 4.13

The truth table below illustrates how the UDP of a two-input multiplexer could be rewritten using shorthand notation. When the *select* input is 0 and the *a* channel is 0, the output is 0, regardless of the value of the input at the *b* channel. When *select* is a 1 and the *b* channel is a 0, the output is a 0 when the *a* channel is a *0, 1,* or *x*. The ? shorthand notation substitutes 0, 1, and *x* in the table row and effectively implements a don't-care condition on the associated input.

```
table
// Shorthand notation:
// ? represents iteration of the table entry over the values 0,1,x.
// i.e., don't care on the input
//      select    a    b    : mux_out
//      0         0    ?    :       0 ; //? = 0,1,x shorthand notation.
//      0         1    ?    :       1 ;
//      1         ?    0    :       0 ;
//      1         ?    1    :       1 ;
//      ?         0    0    :       0 ;
//      ?         1    1    :       1 ;
endtable
```

Hardware devices can exhibit two basic kinds of sequential behavior: level-sensitive behavior (e.g., transparent latch), which is conditioned by an enabling input signal, and edge-sensitive behavior (e.g., a D-type flip-flop), which is synchronized by an input signal. Level-sensitive devices respond to any change of an input while the enabling input is high; edge-sensitive devices ignore their inputs until a synchronizing edge occurs. Sequential hardware devices may have the behavior of one or the other, or a combination of both.

A truth table that describes sequential behavior has input columns followed by a colon (:) and a column for the present state of the device, and another colon and a column for its next state, that is, the state that will be caused by the present inputs. In addition, the output of the UDP must be declared to have type *reg* because the value of the output is produced abstractly, by a table, and must be held in memory during simulation.

End of Example 4.13

MODELING TIP

The output of a sequential user-defined primitive must be declared to have type **reg**.

Example 4.14

A truth-table description of a transparent latch is given below by *latch_rp*. It describes transparent behavior and latching behavior, and also deals with the possibility that, under simulation, the input *enable* might acquire an *x* value.

```
primitive latch_rp (q_out, enable, data);
   output  q_out;
   input   enable, data;
   reg     q_out;

   table
//     enable      data            state       q_out/next_state
        1           1       :       ?       :       1 ;
        1           0       :       ?       :       0 ;
        0           ?       :       ?       :       - ;
// Above entries do not deal with enable = x.
// Ignore event on enable when data = state:
        x           0       :       0       :       - ;
        x           1       :       1       :       - ;
// Note: The table entry '-' denotes no change of the output.
   endtable
endprimitive
```

The transparent latch modeled by *latch_rp* exhibits level-sensitive behavior; that is, the output can change any time that an input changes, depending on the value of the inputs. The value that is scheduled for the output is determined only by the value of the *enable* and *data* inputs. In contrast, a truth table describing edge-sensitive behavior will be activated whenever an input has an event, but whether the output changes depends on whether a synchronizing input has made an appropriate transition. For example, a flip-flop that is sensitive to the rising edge of its clock would have an entry of (01) in the corresponding column of the table. UDPs can describe behavior that is sensitive to either the positive or negative edge (transition) of a clock signal, with built-in semantics for positive (**posedge**) and negative (**negedge**) signal transitions. A falling edge (**negedge**) transition is denoted by the following signal value pairs: (10), (1x), and (x0); rising edges are denoted by (01), (0x), and (x1).

End of Example 4.14

Example 4.15

The UDP *d_prim1* in Figure 4-22 describes the behavior of an edge-sensitive D-type flip-flop. The input signal *clock* synchronizes the transfer of *data* to *q_out*.

The table-entry notation for a sequential behavior uses parentheses to enclose the defining logic values of a signal whose transition affects the output (i.e., the synchronizing input signal). In the table in Figure 4-22, the (01) entry in the column for *clock* denotes a low-to-high transition of the signal *clock*—that is a value change. Note, also, that the row corresponding to the entry of (?0) for *clock* actually denotes 27 input possibilities and replaces 27 rows of entries, as there are two more symbols ? in that row. For example, (?0) represents (00), (10) and (x0). Each of these is combined with three possibilities for the data; each of the resulting nine possibilities is combined with three possibilities for the state. In effect, this row explicitly specifies that the output should not change in any of these situations. Since the model represents the physical behavior of a rising-edge sensitive behavior, the output should not change on a falling edge, or if there is no edge at all (00). Were this row omitted the model would propagate an X value under simulation. Remember, it is desirable that the UDP table be as complete and unambiguous as possible.

A truth table can include both level-sensitive behavior and edge-sensitive behavior to model synchronous behavior with asynchronous set and reset conditions. Because a simulator will search the truth table from top to bottom, the level-sensitive behavior should precede the edge-sensitive behavior in the table.

End of Example 4.15

```
primitive d_prim1 (q_out, clock, data);
  output        q_out;
  input         clock, data;

  reg           q_out;

  table
```

//	clk	data		state		q_out/next_state	
	(01)	0	:	?	:	0 ;	// Rising clock edge
	(01)	1	:	?	:	1 ;	
	(0?)	1	:	1	:	1 ;	
	(?0)	?	:	?	:	- ;	// Falling or steady clock edge
	?	(??)	:	?	:	- ;	// Steady clock, ignore data
							// transitions

```
  endtable
endprimitive
```

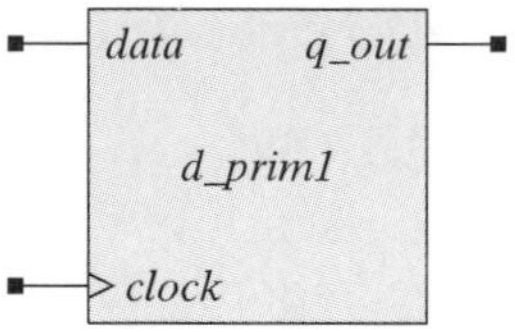

FIGURE 4-22 Truth-table model of a D-type flip-flop.

Example 4.16

A J-K flip-flop having asynchronous *preset* and *clear* with edge-sensitive sequential behavior is described in Figure 4-23. The *preset* and *clear* inputs are active-low, and the output is sensitive to the rising edge of the clock. The signal *clock* synchronizes the changes of *q_out*. Depending on the values of the *preset* and *clear* signals when the clock edge occurs, *q_out* does not change ($j = 0$, $k = 0$), *q_out* gets a value of 0, *q_out* gets a value of 1, ($j = 0$, $k = 1$), or *q_out* is toggled ($j = 1$, $k = 1$).

End of Example 4.16

```
primitive jk_prim (q_out, clk, j, k, preset, clear);
  output  q_out;
  input   clk, j, k, preset, clear;
  reg     q_out;



  table
```

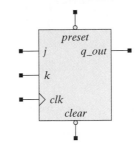

//	clk	j	k	pre	clr		state		q_out/next_state
//	Preset Logic								
	?	?	?	0	1	:	?	:	1 ;
	?	?	?	*	1	:	1	:	1 ;
//	Clear Logic								
	?	?	?	1	0	:	?	:	0 ;
	?	?	?	1	*	:	0	:	0 ;
// Normal Clocking									
//	clk	j	k	pre	clr		state		q_out/next_state
	r	0	0	0	0	:	0	:	1 ;
	r	0	0	1	1	:	?	:	- ;
	r	0	1	1	1	:	?	:	0 ;
	r	1	0	1	1	:	?	:	1 ;
	r	1	1	1	1	:	0	:	1 ;
	r	1	1	1	1	:	1	:	0 ;
	f	?	?	?	?	:	?	:	- ;
// j and k cases									
//	clk	j	k	pre	clr		state		q_out/next_state
	b	*	?	?	?	:	?	:	- ;
	b	?	*	?	?	:	?	:	- ;
/Reduced pessimism.									
	p	0	0	1	1	:	?	:	- ;
	p	0	?	1	?	:	0	:	- ;
	p	?	0	?	1	:	1	:	- ;
	(?0)	?	?	?	?	:	?	:	- ;
	(1x)	0	0	1	1	:	?	:	- ;
	(1x)	0	?	1	?	:	0	:	- ;
	(1x)	?	0	?	1	:	1	:	- ;
	x	*	0	?	1	:	1	:	- ;
	x	0	*	1	?	:	0	:	- ;

```
  endtable
endprimitive
```

FIGURE 4-23   UDP for a J-K flip-flop.

REFERENCES

1. *IEEE Standard Hardware Description Language Based on the Verilog Hardware Description Language*, Language Reference Manual, IEEE Std. 1364-1995. Piscataway, NJ: Institute of Electrical and Electronic Engineers, 1996.
2. Chang H, et al., *Surviving the System on Chip Revolution*. Boston: Kluwer, 1999.
3. Thomas R, Moorby P. *The Verilog Hardware Description Language*, 3rd ed. Boston: Kluwer, 1996.
4. Ciletti MD. *Modeling, Synthesis and Rapid Prototyping with the Verilog HDL*. Upper Saddle River, NJ: Prentice-Hall, 1999.

PROBLEMS

For all of the problems requiring development and verification of a design, a test plan, a testbench, and simulation results are to be provided with the solution. Supporting material, such as a state-transition graph or a truth table should be included too.

1. Using Verilog gate-level primitives, develop and verify a structural model for the circuit shown in Figure P4-1. Use the following name for your testbench, the model, and its ports: *t_Combo_str(), Combo_str (Y, A, B, C, D)*. *Note*: The testbench will have no ports. Exhaustively simulate the circuit and provide graphical output demonstrating that the model is correct. *Note*: If text output is desired, consider using the *$monitor* and *$display* tasks.

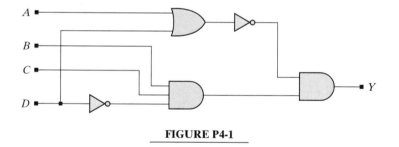

FIGURE P4-1

2. Repeat Problem 4.1 and develop *Combo_prim*, a user-defined primitive implementing a truth-table model of the logic. Then instantiate the UDP in a module, *Combo_UDP*. Verify that the model is correct and that its response matches that of *Combo_str* in Problem 1.
3. Develop a testbench that contains patterns exercising the structural connectivity of the 4-bit units in *Add_rca_16* (see Example 4.2).
4. Using Table 3.1, develop and verify a Verilog model for a circuit that converts a binary coded decimal encoded digit into an Excess-3 encoded digit.

5. Develop a testbench to verify the functionality of *d_prim1* (see Example 4.15). Include a written test plan describing (1) the functional features that will be tested and (2) how they will be tested.

6. Develop a testbench to verify the functionality of *jk_prim* (see Example 4.16). Include a written test plan describing (1) the functional features that will be tested and (2) how they will be tested.

7. Write continuous assignment statements describing the following Boolean functions: $Y1(A, B, C, D) = \Sigma m$ (4,   5,   6,   7,   11,   12,   13), $Y2(A, B, C) = \Sigma m(1, 2, 4, 4)$.

8. Develop and verify a UDP for a negative-edge triggered D-type flip-flop that has active-low reset.

9. Develop a testbench that verifies the functionality of *AOI_str* in Section 4.1.2.

10. Develop a testbench that verifies the functionality of *latch_rp* in Example 4.14.

11. Develop a small set of test patterns that will (1) test a half-adder circuit, (2) test a full-adder circuit, (3) exhaustively test a 4-bit ripple-carry adder, and (4) test a 16-bit ripple-carry adder by verifying that the connectivity between the 4-bit slices are connected correctly, given that the 4-bit slices themselves have been verified.

12. Write a testbench (including a test plan) to verify a gate-level model of a full adder.

13. Write a testbench (including a test plan) to verify a gate-level model of an S-R (set–reset) latch.

14. Write and verify a Verilog module that produces 4-bit output code indicating the number of 1s in an 8-bit input word.

15. Write and verify a structural (gate-level) description of the circuit shown in Figure P4-15.

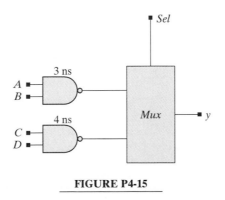

FIGURE P4-15

16. Write and verify a Verilog module of a bidirectional 8-bit ring counter capable of counting in either direction, beginning with first active clock edge after reset.

17. Design and verify a Verilog module of a decade counter—that is, a counter that counts from 0 to 10, then returns a count of 0.

18. The schematic shown in Figure P4-18 is for *Divide_by_11*, a frequency divider that divides *clk* by 11 and asserts its output for one cycle. The unit consists of a chain of toggle-type flip-flops with additional logic to form an output pulse every 11th pulse of *clk*. The asynchronous signal *rst* is active-low and drives *Q* to 1. Develop and verify a model of *Divide_by_11*.

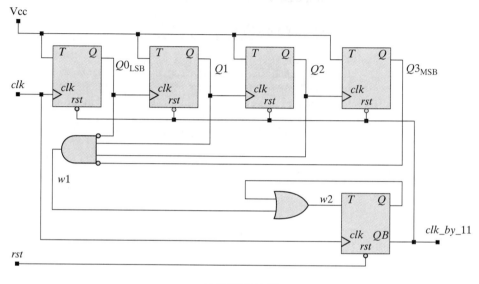

FIGURE P4-18

CHAPTER 5

Logic Design with Behavioral Models of Combinational and Sequential Logic

In Chapter 2 we presented schematics, Boolean equations, and truth table descriptions of combinational logic, and in Chapter 4 we presented their Verilog counterparts. Chapter 3 reviewed fundamental models and manual methods for designing sequential logic. This chapter will present Verilog models based on Boolean equations and introduce a more general and abstract style of modeling for combinational and sequential logic. Algorithmic state machine (ASM) charts will be introduced for developing behavioral models of finite-state machines. More general sequential machines consisting of a datapath and a control unit will be designed by algorithmic state machine and datapath (ASMD) charts, which link a state machine to the datapath that it controls. We will also present some results of synthesizing behavioral models of circuits.

5.1 Behavioral Modeling

Verilog supports structural and behavioral modeling. Structural modeling connects primitive gates and/or functional units to create a specified functionality (for example, an adder) just as parts are connected on a chip or a circuit board. But gate-level models are not necessarily the most convenient or understandable models of a circuit, especially when the design involves more than a few gates. Many modern application-specific integrated circuits (ASICs) have several million gates on a single chip! Also, truth tables become unwieldy when the circuit has several inputs, limiting the utility of

Verilog's user-defined primitives. In today's methodologies for designing ASICs and field-programmable gate arrays (FPGAs), large circuits are partitioned to form an architecture—that is, a configuration of functional units that communicate through their ports—such that each unit is described by a behavioral model of its functionality. Even though the architecture consists of simpler functional units than the whole, designers do not form their gate-level implementation directly. Instead, they write so-called behavioral models, which are automatically synthesized into a gate-level circuit (structure).

Traditionally, designers have designed at the gate (structural) level, using schematic-entry tools to connect gates to form a circuit. But modern design methodology decomposes and represents a design at a meaningful level of abstraction and then synthesizes it into gates. Synthesis tools translate and map the hardware description language (HDL) into a physical technology, such as an ASIC standard-cell library, or a library of programmable parts, such as FPGAs.

Behavioral modeling is the predominant descriptive style used by industry, enabling the design of massive chips. Behavioral modeling describes the functionality of a design—that is, *what* the designed circuit will do, not *how* to build it in hardware. Behavioral models specify the input–output model of a logic circuit and suppress details about its low-level internal structure (architecture) and physical implementation. Propagation delays are not included in the behavioral model of the circuit, but the propagation delays of the cells in the target technology are considered by the synthesis tool when it imposes timing constraints on the physical realization of the logic. Behavioral modeling encourages designers to (1) rapidly create a behavioral prototype of a design (without binding it to hardware detail), (2) verify its functionality, and then (3) use a synthesis tool to optimize and map the design into a selected physical technology, subject to constraints on timing and/or area. If the behavioral model has been written in a synthesis-ready style,[1] the synthesis tool will remove redundant logic, analyze tradeoffs between alternative architectures and/or multilevel equivalent circuits, and ultimately achieve a design that is compatible with area or timing constraints. By focusing the designer's attention on the functionality that is to be implemented, rather than on individual logic gates and their interconnections, behavioral modeling provides the freedom to explore alternatives and tradeoffs before committing a design to production.

Aside from its importance in synthesis, behavioral modeling provides flexibility to a design project by allowing portions of the design to be modeled at different levels of abstraction and completeness. The Verilog language accommodates mixed levels of abstraction, so that portions of the design that are implemented at the gate level (i.e., structurally) can be integrated and simulated concurrently with other portions of the design that are represented by behavioral descriptions.

---

[1]Chapter 6 will consider how to write synthesis-ready models of combinational and sequential logic in Verilog.

5.2   A Brief Look at Data Types for Behavioral Modeling

Before considering behavioral models, we must first understand how information is represented and used in a Verilog model. All computer programs represent information as variables (e.g., integers and real numbers) that can be retrieved, manipulated, and stored in memory. A variable may represent a number used in computation (such as a loop index governing a repetitive sequence of steps), a value of data (such as a binary word), or a computed value (such as the sum of two numbers). In Verilog, a variable can also represent a binary-encoded logic signal in a circuit. A Verilog model describes how the waveforms of variables will evolve in a simulation environment. For example, the *sum* and *c_out* bits of a half adder evolve in the manner specified by the interconnection of primitives in the model. Variables are declared and used according to rules that govern the data types[2] supported by the language.

All variables in Verilog have a predefined type, and there are only two families of data types: nets and registers. *Net variables* act like wires in a physical circuit and establish connectivity between design objects. *Register variables* act like variables in ordinary procedural languages—they store information while the program executes. Not all data types are useful in a synthesis methodology, and we will use mainly the net type **wire** and the register types **reg** and **integer**. A **wire** and a **reg** have a default size of 1 bit. The size of an **integer** is automatically fixed at the word length supported by the host computer, at least 32 bits.

5.3   Boolean-Equation-Based Behavioral Models of Combinational Logic

A Boolean equation describes combinational logic by an expression of operations on variables. Its counterpart in Verilog is the continuous assignment statement.

Example 5.1

The five-input and-or-invert (AOI) circuit that was shown in Figure 4.7 can be described by a single continuous assignment statement that forms the output of the circuit from operations on its inputs, as shown in *AOI_5_CA0* below.

```
module AOI_5_CA0 (y_out, x_in1, x_in2, x_in3, x_in4, x_in5);
  input       x_in1, x_in2, x_in3, x_in4, x_in5;
  output      y_out;

  assign y_out =  ~((x_in1 & x_in2) | (x_in3 & x_in4 & x_in5));

endmodule
```

End of Example 5.1

---

[2]See Appendix C for a discussion of Verilog data types.

The keyword **assign** declares a continuous assignment and binds the Boolean expression on the right-hand side (RHS) of the statement to the variable on the left-hand side (LHS). Verilog has several built-in operators for arithmetic, logical, and machine-oriented operations (e.g., concatenation, reduction, and shifting operators) that can be used in expressions.

The expression in *AOI_5_CA0* in Example 5.1 uses the bitwise inversion ($\sim$) operator,[3] the bitwise AND operator (&), and the bitwise OR operator (|). The assignment is said to be sensitive to the variables in the RHS expression because any time a variable in the RHS changes during simulation, the RHS expression is reevaluated and the result is used to update the LHS. A continuous assignment is said to describe *implicit combinational logic* because the RHS expression of the continuous assignment is equivalent to logic gates that implement the same Boolean function; but note that a continuous assignment can be more compact and understandable than a schematic or a netlist of primitives.

Example 5.2

The five-input AOI circuit can be modified to have an additional input, *enable*, and to have a three-state output, as described by *AOI_5_CA1* below.

```
module AOI_5_CA1 (y_out, x_in1, x_in2, x_in3, x_in4, x_in5, enable);
   input          x_in1, x_in2, x_in3, x_in4, x_in5, enable;
   output         y_out;

   assign y_out = enable ? ~((x_in1 & x_in2) | (x_in3 & x_in4 & x_in5)) : 1'bz;

endmodule
```

The *conditional operator (? :)* acts like a software if-then-else switch that selects between two expressions. In this example, if the value of *enable* is true in *AOI_5_CA1*, the expression to the right of the *?* is evaluated and used to assign value to *y_out*; otherwise, the expression to the right of the *:* is used.[4] This example also illustrates how to write models that include three-state outputs. The value of *y_out* is formed by combinational logic while *enable* is asserted, but has the value *z* otherwise. This functionality corresponds to the equivalent logic circuit shown in Figure 5-1. Note that the continuous assignment statement is an implicit, abstract, and compact representation of the structure described by an equivalent gate-level schematic or netlist of primitives.

End of Example 5.2

A continuous assignment statement establishes an event-scheduling rule between a target net variable and a Boolean expression. A module may contain multiple

---

[3]See Appendix D for a discussion and more examples of Verilog operators.
[4]The syntax of the conditional operator requires that both alternative expressions be specified.

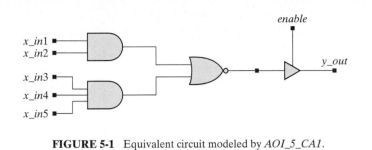

FIGURE 5-1   Equivalent circuit modeled by *AOI_5_CA1*.

continuous assignments, and they are active concurrently with all other continuous as-signments, primitives, behavioral statements, and instantiated modules, just as all of the electrical signals in a circuit arc active concurrently. Continuous assignments can also be written implicitly and efficiently (without the keyword ***assign***) as part of the decla-ration of a wire.

Example 5.3

The model described by *AOI_5_CA2* declares *y_out* to be of type ***wire***, and associates with it a Boolean expression that defines the value of *y_out*.

```
module AOI_5_CA2 (y_out, x_in1, x_in2, x_in3, x_in4, x_in5, enable);
  input       x_in1, x_in2, x_in3, x_in4, x_in5, enable;
  output      y_out;

  wire y_out = enable ? ~((x_in1 & x_in2) | (x_in3 & x_in4 & x_in5)) : 1'bz;

endmodule
```

End of Example 5.3

A continuous assignment statement uses built-in Verilog operators to express how a signal's value is formed abstractly. Each operator has a gate-level counterpart, so such expressions are easily synthesized into physical circuits.

Example 5.4

A continuous assignment statement with a conditional operator provides a convenient way to model the multiplexer circuit shown in Figure 5-2. In *Mux_2_32_CA* an event for signal *select* or for a selected datapath will cause *mux_out* to be updated during simulation.

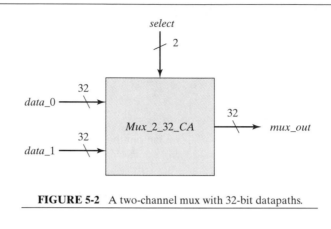

FIGURE 5-2   A two-channel mux with 32-bit datapaths.

```
module Mux_2_ 32_CA ( mux_out, data_1, data_0, select);
  parameter                          word_size = 32;
  output         [word_size -1: 0]  mux_out;
  input          [word_size -1: 0]  data_1, data_0;
  input                             select;

  assign mux_out = select ? data_1 : data_0;
endmodule
```

End of Example 5.4

Note that the size of the 32-bit datapath in *Mux_2_32_CA* is specified by a parameter to make the model scalable, portable, and more useful. Parameters are constant values, and remain fixed during simulation.

5.4   Propagation Delay and Continuous Assignments

Propagation (inertial) delay can be associated with a continuous assignment so that its implicit logic has the same functionality and timing characteristics as its gate-level counterpart.

Example 5.5

The functionality of an AOI gate structure with unit propagation delays on its implicit gates is described below by *AOI_5_CA3*. Each declaration of a wire includes a logic expression assigning value to the wire, and includes a unit time delay. The three continuous assignments are active concurrently during simulation, each having a monitoring mechanism, implemented by the simulator, that detects a change in its RHS expression and schedules a change to update the LHS variable (subject to the propagation delay), just as the equivalent combinational logic would operate under the influence of its inputs.

```
module AOI_5 _CA3 (y_out, x_in1, x_in2, x_in3, x_in4);
  input           x_in1, x_in2, x_in3, x_in4;
  output          y_out;

  wire #1 y1 = x_in1 & x_in2;        // Bitwise and operation
  wire #1 y2 = x_in3 & x_in_4;
  wire #1 y_out = ~ (y1 | y2);       // Complement the result of bitwise OR operation
endmodule
```

End of Example 5.5

Example 5.6

As an alternative to the structural model of a 2-bit comparator that was presented in Example 4.4, the model *compare_2_CA0* given below is described by three (concurrent) continuous assignment statements (implicit combinational logic).[5] The model is equivalent to *compare_2_str* but has no explicit binding to hardware or to primitive gates

```
module compare_2_CA0 (A_lt_B, A_gt_B, A_eq_B, A1, A0, B1, B0);
  input   A1, A0, B1, B0;
  output A_lt_B, A_gt_B, A_eq_B;

  assign A_lt_B = (~A1) & B1 | (~A1) & (~A0) & B0 | (~A0) & B1 & B0;

  assign A_gt_B = A1 & (~B1) | A0 & (~B1) & (~B0) | A1 & A0 & (~B0);

  assign A_eq_B = (~A1) & (~A0) & (~B1) & (~B0) | (~A1) & A0 & (~B1) & B0
                  | A1 & A0 & B1 & B0 | A1 & (~A0) & B1 & (~B0);

endmodule
```

End of Example 5.6

Note that continuous assignment statements suppress detail about the internal structure of the module, and deal only with the Boolean equations that describe the input–output relationships of the comparator. A synthesis tool will create the actual hardware realization of the assignment.

---

[5]The order in which multiple continuous assignments are listed in the source code is arbitrary; that is, the order of the statements has no effect on the results of simulation. The order does not establish a precedence for their evaluation.

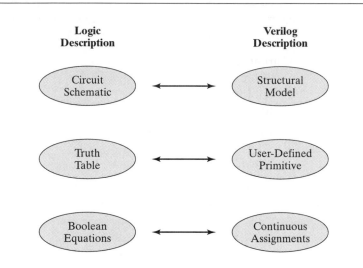

FIGURE 5-3   Verilog counterparts of three common descriptions of combinational logic.

The three Verilog language constructs corresponding to schematic, truth table and Boolean equation descriptions of combinational logic are show in Figure 5-3. All three descriptions describe level-sensitive behavior; that is, variables are updated immediately when an input changes. None of the examples modeled feedback structures, but we will see that a continuous assignment with feedback is a synthesizable model for a hardware latch.

5.5   Latches and Level-Sensitive Circuits in Verilog

The level-sensitive storage mechanism of a latch (See Chapter 3) can be modeled in a variety of ways. First, note that a set of continuous-assignment statements has implicit feedback if a variable in one of the RHS expressions is also the target of an assignment. For example, a pair of cross-coupled NAND gates could be modeled as

```
assign q = set ~& qbar;
assign qbar = rst ~& q;
```

The implied behavior will still be level-sensitive, but it will correspond to the feedback structure of a hardware latch. However, synthesis tools do not accommodate this form of feedback. But they do support the feedback that is implied by a continuous assignment in which the RHS uses a conditional operator.

Example 5.7

The output of a transparent latch follows the data input while the latch is enabled, but otherwise will hold the value it had when the enable input was de-asserted. Example 4.14 presented a truth table model of a transparent latch. Here, *Latch_CA* uses a continuous assignment statement with feedback to model this functionality.

```
module Latch_CA (q_out, data_in, enable);
  output      q_out;
  input       data_in, enable;

  assign q_out = enable ? data_in : q_out;
endmodule
```

Figure 5-4 shows the waveforms produced by simulation of *Latch_CA*. Note how *q_out* follows *data_in* while *enable* is asserted, and latches *q_out* to the value of *data_in* when *enable* is de-asserted. The appearance of *q_out* in the RHS expression and as the LHS target variable implies a structural feedback in hardware, and will be synthesized as a latch.

End of Example 5.7

FIGURE 5-4 Simulation results for a transparent latch modeled by a continuous assignment statement with feedback.

When feedback is used in a continuous-assignment statement with a conditional operator, a synthesis tool will infer the functionality of a latch and its hardware implementation. Chapter 6 will discuss descriptive styles that lead to intentional and accidental synthesis of latches.

Example 5.8

The latch model *Latch_Rbar_CA* below uses a nested conditional operator to add the functionality of an active-low reset to a transparent latch. Simulation of *Latch_Rbar_CA* produces the waveforms shown in Figure 5-5, in which the actions of *enable* and *reset* are apparent.

```
module Latch_Rbar_CA (q_out, data_in, enable, reset);
 output      q_out;
 input       data_in, enable, reset;

 assign q_out = !reset ? 0 : enable ? data_in : q_out;
endmodule
```

End of Example 5.8

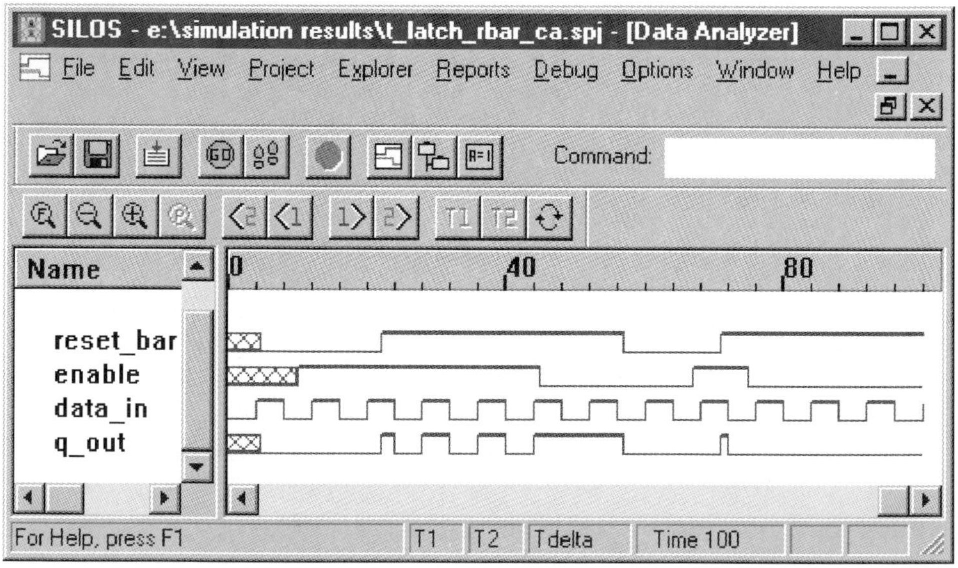

FIGURE 5-5 Simulation results for a transparent latch with active-low *reset* and active-high *enable*.

Verilog supports multiple descriptive styles that can define the same functionality. Continuous assignments are convenient for modeling small Boolean expressions, three-state behavior, and transparent latches. But designers writing large Boolean equation models (continuous assignments) are prone to making mistakes when there are several variables and large expressions. Also, the Boolean expressions might obscure the functionality of the design, even if they are written correctly. So, it is worthwhile to consider other language constructs that offer simpler, but more readable alternatives that describe edge-sensitive as well as level-sensitive behavior.

5.6   Cyclic Behavioral Models of Flip-Flops and Latches

Continuous-assignment statements are limited to modeling level-sensitive behavior—combinational logic and transparent latches. They cannot model an element that has edge-sensitive behavior, such as a flip-flop. Many digital systems operate synchronously, with activity triggered by an edge of a synchronizing signal (commonly called a clock). Verilog uses a cyclic behavior to model edge-sensitive functionality. Like the single-pass behaviors that are used to model signal generators in testbenches (see Example 4.7), cyclic behaviors are abstract—they do not use hardware to specify signal values. Instead, they execute procedural statements to generate the values of variables, just like the statements of an ordinary procedural language (e.g., C) execute to extract, manipulate, and store variables in memory. They are called *cyclic behaviors* because they do not expire after their last procedural statement has executed; instead, they re-execute. The execution of these statements can be unconditional or can be governed by an optional event control expression. Cyclic behaviors are used to model (and synthesize) both level-sensitive and edge-sensitive (synchronous) behavior (e.g., flip-flops).

Example 5.9

The keyword **always** in *df_behav* declares a cyclic behavior corresponding to an edge-triggered flip-flop. At every rising edge of *clk* the behavior's procedural statements execute, computing the value of *q* and storing it in memory. A continuous assignment statement forms *q_bar* from *q* immediately after *q* has changed. The nonblocking, or concurrent, assignment operator ($<=$) will be explained later.

```
module df_behav (q, q_bar, data, set, reset, clk);
input           data, set, clk, reset;
output          q, q_bar;
reg             q;

assign q_bar = ~ q;

always @  (posedge clk)  // Flip-flop with synchronous set/reset
```

```
      begin
        if (reset == 0) q <= 0;
        else if (set ==0) q <= 1;
        else q <= data;
      end
    endmodule
```

End of Example 5.9

In *df_behav* the action of *reset* is synchronous because it has no influence until the procedural statements are evaluated at the active edge of *clk*. The variable *q* retains its residual value until the next active edge of *clk*, as specified by ***posedge** clk*, because *q* was declared as a register variable of type ***reg***.

The operator $(=)$ in a procedural statement is called a procedural-assignment operator. A variable that is assigned value by a procedural-assignment operator in a single-pass or cyclic behavior must be a declared register-type variable (i.e., not a net). Register variables store information during simulation, but do not necessarily imply that the synthesized circuit will have hardware registers.[6]

5.7   Cyclic Behavior and Edge Detection

A cyclic behavior is activated at the beginning of simulation, and it will execute its associated procedural statements subject to timing control imposed by delay control (**#** operator) and event-control expressions (**@** operator).[7] The Verilog keyword ***posedge*** qualifies the event control expression to execute its procedural statements only when a rising edge of the argument signal (e.g., *clk* in Example 5.9) has occurred. Edge semantics for rising (***posedge***) and falling (***negedge***) edges are built into Verilog.

A simulator automatically monitors the variables in an event-control expression, and when the expression changes value, the associated procedural statements execute if the enabling change took place. When all the statements of a cyclic behavior complete execution, the computational activity flow returns to the ***always*** keyword and commences execution again, subject to its event-control expression. If a delay-control operator or an event-control expression is encountered in one of the statements being executed, the activity flow of the behavior is suspended to wait until the indicated time has elapsed or until the event control expression detects the qualifying activity. The conditionals in the procedural statements in the cyclic behavior in *df_behav* (***if*** and ***else if***) test whether the associated expression evaluates to *true*. If it does, the associated statement (or ***begin*** ... ***end*** block statement) is executed.

---

[6]Chapter 6 will discuss whether a register variable in a model synthesizes to a hardware storage element.
[7]A ***wait*** statement will also suspend execution, but it is not synthesized by the leading synthesis tools and will not be discussed here or used in our models.

Example 5.10

The reset action of a flip-flop can be asynchronous. The functionality modeled below by *asynch_df_behav* is sensitive to the rising edge of the clock, but also to the falling edge of *reset* and *set*, with priority given to *reset*. The last clause in the conditional statement executes at a rising edge of *clk* only if the asynchronous inputs are not asserted.

```
module asynch_df_behav (q, q_bar, data, set, clk, reset );
  input        data, set, reset, clk;
  output       q, q_bar;
  reg          q;

  assign       q_bar = ~q;

  always @ (negedge set or negedge reset or posedge clk)
    begin
      if (reset == 0) q <= 0;
        else if (set == 0) q <= 1;
          else q <= data;            // synchronized activity
    end
endmodule
```

End of Example 5.10

Note that *clk* and *clock* are not keywords, *so it is important to place in the last conditional clause of the **if** statement the computational activity associated with the synchronizing signal of a synchronous behavior.* This coding discipline allows a synthesis tool to correctly (1) identify the synchronizing signal (its name and its location in the event-control expression are not predetermined), and (2) infer the need for a flip-flop to hold the value of *q* between the active edges of the synchronizing signal.

The cyclic behavior in *asynch_df_behav* is activated at the beginning of simulation and immediately suspends until its event-control expression changes.[8] The expression is formed as an "event or" of *set*, *reset*, and *clk*.[9] The Verilog language allows a mixture of level-sensitive and edge-qualified variables in an event-control expression, but synthesis tools do not support such models of behavior. See that your description is entirely edge-sensitive or entirely level-sensitive.

Example 5.11

A transparent latch is modeled in *tr_latch* by a cyclic behavior whose level-sensitive event-control expression is sensitive to a change in *enable* or a change in *data*.

---

[8]An event for *clk* occurs at the beginning of simulation if it is assigned a value of 1.
[9]Verilog 2001 (see Appendix H) introduces the option to form the event control-expression more conveniently as a comma-separated sensitivity list.

```
module tr_latch (q_out, enable, data);
  output q_out;
  input enable, data;
  reg q_out;

  always @  (enable or data)
   begin
    if (enable) q_out = data;
   end
endmodule
```

End of Example 5.11

When *enable* is asserted in *tr_latch* the behavior is activated, and *q_out* immediately gets the value of *data*. Then the activity flow returns to the **always** construct and suspends to await the next change of the event-control expression. If *data* changes while *enable* is asserted, the cycle of *q_out* getting *data* repeats. The control flow of the *if* statement has no branch, so while *enable* is de-asserted *q_out* retains the value it had when *enable* was de-asserted. While *enable* is de-asserted the events of *data* reactivate the process, but no assignment is made to *q_out*.

5.8 A Comparison of Styles for Behavioral Modeling

We have already seen how the 2-bit comparator can be described by a gate-level structure (Example 4.4) and by a Boolean equation-based behavioral model (Example 5.6). Next, we compare simpler and more readable alternatives that also use continuous assignments, and then we contrast modeling styles based on (1) continuous assignments, (2) register transfer level logic (RTL), and (3) behavioral algorithms.

5.8.1 Continuous-Assignment Models

A modeling style based on continuous assignments describes level-sensitive behavior. Continuous assignments execute concurrently with each other, with gate-level primitives, and with all of the behaviors in a description.

Example 5 12

The functionality in *compare_2_CA1* is evident from the expressions in the continuous assignment statements. Here, the Verilog concatenation operator ({}) concatenates the bits of the datapaths to form 2-bit vectors. The Boolean value of the RHS expression determines the assignment of 0 or 1 to the LHS variable.[10] Note that the gate-level implementation is not apparent.

```
module compare_2_CA1  (A_lt_B, A_gt_B, A_eq_B, A1, A0, B1, B0);
  input           A1, A0, B1, B0;
  output          A_lt_B, A_gt_B, A_eq_B;

  assign          A_lt_B = ({A1,A0} < {B1,B0});
  assign          A_gt_B = ({A1,A0} > {B1,B0});
  assign          A_eq_B = ({A1,A0} == {B1,B0});
endmodule
```

End of Example 5.12

Another simple, and elegant, implementation of the 2-bit comparator uses continuous-assignment statements and the relational operators, with declared 2-bit vectors *A* and *B*, as shown next.

Example 5.13

The RHS expression of the continuous assignment statements in *compare_2_CA2* are sensitive to *A* and *B*, and evaluate to 1 (true) or 0 (false).

```
module compare_2_CA2 (A_lt_B, A_gt_B, A_eq_B, A, B);
  input [1: 0]    A, B;
  output          A_lt_B, A_gt_B, A_eq_B;

  assign          A_lt_B = (A < B);
  assign          A_gt_B = (A > B);
  assign          A_eq_B = (A == B);
endmodule
```

End of Example 5.13

Suppose now that we want to extend this model to compare two 32-bit words, as shown in Figure 5-6. It is not feasible to write the Boolean equations that compare 32-bit words.

---

[10]In general, a Verilog expression is true if it evaluates to the binary equivalent of a positive integer, and false otherwise.

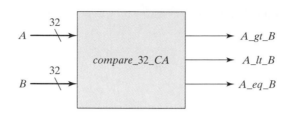

FIGURE 5-6 Block diagram symbol for a 32-bit comparator.

Example 5.14

A 32-bit comparator has the same functionality as a 2-bit comparator. So we modify the model from the previous example by adding a parameter declaration to size the word length of the datapath. The description also declares a comma-separated list of continuous assignments with one statement. The model is readable, understandable, compact, and extendable to datapaths of arbitrary sizes.

```
module compare_32_CA (A_gt_B, A_lt_B, A_eq_B, A, B);
  parameter    word_size = 32;
  input        [word_size-1: 0]   A, B;
  output       A_gt_B, A_lt_B, A_eq_B;

  assign       A_gt_B = (A > B),          // Note: list of multiple assignments
               A_lt_B = (A < B),
               A_eq_B = (A == B);
endmodule
```

End of Example 5.14

5.8.2 Dataflow/RTL Models

Dataflow models of combinational logic describe concurrent operations on signals, usually in a synchronous machine, where computations are initiated at the active edges of a clock and are completed in time to be stored in a register at the next active edge. At each active edge, the hardware registers read and store the data inputs that were formed as a result of the previous clock edge, and then propagate new values to be stored in registers at the next edge. Dataflow models for synchronous machines are also referred to as RTL (register transfer level) models because they describe register activity in a synchronous machine [1, 2]. RTL models are written for a specific architecture—that is, the registers, datapaths, and machine operations and their schedule are known a priori.

A behavioral model of combinational logic can be described by a set of concurrent continuous assignments (see Examples 5.6, and 5.13) or by an equivalent asynchronous (i.e., level-sensitive) cyclic behavior. Cyclic behaviors are declared by the keyword *always*, execute statements in sequential order, and reexecute indefinitely.

Example 5.15

The level-sensitive cyclic behavior in *compare_2_RTL* executes and updates the outputs whenever a bit of either datapath changes.

```
module compare_2_RTL (A_lt_B, A_gt_B, A_eq_B, A1, A0, B1, B0);
    input       A1, A0, B1, B0;
    output      A_lt_B, A_gt_B, A_eq_B;
    reg         A_lt_B, A_gt_B, A_eq_B;

    always @ (A0 or A1 or B0 or B1) begin
      A_lt_B =    ({A1,A0} < {B1,B0});
      A_gt_B =    ({A1,A0} > {B1,B0});
      A_eq_B =    ({A1,A0} == {B1,B0});
    end
endmodule
```

End of Example 5.15

The assignment operator in Example 5.15 is the ordinary procedural assignment operator (=). Consequently, the statements are executed in the listed order, with the storage of value occurring immediately after any statement executes and before the next statement can execute. Because there are no data dependencies between the variables on the LHS of the three procedural assignments in *compare_2_RTL*, the order in which the assignments are listed does not affect the outcome. That is not always the case.

Example 5.16

The shift register shown in Figure 5-7 is described below by a synchronous cyclic behavior with a list of procedural assignments.

```
module shiftreg_PA (E, A, B, C, D, clk, rst);
    output A;
    input  E;
    input  clk, rst;
    reg    A, B, C, D;

    always @ (posedge clk or posedge rst) begin
```

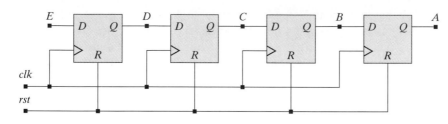

FIGURE 5-7   A 4-bit serial shift register.

```
if (reset) begin A = 0; B = 0; C = 0; D = 0; end
else begin
  A = B;
  B = C;
  C = D;
  D = E;
 end
end
endmodule
```

Now consider what happens if the order of the procedural assignments in the model is reversed, as in *shiftreg_PA_rev* below. The list of statements executes in sequential order, from top to bottom. The effect of the assignment made by the first procedural statement is immediate. So *D* changes, and the updated value is used in the second statement, and so forth. The statements execute sequentially, but at the same time step of the simulator. The list of four statements is equivalent to a single statement that assigns *E* to *A*. Synthesis tools recognize this form of expression substitution, and the circuit that is synthesized consists of a single flip flop, shown in Figure 5-8.

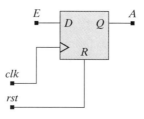

FIGURE 5-8   Circuit synthesized as a result of expression substitution in an incorrect model of a 4-bit serial shift register.

```
module shiftreg_PA_rev (A, E, clk, rst);
  output A;
  input  E;
  input  clk, rst;
  reg    A, B, C, D;

  always @ (posedge clk or posedge rst) begin
    if (rst) begin A = 0; B = 0; C = 0; D = 0; end
    else begin
      D = E;
      C = D;
      B = C;
      A = B;
    end
  end
endmodule
```

End of Example 5.16

Procedural assignments are called *blocked* assignments because a statement that makes a procedural assignment must complete execution (i.e., write results to memory) before the next statement in the behavior can execute. The statements that follow a procedural assignment are blocked from executing until the statement with the procedural assuagement completes execution. This sets the stage for expression substitution. *Failure to appreciate the effects of expression substitution can lead to incorrect models.*

An alternative Verilog dataflow model uses concurrent procedural assignments, also called non-blocking assignments, in a cyclic behavior. Nonblocking assignments are made with the nonblocking assignment operator ($<=$), instead of the ordinary (procedural, sequential) assignment operator ($=$). Nonblocking assignment statements effectively execute concurrently (in parallel) rather than sequentially, so the order in which they are listed has no effect. Moreover, a simulator must implement a sampling mechanism by which all of the variables referenced by the RHS of the statements with nonblocking assignments are sampled, held in memory, and used to update the LHS variables concurrently.[11] Consequently, changes to the listed order of the nonblocking assignments do not affect the outcome of the assignments to the LHS variable, because the assignments are based on the values that were held by the RHS variables immediately before the statements executed.

---

[11]It is advisable to avoid having multiple behaviors assign value to the same variable, because software race conditions make the outcome indeterminate. See "Additional Features of Verilog" at the companion website.

Example 5.17

An equivalent model of the 4-bit serial shift register shown in Figure 5-7 is described below with nonblocking assignment operators ($<=$).

```
module shiftreg_nb (A, E, clk, rst);
  output  A;
  input   E;
  input   clk, rst;
  reg     A, B, C, D;

  always @ (posedge clk or posedge rst) begin
    if (rst) begin A <= 0; B <= 0; C <= 0; D <= 0; end
    else begin
      A <= B;              //        D <= E;
      B <= C;              //        C <= D;
      C <= D;              //        B <= D;
      D <= E;              //        A <= B;
    end
  end
endmodule
```

End of Example 5.17

The commented (//) nonblocking assignments in *shiftreg_nb* have reversed order and would lead to the same results in simulation, and will synthesize to the same structure. The statements in a list of nonblocking assignments execute concurrently, without dependence on their relative order. This style describes the concurrency that is found in actual hardware and the register transfers that occur within synchronous machines.

When a cyclic behavior executes nonblocking assignments, the simulator evaluates each of the RHS expressions before assigning values to their LHS targets. This, in general, prevents any interaction between the assignments and eliminates dependencies on their relative order. This is not the case for ordinary (i.e., sequential) procedural assignments (i.e., those using the = operator), because such statements execute in sequence, and only after the immediately preceding statement has completed execution. If the functionality being modeled by a cyclic behavior does not depend on the sequence in which the statements are written, either blocking or nonblocking assignments can be used (see Example 5.15). However, if we are modeling logic that includes edge-driven register transfers, it is strongly recommended that the edge-sensitive (synchronous) operations be described by nonblocking assignments and that combinational logic be described with blocked assignments. This practice will prevent race conditions between combinational logic and register operations.

5.8.3   Algorithm-Based Models

A behavioral model described by a circuit's input–output algorithm is more abstract than an RTL description. The algorithm prescribes a sequence of procedural assignments within a cyclic behavior. The outcome of executing the statements determines the values of storage variables and, ultimately, the output of the machine. The algorithm described by the model does not have explicit binding to hardware, and it does not have an implied architecture of registers, datapaths, and computational resources. This style is most challenging for a synthesis tool because it must perform what is referred to as "architectural synthesis," which extracts the resources (e.g., determines actual requirements for processors, datapaths, and hardware memory) and scheduling requirements that support the algorithm and then maps the description into an RTL model whose logic can be synthesized.

Not all algorithms can be implemented in hardware. Nonetheless, this descriptive style is useful and attractive, because it is abstract, and eliminates the need for an a priori architecture. Also, the description can be very readable and understandable. The key distinction to remember is that the assignment statements in a dataflow (RTL) model execute concurrently (in parallel) and operate on explicitly declared registers in the context of a specified architecture; the statements in an algorithmic model execute sequentially, without an explicit architecture.

Example 5.18

The cyclic behavior in *compare_2_algo* is activated whenever any bit of *A* or *B* changes. The algorithm first initializes all register variables to 0, a precaution that will prevent the synthesis of unwanted latches. Then the algorithm traverses a decision tree to determine which of the three outputs to assert. The nonasserted outputs will retain the value that was assigned to them at the beginning of the sequence.

```
module compare_2_algo (A_lt_B, A_gt_B, A_eq_B, A, B);
  output        A_lt_B, A_gt_B, A_eq_B;
  input [1: 0]  A, B;

  reg           A_lt_B, A_gt_B, A_eq_B;

  always @  (A or B)      // Level-sensitive behavior
    begin
     A_lt_B = 0;
     A_gt_B = 0;
     A_eq_B = 0;
     if (A == B)          A_eq_B = 1;      // Note: parentheses are required
     else if (A > B)      A_gt_B = 1;
     else                 A_lt_B = 1;
    end
endmodule
```

End of Example 5.18

Figure 5-9 shows the gate-level schematic obtained by synthesizing[12] *compare_2_algo* and targeting the implementation to generic gates. Note that the algorithm has register variables to support its execution, but does not need hardware memory because it synthesizes to combinational logic.

5.8.4   Port Names: A Matter of Style

Design teams follow elaborate rules that govern the style of their Verilog models. This is done to ensure that only constructs supported by synthesis tools are used. Other rules govern the use of upper- and lower-case text, and naming conventions for signals, modules, functions, tasks, and ports, with the aim of increasing the readability and the re-usability of the code [3]. Signals should be given names that describe their use (e.g., *clock*), and modules, functions and tasks should be given names that describe the encapsulated functionality (e.g., *comparator*). The examples in the rest of this book will generally follow a particular port-naming convention. The ports will be ordered in the following sequence: (datapath bidirectional signals, bidirectional control signals, datapath outputs, control outputs, datapath inputs, control inputs, synchronizing signals).

5.8.5   Simulation with Behavioral Models

An event at the input of a primitive causes the simulator to schedule an updating event for its output. Likewise, an event in the RHS expression of a continuous-assignment statement causes the scheduling of an event for the assignment's target variable. In both cases, the scheduling is governed by any propagation delay associated with the primitive or continuous assignment, which has the effect of scheduling the output/target event to occur at a future time step of the simulator, rather than in the current time step. Simulators behave differently, though, when a cyclic process is activated. Their associated statements execute sequentially, in the same time step, until the simulator

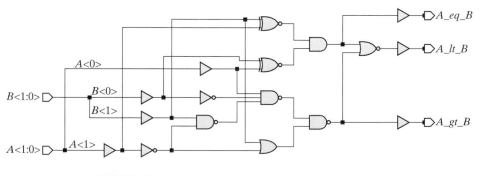

FIGURE 5-9   Synthesis results derived from *compare_2_algo*.

---

[12]With Synopsys' Design Compiler. Note that the tool represents vector ranges by the braces <> rather than [ ].

encounters either (1) a delay control operator (#), (2) an event control operator (@), (3) a *wait* construct, or (4) the last statement of the behavior. The first three have the effect of suspending the execution of the behavioral statement until a condition is satisfied; the last possibility causes the activity to restart from the first statement of the behavior. Models for primitives and continuous assignments cannot suspend themselves. They execute immediately. Cyclic behaviors can suspend themselves. When they do, their activity can cause other behaviors, primitives, and continuous assignments to be activated. But until an active behavior is suspended, the rest of the world waits for it to suspend. A consequence of this is that a cyclic or single-pass behavior that has a loop that does not include a mechanism for suspension of its activity will execute endlessly, and consume the attention of the simulator. Good modeling will prevent this from happening; otherwise, reach for the *off* button.

If multiple behaviors are activated at the same time step, the order in which the simulator executes them is indeterminate. Care must be taken to avoid having such behaviors assign value to the same register, because the outcome of the assignments will be indeterminate. Synthesis tools will warn you of such features in your model.

5.9   Behavioral Models of Multiplexers, Encoders, and Decoders

In Chapter 3, we examined some of the basic building blocks of combinational logic: multiplexers, encoders, and decoders. Here we present their Verilog models to illustrate alternative level-sensitive behavioral descriptions, and the results of synthesizing them into an ASIC library.[13]

Example 5.19

Mux_4_32_case is a behavioral model of the four-channel, 32-bit, multiplexer shown in Figure 5.10 with a three-state output. The *default* case item covers cases that might occur in simulation, and it is a recommended practice to avoid unintentional synthesis of hardware latches if a case statement is not fully decoded for all possibilities that use 0 and 1. If the case items are not completely decoded, then the default assignment would be treated as a don't-care condition in synthesis and could lead to a smaller circuit.

```
module Mux_4_32_case
  (mux_out, data_3, data_2, data_1, data_0, select, enable);
output   [31: 0]  mux_out;
input    [31: 0]  data_3, data_2, data_1, data_0;
input    [1: 0]   select;
```

---

[13]Styles for writing synthesis-friendly models of combinational and sequential logic will be presented in Chapter 6.

```
input           enable;
reg    [31: 0]  mux_int;

assign mux_out = enable ? mux_int : 32'bz;

always @ ( data_3 or data_2 or data_1 or data_0 or select)
  case (select)
     0:           mux_int = data_0;
     1:           mux_int = data_1;
     2:           mux_int = data_2;
     3:           mux_int = data_3;
     default:     mux_int = 32'bx;              // May execute in simulation
  endcase
endmodule
```

End of Example 5.19

The Verilog *case* statement is similar to its counterpart in other languages (e.g., the switch statement in C). It searches from top to bottom to find a match between the case expression and a case item, expressed as a value in Verilog's four-value logic system. The *case* statement executes the first match found, and does not consider any remaining possibilities.

The keyword *always* in *Mux_4_32_case* declares a behavior, or process (computational activity flow), that begins execution when the event-control expression *(data_3 or data_2 or data_1 or data_0 or select)* changes under simulation. The behavior has a simple interpretation: whenever a datapath input or the select bus changes value,

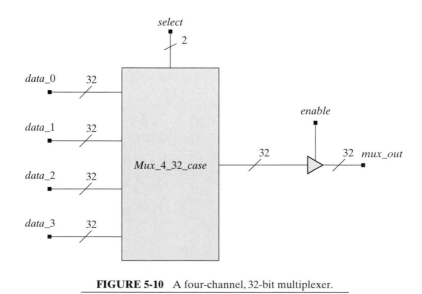

FIGURE 5-10 A four-channel, 32-bit multiplexer.

decode and update the value of an internal storage variable, *mux_int*. A continuous-assignment statement is included in *Mux_4_ 32_case* to describe a three-state output under the active-high control of *enable*.

The **@** operator in *Mux_4_32_case* denotes event control, meaning that the procedural statement(s) that follow the event-control expression do not execute until an activating event occurs. When such an event occurs, the statements execute in sequence, top to bottom. When the last statement completes execution, the computational activity returns to the location of the keyword ***always***, where the event control operator **@** suspends the behavior until the next sensitizing event occurs. Then the cycle repeats. A cyclic behavior becomes active in simulation at time 0, when the simulation begins, but in this example the activity immediately suspends until the event-control expression changes. Then the ***case*** statement executes, and assigns value to *mux_int*, and immediately returns control to the event-control operator.

Example 5.20

An alternative model uses nested conditional statements (***if***) to model a multiplexer. The model *Mux_4_32_if* also includes a continuous assignment that forms a three-state output.

```
module Mux_4_32_if
  (mux_out, data_3, data_2, data_1, data_0, select, enable);
  output [31: 0]  mux_out;
  input  [31: 0]  data_3, data_2, data_1, data_0;
  input  [1: 0]   select;
  input           enable;
  reg    [31: 0]  mux_int;

  assign mux_out = enable ? mux_int : 32'bz;

  always @ ( data_3 or data_2 or data_1 or data_0 or select)
    if (select == 0) mux_int = data_0; else
      if (select == 1) mux_int = data_1; else
        if (select == 2) mux_int = data_2; else
          if (select == 3) mux_int = data_3; else mux_int = 32'bx;
endmodule
```

End of Example 5.20

Example 5.21

Nested conditional assignments, using the *? :* operator, are used in *Mux_4_32_CA* to model the same functionality as *Mux_4_32_if*.

```
module Mux_4_32_CA (mux_out, data_3, data_2, data_1, data_0, select, enable);
  output [31: 0]  mux_out;
  input  [31: 0]  data_3, data_2, data_1, data_0;
  input  [1: 0]   select;
  input           enable;
  wire   [31: 0]  mux_int;
  assign mux_out = enable ? mux_int : 32'bz;
  assign mux_int = (select == 0) ? data_0 :
                        (select == 1) ? data_1:
                              (select == 2) ? data_2:
                                    (select == 3) ? data_3: 32'bx;
endmodule
```

End of Example 5.21

The combinational encoders and decoders discussed in Chapter 3 can be modeled conveniently with cyclic behaviors.

Example 5.22

Two implementations of an 8:3 encoder are shown below. Neither decodes fully all possible patterns of *Data*, but both cover the remaining outcomes with a ***default*** assignment. The result of synthesis, shown in Figure 5-11, is combinational. This model is

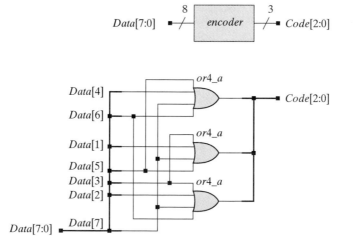

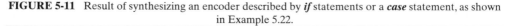

FIGURE 5-11 Result of synthesizing an encoder described by *if* statements or a *case* statement, as shown in Example 5.22.

intended for applications in which only the indicated words of *Data* occur in operation. The default assignments will be don't-cares in synthesis, and are needed to prevent synthesis of a circuit having latched outputs.

```
module encoder (Code, Data);
  output      [2: 0] Code;
  input       [7: 0] Data;
  reg         [2: 0] Code;

  always @  (Data)
   begin
    if (Data == 8'b00000001) Code = 0; else
    if (Data == 8'b00000010) Code = 1; else
    if (Data == 8'b00000100) Code = 2; else
    if (Data == 8'b00001000) Code = 3; else
    if (Data == 8'b00010000) Code = 4; else
    if (Data == 8'b00100000) Code = 5; else
    if (Data == 8'b01000000) Code = 6; else
    if (Data == 8'b10000000) Code = 7; else Code = 3'bx;
   end

/* Alternative description is given below

  always @  (Data)
   case (Data)
       8'b00000001  :  Code = 0;
       8'b00000010  :  Code = 1;
       8'b00000100  :  Code = 2;
       8'b00001000  :  Code = 3;
       8'b00010000  :  Code = 4;
       8'b00100000  :  Code = 5;
       8'b01000000  :  Code = 6;
       8'b10000000  :  Code = 7;
       default      :  Code = 3'bx;
   endcase
   */
endmodule
```

End of Example 5.22

Example 5.23

Alternative behaviors describing an 8:3 priority encoder are shown below.[14] The result of synthesizing the circuit is shown in Figure 5-12. Note that the conditional (*if*) statement has an implied priority of execution, and that the **casex** statement combined with *x* in the case items implies priority also. The **casex** statement ignores *x* and *z* in bits of the *case item* (e.g., *Data[6]*) and the *case expression (Data)*—they are treated as don't-cares. The default assignments in both styles provide flexibility to the logic optimizer of a synthesis tool.

```
module priority (Code, valid_data, Data);
  output        [2: 0]   Code;
  output                 valid_data;
  input         [7: 0]   Data;
  reg           [2: 0]   Code;

  assign                 valid_data = |Data; // "reduction or" operator
  always @  (Data)
   begin
     if (Data[7]) Code = 7; else
     if (Data[6]) Code = 6; else
     if (Data[5]) Code = 5; else
     if (Data[4]) Code = 4; else
     if (Data[3]) Code = 3; else
     if (Data[2]) Code = 2; else
     if (Data[1]) Code = 1; else
     if (Data[0]) Code = 0; else
                  Code = 3'bx;

   end
/*// Alternative description is given below

always @  (Data)
  casex (Data)
     8'b1xxxxxxx  :  Code = 7;
     8'b01xxxxxx  :  Code = 6;
     8'b001xxxxx  :  Code = 5;
     8'b0001xxxx  :  Code = 4;
     8'b00001xxx  :  Code = 3;
     8'b000001xx  :  Code = 2;
     8'b0000001x  :  Code = 1;
     8'b00000001  :  Code = 0;
     default      :  Code = 3'bx;
  endcase
*/
endmodule
```

End of Example 5.23

---

[14]The *reduction or* operator is used to form the logic for *valid_data*. The operator forms the *or* of the bits in a word.

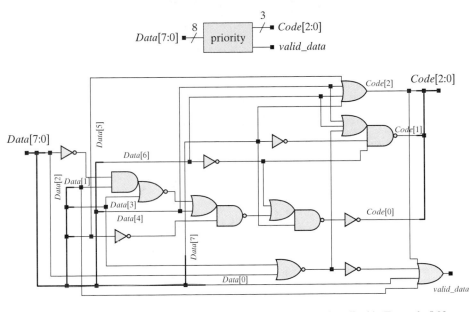

FIGURE 5-12 Circuit synthesized for the 8:3 priority encoder described in Example 5.23.

Example 5.24

A 3:8 decoder is described by the alternative behaviors shown below. The decoders synthesize to the circuit in Figure 5-13.

```verilog
module decoder (Data, Code);
output      [7: 0] Data;
input       [2: 0] Code;
reg         [7: 0] Data;

always @  (Code)
 begin
  if (Code == 0) Data = 8'b00000001; else
  if (Code == 1) Data = 8'b00000010; else
  if (Code == 2) Data = 8'b00000100; else
  if (Code == 3) Data = 8'b00001000; else
  if (Code == 4) Data = 8'b00010000; else
  if (Code == 5) Data = 8'b00100000; else
  if (Code == 6) Data = 8'b01000000; else
  if (Code == 7) Data = 8'b10000000; else
                 Data = 8'bx;
 end

/* Alternative description is given below
always @  (Code)
```

```
case (Code)
  0       :  Data = 8'b00000001;
  1       :  Data = 8'b00000010;
  2       :  Data = 8'b00000100;
  3       :  Data = 8'b00001000;
  4       :  Data = 8'b00010000;
  5       :  Data = 8'b00100000;
  6       :  Data = 8'b01000000;
  7       :  Data = 8'b10000000;
  default :  Data = 8'bx;
  endcase
*/
endmodule
```

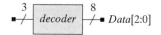

End of Example 5.24

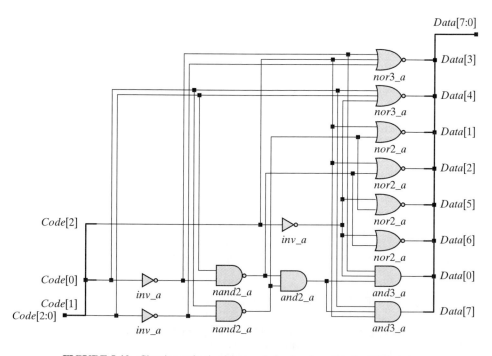

FIGURE 5-13 Circuit synthesized from a behavioral model of a 3:8 decoder.

Example 5.25

The seven-segment light-emitting diode (LED) display depicted in Figure 5-14 is a useful circuit in many applications using prototyping boards. Module *Seven_Seg_Display* accepts 4-bit words representing binary coded decimal (BCD) digits and displays their decimal value. The display has active-low illumination outputs,[15] and can be implemented with combinational logic.[16] The description synthesizes into a combinational circuit. Several of the input codes are unused and should not occur under ordinary operation. One possibility is to assign don't-cares to those codes. However, this would display an output if such an input code occurred. Instead, the default assignment blanks the display for all unused codes. This prevents a bogus display condition, and, as we will see in Chapter 6, prevents the synthesis tool from synthesizing a latched output. Why? If the default assignment is omitted, an event of an input that is not decoded will be detected by the event control expression of the cyclic behavior, but will not cause *Display* to be an assigned value. The implication is that the output should remain at whatever value it had before the input event occurred; that is, it should behave like a latch! Consequently, the displayed value of *Display* would not correspond to the BCD.

```
module Seven_Seg_Display (Display, BCD);
  output      [6: 0]   Display;
  input       [3: 0]   BCD;
  reg         [6: 0]   Display;
  //                       abc_defg
  parameter   BLANK = 7'b111_1111;
  parameter   ZERO  = 7'b000_0001;      // h01
  parameter   ONE   = 7'b100_1111;      // h4f
  parameter   TWO   = 7'b001_0010;      // h12
  parameter   THREE = 7'b000_0110;      // h06
  parameter   FOUR  = 7'b100_1100;      // h4c
  parameter   FIVE  = 7'b010_0100;      // h24
  parameter   SIX   = 7'b010_0000;      // h20
  parameter   SEVEN = 7'b000_1111;      // h0f
  parameter   EIGHT = 7'b000_0000;      // h00
  parameter   NINE  = 7'b000_0100;      // h04

  always @ (BCD)

    case (BCD)
      0:        Display = ZERO;
      1:        Display = ONE;
      2:        Display = TWO;
```

---

[15] An active-low signal is asserted if its value is 0.
[16] The underscore character is used in the parameters of *Seven_seg_display* to make the representation of a number more readable.

```
    3:              Display = THREE;
    4:              Display = FOUR;
    5:              Display = FIVE;
    6:              Display = SIX;
    7:              Display = SEVEN;
    8:              Display = EIGHT;
    9:              Display = NINE;
 default:          Display = BLANK;
 endcase
endmodule
```

End of Example 5.25

5.10   Dataflow Models of a Linear-Feedback Shift Register

RTL models are popular in industry because they are easily synthesized by modern tools for electronic design automation (EDA). The next example illustrates an RTL model of a synchronous circuit, an autonomous linear-feedback shift register that executes concurrent transformations on a datapath under the synchronizing control of its only input, a clock signal.

Example 5.26

Linear-feedback shift registers (LFSRs) are commonly used in data-compression circuits implementing a signature analysis technique called cyclic-redundancy check (CRC) [4]. Autonomous LFSRs are used in applications requiring pseudo-random binary numbers.[17] For example, an autonomous LFSR can be a random pattern generator

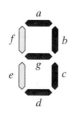

FIGURE 5-14   A seven-segment LED display.

---

[17]LFSRs are also used as fast counters when only the terminal count is needed.

Reset

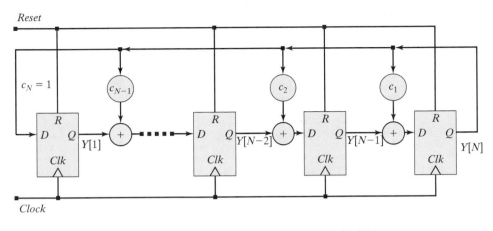

Clock

FIGURE 5-15   LFSR with modulo-2 (exclusive-or) addition.

providing stimulus patterns to a circuit. The response to these patterns can be compared to the circuit's expected response and thereby reveal the presence of an internal fault. The autonomous LFSR shown in Figure 5-15 has binary tap coefficients $C_1, ..., C_N$ that determine whether $Y(N)$ is fed back to a given stage of the register. The structure shown has $C_N = 1$ because $Y[N]$ is connected directly to the input of the leftmost stage. In general, if $C_{N-j+1} = 1$, then the input to stage j is formed as the *exclusive-or* of $Y[j - 1]$ and $Y[N]$, for $j = 2, ... N$. Otherwise, the input to stage j is the output of stage $j - 1$—$Y[j] <= Y[j - 1]$. The vector of tap coefficients determines the coefficients of the characteristic polynomial of the LFSR, which characterize its cyclic nature [2]. The characteristic polynomial determines the period of the register (the number of cycles before a pattern repeats).

The Verilog code below describes an eight-cell autonomous LFSR with a synchronous (edge-sensitive) cyclic behavior using an RTL style of design. Each bit of the register is assigned a value concurrently with the other bits; the order of the listed nonblocking assignments is of no consequence. The movement of data through the register under simulation is shown in binary and hexadecimal format in Figure 5-16 for the initial state and three cycles of the clock. Note that this model is not fully parameterized, because the register transfers are correct only if *Length* $= 8$.

```
module Auto_LFSR_RTL (Y, Clock, Reset);
  parameter             Length = 8;
  parameter             initial_state = 8'b1001_0001;        // 91h
  parameter [1: Length] Tap_Coefficient = 8'b1100_1111;

  input             Clock, Reset;
  output  [1: Length]   Y;
  reg     [1: Length]   Y;

  always @ (posedge Clock)
    if (Reset == 0) Y <= initial_state;                 // Active-low reset to initial state
```

```
        else begin
          Y[1] <= Y[8];
          Y[2] <= Tap_Coefficient[7] ? Y[1] ^ Y[8] : Y[1];
          Y[3] <= Tap_Coefficient[6] ? Y[2] ^ Y[8] : Y[2];
          Y[4] <= Tap_Coefficient[5] ? Y[3] ^ Y[8] : Y[3];
          Y[5] <= Tap_Coefficient[4] ? Y[4] ^ Y[8] : Y[4];
          Y[6] <= Tap_Coefficient[3] ? Y[5] ^ Y[8] : Y[5];
          Y[7] <= Tap_Coefficient[2] ? Y[6] ^ Y[8] : Y[6];
          Y[8] <= Tap_Coefficient[1] ? Y[7] ^ Y[8] : Y[7];
        end
      endmodule
```

End of Example 5.26

5.11   Modeling Digital Machines with Repetitive Algorithms

An algorithm for modeling the behavior of a digital machine may execute some or all
of its steps repeatedly in a given machine cycle, depending on whether the steps executed unconditionally or not. For example, an algorithm that sequentially shifts the bits
of an LFSR can be described by a *for* loop in Verilog.

Example 5.27

The LFSR in Example 5.26 is modeled again below by *Auto_LFSR_ALGO*, an algorithm-based behavioral model that uses a *for* loop to sequence through the concurrent

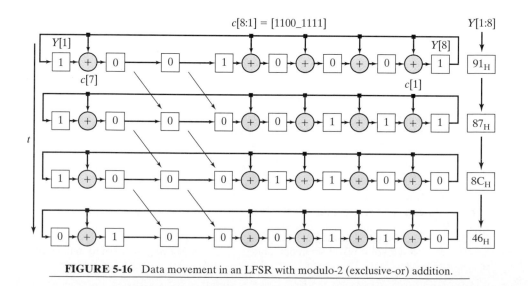

FIGURE 5-16   Data movement in an LFSR with modulo-2 (exclusive-or) addition.

(nonblocking) register assignments one at a time, beginning with the cell to the right of the MSB. The machine's activity in one clock cycle is determined by seven iterations of the loop, followed by a final assignment statement that updates the cell of the MSB. The performance of this machine is identical to that of the machine in Example 5.26.

```
module Auto_LFSR_ALGO (Y, Clock, Reset);
  parameter      Length = 8;
  parameter      initial_state = 8'b1001_0001;
  parameter      [1: Length] Tap_Coefficient = 8'b1100_1111;
  input          Clock, Reset;
  output         [1: Length] Y;
  integer        Cell_ptr;
  reg            Y;

  always @ (posedge Clock)
    begin
     if (Reset == 0) Y <= initial_state;                  // Arbitrary initial state, 91h
     else begin  for (Cell_ptr = 2; Cell_ptr <=Length; Cell_ptr = Cell_ptr +1)
       if (Tap_Coefficient [Length - Cell_ptr + 1] == 1)
       Y[Cell_ptr] <= Y[Cell_ptr - 1]^ Y [Length];
       else
       Y[Cell_ptr] <= Y[Cell_ptr - 1];
       Y[1] <= Y[Length];
     end
    end
endmodule
```

A for loop has the form:

```
for(initial_statement;   control_expression;   index_statement)
statement_for_execution;
```

End of Example 5.27

At the beginning of execution of a ***for*** loop, *initial_statement* executes once, usually to initialize a register variable (i.e., an ***integer*** or a ***reg***) that controls the loop. If *control_expression* is true, the *statement_for_execution* will execute.[18] After the *statement_for_execution* has executed, *the index_statement* will execute (usually to increment a counter). Then the activity flow will return to the beginning of the ***for*** statement and check the value of the *control_expression* again. If *control_expression* is false, the loop terminates and the activity flow proceeds to whatever statement immediately follows the *statement_for_execution*. (*Note*: The value of the register variable governed by *control_expression* in the ***for*** loop may be changed in the body of the loop during execution.)

---

[18]*Statement_for_execution* can be a single statement or a block statement (i.e., **begin** ... **end**).

Verilog has three additional loop constructs for describing repetitive algorithms: **repeat, while**, and **forever**. The **repeat** loop (see Example 5.28) executes an associated statement or block of statements a specified number of times. When the activity flow within a behavior reaches the **repeat** keyword, an expression is evaluated once to determine the number of times that the statement is to be executed. If the expression evaluates to *x* or *z*, the result will be treated as 0 and the statement will not be executed; that is, the execution skips to the next statement in the behavior. Otherwise, the execution repeats for the specified number of times, unless it is prematurely terminated by a **disable** statement within the activity flow (see Example 5.33).

Example 5.28

A **repeat** loop is used in the fragment of code below to initialize a memory array.

```
...
word_address = 0;
repeat (memory_size)
  begin
    memory [ word_address] = 0;
    word_address = word_address + 1;
  end
...
```

End of Example 5.28

Example 5.29

In this example the **for** loop is used to assign values to bits within a register after it has been initialized to *x*. The results of executing the loop are shown in Figure 5-17.

```
reg [15: 0] demo_register;
integer K;

...
for (K = 4; K; K = K - 1)
  begin
    demo_register [K + 10] = 0;
    demo_register [K + 2] = 1;
  end
...
```

At the beginning of execution the statement $K = 4$ executes and assigns the value 4 to K. Thus, the *control_expression*, K, is a "TRUE" value. The assignments to *demo_register* are made and then K is decremented. This process continues until a

15	14	13	12	11	10	9	8	7	6	5	4	3	2	1	0
x	0	0	0	0	x	x	x	x	1	1	1	1	x	x	x

FIGURE 5-17   Register contents after execution of the *for* loop.

decrementation assigns the value $K = 0$. This produces a false value for the expression that controls the loop (i.e., K) and control passes to whatever statement follows the for loop's *statement_for_execution*.

End of Example 5.29

Example 5.30

A majority circuit asserts its output if a majority of the bits of an input word are asserted. The description in *Majority_4b* is suitable for a 4-bit datapath, and uses a ***case*** statement to decode the bit patterns. However, this model is hardwired and becomes cumbersome for long word lengths. The parameterized alternative, *Majority*, uses a ***for*** loop to count the asserted bits in *Data*. A final procedural assignment asserts Y after the loop has competed execution, provided that *count* exceeds the value defined by the parameter *majority*. (The parameters in *Majority* provide flexibility in sizing *Data* and *count*, and in setting an assertion threshold, *majority*. Figure 5-18 shows a segment of simulation results for *Majority*.

```
module Majority_4b (Y, A, B, C, D);
  input    A, B, C, D;
  output  Y;
  reg      Y;
  always @ (A or B or C or D) begin
    case ({A, B,C, D})
      7, 11, 13, 14, 15:    Y = 1;
      default             Y = 0;
    endcase
  end
endmodule

module Majority (Y, Data);
  parameter    size = 8;
  parameter    max = 3;
```

```
parameter        majority = 5;
input            [size-1: 0]        Data;
output                              Y;
reg                                 Y;
reg              [max-1: 0]         count;
integer                             k;

always @ (Data) begin
  count = 0;
  for (k = 0; k < size; k = k + 1) begin
    if (Data[k] == 1) count = count + 1;
  end
  Y = (count >= majority);
  end
endmodule
```

A Verilog *while* loop has the form:

```
while (expression) statement;
```

End of Example 5.30

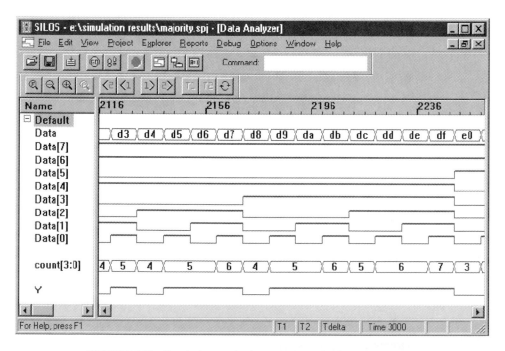

FIGURE 5-18 Simulation results for a parameterized majority circuit.

When the *while* statement is encountered during the activity flow of a cyclic or single-pass behavior, *statement*[19] executes repeatedly while a Boolean *expression* is true. When the *expression* is false the activity flow skips to whatever statement follows *statement*. For example, the statement below increments a synchronous counter while *enable* is asserted.

while (enable) **begin @ (posedge** clock) count <= count + 1; **end**

5.11.1 Intellectual Property Reuse and Parameterized Models

Models have increased value if they are written to be extendable to more than one application. Use parameters to specify bus widths, word length, and other details that customize a model to an application.

Example 5.31

The model *Auto_LFSR_Param* describes the same functionality as *Auto_LFSR_RTL* and *Auto_LFSR_ALGO* (see Example 5.26 and Example 5.27), but uses a parameterized *for* loop, conditional operators, and concurrent assignments to the register cells. Unlike *Auto_LFSR_RTL*, which was hard-wired to eight cells, *Auto_LFSR_param* is easily extended to an arbitrary length by changing only its parameters.[20]

```
module Auto_LFSR_Param (Y, Clock, Reset);
   parameter              Length = 8;
   parameter              initial_state = 8'b1001_0001; //    Arbitrary initial state
   parameter [1: Length] Tap_Coefficient = 8'b1100_1111;

   input                  Clock, Reset;
   output [1: Length]     Y;
   reg [1: Length]        Y;
   integer                k;

   always @ (posedge Clock)
     if (Reset==0) Y <= initial_state;
       else begin
         for (k = 2; k <= Length; k = k + 1)
           Y[k] <= Tap_Coefficient[Length-k+1] ? Y[k-1] ^ Y[Length] : Y[k-1];
           Y[1] <= Y[Length];
       end
endmodule
```

End of Example 5.31

---

[19]Statement can be a single statement or a block statement (i.e., **begin** ... **end**).
[20]For example, a testbench could assign a different initial state.

Example 5.32

The algorithm in *count_of_1s* uses the Verilog right-shift operator ($>>$) in counting the number of 1s that are within a register, by counting the number of times that a 1 is observed in the LSB as the word is shifted repeatedly to the right.[21] The right-shift operator fills a 0 in the MSB position that is emptied by the shift operation.

```
begin: count_of_1s         // count_of_1s declares a named block of statements
  reg [7: 0] temp_reg;

  count = 0;
  temp_reg = reg_a;        // load a data word
   while (temp_reg)
    begin
      if (temp_reg[0]) count = count + 1;
      temp_reg = temp_reg >> 1;
    end
end
```

An alternative description is shown below:

```
begin: count_of_1s
  reg [7: 0] temp_reg;

  count = 0;
  temp_reg = reg_a;        // load a data word
   while (temp_reg)
    begin
      count = count + temp_reg[0];
      temp_reg = temp_reg >> 1;
    end
end
```

The right-shift operation will eventually cause *temp_reg* to have a value of 0, thereby causing the loop to terminate.

End of Example 5.32

---

[21]In general, the operator can be accompanied by an integer value to shift a word by a specified number of positions, e.g., the statement, $word< = word >> 3$; will shift **word** by three bits to the right, and fill in with 0s from the left. The left-shift operator ($<<$) has similar effect, but in the reverse direction as the right-shift operator.

5.11.2 Clock Generators

Clock generators are used in testbenches to provide a clock signal for testing the model of a synchronous circuit. A flexible clock generator will be parameterized for a variety of applications. The *forever* loop causes unconditional repetitive execution of statements, subject to the *disable* statement, and is a convenient construct for describing clocks.

Example 5.33

The code below produces the symmetric waveforms in Figure 5-19 under simulation. The loop mechanism *forever* executes until the simulation terminates. This example also illustrates how the activity of an *initial* behavior may continue for the duration of a simulation, without expiring. The *disable* statement terminates execution after 3500 time steps by disabling the named block *clock_loop*.

```
parameter half_cycle = 50;
parameter stop_time = 350;
initial
  begin: clock_loop       // Note: clock_loop is a named block of statements
    clock = 0;
    forever
      begin
        #half_cycle clock = 1;
        #half_cycle clock = 0;
      end
  end

initial
  #350 disable clock_loop;
```

End of Example 5.33

In many situations, loops can be constructed using any of the four basic looping mechanisms of Verilog, but be aware that some EDA synthesis tools will synthesize only the *for* loop. Also, note that *always* and *forever* are not the same construct, though both are cyclic. First, the *always* construct declares a concurrent behavior. The *forever* loop is a computational activity flow and is used only within a behavior. Its execution is not necessarily concurrent with any other activity flow. The second significant distinction is that *forever* loops can be nested; on the other hand, cyclic and single-pass behaviors may not be nested. Finally, a *forever* loop executes only when it is reached within a sequential activity flow. An *always* behavior becomes active and can execute at the beginning of simulation.

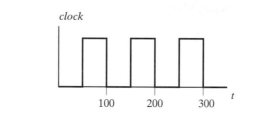

FIGURE 5-19   Clock waveforms implemented with a *forever* loop.

The *disable* statement is used to prematurely terminate a named block of procedural statements. The effect of executing the *disable* is to transfer the activity flow to the statement that immediately follows the named block or task in which *disable* was encountered during simulation.

Example 5.34

The *find_first_one* module below finds the location of the first 1 in a 16-bit word. When *disable* executes, the activity flow exits the *for* loop, proceeds to *end*, and then returns to the *always* to await the next event on *trigger*. At that time *index_value* holds the value at which *A_word* is one.

```
module find_first_one (index_value, A_word, trigger);
  output [3: 0]   index_value;
  input [15: 0]   A_word;
  input           trigger;
  reg    [3: 0]   index_value;
  always @  (trigger)
    begin: search_for_1
    index_value = 0;
    for (index_value = 0; index_value <= 15; index_value = index_value + 1)
      if (A_word[index_value] == 1) disable search_for_1;
    end
endmodule
```

End of Example 5.34

5.12 Machines with Multicycle Operations

Some digital machines have repetitive operations distributed over multiple clock cycles. This activity can be modeled in Verilog by a synchronous cyclic behavior that has as many nested edge-sensitive event-control expressions as are needed to complete the operations.

Example 5.35

A machine that is to form the sum of four successive samples of a datapath could store the samples in registers and then use multiple adders to form the sum, or it could use one adder to accumulate the sum sequentially. The implicit state machine *add_4cycle* adds four successive samples on a data bus.

```
module add_4cycle (sum, data, clk, reset);
  output        [5: 0]   sum;
  input         [3: 0]   data;
  input                  clk, reset;
  reg                    sum;

  always @ (posedge clk) begin:  add_loop
    if (reset) disable add_loop;                      else sum <= data;
      @ (posedge clk) if (reset) disable add_loop;    else sum <= sum + data;
        @ (posedge clk) if (reset) disable add_loop;  else sum <= sum + data;
          @ (posedge clk) if (reset) disable add_loop; else sum <= sum + data;
  end
endmodule
```

The behavior in *add_4cycle* contains four event control expressions. The sum is initialized to the first sample of data in the first clock cycle. Four samples of data are accumulated after four clock cycles, before the activity flow returns to the first event control expression to await a new sequence of samples of data. Note that the ***disable*** statement is included within the reset statement in each clock cycle to ensure that the machine reinitializes properly, regardless of when reset is asserted [5]. A hardware realization of *add_4cycle* is shown in Figure 5-20. It synthesizes a state machine[22] to control the four-cycle operation and uses only one adder.

End of Example 5.35

---

[22]See Appendix G for a description of ASIC flip-flop standard cells used in the examples.

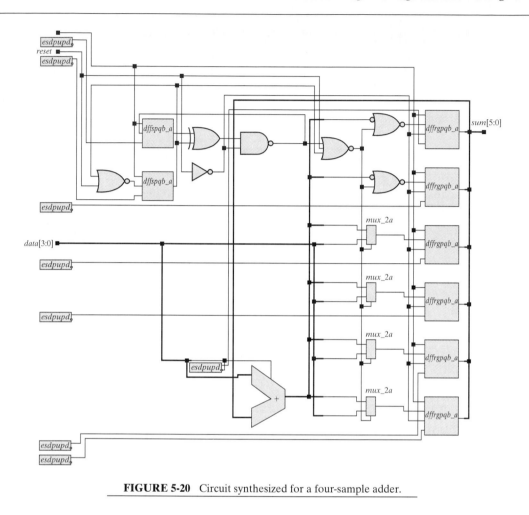

FIGURE 5-20   Circuit synthesized for a four-sample adder.

5.13   Design Documentation with Functions and Tasks: Legacy or Lunacy?

Verilog models are a legacy of their author. Whether a model is useful to anyone else depends on the correctness and clarity of the description. Even a correct model has limited utility if its credibility is compromised by poor documentation and style. Verilog has two types of subprograms that can improve the clarity of a description by encapsulating and organizing code into tasks and functions. Tasks create a hierarchical organization of the procedural statements within a Verilog behavior; functions substitute for an expression. Tasks and functions let designers manage a smaller segment of code. Both constructs facilitate a readable style of code, with a single identifier convey-ing the meaning of many lines of code. Encapsulation of Verilog code into tasks or functions hides the details of an implementation from the outside world. Overall, tasks and functions improve the readability, portability, and maintainability of a model.

5.13.1 Tasks

Tasks are declared within a module, and they may be referenced only from within a cyclic or single-pass behavior. A task can have parameters passed to it, and the results of executing the task can be passed back to the environment. When a task is called, copies of the parameters in the environment are associated with the inputs, outputs, and inouts within the task according to the order in which the inputs, outputs, and in-outs are declared. The variables in the environment are visible to the task. Additional, local, variables may be declared within a task. A word of caution: a task can call itself, but the memory supporting the variables of a task is shared by all calls. The original standard (1995) language does not support recursion, so anticipate side effects.[23]

A task must be named, and may include declarations of any number or combina-tion of the following: *parameter, input, output, inout, reg, integer, real, time, realtime,* and *event*. The variable types *real, time* and *realtime* are additional members of the register family of types (see Appendix D). The keyword *event* declares an abstract event. Abstract events are used in high-level modeling, but we will not use them in our examples because they are not supported by synthesis tools. All of the declarations of variables are local to the task. The arguments of the task retain the type they hold in the environment that invokes the task. For example, if a wire bus is passed to the task, it may not have its value altered by an assignment statement within the task. All the ar-guments to the task are passed by a value—not by a pointer to the value. When a task is invoked, its formal and actual arguments are associated in the order in which the task's ports have been declared.

Example 5.36

The module *adder_task* contains a user-defined task that adds two 4-bit words and a carry bit. The circuit produced by the synthesis tool is in Figure 5-21.

```
module adder_task (c_out, sum, c_in, data_a, data_b, clk, reset);
  output          [3: 0]    sum;
  output                    c_out;
  input           [3: 0]    data_a, data_b;
  input                     clk, reset;
```

---

[23]Verilog-2001 adds *automatic* tasks and functions, which allocate unique storage to each call of a task or function, thereby supporting recursion.

```
input                              c_in;
reg                                sum;
reg                                c_out;

always @  (posedge clk or posedge reset)
  if (reset)      {c_out, sum} <= 0; else
  add_values (c_out, sum, data_a, data_b, c_in);

task add_values;
  output              [3: 0]    sum;
  output                        c_out;
  input               [3: 0]    data_a, data_b;
  input                         c_in;

  begin
   {c_out, sum} <= data_a + (data_b + c_in);
  end
 endtask
endmodule
```

End of Example 5.36

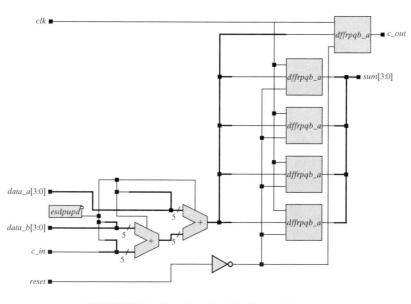

FIGURE 5-21   Circuit synthesized from *adder_task*.

5.13.2 Functions

Verilog functions are declared within a parent module and can be referenced in any valid expression—for example, in the RHS of a continuous-assignment statement. A function is also implemented by an expression and returns a value at the location of the function's identifier. Functions may implement only combinational behavior, that is, they compute a value on the basis of the present value of the parameters that are passed to the function. Consequently, they may not contain timing controls (no delay control [#], event control [@], or *wait* statements), and may not invoke a task. However, they may call other functions, but not recursively.

A function may contain a declaration of inputs and local variables. The value of a function is returned by its name when the expression calling the function is executed. Consequently, a function may not have any declared output or inout port (argument). It must have at least one input argument. The execution, or evaluation, of a function takes place in zero time, that is, in the same time step that the calling expression is evaluated by the host simulator. The definition of a function implicitly defines an internal register variable with the same name, range, and type as the function itself; this variable must be assigned value within the function body.

Example 5.37

The function *aligned_word* in *word_aligner* shifts (<< is the left-shift operator) a word to the left until the most significant bit is a 1. The input to *word_aligner* is an 8-bit word, and the output is also an 8-bit word.

```
module word_aligner (word_out, word_in);
  output  [7: 0]   word_out;
  input   [7: 0]   word_in;

  assign word_out = aligned_word(word_in);

  function  [7: 0] aligned_word;
    input     [7: 0]        word;
    begin
      aligned_word = word_in;
      if (aligned_word != 0)
        while (aligned_word[7] == 0) aligned_word = aligned_word << 1;
    end
  endfunction
endmodule
```

End of Example 5.37

Example 5.38

The Verilog model *arithmetic_unit* uses functions with descriptive names to make the source code more readable. The combinational circuit synthesized from *arithmetic_unit* is shown in Figure 5-22.

```
module arithmetic_unit (result_1, result_2, operand_1, operand_2,);
   output                [4: 0] result_1;
   output                [3: 0] result_2;
   input                 [3: 0] operand_1, operand_2;


   assign result_1 = sum_of_operands (operand_1, operand_2);
   assign result_2 = largest_operand (operand_1, operand_2);

   function [4: 0] sum_of_operands;
   input [3: 0] operand_1, operand_2;

   sum_of_operands = operand_1 + operand_2;
   endfunction

   function [3: 0] largest_operand;
   input [3: 0] operand_1, operand_2;

   largest_operand = (operand_1 >= operand_2) ? operand_1 : operand_2;
   endfunction
endmodule
```

End of Example 5.38

Functions and tasks are both used to improve the readability of a Verilog model and to exploit re-usable code. Functions are equivalent to combinational logic, and cannot be used to replace code that contains event control (@) or delay control (#) operators. Tasks are more general than functions, and may contain timing controls. Tasks that are to be synthesized may contain event-control operators, but not delay-control operators.

5.14 Algorithmic State Machine Charts for Behavioral Modeling

Many sequential machines implement algorithms (i.e., multistep sequential computations) in hardware. A machine's activity consists of a synchronous sequence of operations on the registers of its datapaths, usually under the direction of a controlling state machine. State-transition graphs (STGs) indicate the transitions that result from inputs that are applied when a state machine is in a particular state, but STGs do not directly

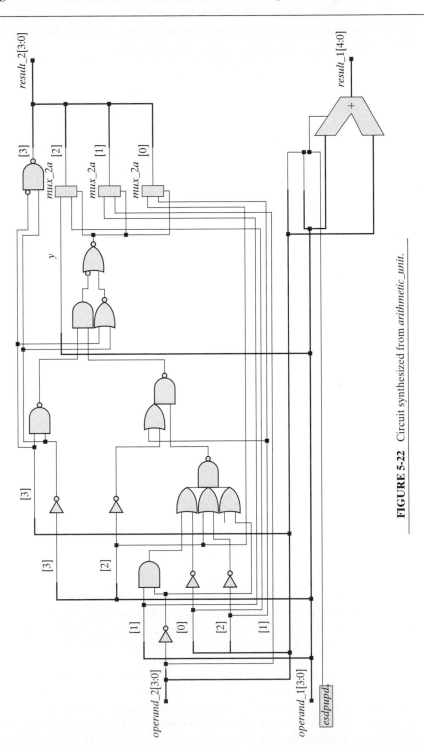
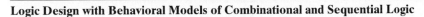

FIGURE 5-22   Circuit synthesized from *arithmetic_unit.*

display the evolution of states under the application of input data. Fortunately, there is an alternative format for describing a sequential machine.

Algorithmic state machine (ASM) charts are an abstraction of the functionality of a sequential machine, and are a key tool for modeling their behavior [1, 2, 6, 7, 8]. They are similar to software flowcharts, but display the time sequence of computational activity (e.g., register operations) as well as the sequential steps that occur under the influence of the machine's inputs. An ASM chart focuses on the activity of the machine, rather than on the contents of all the storage elements. Sometimes it is more convenient, and even essential, to describe the state of a machine by the activity that unfolds during its operation, rather than the data that are produced by the machine. For example, instead of describing a 16-bit counter by its contents we can view it as a datapath unit and describe its activity (e.g., counting, waiting, etc.).

ASM charts can be very helpful in describing the behavior of sequential machines, and in designing a state machine to control a datapath. We will introduce ASM charts in this chapter and make extensive use of them in designing sequential machines and datapath controllers in Chapter 6 and Chapter 7.

An ASM chart is organized into blocks having an internal structure formed from the three fundamental elements shown in Figure 5.23(a): a state box, a decision box, and a conditional box [2]. State boxes are rectangles, conditional boxes are rectangles with round corners, and decision boxes are diamond-shaped. The basic unit of an ASM chart is an ASM block, shown in Figure 5.23(b). A block contains one state box and an optional configuration of decision diamonds and conditional boxes placed on directed paths leaving the block. An ASM chart is composed of ASM blocks; the state box represents the state of the machine between synchronizing clock events. The blocks of an ASM chart are equivalent to the states of a sequential machine. Given an ASM chart, equivalent information can be expressed by a state-transition graph, but with less clarity about the activity of the machine.

Both types of state machines (Mealy and Moore) can be represented by ASM charts. The outputs of a Moore-type machine are usually listed inside a state box. The values of the variables in the decision boxes determine the possible paths through the block under the action of the inputs. The ASM chart for a vehicle speed controller (see Figure 5-23c) has a Mealy-type output indicating that the tail lights of the vehicle are illuminated while the brake is applied.

Conditional outputs (Mealy outputs) are placed in a conditional box on an ASM chart. These boxes are sometimes annotated with the register operations that occur with the state transition in more general machines that have datapath registers as well as a state register, but we will avoid that practice in favor of the ASMD charts that will be discussed below. The decision boxes along a path in an ASM chart imply a priority decoding of the decision variables. For example, in Figure 5-23(c) the brake has priority over the accelerator. Only paths leading to a change in state are shown, and if a variable does not appear in a decision box on a path leaving a state, then it is understood that the path is independent of the value of the variable. (The accelerator is not decoded in state S_high in Figure 5-23c.) ASM charts can become cluttered, so we sometimes place only the asserted value of a decision variable on the corresponding path and do not label paths with de-asserted decision variables unless the omission would lead to confusion. We also may omit showing default transitions and assertions that

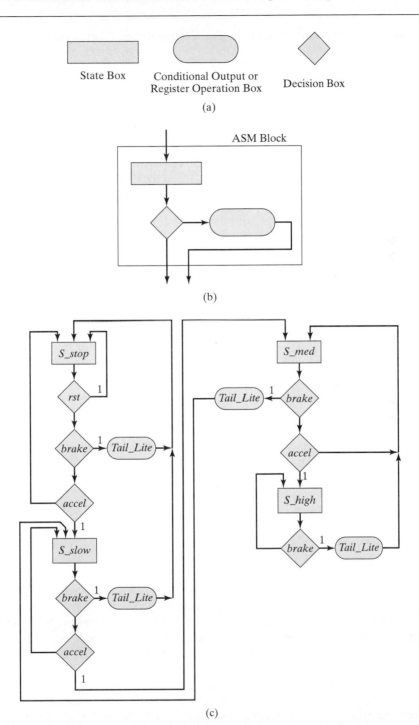

(a)

(b)

(c)

FIGURE 5-23 Algorithmic state machine charts: (a) symbols, (b) an ASM block, and (c) an ASM chart for a vehicle speed controller.

return to the same state and paths that return to the reset state when a reset signal is asserted.

5.15 ASMD Charts

One important use of a state machine is to control register operations on a datapath in a sequential machine that has been partitioned into a controller and a datapath. The controller is described by an ASM chart, and we modify the chart to link it to the data-path that is controlled by the machine. The chart is modified by annotating each of its paths to indicate the concurrent register operations that occur in the associated data-path unit when the state of the controller makes a transition along the path. ASM charts that have been linked to a datapath in this manner are called algorithmic state machine and datapath (ASMD) charts. ASMD charts are motivated by the finite-state machine–datapath paradigm (FSMD) that was introduced in other works as a univer-sal model that represents all hardware design [9].

   ASMD charts help clarify the design of a sequential machine by separating the design of its datapath from the design of the controller, while maintaining a clear rela-tionship between the two units. Register operations that occur concurrently with state transitions are annotated on a path of the chart, rather than in conditional boxes on the path, or in state boxes, because these registers are not part of the controller. The out-puts generated by the controller are the signals that control the registers of the data-path and cause the register operations that annotate the ASMD chart.

   In the examples that follow, we will adhere to a practice of indicating an asyn-chronous reset signal by a labeled path entering a reset state, but not emanating from another state. A synchronous reset signal will be denoted by a decision diamond placed on the path leaving the reset state. The diamond will have an exit path that returns to the reset state if the reset signal is asserted. It will not be shown at other states.

Example 5.39

The architecture and ASMD chart in Figure 5-24 describe the behavior of *pipe_2stage*, a two-stage pipeline that acts as a 2:1 decimator with a parallel input and output. Dec-imators are used in digital signal processors to move data from a high-clock-rate data-path to a lower-clock-rate datapath. They are also used to convert data from a parallel format to a serial format. In the example shown here, entire words of data can be trans-ferred into the pipeline at twice the rate at which the content of the pipeline must be dumped into a holding register or consumed by some processor. The content of the holding register, *R0*, can be shifted out serially, to accomplish an overall parallel-to-serial conversion of the data stream.

   The ASMD chart in Figure 5-24(b) indicates that the machine has synchronous reset to *S_idle*, where it waits until *rst* is de-asserted and *En* is asserted. Note that tran-sitions that would occur from the other states to *S_idle* under the action of *rst* are not

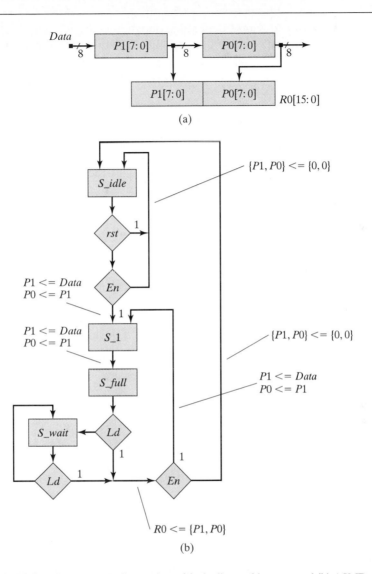

FIGURE 5-24 Two-stage pipeline register: (a) pipeline architecture, and (b) ASMD chart.

shown. With *En* asserted, the machine transitions from *S_idle* to *S_1*, accompanied by concurrent register operations that load the MSByte of the pipe with *Data* and move the content of *P1* to the LSByte (*P0*). At the next clock the state goes to *S_full*, and now the pipe is full. If *Ld* is asserted at the next clock, the machine moves to *S_1* while dumping the pipe into a holding register *R0*. If *Ld* is not asserted, the machine enters *S_wait* and remains there until *Ld* is asserted, at which time it dumps the pipe and returns to *S_1* or to *S_idle*, depending on whether *En* is asserted too. The data rate at *R0* is one-half the rate at which data is supplied to the unit from an external datapath.

End of Example 5.39

Note that the ASMD chart in Figure 5-24 is not fully developed. The conditional outputs that interface the controller with the datapath and cause the indicated operations on the datapath must be added to complete the chart.[24] Thus, the design of a datapath controller (1) begins with an understanding of the sequential register operations that must execute on a given datapath architecture, (2) defines an ASM chart describing a state machine that is controlled by primary input signals and/or status signals from the datapath, (3) forms an ASMD chart by annotating the arcs of the ASM chart with the datapath operations associated with the state transitions of the controller, (4) annotates the state of the controller with unconditional output signals, and (5) includes conditional boxes for the signals that are generated by the controller to control the datapath. If signals report the status of the datapath to the controller, these are placed in decision diamonds too, to indicate that there is feedback linkage between the machines. This decomposition of effort leads to separately verifiable models for the controller and the datapath. The final step in the design process is to integrate the verified models within a parent module and to verify the functionality of the overall machine. We will address this methodology in more detail in Chapter 7.

The register operations of ASMD charts are usually written in register transfer notation (RTN), a set of symbols and semantics that compactly specifies the instruction set of a computer [2, 6, 7]. We will describe those operations by Verilog's operators, which correspond to common hardware operations. Note the concatenation and nonblocking assignment operators in Figure 5-24. Datapath register operations made with a nonblocking assignment operator are concurrent, so the register transfers denoted by *R0 <= {P1, P0}* and *{P1, P0} <= 0* are concurrent and do not race.

5.16 Behavioral Models of Counters, Shift Registers, and Register Files

The storage elements of counters and registers usually have the same synchronizing and control signals.[25] A counter generates a sequence of related binary words; a register stores data that can be retrieved and/or overwritten under the control of a host

---

[24]See Problem 29 at the end of the chapter.

[25]An exception is a ripple counter, which connects the output of a stage to the clock input of an adjacent stage [5].

processor. The cells of a shift register exchange contents in a systematic and synchronous manner. Register files are a collection of registers that share the same synchronizing and control signals. Behavioral descriptions of a wide variety of counters, shift registers, and register files are routinely synthesized by modern synthesis tools.

5.16.1 Counters

Example 5.40

Suppose a 3-bit counter has the features that it can count up or down or hold the count. The counter could be modeled by choosing a state consisting of the content of the register holding the count, but instead we choose to associate the state with the activity of the machine, which consists of idling, incrementing, or decrementing. The latter approach allows us to model the counter independently of its word length. We let the mode of counting be determined by a 2-bit input word, *up_dwn*, with options to count up, count down, or hold the count, and included an active low asynchronous reset of the counter. Two versions of the ASM chart for the machine are shown in Figure 5-25—one, Figure 5-25(a), focuses attention on the state transitions of the machine; the other, Figure 5-25(b), shows the conditional output/operation boxes annotated with register operations using the Verilog operator $<=$ to effect concurrent (nonblocking) register assignments. Once the machine's operation is verified in (a), the concurrent register operations that are linked to the state transitions can be included, as shown by the conditional output boxes in Figure 5-25(b).

The activity of the counter has three states: idling (*S_idle*), incrementing (*S_incr*), and decrementing (*S_decr*). The asynchronous active-low reset signal, *reset_* drives the state to *S_idle* and its action is not confined to the active edges of the clock. The signal *reset_* is shown only at *S_idle*, to indicate that *S_idle* is reached from any state when *reset_* is asserted. The machine enters *S_idle* asynchronously from any state under the action of *reset_* and enters synchronously from *S_decr* and *S_incr* if *up_dwn* is 0 or 3. Otherwise, the count is either incremented or decremented.

Note that the ASM charts in Figure 5-25 are independent of the word length of the counter and that they capture the functionality of the machine. They can be adapted to a variety of applications.

A controller and datapath implementation of a sequential machine based on Figure 5-25 would require a 4-bit register to hold *count* and a separate 2-bit register to hold the state. A closer look at the machine suggests even further simplification. The counter can be viewed as having a single (equivalent) state, *S_running*, and there is no need for a state register, only a datapath register for *count*. Two ASMD charts are shown in Figure 5-26, one (a) for a machine with asynchronous reset, and the other (b) for a machine with synchronous reset. The action of *reset_* is to drive the state to *S_running* and flush the register holding *count*. *reset_* is shown as a synchronous entry into *S_running* in Figure 5-26(a). A decision diamond for *reset_* is shown in Figure 5-26(b)

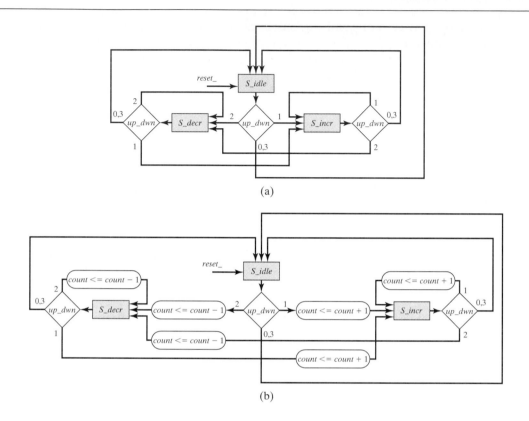

FIGURE 5-25 ASM charts for a behavioral model of an up–down counter having synchronous reset: (a) without conditional output boxes, and (b) with conditional output boxes for register operations outputs generated by the state machine.

on the path leaving *S_running* to remind us that the machine ignores *up_dwn* if *reset_* is asserted.

The Verilog model of the counter can be derived from the ASMD chart by noting that at every clock edge the machine either increments the count, decrements the count, or leaves the count unchanged. The cyclic behavior shown in *Up_Down_Implicit1* describes the decision tree for the state changes and the operations on the datapath register. It suppresses the details of the control signals that will control the hardware datapath.

```
module Up_Down_Implicit1 (count, up_dwn, clock, reset_);
 output    [2: 0]   count;
 input     [1: 0]   up_dwn;
 input              clock, reset_;

 reg       [2: 0]   count;
```

```
always @ (negedge clock or negedge reset_)
    if (reset_ == 0)                              count <= 3'b0; else
    if (up_dwn == 2'b00 || up_dwn == 2'b11) count <= count; else
    if (up_dwn == 2'b01)                     count <= count + 1; else
    if (up_dwn == 2'b10)                     count <= count −1;
endmodule
```

End of Example 5.40

Example 5.41

A ring counter asserts a single bit that circulates through the counter in a synchronous manner. The movement of data in an 8-bit ring counter is illustrated in Figure 5-27. Given an external synchronizing signal, *clock*, the behavior described by *ring_counter* ensures the synchronous movement of the asserted bit through the register and the automatic restarting of the count at *count[0]* after *count[7]* is asserted at the end of a cycle. Note that the activity of the machine is the same in every clock cycle, and that *ring_counter* is an implicit state machine. The synthesized circuit is shown in Figure 5-28. The D-type flip-flops in the implementation are active on the rising edge of the clock, have gated data (i.e., the datapath of the flip-flop is driven by the output of a multiplexer whose control signal selects between the output of the flip-flop and the external datapath), and have an asynchronous active-low reset.

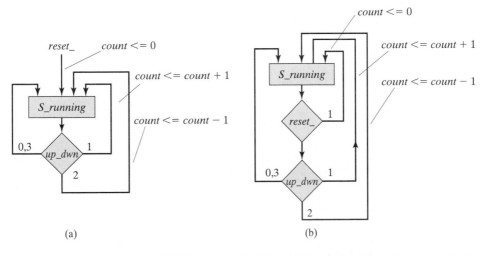

(a)   (b)

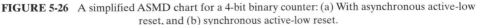

FIGURE 5-26 A simplified ASMD chart for a 4-bit binary counter: (a) With asynchronous active-low reset, and (b) synchronous active-low reset.

```
module ring_counter (count, enable, clock, reset);
    output      [7: 0]    count;
    input                 enable, reset, clock;
    reg         [7: 0]    count;

    always @  (posedge reset or posedge clock)
        if (reset == 1'b1)      count <= 8'b0000_0001; else
            if (enable == 1'b1)   count <= {count[6: 0], count[7]};    // Concatenation
                                                                          operator
endmodule
```

End of Example 5.41

Example 5.42

Our last example of a counter is a 3-bit up–down counter, but modified to include two additional features: a signal, *counter_on*, which enables the counter, and a signal, *load*, which loads an initial count from an external datapath. The description of the counter exploits Verilog's built-in arithmetic and implements the counter with an *if* statement.

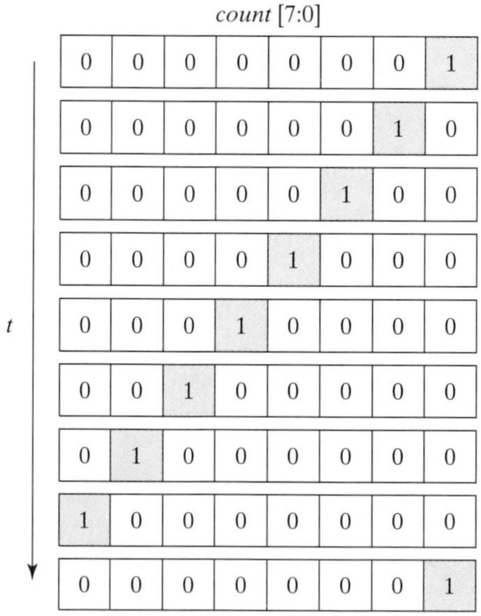

FIGURE 5-27   Data movement in a 8-bit ring counter.

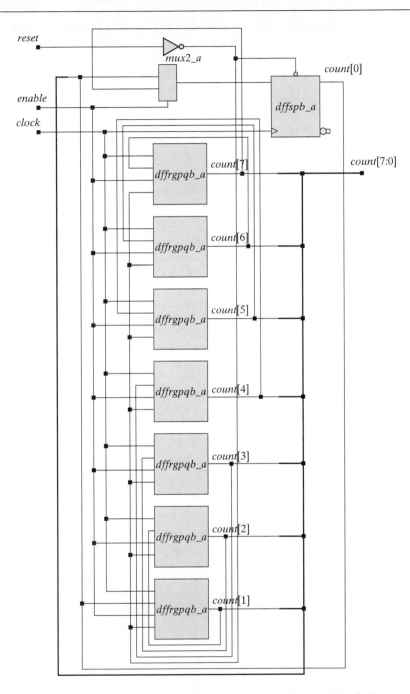

FIGURE 5-28 Ring counter synthesized from a Verilog behavioral description.

The synthesized circuit and block diagram are shown in Figure 5-29. In this implementation, the library cell *dffrgpqb_a* is a D-type flip-flop active on the rising edge, having internally gated data[26] and asynchronous active-low reset.

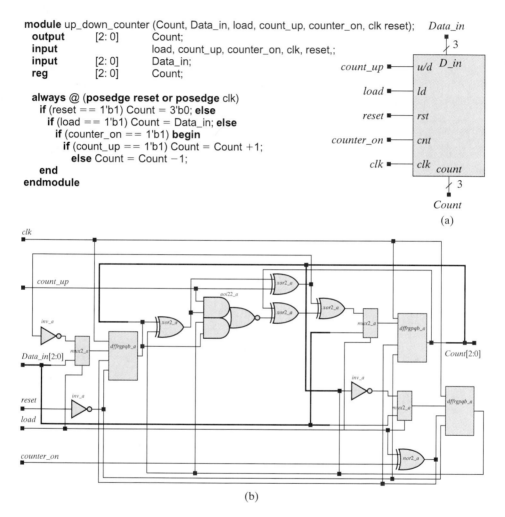

```
module up_down_counter (Count, Data_in, load, count_up, counter_on, clk reset);
    output      [2: 0]      Count;
    input                   load, count_up, counter_on, clk, reset,;
    input       [2: 0]      Data_in;
    reg         [2: 0]      Count;

    always @ (posedge reset or posedge clk)
        if (reset == 1'b1) Count = 3'b0; else
        if (load == 1'b1) Count = Data_in; else
            if (counter_on == 1'b1) begin
                if (count_up == 1'b1) Count = Count +1;
                else Count = Count −1;
        end
endmodule
```

(a)

(b)

FIGURE 5-29 3-bit up–down counter with additional features that load an initial count and enable the counting activity: (a) block diagram symbol, and (b) the circuit synthesized for the counter.

End of Example 5.42

---

[26]Cell libraries include such flip-flops because the physical layout of the mask for the integrated unit requires less area than connected, but distinct, units that accomplish the same functionality. The integrated unit will also have superior performance (smaller input–output propagation delays).

5.16.2 Shift Registers

Example 5.43

Shift_reg4 below declares an internal 4-bit register, *Data_reg*, which creates *Data_out* by a continuous assignment to the least significant bit (LSB) of the register and forms the register contents synchronously from a concatenation of the scalar *Data_in* with the three leftmost bits of the register. Notice that the register variable, *Data_reg*, is referenced by concatenation in a nonblocking assignment before it is assigned value in a synchronous behavior. This implies the need for memory, and synthesizes to the flip-flop structure shown in Figure 5-30. Also, recall that the values on the RHS of the nonblocking assignments are the values of the variables immediately before the active edge of the clock, and the values on the LHS are the values formed after the edge.

```
module Shift_reg4 (Data_out, Data_in, clock, reset);
  output              Data_out;
  input               Data_in, clock, reset;
  reg   [3: 0]        Data_reg;

  assign              Data_out = Data_reg[0];
  always @  (negedge reset or posedge clock)
    begin
      if (reset == 1'b0)    Data_reg <= 4'b0;
      else                  Data_reg <= {Data_in, Data_reg[3:1]};
    end
endmodule
```

End of Example 5.43

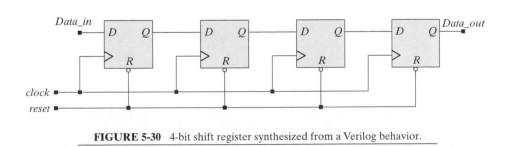

FIGURE 5-30 4-bit shift register synthesized from a Verilog behavior.

Example 5.44

In this example, a register with reset and parallel load is synthesized from the Verilog description of *Par_load_reg4*. The structure of the synthesized result is shown in Figure 5-31. The muxes and flip-flops are implemented as library cells.

```verilog
module Par_load_reg4 (Data_out, Data_in, load, clock, reset);
    input       [3: 0]    Data_in;
    input                 load, clock, reset;
    output      [3: 0]    Data_out;           // Port size
    reg                   Data_out;           // Data type

    always @  (posedge reset or posedge clock)
      begin
        if (reset == 1'b1)            Data_out <= 4'b0;
        else if (load == 1'b1)        Data_out <= Data_in;
      end
endmodule
```

End of Example 5.44

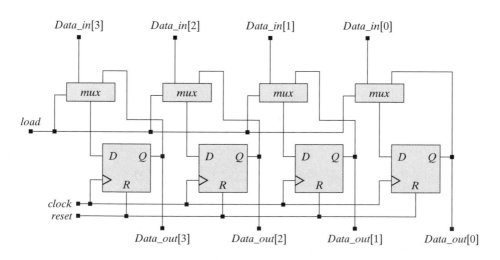

FIGURE 5-31   4-bit register with parallel load, synthesized from a Verilog behavior.

Example 5.45

Barrel shifters are used in digital signal processors to avoid overflow problems by scaling the input and output of a datapath operation. Scaling is accomplished by shifting the bits of a word to the left or to the right. Shifting a word to the right effectively divides the word by a power of 2, and shifting the word to the left multiplies the word by a power of 2. Words are shifted to the right to prevent overflow that might result from an arithmetic operation, and then the final result is shifted to the left. The shifting action of a barrel shifter can be implemented with combinational logic, but the model presented below uses registered logic and circulates the word through a storage register, by exploiting concatenation, as depicted in Figure 5-32. The top register shows the pattern before the shift, and the bottom register shows the pattern that results from the shift. The circuit synthesized from *barrel_shifter* is shown in Figure 5-32.

```verilog
module barrel_shifter (Data_out, Data_in, load, clock, reset);
  output      [7: 0]    Data_out;
  input       [7: 0]    Data_in;
  input                 load, clock, reset;
  reg         [7: 0]    Data_out;

  always @  (posedge reset or posedge clock)
    begin
      if (reset == 1'b1)      Data_out <= 8'b0;
      else if (load == 1'b1)  Data_out <= Data_in;
      else                    Data_out <= {Data_out[6: 0], Data_out[7]};
    end
endmodule
```

End of Example 5.45

Example 5.46

A 4-bit universal shift register is an important unit of digital machines that employ a bit-slice architecture, with multiple identical slices of a 4-bit shift register chained together with additional logic to form a wider and more versatile datapath [8]. Its features include synchronous reset, parallel inputs, parallel outputs, bidirectional serial input from either the LSB or the most significant bit (MSB), and bidirectional serial output to either the LSB or the MSB. In the serial-in, serial-out mode the machine can delay an input signal for 4 clock ticks, and act as a uni-directional shift register. In parallel-in, serial-out mode it operates as a parallel-to-serial converter, and in the serial-in, parallel-out mode it operates as a serial-to-parallel converter. Its parallel-in, parallel-out mode, combined with shift operations, allows it to perform any of the operations of less versatile unidirectional shift registers.

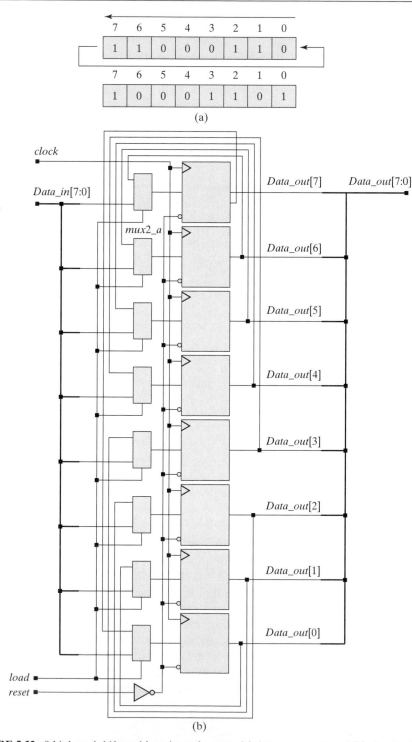

(b)

FIGURE 5-32 8-bit barrel shifter with registered output: (a) data movement, and (b) circuit synthesized from a Verilog behavioral model, *barrel shifter*.

```
module Universal_Shift_Reg  (Data_Out, MSB_Out, LSB_Out, Data_In,
  MSB_In, LSB_In, s1, s0, clk, rst);
output  [3: 0]  Data_Out;
output          MSB_Out, LSB_Out;
input   [3: 0]  Data_In;
input           MSB_In, LSB_In;
input           s1, s0, clk, rst;
reg             Data_Out;

assign MSB_Out = Data_Out[3];
assign LSB_Out = Data_Out[0];

always @ (posedge clk) begin
 if (rst) Data_Out <= 0;
 else  case ({s1, s0})
   0:  Data_Out <= Data_Out;              // Hold
   1:  Data_Out <= {MSB_In, Data_Out[3:1]};   // Serial shift from MSB
   2:  Data_Out <= {Data_Out[2: 0], LSB_In};  // Serial shift from LSB
   3:  Data_Out <= Data_In;               // Parallel Load
  endcase
 end
endmodule
```

We can anticipate that the gate-level machine will consist of four D-type flip-flops with steering logic to manage the datapaths supporting the specified features. The block diagram symbol and simulation results verifying the functionality of the model are shown in Figure 5-33. The waveforms for *Data_Out* illustrate the right-shift, left-shift, and load operations.

End of Example 5.46

5.16.3 Register Files and Arrays of Registers (Memories)

A register file consists of a small number of registers and is integrated with additional logic. It supports write and nondestructive read operations. Usually implemented by D-type flip-flops, register files are not used for mass storage because they occupy significantly more silicon area than compiled memory. A common application combines a register file in tandem with an arithmetic and logic unit (ALU), as shown in Figure 5-34. The dual-channel outputs of the register file form the datapaths to the ALU, and the output of the ALU is stored in the register file at a designated location. A host processor provides the addresses for the operations and controls the sequence of reading and writing to prevent a simultaneous read and write affecting the same location.[27]

---

[27]More complex register files have logic that allows a read operation to return the value currently written.

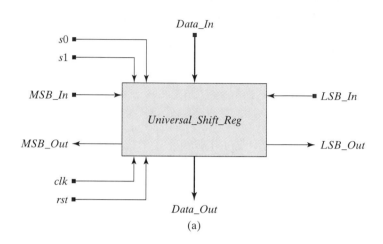

(a)

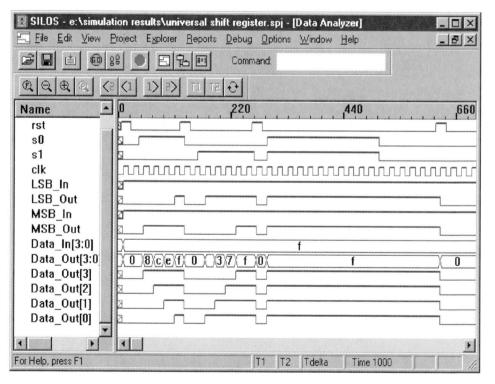

(b)

FIGURE 5-33 4-bit universal shift register: (a) block diagram symbol, and (b) simulation results verifying
the 4-bit universal shift register.

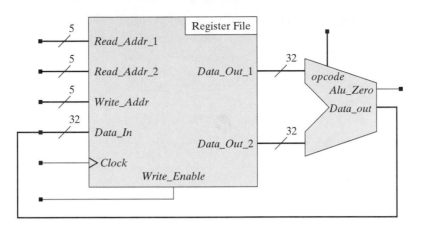

FIGURE 5-34   A 32-word register file in tandem with an ALU with a 32-bit datapath.

Example 5.47

The single-input, dual-output register file modeled on the next page as an implicit state machine by *Register_File* introduces the concept of a Verilog memory, a provision for declaring an array of words.[28] By appending an additional array range ([31: 0]) to the declaration of *Reg_File* in the module *Register_File*, we declare 32 words of memory, with each word having 32 bits. The dual output datapaths are formed by continuous-assignment statements using the 5-bit address provided by the host processor. The writing operation occurs synchronously under the control of *Write_Enable*, and the processor must ensure that data is not read while *Write_Enable* is asserted and *Clock* has a rising edge. Decoders are synthesized automatically by a synthesis tool, and are inside the register file, where they decode the addresses to locate a specific register.

```
module Register_File  (Data_Out_1, Data_Out_2, Data_in, Read_Addr_1,
        Read_Addr_2, Write_Addr, Write_Enable, Clock);
  output [31: 0]  Data_Out_1, Data_Out_2;
  input  [31: 0]  Data_in;
  input  [4: 0]   Read_Addr_1, Read_Addr_2, Write_Addr;
  input           Write_Enable, Clock;
  reg    [31: 0]  Reg_File [31: 0];  // 32bit x32 word memory declaration
```

----

[28]Verilog does not support two-dimensional arrays in which any cell can be addressed. A word in a Verilog memory can be addressed directly. A cell (bit) in a word can be addressed indirectly by first loading the word into a buffer register and then addressing the bit of the word.

```
    assign Data_Out_1 = Reg_File[Read_Addr_1];
    assign Data_Out_2 = Reg_File[Read_Addr_2];

    always @ (posedge Clock) begin
      if (Write_Enable) Reg_File [Write_Addr] <= Data_in;
    end
  endmodule
```

End of Example 5.47

5.17   Switch Debounce, Metastability, and Synchronizers for Asynchronous Signals

Sequential circuits use flip-flops and latches as storage elements, but both devices are subject to a condition called metastability. A hardware latch can enter the metastable state if a pulse at one of its inputs is too short, or if both inputs are asserted either simultaneously or within a sufficiently small interval of each other. A transparent latch can go into a metastable state too, if the data are unstable around the edge of the enable input. D-type flip-flops (edge-triggered) are formed from two cascaded transparent latches with complementary clocks. A D-type flip-flop can enter the metastable state if the data are unstable in the setup interval preceding the clock edge or if the clock pulse is too narrow. Given the vulnerability of storage devices to metastability, it is important that systems be designed to minimize the impact of signals that could cause the system to be upset by this condition.

Many physical systems that are intended to operate synchronously have asynchronous input signals. A signal is asynchronous if it is not controlled by a clock or if it is synchronized by a clock in a different domain. In both cases, a signal transition can occur in a random manner with respect to the active edge of the clock that is controlling sequential devices. Traffic lights, computer keyboards, and elevator buttons have inputs that arrive randomly. If they happen to arrive during the setup interval of a flip-flop they could cause the flip-flop to enter a metastable state and remain there for an indefinite time, upsetting the operation of the system.

If a mechanical switch generates an input that drives a flip-flop of a circuit, the input signal could oscillate during the setup interval of the flip-flop and cause it to enter a metastable state [7–10]. Figure 5-35 illustrates a simple push-button switch configuration in which the data line to the flip-flop is normally pulled down. When the spring-loaded button is pushed down a connection is made to pull the line up to Vdd. The mechanical contact will vibrate momentarily, for a few milliseconds, creating an unstable signal on the line. There are various ways to deal with switch bounce, depending on the application. For example, the push button switches on a student prototyping board for an FPGA have a resistor-capacitor (RC) lowpass filter and a buffer placed between the switch and the chip [11].

As an alternative remedy, the circuit shown in Figure 5-36 uses a single pole-double throw switch to eliminate the effect of the bounce. With the switch initially in the upper position, the line driving the upper NAND gate is pulled down to ground. The connection between the arm of the switch and the circuit is broken momentarily when the arm moves from the top position to the bottom contact. The arm of the switch still bounces when it arrives at the bottom contact, but as long as it does not bounce back to the top contact the signal at the bottom input to the NAND latch can oscillate and not affect the circuit because the top output of the latch has already made a transition from 1 to 0, thereby blocking the activity at the bottom switch contact from affecting the circuit. A similar action occurs when the switch is thrown to the top position. Keyboards commonly have debounce circuitry built into the keys.

A flip-flop may enter a metastable state if the data input changes within a finite interval before or after the clock transition. The output of the device has an output between a 0 and 1, and cannot be decoded with certainty. The physical situation is illustrated in Figure 5-37, where a ball must roll over a pinnacle before making a state transition. If it should be pushed with only enough energy to reach, but not pass, the pinnacle, it would reside there indefinitely. A circuit in the metastable state will remain there for an unpredictable time before returning to the state it had before the clock, or making a transition to the opposite state. Asynchronous inputs are problematic, because their transitions are unpredictable.

Metastability cannot be prevented, but its effect can be reduced. Experimental results have shown that the mean time between failures of a circuit with an asynchronous input is exponentially related to the length of time available for recovery from the metastable condition. Thus, high-speed digital circuits rely on synchronizers to create a

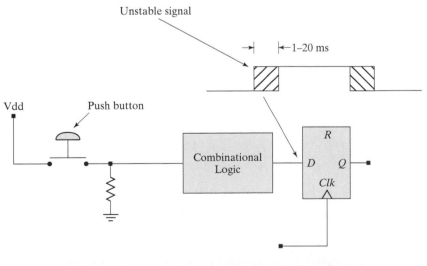

FIGURE 5-35 A push-button input device with closure bounce.

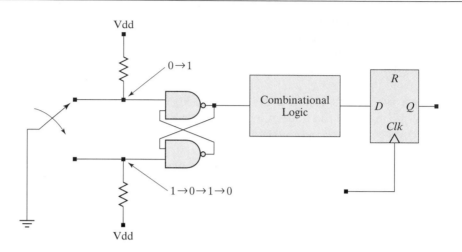

FIGURE 5-36   A NAND latch configuration for eliminating the effects of switch closure bounce.

time buffer for recovering from a metastable event, thereby reducing the possibility that metastability will cause a circuit to malfunction.[29]

The first rule of synchronization is that an asynchronous signal should never be synchronized by more than one synchronizer. To do so would risk having the outputs of multiple synchronizers produce different synchronized signals in the event that one or more of them is driven into the metastable condition.

There are two basic types of synchronizer circuits, depending on whether the asynchronous input pulse has a width that is larger or smaller than the period of the clock. In the former case, a synchronizer consists of a multistage shift register placed between the asynchronous input and the circuit. Multiple stages are used because the clock periods are ever-shrinking as technology advances, making it more likely that the

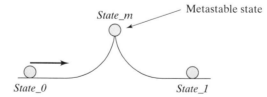

FIGURE 5-37   An illustration of how metastability can happen in a physical system.

---

[29]For an extensive treatment of synchronizers, see reference 8.

metastability of a single flip-flop might not be resolved in a single period. The circuit in Figure 5-38(a) treats the situation in which the width of the asynchronous input pulse is larger than the clock period. Two flip-flops are placed between *Asynch_in* and the circuit; the second flip-flop operates synchronously, and the flip-flop driven by *Asynch_in* guards against metastability. Consider how the circuit operates: if the asynchronous input signal reaches a stable condition outside the setup interval, it will be clocked through with a latency of two cycles. On the other hand, if *Asynch_in* is unstable during the setup interval (due to bounce or to a late-arriving input) there are two possibilities. Assuming that the circuit was in a reset condition, if the unstable input is sampled as a 0, but ultimately settles to a 1, the 1 will appear at the output with

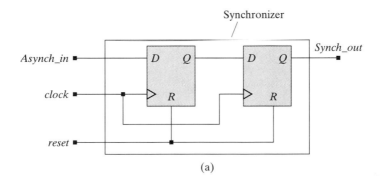

(a)

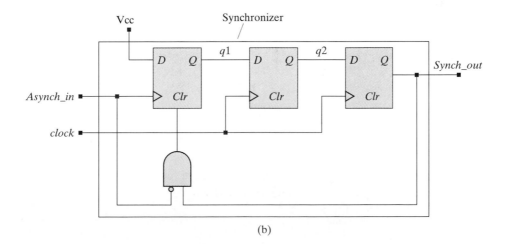

(b)

FIGURE 5-38 Synchronizer circuits for asynchronous input signals: (a) circuit for use when the width of the asynchronous input pulse is greater than the period of the clock, (b) circuit for use when the width of the asynchronous input pulse is less than the period of the clock, (c) waveforms in the circuit of (b) when the asynchronous pulse does not cause a metastable condition, and (d) waveforms of the circuit in (b) when the asynchronous input signal causes a metastable condition.

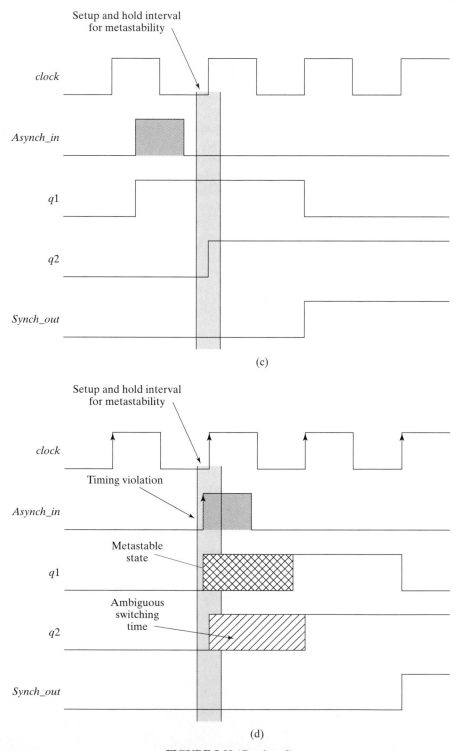

(c)

(d)

FIGURE 5-38 (Continued)

a latency of three cycles. If the signal settles to 0, it will have arrived with a latency of two cycles. Thus, the maximum latency is $n + 1$, where n is the number of stages in the synchronizer chain. The second flip-flop does not see the instability; the only effect is at the entry stage, which could lead to one additional cycle of latency. Latency is tolerable because the asynchronous input itself does not have a predictable arrival time; but ambiguous output transition times resulting from metastability are problematic because they can cause upset of the system.

If the width of the asynchronous input pulse may be less than the period of the clock, the circuit in Figure 5-38(b) is used, with an additional cost in hardware. Note that the first flip-flop has Vcc connected to its data input and has *Asych_in* connected to its clock input and that the remaining two flip-flops are triggered by the clock of the system. A short pulse at *Asynch_in* will drive $q1$ to 1; this value will propagate to *Synch_out* after the next two clock edges. When *Synch_out* becomes 1, the signal at *Clr* is asserted, assuming that *Asynch_in* returns to 0 before *Synch_out* becomes 1. The flip-flop driving $q2$ guards against metastability in the first stage of the chain.

Synchronizers are also used when a signal must cross a boundary between two clock domains (see Example 9.9). If *clock_1* in Figure 5-39 is slower than *clock_2*, the synchronizer in Figure 5-38(a) should be used to synchronize the interface signals controlling data transfer between the domains, otherwise the synchronizer in Figure 5-38(b) should be used. Care must be taken to anticipate that more than one active edge of *clock_2* will occur while *asynch_in* is asserted if *asynch_in* is synchronized to *clock_1*, that is it has duration $T_{clock\_1} > T_{clock\_2}$.[30]

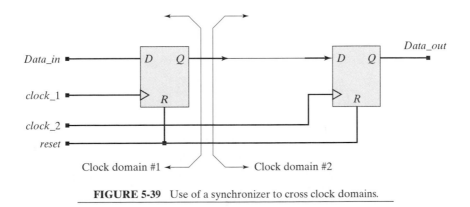

FIGURE 5-39 Use of a synchronizer to cross clock domains.

---

[30]See Chapter 9.

5.18   Design Example: Keypad Scanner and Encoder

Keypad scanners are used to enter data manually in digital telephones, computer keyboards, and other digital systems. Telephones, computers, and other equipment have a keypad. A keypad scanner responds to a pressed key and forms a code that uniquely identifies the key that is pressed. It must take into account the asynchronous nature of the inputs and deal with switch debounce. Also, it must not interpret a key to be pressed repeatedly if it is pressed once and held down.

Let us consider a scheme for designing a scanner/decoder for the hexadecimal keypad circuit shown in Figure 5-40. Each row of the keypad is connected to ground by a pull-down resistor. When a button is pressed, a connection is established between a row and a column at the location of the button; this connection will pull the row line up to the value of the column line at the location of the pressed key. If that column line is connected to the supply voltage, the row that is connected to that column by the pressed button will be pulled to the supply too; otherwise the row line is pulled down to 0. The keypad code generator unit has control over the column lines and will

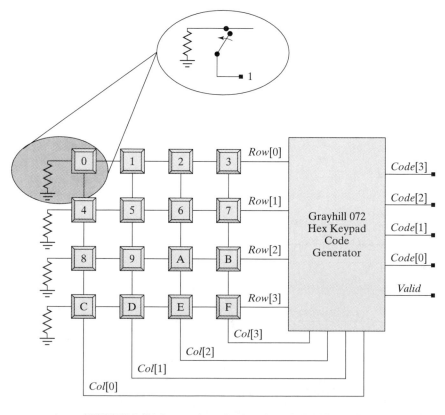

FIGURE 5-40   Scanner/encoder for a hexadecimal keypad.

exercise that control to systematically assert voltage on the column lines to detect the location of a pressed button.

A keypad code generator must implement a decoding scheme that (1) detects whether a button is pressed, (2) identifies the button that is pressed, and (3) generates an output consisting of the unique code of the button. The scanner/encoder will be implemented as a synchronous sequential machine with the button codes shown in Table 5-1. The outputs of the machine are the column lines, the code lines, and a signal, *Valid*, that indicates whether the value of code is valid. We will use the synchronizer in Figure 5-38(a) for the asynchronous input.

To detect that a button is pressed, the machine can assert a 1 simultaneously on all of the column lines until detecting that a row line has been pulled up (by sensing that the OR of the row lines is a 1. The identity of the row line that is asserted is still not known. Then the scanner applies a 1 sequentially to each of the column lines, one at a time, until a row line is detected to be asserted. The location of the asserted column and row corresponds to the key that is pressed. This information is encoded uniquely, for each key. We will assume that only one switch is closed at a time.

The behavior of the keypad scanner/encoder is represented by the ASM chart[31] shown in Figure 5-41. The machine resides in state S_0, with all column lines asserted, until one row line or more is asserted. A signal formed from OR-ing the rows launches the activity of the machine. In S_1 only column 0 is asserted. If a row is also asserted the output *Valid* is asserted for one clock and the machine moves to S_5, where it remains with all of the column lines asserted until the row is de-asserted.[32] Then the

Table 5.1   Keypad codes for a hexadecimal scanner.

Key	Row[3:0]	Col[3:0]	Code
0	0 0 0 1	0 0 0 1	0 0 0 0
1	0 0 0 1	0 0 1 0	0 0 0 1
2	0 0 0 1	0 1 0 0	0 0 1 0
3	0 0 0 1	1 0 0 0	0 0 1 1
4	0 0 1 0	0 0 0 1	0 1 0 0
5	0 0 1 0	0 0 1 0	0 1 0 1
6	0 0 1 0	0 1 0 0	0 1 1 0
7	0 0 1 0	1 0 0 0	0 1 1 1
8	0 1 0 0	0 0 0 1	1 0 0 0
9	0 1 0 0	0 0 1 0	1 0 0 1
A	0 1 0 0	0 1 0 0	1 0 1 0
B	0 1 0 0	1 0 0 0	1 0 1 1
C	1 0 0 0	0 0 0 1	1 1 0 0
D	1 0 0 0	0 0 1 0	1 1 0 1
E	1 0 0 0	0 1 0 0	1 1 1 0
F	1 0 0 0	1 0 0 0	1 1 1 1

---

[31]The asynchronous reset is shown at only S_0, but it is understood that asynchronous reset drives the state to S_0 from any state in the chart.

[32]*Valid* can be used to control the writing of data to a storage unit, such as a first in, first out (FIFO) memory. See Problem 7.11.

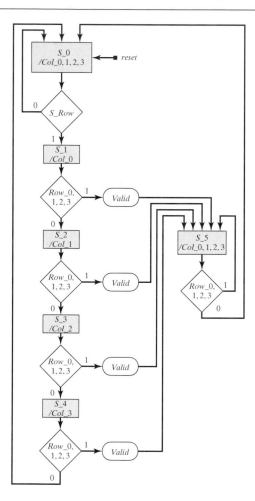

FIGURE 5-41   ASM chart of keypad scanner circuit.

machine returns to *S_0* for one cycle. Note that all of the columns are asserted in *S_5*, although only the column corresponding to the key need be asserted. This eliminates the need to add two more states to the machine as it moves from *S_2*, *S_3* and *S_4* and awaits de-assertion of the row.

The ASM chart implies that the decoding of the columns occurs in a priority manner, beginning with *Column_0*. If more than one column is asserted the first one decoded determines the code. The decision diamond leaving *S_0* tests the synchronized row signal. The decision diamonds leaving the other states can test the unsynchronized signal because in those states the signal from the key has already settled.

The Verilog models for the keypad scanner and its testbench are presented below, along with supporting modules. The scanner is to be tested within the Verilog environment, not on a physical prototyping board with a physical keypad. Therefore, the testbench shown in Figure 5-42 must include (1) a signal generator that will simulate the assertion of a key, and (2) a module, *Row_Signal*, that will assert a row line corresponding to the asserted key, and (3) *Hex_Keypad_Grayhill_072*,[33] the unit under test (UUT). After the model of the keypad scanner has been verified, it can serve as a user interface in simulating other systems, and can also be used in a physical environment with confidence that it should function correctly, which greatly reduces the scope of a search for the source of an error in the operation of a prototype.

The signal generator for key assertions is embedded in the testbench, where a single-pass behavior assigns values to *key* within a **for** loop, and a level-sensitive behavior reacts to changes in *key* by generating an ASCII string identifying the pressed key.[34] This enhances the display of waveforms produced by the simulator. The module *Row_Signal* detects assertion of a key and determines the row in which it is located. Its role is to replace the physical keypad driving the encoding unit. *Synchronizer* is a two-stage shift register whose output is a synchronized version of the OR of the asynchronous row signals.

Test bench for Hex_Keypad_Grayhill_072

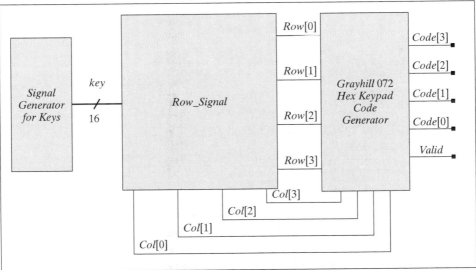

FIGURE 5-42 Testbench organization for the Grayhill 072 hexadecimal keypad scanner/encoder.

---

[33]See www.grayhill.com.
[34]Verilog does not have a data type for strings. Strings must be stored in a declared register, at 8 bits per character.

It asserts when any key is pushed. When the output of the synchronizer changes, the code generator unit determines which of the possible keys was pressed. It contains a state machine that sequentially asserts columns, and a level-sensitive behavior that parses row and column data to assert a hexadecimal code. The testbench instantiates *Hex_Keypad_Grayhill_072, Row_Signal*, and *Synchronizer*. Note that the code generator is synchronized by the positive edge of the clock, so *Synchronizer* is sensitive to the falling edge. This eliminates a potential race condition in the hardware.

The simulation results are shown in Figure 5-43, with pressed key values displayed in text format; the signals *A_Row* and *S_Row* are the input and output, respectively, of the synchronizer. The transitions of *S_Row* exhibit a latency of 1 cycle compared to those of *A_Row*. Notice that *valid* signals completion of the operations by asserting for one cycle after the key has been scanned and encoded.

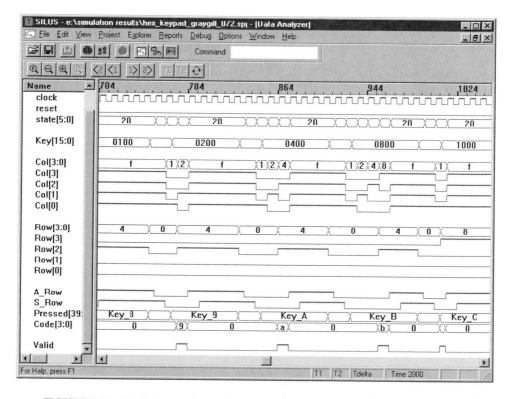

FIGURE 5-43 Simulation results for the Grayhill 072 hexadecimal keypad scanner/encoder.

```
// Decode the asserted Row and Col

//        Grayhill 072 Hex Keypad
//                Col[0]    Col[1]    Col[2]    Col[3]
//        Row[0]  0         1         2         3
//        Row[1]  4         5         6         7
//        Row[2]  8         9         A         B
//        Row[3]  C         D         E         F

module Hex_Keypad_Grayhill_072 (Code, Col, Valid, Row, S_Row, clock, reset);
  output   [3: 0]  Code;
  output           Valid;
  output   [3: 0]  Col;
  input    [3: 0]  Row;
  input            S_Row;
  input            clock, reset;
  reg              Col, Code;
  reg      [5: 0]  state, next_state;

// One-hot
  parameter     S_0 = 6'b000001, S_1 = 6'b000010, S_2 = 6'b000100;
  parameter     S_3 = 6'b001000, S_4 = 6'b010000, S_5 = 6'b100000;

  assign Valid = ((state == S_1) || (state == S_2) || (state == S_3) || (state == S_4))
         && Row;

// Does not matter if the row signal is not the debounced version.
// Assumed to settle before it is used at the clock edge

  always @ (Row or Col)
    case ({Row, Col})
        8'b0001_0001:   Code = 0;
        8'b0001_0010:   Code = 1;
        8'b0001_0100:   Code = 2;
        8'b0001_1000:   Code = 3;

        8'b0010_0001:   Code = 4;
        8'b0010_0010:   Code = 5;
        8'b0010_0100:   Code = 6;
        8'b0010_1000:   Code = 7;

        8'b0100_0001:   Code = 8;
        8'b0100_0010:   Code = 9;
        8'b0100_0100:   Code = 10;          // A
        8'b0100_1000:   Code = 11;          // B

        8'b1000_0001:   Code = 12;          // C
        8'b1000_0010:   Code = 13;          // D
        8'b1000_0100:   Code = 14;          // E
        8'b1000_1000:   Code = 15;          // F
```

```verilog
      default:            Code = 0;                    // Arbitrary choice
    endcase

  always @ (posedge clock or posedge reset)
   if (reset) state <= S_0;
   else state <= next_state;

  always @ (state or S_Row or Row) // Next-state logic
   begin next_state = state; Col = 0;
    case (state)
     // Assert all rows
     S_0: begin Col = 15; if (S_Row) next_state = S_1; end
     // Assert col 0
     S_1: begin Col = 1; if (Row) next_state = S_5; else next_state = S_2; end
     // Assert col 1
     S_2: begin Col = 2; if (Row) next_state = S_5; else next_state = S_3; end
     // Assert col2
     S_3: begin Col = 4; if (Row) next_state = S_5; else next_state = S_4; end
     // Assert col 3
     S_4: begin Col = 8; if (Row) next_state = S_5; else next_state = S_0; end
     // Assert all rows
     S_5: begin Col = 15; if (Row == 0) next_state = S_0; end
    endcase
   end
endmodule

module Synchronizer (S_Row, Row, clock, reset);
  output        S_Row;
  input [3: 0]  Row;
  input         clock, reset;
  reg           A_Row, S_Row;
  // Two stage pipeline synchronizer
  always @ (negedge clock or posedge reset)
  begin
   if (reset)  begin        A_Row <= 0;
                            S_Row <= 0;
   end
   else begin               A_Row <= (Row[0] || Row[1] || Row[2] || Row[3]);
                            S_Row <= A_Row;
   end
  end
endmodule

module Row_Signal (Row, Key, Col);            // Scans for row of the asserted key
  output [3: 0]  Row;
  input  [15: 0] Key;
  input  [3: 0]  Col;
  reg            Row;
```

```verilog
        always @ (Key or Col) begin            // Combinational logic for  key assertion
          Row[0] = Key[0]  && Col[0] || Key[1]  && Col[1] || Key[2]  && Col[2] || Key[3]
          && Col[3];
          Row[1] = Key[4]  && Col[0] || Key[5]  && Col[1] || Key[6]  && Col[2] || Key[7]
          && Col[3];
          Row[2] = Key[8]  && Col[0] || Key[9]  && Col[1] || Key[10] && Col[2] || Key[11]
          && Col[3];
          Row[3] = Key[12] && Col[0] || Key[13] && Col[1] || Key[14] && Col[2] || Key[15]
          && Col[3];
        end
      endmodule

//////////////////////////// Test Bench ////////////////////////////
      module test_Hex_Keypad_Grayhill_072 ();
        wire [3: 0]    Code;
        wire           Valid;
        wire [3: 0]    Col;
        wire [3: 0]    Row;
        reg            clock, reset;
        reg [15: 0]    Key;
        integer        j, k;
        reg[39: 0]     Pressed;
        parameter      [39: 0] Key_0 = "Key_0";
        parameter      [39: 0] Key_1 = "Key_1";    // "one-hot" code for pressed key
        parameter      [39: 0] Key_2 = "Key_2";
        parameter      [39: 0] Key_3 = "Key_3";
        parameter      [39: 0] Key_4 = "Key_4";
        parameter      [39: 0] Key_5 = "Key_5";
        parameter      [39: 0] Key_6 = "Key_6";
        parameter      [39: 0] Key_7 = "Key_7";
        parameter      [39: 0] Key_8 = "Key_8";
        parameter      [39: 0] Key_9 = "Key_9";
        parameter      [39: 0] Key_A = "Key_A";
        parameter      [39: 0] Key_B = "Key_B";
        parameter      [39: 0] Key_C = "Key_C";
        parameter      [39: 0] Key_D = "Key_D";
        parameter      [39: 0] Key_E = "Key_E";
        parameter      [39: 0] Key_F = "Key_F";
        parameter      [39: 0] None = "None";

        always @ (Key) begin
          case (Key)
            16'h0000:   Pressed = None;
            16'h0001:   Pressed = Key_0;
            16'h0002:   Pressed = Key_1;
            16'h0004:   Pressed = Key_2;
            16'h0008:   Pressed = Key_3;

            16'h0010:   Pressed = Key_4;
            16'h0020:   Pressed = Key_5;
```

```
    16'h0040:    Pressed = Key_6;
    16'h0080:    Pressed = Key_7;

    16'h0100:    Pressed = Key_8;
    16'h0200:    Pressed = Key_9;
    16'h0400:    Pressed = Key_A;
    16'h0800:    Pressed = Key_B;

    16'h1000:    Pressed = Key_C;
    16'h2000:    Pressed = Key_D;
    16'h4000:    Pressed = Key_E;
    16'h8000:    Pressed = Key_F;

    default: Pressed = None;
   endcase
 end


 Hex_Keypad_Grayhill_072 M1(Code, Col, Valid, Row, S_Row, clock, reset);
 Row_Signal M2(Row, Key, Col);
 Synchronizer M3(S_Row, Row, clock, reset);

   initial #2000 $finish;
   initial begin clock = 0; forever #5 clock = ~clock; end
   initial begin reset = 1; #10 reset = 0; end
   initial  begin for (k = 0; k <= 1; k = k+1) begin Key = 0; #25 for (j = 0; j <= 16; j
   = j+1) begin
     #20 Key[j] = 1; #60 Key = 0; end end end
 endmodule
```

REFERENCES

1. Lee S. *Design of Computers and Other Complex Digital Devices*. Upper Saddle River, NJ: Prentice-Hall, 2000.
2. Mano MM, Kime CR. *Logic and Computer Design Fundamentals*. Upper Saddle River, NJ: Prentice-Hall, 1997.
3. "Verilog HDL Coding—Semiconductor Reuse Standard." Chandler, AZ: Motorola, 1999.
4. Abramovici M, et al. *Digital Systems Testing and Testable Design*. Rockville, MD: Computer Science Press, 1990.
5. Ciletti MD. *Modeling, Synthesis and Rapid Prototyping with the Verilog HDL*. Upper Saddle River, NJ: Prentice-Hall, 1999.
6. Clare CR. *Designing Logic Systems Using State Machines*. New York: McGraw-Hill, 1971.
7. Heuring VP, Jordan HF. *Computer Systems Design and Architecture*. Menlo Park, CA: Addison-Wesley Longman, 1997.
8. Wakerly JF. *Digital Design Principles and Practices*, 3rd ed. Upper Saddle River, NJ: Prentice-Hall, 2000.
9. Gajski D, et al. "Essential Issues in Codesign." In: Staunstrup J, Wolf W, eds. *Hardware/Software Co-Design: Principles and Practices*. Boston: Kluwer, 1997.

10. Katz RH. *Contemporary Logic Design*. Redwood City, CA: Benjamin Cummings, 1994.

11. Digilent Inc. documentation (www.digilent.cc).

PROBLEMS

Note: For all of the problems requiring development and verification of a design, a carefully written test plan, a documented testbench, and simulation results are to be provided with the solution. As a minimum, the test plan should describe (1) the functional features that will be tested, and (2) how they will be tested.

1. Using a single continuous assignment, develop and verify a behavioral model implementing a Boolean equation describing the logic of the circuit below. Use the following names for the testbench, the model, and its ports: *t_Combo_CA()*, and *Combo_CA (Y, A, B, C, D)*, respectively. *Note*: The testbench will have no ports. Exhaustively simulate the circuit and provide graphical and text output demonstrating that the model is correct.

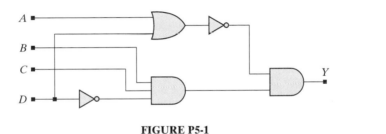

FIGURE P5-1

2. Develop a testbench *t_Combo_all*, a testbench in which *Combo_str*, *Combo_UDP* and *Combo_CA*, (see Problems 4.1 and 4.2) are all instantiated. In the testbench environment, name their outputs as *Y_str, Y_UDP*, and *Y_CA*, respectively. Simulate the models and produce graphical output showing their waveforms. Discuss the results.

3. Repeat Problem 2 after assigning unit delay to all primitives and continuous assignments. Discuss the results.

4. Write a structural model that has the same functionality as *AOI_5_CA1*, (see Example 5.2) and then write a testbench that confirms that the two models have the same behavior.

5. Write a testbench to verify a gate-level model of an SR latch.

6. Write a testbench and verify that *tr_latch* (Example 5.11) correctly models a transparent latch.

7. Develop a testbench to verify the functionality of the user-defined primitive *AOI_UDP* (See Example 4.11). The testbench is to simulate the primitive and compare its output to that of a continuous assignment statement that describes the same functionality. The result of the comparison is to be reported by an error signal.

8. Using a cyclic behavior (*always*), write and verify a model of a transparent latch having active high *enable*, and active low *set* and *reset*. The action of *reset* is to drive the output of the latch to 0.

9. Write and verify a behavioral model of J-K flip-flop with active-low asynchronous reset.
10. Write and verify a Verilog model that will assert its output if a 4-bit input word is not a valid binary coded decimal code.
11. Explain why the code fragment shown below will execute endlessly, and recommend an alternative description.

> **reg** [3: 0] K
> **for** (K=0; K<=15; K = K+1) **begin**
> ...
> **end**

12. Using continuous assignment statements, develop and verify a model for *compare_4_32_CA*, a circuit that compares four 32-bit unsigned binary words and asserts output(s) indicating which words have the largest value and which words have the smallest value.
13. Using a level-sensitive cyclic behavior and a suitable algorithm, develop and verify a model for *compare_4_32_ALGO*, a circuit that compares four 32-bit words and asserts output(s) indicating which words have the largest value and which words have the smallest value.
14. Verify the functionality of *Universal_Shift_Reg* in Example 5.45.
15. Write a Verilog description of the circuit shown in Figure P5-15 and verify that the circuit's output, *P_odd*, is asserted if successive samples of *D_in* have an odd number of 1s.

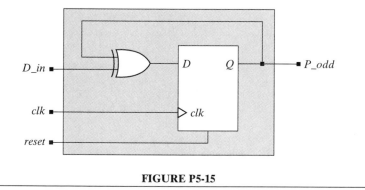

FIGURE P5-15

16. Develop and verify a Verilog model of a 4-bit binary synchronous counter with the following specifications: negative edge-triggered synchronization, synchronous load and reset, parallel load of data, active-low enabled counting.
17. Modify the counter of the previous problem to have an additional output (RCO [ripple-carry output]) that asserts while the counter is at 1111_2. Cascade two such counters and demonstrate that the unit now works as an 8-bit counter.
18. Develop and verify a Verilog model of a 4-bit Johnson counter.

19. Develop and verify a Verilog model of a 4-bit BCD counter.
20. Develop and verify a Verilog model of a modulo-6 counter.
21. Write a testbench and verify the functionality of *up_down_counter* in Example 5.41.
22. Write and verify a Verilog model of *up_down_counter_par_load*, by modifying the counter in Example 5.41 to have a parallel load capability.
23. Write a parameterized and portable Verilog model of an 8-bit ring counter whose movement is from its MSB to its LSB.
24. Write and verify a Verilog mode for a "jerky" ring counter having the register sequence shown in Figure P5-24(a); repeat for the counter in Figure P5-24(b).

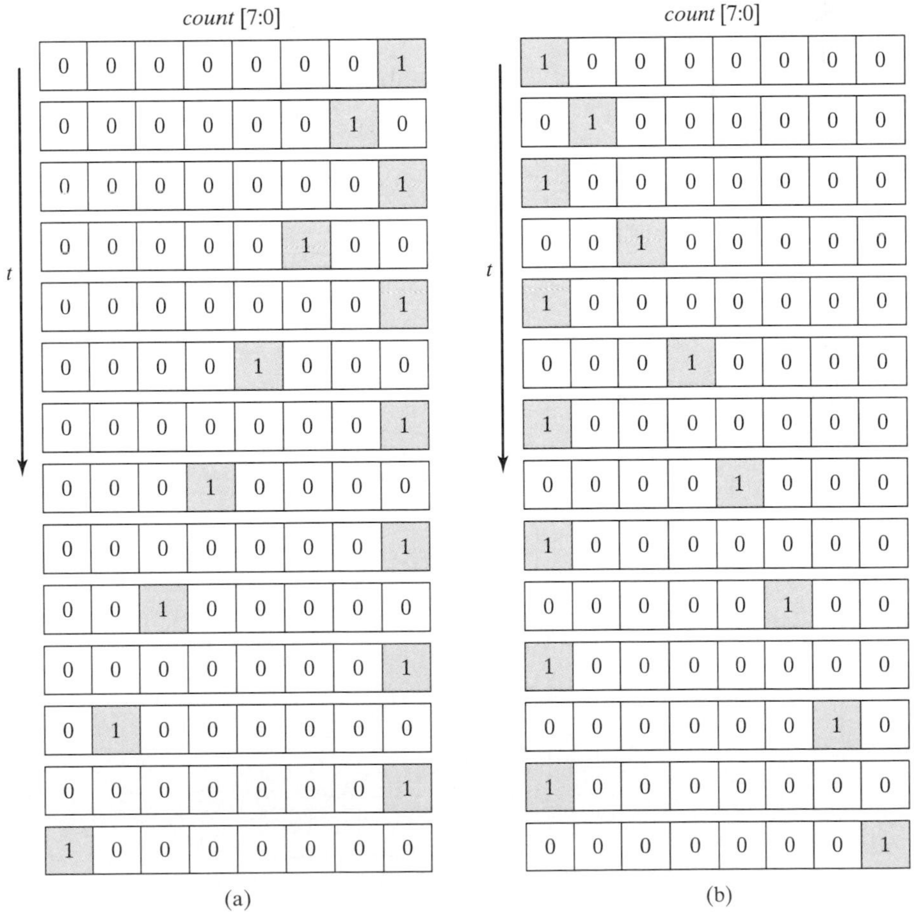

count [7:0]

(a)

0	0	0	0	0	0	0	1
0	0	0	0	0	0	1	0
0	0	0	0	0	0	0	1
0	0	0	0	0	1	0	0
0	0	0	0	0	0	0	1
0	0	0	0	1	0	0	0
0	0	0	0	0	0	0	1
0	0	0	1	0	0	0	0
0	0	0	0	0	0	0	1
0	0	1	0	0	0	0	0
0	0	0	0	0	0	0	1
0	1	0	0	0	0	0	0
0	0	0	0	0	0	0	1
1	0	0	0	0	0	0	0

count [7:0]

(b)

1	0	0	0	0	0	0	0
0	1	0	0	0	0	0	0
1	0	0	0	0	0	0	0
0	0	1	0	0	0	0	0
1	0	0	0	0	0	0	0
0	0	0	1	0	0	0	0
1	0	0	0	0	0	0	0
0	0	0	0	1	0	0	0
1	0	0	0	0	0	0	0
0	0	0	0	0	1	0	0
1	0	0	0	0	0	0	0
0	0	0	0	0	0	1	0
1	0	0	0	0	0	0	0
0	0	0	0	0	0	0	1

FIGURE P5-24

25. Write and verify a Verilog model for a counter with the sequence shown in Figure P5-25.

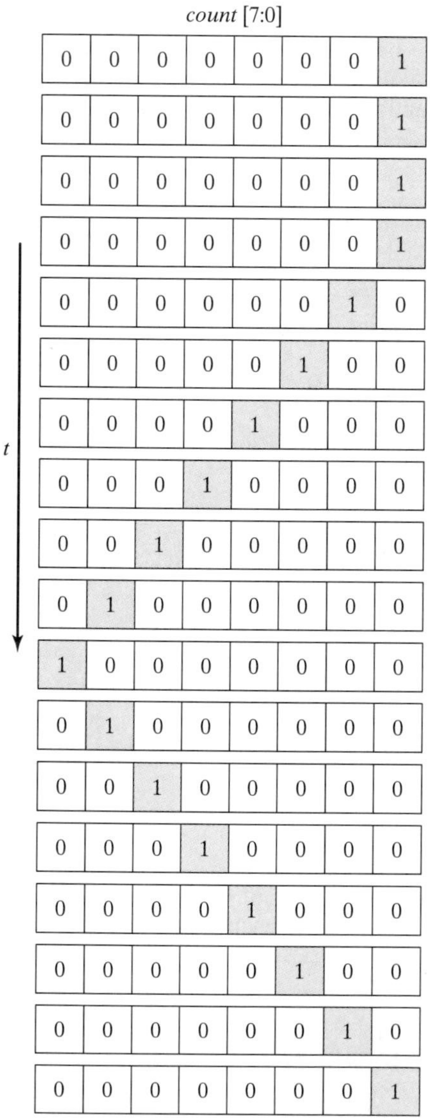

count [7:0]

0	0	0	0	0	0	0	1
0	0	0	0	0	0	0	1
0	0	0	0	0	0	0	1
0	0	0	0	0	0	0	1
0	0	0	0	0	0	1	0
0	0	0	0	0	1	0	0
0	0	0	0	1	0	0	0
0	0	0	1	0	0	0	0
0	0	1	0	0	0	0	0
0	1	0	0	0	0	0	0
1	0	0	0	0	0	0	0
0	1	0	0	0	0	0	0
0	0	1	0	0	0	0	0
0	0	0	1	0	0	0	0
0	0	0	0	1	0	0	0
0	0	0	0	0	1	0	0
0	0	0	0	0	0	1	0
0	0	0	0	0	0	0	1

t

FIGURE P5-25

26. For an 8-bit datapath, develop a Verilog model for the sequential machine represented by the ASMD chart in Figure P5-26. *Note*: The machine has synchronous reset, and the action of *rst* drives the state to *S_idle* from every state. The register operations and state transitions are to be synchronized to the rising edge of the clock.

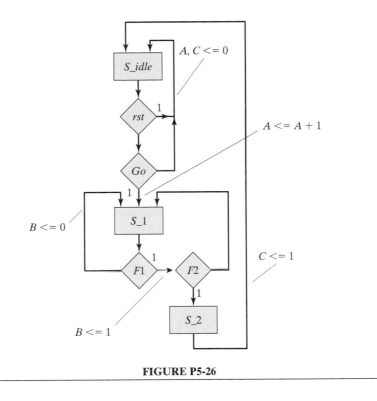

FIGURE P5-26

27. Develop and verify an 8-bit ALU having input datapaths *a* and *b*, output data path {*c_out, sum*}, and an operand *Oper*, and having the functionality indicated in Figure P5-27.

Operand	Function
Add	$a + b + c\_in$
Subtract	$a + \sim b + c\_in$
Subtract_a	$b + \sim a + \sim c\_in$
Or_ab	$\{1'b0, a \mid b\}$
And_ab	$\{1'b0, a \,\&\, b\}$
not_ab	$\{1'b0, (\sim a) \,\&\, b\}$
exor	$\{1'b0, a \wedge b\}$
exnor	$\{1'b0, a \sim^\wedge b\}$

FIGURE P5-27

28. Combine the ALU from Problem 27 with an 8-bit version of a register file to form an architecture like that shown in Figure 5-34. Develop a testbench to verify each functional unit and the overall structure.

29. Develop and verify a Verilog model of the two-stage pipeline machine described by the ASMD chart in Figure 5-24. Partition the design into a controller and a datapath unit. Modify the ASMD chart to show conditional output boxes whose asserted signals cause the register operations in the datapath.

30. Modify the machine described by the ASMD chart in Figure 5-24 by replacing the registers *P0* and *P1* by an appropriately sized shift register, whose contents are shifted under the actions of *En* and *Ld*. Develop and verify a Verilog model of the machine.

31. A designer contends that the reset signal can be removed from the hexadecimal keypad machine presented in Section 5.18 if logic is added to direct unused states to *S_5*. Discuss the validity of this claim. If it is true, what are the trade-offs between the two circuits?

32. Modify the ASM chart for the keypad scanner circuit (see Section 5.18) to require that a key be held for 25 clock cycles before it is interpreted to be a valid keystroke. Write and verify a Verilog model of the revised design.

33. Write a Verilog model of *Clock_Prog*, a programmable clock generator with an output port (*clk*) and the parameters *Latency, Offset*, and *Pulse_Width* as shown in Figure P5-33. The default (declared) values of the parameters are 100, 50, and 50, respectively. The testbench is to be named *t_Clock_Prog*. Include the following "annotation module" in your project. The role of this module is to override the default parameters in the clock generator with those that are to be used in a given application. It uses hierarchical dereferencing, where M1 is the instance name of the UUT in *t_Clock_Prog*. The testbench must demonstrate that this works. The example below replaces the default values of *Latency, Offset*, and *Pulsewidth* by 10, 5, and 5, respectively.

```
module annotate_Clock_Prog ();
  defparam t_Clock_Prog.M1.Latency = 10;
  defparam t_Clock_Prog.M1.Offset = 5;
  defparam t_Clock_Prog.M1.Pulse_Width = 5;
endmodule
```

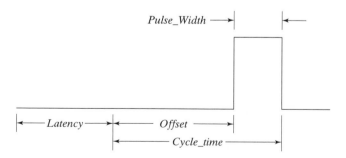

FIGURE P5-33 Parameterized waveform of a programmable clock generator.

34. Write and verify a Verilog module of a programmable 8-bit testbench pattern generator, *Pattern_Gen_8B*, having an output word, *Stim_Pattern*, and an input, *enable*, which activates the generator. Once the generator is activated, the bit patterns are to be generated exhaustively for a prescribed number of cycles, provided that *enable* is asserted. The module is to have the following parameters:

    Offset: An initial offset between the time that enable is asserted and the time that the first pattern of *Stim_Pattern* is asserted.
    Stim_size: The width of *Stim_Pattern*.
    Cycles: The number of times the entire pattern set is to be repeated.
    Period: The interval between successive patterns.

35. Hamming codes are used in computer memory systems to detect and correct errors in data [8]. Each information word is augmented by additional bits that distinguish valid encoded words from corrupted words. A Hamming code with a minimum distance of 4 bits can correct a single bit error in the encoded word and detect any 2-bit error. Figure P5-35 lists information bytes (*D_word*) and the parity bits that are appended to the bytes to form an encoded word (*H_word*). (a) Develop a Verilog module, *Hamming_Encoder_MD3*, that will produce a 7 bit value *H_word* from a 4-bit value, *D_word*. (b) Develop a testbench to verify the functionality of *Hamming_Encoder_MD3*. (c) Browse the literature on parity circuits to find a decoder that will correct a single-bit error

H_word	
D_word	Parity Bits
0000	000
0001	011
0010	101
0011	110
0100	110
0101	101
0110	011
0111	000
1000	111
1001	100
1010	010
1011	001
1100	001
1101	010
1110	100
1111	111

FIGURE 5-P35

in *H_word*. (d) Develop and verify a Verilog module, *Hamming_Decoder_MD3*, that will detect any 2-bit error in *H_word* and correct any single-bit error. (e) Develop a testbench that will simulate the encoder, introduce random 1 and 2-bit errors in *H_word*, and take action to correct a single bit error and signal the presence of a 2-bit (uncorrectable) error. (f) Demonstrate the decoder's behavior if a 3-bit (burst) error corrupts *H_word*.

36. Using library cells with known setup and hold timing parameters, develop and verify a model of the synchronizer circuit in Figure 5-38(b). Explore its operation with *Asynch_in* being (a) a long pulse, and (b) a short pulse relative to the clock. How does it behave if an asynchronous pulse arrives within two or more successive clock cycles?

37. Using continuous assignments, develop and verify a model of a transparent latch having active high enable, active low reset, and active high set. Use the text below.

> **module** Latch_RbarS_CA (q_out, data_in, enable, reset_bar, set);
> ...
> **endmodule**

38. Using *d_prim1* (See Example 4.15) develop and verify a model of a 4-bit shift register with parallel load.

39. The shifting operators in Verilog describe logical shifts operations. The left-shift operator (<<) pushes a 0 into the LSB of the word, and the right-shift operator (>>) pushes a 0 into the MSB of the word. In contrast, an arithmetic shift operator would push a 1 into the MSB of a register as it shifts its contents to the right. Arithmetic shifts are commonly executed in hardware units for division, to extend the sign bit of a number that is represented in a 2s complement format. Develop and verify a parameterized Verilog model of a sequential machine that will execute a arithmetic right shift and a logical left shift. Use the module header below.

> **module** Arithmetic_Right_Shift (data_out, data_in, clk, reset);
> **endmodule**

The unit is to be active on the positive edge of a signal *clk*, with an active-high reset. The datapath of the unit is to be parameterized by *word length*.

CHAPTER 6

Synthesis of Combinational and Sequential Logic

Chapter 1 identified the major steps of a design flow for ASICs, and Chapters 2 and 3 reviewed basic concepts and Karnaugh maps for designing combinational and sequential circuits by manual methods. Chapter 4 introduced the Verilog HDL for describing the functionality of a logic circuit. Chapter 5 developed behavioral models of several combinational and sequential logic circuits and showed the results of synthesizing them with modern design tools.

The design flow for an application-specific integrated circuit (ASIC) depends on software tools to manage and manipulate the databases that describe large, complex circuits. Among these tools, the synthesis engine plays a strategic role by automating the task of minimizing a set of Boolean functions and mapping the result into a hardware implementation that meets design objectives. Manual methods relying on Karnaugh maps cannot accommodate large circuits, and their use is error-prone, tedious, and time consuming. Automated software tools can optimize logic quickly, and without error. To use them effectively, a designer must understand a hardware description language (HDL) and be skilled at writing descriptions that conform to the constraints imposed by the tool. Just as understanding Karnaugh maps is the key to manual design methods, *understanding how to write synthesis-friendly Verilog models is the key to automated design methods.*

Synthesis tools perform many tasks, but the following steps are critical: (1) detect and eliminate redundant logic, (2) detect combinational feedback loops, (3) exploit don't-care conditions, (4) detect unused states, (5) detect and collapse equivalent states, (6) make state assignments, and (7) synthesize optimal, multilevel realizations of logic subject to constraints on area and/or speed in a physical technology. This last step involves both optimization and technology mapping. The steps that were performed manually in Chapters 2 and 3 will be executed automatically by a synthesis tool. This will shorten the design cycle, reduce the burden placed on the designer, and increase the likelihood that the design will be correct.

HDLs are the entry point for a modern synthesis-oriented design flow for ASICs and field-programmable gate arrays (FPGAs). A designer must understand how to use language constructs to describe combinational and sequential logic, and must know how to write synthesis-friendly descriptions. In this chapter we will present several examples that demonstrate how to write synthesis-ready models of combinational and sequential logic (that is, models that can be used with a synthesis tool to produce a gate-level realization of the described functionality). The examples will help the reader anticipate the results of synthesis—that is, know what circuit will be created from the description.

6.1   Introduction to Synthesis

Circuit design begins with specification of the circuit's functionality—what the circuit does—and ends with physical hardware that implements the functionality reliably, with sufficient performance and acceptable cost. Models of circuits can be classified according to levels of abstraction and views [1]. There are three common levels of abstraction: architectural, logical, and physical. A description at the architectural level of abstraction implies the operations that must be executed by the circuit to transform a sequence of inputs into a specified sequence of outputs, but does not associate the operations with specific clock cycles. These operations will ultimately be implemented by distinct, interconnected, and synchronized functional units, hence our use of the term *architecture*. The design challenge here is to extract from the architectural description a structure of computational resources that implements the functionality of the machine.

A model at the logical level of abstraction describes a set of variables and a set of Boolean functions that the circuit must implement. An architecture of register resources and functional units and the sequential activity of a logic-level model are part of its description. The design task here is to translate the Boolean descriptions into an optimized netlist of combinational gates and storage registers that will implement the circuit at a satisfactory level of performance. Synthesis tools are used for this task. A geometrical model describes the shapes that define the doping regions of semiconductor materials used to fabricate transistors. HDLs do not treat geometric models.

Design may begin at a high level of abstraction, but ultimately ends in physical reality. Along the path between the two, an HDL can facilitate the design process by providing different views of the circuit. There are three common views: behavioral, structural, and physical. A behavioral view of an architectural model could be an algorithm that specifies a sequence of data transformations. A structural view of the same model might consist of a structure of datapath elements (registers, memory, adders, and a controller) that implement the algorithm. State-transition graphs, next state/output tables, and algorithmic state machine (ASM) charts are behavioral views of a logic-level model of a circuit. A structural view of a logic-level model would consist of a schematic of gates that implement the functionality of the behavior described by the ASM chart in the behavioral view of models. Physical descriptions, the actual geometric patterns of the physical devices that implement the circuit, will not be considered here.

Synthesis creates a sequence of transformations between views of a circuit, from a higher level of abstraction to a lower one, with each step leading to a more detailed description of the physical reality. Figure 6-1 shows a modified Y-chart [2] depicting an axis for the behavioral, structural, and physical views of a circuit. We have annotated the behavioral and structural axis to show Verilog constructs that can be used to create a view. The chart also shows a sequence of transformations, in which: (1) behavioral synthesis transforms an algorithm (behavioral view) to an architecture of registers and a schedule of operations that occur in specified clock cycles (structural view), (2) a Verilog model of this architecture is formed as a data flow/register transfer level (RTL) description (behavioral view), and (3) logic synthesis translates the data flow/RTL description into a Boolean representation and synthesizes it into a netlist (structural view). The Y-chart in Figure 6-1 has been annotated to represent these transformations, and to indicate which Verilog constructs (e.g., continuous assignments) describe the design at each stage of the activity.

6.1.1 Logic Synthesis

Logic synthesis generates a structural view from a logic level view (description) of a circuit [1]. The resulting structural view is a netlist of structural primitives. The logic-level view is a set of Boolean equations described by a set of continuous-assignment statements in Verilog or an equivalent level-sensitive behavior. Logic synthesis includes

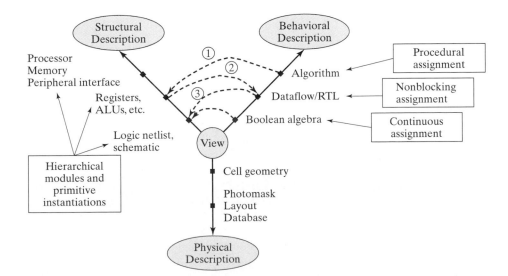

FIGURE 6-1 A Y-chart representation of Verilog constructs supporting synthesis activity in three views of a circuit: behavioral, structural, and physical.

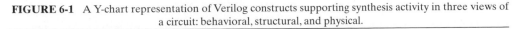

transforming a given netlist of primitives into an optimized netlist of Verilog primitives and mapping a generic optimal netlist into an equivalent circuit composed of physical resources in a target technology, such as the cells in an ASIC cell library. A logic synthesis tool has the general organization shown in Figure 6-2. The tool forms a hardware realization (technology implementation) from a Verilog RTL model or primitive netlist.

The translation engine of a synthesis tool reads and translates a Verilog-based description of the input–output behavior of a circuit into an intermediate internal representation of Boolean equations describing combinational logic and other representations of storage elements and synchronizing signals. Techniques for simultaneously optimizing the set of equations will remove redundant logic, exploit don't-care conditions, and share internal logic subexpressions as much as possible to produce an optimized, generic (technology-independent) multilevel logic implementation.

In general, there may be multiple realizations of a multiinput, multioutput combinational logic circuit, but the transformations made in synthesis are guaranteed to maintain the input–output equivalence of the circuit and produce a testable circuit [3].[1] The optimization process is based on an iterative search, not the solution of an analytical model, so the result is not necessarily the global optimum that could be found over the domain of possible circuits that have equivalent input–output behavior. Logic optimization is followed by performance optimization, which seeks a circuit that has optimal performance in the physical technology.

The translation engine creates an internal product-of-sums (POS) of a Boolean expression by factoring the sum-of-products (SOP) form into expressions whose Boolean product generates the SOP form. When the POS representation for two or more outputs contain a common subexpression, the synthesis tool may minimize the internal logic needed to realize the circuit by generating the common subexpression once and sharing it (through fanout) among output variables.

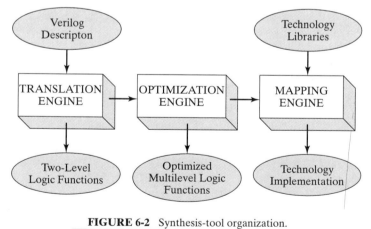

FIGURE 6-2 Synthesis-tool organization.

---

[1]We will consider fault simulation and test generation in Chapter 11.

Techniques for simultaneously optimizing the set of equations will remove redundant logic, exploit don't-care conditions, and share internal logic subexpressions [3] as much as possible to produce an optimized, generic (technology-independent) multilevel logic implementation. For example, a "?" entry in the input–output table of a combinational user-defined primitive (UDP) represents a don't-care condition that can be used in Boolean minimization. However, this may lead to a mismatch between the seed code's behavior and that of the synthesis product.

The functional relationships between the inputs and outputs of a combinational Boolean circuit can be expressed as a set of two-level Boolean equations in either SOP or POS form, which must be optimized by the tool. A set of Boolean equations describing a multiinput, multioutput (MIMO) combinational logic circuit can always be optimized to obtain a set of Boolean equations whose input–output behavior is equivalent while containing the fewest literals [4]. Software tools exist to perform this optimization and to cover the resulting Boolean equations by the resources of a technology library. Consequently, a Verilog description consisting only of a netlist of combinational primitives without feedback can always be synthesized. Some vendors, however, choose not to implement synthesis of a UDP.

Espresso [5] is a commonly used software system developed at the University of California at Berkeley for minimizing the number of cubes in a single Boolean function. It performs several transformations on a circuit to arrive at its optimal representation. For example, the transformation *Expand* replaces cubes with prime implicants that have fewer literals. *Irredundant* extracts from a cover of the function a minimal subset that also covers the function (i.e., redundant logic has been removed). *Reduce* transforms an irredundant cover into a new cover of the same function.

Espresso minimizes a single Boolean function of several Boolean variables, but it does not provide a solution to the problem of optimizing a MIMO combinational logic circuit. In general, the optimization of a set of Boolean equations is not obtained by applying Espresso separately to the individual functions in the set. Instead, a multilevel optimization program, such as misII [3], must be used to simultaneously optimize the set of equations as an aggregate.

Logic synthesis treats a set of individual Boolean input–output equations as a multilevel circuit (see Figure 6-3). By removing redundant logic, sharing internal logic, and exploiting input and output don't-care conditions, a logic synthesis tool optimizes a multilevel set of Boolean equations and achieves a better realization (i.e., area-efficient) than could be obtained by merely optimizing the individual input–output equations.

Like Espresso, the misII multilevel logic optimization program performs several transformations on a logic circuit while searching for an optimal description of a digital circuit. Four transformations play a key role in the misII algorithm for logic synthesis: decomposition, factoring, substitution, and elimination.

The operation of *decomposition* transforms the circuit by expressing a single Boolean function (i.e., the Boolean expression representing the logic value of a node in the circuit) in terms of new nodes.

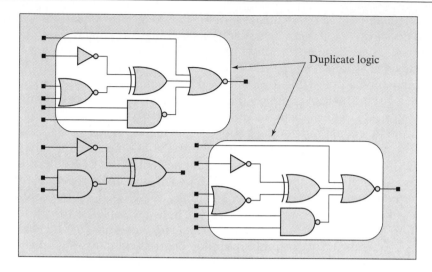

FIGURE 6-3 Multilevel combinational logic.

Example 6.1

Figure 6-4 shows the schematic of a function, *F*, that is to be decomposed in terms of new nodes *X* and *Y*. The original form of *F* is described by the Boolean equation:

$$F = abc + abd + a'b'c' + b'c'd'$$

The Espresso operation of *decomposition* expresses *F* in terms of two additional internal nodes, *X* and *Y*, to form the circuit shown in Figure 6-5. These internal nodes could then be re-used to form other expressions and thereby achieve a reduction in hardware area.

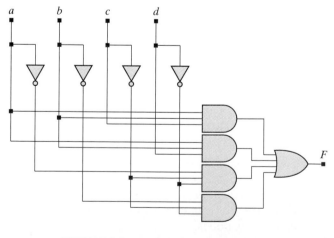

FIGURE 6-4 Circuit before decomposition.

$$F = XY + X'Y'$$
$$X = ab$$
$$Y = c + d$$

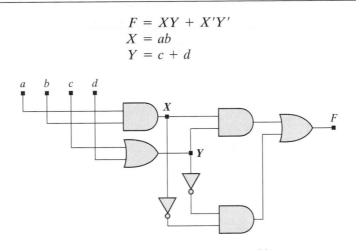

FIGURE 6-5 Circuit after decomposition.

End of Example 6.1

In contrast to *decomposition*, which represents an *individual* function in terms of intermediate nodes,[2] the operation of *extraction* expresses a *set* of functions in terms of intermediate nodes by expressing each function in terms of its factors and then detecting which factors are shared among functions.

Example 6.2

Figure 6-6 shows a directed acyclic graph (DAG) [1] representing the set of functions, *F, G*, and *H*, which is to be decomposed in terms of new nodes *X* and *Y*, with

$$F = (a + b)cd + e$$
$$G = (a + b)e'$$
$$H = cde$$

and *X* and *Y* are given by:

$$X = a + b$$
$$Y = cd$$

The nodes of the graph represent Boolean operations on the data that are associated with the edges that enter the node. The *extraction* process finds those members of the set of functions with the factor $(a + b)$ and the factor cd. The factors are extracted from those functions and replaced by the new internal nodes *X* and *Y* to produce the new DAG shown in Figure 6-7.

End of Example 6.2

[2]The nodes of the directed acyclic graph represent operations that are performed on data (e.g., addition).

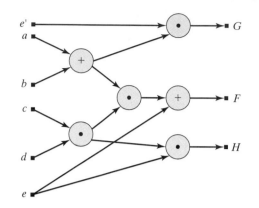

FIGURE 6-6   DAG of a set of functions before extraction.

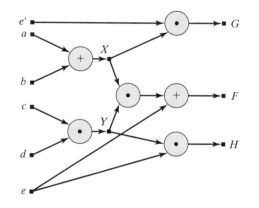

FIGURE 6-7   Directed acyclic graph of a set of functions after extraction.

The optimization process seeks a set of intermediate nodes to optimize the circuit's delay and area. This step may lead to significant reduction in the overall silicon area, because the intermediate nodes correspond to factors that are common to more than one Boolean function, and therefore can be shared to eliminate replicated logic. The task of finding the common factors among a set of functions is called *factoring*. Factoring produces a set of functions in a products of sum form. It creates a structural transformation of the circuit from a two-level realization to an equivalent multilevel realization that uses less area, but is possibly slower.

Example 6.3

Figure 6-8 shows a DAG representing a function, F, that is factored to identify its Boolean factors in product of sums form. The function represented by the DAG is described by the Boolean equation:

$$F = ac + ad + bc + bd + e$$

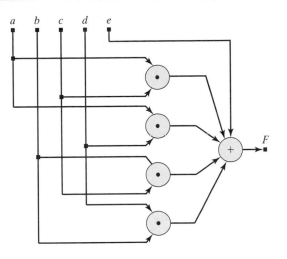

FIGURE 6-8 The DAG of a function before factoring.

The factored form of F is

$$F = (a + b)(c + d) + e$$

Factorization seeks the factored representation of a function with the fewest number of literals. The DAG of the factored form of F is shown in Figure 6-9.

End of Example 6.3

The *substitution* process expresses a Boolean function in terms of its inputs and another function. Since both functions need to be implemented, this step provides a potential reduction of replicated logic.

Example 6.4

The DAG in Figure 6-10(a) represents the function F before the function G is substituted into it, where

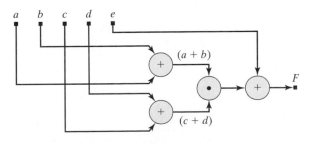

FIGURE 6-9 The DAG of the factored form of a function.

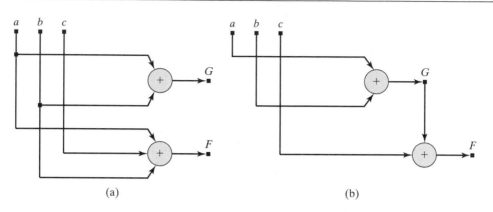

FIGURE 6-10   DAG of a function before (a) and after (b) substitution.

$$G = a + b$$
$$F = a + b + c$$

After substitution, F has the form:

$$F = G + c$$

and the DAG shown in Figure 6-10(b).

End of Example 6.4

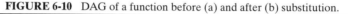

Sometimes the process of *decomposition* has to be undone in the search for an optimal realization. *Elimination* removes (collapses) a node in a function and reduces the structure of the circuit. This step is also referred to as *flattening* the circuit. The transformation would ultimately eliminate the area-efficient internal multilevel structure and create a faster two-level structure.

Example 6.5

The DAG in Figure 6-11(a) represents the function F before the function G is eliminated from it, where before elimination

$$F = Ga + G'b$$
$$G = c + d$$

and after elimination:

$$F = ac + ad + bc'd'$$

The new DAG for F is shown in Figure 6-11(b).

End of Example 6.5

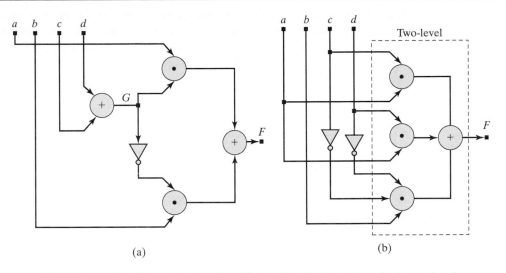

FIGURE 6-11 DAG of a function before (a) and after (b) elimination of an internal node.

In general, there may be multiple realizations of a multiinput, multioutput combinational logic circuit, but the transformations in misII are guaranteed to maintain the input–output equivalence of the circuit and produce a testable circuit. Multilevel networks may be used because two-level networks might require higher fan-in than is practical. Multilevel circuits present an opportunity to share logic, but can be slower than a two-level counterpart.

6.1.2 RTL Synthesis

RTL synthesis begins with an architecture and converts language-based RTL statements into a set of Boolean equations that can be optimized by a logic synthesis tool. This level of synthesis transforms a logic-level view described in terms of operations on registers in the context of a fixed architecture, and creates equivalent Boolean equations and synthesizes an optimal realization of the given architecture.

RTL synthesis begins with the assumption that a set of hardware resources is available and that the scheduling and allocation of resources have been determined,[3] subject to the constraints imposed by the resources of the architecture. The RTL description represents either a finite-state machine or a more general sequential machine that makes register transfers within the boundaries of a predefined clock cycle. RTL descriptions in Verilog use language operators and make synchronous concurrent assignments to register variables (i.e., nonblocking assignments). The Verilog language operators represent a variety of register transfer operations, and are easily synthesized. Logic synthesis tools operate in this domain (i.e., they generally lack the ability

---

[3]Resource allocation and scheduling are discussed in Chapter 9.

to analyze tradeoffs between scheduling and allocation of resources, and use, instead, the implicit solution imposed by the writer of the description). Nonetheless, these tools have a broad and significant scope of use. The synthesis engine must minimize and optimally encode the state of the RTL-described machine, optimize the associated combinational logic, and map the result into the target technology.

6.1.3   High-Level Synthesis

High-level synthesis, also called "behavioral synthesis" or "architectural synthesis," has the goal of finding an architecture whose resources can be scheduled and allocated to implement an algorithm, such as an algorithm for digital signal processing (DSP).[4] The algorithm to be synthesized describes only the functionality of the circuit; it does not explicitly declare a structure of registers and datapaths. Thus, many different architectures may implement the same functional specification.

The starting point for high-level synthesis is an input–output algorithm, with no details about the implementation. A behavioral synthesis tool executes two main steps in order to create an architecture of datapath elements, control units, and memory: resource *allocation* and resource *scheduling*. Dataflow graphs display dependencies between data, and the allocation step identifies the operators used in an algorithm (e.g., +) and infers the need for memory resources to hold data implied by the sequential activity of the algorithm. Allocation binds these operators and memory resources to datapath resources (e.g., multiplication operators can be bound to a multiplier cell).

In scheduling, the operations in the behavioral description are assigned to specific clock cycles (implicit states) to implement the ordered sequential activity flow of the algorithm. For example, Figure 6-12(a) [2] shows three procedural assignments that execute sequentially in a Verilog cyclic behavior within a hypothetical algorithm. A compiler must form parse trees from the statements, and then extract a dataflow graph from the set of parse trees. Figure 6-12b shows a DFG and a sequential schedule for its operators. The statements imply a time-ordered sequential activity, and are allocated to the clock cycles.

Scheduling assigns the operations of the behavioral description to clock cycles. The dataflow graph shown in 6-12(b) can be used to infer memory read/writes and schedule activity in clock cycles using the available resources. This ultimately determines the number of computational units (operators) that will be used in a given clock cycle and shared between cycles. Because multiple architectures might have the same functionality, a behavioral synthesis tool must explore architectural alternatives and consider a number of tradeoffs and/or constraints (e.g., data rates, input/output channels, clock period, pipelining, datapath width, latency, throughput, speed, area, and power). High-level synthesis is an area of active research, although electronic design automation (EDA) tool vendors now offer tools for behavioral synthesis.[5] Coding styles and other details of these tools will not be presented here. Also, note that system-level

---

[4]We will consider algorithm synthesis in more detail in Chapter 9.
[5]For example, Synopsys's Behavioral Compiler.

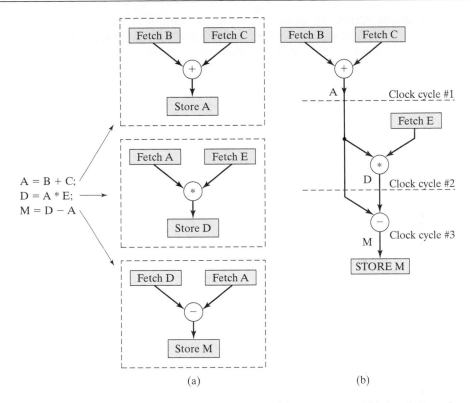

FIGURE 6-12   Representation of a behavioral model: (a) parse trees, and (b) data flow graph.

design languages are emerging, but the industry is still in a state of flux, with contending groups advocating System C, Spec C, and Superlog (see www.systemc.org and www.co-design.com).

6.2   Synthesis of Combinational Logic

There are many ways to describe combinational logic with the Verilog HDL, but some are not supported by synthesis tools.[6] Synthesizable combinational logic can be described by (1) a netlist of structural primitives, (2) a set of continuous-assignment statements, and (3) a level-sensitive cyclic behavior. User-defined primitives and ***assign*** ··· ***deassign*** procedural continuous assignments can also describe combinational logic, but most EDA vendors have chosen to not support these options.

---

[6]See "Additional Features of Verilog" at the companion web site for Verilog constructs that are commonly supported by synthesis tools, and constructs that are not supported.

A design that is expressed as a netlist of primitives should be synthesized to re-move any redundant logic before mapping the design into a technology. This provides a measure of safety to the design, because most designers have difficulty discovering and removing redundant logic from any but the simplest circuits. Synthesizing a netlist ensures that the logic has been minimized correctly.

Example 6.6

Figure 6-13(a) shows the preoptimized schematic of the circuit described by the netlist of primitives in *boole_opt*. The synthesized circuit shown in Figure 6-13(b) has a more efficient gate-level implementation (less area) than the generic circuit would have if parts from a cell library are substituted for the primitives.

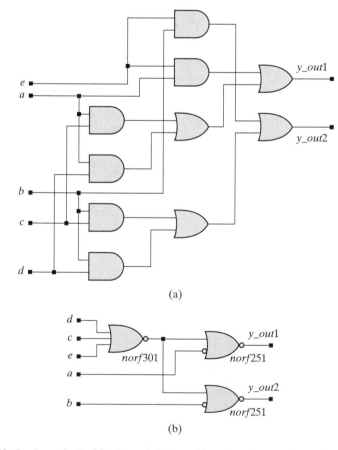

FIGURE 6-13 Logic synthesis: (a) schematic for a netlist of primitives, and (b) the circuit synthesized from the netlist.

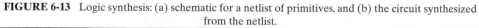

```
module boole_opt(y_out1, y_out2, a, b, c, d, e);
  output       y_out1, y_out2;
  input        a, b, c, d, e;

  and          (y1, a, c);
  and          (y2, a, d);
  and          (y3, a, e);
  or           (y4, y1, y2);
  or           (y_out1, y3, y4);
  and          (y5, b, c);
  and          (y6, b, d);
  and          (y7, b, e);
  or           (y8, y5, y6);
  or           (y_out2, y7, y8);
endmodule
```

End of Example 6.6

Continuous-assignment statements are synthesizable. The expression that assigns value to a net variable in a continuous assignment statement will be translated by the synthesis tool to an equivalent Boolean equation, which can be optimized and synthesized in physical hardware simultaneously with other continuous assignments in the module.[7] The designer must verify that the continuous assignment correctly describes the logic, and that the synthesized multilevel logic circuit has the correct functionality.

Example 6.7

The continuous assignment in *or_nand* synthesizes to the circuit in Figure 6-14.

```
module or_nand (y, enable, x1, x2, x3, x4);
  output    y;
  input     enable, x1, x2, x3, x4;

  assign y = ~(enable & (x1 | x2) & (x3 | x4));
endmodule
```

End of Example 6.7

A level-sensitive cyclic behavior will synthesize to combinational logic if it assigns a value to each output for every possible value of its inputs. This implies that the event control expression of the behavior must be sensitive to every input, and that every path of the activity flow must assign value to every output.

---

[7]Remember, simultaneous optimization of Boolean equations discovers and exploits logic that can be shared among them.

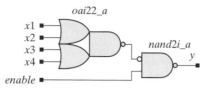

FIGURE 6-14 Combinational logic synthesis: circuit synthesized from a continuous assignment.

Example 6.8

Chapters 4 and 5 presented examples that described a 2-bit comparator by Boolean equations and a netlist of primitives (see Example 4.4), and a set of continuous assignments (see Examples 5.6, 5.12, and 5.13). The descriptions synthesize equivalent circuits. The model below, *comparator*, describes the functionality of a 2-bit comparator by a level-sensitive behavior. Its algorithm exploits the fact that the data words are identical if all of their bits match in each position. Otherwise, the most significant bit at which the words differ determines their relative magnitude. This algorithm is not as simple or as elegant as the ones in Chapters 4 and 5, but it serves to illustrate the power of synthesis tools to correctly synthesize combinational logic from a level-sensitive cyclic behavior containing a loop construct.[8] The synthesized circuit is shown in Figure 6-15.

```verilog
module comparator (a_gt_b, a_lt_b, a_eq_b, a, b);  // Alternative algorithm
  parameter      size = 2;
  output                           a_gt_b, a_lt_b, a_eq_b;
  input          [size: 1]         a, b;
  reg                              a_gt_b, a_lt_b, a_eq_b;
  integer                          k;

  always @ ( a or b) begin: compare_loop
    for (k = size; k > 0; k = k-1) begin
      if (a[k] ! = b[k]) begin
        a_gt_b = a[k];
        a_lt_b = ~a[k];
        a_eq_b = 0;
        disable compare_loop;
      end                 // if
    end                   // for loop
    a_gt_b = 0;
    a_lt_b = 0;
    a_eq_b = 1;
  end                     // compare_loop
endmodule
```

---

[8]A cyclic behavior executes repeatedly, subject to embedded timing controls.

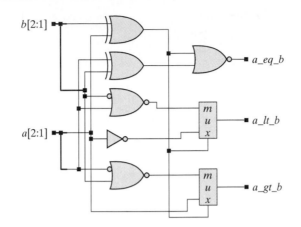

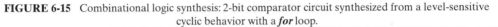

FIGURE 6-15 Combinational logic synthesis: 2-bit comparator circuit synthesized from a level-sensitive cyclic behavior with a *for* loop.

End of Example 6.8

A synthesizable Verilog model of combinational logic describes the functionality of the circuit and is written independently of the technology in which the physical circuit will be realized. Timing imposes a constraint on the speed of a circuit, but not on its functionality. The speed at which ASIC parts can operate is proportional to the physical geometry of the part. Faster parts require more area. Technology-dependent timing constructs, such as gate propagation delays, are not to be included in the model of functionality. Synthesis tools may require that the logic be free of feedback loops (e.g., no cross-coupled NAND gates). Functions and tasks will synthesize to combinational logic if they do not contain incomplete case statements or conditionals (*if*), and do not contain embedded timing controls (**#, @**, or *wait*).

A synthesis tool synthesizes combinational logic to implement the expression of the *case* construct, and the expressions associated with a conditional operator. If a multiplexed datapath has control logic other than a single select bus, the synthesis tool will create additional combinational logic on the control line of the mux to govern the activity of the multiplexer.

Example 6.9

The continuous assignment in *mux_logic* has logic determining whether *sig_a* or *sig_b* is selected. The description synthesizes to the circuit shown in Figure 6-16, which has logic at the control line of the synthesized mux.

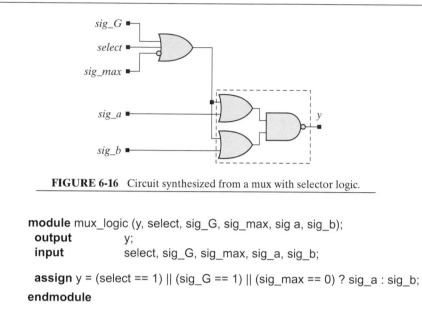

FIGURE 6-16   Circuit synthesized from a mux with selector logic.

```
module mux_logic (y, select, sig_G, sig_max, sig a, sig_b);
    output        y;
    input         select, sig_G, sig_max, sig_a, sig_b;

    assign y = (select == 1) || (sig_G == 1) || (sig_max == 0) ? sig_a : sig_b;
endmodule
```

End of Example 6.9

6.2.1   Synthesis of Priority Structures

A *case* statement implicitly attaches higher priority to the first item that it decodes than to the last one, and an *if* statement implies higher priority to the first branch than to the remaining branches. A synthesis tool will determine whether the case items of a *case* statement are mutually exclusive. If they are mutually exclusive, the synthesis tool will treat them as though they had equal priority and will synthesize a mux rather than a priority structure. Even when the list of case items is not mutually exclusive, a synthesis tool might allow the user to direct that they be treated without priority (e.g., Synopsys *parallel_case* directive). This would be useful if only one case item could be selected at a time in actual operation. An *if* statement will synthesize to a mux structure when the branching is specified by mutually exclusive conditions, as in Example 6.9, but when the branching is not mutually exclusive the synthesis tool will create a priority structure.

Example 6.10

The conditional activity flow within *mux_4pri* is not governed by mutually exclusive conditions. This results in synthesis of an implied priority for datapath *a* because *sel_a* decodes *a* independently of *sel_b* or *sel_c*, shown in Figure 6-17. The event control expression of the cyclic behavior monitors all of the signals that are referenced within the behavior.

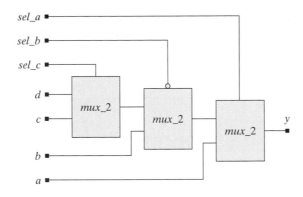

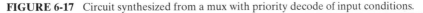

FIGURE 6-17 Circuit synthesized from a mux with priority decode of input conditions.

```
module mux_4pri (y, a, b, c, d, sel_a, sel_b, sel_c);
  output   y;
  input    a, b, c, d, sel_a, sel_b, sel_c;
  reg      y;

    always @ (sel_a or sel_b or sel_c or a or b or c or d)
    begin
      if   (sel_a == 1)        y = a; else
      if   (sel_b == 0)        y = b; else
      if   (sel_c == 1)        y = c; else
                               y = d;

  end
endmodule
```

End of Example 6.10

6.2.2 Exploiting Logical Don't-Care Conditions

When *case*, conditional branch (*if*), or conditional assignment (*? ... :*) statements are used in a Verilog behavioral description of combinational logic, the behavioral model and the synthesized netlist should produce the same simulation results (except for time-dependent behavior) if the seed code has *default* assignments that are purely 0 or 1 values (i.e., the *default* does not explicitly assign an **x** or a **z** value). Simulation results may differ if the default or branch statement makes an explicit assignment of an **x** or a **z**. A synthesis tool will treat *casex* and *casez* statements as *case* statements. Those *case* items that decode to explicit assignment of **x** or **z** will be treated as don't-care conditions for the purpose of logic minimization of the equivalent Boolean expressions, (i.e., it does not matter what value is assigned to the object of the *case* assignment under those input conditions).[9] The physical hardware will propagate either a 0 or a 1, while the HDL model will propagate an **x** in simulation. This may lead to a mismatch between the results obtained by simulating the seed code and the synthesis product.

Synthesis Tip

An assignment to **x** in a *case* or an *if* statement will be treated as a don't-care condition in synthesis.

Example 6.11

The counter and seven-segment display shown in Figure 6-18(a) are modeled below, in Verilog, by *Latched_Seven_Seg_Display*, with active low outputs.[10] The counter increments while *Enable* is asserted. The display is blank while *Blanking* is asserted. If *Blanking* is not asserted the display shows only even-valued numbers, and remains latched if BCD is the code of an odd number or an invalid number. The waveforms for *Display_L* and *Display_R* are shown in Figure 6-18(b), with values corresponding to the codes for active-low seven-segment displays. The latching action is implemented by not having a *default* item in the *case* statement. When the level-sensitive behavior is activated by a change in *count*, the values of *Display_L* and *Display_R* change only when *count* decodes to an even number; otherwise *Display_L* and *Display_R* remain unchanged (latched).

```
module Latched_Seven_Seg_Display
(Display_L, Display_R, Blanking, Enable, clock, reset);
output        [6: 0]    Display_L, Display_R;
input                   Blanking, Enable, clock, reset;
reg           [6: 0]    Display_L, Display_R;
reg           [3: 0]    count;
//                      abc_defg
parameter     BLANK  = 7'b111_1111;
parameter     ZERO   = 7'b000_0001;        // h01
parameter     ONE    = 7'b100_1111;        // h4f
```

---

[9]See Examples 5.22 (8-bit encoder), 5.23 (8:3 priority encoder), and 5.24 (3:8 decoder) in Chapter 5 for circuits synthesized from level-sensitive cyclic behaviors using default assignments in *case* and *if* statements.
[10]The underscore character is used in the parameters of *Latched_Seven_Seg_Display* to make the representation of a number more readable.

```
     parameter     TWO    = 7'b001_0010;        // h12
     parameter     THREE  = 7'b000_0110;        // h06
     parameter     FOUR   = 7'b100_1100;        // h4c
     parameter     FIVE   = 7'b010_0100;        // h24
     parameter     SIX    = 7'b010_0000;        // h20
     parameter     SEVEN  = 7'b000_1111;        // h0f
     parameter     EIGHT  = 7'b000_0000;        // h00
     parameter     NINE   = 7'b000_0100;        // h04

  always @ (posedge clock)
    if (reset) count < = 0;
    else if (Enable) count < = count +1;
  always @ (count or Blanking)
    if (Blanking) begin Display_L = BLANK; Display_R = BLANK; end  else
    case (count)
      0:            begin Display_L = ZERO; Display_R = ZERO; end
      2:            begin Display_L = ZERO; Display_R = TWO; end
      4:            begin Display_L = ZERO; Display_R = FOUR; end
      6:            begin Display_L = ZERO; Display_R = SIX; end
      8:            begin Display_L = ZERO; Display_R = EIGHT; end
      10:           begin Display_L = ONE; Display_R = ZERO; end
      12:           begin Display_L = ONE; Display_R = TWO; end
      14:           begin Display_L = ONE; Display_R = FOUR; end
      //default:    begin Display_L = BLANK; Display_R = BLANK; end
    endcase
  endmodule
```

End of Example 6.11

The absence of ***default*** assignments in the level-sensitive cyclic behavior in *Latched_Seven_Seg_Display* latches the output and will cause latches to be implemented by a synthesis tool. When default assignments are unconstrained, the synthesis tool can exploit them as don't-care conditions and reduce the logic needed to implement the circuit. The next example demonstrates how to exploit the don't-cares and how to form three-state outputs.

Synthesis Tip

If a conditional operator assigns the value **z** to the right-hand side expression of a continuous assignment in a level-sensitive behavior, the statement will synthesize to a three-state device driven by combinational logic.

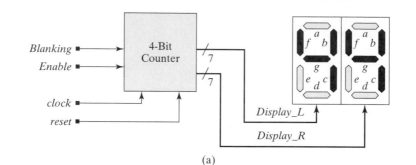

(a)

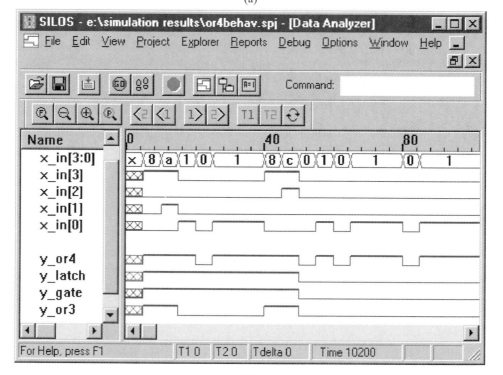

(b)

FIGURE 6-18 Seven-segment LED display: (a) counter and display units, and (b) simulation results showing the action of *reset, Blanking*, and *Enable*. The values of *Display_L* and *Display_R* correspond to the 7-bit codes of the active-low seven-segment display units for the values of *count*.

Example 6.12

The level-sensitive cyclic behavior in *alu_with_z1* describes the combinational logic of a simple ALU, and a continuous assignment describes a three-state output. The **default** assignments of the **case** statement constrain the output of the ALU to be 0. A second version, *alu_with_z2*, which is not shown, has the same description, except the **default** assignments of the **case** statement are don't-cares (4'bx) instead of 0. The synthesized circuits are shown in Figure 6-19. Both have three-state output inverters (which take

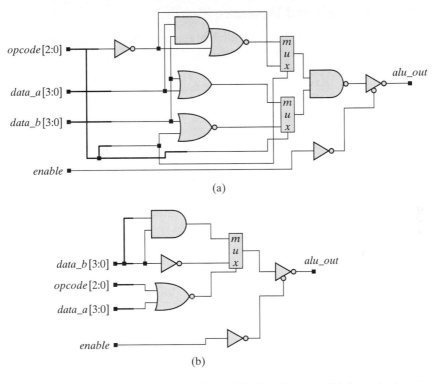

(a)

(b)

FIGURE 6-19 Circuits synthesized from (a) *alu_with_z1* and (b) *alu_with_z2*.

less area than buffers), but the circuit for *alu_with_z2* has a simpler realization, because the synthesis tool is able to exploit the don't-cares implied by the ***default*** assignment of 4'*bx* in the ***case*** statement.

```
module alu_with_z1 (alu_out, data_a, data_b, enable, opcode);
   input         [2: 0]    opcode;
   input         [3: 0]    data_a, data_b;
   input                   enable;
   output                  alu_out;              // scalar for illustration
   reg           [3: 0]    alu_reg;

   assign alu_out = (enable == 1) ? alu_reg : 4'bz;

   always @ (opcode or data_a or data_b)
     case (opcode)
       3'b001:          alu_reg = data_a | data_b;
       3'b010:          alu_reg = data_a ^ data_b;
       3'b110:          alu_reg = ~data_b;
       default:         alu_reg = 4'b0;    // alu_with_z2 has default: alu_reg = 4'bx;
     endcase
endmodule
```

End of Example 6.12

6.2.3   ASIC Cells and Resource Sharing

An ASIC cell library usually contains cells that are more complex than combinational primitive gates.[11] For example, most libraries will contain a model for a full-adder cell. Whether the synthesis tool exploits the available model or builds another circuit depends on the designer's Verilog description. The tool must share resources as much as possible to minimize needless duplication of circuitry.

Synthesis Tip

Use parentheses to control operator grouping and reduce the size of a circuit.

Example 6.13

A synthesis tool mapped the addition operator in the Verilog description below to a full-adder ASIC library cell to create the circuit shown in Figure 6-20(a). An alternative implementation shown in Figure 6.20(b) builds two different 5-bit adder blocks out of basic library cells in a structure that depends on the speed goal for the design (details not shown). The *esdpupd* device provides 0 (or 1) where needed. The leftmost adder forms *A[3: 0] + B[3: 0]*. The 5-bit result is one of the inputs to the right-most adder block. That block has a second 5-bit input formed by the *C_in bit* and four 0s. The carry-in of each block is hard-wired to ground.

```
module badd_4 (Sum, C_out, A, B, C_in);
    output    [3: 0]   Sum;
    output             C_out;
    input     [3: 0]   A, B;
    input              C_in;

    assign {C_out, Sum} = A + B + C_in;
endmodule
```

End ofExample 6.13

A synthesis tool must recognize whether the physical resources required to implement complex (large area) behaviors can be shared. If the data flows within the behavior do not conflict, the resource can be shared between one or more paths. For example, the addition operators in the continuous assignment below are in mutually exclusive datapaths and can be shared in hardware.

---

[11]The fabrication masks of the complex cells are fixed and create a more efficient and faster implementation of the functionality than an aggregate of simpler cells with equivalent functionality.

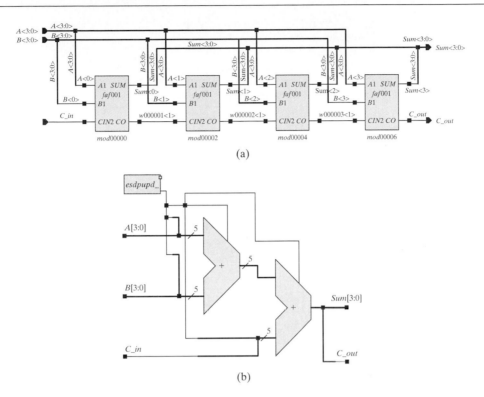

(a)

(b)

FIGURE 6-20 Result of synthesizing the + operator (a) using library full-adder cells, and (b) using 5-bit
adder blocks with hard-wired inputs to accommodate 4-bit datapaths.

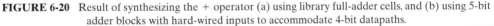

assign y_out = sel ? data_a + accum : data_a + data_b;

Consequently, the operators can be implemented by a shared adder whose input datapaths are multiplexed. This feature is vendor-dependent. If the tool does not automatically implement resource sharing, the description must be written to force the sharing.

Example 6.14

The use of parentheses in the description in *res_share* forces the synthesis tool to multiplex the datapaths and produce the circuit shown in Figure 6-21.

```
module res_share (y_out, sel, data_a, data_b, accum);
   output        [4: 0]   y_out;
   input         [3: 0]   data_a, data_b, accum;
   input                  sel;

   assign y_out = data_a + (sel ? accum : data_b);
endmodule
```

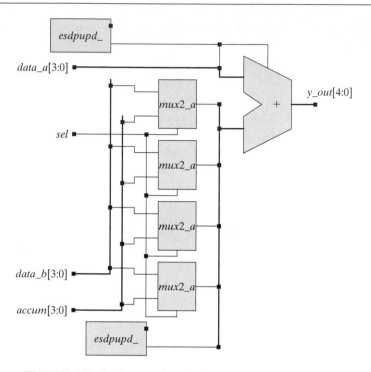

FIGURE 6-21   Implementation of a datapath with shared resources.

Failure to include the parentheses in the expression for *y_out* in *res_share* will lead to synthesis of a circuit that uses two adders. The most efficient implementation multiplexes the datapaths and shares the adder between them, rather than multiplex the outputs of separate adders. The important design tradeoff is that the mux will occupy significantly less area than the adder that it replaces.

End of Example 6.14

As a general guideline, if arithmetic functions are to be inferred from language-based operators, the operators should be grouped as much as possible within a single cyclic behavior to allow the synthesis engine to share the hardware resources that will be used to implement the function.

6.3   Synthesis of Sequential Logic with Latches

Latches are synthesized in two ways: intentionally and accidentally. If a designer does not understand how a synthesis tool infers latch-based logic from a Verilog description, surprises will occur when what was intended to be combinational logic is synthesized as latched logic.

We have discussed three common ways to describe synthesizable combinational logic: (1) netlist of primitives, (2) Boolean equations described by continuous-assignment statements, and (3) a level-sensitive cyclic behavior. Can any or all of these styles lead to a circuit with latches?

Synthesis Tip

A feedback-free netlist of combinational primitives will synthesize into latch-free combinational logic.

Feedback-free netlists of combinational primitives will synthesize into latch-free combinational logic. Synthesis tools may not allow a netlist of primitives to have feedback (e.g., cross-coupled NAND gates), so such a description will be flagged as an error condition and the code will not synthesize into anything.

A set of continuous assignments describing combinational logic must not have feedback among them. For example, the pair of continuous assignments in Section 5.5 describe a NAND-latch, but this model will not synthesize because it has structural feedback.[12]

Synthesis Tip

A set of feedback-free continuous assignments will synthesize into latch-free combinational logic.

A continuous assignment that uses the conditional operator (**? :**) with feedback, (i.e., the target variable of the assignment appears in an expression of the operator) will synthesize to a latch. This is a common and preferred way to intentionally model a latch, such as static random access memory (SRAM),[13] and it will synthesize.

Synthesis Tip

A continuous assignment using a conditional operator with feedback will synthesize into a latch.

Example 6.15

A cell of an SRAM memory can be modeled by the following continuous assignment statement having feedback:

---

[12]Synthesis tools may not infer latches from structural feedback loops.
[13]We will consider SRAMs in more detail in Chapter 8.

    assign data_out = (CS_b == 0) ? (WE_b == 0) ? data_in : data_out : 1'bz;

The signals CS_b and WE_b implement the active-low chip-select and write-enable functions of the cell. If the chip is selected and $WE\_b == 0$, $data\_out$ follows $data\_in$ (transparent mode), but when WE_b switches to 1, $data\_out = data\_out$ (latched mode). Synthesis tools infer the behavior of a latch from this statement because the output of the device is not affected by $data\_in$ while $WE\_b=1$, and $data\_out$ holds the residual value that it had at the moment WE_b switched to 1. If $CS\_b == 1$, the cell is in the three-state, high-impedance condition.

End of Example 6.15

6.3.1 Accidental Synthesis of Latches

Example 6.16

The model *or4_behav* describes a four-input OR gate. The algorithm within the cyclic behavior initializes the output to 0, then tests the inputs sequentially. If an input is 1, the output is set to 1, and the sequence terminates. The description synthesizes to combinational logic. The loop-index variable, k, is eliminated by the synthesis process, and has no hardware counterpart. Note that the output variable, y, was declared as a register variable but does not synthesize into a storage element.[14]

```
module or4_behav (y, x_in);

    parameter        word_length = 4;
    output                              y;
    input            [word_length - 1: 0]   x_in;
    reg                                 y;
    integer                             k;

    always @  x_in
      begin: check_for_1
        y = 0;
        for (k = 0; k <= word_length -1; k = k+1)
          if (x_in[k] == 1)
            begin
              y = 1;
              disable check_for_1;
            end
      end
endmodule
```

    Now consider the description *or4_behav_latch*, whose event control expression is not sensitive to *x_in[0]*. This leads to synthesis of a latched output—the latch implements

---

[14]Users of Verilog who do not realize that *all* the variables that are assigned value by a procedural assignment are register variables will be mystified by the results of synthesis, if they assume that every register variable will synthesize to a flip-flop.

the functionality that is implied by the circuit assigning value to the output only when the cyclic behavior is activated solely by a change in *x_in[3:1]*.

```
module or4_behav_latch (y, x_in);
  parameter      word_length = 4;
  output                                      y;
  input          [word_length - 1: 0]        x_in;
  reg                                         y;
  integer                                     k;

  always @  (x_in[3:1])
    begin: check_for_1
      y = 0;
      for (k = 0; k <= word_length -1; k = k+1)
        if (x_in[k] == 1)
          begin
            y = 1;
            disable check_for_1;
          end
    end
endmodule
```

The circuits that implement the functionality of *or4_behav* and *or4_behav_latch* are shown in Figure 6-22(a, b). The value of *x_in[0]* is latched in *or4_behav_latch*, and the latch is controlled by OR-ing *x_in[1]*, *x_in[2]*, and *x_in[3]*. If any of these inputs is 1, the value of *x_in[0]* is passed to the OR gate forming *y*. But if these inputs all switch to 0, the value of *x_in[0]* is latched and the value of *y* is independent of changes in *x_in[0]*. The simulation results in Figure 6-22(c) show the waveforms produced by simulation of *or4_behav*, *or4_behav_latch*, *y_gate* (the response of the gate-level circuit in Figure 6-21(b), and *y_or3_behav*, a three-input OR gate driven by *x_in[1]*, *x_in[2]*, and *x_in[3]*. The waveforms demonstrate the latching behavior of *or4_behav_latch*.[15]

Level-sensitive cyclic behaviors will synthesize into combinational logic if the description does not imply the need for storage. If storage is implied by the model, then a latch will be introduced into the implementation. To avoid latches, all of the variables that are assigned value by the behavior must be assigned value under all events that affect the right-hand side expressions of the assignments that implement the logic. Failure to do so will produce a design with unwanted latches. Consequently, all inputs to a level-sensitive behavior that is to implement combinational logic must be included in the event control expression. The right-hand side operands of assignments within the behavior are inputs. Likewise, any control signals whose transitions affect the assignments to the target register variables in the behavior are considered to be inputs to the behavior.

Remember: If the output of combinational logic is not completely specified for all cases of the inputs, then a latch will be inferred. Signals that appear as operands on the right-hand side of any assignment in a level-sensitive cyclic behavior may not appear on the left-hand side of the expression. If this rule is not observed, the behavior has implicit feedback and will not synthesize into combinational logic.

---

[15]The testbench for this circuit, *t_or4_behav*, contains a more elaborate test than that shown in Figure 6.22.

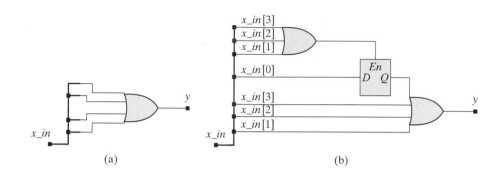

(a)   (b)

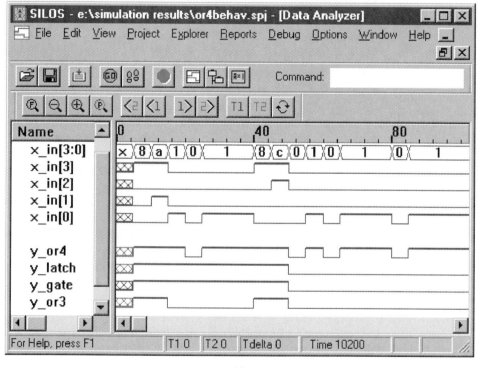

(c)

FIGURE 6-22 Four-input OR gate: (a) circuit synthesized from a level-sensitive cyclic behavior, (b) synthesized from a latch-inducing model, and (c) simulation results for both circuits.

End of Example 6.16

> ### Synthesis Tip
>
> A Verilog description of combinational logic must assign value to the outputs for all possible values of the inputs.

Verilog *case* statements and conditionals (*if*) that do not include all possible cases or conditions are *incompletely specified* and may lead to synthesis of unwanted latches in the design. When a *case* statement (in a level-sensitive cyclic behavior) does not specify an output for all of the possible inputs, the synthesis tool infers an implicit latch (i.e., the description implies that the output should retain its residual value under the conditions that were left unspecified). Caution must be taken to ensure that *case* and conditional branching (*if*) statements are complete, either explicitly or by default. If an expression associated with a conditional operator in a continuous assignment assigns the target variable (left-hand side [LHS]) to itself, the statement will synthesize a latch, but an incomplete conditional operator will cause a syntax error.[16]

Example 6.17

When a *case* statement is incompletely decoded, a synthesis tool will infer the need for a latch to hold the residual output when the select bits take the unspecified values. The latch is enabled by the event-or of the cases under which the assignment is explicitly made. In this example, the latch[17] is enabled *when {sel_a, sel_b} == 2'b10 or {sel_a, sel_b} == 2'b01*. Figure 6-23(a) shows a generic implementation, and Figure 6-23(b) shows an implementation using an actual cell library. The latter uses the library's 2-channel mux, and a model (*esdpupd*) for an electrostatic discharge pull-up/down device, which disables the hardware latch's active-low reset, rather than letting it float. Its internal structure consists of an instantiation of the *pullup* and *pulldown* primitives. Its outputs are the outputs of the two primitives. In this example the output of the *pullup* primitive is connected to the latch's reset input.

```
module mux_latch (y_out, sel_a, sel_b, data_a, data_b);
    output      y_out;
    input       sel_a, sel_b, data_a, data_b;
    reg         y_out;

    always @ ( sel_a or sel_b or data_a or data_b)
      case ({sel_a, sel_b})
        2'b10: y_out = data_a;
        2'b01: y_out = data_b;
      endcase
endmodule
```

---

[16]The syntax of the conditional operator (**?** ... **:** ...) requires a statement for the true condition and the false condition.

[17]Appendix G describes the various latches and flip-flops shown in synthesized circuits.

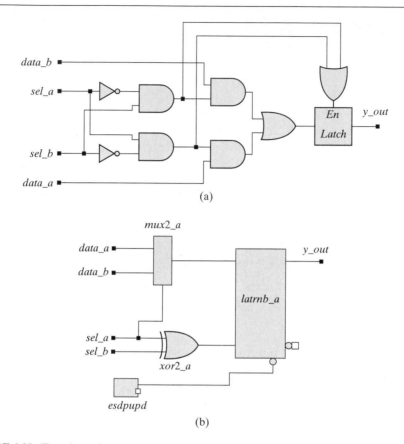

(a)

(b)

FIGURE 6-23   Two-channel mux with latched output synthesized from an incompletely specified case
statement using (a) generic parts, and (b) parts from a cell library.

End of Example 6.17

6.3.2   Intentional Synthesis of Latches

A synthesis tool infers the need for a latch when a register variable in a level-sensitive
behavior is assigned value in some threads of activity,[18] but not in others (e.g., an incomplete *if* statement in a behavior). Threads of execution that do not explicitly assign value
to a register variable imply the need for the variable to retain the value it had before the

---

[18]A thread of activity is a sequence of executed statements determined, for example, by the evaluation of
conditional branch statements (*if . . . else*).

behavior was activated. In general, a level-sensitive behavior may contain several such variables, and all of them will be synthesized as latches. The synthesis tool must identify the datapaths through the latches and identify their control signals. The control signal of a given latch will be the signal whose value controls the branching of the activity flow to the statements that do not assign value to the associated register variable. If the activity flow assigns value to a given register variable in all possible threads of the activity, a latch will be inferred only if a path assigns a variable its own value (i.e., self-feedback); otherwise, in the absence of feedback, the behavior does not imply a latch.

In synthesis, latches implement incompletely specified assignments to register variables in *case* and conditional branch (*if*) statements in a level-sensitive cyclic behavior. If a *case* statement has a *default* assignment with feedback (i.e., the variable is explicitly assigned to itself), the synthesis tool will form a mux structure with feedback. Likewise, if an *if* statement in a level-sensitive behavior assigns a variable to itself, the result will be a mux structure with feedback. We'll see that if the behavior is edge-sensitive, incomplete **case** and conditional statements synthesize register variables to flip-flops; if the statements are completed with feedback, the result is a register whose output is fed back through a mux at its datapath. (If the cell library has a cell with a gated datapath, the tool will select that part).

The functionality of a latch is also inferred when the conditional operator ($? \ldots : \ldots$) is implemented with feedback, but the actual implementation chosen by a synthesis tool depends on the context. If the conditional operator is used in a continuous assignment, the result will be a mux with feedback. If the conditional operator is used in an edge-sensitive cyclic behavior, the result will be a register with a gated data path in a feedback configuration with the output of the register. If it is used in a level-sensitive cyclic behavior, the results will be a hardware latch.

Example 6.18

The description given by *latch_if1* below assigns *data_out* with feedback in an *if* statement in a level-sensitive cyclic behavior. The result of synthesizing the latch circuit is shown in Figure 6-24, which has the structure of a mux with feedback.

```
module latch_if1(data_out, data_in, latch_enable);
   output      [3: 0]    data_out;
   input       [3: 0]    data_in;
   input                 latch_enable;
   reg         [3: 0]    data_out;

   always @ (latch_enable or data_in)
     if (latch_enable) data_out = data_in;
        else data_out = data_out;
endmodule
```

End of Example 6.18

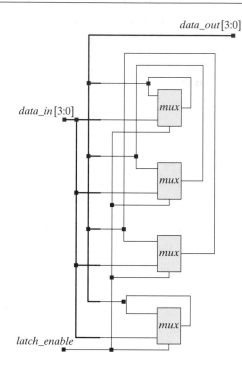

FIGURE 6-24   Latched circuit synthesized from *latch_if1*.

Synthesis Tip

An *if* statement in a level-sensitive behavior will synthesize to a latch if the statement assigns value to a register variable in some, but not all, branches (i.e., the statement is incomplete).

Example 6.19

The event control expression of the level-sensitive behavior in *latch_if2* is sensitive to both *latch_enable* and *data_in*, but the *if* statement is incomplete. This descriptive style maps preferentially to a hardware latch, shown in Figure 6-25, rather than a feedback-mux configuration.

```
module latch_if2 (data_out, data_in, latch_enable);
  output [3: 0]    data_out;
  input  [3: 0]    data_in;
  input            latch_enable;
  reg    [3: 0]    data_out;

  always @  (latch_enable or data_in)
    if (latch_enable) data_out = data_in;        // Incompletely specified
endmodule
```

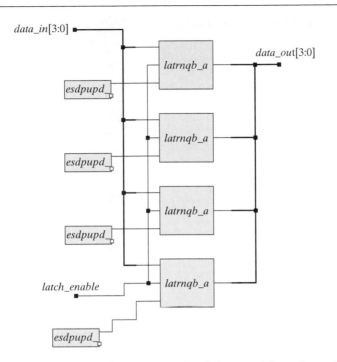

data_in[3:0]

data_out[3:0]

latrnqb_a

*esdpupd_*

latrnqb_a

*esdpupd_*

latrnqb_a

*esdpupd_*

latch_enable

latrnqb_a

*esdpupd_*

FIGURE 6-25 Latches synthesized from *latch_if2*, a description containing an incompletely specified conditional branch (*if*) statement.

End of Example 6.19

The synthesis results in Figures 6-24 and 6-25 illustrate how a slight change in the behavioral description can influence the structure of the synthesized circuit. The structure in Figure 6-24 synthesized from a complete conditional branch with feedback, whereas the description in Figure 6-25 synthesized from an incomplete conditional branch. The circuits are equivalent in simulation, but will have different area/speed tradeoffs in hardware. An *if* statement that is completed with feedback is equivalent to the following conditional assignment statement:

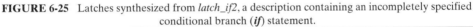

assign data_out [3:0] = latch_enable ? data_in [3: 0] : data_out[3: 0];

This statement will synthesize to the same structure, and it is commonly used to describe a latch. Remember, the conditional operator must be completed with two expressions, one for the true condition, and the other for the false condition.

Example 6.20

An *sn54170* register file consists of an array of four 4-bit words. Two address busses, *wr_sel* and *rd_sel* provide addresses for write and read operations. Two enable lines, *wr_enb* and *rd_enb* control transparent-low latches. The level-sensitive behavior in *sn54170* has an incomplete branching statement. The synthesized circuit in Figure 6-26 has an array of latches holding *latched_data*.

```
module sn54170 (data_out, data_in, wr_sel, rd_sel, wr_enb, rd_enb);
  output        [3: 0]    data_out;
  input                   wr_enb, rd_enb;
  input         [1: 0]    wr_sel, rd_sel;
  input         [3: 0]    data_in;

  reg           [3: 0]    latched_data    [3: 0];

  always @ (wr_enb or wr_sel or data_in) begin
    if (!wr_enb) latched_data[wr_sel] = data_in;
  end

  assign data_out = (rd_enb) ? 4'b1111 : latched_data[rd_sel];

endmodule
```

End of Example 6.20

6.4   Synthesis of Three-State Devices and Bus Interfaces

Three-state devices allow buses to be shared among multiple devices. The preferred style for inferring a three-state bus driver uses a continuous assignment statement that has one branch set to a three-state logic value (z).

Example 6.21

The circuit shown in Figure 6-27 is a typical configuration for integrating some core logic with a three-state unidirectional interface to a bus. Because an expression of the conditional assignment assigns $32'bz$ to *data_to_bus*, a synthesis tool will infer a 32-bit wide output with three-state drivers from the style used in the Verilog description of *Uni_dir_bus*.

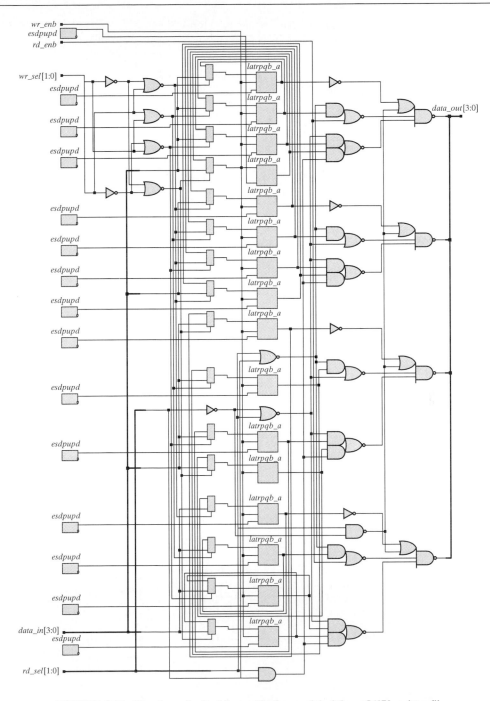

FIGURE 6-26 Circuit synthesized from a Verilog model of the *sn54170* register file.

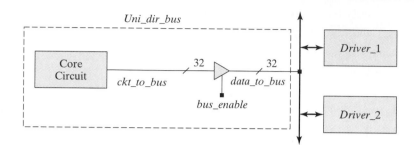

FIGURE 6-27   Unidirectional interface to a bus.

```
module Uni_dir_bus ( data_to_bus, bus_enable);
  input                    bus_enable;
  output        [31: 0]    data_to_bus;
  reg           [31: 0]    ckt_to_bus;

assign data_to_bus = (bus_enabled) ? ckt_to_bus : 32'bz;

// Description of core circuit goes here to drive ckt_to_bus

endmodule
```

End of Example 6.21

Example 6.22

The circuit shown in Figure 6-28 has a bidirectional interface with an external bus. The port of the circuit is declared to be bidirectional (***inout***), and a pair of continuous assignments is used to model the inbound and outbound datapaths.

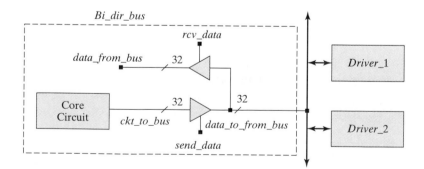

FIGURE 6-28   Bidirectional interface to a bidirectional bus.

```
module Bi_dir_bus (data_to_from_bus, send_data, rcv_data);
  inout   [31: 0]   data_to_from_bus;
  input             send_data, rcv_data;
  wire    [31: 0]   ckt_to_bus;
  wire    [31: 0]   data_to_from_bus, data_from_bus;

  assign data_from_bus = (rcv_data) ? data_to_from_bus : 32'bz;
  assign data_to_from_bus = (send_data) ? reg_to_bus : data_to_from_bus;

  // Behavior using data_from_bus and generating
  // ckt_to_bus goes here

endmodule
```

End of Example 6.22

6.5   Synthesis of Sequential Logic with Flip-Flops

Flip-flops are synthesized only from edge-sensitive cyclic behaviors, but not every register variable that is assigned value in an edge-sensitive behavior synthesizes to a flip-flop. What determines the outcome of synthesis from an edge-sensitive behavior? This section will address the following related questions: How does a synthesis tool infer the need for a flip-flop? Or when does a register variable that is assigned value within an edge-sensitive behavior automatically synthesize to a flip-flop? A flip-flop is synchronized by a clock signal—how does the synthesis tool distinguish the synchronizing signal from other signals, or must it be identified by a special word, like "clock"? If a design has multiple flip-flops, how can the model be written to ensure that they function concurrently?

A register variable in an edge-sensitive behavior will be synthesized as a flip-flop (1) if it is referenced outside the scope of the behavior, (2) if it is referenced within the behavior before it is assigned value, or (3) if it is assigned value in only some of the branches of the activity within the behavior. All of these situations imply the need for memory, or residual value. The fact that these conditions occur in an edge-sensitive behavior dictates that the memory be a flip-flop, rather than a latch.

Recall that an incomplete conditional statement (i.e., a *if* ... *else* statement or a *case* statement) in a level-sensitive cyclic behavior will synthesize to a latch. However, if the behavior is *edge*-sensitive, these types of statements will *not* create latches, but they will synthesize logic that implements a "clock enable," because the *incomplete* statements imply that the affected variables should not change under the conditions implied by the logic, even though the clock makes a transition.

The sequence in which signals are decoded in the statement that follows the event-control expression of an edge-sensitive cyclic behavior determines which of the edge-sensitive signals are control signals and which is the clock (i.e., the synchronizing signal). If the event-control expression is sensitive to the edge of more than one signal, an *if* statement must be the first statement in the behavior. The control signals that appear in the event-control expression must be decoded explicitly in the branches of the *if* statement (e.g., decode the reset condition first). The synchronizing signal is not tested explicitly in the body of the *if* statement, but, by default, the last branch must describe the synchronous activity, independently of the actual names given to the signals.

Synthesis Tip

A variable that is referenced within an edge-sensitive behavior before it is assigned value in the behavior will be synthesized as the output of a flip-flop.

Example 6.23

The nonblocking assignments to *data_a* and *data_b* in *swap_synch* describe a synchronous data swapping mechanism. The statements assigning value to *data_a* and *data_b* execute nonblocking assignments concurrently, so both variables are sampled (referenced) before receiving value; both are synthesized as the output of a flip-flop in Figure 6-29. Note that *set1* and *set2* are explicitly decoded first. The last clause of the *if* statement assigns values to *data_a* and *data_b*. Those (nonblocking) assignments are synchronized to the rising edge of *clk*, which is not referenced explicitly in the *if* statement.

```
module swap_synch (data_a, data_b set1, set2, clk,);
  output          data_a, data_b;
  input           clk, set1, set2, swap;
  reg             data_a, data_b;

  always @ (posedge clk)
    begin
      if (set1) begin data_a <= 1; data_b <= 0; end else
        if (set2) begin data_a <= 0; data_b < = 1; end
          else
            begin
              data_b <= data_a;
              data_a <= data_b;
            end
    end
endmodule
```

End of Example 6.23

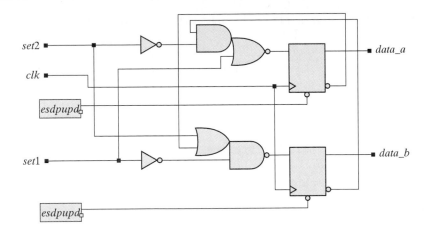

FIGURE 6-29 Circuit synthesized for *swap_synch*, with variables referenced before being assigned value.

Example 6.24

The functionality of a 4-bit parallel-load data register is described by *D_reg4_a*. The positive edge of *reset* appears in the event control expression and appears in the first clause of the *if* statement; the positive edge of *clock* also appears in the event-control expression, but is not explicitly decoded by the branch statement that follows the event-control expression. This enables the synthesis tool to correctly infer the need for a resettable flip-flop, active on the positive edge of *clock*. The value of *Data_out* is synchronized to the positive edge of *clock*, so the synthesis tool creates the 4-bit array of flip-flops shown in Figure 6-30.

```
module D_reg4_a  (Data_out, clock, reset, Data_in);
   output        [3: 0]    Data_out;
   input         [3: 0]    Data_in;
   input                   clock, reset;
   reg           [3: 0]    Data_out;

   always @  (posedge clock or posedge reset)
     begin
      if (reset == 1'b1) Data_out <= 4'b0;
        else Data_out <= Data_in;
        end
   endmodule
```

End of Example 6.24

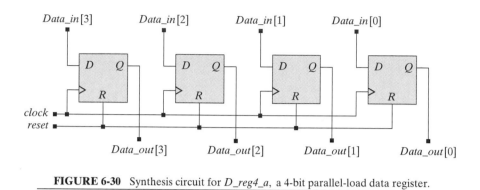

FIGURE 6-30 Synthesis circuit for *D_reg4_a*, a 4-bit parallel-load data register.

In general, the event-control expression of a cyclic behavior describing sequential logic must be synchronized to a single edge (***posedge*** or ***negedge***, but not both) of a single clock (synchronizing signal). Multiple behaviors need not have the same synchronizing signal, or be synchronized by the same edge of the same signal, but the optimization process requires that all of the synchronizing signals (clocks) have the same period. Otherwise, it would not be possible to optimize the performance of the logic.

Synthesis Tip

A variable that is assigned value by a cyclic behavior before it is referenced within the behavior, but is not referenced outside the behavior, will be eliminated by the synthesis process.

It is important to realize that not every register variable that is assigned value in a cyclic behavior synthesizes to a hardware storage device. The 4-input OR gate that was described by a Verilog behavior (*or4_behav*) in Example 6.16 has a register variable, k, that is used within the behavior, but is not referenced outside the behavior. It is not referenced before it is assigned value. The variable k merely supports the algorithm and does not require hardware memory—it does not have a life beyond the computation performed by the behavior. It is eliminated by synthesis. The behavior correctly synthesizes to a hardware OR gate, without a memory element.

Synthesis Tip

A variable that is assigned value by an edge-sensitive behavior and is referenced outside the behavior will be synthesized as the output of a flip-flop.

Example 6.25

The behavior in *empty_circuit* assigns value to the register variable *D_out*, but *D_out* is not referenced outside the scope of the behavior. Consequently, a synthesis tool will eliminate *D_out*. If *empty_circuit* is modified to declare *D_out* as an output port, *D_out* will be synthesized as the output of a flip-flop.

```
module empty_circuit (D_in, clk);
  input    D_in;
  input    clk;
  reg      D_out;

  always @ (posedge clk)  begin
    D_out < = D_in;
  end
endmodule
```

End of Example 6.25

6.6   Synthesis of Explicit State Machines

Explicit state machines have an explicitly declared state register and explicit logic that governs the evolution of the state under the influence of the inputs. In this section, we will present a recommended style for describing explicit state machines, will use that style to write Verilog models of the explicit state machines that were designed by manual methods in Chapter 3, and then synthesize their implementation in hardware. Explicit machines can be described by two behaviors, an edge-sensitive behavior that synchronizes the evolution of the state and a level-sensitive behavior that describes the next state and output logic.

6.6.1   Synthesis of a BCD-to-Excess-3 Code Converter

The circuit for a Mealy-type binary coded decimal (BCD)-to-Excess-3 code converter was designed by manual methods in Example 3.2 to illustrate the basic steps of (1) forming a state-transition graph, (2) defining a state table, (3) choosing a state assignment, (4) encoding the next state and output table, (5) developing and minimizing the Karnaugh maps for the encoded state bits and output, and (6) creating the circuit's implementation (schematic). Here we will revisit the code converter to model and synthesize it in Verilog.

Beginning with the state-transition graph (see Figure 3-19), we write the behavioral model, *BCD_to_Excess_3b*, and verify that it has the same functionality as the previous design. Note that a significant amount of work is required merely to change the state assignment in the original design, but the behavioral model requires only a change to the declared parameters defining the state codes. A synthesis tool will automatically reflect the change in the resulting gate-level structure.

> **Synthesis Tip**
>
> Use two cyclic behaviors to describe an explicit state machine: a level-sensitive behavior to describe the combinational logic for the next state and outputs and an edge-sensitive behavior to synchronize the state transitions.

The model, *BCD_to_Excess_3b*, has two cyclic behaviors.[19] An edge-sensitive behavior describes the state transitions and a level-sensitive behavior describes the next-state and output logic. Note that the assignments in the edge-sensitive behavior are nonblocking and that those in the level-sensitive behavior are blocked (procedural) assignments. The Verilog language specifies that nonblocking assignments and blocking assignments that are scheduled to occur in the same time step of simulation execute in a particular order. The nonblocking assignments are sampled first, at the beginning of the time step (before any assignments are made), then the blocked assignments are executed. After the blocked assignments execute, the nonblocking assignments are completed by assigning to the left-hand side of the statements the values that were determined by the sampling at the beginning of the time step. This mechanism ensures that nonblocking assignments execute concurrently, independent of their order, and that race conditions cannot propagate through blocked assignments and thereby affect the nonblocking assignments. Nonblocking assignments describe concurrent synchronous register transfers in hardware.

Synthesis Tip

Use the procedural assignment operator ($=$) in the level-sensitive cyclic behaviors describing the combinational logic of a finite-state machine.

Matching simulation results between a behavioral model and a synthesized circuit does not guarantee that an implementation of the circuit is correct. The waveforms in Figure 6-31 were obtained by simulating *BCD_to_Excess_3b*; they match those of the manually designed gate-level model in Figure 3-23. However, note that *BCD_to_Excess_3b* does not include a ***default*** assignment in its ***case*** statement. This leads to latches in the synthesized circuit shown in Figure 6-32(a). On the other hand, with *don't-care* default assignments to *next_state* and *B_out*, *BCD_to_Excess_3c* synthesizes to the circuit in Figure 6-32(b), without latches. Figure 6-33(c) shows the simulation results for both circuits. The waveforms of both circuits match those of the manual design (Figure 3-23) because the testbench exercises the circuit over only the allowable input sequences. The don't-care assignments of *BCD_to_Excess_3c* give greater flexibility to the synthesis tool than the implied latch structure of *BCD_to_Excess_3b*, thus it is advisable to include ***default*** assignments in all ***case*** statements. Additionally, the latches in *BCD_to_Excess_3b* waste hardware and silicon area.

Synthesis Tip

Use the nonblocking assignment operator ($<=$) in the edge-sensitive cyclic behaviors describing the state transitions of a finite-state machine and the register transfers of the datapath of a sequential machine.

---

[19]It is sometimes convenient to use continuous assignments to describe the output combinational logic.

```verilog
module BCD_to_Excess_3b (B_out, B_in, clk, reset_b);

output B_out;
input            B_in, clk, reset_b;
parameter        S_0 = 3'b000,              // State assignment
                 S_1 = 3'b001,
                 S_2 = 3'b101,
                 S_3 = 3'b111,
                 S_4 = 3'b011,
                 S_5 = 3'b110,
                 S_6 = 3'b010,
                 dont_care_state = 3'bx,
                 dont_care_out = 1'bx;
reg      [2: 0]  state, next_state;
reg              B_out;

always @ (posedge clk or negedge reset_b)
  if (reset_b == 0) state <= S_0; else state <= next_state;

always @ (state or B_in) begin
  B_out = 0;
  case (state)
    S_0:  if (B_in == 0) begin next_state = S_1; B_out = 1; end
          else if (B_in == 1) begin next_state = S_2; end

    S_1:  if (B_in == 0) begin next_state = S_3; B_out = 1; end

          else if (B_in == 1) begin next_state = S_4; end

    S_2:  begin next_state = S_4; B_out = B_in; end

    S_3:  begin next_state = S_5; B_out = B_in; end

    S_4:  if (B_in == 0) begin next_state = S_5; B_out = 1; end
          else if (B_in == 1) begin next_state = S_6; end

    S_5:  begin next_state = S_0; B_out = B_in; end

    S_6:  begin next_state = S_0; B_out = 1; end
    /*  Omitted for BCD_to_Excess_3b version
        Included for BCD_to_Excess_3c version
    default: begin  next_state = dont_care_state; B_out = dont_care_out; end

    */
  endcase
  end
endmodule
```

Synthesis Tip

Decode all possible states in a level-sensitive behavior describing the combinational next state and output logic of an explicit state machine.

If all possible states listed as a case item in a ***case*** statement are not decoded in the level-sensitive behavior describing the next state and output logic of a state machine, the combinational logic describing the next state and the output will be synthesized as the outputs of latches, the circuit may have more hardware than needed, and it may not function as intended.

Synthesis tools impose some additional restrictions on how state machines can be modeled. The state register of an explicit state machine must be assigned value as an aggregate, i.e., bit select and part select assignments to the state register variable are not allowed by a synthesis tool. The entire register must be assigned value. Asynchronous control signals (e.g., set and reset) must be scalars in the event-control expression of the behavior. Lastly, for synthesis, the value that is assigned to the state register must be either a constant (e.g., *state_reg = start_state*) or a variable that evaluates to a constant after static evaluation (i.e., the state-transition diagram must specify a fixed relationship). The description of *BCD_to_Excess_3b* satisfies these constraints.

A behavior describing the synchronous activity of an explicit state machine may contain only one clock-synchronized event-control expression. This rule applies whether the same or some other behavior describes the machine's next state and output. The description of an explicit state machine will also include an explicitly declared state register variable of type ***reg***. Only one such register may be identified for a machine, which implies that each assignment to the state register must assign value to the whole register, rather than to a bit select or a part select. The constraints on procedural assignments to the same register ensure that it is possible to associate a fixed-state transition diagram with the behavior.

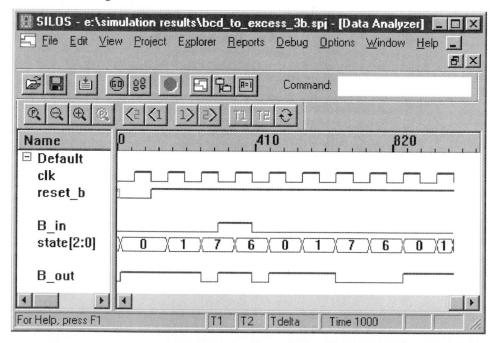

FIGURE 6-31 Results obtained from simulation of *BCD_to_Excess_3b*, a Verilog behavioral model of a
BCD-to-Excess-3 code converter.

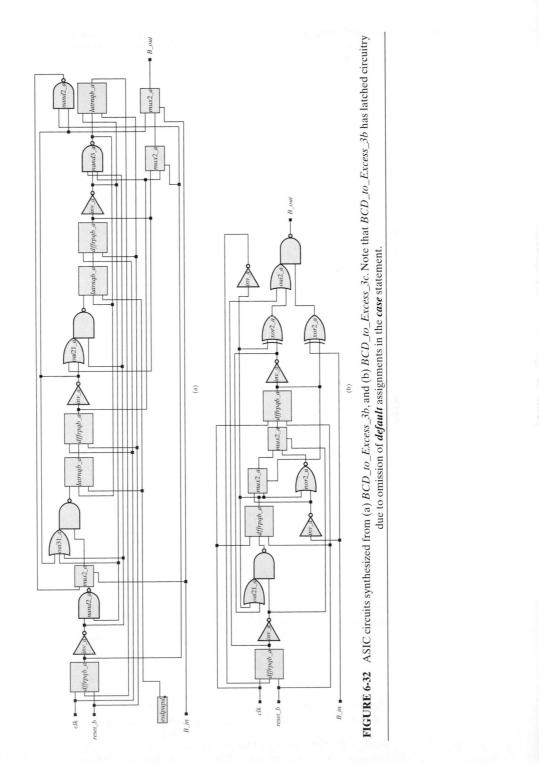

FIGURE 6-32 ASIC circuits synthesized from (a) *BCD_to_Excess_3b*, and (b) *BCD_to_Excess_3c*. Note that *BCD_to_Excess_3b* has latched circuitry due to omission of *default* assignments in the *case* statement.

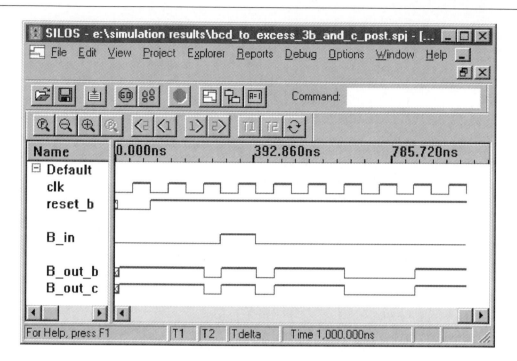

FIGURE 6-33 Post-synthesis simulation of the ASIC circuits synthesized from *BCD_to_Excess_3b* and *BCD_to_Excess_3c*.

6.6.2 Synthesis of a Mealy-Type NRZ-to-Manchester Line Code Converter

A serial line converter that converts a non–return-to-zero (NRZ) bit stream into a Manchester encoded bit stream was designed by manual methods as a Mealy-type finite state machine in Chapter 3 (See Section 3.7.1). The same state machine is described here by a Verilog behavioral model, *NRZ_2_Manchester_Mealy*, shown below.

```
module NRZ_2_Manchester_Mealy (B_out, B_in, clock, reset_b);
    output          B_out;
    input           B_in;
    input           clock, reset_b;
    reg [1: 0]      state, next_state;
    reg             B_out;
    parameter       S_0 = 0,
                    S_1 = 1,
                    S_2 = 2,
                    dont_care_state = 2'bx,
                    dont_care_out = 1'bx;

always @ (negedge clock or negedge reset_b)
    if (reset_b == 0) state <= S_0; else state <= next_state;

always @ (state or B_in ) begin
    B_out = 0;
```

```
        case (state)                                        // Partially decoded
          S_0: if (B_in == 0) next_state = S_1;
               else if (B_in == 1) begin next_state = S_2; B_out = 1; end
          S_1: begin next_state = S_0; B_out = 1;  end
          S_2: begin next_state = S_0;  end
          default:  begin next_state = dont_care_state; B_out = dont_care_out; end
        endcase
      end
    endmodule
```

The simulation results shown in Figure 6-34 match those shown in Figure 3-30 for the gate level (manual) design of the Mealy-type code converter.[20] We note again that *B_in* is switching on active edges of *clock_1* in Figure 6-34, which coincides with alternate active edges of *clock_2*. For those edges, *B_in* changes at the same time as the state. As a general rule, avoid having the inputs change at the same time that the state changes unless it happens that the inputs are treated as don't-cares at those edges, as they are in this example. The netlist[21] and schematic of the circuit synthesized from *NRZ_2_Manchester_Mealy* are shown in Figure 6-35.

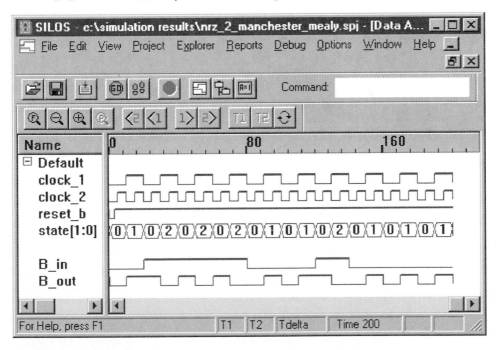

FIGURE 6-34 Results obtained from simulation of *NRZ_2_Manchester_Mealy*, a behavioral model of a Mealy-type FSM of an NRZ-to-Manchester serial line converter.

---

[20]Problem 2 at the end of this chapter addresses the postsynthesis verification step in which the functionality of the synthesized circuit is shown to match that of the behavioral model.

[21]The synthesis tool (Synopsys) generates names of wires and module instances using a more general naming convention supported by Verilog's escaped identifiers. These identifiers begin with a backslash (\) and end with white space; any printable ASCII character can be used in an escaped identifier. Here the instance names of the flip-flops correspond to the bits of the machine's state.

```
module NRZ_2_Manchester_Mealy (B_out, B_in, clock, reset_);
input B_in, clock, reset;
output B_out;
wire \next_state<1>, \next_state<0>, \state<1>, \state<0>, n80, n81, n82, n83;
  buff101 U26 (.A1(n81), .O(n80));
  norf201 U27 (.A1(n81), .B1(n82), .O(\next_state<1>));
  norf201 U28 (.A1(B_in), .B1(n80), .O(\next_state<0>));
  blf00101 U29 (.A1(n83), .B2(\state<1>), .(C2(n82), .O(B_out));
  nanf251 U30 (.A1(\state<1>), .B2(n83), .O(n81));
  invf101 U31 (.A1(B_in), .O(n82));
  invf101 U32 (.A1(\state<0>), .O(n83));
  dfrf301 \state_reg<1> (.DATA1(\next_state<1>), .CLK2(clock), .RST3(
      reset), .Q(\state<1>));
  dfrf301 \state_reg<0> (.DATA1(\next_state<0>), .CLK2(clock), .RST3(
      reset), .Q(\state<0>));
endmodule
```

(a)

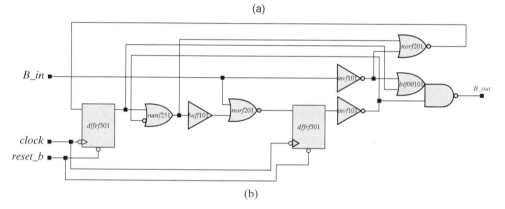

(b)

FIGURE 6-35 ASIC circuit synthesized from *NRZ_2_Manchester_Mealy*: (a) netlist (shown with port
connections made by names of formal and actual signals, and (b) schematic.

6.6.3 Synthesis of a Moore-Type NRZ-to-Manchester Line Code Converter

The explicit finite-state machine describing the Moore-type NRZ-to-Manchester serial line code converter that was designed in Chapter 3 (see the state-transition graph in
Figure 3-31) has the Verilog behavioral description below. The waveforms in Figure 6-36,
produced by simulating *NRZ_2_Manchester_Moore*, are identical to those of the gate-
level model in Figure 3-35. The circuit synthesized from *NRZ_2_Manchester_Moore* is
shown in Figure 6-37.[22]

```
module NRZ_2_Manchester_Moore (B_out, B_in, clock, reset_b);
  output         B_out;
  input          B_in;
  input          clock, reset_b;
  reg [1: 0]     state, next_state;
  reg            B_out;
```

---

[22]Problem 3 at the end of this chapter addresses the postsynthesis verification step in which the functionality
of the synthesized circuit is shown to match that of the behavioral model.

```
parameter      S_0 = 0,
               S_1 = 1,
               S_2 = 2,
               S_3 = 3;

always @ (negedge clock or negedge reset_b)
  if (reset_b == 0) state <= S_0; else state < = next_state;

always @ (state or B_in ) begin
  B_out = 0;
  case (state)                                        // Fully decoded
    S_0: begin if (B_in == 0) next_state = S_1; else next_state = S_3; end
    S_1: begin next_state = S_2; end
    S_2: begin B_out = 1; if (B_in == 0) next_state = S_1; else next_state = S_3;
    end
    S_3: begin B_out = 1; next_state = S_0; end
  endcase
  end
endmodule
```

6.6.4   Synthesis of a Sequence Recognizer

A sequence recognizer asserts an output, *D_out*, when a given pattern of consecutive bits has been received in its serial input stream, *D_in* [6]. The data is typically synchronized by the opposite edge of the clock whose active edge controls the state transitions of the machine (i.e., the synchronization is said to be antiphase). Thus, data would be applied on the rising edge of the clock if the state transitions are to occur on the falling

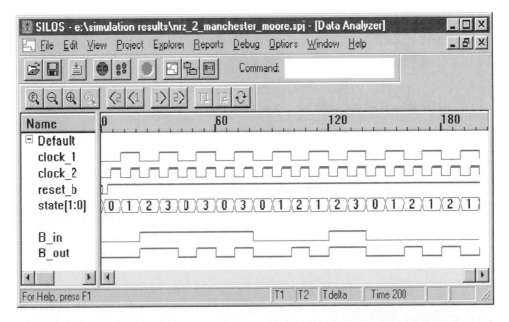

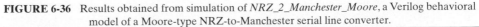

FIGURE 6-36   Results obtained from simulation of *NRZ_2_Manchester_Moore*, a Verilog behavioral model of a Moore-type NRZ-to-Manchester serial line converter.

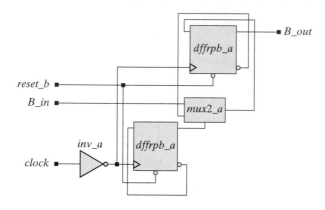

FIGURE 6-37 ASIC circuit synthesized from *NRZ_2_Manchester_Moore*.

edge of the clock, and vice versa. Sequence recognizers can be realized as explicit finite-state machines of the Mealy or Moore type.

We will follow two conventions to describe sequence recognizers. The first convention clarifies the semantics of how the machine receives input bits. It states that the output of a Mealy machine is valid immediately *before* the active edge of the clock controlling the machine, and successive values are received in successive clock cycles.[23] This has implications for interpreting when the output is valid. The output immediately before the active edge of the clock is valid, reflecting the sampled value of the input and the state of the machine before the edge.

The second convention distinguishes between resetting and nonresetting machines. A nonresetting machine continues to assert its output if the input bit pattern is overlapping (i.e., the specified sequence of m bits has a nonempty intersection with a pattern formed from bits that immediately follow the mth bit). For example, overlapping sequences of 1111_2 are present in the bit stream 001111110_2. A resetting machine that detects a pattern of length m embedded within a longer pattern must de-assert when the $m + 1$th bit arrives, independently of its value. It then begins a new cycle of detecting the next m bits.

Example 6.26

The sequence recognizer in Figure 6-38(a) is to sample the serial input, D_in, on the falling edge of the clock and assert D_out if three successive samples are 1. The machine is to have a synchronous reset action and an enable signal, En, that controls whether the machine is active or not. The ASM chart in Figure 6-38(b) describes a nonresetting Mealy-type finite-state machine (FSM) implementing the desired behavior, and the chart in Figure 6-38(c) describes a Moore machine version.

---

[23]The data must be stable prior to the active edge of the clock for at least the setup time of any flip-flop driven by D_in.

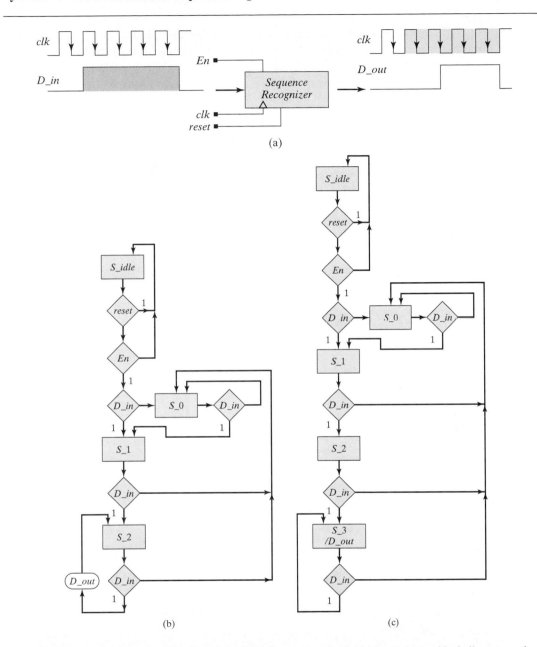

FIGURE 6-38 Sequence recognizer for detecting three successive 1s: (a) input–output block diagram and waveforms for a Moore-type output, (b) ASM chart for a Mealy-type explicit FSM implementation, and (c) ASM chart for a Moore-type explicit FSM realization.

End of Example 6.26

In the Mealy version of the machine, the signal *reset* places the state of the machine in *S_idle*, where it resides until *En* is asserted.[24] After *En* is asserted the machine makes transitions to its other states, depending solely on *D_in*. Two successive samples of 1 will cause a transition to *S_2*, where *D_out* is asserted as long as *D_in* is held at 1. The structure of the ASM chart specifies that the Mealy machine will remain in *S_2* until a 0 is received, (i.e., the machine is nonresetting); similarly, the Moore machine will remain in *S_3*. Note that the Moore machine has an extra state, because *D_out* in the Moore machine does not anticipate *D_in* and asserts *D_out* in the state reached *after* the third active edge of the clock (the Mealy machine anticipates *D_in* and asserts *D_out before* the third clock transition).

The Verilog models of the explicit state machines (given below) implement their output combinational logic with a continuous-assignment statement (see Chapter 4). To illustrate their response to two different formats for a serial line code converter (see Section 3.7), the testbench includes two instantiations of each machine. One machine

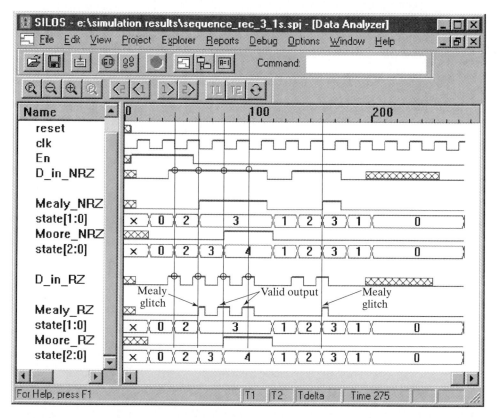

FIGURE 6-39 Sequence recognizer: simulation results for behavioral models of Mealy and Moore machines that detect three successive 1s in a serial bit stream encoded in NRZ and RZ formats.

---

[24]The structure of the ASM chart at *S_idle* implies that the reset action is synchronous, because *reset* is tested at only the active edges of the clock. For simplicity, the ASM chart omits showing that *reset* will cause a transition from any state to *S_idle*.

has a bitstream for *D_in* encoded in an NRZ format, and the other has a return-to-zero (RZ) format [7].

Simulation results are shown in Figure 6-39. First, compare the waveforms of *Mealy_NRZ* and *Moore_NRZ*. Note that *Mealy_NRZ* asserts in *S_2* (after two clocks) while *D_in* is *1*, and anticipates the third clock edge marking the end of the recognized pattern, but *Moore_NRZ* does not assert until after the third clock edge.

The importance of the convention stating that valid outputs are determined by the value of the inputs immediately prior to the active edge of the clock is illustrated by the waveform *Mealy_RZ*. Note that the Mealy machine has an invalid assertion of its output when the input has an RZ format, an apparent glitch. This assertion occurs immediately after the second clock and persists until *D_in* is possibly de-asserted in the next bit-time of the input. The valid output is 0, which is the value of *Mealy_RZ* immediately *before* the second clock. The value of *Mealy_RZ* immediately before the third clock is 1, which is a valid output. The processor that communicates with this machine would be responsible for interpreting *Mealy_RZ* correctly by detecting the value of the output immediately *before* the active edge of the clock.

The testbench also demonstrates the behavior of the machine if the input has a value of $1'bx$ in simulation. The Verilog code was written to direct the machine to return to *S_idle* if the input is not a 0 or a 1. This situation cannot occur in the physical machine, but the state could be assigned *x* in simulation. The circuits synthesized from the Mealy and Moore machines are shown in Figure 6-40.

```
module Seq_Rec_3_1s_Mealy (D_out, D_in, En, clk, reset);
  output        D_out;
  input         D_in, En;
  input         clk, reset;

  parameter     S_idle = 0;                         // Binary code
  parameter     S_0 =    1;
  parameter     S_1 =    2;
  parameter     S_2 =    3;
  reg           [1: 0]   state, next_state;

  always @ (negedge clk)
  if (reset == 1) state <= S_idle; else state <= next_state;

  always @ (state or D_in) begin
    case (state)                                    // Partially decoded

      S_idle:   if ((En == 1) && (D_in == 1))       next_state = S_1; else
                if ((En  == 1) && (D_in == 0))      next_state = S_0;
                else                                next_state = S_idle;

      S_0:      if (D_in == 0)                       next_state = S_0; else
                if (D_in == 1)                       next_state = S_1;
                else                                next_state = S_idle;

      S_1:      if (D_in == 0)                       next_state = S_0; else
                if (D_in == 1)                       next_state = S_2;
                else                                next_state = S_idle;
```

```
          S_2:          if (D_in == 0)                    next_state = S_0; else
                        if (D_in == 1)                    next_state = S_2;
                        else                              next_state = S_idle;

        default:                                          next_state = S_idle;
      endcase
    end

    assign D_out = ((state == S_2) && (D_in == 1 ));      // Mealy output
  endmodule

  module Seq_Rec_3_1s_Moore (D_out, D_in, En, clk, reset);
    output          D_out;
    input           D_in, En;
    input           clk, reset;

    parameter       S_idle = 0;             // One-Hot
    parameter       S_0 =    1;
    parameter       S_1 =    2;
    parameter       S_2 =    3;
    parameter       S_3 =    4;

    reg             [2: 0]     state, next_state;

    always @ (negedge clk)
      if (reset == 1) state <= S_idle; else state <= next_state;

    always @ (state or D_in) begin
      case (state)

        S_idle:       if ((En == 1) && (D_in == 1))       next_state = S_1; else
                      if ((En  == 1) && (D_in == 0))      next_state = S_0;
                      else                                next_state = S_idle;

        S_0:          if (D_in == 0)                      next_state = S_0; else
                      if (D_in == 1)                      next_state = S_1;
                      else                                next_state = S_idle;

        S_1:          if (D_in == 0)                      next_state = S_0; else
                      if (D_in == 1)                      next_state = S_2;
                      else                                next_state = S_idle;

        S_2, S_3:     if (D_in == 0)                      next_state = S_0; else
                      if (D_in == 1)                      next_state = S_3;
                      else                                next_state = S_idle;

        default:                                          next_state = S_idle;
      endcase
    end

    assign D_out = (state == S_3);                        // Moore output
  endmodule
```

```
module t_Seq_Rec_3_1s ();
reg D_in_NRZ, D_in_RZ, En, clk, reset;

wire Mealy_NRZ;
wire Mealy_RZ;
wire Moore_NRZ;
wire Moore_RZ;

Seq_Rec_3_1s_Mealy M0 (Mealy_NRZ, D_in_NRZ, En, clk, reset);
Seq_Rec_3_1s_Mealy M1 (Mealy_RZ, D_in_RZ, En, clk, reset);
Seq_Rec_3_1s_Moore M2 (Moore_NRZ, D_in_NRZ, En, clk, reset);
Seq_Rec_3_1s_Moore M3 (Moore_RZ, D_in_RZ, En, clk, reset);

initial #275 $finish;

initial begin #5 reset = 1; #1 reset = 0; end
initial begin
  clk = 0; forever #10 clk = ~clk;
end
initial begin
  #5 En = 1;
  #50 En = 0;
end

initial fork
  begin #10 D_in_NRZ = 0;   #25 D_in_NRZ = 1;   #80 D_in_NRZ = 0; end
  begin #135 D_in_NRZ = 1; #40 D_in_NRZ = 0; end
  begin #195 D_in_NRZ = 1'bx; #60 D_in_NRZ = 0; end
join

initial fork
  #10 D_in_RZ = 0;
  #35 D_in_RZ = 1;   #45 D_in_RZ = 0;
  #55 D_in_RZ = 1;   #65 D_in_RZ = 0;
  #75 D_in_RZ = 1;   #85 D_in_RZ = 0;
  #95 D_in_RZ = 1;   #105 D_in_RZ = 0;
  #135 D_in_RZ = 1; #145 D_in_RZ = 0; #155 D_in_RZ = 1; #165 D_in_RZ = 0;
  #195 D_in_RZ = 1'bx; #250 D_in_RZ = 0;
join
endmodule
```

The data bits of the sequence recognizer in Figure 6-38(a) were used to control explicit state machines. This led to implementations of Mealy and Moore sequence recognizers having extra logic that forces the machines to *S_idle* if they enter an unused state, depending on the state assignment scheme. An alternative approach is to consider the sequence recognizer as a datapath unit in which the input bits are shifted through a register, with simple logic detecting whether the contents of the register match the pattern of 1s. The basic cores of two such implicit state machines are shown in Figire 6-41.

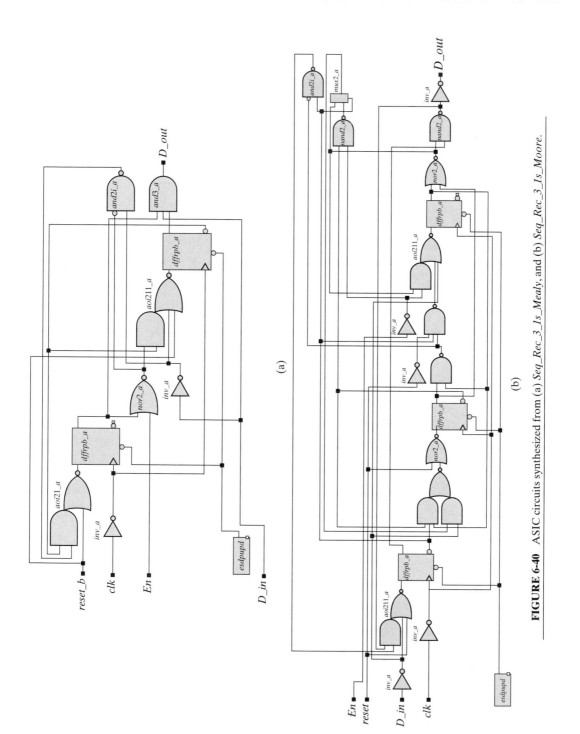

FIGURE 6-40 ASIC circuits synthesized from (a) *Seq_Rec_3_1s_Mealy*, and (b) *Seq_Rec_3_1s_Moore*.

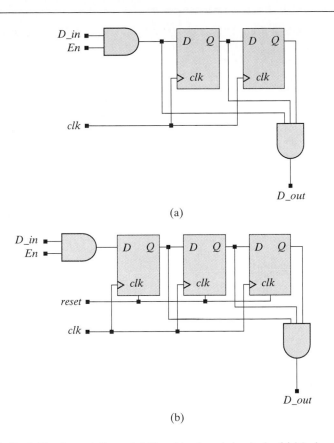

(a)

(b)

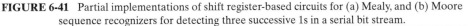

FIGURE 6-41   Partial implementations of shift register-based circuits for (a) Mealy, and (b) Moore sequence recognizers for detecting three successive 1s in a serial bit stream.

Both machines gate the datapath to the shift register, which is a slight difference from the machines described by *Seq_Rec_3_1s_Mealy* and *Seq_Rec_3_1s_Moore*, which ignore *En* after their state has moved from *S_idle*. The Mealy-type machine in Figure 6-41 also gates *D_in* with the content of the register, has fewer states, and requires one less flip-flop than the Moore version.

The circuits shown in Figure 6-41 lack the logic required to fix (recirculate) the contents of the register when *En* is not asserted. Verilog models for the full implementations with shift registers are given below. They synthesize to the circuits shown in Figure 6-42, which are considerably simpler than those in Figure 6-40, where the machine's state is decoded to determine assertions of the output. In Figure 6-42 the data is stored and decoded directly. Thus, *an explicit state machine implementation of a sequence recognizer is not necessarily the most efficient implementation*. Note that the simulations results shown in Figure 6-42(c) match those in Figure 6-39.

```
module Seq_Rec_3_1s_Mealy_Shft_Reg (D_out, D_in, En, clk, reset);
  output          D_out;
  input           D_in, En;
  input           clk, reset;
  parameter       Empty = 2'b00;
  reg     [1: 0]  Data;

  always @ (negedge clk)
  if (reset == 1) Data <= Empty; else if (En  == 1) Data <= {D_in, Data[1]};

  assign D_out = ((Data == 2'b11) && (D_in == 1 ));   // Mealy output
endmodule

module Seq_Rec_3_1s_Moore_Shft_Reg (D_out, D_in, En, clk, reset);
  output          D_out;
  input           D_in, En;
  input           clk, reset;
  parameter       Empty = 2'b00;
  reg     [2: 0]  Data;

  always @ (negedge clk)
   if (reset == 1) Data <= Empty; else if (En == 1) Data <= {D_in, Data[2:1]};

  assign D_out = (Data == 3'b111);    // Moore output
endmodule
```

6.7   Registered Logic

Variables whose values are assigned synchronously with a clock signal are said to be *registered*. Registered signals are updated at the active edges of the clock and are stable otherwise (i.e., they cannot glitch). The outputs of a Moore-type state machine are not registered, but they cannot glitch with changes at the machine's input. Their settling time will be longer than if they were registered.

Example 6.27

The output of *mux_reg* below is synchronized by the rising edge of *clock*, so the synthesis tool implements the combinational logic of a 4-channel mux with 8-bit datapaths, but registers the outputs of the mux in a bank of D-type flip-flops, as shown in Figure 6-43.

Figure 6-44   shows structures for registering the outputs of Mealy and Moore-type state machines. The output of the storage register in the structures in Figure 6-44(a) and 6-44(b) lags the combinational values by one clock cycle (i.e., the output of the register corresponds to the state of the machine in the previous cycle). The structures

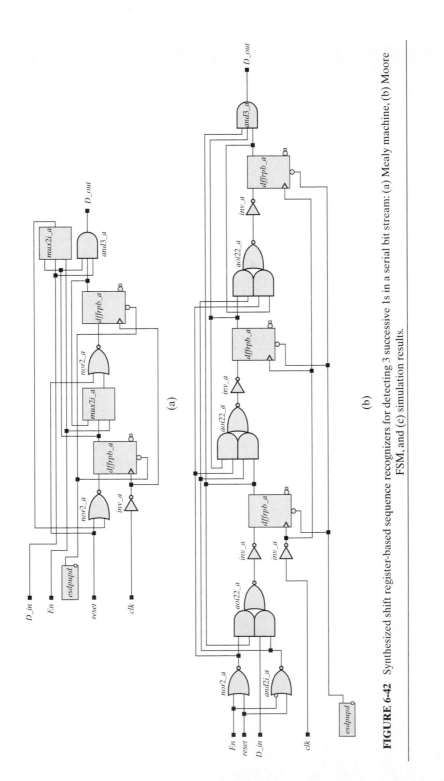

FIGURE 6-42 Synthesized shift register-based sequence recognizers for detecting 3 successive 1s in a serial bit stream: (a) Mealy machine, (b) Moore FSM, and (c) simulation results.

293

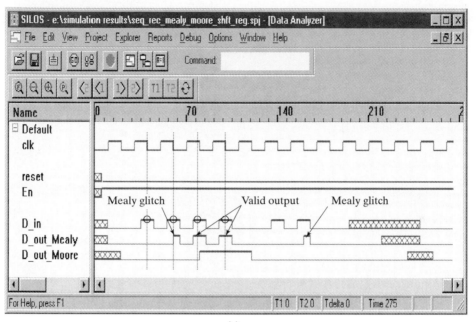

(c)

FIGURE 6-42 Continued

```
module mux_reg (y, a, b, c, d, select, clock);
   output [7: 0]   y;
   input   [7: 0]  a, b, c, d;
   input   [1: 0]  select;
   input           clock;
   reg             y;

   always @ (posedge clock)
      case (select)
         0: y <= a;        // non-blocking
         1: y <= b;
         2: y <= c;
         3: y <= d;
         default y <= 8'bx;
      endcase
endmodule
```

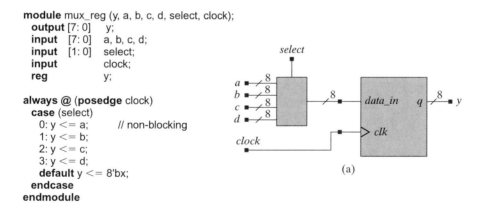

FIGURE 6-43   Multiplexer with registered output: (a) structural block diagram, and (b) synthesized circuit.

End of Example 6.27

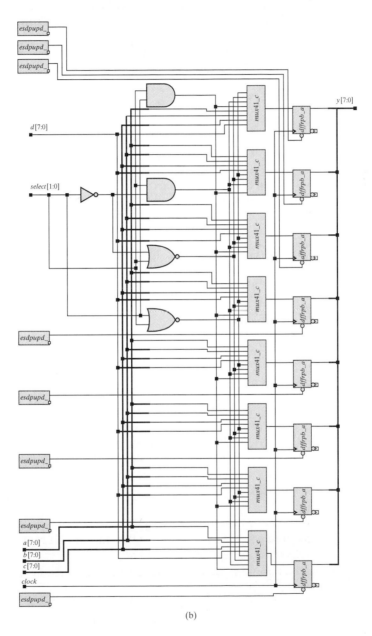

(b)

FIGURE 6-43 Continued

in Figure 6-44(c) and 6-44(d) can be used if it is important that the registered outputs be formed in the same cycle as the state. The registered Mealy outputs are formed from the *next* state and inputs at the time of the active edge of the clock; the value stored in the output register will correspond to the state reached at the clock transition and the inputs that caused the state transition. The registered Moore outputs are formed from the *next* state at the time of the active edge of the clock. The value stored in the output register will correspond to the state that is stored in the state register. The output is a registered Moore-type output.

Example 6.28

The sequence recognizers in Example 6.26 did not have registered outputs. The simulation results that were shown in Figure 6-42 had invalid assertions (glitches) of the output. The output of both machines can be registered. Include the following code in *Seq_Rec_3_1s_Mealy*[25]:

```
reg D_out_reg;
always @ (negedge clk)
  if (reset == 1) D_out_reg <= 0;
  else D_out_reg <= ((state == S_2) && (next_state == S_2) && (D_in == 1 ));
```

Notice that the clause (*state==S_2*) is included in the logic to prevent a premature assertion while the state of the machine is *S_1* (see the ASM chart in Figure 6-38b). Include the following code in *Seq_Rec_3_1s_Moore*:

```
reg D_out_reg;
always @ (negedge clk)
  if (reset == 1) D_out_reg <= 0; else D_out_reg <= (next_state == S_3);
```

The waveforms shown in Figure 6-45 show both registered and unregistered outputs for NRZ and return-to-zero (RZ) formatted serial inputs to the machine. Note that the unregistered output of the Mealy machine changes with the input, but the registered output does not, and that the value of the registered Mealy output corresponds to the value implied by the input and next state at the active (falling) edge of the clock. The unregistered output anticipates the clock; the output of the registered machine does not. The waveforms of the registered and unregistered Moore outputs are identical. In the case of the unregistered machine, the output is formed by combinational logic; in the case of the registered machine the output is the output of a register.

End of Example 6.28

---

[25]The port declarations of each machine must be modified to include the registered output.

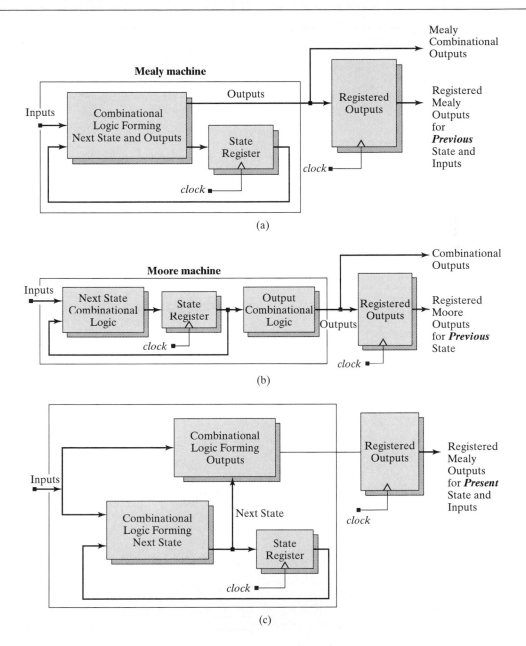

FIGURE 6-44 Registered outputs in (a) a Mealy machine, and (b) a Moore machine, (c) Mealy machine registered on the next-state function, and (d) Moore machine registered on the next-state function.

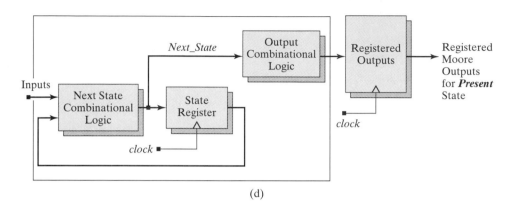

(d)

FIGURE 6-44 Continued

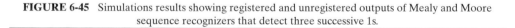

FIGURE 6-45 Simulations results showing registered and unregistered outputs of Mealy and Moore sequence recognizers that detect three successive 1s.

6.8   State Encoding

The method for designing a sequential machine requires that a set of flip-flops be chosen to represent the states of the machine and that a unique binary code be assigned to each state. State encoding determines the number of flip-flops that are required to hold the state and influences the complexity of the combinational logic used to realize the next state and output of a synthesized state machine. The task of assigning a code to the states of a machine is called *state assignment*, or *state encoding*. Because the number of different state assignments grows exponentially with the dimension of the state, it is not feasible to enumerate the possible state assignments of any but the simplest machines. In the quest for an optimal assignment, computer-based heuristic methods are used instead. Various algorithms are embedded within a synthesis tool to search for a good state assignment. Alternatively, the designer can assign state codes manually.

There are some general guidelines for manually assigning a state code: (1) if two states have the same next state for a given input, give them logically adjacent state assignments, (2) assign logically adjacent state codes to the next states of a given state, and (3) assign logically adjacent state codes to states that have the same output for a given input. These guidelines increase the possibility, but do not guarantee, that the combinational logic required to implement the output and next-state functions will be minimal [6].

The number of flip flops must be sufficient to represent the number of states as a binary number (e.g., a machine with eight states would require at least three flip-flops, and the state could be represented as a binary number. Other encodings are possible. The designer can choose the state assignment or allow the tool to optimize the assignment for the performance constraints. Alternatively, the design of the state machine can be taken out of the hands of the general synthesis tool and passed to a special state machine design tool and companion optimizer. Table 6-1 lists several common state-assignment codes.

If the designer assigns the state code, the synthesis tool will treat the state machine like it treats random logic, and make no attempt to find an optimal assignment. A state machine with N states will require at least $\log_2 N$ flip-flops to store the encoded representation of the state, but it could have more. For example, a machine with 64 states will require at least 8 flip-flops to encode the state. A BCD format simply adds one to a code to obtain the next code in the sequence. It uses the minimal number of flip-flops, but does not necessarily lead to an optimal realization of the combinational logic used to decode the next state and output of the machine. If a machine has more than 16 states, a binary code will result in a relatively large amount of next-state logic; the machine's speed will also be slower than alternative encodings. A Gray code uses the same number of bits as a binary code, but has the feature that two adjacent codes differ by only one bit, which can reduce the electrical noise in a circuit. A Johnson code has the same property, but uses more bits.

A code that changes by only one bit between adjacent codes will reduce the simultaneous switching of adjacent physical signal lines in a circuit, thereby minimizing the possibility of electrical crosstalk. These codes also minimize transitions through intermediate

TABLE 6-1 Commonly used state-assignment codes.

#	Binary	One-Hot	Gray	Johnson
0	0000	0000000000000001	0000	00000000
1	0001	0000000000000010	0001	00000001
2	0010	0000000000000100	0011	00000011
3	0011	0000000000001000	0010	00000111
4	0100	0000000000010000	0110	00001111
5	0101	0000000000100000	0111	00011111
6	0110	0000000001000000	0101	00111111
7	0111	0000000010000000	0100	01111111
8	1000	0000000100000000	1100	11111111
9	1001	0000001000000000	1101	11111110
10	1010	0000010000000000	1111	11111100
11	1011	0000100000000000	1110	11111000
12	1100	0001000000000000	1010	11110000
13	1101	0010000000000000	1011	11100000
14	1110	0100000000000000	1001	11000000
15	1111	1000000000000000	1000	10000000

states, when state changes occur in the operation of the actual hardware. The problem of intermediate transitions arises because flip-flops in the state register do not change simultaneously. When more than one bit changes to make a state transition and the bits do not switch simultaneously, an intermediate state is present momentarily in the state register. This could have undesirable consequences.

A popular design methodology called *one-hot encoding* (for active-high logic, one-cold for active-low logic) uses more than the minimum number of flip-flops; in fact, it uses one for each state. The decoding logic in a one-hot machine uses fewer gates because the machine has to decode only a single bit of a register rather than a vector pattern.

One-hot state encoding uses more flip-flops than other forms of encoding, but it usually leads to simpler (fewer levels) decoding logic for the next state and the output of the machine. The decoding logic for one-hot machines does not become more complex as more states are added to the design. Thus, the speed at which the machine can operate is not limited by the time to decode the state. One-hot machines can be faster, and the silicon area required by the extra flip-flops can be offset by the area saved by using simplified decoding logic. It is also quite easy to modify a one-hot design, because adding or removing a state does not affect the encoding of the other states. The design effort is reduced, too, because there is no need to encode a state-transition table. The state-transition graph (STG) is sufficient.

One-hot encoding used with a *case* statement might not produce the same results as one-hot encoding with an *if* statement. A *case* statement implicitly references all of the bits in the case expression, so that a one-hot encoding with an *if* statement that tests individual bits might provide simpler decoding logic.

One-hot encoding usually does not correspond to the optimal state assignment, but it has overriding merit in some applications. For example, programmable logic, such as a field-programmable gate array (FPGA), will have a fixed amount of flip-flop and combinational logic resources. Saving them does not necessarily provide a benefit.

In the Xilinx architecture discussed in Chapter 8, a configurable logic block (CLB) has three lookup tables for implementing combinational logic. An application that requires more decoding logic than that available in a single CLB will have to use additional CLBs. An attractive alternative might be to use one-hot state assignment to reduce the number of CLBs used by the machine and to reduce the need to use interconnect resources between CLBs. One-hot encoding is more reliable than binary encoding because fewer bits make transitions. Be aware, though, that in large machines, one-hot encoding will have several unused states, in addition to requiring more registers than alternative encodings.[26] Gray encoding is recommended for machines with more than 32 states because it requires fewer flip-flops than one-hot encoding and is more reliable than binary encoding because fewer bits change simultaneously.

A word of caution: If a state assignment does not exhaust the possibilities of a code, then additional logic will be required to detect and recover from transitions into unused states. Such transitions should not occur, but noise could cause the state to change to an unexpected value. It is essential that the machine be able to recover from such a state and resume operation. This additional logic will have an impact on the overall area required to realize the design.

6.9   Synthesis of Implicit State Machines, Registers, and Counters

An implicit state machine has one or more clock-synchronized (i.e., edge-sensitive) event-control expressions in a behavior. The synchronous behavior of an explicit finite-state machine (FSM) can contain only one such event-control expression, but an implicit FSM can contain multiple edge-sensitive event-control expressions in the same behavior. The clock edges of an implicit FSM define the boundaries of the machine's state transitions (i.e., the machine is in a fixed state between clock transitions). It is essential that the multiple event-control expressions of an implicit FSM be synchronized to the same edge of the same clock, either *posedge* or *negedge* but not both.

6.9.1   Implicit State Machines

The description of an implicit FSM does not represent the value of its state by an explicitly declared register variable (*reg*). Instead, the state is defined implicitly by the evolution of activity within a cyclic (*always* ...) behavior. Implicit FSMs may contain multiple clock-synchronized event-control expressions within the same behavior, and are considered to be a more general style of design than explicit FSMs. These machines have a limitation—*each state may be entered from only one other state,* because states are determined by the evolution of the behavior from clock cycle to clock cycle, and a clock cycle can be entered only from the immediately preceding clock cycle. Thus, the ASM charts of the sequence recognizers in Figure 6-38 cannot be implemented as implicit state machines, but the counters and registers described in Chapter 5 can all be described as single-cycle implicit state machines. Likewise, the Verilog models of the

---

[26]This is not an issue in register-rich FPGAs.

shift-register-based sequence recognizers in Figure 6-41 are implicit state machines. *Any sequential machine with an identical activity flow in every cycle is a one-cycle implicit state machine, and its activity can be described by one state, running.* The simplest example of such a machine is a D-type flip-flop. Typically, an implicit state machine can be described with fewer statements than a corresponding explicit machine, which must have an elaborate, explicit, STG description. The STG of an implicit machine is implicit, and could be constructed from the behavioral description, if necessary.

Synthesis tools infer the existence of an implicit FSM when a cyclic (***always***) behavior has more than one embedded, clock-synchronized, event-control expression. The multiple event-control expressions within an implicit FSM separate the activity of the behavior into distinct clock cycles of the machine. For example, the behavior below has register assignments to *reg_a* and *reg_c* in the first clock cycle, and to *reg_g* and *reg_m* in the second clock cycle. Both cycles must execute before the activity flow returns to the beginning of the behavior. Note that the event-control expressions that are imbedded within the behavior are not accompanied by the ***always*** keyword, which declares a behavior and cannot be nested. The role of these embedded event-control expressions is to suspend execution of the simulation until the active edge of the clock.

```
always @ (posedge clk)          // Synchronized event before first assignment
  begin
   reg_a <= reg_b;              // Executes in first clock cycle
   reg_c <= reg_d;             // Executes in first clock cycle.
   @ (posedge clk)             // Begins second clock cycle.
    begin
     reg_g <= reg_f;           // Executes in second clock cycle.
     reg_m <= reg_r;          // Executes in second clock cycle.
    end
  end
```

The states of an implicit FSM are not enumerated a priori. Each edge-sensitive transition determines a state transition. The synthesis tool will use this information to determine the size of a physical register that will be synthesized to represent the state (the synthesized circuit will contain registers designated as "multiple wait states"). The tool will also extract and optimize the combinational logic that governs the state transitions in the physical machine.

6.9.2   Synthesis of Counters

Synthesis tools easily synthesize a variety of counters and shift registers as single-cycle implicit state machines. Several were presented in Chapter 5. Even a ripple counter can be described and synthesized as a cascade of individual implicit state machines (one for each cell).

Example 6.29

A 4-bit ripple counter can be implemented with toggle (T-type) flip-flops. This type of counter has limited practical application because it takes excessive time to propagate changes through the cascaded chain of flip-flops, especially for long counters. The output count is also subject to glitches during the transitions. The Verilog description of *ripple_counter* uses four behaviors to model the rippling effect, with successive stages of the counter triggered by the output of their immediately previous stage. The toggling action is controlled by the input *toggle*. No ASM chart is developed because devices trigger on signals other than the clock. The circuit simulates correctly and synthesizes. The wires *c0*, *c1*, and *c2* are required because the event-control expression must be a simple variable (not a bit-select) to comply with the style sheet for the synthesis tool that produced the result. The structure of the counter and the synthesized result are shown in Figures 6-46 and 6-47, respectively.

```
module ripple_counter (count, toggle, clock, reset,);
    output      [3: 0]    count;
    input                 toggle, clock, reset;
    reg         [3: 0]    count;
    wire                  c0, c1, c2;

    assign c0 = count[0];
    assign c1 = count[1];
    assign c2 = count[2];

    always @ (posedge reset or posedge clock)
        if (reset == 1'b1) count[0] <= 1'b0; else
        if (toggle == 1'b1) count[0] <= ~count[0];

    always @ (posedge reset or negedge c0)
        if (reset == 1'b1) count[1] <= 1'b0; else
        if (toggle == 1'b1) count[1] <= ~count[1];

    always @ (posedge reset or negedge c1)
        if (reset == 1'b1) count[2] <= 1'b0; else
        if (toggle == 1'b1) count[2] <= ~count[2];

    always @ (posedge reset or negedge c2)
        if (reset == 1'b1) count[3] <= 1'b0; else
        if (toggle == 1'b1) count[3] <= ~count[3];
endmodule
```

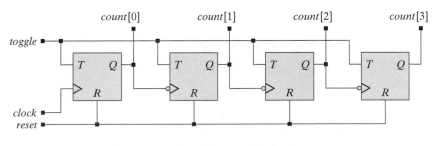

FIGURE 6-46 Structure of a 4-bit ripple counter.

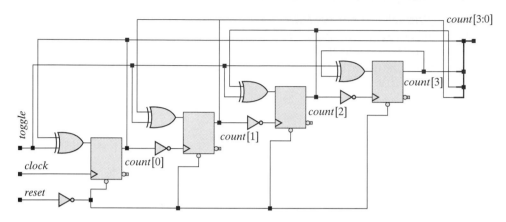

FIGURE 6-47   Synthesized circuit for a 4-bit ripple counter.

End of Example 6.29

Example 6.30

The ring counter presented in Example 5.40 is a single-cycle implicit state machine. Its description is greatly simplified compared to an alternative machine based on an elaboration of all possible states of an 8-bit register storing an explicit state.

End of Example 6.30

6.9.3   Synthesis of Registers

Storage elements in a sequential machine can be implemented with flip-flops or with latches, depending on the clocking scheme used by the machine. We will use the term *register* to mean a memory structure formed by a group of D-type flip-flops with a common clock.[27]

Example 6.31

The shift register described by *shifter_1* below includes combinational logic forming the register variable *new_signal*. Since *new_signal* receives value within a synchronous behavior and is referenced outside the behavior, it will be synthesized as the output of a flip-flop, with the structure shown in Figure 6-48.

```
module shifter_1 (sig_d, new_signal, Data_in, clock, reset);
  output      sig_d, new_signal;
  input       Data_in, clock, reset;
```

[27]Asynchronous registers using latches will be discussed in Chapter 7.

```
      reg                  sig_a, sig_b, sig_c, sig_d, new_signal;
    always @ (posedge reset or posedge clock)
      begin if (reset == 1'b1)
        begin
          sig_a <= 0;
          sig_b <= 0;
          sig_c <= 0;
          sig_d <= 0;
          new_signal <= 0;
        end
      else
        begin
          sig_a <= Data_in;
          sig_b <= sig_a;
          sig_c <= sig_b;
          sig_d <= sig_c;
          new_signal <= (~ sig_a) & sig_b;
        end
      end
    endmodule
```

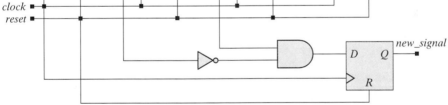
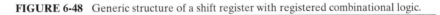

FIGURE 6-48 Generic structure of a shift register with registered combinational logic.

End of Example 6.31

Example 6.32

In *shifter_2*, *new_signal* is formed outside of the behavior in a continuous assignment, and is synthesized as the combinational logic shown in Figure 6-49.

```
    module shifter_2 (sig_d, new_signal, Data_in, clock, reset);
      output            sig_d, new_signal;
      input             Data_in, clock, reset;

      reg               sig_a, sig_b, sig_c, sig_d, new_signal;

    always @ (posedge reset or posedge clock)
```

```
            begin
             if (reset == 1'b1)
               begin
                 sig_a <= 0;
                 sig_b <= 0;
                 sig_c <= 0;
                 sig_d <= 0;
               end
             else
               begin
                 sig_a <= shift_input;
                 sig_b <= sig_a;
                 sig_c <= sig_b;
                 sig_d <= sig_c;
               end
            end
            assign new_signal = (~ sig_a) & sig_b;
            endmodule
```

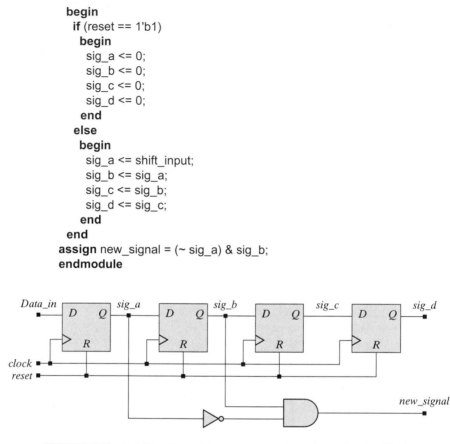

FIGURE 6-49   A shift register with separate, unregistered combinational logic.

End of Example 6.32

Example 6.33

An accumulator is an important part of an arithmetic and logic unit (ALU) of a digital machine. Here, an accumulator forms a running sum of the samples of an input. Two versions, *Add_Accum_1* and *Add_Accum_2* are shown below. *Add_Accum_1* forms *overflow_1* one cycle after storing the results of an overflow condition, as shown in the simulation results in Fig. 6-50. *Add_Accum_2* forms *overflow_2* as an unregistered Mealy output. The machines synthesize differently, too, as shown in Figure 6-50(b), where *overflow_1* is formed in *Add_Accum_1 as a registered version of overflow_2*, which is formed in *Add_Accum_2*.

```
            module Add_Accum_1 (accum, overflow, data, enable, clk, reset_b);
            output      [3: 0]    accum;
            output                overflow;
```

```
    input       [3: 0]    data;
    input                 enable, clk, reset_b;
    reg                   accum, overflow;

    always @ (posedge clk or negedge reset_b)
      if (reset_b == 0) begin accum <= 0; overflow <= 0; end
      else if (enable) {overflow, accum} <= accum + data;
endmodule

module Add_Accum_2 (accum, overflow, data, enable, clk, reset_b);
    output      [3: 0]    accum;
    output                overflow;
    input       [3: 0]    data;
    input                 enable, clk, reset_b;
    reg                   accum;
    wire        [3:0]     sum;
    assign                {overflow, sum} = accum + data;

    always @ (posedge clk or negedge reset_b)
      if (reset_b == 0) accum <= 0;
      else if (enable) accum <= sum;
endmodule
```

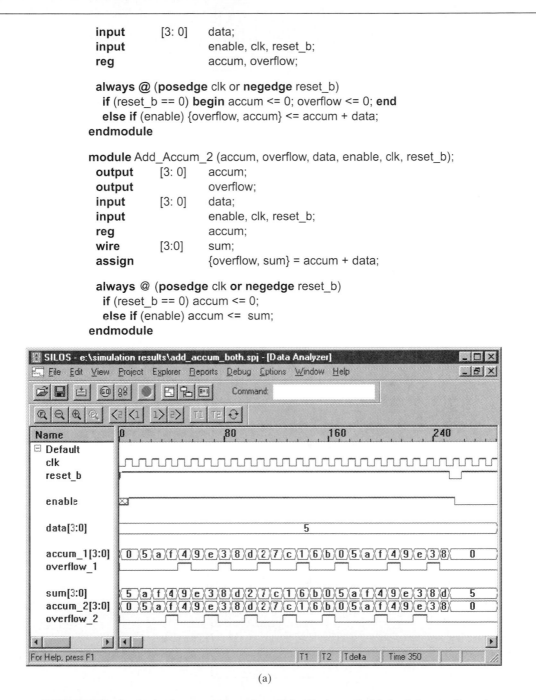

(a)

FIGURE 6-50 Synthesis of an accumulator for a 4-bit wide datapath: (a) simulation results, and (b) synthesized circuit.

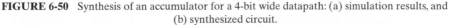

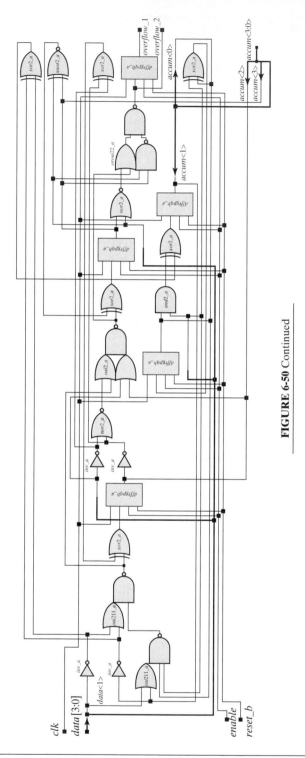

FIGURE 6-50 Continued

End of Example 6.33

6.10   Resets

Every sequential module in a design should have a reset signal. Otherwise, the initial state of the machine and its operation cannot be controlled. If an initial state cannot be assigned, the machine cannot be tested for manufacturing defects. Special care must be taken in describing the reset action of an implicit state machine that contains more than one event-control expression. Such machines must be disabled by an external agent. The first statement in the behavior that is associated with a reset signal must be a conditional statement that terminates execution of the behavior if the reset signal is asserted. Be careful: The *disable* statement must ensure that the machine begins executing at the top of the behavior when the reset is de-asserted. Incomplete resets will cause extra logic to be synthesized. Worse yet, the machine will reset to a different state, depending on when the reset is asserted.

Asynchronous reset signals can glitch, so it is recommended that asynchronous reset inputs be synchronized.[28] This can be done with a separate synchronizer, or by writing the Verilog model to have an edge-sensitive synchronous behavior without including the reset signal in the event control expression.

Example 6.34

A Moore-type sequence recognizer that asserts *D_out* after two successive samples of D_in are both either 1 or 0 is described by an implicit-state machine, and illustrates the care that must be taken with reset signals. The Verilog model, Seq_Rec_Moore_imp, uses a two-stage shift register to hold samples of the input bit stream, and generates the waveforms in Figure 6-51. Additional logic, in the form of a state machine, must prevent the output from asserting prematurely (i.e., before two samples have been received). A variable, flag, will be set after two samples have been received. We will illustrate start-up of the machine and explore the consequences of partially resetting the shift register.

If only the last stage of the shift register is flushed under the action of *reset*, the description synthesizes to the circuit shown in Figure 6-52. The synthesis tool forms a state register, *multiple_wait_state*. At the first active edge of *clock*, the machine flushes *last_bit* or loads *this_bit*, depending on *reset*. If *reset* is asserted it terminates the activity of *machine* until *reset* is de-asserted. Then *this_bit* is loaded with the first sample of the bit stream, and the machine enters an endless loop in which data is shifted through the pipeline. *flag* is asserted at the second active edge of *clock*, and it remains asserted until the activity of the named block *machine* is terminated by *reset*. Signal *flag* is used in the continuous assignment to *D_out*, ensuring that the machine will not assert prematurely after a reset condition. The entire behavior must be encapsulated as the named block *wrapper_for_synthesis* to enable the synthesis tool to create an implementation of the circuit.

---

[28]FPGAs commonly conserve routing resources by having a global set/reset that is automatically wired to all of its sequential devices. These signal paths may be slower than routed signals in more advanced technologies (e.g., the Xilinx Virtex parts).

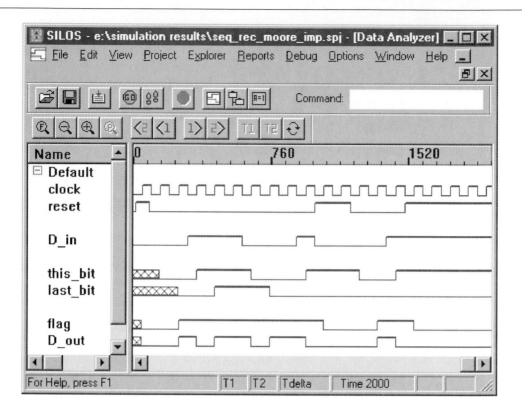

FIGURE 6-51 Simulation results for *Seq_Rec_Moore_imp.*

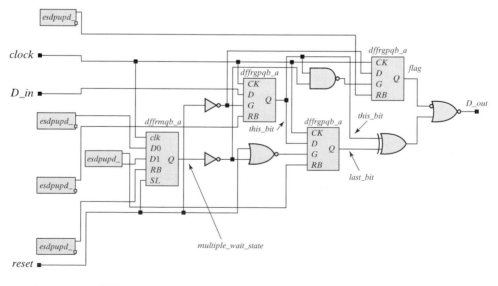

FIGURE 6-52 Circuit synthesized from *Seq_Rec_Moore_imp.*

```
module Seq_Rec_Moore_imp (D_out, D_in, clock, reset);
  output        D_out;
  input         D_in;
  input         clock, reset;
  reg           last_bit, this_bit, flag;
  wire          D_out;

  always begin: wrapper_for_synthesis
    @ (posedge clock /* or posedge reset*/) begin: machine
      if (reset == 1) begin
        last_bit <= 0;
        //  this_bit <= 0;
        // flag <= 0;
        disable machine; end
      else begin
        //  last_bit <= this_bit;
        this_bit <= D_in;
        forever
          @ (posedge clock  /* or posedge reset */) begin
          if (reset == 1) begin
            // last_bit <= 0;
            // this_bit <= 0;
            flag <= 0;
            disable machine; end
          else begin
            last_bit <= this_bit;
            this_bit <= D_in;
            flag <= 1; end              // second edge
        end
      end
    end // machine
  end  // wrapper_for_synthesis

  assign D_out = (flag && (this_bit == last_bit));
endmodule
```

In Figure 6-52 two gated-input flip-flops (*dffrgpqb_a*) form a pipeline for *last_bit* and *this_bit*. When the gate input (*G*) of this type of flip-flop is low, the *Q* output is connected to the *D* input through internal feedback, while ignoring the external *D* input. Otherwise, the external *D* input is the input. A third gated-input flip-flop holds *flag*, which is gated together with the difference of *last_bit* and *this_bit* to form *out_bit*. The active-low input (*RB*) of all of the flip-flops is disabled by the *esdpupd* device. The synthesis tool inserts a D-type flip-flop with multiplexed input (*dffrmpqb_a*) to hold *multiple_wait_state* (created by the synthesis tool) indicating whether two samples have been received or not. The active-low *RB* (reset) input of the flip-flop is disabled (for synchronous operation), and the active-low *SL* (set) is wired to *reset*. The *D0* and *D1* inputs are wired to power and ground, respectively, through *esdpupd* and the active-low *SL* input, which is connected to *reset*. When *SL* is low (*reset* is not asserted) *D0* is selected, and when *SL* is high (*reset* is asserted) *D1* is selected.

Now consider the action of *reset*. While *reset* is asserted, its inverted value causes *this_bit* and *last_bit* to hold their value (through internal feedback); it also drives the NAND-gate at the input to *flag* to get the value of its external input, which is held to 0. Thus, the reset conditions specified by the behavioral description are met for *this_bit*, *last_bit*, and *flag*.

At the first clock after *reset* is de-asserted, the *multiple_wait_state* gets 1, setting up the datapath from *this_bit* to *last_bit* on subsequent clocks. Also, *this_bit* gets *in_bit* after *reset* is de-asserted.

Note that the first reset statement within the loop in S*eq_Rec_Moore_imp* sets only *flag* to 0, not *this_bit* and *last_bit*. This implies that *this_bit* and *last_bit* are to remain unchanged by *reset* (i.e., the pipeline is not flushed). If the comments in the model are removed to cause the reset action to flush the pipeline, the description synthesizes to the simpler realization shown in Fig. 6-53. *When registers are not flushed on reset, additional logic is required to feed their outputs back to their inputs to retain their state under the action of the clock.* This additional logic can be avoided by driving the register to a known value on reset. In this example, the flag register prevents undesirable consequences of not flushing the registers and not fully loading the pipeline, but this style leads to needless logic of extra muxes and/or more complicated flip-flop cells.

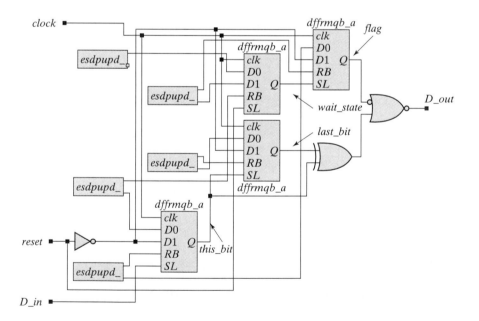

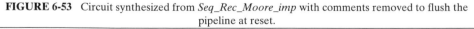

FIGURE 6-53   Circuit synthesized from *Seq_Rec_Moore_imp* with comments removed to flush the pipeline at reset.

End of Example 6.34

6.11   Synthesis of Gated Clocks and Clock Enables

Designers intentionally avoid using gated clocks because they can lead to problematic timing behavior of the host circuit. On the other hand, low-power designs deliberately disable clocks to reduce or eliminate power wasted by useless switching of transistors. Improperly gated clocks can add skew to the clock path and cause the clock signal to violate a flip-flop's constraint on the minimum width of the clock pulse. The recommended way to write a Verilog description that synthesizes a gated clock is shown below.

```
module best_gated_clock (clock, reset_, data_gate, data, Q);
  input        clock, reset_, data, data_gate;
  output       Q;
  reg          Q;

  always @ (posedge clock or negedge reset_)
    if (reset_ == 0) Q <= 0; else if (data_gate) Q <= data;       // Infers storage
endmodule
```

The description multiplexes the data with the output of a flip-flop. When the signal *data_gate* is asserted, the data is presented to the input of the flip-flop. When *data_gate* is de-asserted, the output of the flip-flop is unchanged. This description synthesizes into the circuit shown in Figure 6-54. The circuit is synchronized by *clock*, but *data_gate* gates the action of clock. Cell libraries may contain an encapsulation of this structure as a library cell for a flip-flop. Note that the synthesis tool selected an inverter and a negative-edge-sensitive flip-flop based on the availability of cells in the library.

Whether a synthesis tool infers a clock enable circuit from a Verilog description depends on the style of coding. For example, the cyclic behavior given below implies that *q_out* gets value conditionally, depending on *enable*. This will synthesize logic to implement a clock enable.

```
always @ (posedge clk)

if (enable == 1) q_out <= Data_in;
```

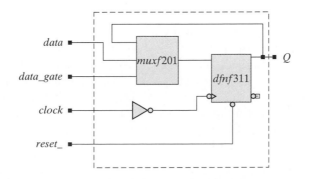

FIGURE 6-54   Synthesis result for the recommended structure of a gated clock.

6.12   Anticipating the Results of Synthesis

The results of synthesis should not be taken at face value. Instead, it is advisable to anticipate what the synthesis process will produce and then examine the results against those expectations. There are more details about synthesis than can be covered here, but this section will cover some of the basic rules that will help the designer anticipate the results of synthesis and write Verilog descriptions that infer the desired result [8]. Each vendor's tool operates differently, so it is advisable to experiment with a synthesis tool to learn how it handles particular styles of coding.

6.12.1 Synthesis of Data Types

Nets that are primary inputs or outputs will be retained in the design, but internal nets may be eliminated by the action of the synthesis tool. Integers are stored as 32-bit data objects, so use sized numbers (e.g., 8'b0110_1110) to reduce the size of the register required to hold a parameter. Do not use explicit values of x or z in logical tests (e.g., A==4'bx). They have no hardware counterpart.

6.12.2 Operator Grouping

All of the predefined Verilog operators may be used in expressions forming a binary or Boolean value. Some operators may be treated in a special way by the technology mapper that is part of a synthesis tool. For example, the Verilog operators + , − , < , > , and =   may be mapped directly to a library element if it is available. Otherwise, the synthesis tool will convert the operator into an equivalent set of Boolean equations that will be optimized. Be aware that the operands of some Verilog operators must be restricted for successful synthesis. Shift operators (<< , >>) within a behavior are synthesizable, provided that the shift is by a constant number of bits. The reduction, bitwise, and logical connective operators (see Appendix D) are each equivalent to operations performed by a logic gate. Thus, these operators are translated into a set of equivalent Boolean equations and synthesized into combinational logic. The synthesis engine will optimize these equations then map the generic description into the target technology library.

The conditional operator (*? ...:*) synthesizes into library muxes or into gates that implement the functionality of a mux. The expression to the left of *?* is formed as control logic for the mux. A conditional operator must be complete—an expression must be given for both the true and false conditions. When an expression has multiple operators, the architecture of the synthesized result will reflect the parsing of the compiler (i.e., left-to-right) and the precedence of the operators. The designer can influence the outcome by using parentheses to form sub-expressions.

Example 6.35

The continuous assignment to *sum1* in *operator_group* is equivalent to the continuous assignment to *sum2*, but *sum2* will synthesize to a faster circuit.

```
module operator_group (sum1, sum2, a, b, c, d);
   output  [4: 0]   sum1, sum2;
   input   [3: 0]   a, b, c, d;

   assign sum1 = a + b + c + d;
   assign sum2 = (a + b) + (c + d);

endmodule
```

End of Example 6.35

The structures of the synthesized circuits are shown in Figure 6-55. The architectural improvement created by the grouping of terms leads to tradeoffs in the synthesized result. The logic for *sum1* has three levels, compared to two for *sum2*, so *sum2* will be approximately 30% faster. The longest path forming *sum1* goes through three adders. If power is a consideration, input *d* forming *sum1* could be used for the signal that changes more frequently. It can also be used to accommodate a late-arriving signal by having to pass through only one adder. Within these overall structures, the synthesis tool can also optimize the implementation of the individual adders.

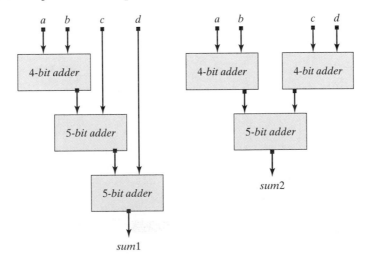

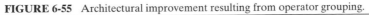

FIGURE 6-55 Architectural improvement resulting from operator grouping.

6.12.3 Expression Substitution

Synthesis tools perform expression substitution to determine the outcome of a sequence of procedural (blocking) assignments in a behavior. The designer can often write an alternative and more readable description of the same functionality. If it is necessary for procedural assignments to be used, be aware that expression substitution will affect the result.

Example 6.36

The assignments in *multiple_reg_assign* execute sequentially, with immediate changes to the target register variables, so *data_a + data_b* is substituted into the expression for *data_out2* and used in the subsequent assignment to *data_out1*. Figure 6-56(a) shows the effective data flow implemented by the functionality. The behavior of *expression_sub* is equivalent to the behavior of *multiple_reg_assign*, but the former style makes the effect of expression substitution more apparent. Both versions synthesize to the circuit in Figure 6-56(b). A recommended style is given by *expression_sub_nb_equiv*, which implements equivalent logic with the nonblocking operator ($<=$).

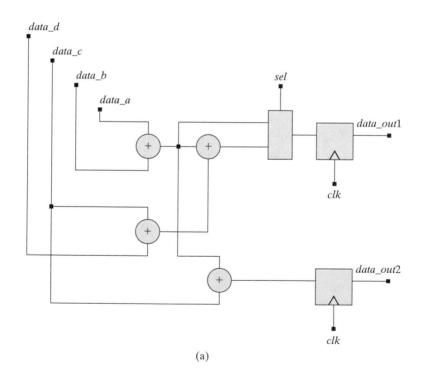

(a)

FIGURE 6-56 Dataflow structure and synthesized circuit resulting from (a) *multiple_reg_assign*, and (b) *expression_sub*.

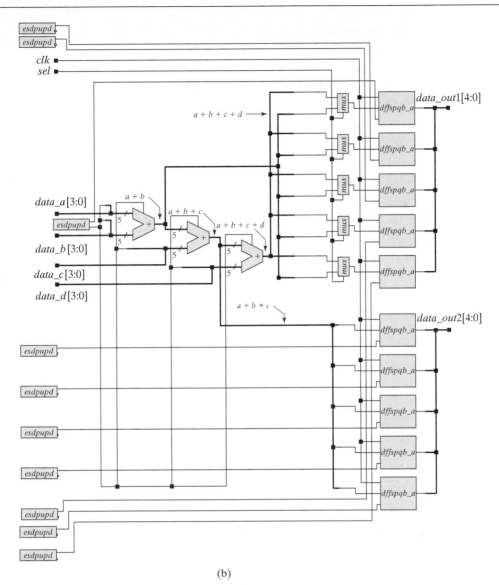

(b)

FIGURE 6-56 Continued

```
module multiple_reg_assign
    (data_out1, data_out2, data_a, data_b, data_c, data_d, sel, clk);

    output      [4: 0]    data_out1, data_out2;
    input       [3: 0]    data_a, data_b, data_c, data_d;
    input                 clk;

    reg                   data_out1, data_out2;
```

```verilog
      always @ (posedge clk)
        begin
          data_out1 = data_a + data_b ;
          data_out2 = data_out1 + data_c;
          if (sel == 1'b0)
            data_out1 = data_out2 + data_d;
        end
      endmodule

module  expression_sub
  (data_out1, data_out2, data_a, data_b, data_c, data_d, sel, clk);

  output        [4: 0]    data_out1, data_out2;
  input         [3: 0]    data_a, data_b, data_c, data_d;
  input                   sel, clk;
  reg                     data_out1, data_out2;

      always @ (posedge clk)
        begin
          data_out2 = data_a + data_b + data_c;
          if (sel == 1'b0)
            data_out1 = data_a + data_b + data_c + data_d;
          else
            data_out1 = data_a + data_b;
        end
      endmodule

module  expression_sub_nb
  (data_out1nb, data_out2nb, data_a, data_b, data_c, data_d, sel, clk);

  output        [4: 0]    data_out1nb, data_out2nb;
  input         [3: 0]    data_a, data_b, data_c, data_d;
  input                   sel, clk;
  reg           [4: 0]    data_out1nb, data_out2nb;

      always @ (posedge clk)
        begin
          data_out2nb <= data_a + data_b + data_c;
          if (sel == 1'b0)
            data_out1nb <= data_a + data_b + data_c + data_d;
          else
            data_out1nb <= data_a + data_b;
        end
      endmodule
```

End of Example 6.36

6.13   Synthesis of Loops

A loop in a cyclic behavior is said to be *static*, or *data-independent*, if the number of its iterations can be determined by the compiler *before* simulation (i.e., the number of iterations is fixed and independent of the data). A loop is said to be data-dependent if the number of iterations depends on some variable during operation. In addition to having a dependency on data, a loop may have a dependency on embedded timing controls (i.e., an event-control expression. Figure 6-57 shows possible loop structures. In principle, static loops can be synthesized using *repeat*, *for*, *while*, and *forever* loop constructs, but a given vendor might choose to confine the descriptive style of a static loop to a particular construct. The most likely form is that of a *for* loop. Non-static loops that do not have internal timing controls are problematic—they cannot be synthesized.

6.13.1   Static Loops without Embedded Timing Controls

If a loop has no internal timing controls and no data dependencies, its computational activity is implicitly combinational. The mechanism of the loop is artificial—the

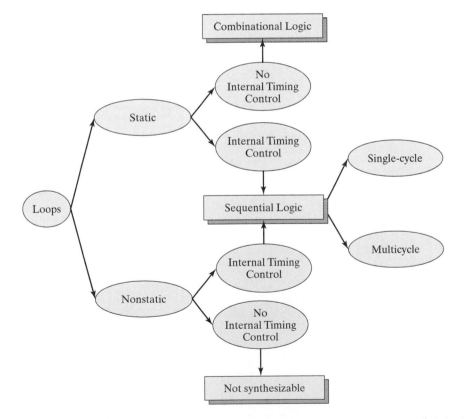

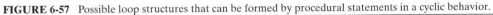

FIGURE 6-57   Possible loop structures that can be formed by procedural statements in a cyclic behavior.

computations of the loop can be performed without memory, instantaneously. The iterative computational sequence has a noniterative counterpart that can be obtained by unrolling the loop, and the operations in the unrolled loop can occur at a single time step of the simulator.

Example 6.37

The loop in *for_and_loop_comb* does not depend on the data and does not have embedded event controls. It iterates for a fixed, predetermined number of steps and terminates. The description synthesizes to the anticipated combinational circuit in Figure 6-58.

```
module for_and_loop_comb (out, a, b);
  output [3: 0]    out;
  input            [3: 0]    a, b;

  reg              [2: 0]    i;
  reg              [3: 0]    out;
  wire             [3: 0]    a, b;

  always @ (a or b)
    begin
    for (i = 0; i <= 3; i = i+1)
      out[i] = a[i] & b[i];
    end
endmodule
```

The unrolled loop in Example 6.36 is equivalent to the following assignments:

```
out[0] = a[0] & b[0];
out[1] = a[1] & b[1];
out[2] = a[2] & b[2];
out[3] = a[3] & b[3];
```

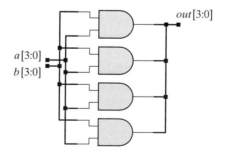

FIGURE 6-58   Synthesis of bitwise-and operations in a static *for* loop.

which correspond to a bitwise-and of two 4-bit datapaths. There are no dependencies between the assignments, so the order in which the statements are evaluated does not affect the outcome of the evaluations.

End of Example 6.37

A tool that supports synthesis of a static ***repeat*** loop with no internal timing controls will replace it by equivalent, synthesized combinational logic. The behavior of a ***for*** loop with static range is equivalent to a ***repeat*** loop with the same range, so some tools support only the ***for*** loop.

Example 6.38

Suppose a sequential machine has the task of receiving a word of data in parallel format and then asserting an output that encodes the number of 1s in the word. If the hardware is fast enough, the functionality can be implemented by combinational logic and can execute in one clock cycle. The loop in the Verilog model *count_ones_a* below has no internal timing controls and is static—it does not depend on *data*. *The order in which the statements in the cyclic behavior execute is important*, and the algorithm in the model uses the blocked assignment operator (=).

Note that *bit_count* asserts after the loop has executed, within the same clock cycle in which the loop executes. Simulation results are shown in Figure-59(a). Be careful in interpreting the results, for the loop executes in one time step of simulation, virtually instantaneously. Consequently, the displayed values of *temp*, *count*, and *bit_count* are final values that result after any values generated in intermediate simulation cycles have been overwritten. This version of the machine synthesizes into clock-compatible combinational logic (i.e., the outputs are stable within one cycle of the clock) with registered outputs.

The synthesized circuit is shown in Figure 6-59(b). The contents of register variables *index* and *temp* do not have a lifetime outside of the cyclic behavior in which they are assigned value (i.e., they are not referenced elsewhere). Both variables are eliminated by the synthesis tool. The only synthesized register is *bit_count*. Signal *bit_count* is registered because its value is assigned within an edge-sensitive cyclic behavior; *bit_count* is also an output port. The reset action is synchronous, and the reset inputs of the selected D-type flip-flops are hard wired to 1 to disable them.

```
module count_ones_a (bit_count, data, clk, reset);
    parameter                          data_width = 4;
    parameter                          count_width = 3;
    output     [count_width-1: 0]      bit_count;
    input      [data_width-1: 0]       data;
    input                              clk, reset;
    reg        [count_width-1: 0]      count, bit_count, index;
    reg        [data_width-1: 0]       temp;
```

```
always @ (posedge clk)
  if (reset) begin count = 0; bit_count = 0; end
  else begin
    count = 0;
    bit_count = 0;
    temp = data;
    for (index = 0; index < data_width; index = index + 1) begin
      count = count + temp[0];
      temp = temp > 1;
    end
    bit_count = count;
  end
endmodule
```

End of Example 6.38

6.13.2 Static Loops With Embedded Timing Controls

If a static loop has an embedded edge-sensitive event-control expression, the computational activity of the loop is synchronized and distributed over one or more cycles of the clock. As a result, the behavior is that of an implicit state machine in which each iteration of the loop occurs after a clock edge. The behavior may include additional computational activity that is placed in the cycle that immediately follows the loop's expiration.

Example 6.39

As an alternative to the static loop without embedded timing controls, we will now consider three equivalent versions of the machine that counts the 1s in a word of data, with each machine using a different static loop structure, and having embedded timing controls. Machines *count_ones_b0, count_ones_b1*, and *count_ones_b2* use *forever, while*, and *for* loops, respectively. Each loop structure has an embedded event-control expression that is synchronized by an external clock signal. The loops execute for a fixed number of clock cycles, independently of the data. The loops can be unrolled and controlled by an FSM (sequential logic) whose state transitions correspond to the iterations of the loop. The simulation results shown in Figure 6-60 for a 4-bit data path demonstrate that the machines have identical functionality. Signals *bit_count_0, bit_count_1*, and *bit_count_2* are the outputs of the machines *count_ones_b0, count_ones_b1*, and *count_ones_b2*, respectively. Only the styles in *count_ones_b0* and *count_ones_b1* were supported by the synthesis tool.[29] Only *count_ones_b2* has the same functionality if nonblocking assignments replace the procedural assignments in the models (see Problem 45 at the end of this chapter).

---

[29]Synopsys Design Compiler™ was used to obtain the synthesized circuits. It is common for electronic design automation tools to support a restricted descriptive style.

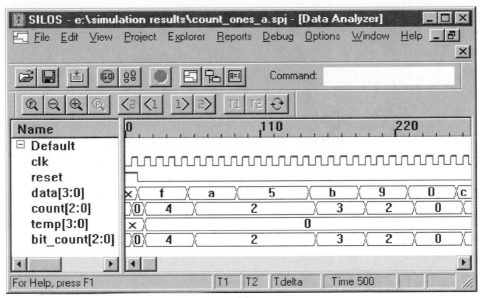

(a)

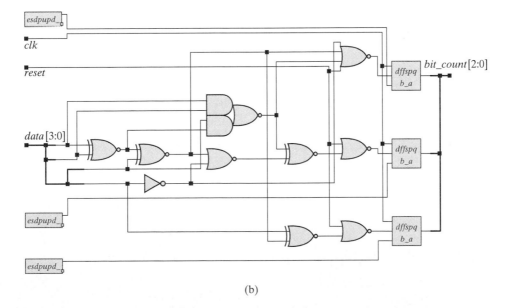

(b)

FIGURE 6-59 Results for (a) simulation of *count_ones_a*, and (b) circuit synthesized from *count_ones_a*.

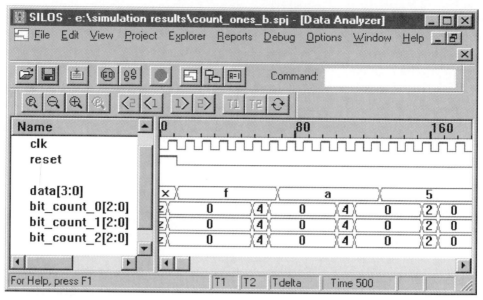

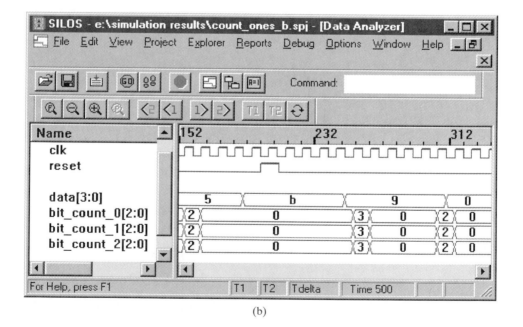

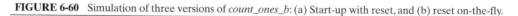

FIGURE 6-60 Simulation of three versions of *count_ones_b*: (a) Start-up with reset, and (b) reset on-the-fly.

```verilog
module count_ones_b0 (bit_count, data, clk, reset);
   parameter                          data_width = 4;
   parameter                          count_width = 3;
   output        [count_width-1: 0]   bit_count;
   input         [data_width-1: 0]    data;
   input                              clk, reset;
   reg           [count_width-1: 0]   count, bit_count;
   reg           [data_width-1: 0]    temp;
   integer                            index;

   always begin: wrapper_for_synthesis
    @ (posedge clk) begin: machine
     if (reset) begin bit_count = 0;  disable machine; end
     else
       count = 0; bit_count = 0; index = 0; temp = data;
       forever  @ (posedge clk)
         if (reset) begin bit_count = 0; disable machine; end
         else if (index < data_width-1) begin
           count = count + temp[0];
           temp = temp >> 1;
           index = index + 1;
         end
         else begin
           bit_count = count + temp[0];
           disable machine;
         end
    end // machine
   end // wrapper_for_synthesis
endmodule

module count_ones_b1 (bit_count, data, clk, reset);
   parameter                          data_width = 4;
   parameter                          count_width = 3;
   output        [count_width-1: 0]   bit_count;
   input         [data_width-1: 0]    data;
   input                              clk, reset;
   reg           [count_width-1: 0]   count, bit_count;
   reg           [data_width-1: 0]    temp;
   integer                            index;

   always begin: wrapper_for_synthesis
    @ (posedge clk) begin: machine
     if (reset) begin bit_count = 0; disable machine; end
     else  begin
       count = 0; bit_count = 0; index = 0; temp = data;
       while (index < data_width) begin
         if (reset) begin bit_count = 0; disable machine; end
```

```verilog
        else if ((index < data_width) && (temp[0] ))
          count = count + 1;
          temp = temp >> 1;
          index = index +1;
          @ (posedge clk);
        end
      if (reset) begin bit_count = 0; disable machine; end
      else bit_count = count;
        disable machine;
      end
    end // machine
   end // wrapper_for_synthesis
 endmodule

  module count_ones_b2 (bit_count, data, clk, reset);
    parameter                          data_width = 4;
    parameter                          count_width = 3;
    output         [count_width-1: 0]  bit_count;
    input          [data_width-1: 0]   data;
    input                              clk, reset;
    reg            [count_width-1: 0]  count, bit_count;
    reg            [data_width-1: 0]   temp;
    integer                            index;

   always begin: machine
     for (index = 0; index <= data_width; index = index +1)   begin
       @ (posedge clk)
         if (reset) begin bit_count = 0; disable machine; end
                 else if (index == 0) begin count = 0; bit_count = 0; temp = data; end
         else if (index < data_width) begin
            count = count + temp[0]; temp = temp >> 1; end
         else bit_count = count + temp[0];
       end
     end // machine
   endmodule
```

End of Example 6.39

6.13.3 Nonstatic Loops without Embedded Timing Controls

The number of iterations to be executed by a loop having a data dependency cannot be determined before simulation. If the loop does not have embedded timing control, the behavior can be simulated, *but it cannot be synthesized*. Under the action of a simulator, the behavior is virtually sequential and can be simulated. But hardware cannot execute the computation of the loop in a single cycle of the clock. We will demonstrate this by the following example.

Example 6.40

The computational activity of counting the 1s in a word of data is wasted after the last 1 is found. The data-dependent loop in *count_ones_c* implements a more efficient machine than *count_ones_b*. At the first active edge of the clock after reset, the machine loads *data* into *temp*, and counts the number of 1s in *temp* by repeatedly adding the value of the LSB of *temp* to *count*, and shifting the word. This continues as long as the word is not empty of 1s (i.e., until the reduction-or of *temp*, | *temp*, is false). So the data word 0001_2 will execute in fewer iterations than 1000_2 because the reduction-or of the word that results from the first right-shift in *count_ones_c* is empty of 1s. The computational activity occurs in a single cycle of the clock, so the efficiency would be apparent in simulation with long words of data. The simulation results in Figure 6-61 show the value of *index* at the end of the loop. Note that the final results (i.e., those at the end of the clock cycle of computation) are displayed. Although this behavior is attractive for simulation, it cannot be synthesized. The task of counting the 1s in a word is fundamentally combinational, but combinational logic cannot perform the sequential steps of the loop in one cycle of the clock, and at the same time terminate the activity if the word becomes empty of 1s. The loop cannot be unrolled statically because its length is data-dependent.

```verilog
module count_ones_c (bit_count, data, clk, reset);
  parameter                          data_width = 4;
  parameter                          count_width = 3;
  output        [count_width-1: 0]   bit_count;
  input         [data_width-1: 0]    data;
  input                              clk, reset;
  reg           [count_width-1: 0]   count, bit_count, index;
  reg           [data_width-1: 0]    temp;

  always @ (posedge clk)
    if (reset) begin count = 0; bit_count = 0; end
    else begin
      count = 0;
      temp = data;
      for (index = 0; | temp; index = index + 1) begin
        if (temp[0] ) count = count + 1;
        temp = temp >> 1;
      end
      bit_count = count;
    end
endmodule
```

End of Example 6.40

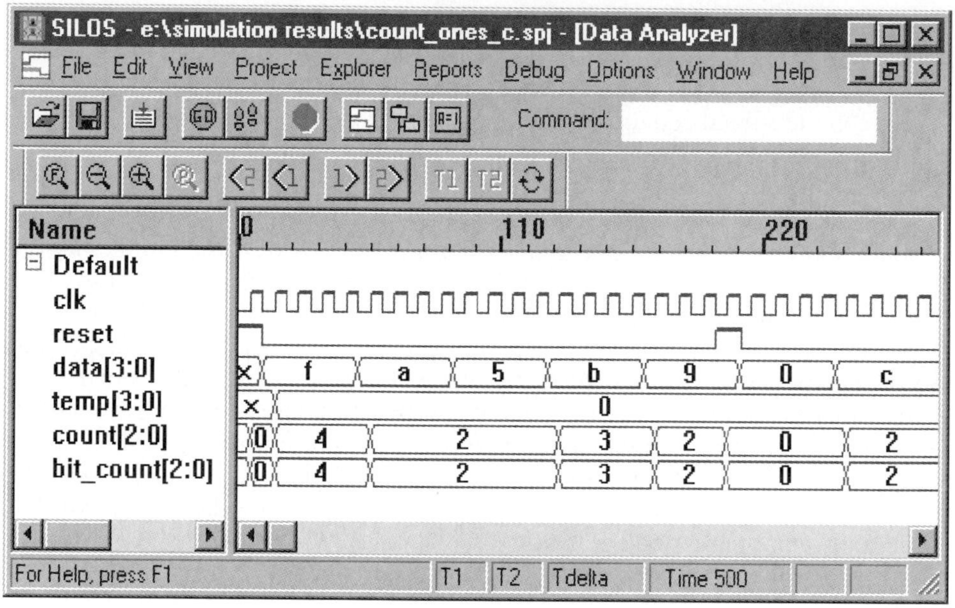

FIGURE 6-61   Simulation of *count_ones_c*, which has a data-dependent loop, and cannot be synthesized.

6.13.4 Nonstatic Loops with Embedded Timing Controls

A nonstatic loop may implement a multicycle operation. The data dependency alone is not a barrier to synthesis because the activity of the loop can be distributed over multiple cycles of the clock. However, the iterations of a non-static loop must be separated by a synchronizing edge-sensitive event control expression in order to be synthesized.

Example 6.41

The cyclic behavior in *count_ones_d* has edge-sensitive timing controls within a nonstatic *while* loop. The sequential activity of the loop is distributed over multiple cycles of the clock. First, the data is loaded into a shift register. Then it shifts the data through the register on successive clock cycles. After all of the data have been shifted, one more cycle elapses before *bit_count* is ready. Simulation results are presented in Figure 6-62.[30] Note that when $data = 3_H = 0011_2$, the loop terminates after the second cycle.

```
module count_ones_d (bit_count, data, clk, reset);
    parameter                                     data_width = 4;
    parameter                                     count_width = 3;
    output        [count_width-1: 0]              bit_count;
    input         [data_width-1: 0]               data;
    input                                         clk, reset;
```

---

[30]See Problem 6.13 for an exercise requiring synthesis of ***count_ones_d***.

```
reg               [count_width-1: 0]        count, bit_count;
reg               [data_width-1: 0]         temp;

always begin: wrapper_for_synthesis
@ (posedge clk)
 if (reset) begin count = 0; bit_count = 0; end
  else begin: bit_counter
   count = 0;
   temp = data;
   while (temp)
   @(posedge clk)
    if (reset) begin
    count = 2'b0;
    disable bit_counter; end
    else begin
     count = count + temp[0];
     temp = temp >> 1;
    end
   @(posedge clk)
    if (reset) begin
     count = 0;
     disable bit_counter; end
    else bit_count = count;
   end // bit_counter
  end
endmodule
```

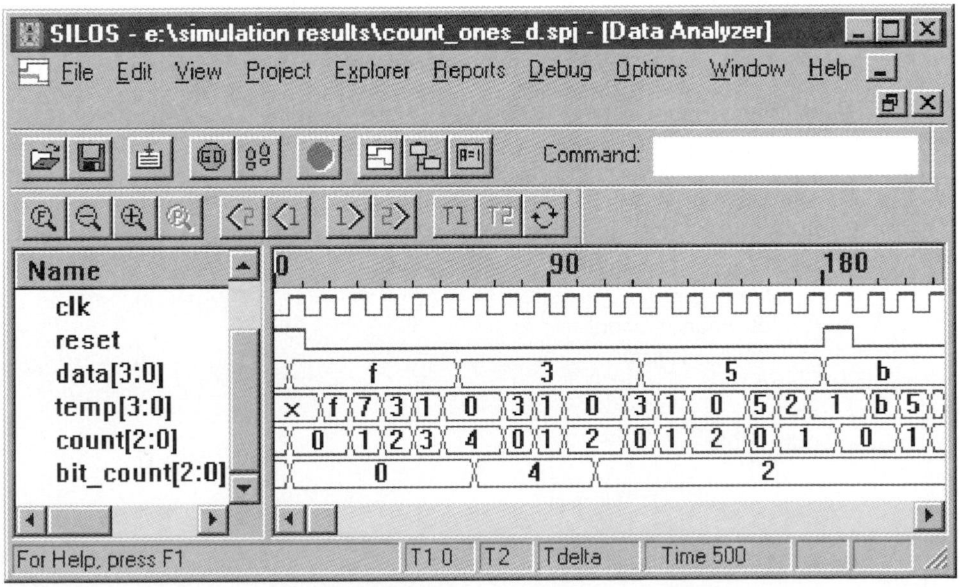

FIGURE 6-62 Simulation of *count_ones_d*.

The next version of a machine that counts the 1s in a word of data, *count_ones_SD*, adds signals *start* and *done* to the port structure, and eliminates the data dependency in the loop. *start* launches counting.[31]

```verilog
module count_ones_SD (bit_count, done, data, start, clk, reset);
  parameter                              data_width = 4;
  parameter                    .         count_width = 3;
  output       [count_width-1: 0]        bit_count;
  output                                 done;
  input        [data_width-1: 0]         data;
  input                                  start, clk, reset;
  reg          [count_width-1: 0]        count, bit_count, index;
  reg          [data_width-1: 0]         temp;
  reg                                    done, start;

  always @ (posedge clk) begin: bit_counter
    if (reset) begin count = 0; bit_count = 0; done = 0; end
    else if start begin
      done = 0;
      count = 0;
      bit_count = 0;
      temp = data;
      for (index = 0; index < data_width; index = index + 1)

        @ (posedge clk)
        if (reset) begin
          count = 0;
          bit_count = 0;
          done = 0;
          disable bit_counter; end
        else begin
          count = count + temp[0];
          temp = temp >> 1;
        end

      @ (posedge clk)        // Required for final register transfer
      if (reset) begin  count = 0; bit_count = 0; done = 0;
        disable bit_counter; end
      else begin
        bit_count = count;
        done = 1; end
    end
  end
endmodule
```

---

[31]See Problem 14 at the end of this chapter for an exercise requiring the use of alternative loop structures in *count_ones_SD*.

Note in the simulation results in Figure 6-63 that *start* asserts for one cycle. Signal *index* increments at each edge of the clock until all bits have been counted. Then *done* is asserted. When *reset* is asserted in the middle of a counting sequence, the machine re-initializes the registers and restarts the sequence.

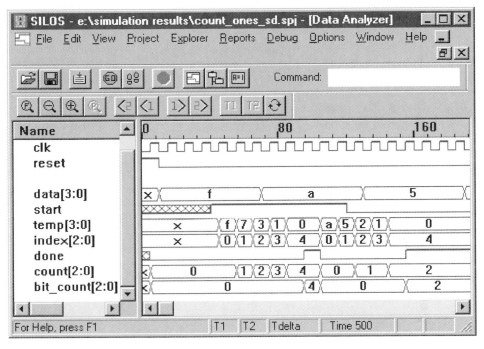

FIGURE 6-63   Simulation of *count_ones_SD*.

End of Example 6.41

6.13.5 State-Machine Replacements for Unsynthesizable Loops

Synthesis tools do not support nonstatic loops that do not have embedded timing controls. Such machines are not directly synthesizable, but their loop structures can be replaced by equivalent synthesizable sequential behavior.

Example 6.42

The ASMD chart in Figure 6-64 describes a state machine that counts the 1s in a word and terminates activity as soon as possible. The machine remains in its reset state, *S_idle*, until an external agent asserts *start*. This action asserts the Mealy output, *load_temp*, which will cause *data* to be loaded into register *temp* when the state makes a transition to *S_counting* at the next active edge of *clk*. The machine remains in

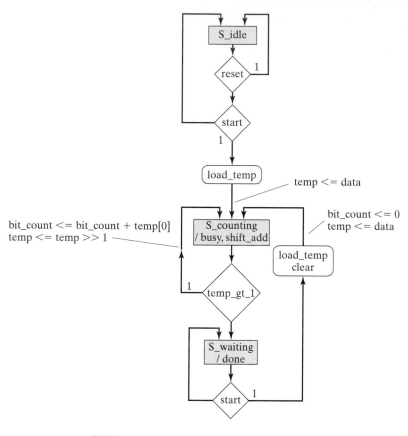

FIGURE 6-64 ASMD chart of *count_ones_SM*.

S_counting while *temp* contains a 1. At each subsequent clock *temp* is shifted toward its LSB, and *temp[0]* is added to *bit_count*. When *temp* finally has a 1 in only the LSB, the machine's state moves to *S_waiting*, where *done* is asserted as a Moore output. The state remains in *S_waiting* until *start* is de-asserted. The branches of the algorithmic state machine and datapath (ASMD) chart show the control signals that are generated by the controller and are annotated with the register operations of the machine. Implicitly, those registers must remain unchanged for state transitions that traverse the other branches.

The Verilog sequential machine *count_ones_SM* avoids the problem of having to synthesize a nonstatic loop. The waveforms of the behavioral model, shown in Figure 6-65, demonstrate that the machine terminates as soon as *temp* is detected to be empty of 1s. The circuit synthesized from *count_ones_SM* is shown in Figure 6-66. We partitioned the machine into a datapath and a controller and encapsulated the counter in a separate module to reduce the complexity of the drawings and expose the individual functional units of the design. Note that the signals *busy* and *shift_add* are hard wired together as outputs of the controller. This is a consequence of the ASMD

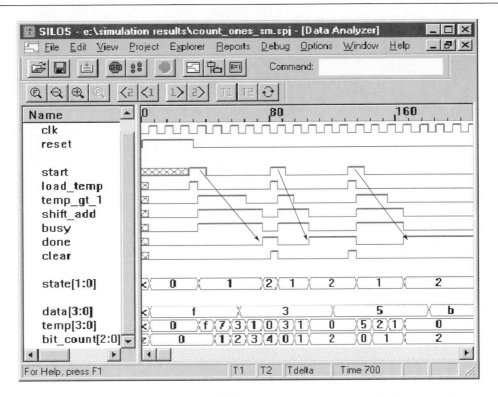

FIGURE 6-65 Simulation of *count_ones_SM* demonstrating behavior equivalent to a nonstatic loop.

chart's assertion of *busy* and *shift_add* in only state *S_counting*. Both signals were pre-served in the synthesis process, even though they are identical, because both are ports of the controller. The machine can be modified to remove one of the signals from the port. Alternatively, the module boundaries could be removed, which would allow the synthesis tool freedom to eliminate any redundant logic.

```
module count_ones_SM (bit_count, busy, done, data, start, clk, reset);
    parameter                      counter_size = 3;
    parameter                      word_size = 4;

    output   [counter_size -1 : 0]  bit_count;
    output                          busy, done;
    input    [word_size-1: 0]       data;
    input                           start, clk, reset;
    wire                            load_temp, shift_add, clear;
    wire                            temp_0, temp_gt_1;

    controller M0 (load_temp, shift_add, clear, busy, done, start, temp_gt_1, clk,
        reset);
    datapath M1 (temp_gt_1, temp_0, data, load_temp, shift_add, clk, reset);
    bit_counter_unit M2 (bit_count, temp_0, clear, clk, reset);

endmodule
```

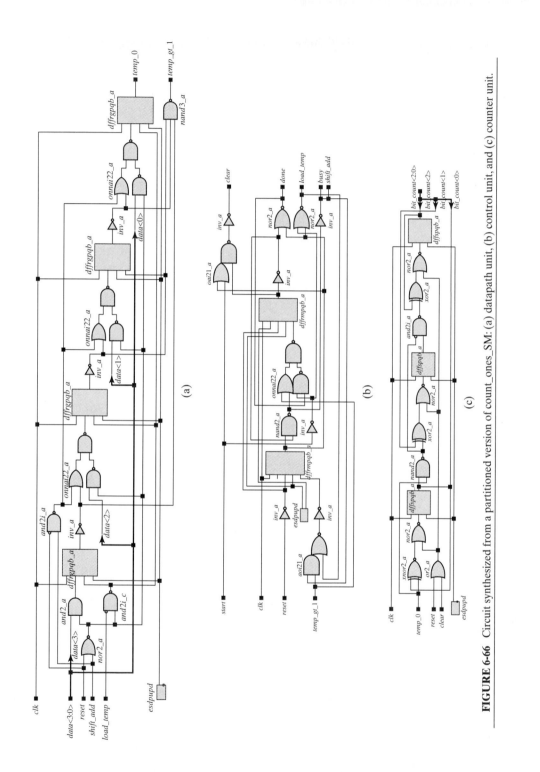

FIGURE 6-66 Circuit synthesized from a partitioned version of count_ones_SM: (a) datapath unit, (b) control unit, and (c) counter unit.

```verilog
module controller (load_temp, shift_add, clear, busy, done, start, temp_gt_1, clk,
    reset);
    parameter                       state_size = 2;
    parameter                       S_idle = 0;
    parameter                       S_counting = 1;
    parameter                       S_waiting = 2;

    output                          load_temp, shift_add, clear, busy,
                                    done;
    input                           start, temp_gt_1, clk, reset;
    reg                             bit_count;
    reg     [state_size-1 : 0]      state, next_state;
    reg                             load_temp, shift_add, busy, done,
                                    clear;

    always @ (state or start or temp_gt_1) begin
      load_temp = 0;
      shift_add = 0;
      done = 0;
      busy = 0;
      clear = 0;
      next_state = S_idle;

      case (state)
        S_idle:     if (start) begin next_state = S_counting; load_temp = 1; end

        S_counting: begin busy = 1; if (temp_gt_1) begin next_state = S_counting;
                      shift_add = 1; end
                    else begin next_state = S_waiting; shift_add = 1;  end
                    end

        S_waiting:  begin
                      done = 1;
                      if (start) begin next_state = S_counting; load_temp = 1; clear =
                      1; end
                       else next_state = S_waiting;
                    end

        default:    begin clear = 1; next_state = S_idle; end
      endcase
    end

    always @ (posedge clk) // state transitions
      if (reset)
        state <= S_idle;
      else state <= next_state;
endmodule

module datapath (temp_gt_1, temp_0, data, load_temp, shift_add, clk, reset);
    parameter                       word_size = 4;
```

```
        output                                      temp_gt_1, temp_0;
        input            [word_size-1: 0]           data;
        input                                        load_temp, shift_add, clk, reset;

        reg              [word_size-1: 0]           temp;
        wire                                         temp_gt_1 = (temp > 1);
        wire                                         temp_0 = temp[0];

        always @ (posedge clk)                      // state and register transfers
          if (reset) begin
            temp <= 0; end
          else begin
            if (load_temp) temp <= data;
            if (shift_add) begin temp <= temp > 1; end
          end
      endmodule

      module bit_counter_unit (bit_count, temp_0, clear, clk, reset);
        parameter                                   counter_size = 3;
        output           [counter_size -1 : 0]      bit_count;
        input                                        temp_0;
        input                                        clear, clk, reset;
        reg                                          bit_count;

        always @ (posedge clk)                      // state and register transfers
          if (reset || clear)
            bit_count <= 0;
          else bit_count <= bit_count + temp_0;
      endmodule
```

End of Example 6.42

Note that in the previous example all of the outputs in the level-sensitive behavior were assigned value at the beginning of the behavior. Then only changes were assigned in subsequent states. This style makes the code more readable and prevents synthesis of unwanted latches, which will happen if a variable in a level-sensitive behavior is assigned value in some, but not all, of the paths of the activity flow through the behavior.

Example 6.43

Our final example of a machine that counts the 1s in a word of data is described by an implicit state machine. Its behavior (see Figure 6-67) is equivalent to that of a nonstatic loop, but it is synthesizable (see Problem 42 at the end of this chapter).

```
      module count_ones_IMP (bit_count, start, done, data, data_ready, clk, reset);
        parameter                                   word_size = 4;
        parameter                                   counter_size = 3;
        parameter                                   state_size = 2;
```

```
output        [counter_size -1 : 0]    bit_count;
output                                 start, done;
input         [word_size-1: 0]         data;
input                                  data_ready, clk, reset;

reg                                    bit_count;
reg           [state_size-1 : 0]       state, next_state;
reg                                    start, done, clear;
reg           [word_size-1: 0]         temp;

always @ (posedge clk) if (reset)
  begin temp<= 0; bit_count <= 0; done <= 0; start <= 0; end
else if (data_ready && data && !temp)
  begin temp <= data; bit_count <= 0; done <= 0; start <= 1; end
else if (data_ready && (!data) && done)
  begin bit_count <= 0; done <= 1; end
else if (temp == 1)
  begin bit_count <= bit_count + temp[0]; temp <= temp > 1; done <= 1; end
else if (temp && !done)
  begin start <= 0; temp <= temp > 1; bit_count <= bit_count + temp[0]; end
endmodule
```

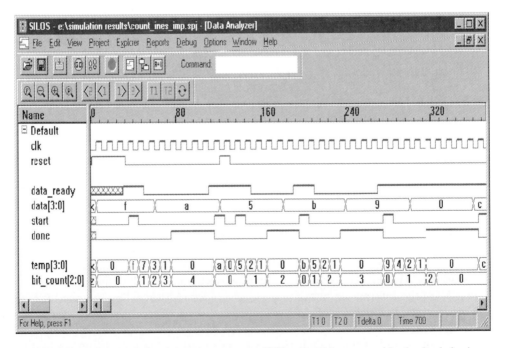

FIGURE 6-67 Simulation results for *count_ones_IMP*, an implicit state machine having behavior equivalent to a nonstatic loop.

End of Example 6.43

6.14   Design Traps to Avoid

In general, avoid referencing the same variable in more than one cyclic (*always*) behavior. When variables are referenced in more than one behavior, there can be races in
the software, and the postsynthesis simulated behavior may not match the presynthesis
behavior. Never assign value to the same variable in multiple behaviors.

6.15   Divide and Conquer: Partitioning a Design

VLSI circuits commonly contain more gates (several million) than a synthesis tool can
accommodate. It is standard practice to partition such circuits hierarchically into small
functional units, which have a manageable complexity. This partitioning is done top-
down, across one or more layers of hierarchy. Synthesis tools commonly synthesize circuits with about 10,000 to 50,000 gates, but beyond that the return diminishes as the
run-time becomes large. Following decomposition, the lowest-level modules of the design can be synthesized individually. It is easier to modify the lower-level modules to
make the design more amenable to synthesis, then to work with large modules.

   Structural modeling composes a circuit by connecting primitive gates to create a
specified functionality (e.g., an adder) just as parts are connected on a chip or a circuit
board. But gate-level models are not necessarily the most convenient or understandable models of a circuit, especially when the design involves more than a few gates.
Many modern ASICs have several million gates on a single chip! Also, truth tables become unwieldy when the circuit has several inputs, limiting the utility of user-defined
primitives. Architectural partitioning forms a structural model, but the functional units
in the architecture have much more complexity than basic combinational logic gates,
and are modeled behaviorally.

   Partitioning is not done randomly. Skill and experience play a role, and the designer's choice of hierarchical boundaries can have a strong impact on the quality of
the synthesis product, and the cost of the effort. Partitioning the design into smaller
functional units can improve the readability of the description, improve the synthesis
result, shorten the optimization cycle, and simplify the synthesis process.

   In general, a design should be partitioned along functional lines into smaller
functional units, each with a common clock domain, and each of which is to be verified
separately. The hierarchy of the design should separate the clock domains, thereby
clarifying the interaction between multiple clocks and revealing the need for synchronizer circuits. The logic in each clock domain can be verified separately, before integrating
the system.

   Functionally related logic should be grouped within a partition, so that the synthesis tool will be able to exploit opportunities for sharing logic, with a minimum of
routing between blocks. If a module is used in multiple places in the design, it should be
optimized separately for area and then instantiated as needed. This strategy will result
in an overall design that is very efficient in its use of area.

It is also recommended that a module contain no more than one state machine. This will allow the synthesis tool to optimize the logic for a machine without the influence of extraneous logic. Logic in different clock domains (e.g., with interacting state machines) should be encapsulated in separate blocks of the partition. Synchronizers should be used where signals cross between the domains.

The partition of a design should group registers and their logic, so that their control logic might be implemented efficiently. Otherwise, splitting registers and logic across boundaries of the partition might lead to extra/duplicate control logic. Place the combinational logic driving the datapath of a register in the same module as the destination register. Likewise, any glue logic between module boundaries should be included within a module. If glue logic sits outside the modules, it cannot be absorbed by either of them.

Module boundaries are preserved in synthesis (i.e., optimized separately) so combinational logic should not be distributed between modules. Placing the logic in a single module will allow the synthesis tool to achieve the maximum exploitation of common logic. Do not include clocks trees, input–output pads, and test registers (see Chapter 11) in a design that is to be synthesized. Add them to the design after synthesis.

REFERENCES

1. De Micheli G. *Synthesis and Optimization of Digital Circuits*. New York: McGraw-Hill, 1994.
2. Gajski D, et al. *High-Level Synthesis*. Boston: Kluwer, 1992.
3. Bartlett K, et al. "Multilevel Logic Minimization Using Implicit Don't-Cares." *IEEE Transactions on Computer Aided Design of Integrated Circuits, CAD-5*, 723–740, 1986.
4. Brayton RK, et al. "MIS: A Multiple-Level Interactive Logic Optimization System." *IEEE Transactions on Computer-Aided Design of Integrated Circuits and Systems, CAD-6*, 1062–1081, 1987.
5. Brayton RK, et al. *Logic Minimization Algorithms for VLSI Synthesis*. Boston: Kluwer, 1984.
6. Katz RH. *Contemporary Logic Design*. Redwood City, CA: Benjamin Cummings, 1994.
7. Wakerly JF. *Digital Design Principles and Practices*, 3rd ed. Upper Saddle River, NJ: Prentice-Hall, 2000.
8. Ciletti MD. *Modeling, Synthesis and Rapid Prototyping with the Verilog HDL*. Upper Saddle River, NJ: Prentice-Hall, 1999.

PROBLEMS

1. Synthesize the universal shift register that was presented in Example 5.45. Verify that the waveforms produced by the synthesized circuit match those of the behavioral model.
2. Synthesize *NRZ_2_Manchester_Mealy* (see Section 6.6.2) and verify that the postsynthesis simulation results match those shown in Figure 6-34 for the behavioral model. Note the time delays that result from the physical cells.

3. Synthesize *NRZ_2_Manchester_Moore* (see Section 6.6.3) and verify that the postsynthesis simulation results match those shown in Figure 6-37 for the behavioral model. Note the time delays that result from the physical cells.

4. The sequence recognizers described by the ASMD charts in Figure 6-38 are *nonresetting*—they assert after three successive 1s are received and continue to assert until a 0 is detected. Develop ASMD charts for resetting Mealy and Moore machines that will detect three successive 1 in a serial bit stream, assert their output, and then return to the reset state, *S_idle*, where the outputs are de-asserted, before processing additional bits. Develop and verify a Verilog model of each machine. Synthesize the machines and verify that the functionality of each synthesized machine matches that of its behavioral model.

5. Develop ASMD charts for resetting Mealy and Moore machines that will detect the pattern 10101010_2 in a serial bit stream (with the LSB arriving first), assert their output, and then return to the reset state, *S_idle*, where the outputs are de-asserted, before processing additional bits. Develop and verify a Verilog model of each machine. Synthesize the machines and verify that the functionality of each synthesized machine matches that of its behavioral models.

6. Develop ASM charts for nonresetting Mealy and Moore machines that will detect the pattern 0010_2 in a serial bit stream, with the LSB arriving first. Synthesize the machines and verify that the functionality of each synthesized machine matches that of its behavioral model.

7. Develop ASM charts for nonresetting Mealy and Moore machines that will detect the patterns 0111_2 or 1000_2 in a serial bit stream, with the LSB arriving first. Synthesize the machines and verify that the functionality of each synthesized machine matches that of its behavioral model.

8. Develop an ASMD chart for a nonresetting sequential machine that implements the majority function. The machine is to assert *D_out* if the serial input, *D_in*, contains two or more 1s in the last 3 bits. Synthesize the machines and verify that the functionality of each synthesized machine matches that of its behavioral model.

9. Critique the following code.

```
module clock_Prog (clk, Pulse_Width, Latency, Offset);
  input Pulse_Width, Latency, Offset;
  output clk;
  reg clk, Pulse_Width, Latency, Offset;
  parameter Pulse_Width = 5;
  parameter Latency = 5;
  parameter Offset = 10;
  parameter a_cycle = Pulse_Width;
  //parameter max_time=1000;

  initial
    clk = 0;
  always begin
    #a_cycle clk = ~clk;
    end
    //initial
    // #max_time $finish;
endmodule
```

10. Explain whether the circuit described below implements combinational logic with, or without, a latched output.

```
module or4_something(y, x_in);
    parameter        word_length = 4;
    output                          y;
    input            [word_length - 1: 0]        x_in;

    reg                             y;
    integer                         k;

    always @ x_in begin
      y = 0;
      if (x_in[0] == 1) y = 1;
      else if (x_in[1] == 1) y = 1;
      else if (x_in[2] == 1) y = 1;
    end
endmodule
```

11. The machine *Seq_Rec_3_1s_Mealy* in Example 6.26 uses a one-hot code for the state assignment. Synthesize an alternative machine using a simple binary code. Compare the new machine to the original machine. Discuss the tradeoffs.

12. The machine *Seq_Rec_3_1s_Mealy* in Example 6.26 partially decodes the possible one-hot state codes and assigns *next_state = S_idle* by default. Synthesize an alternative machine using the *dont_care_state* default assignment. Compare the result to the original machine. Discuss the tradeoffs.

13. Synthesize *count_ones_b0, count_ones_b1* and count_ones_d (see Examples 6.38 and 6.41) and compare the results.

14. Use (a) **while**, and (b) **forever** loop to develop synthesizable implementations of *count_ones_SD* (see Example 6.41). Will any of these synthesize?

15. Develop, verify, and synthesize *count_ones_DPC*, a sequential machine that counts the 1s in a word of *data*. View the machine's datapath as consisting of two storage registers, *bit_count* and *temp*, and a controller that manipulates the datapath to form *bit_count* and *temp*. The machine is to suspend execution as soon as the last 1 is processed in the word.

16. Develop, verify, and synthesize *count_gray_bin*, a 4-bit counter that can count in Gray or in binary code, depending on an input mode.

17. Design and synthesize a sequential machine whose inputs are *clk* and *reset*, and whose outputs are *clk_by_6* and *clk_by_10*, (i.e., clock divider outputs that divide *clk* by 6 and 10, respectively).

18. Determine whether a synthesis tool detects and removes equivalent states from the sequential machine described by the STG in Figure P6-18.

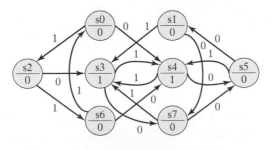

FIGURE P6-18

19. Synthesize *pipe_2stage*, described by the ASMD chart in Example 5.39.
20. Synthesize a circuit that will detect an illegal BCD-encoded word.
21. Digital switches are used in telephone central offices to sample local analog voice signals, convert them to digital signals, and multiplex them with other similar signals for digital transmission on the global phone network [7]. Figure P6-21(a) shows a configuration in which analog signals are sampled at a rate of 8000 samples per second, with each sample represented as an 8-bit word. To achieve a substantial savings in copper wire and other circuitry, thirty-two 64-Kbps voice channels are multiplexed onto a serial channel having a bandwidth

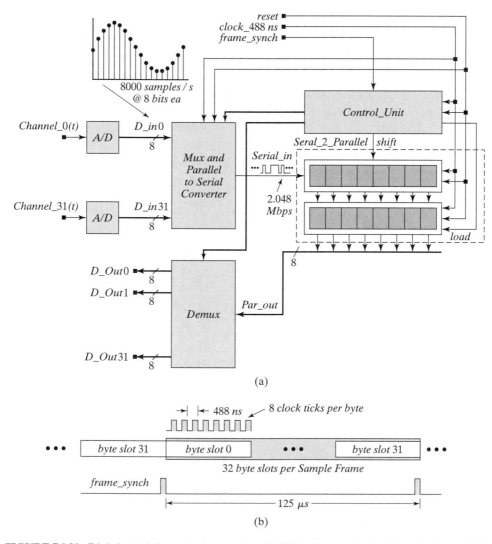

(a)

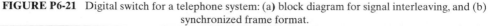

(b)

FIGURE P6-21 Digital switch for a telephone system: (a) block diagram for signal interleaving, and (b) synchronized frame format.

of 2.048 Mbps. Each bit of a sample has a time of 488 ns, and the words of the sampled signals are interleaved in a 32-byte frame, with each frame having a period of 125 μs, as shown in Figure P6-21(b). A frame is synchronized by a pulse of *frame_synch*, which occurs in the bit time immediately before the first byte slot of the frame. An 8-bit shift register receives the digitally encoded bits of *Serial_in*, from least significant bit to most significant bit, in sequence, then the byte is loaded into a holding register that drives a demultiplexer. Develop, verify (before and after synthesis), and synthesize a Verilog module that encapsulates the functionality shown in Figure P6-21(a), where the outputs of the A/D converters are inputs to a module that interleaves the sample bytes, with separate submodules for the control unit, the mux, the demux, the parallel to serial converter, and the serial to parallel converter. Define additional interface signals as needed to complete the design. Model the multiplexer so that its outputs will be registered. Carefully document your work.

22. Synthesize a Verilog description of the combinational logic described by the following Boolean function. Compare the schematic of the synthesized circuit to (a) that of the original circuit and to (b) a simplified version of the function obtained by using Karnaugh maps.

$f(a, b, c, d) = \Sigma\, m(0, 2, 5, 7, 8, 10, 13, 15)$

23. Develop a behavioral model that implements the functionality described by *Divide_by_11* (see Problem 4.19). Synthesize the circuit and compare the result to the circuit synthesized from the structural model shown in Figure P4-19.

24. Under what conditions will a synthesis tool create combinational logic?

25. Under what conditions will a synthesis tool create a circuit that implements a transparent latch?

26. Under what conditions will a synthesis tool create an edge-triggered sequential circuit?

27. Discuss how to describe a synchronous reset condition using Verilog.

28. Synthesize and verify a cell-based implementation of the circuit described by *compare_4_32_CA* (see Problem 12 in Chapter 5).

29. Synthesize and verify a cell-based implementation of the circuit described by *compare_4_32_ALGO* (see Problem 13 in Chapter 5).

30. Synthesize and verify a cell-based implementation of the ring counter described in Problem 24a in Chapter 5.

31. Synthesize and verify a cell-based implementation of the ring counter described in Problem 24b in Chapter 5.

32. Synthesize and verify a cell-based implementation of the sequential machine described in Problem 26 in Chapter 5.

33. Synthesize and verify a cell-based implementation of the 8-bit ALU described in Problem 27 of Chapter 5.

34. Synthesize and verify a cell-based implementation of the sequential machine described in Problem 28 in Chapter 5.

35. Synthesize and verify a cell-based implementation of the sequential machine described by the ASMD chart in Figure 5.24.

36. Synthesize and verify a cell-based implementation of the keypad scanner described in Problem 32 in Chapter 5.

37. Synthesize and verify a cell-based implementation of the programmable pattern generator described in Problem 33 in Chapter 5.

38. Synthesize and verify a cell-based implementation of *Hamming_Encoder_MD3*, described in Problem 35 in Chapter 5.

39. Synthesize and verify a cell-based implementation of the synchronizer circuit described in Problem 36 in Chapter 5.

40. Synthesize and verify a cell-based implementation of the transparent latch described in Problem 37 in Chapter 5.

41. Develop, verify, and synthesize *Binary_Counter_Imp*, an implicit-state machine implementing a 4-bit counter by executing a register transfer operation ($count <= count + 1$) in every clock cycle, depending on an input signal *enable*.

42. Synthesize *count_ones_IMP_gates*, and verify that the functionality of the synthesized circuit matches that of the behavioral model, *count_ones_IMP*, given in Example 6.42.

43. A token ring local area network (LAN) consists of a set of computers organized in a ring topology, with each machine connected to a bus by a LAN adapter (see Figure P7-3). LAN-based machines communicate by transmitting and receiving packets of bits containing encoded source and destination addresses, and a message. The LAN adaptor executes handshaking protocols by decoding the address in a received packet and determining whether its host machine is the recipient of the message or whether to pass the message on to the next machine on the LAN. It also transmits packets from the host machine to a target machine. The protocol for the network specifies how the start of a packet will be recognized, and how the source and destination addresses are encoded. The adaptor must recognize the start sequence, decode the address, and acknowledge that it is the recipient.

The protocol for the token ring shown in Figure P6-43 marks the start of a packet by two successive 0s. Suppose a LAN adapter is hardwired to recognize either the address 100_2 or 010_2. The adapter is to assert a Moore-type output, *P_IN*, if a packet arrives and decodes to the adaptor's addresses. Develop an ASM chart for the packet detection and address decoding circuit, write and verify a Verilog model of the machine, and synthesize the machine. Note whether the synthesis process has detected and/or removed any equivalent states. Explore alternative state assignments.

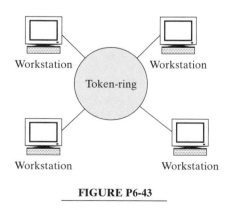

FIGURE P6-43

44. The rotor of a conventional motor rotates continuously when power is applied. The rotor of a stepper motor can be controlled to move a prescribed number of steps or to rotate continuously at a prescribed speed. Stepper motors are widely used in computer numerical control, where they control precision machine tools, and in personal computers, where they control floppy disk drives and control the paper feed of line printers, and in numerous other applications in which accurate positioning without feedback control is a requirement. A simple stepper motor has the configuration shown in Figure P6-44, in which a set of fixed stator coils form electromagnets located symmetrically about the perimeter of the motor assembly. A permanently magnetized rotor[32] is free to rotate on a central axis under the influence of the magnetic fields that are created when the coils are energized. The angular position of the rotor is controlled by sequentially energizing the coils to create a rotating magnetic field that exerts force on the poles of the rotor magnets. The stator coils can be energized sequentially to make the rotor move to a desired angular position or to rotate continuously at a selected speed. The angular acceleration of the rotor can be controlled by varying the spacing between the pulses that energize the coils. The motor in Figure P6-44 has only four stator coils, so the rotor can be held stable in any of eight different positions (states), with an angular resolution of 45 degrees, (the motor steps in increments of 45 degrees). Increasing the number of coils increases the resolution of the position of the rotor.[33] The coils are energized individually, to hold the rotor at the position of the energized stator, or in an

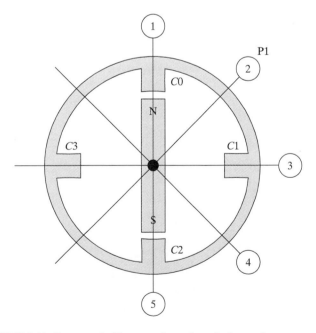

FIGURE P6-44 Rotor and stNator configuration of a four-pole stepper motor.

---

[32]Another class of stepper motors has a variable reluctance.
[33]Commercial motors have angular resolution ranging from 0.72 to 90 degrees.

adjacent/pairwise fashion, to orient the stator at an angle midway between two adjacent stators. For example, if coils *C0* and *C1* are both energized, the rotor will move from the position of *C0* to the position P1. Successively energizing (1) *C0*, (2) *C0* and *C1*, (3) *C1*, (4) *C1* and *C2*, and (5) *C2* will rotate the rotor clockwise to the position of *C2*. The speed at which the motor rotates is determined by the time interval between the inputs. Design and synthesize a Moore-type state machine to control the direction, speed, and angular acceleration of the motor's rotor, with inputs that program the machine to (1) advance by a specified number of steps in a given direction, (2) spin at a specified rate, and (3) accelerate from rest to a specified angular velocity (with stepwise increments in speed). Pulse rates from 1500 to 2500 pulses per second are allowed.

45. Explain the consequences of replacing procedural assignments by non blocking assignments in Example 6.39 and Example 6.41.

CHAPTER 7

Design and Synthesis of Datapath Controllers

Digital systems range from those that are control-dominated to those that are data-dominated. Control-dominated systems are reactive systems, responding to external events; data-dominated systems are shaped by the requirements of high-throughput data computation and transport, as in telecommunications and signal processing [1]. Sequential machines are commonly classified and partitioned into datapath units and control units.

Most datapaths include arithmetic units, such as arithmetic and logic units (ALUs), adders, multipliers, shifters, and digital signal processors, but some do not, such as graphics coprocessors. Datapath units consist of computational resources (e.g., ALUs and storage registers), logic for moving data through the system and between the computation units and the internal registers, and datapaths for moving data to and from the external environment. The datapath unit in Figure 7-1 is controlled by a finite-state machine (FSM) that coordinates the execution of instructions that perform operations on the datapath. Datapath units are characterized by repetitive operations on different sets of data, as in signal processing, image processing, and multimedia [2]. Architectures that are dominated by control units will generally have a significant amount of random (irregular) logic, together with some regular structures, like multiplexers for steering signals, and comparators.

7.1 Partitioned Sequential Machines

Partitioning a sequential machine into a datapath and a controller clarifies the architecture and simplifies the design of the system. The process by which the machine is designed is said to be application-driven, because the sequence of operations that must

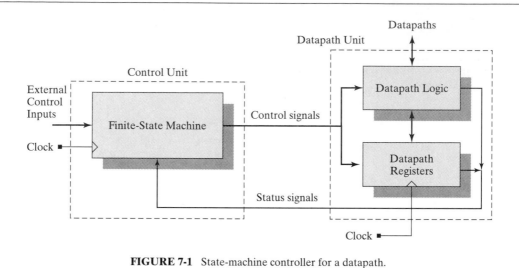

FIGURE 7-1   State-machine controller for a datapath.

be performed by the datapath unit in a particular application determines the resources composing its architecture, the set of instructions that must be executed by the datapath, and, ultimately, the FSM that controls the datapath.

The sequence of steps in an application-driven design process is illustrated in Figure 7-2. Once the architecture of the datapath unit has been selected to support the instruction set of an application, sequences of operations (control states) that support the instruction set can be identified. The control states are used to schedule assertions of the signals that control the movement and manipulation of data as the machine executes instructions. Then an FSM can be designed to generate the control signals. In this section we will illustrate the design of datapath controllers for some simple functional units, to prepare for the design of a stored-program reduced instruction-set computer in the next section.

Control units orchestrate, coordinate, and synchronize the operations of datapath units. The control unit of a machine generates the signals that load, read, and shift the contents of storage registers; fetch instructions and data from memory; store data in memory; steer signals through muxes; control three-state devices; and control the operations of ALUs and other complex datapath units. In synchronous machines, a common clock synchronizes the activities of the controller and datapath functional units. Note that the control unit in Figure 7-1 is implemented as an FSM, and is itself controlled by external input signals and by status signals from the datapath unit. The FSM produces the signals that control the operation of the datapath unit.

Datapath units are commonly described by dataflow graphs; control units are commonly modeled by state transition graphs and/or algorithmic-state machine (ASM) charts for FSMs. Partitioned sequential machines can be modeled by an FSM and datapath (FSMD), a combined control-dataflow graph, which expresses datapath operations in the context of a state-transition graph (STG). We favor using an ASM and datapath (ASMD) chart, which likewise links an ASM chart for a control unit to the operations of the datapath that it controls.

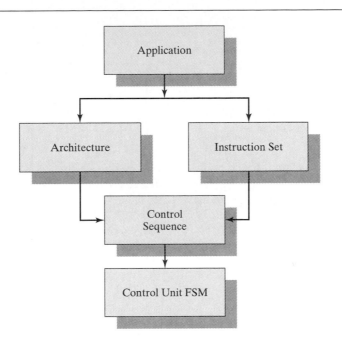

FIGURE 7-2   Application-driven architecture, instruction set, and control sequence for a datapath controller.

7.2   Design Example: Binary Counter

Consider a synchronous 4-bit binary counter that is to be incremented by a count of 1 at each active edge of the clock, and whose count is to wrap around to 0 when the count reaches 1111_2. We could describe the counter by an implicit state machine, *Binary_Counter_Imp*, executing a register transfer operation (*count* $<=$ *count + 1*) conditionally, in every clock cycle, depending on *enable*, and then synthesize a hardware realization directly.[1] Other approaches are possible. One is to partition the machine into an architecture of separate datapath and control units, as shown in Figure 7-3 for *Binary_Counter_Arch*.

The functional elements of the architecture of the datapath unit consist of (1) a 4-bit register to hold *count*, (2) a mux that steers either *count* or the sum of *count* and 0001_2 to the input of the register, and (3) a 4-bit adder to increment *count*. The signal *enable* must be asserted for counting to occur, and the signal *rst* overrides all activity and drives *count* to a value of 0000_2. The input *rst* must be de-asserted and *enable* must be asserted for the machine to begin counting and to continue counting. The control unit for this simple machine passes *enable* directly to the datapath unit.

Now view the counter itself as an explicit-state machine, *Binary_Counter_STG*, having state *count* (the contents of the counter), and inputs *enable, clk*, and *rst*. The

[1]See Problem 41 in Chapter 6.

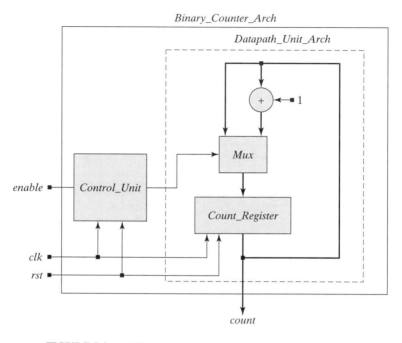

FIGURE 7-3 Architecture for a synchronous 4-bit binary counter.

simplified STG of the machine is shown in Figure 7-4, with *count* entered as the state label within each node. (The reset-directed arcs are not shown, nor are the arcs that return to the current state if *enable* is not asserted.) The STG can be used to develop an explicit state machine with two cyclic behaviors, one defining the next-state/output combinational logic, and the other synchronizing the state transitions. Note that this approach to the design of the machine can become unwieldy, because the size of the graph increases with the width of the datapath. It is often true that the number of states of the datapath registers is enormous compared to the number of states of the control unit. Partitioning the design eliminates the need to consider the state of the datapath registers, except to generate status signals that are fed back to the control unit.

The preceding three views of a 4-bit binary counter illustrate how partitioning a sequential machine into a datapath and a controller can reduce the size of the state that needs to be considered in the design and simplify the control unit of the machine. In this example, the implicit state machine has the simplest description; it suppresses structural detail, leaving it to the synthesis tool. The partitioned machine has the most structural detail, a simple controller, and a datapath register whose state does not influence the design; the STG-based approach required a detailed STG and led to a state machine with 16 states, because the state of the machine was the state of the register holding *count*. The relative complexity of the Verilog models and the synthesized hardware are tradeoffs between the equivalent, but alternative machines.[2]

---

[2]See Problems 10 and 11 at the end of this chapter.

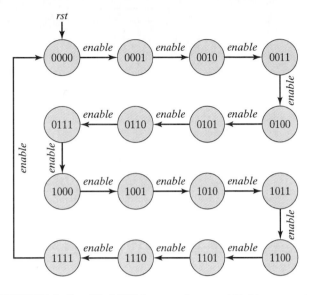

FIGURE 7-4 Simplified STG for a synchronous 4-bit binary counter.

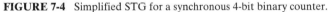

Another view of the binary counter is based on the counter's *activity*. The machine described by the ASM chart in Figure 7-5, *Binary_Counter_ASM*, has one state, *S_running*.[3] At every clock, *enable* is tested and a transition is made back to *S_running*; if *enable* is asserted, a conditional register operation that increments the counter is executed concurrently with the state transition. Even though the ASM chart is simple, it describes the activity of more complex counters and other single-cycle machines that have a different relationship governing the register operations. For example, the function *next_count* could describe a Johnson counter.[4] Note that the description does not require an explicit state register because it remains in the same state. In fact, it reduces to the implicit state machine described above.

A fifth approach to designing the binary counter partitions the machine into a control unit and a datapath unit, but designs a register transfer level (RTL) behavioral model for the datapath unit, rather than a structural model (as we did in Figure 7-3). This style separates the design of the control unit from the design (and synthesis) of the datapath unit and simplifies the description of the datapath unit. It separates the unit that determines *what* happens from the unit that determines *when* it happens. This style might seem like overkill for this counter (and it is), but the style is critical to successful design and synthesis of more complex machines, and is easily implemented in Verilog. Figure 7-6 shows an ASMD chart for *Binary_Counter_Part_RTL*, a counter partitioned into a datapath unit and a control unit, with signal *enable_DP* linking the controller to the datapath. The model's control units passes *enable* through to the datapath unit. The state machine of the controller consists of only the pass-through logic.

---

[3]This machine is actually the same as *Binary_Counter_Imp* mentioned above.
[4]See Problem 18 in Chapter 5.

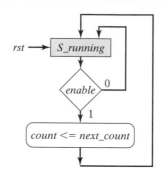

FIGURE 7-5   ASM chart for a synchronous 4-bit binary counter.

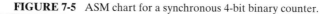

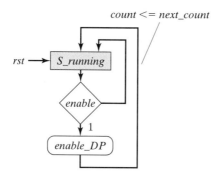

FIGURE 7-6   ASMD chart for a datapath unit, a synchronous 4-bit binary counter, controlled by a state machine.

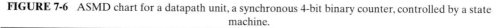

Example 7.1

The Verilog description *Binary_Counter_Part_RTL* has two nested modules: *Control_Unit*, and *Datapath_Unit*. The datapath has been described with flexibility to implement codes for other counters, independently of the control unit. The simulation results in Figure 7-7 show that counting begins at the first rising edge of *clk* after *enable* is asserted, that counting continues while *enable* is asserted, and that *enable_DB* replicates *enable*.

```
module Binary_Counter_Part_RTL (count, enable, clk, rst);
  parameter           size = 4;
  output   [size -1: 0]   count;
  input                enable;
  input                clk, rst;
  wire                 enable_DP;

Control_Unit  M0 (enable_DP, enable, clk, rst);
Datapath_Unit M1 (count, enable_DP, clk, rst);
endmodule
```

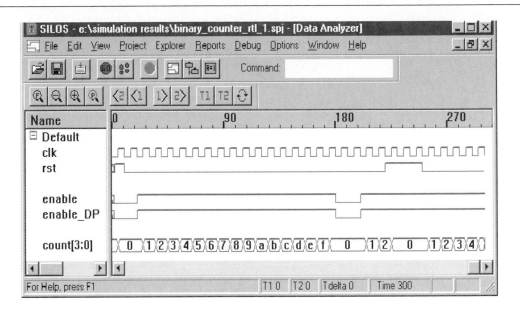

FIGURE 7-7 Simulation results for *Binary_Counter_RTL*, a synchronous 4-bit binary counter controlled by a state machine.

```
module Control_Unit  (enable_DP, enable, clk, rst);
  output       enable_DP;
  input        enable;
  input        clk, rst;                 // Not needed
  wire         enable_DP = enable;       // pass through

endmodule

module Datapath_Unit (count, enable, clk, rst);
  parameter    size = 4;
  output       [size-1: 0] count;
  input        enable;
  input        clk, rst;
  reg          count;
  wire         [size-1: 0] next_count;

  always @ (posedge clk)
   if (rst == 1) count <= 0;
     else if (enable == 1) count <= next_count(count);

  function     [size-1: 0]       next_count;
    input      [size-1: 0]       count;
    begin
     next_count = count + 1;
    end
  endfunction
endmodule
```

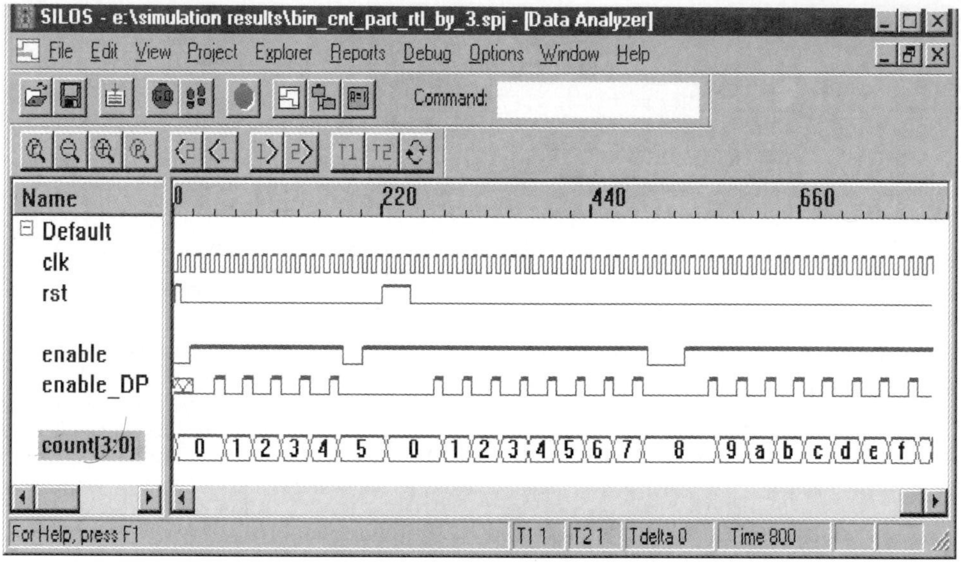

FIGURE 7-8 Simulation results for *Binary_Counter_Part_RTL_by_3* with a control unit to increment the datapath counter every third cycle.

Next, we will redesign the counter to form *Binary_Counter_Part_RTL_by_3*, which increments its count every third clock cycle. Only the control unit must change. One approach is to model the control unit by the implicit Moore machine shown below. The simulated activity of the machine is shown in Figure 7-8. *Caution*: The machine has an interesting feature that calls for a more careful approach.[5]

```verilog
module Control_Unit_by_3  (enable_DP, enable, clk, rst);
  output        enable_DP;
  input         enable;
  input         clk, rst;            // Not needed

  reg           enable_DP;

  always begin: Cycle_by_3
    @ (posedge clk) enable_DP  <= 0;
    if ((rst == 1) || (enable != 1)) disable Cycle_by_3; else
      @ (posedge clk)
        if ((rst == 1) || (enable != 1)) disable Cycle_by_3; else
          @ (posedge clk)
            if ((rst == 1) || (enable != 1)) disable Cycle_by_3;
              else enable_DP <= 1;
  end // Cycle_by_3
endmodule
```

End of Example 7.1

---

[5]See Problem 12 at the end of this chapter.

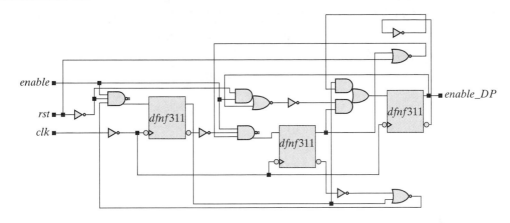

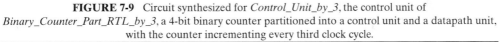

FIGURE 7-9 Circuit synthesized for *Control_Unit_by_3*, the control unit of *Binary_Counter_Part_RTL_by_3*, a 4-bit binary counter partitioned into a control unit and a datapath unit, with the counter incrementing every third clock cycle.

The Verilog module *Binary_Counter_Part_RTL_by_3*, with the modified control unit, is synthesizable. We anticipate that two flip-flops will be needed to implement the implicit-state machine of the control unit, because the state evolves through three embedded clock cycles. One flip-flop will be needed to register *enable_DP*, and four flip-flops will be needed to implement the register holding *count* in the datapath unit. The synthesis results in Figure 7-9 confirm this use of resources for the control unit. However, the postsynthesis behavior of *Binary_Counter_Part_RTL_by_3* is problematic.[6]

7.3 Design and Synthesis of a RISC Stored-Program Machine

Reduced instruction-set computers (RISC) are designed to have a small set of instructions that execute in short clock cycles, with a small number of cycles per instruction. RISC machines are optimized to achieve efficient pipelining of their instruction streams [2]. In this section we will model a simple RISC machine. Our companion website (www.prenhall.com/ciletti) includes the machine's source code and an assembler that can be used to develop programs for student projects. The machine also serves as a starting point for developing architectural variants and a more robust instruction set.

Designers make high-level tradeoffs in selecting an architecture that serves an application. Once an architecture has been selected, a circuit that has sufficient performance (speed) must be synthesized. Hardware description languages (HDLs) play a key role in this process by modeling the system and serving as a descriptive medium that can be used by a synthesis tool.

As an example, the overall architecture of a simple RISC is shown in Figure 7-10. *RISC_SPM* is a stored-program RISC-architecture machine [3, 4]—its instructions are contained in a program stored in memory.

---

[6]See Problem 12 at the end of this chapter.

RISC_SPM

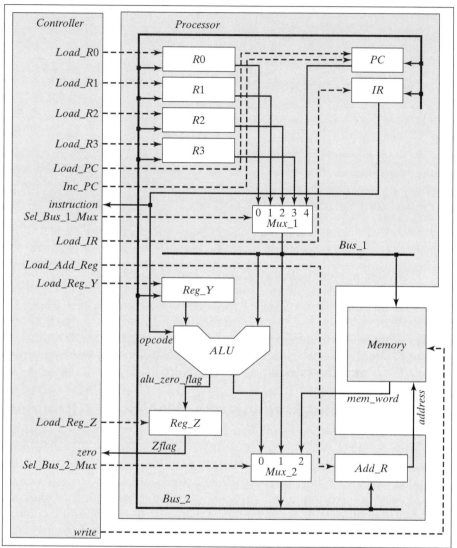

FIGURE 7-10 Architecture of *RISC_SPM*, an RISC stored-program machine (SPM).

The machine consists of three functional units: a processor, a controller, and memory. Program instructions and data are stored in memory. In program-directed operation, instructions are fetched synchronously from memory, decoded, and executed to (1) operate on data within the arithmetic and logic unit (*ALU*), (2) change the contents of storage registers, (3) change the contents of the program counter (*PC*), instruction register (*IR*) and the address register (*ADD_R*), (4) change the contents of memory, (5) retrieve data and instructions from memory, and (6) control the movement of data on the system busses. The instruction register contains the instruction that is currently being executed; the program counter contains the address of the next instruction to be executed; and the address register holds the address of the memory location that will be addressed next by a read or write operation.

7.3.1   RISC SPM: Processor

The processor includes registers, datapaths, control lines, and an ALU capable of performing arithmetic and logic operations on its operands, subject to the opcode held in the instruction register. A multiplexer, *Mux_1*, determines the source of data that is bound for *Bus_1*, and a second mux, *Mux_2*, determines the source of data bound for *Bus_2*. The input datapaths to *Mux_1* are from four internal general-purpose registers (*R0, R1, R2, R3*), and from the PC. The contents of *Bus_1* can be steered to the ALU, to memory, or to *Bus_2* (via *Mux_2*). The input datapaths to *Mux_2* are from the *ALU*, *Mux_1*, and the memory unit. Thus, an instruction can be fetched from memory, placed on *Bus_2*, and loaded into the instruction register. A word of data can be fetched from memory, and steered to a general-purpose register or to the operand register (*Reg_Y*) prior to an operation of the ALU. The result of an *ALU* operation can be placed on *Bus_2*, loaded into a register, and subsequently transferred to memory. A dedicated register (*Reg_Z*) holds a flag indicating that the result of an *ALU* operation is 0.[7]

7.3.2   RISC SPM: ALU

For the purposes of this example, the ALU has two operand datapaths, *data_1* and *data_2*, and its instruction set is limited to the following instructions:

Instruction	Action
ADD	Adds the datapaths to form *data_1* + *data_2*
SUB	Subtracts the datapaths to form *data_1* − *data_2*
AND	Takes the bitwise-and of the datapaths, *data_1* & *data_2*
NOT	Takes the bitwise Boolean complement of *data_1*

7.3.3   RISC SPM: Controller

The timing of all activity is determined by the controller. The controller must steer data to the proper destination, according to the instruction being executed. Thus, the design of the controller is strongly dependent on the specification of the machine's ALU and datapath resources and the clocking scheme available. In this example, a single clock will be used, and execution of an instruction is initiated on a single edge of the clock

---

[7]This can be used to monitor a loop index.

(e.g., the rising edge). The controller monitors the state of the processing unit and the instruction to be executed and determines the value of the control signals. The controller's input signals are the instruction word and the zero flag from the *ALU*. The signals produced by the controller are identified as follows:

Control Signal	Action
Load_Add_Reg	Loads the address register
Load _PC	Loads *Bus_2* to the program counter
Load_IR	Loads *Bus_2* to the instruction register
Inc_PC	Increments the program counter
Sel_Bus_1_Mux	Selects among the *Program_Counter, R0, R1, R2,* and *R3* to drive *Bus_1*
Sel_Bus_2_Mux	Selects among *Alu_out, Bus_1,* and memory to drive *Bus_2*
Load_R0	Loads general-purpose register *R0*
Load_R1	Loads general-purpose register *R1*
Load_R2	Loads general-purpose register *R2*
Load_R3	Loads general-purpose register *R3*
Load_Reg_Y	Loads *Bus_2* to the register *Reg_Y*
Load Reg_Z	Stores output of *ALU* in register *Reg_Z*
write	Loads *Bus_1* into the *SRAM* memory at the location specified by the address register

The control unit (1) determines when to load registers, (2) selects the path of data through the multiplexers, (3) determines when data should be written to memory, and (4) controls the three-state busses in the architecture.

7.3.4 RISC SPM: Instruction Set

The machine is controlled by a machine language program consisting of a set of instructions stored in memory. So, in addition to depending on the machine's architecture, the design of the controller depends on the processor's instruction set (i.e., the instructions that can be executed by a program). A machine language program consists of a stored sequence of 8-bit words (bytes). The format of an instruction of *RISC_SPM* can be long or short, depending on the operation.

Short instructions have the format shown in Figure 7-11(a). Each short instruction requires 1 byte of memory. The word has a 4-bit opcode, a 2-bit source register address, and a 2-bit destination register address. A long instruction requires 2 bytes of memory. The first word of a long instruction contains a 4-bit opcode. The remaining 4 bits of the word can be used to specify addresses of a pair of source and destination registers, depending on the instruction. The second word contains the address of the memory word that holds an operand required by the instruction. Figure 7-11(b) shows the 2-byte format of a long instruction.

opcode				source		destination	
0	0	1	0	0	1	1	0

opcode				source		destination	
0	1	1	0	1	0	don't care	don't care
address							
0	0	0	1	1	1	0	1

(a) (b)

FIGURE 7-11 Instruction format of (a) a short instruction, and (b) a long instruction.

The instruction mnemonics and their actions are listed below.

Single-Byte Instruction	Action
NOP	No operation is performed; all registers retain their values. The addresses of the source and destination register are don't-cares, they have no effect.
ADD	Adds the contents of the source and destination registers and stores the result into the destination register.
AND	Forms the bitwise-and of the contents of the source and destination registers and stores the result into the destination register.
NOT	Forms the bitwise complement of the content of the source register and stores the result into the destination register.
SUB	Subtracts the content of the source register from the destination register and stores the result into the destination register.

Two-Byte Instruction	Action
RD	Fetches a memory word from the location specified by the second byte and loads the result into the destination register. The source register bits are don't-cares (i.e., unused).
WR	Writes the contents of the source register to the word in memory specified by the address held in the second byte. The destination register bits are don't-cares (i.e., unused).
BR	Branches the activity flow by loading the program counter with the word at the location (address) specified by the second byte of the instruction. The source and destination bits are don't-cares (i.e., unused).
BRZ	Branches the activity flow by loading the program counter with the word at the location (address) specified by the second byte of the instruction if the zero flag register is asserted.

The *RISC_SPM* instruction set is summarized in Table 7-1.

The program counter holds the address of the next instruction to be executed. When the external reset is asserted, the program counter is loaded with 0, indicating that the bottom of memory holds the next instruction that will be fetched. Under the action of the clock, for single-cycle instructions, the instruction at the address in the program counter is loaded into the instruction register and the program counter is incremented. An instruction decoder determines the resulting action on the datapaths and the ALU. A long instruction is held in 2 bytes, and an additional clock cycle is required to

TABLE 7-1 Instruction set for the *RISC_SPM* machine.

Instr	Instruction Word			Action
	opcode	src	dest	
NOP	0000	??	??	none
ADD	0001	src	dest	dest <= src + dest
SUB	0010	src	dest	dest <= dest − src
AND	0011	src	dest	dest <= src && dest
NOT	0100	src	dest	dest <= ~ src
RD*	0101	??	dest	dest <= memory [Add_R]
WR*	0110	src	??	memory[Add_R] < = src
BR*	0111	??	??	PC <= memory[Add_R]
BRZ*	1000	??	??	PC <= memory [Add_R]
HALT	1111	??	??	Halts execution until reset

* Requires a second word of data; ? denotes a don't-care.

execute the instruction. In the second cycle of execution, the second byte is fetched from memory at the address held in the program counter, then the instruction is completed. Intermediate contents of the ALU may be meaningless when two-cycle operations are being executed.

7.3.5 RISC SPM: Controller Design

The machine's controller will be designed as an FSM. Its states must be specified, given the architecture, instruction set, and clocking scheme used in the design. This can be accomplished by identifying what steps must occur to execute each instruction. We will use an ASM chart to describe the activity within the machine, *RISC_SPM*, and to present a clear picture of how the machine operates under the command of its instructions.

The machine has three phases of operation: *fetch, decode*, and *execute*. Fetching retrieves an instruction from memory, decoding decodes the instruction, manipulates datapaths, and loads registers; execution generates the results of the instruction. The fetch phase will require two clock cycles—one to load the address register and one to retrieve the addressed word from memory. The decode phase is accomplished in one cycle. The execution phase may require zero, one, or two more cycles, depending on the instruction. The *NOT* instruction can execute in the same cycle that the instruction is decoded; single-byte instructions, such as *ADD*, take one cycle to execute, during which the results of the operation are loaded into the destination register. The source register can be loaded during the decode phase. The execution phase of a 2-byte instruction will take two cycles: (for example *RD*), one to load the address register with the second byte, and one to retrieve the word from the memory location addressed by the second byte and load it into the destination register. The controller for *RISC_SPM* has the 11 states listed below, with the control actions that must occur in each state.

S_idle	State entered after reset is asserted. No action.
S_fet1	Load the address register with the contents of the program counter. (*Note*: *PC* is initialized to the starting address by the reset action.) The state is entered at the first active clock after reset is de-asserted, and is revisited after a *NOP* instruction is decoded.
S_fet2	Load the instruction register with the word addressed by the address register, and increment the program counter to point to the next location in memory, in anticipation of the next instruction or data fetch.
S_dec	Decode the instruction register and assert signals to control datapaths and register transfers.
S_ex1	Execute the *ALU* operation for a single-byte instruction, conditionally assert the zero flag, and load the destination register.
S_rd1	Load the address register with the second byte of a *RD* instruction, and increment the *PC*.
S_rd2	Load the destination register with the memory word addressed by the byte loaded in *S_rd1*.
S_wr1	Load the address register with the second byte of a *WR* instruction, and increment the *PC*.
S_wr2	Load the destination register with the memory word addressed by the byte loaded in *S_wr1*.
S_br1	Load the address register with the second byte of a *BR* instruction, and increment the *PC*.
S_br2	Load the program counter with the memory word addressed by the byte loaded in *S_br1*.
S_halt	Default state to trap failure to decode a valid instruction.

The partitioned ASM chart for the controller of *RISC_SPM* is shown in Figure 7-12, with the states numbered for clarity. Once the ASM charts have been built, the designer can write the Verilog description of the entire machine, for the given architectural partition. This process unfolds in stages. First, the functional units are declared according to the partition of the machine. Then their ports and variables are declared and checked for syntax. Then the individual units are described, debugged, and verified. The last step is to integrate the design and verify that it has correct functionality.

The top-level Verilog module *RISC_SPM* integrates the modules of the architecture of Figure 7-10 and will be presented first. Three modules are instantiated: *Processing_Unit*, *Control_Unit*, and *Memory_Unit*, with instance names *M0_Processor*, *M1_Controller*, and *M2_Mem*, respectively. The parameters declared at this level of the hierarchy size the datapaths between the three structural/functional units.

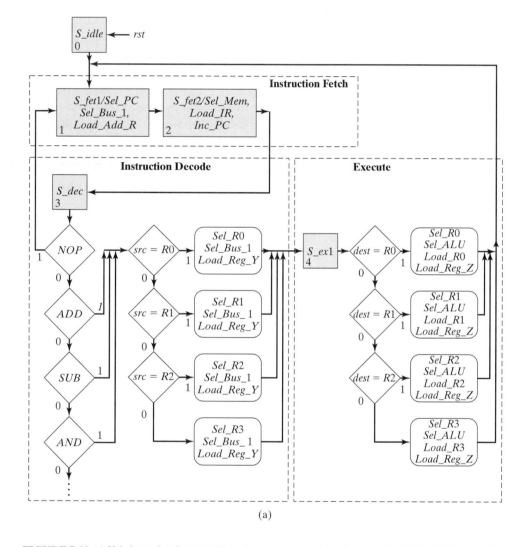

(a)

FIGURE 7-12   ASM charts for the controller of a processor that implements the *RISC_SPM* instruction
set: (a) *NOP, ADD, SUB, AND*, (b) *RD*, (c) *WR*, (d) *BR, BRZ*, and (e) *NOT*.

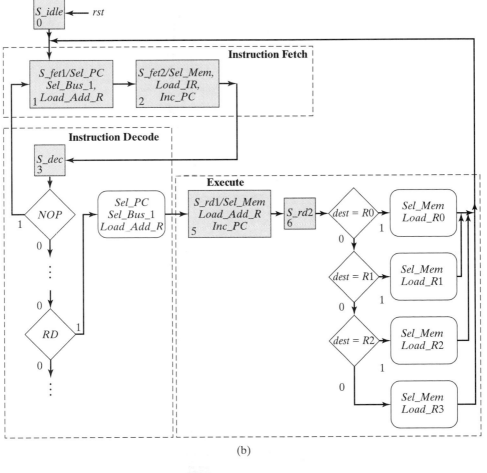

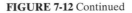

(b)

FIGURE 7-12 Continued

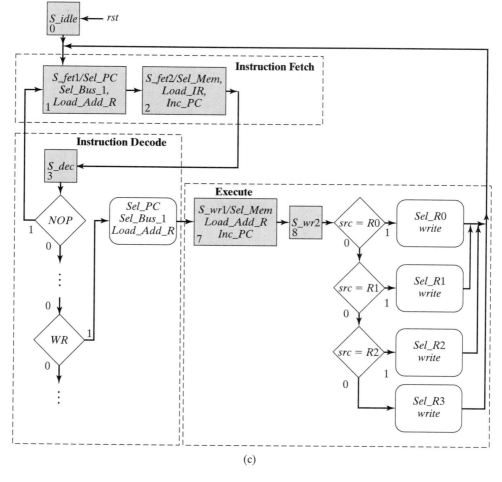

(c)

FIGURE 7-12 Continued

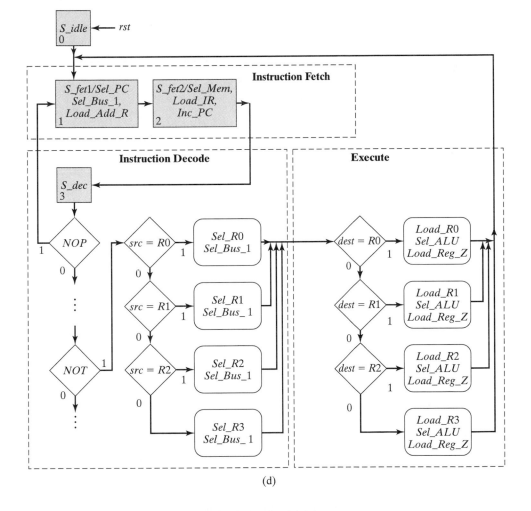

(d)

FIGURE 7-12 Continued

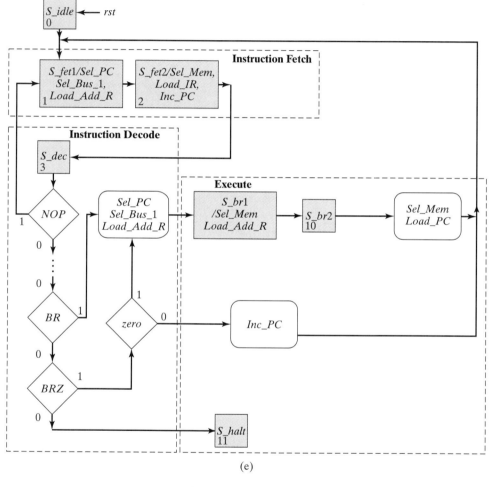

(e)

FIGURE 7-12 Continued

```
module RISC_SPM (clk, rst);
 parameter word_size = 8;
 parameter Sel1_size = 3;
 parameter Sel2_size = 2;
 wire [Sel1_size-1: 0] Sel_Bus_1_Mux;
 wire [Sel2_size-1: 0] Sel_Bus_2_Mux;
 input clk, rst;

 // Data Nets
 wire zero;
 wire [word_size-1: 0] instruction, address, Bus_1, mem_word;

 // Control Nets
 wire Load_R0, Load_R1, Load_R2, Load_R3, Load_PC, Inc_PC, Load_IR;
 wire Load_Add_R, Load_Reg_Y, Load_Reg_Z;
 wire write;

 Processing_Unit M0_Processor
  (instruction, zero, address, Bus_1, mem_word, Load_R0, Load_R1,
  Load_R2, Load_R3, Load_PC, Inc_PC, Sel  Bus_1_Mux, Load_IR,
  Load_Add_R, Load_Reg_Y, Load_Reg_Z,  Sel_Bus_2_Mux, clk, rst);

 Control_Unit M1_Controller (Load_R0, Load_R1, Load_R2, Load_R3, Load_PC,
 Inc_PC, Sel_Bus_1_Mux, Sel_Bus_2_Mux , Load_IR, Load_Add_R,
 Load_Reg_Y, Load_Reg_Z, write, instruction, zero, clk, rst);

 Memory_Unit M2_MEM (
  .data_out(mem_word),
  .data_in(Bus_1),
  .address(address),
  .clk(clk),
  .write(write) );
endmodule
```

The Verilog model of the machine's processor will describe the architecture, register operations, and datapath operations that are represented by the functional units shown in Figure 7-10. The processor instantiates several other modules, which must be declared too.

```
module Processing_Unit (instruction, Zflag, address, Bus_1, mem_word,
 Load_R0, Load_R1, Load_R2, Load_R3, Load_PC, Inc_PC, Sel_Bus_1_Mux,
 Load_IR, Load_Add_R, Load_Reg_Y, Load_Reg_Z,
 Sel_Bus_2_Mux, clk, rst);

 parameter word_size = 8;
 parameter op_size = 4;
 parameter Sel1_size = 3;
 parameter Sel2_size = 2;

 output [word_size-1: 0] instruction, address, Bus_1;
 output                  Zflag;
```

```
      input [word_size-1: 0]   mem_word;
      input                    Load_R0, Load_R1, Load_R2, Load_R3, Load_PC,
                               Inc_PC;
      input [Sel1_size-1: 0]   Sel_Bus_1_Mux;
      input [Sel2_size-1: 0]   Sel_Bus_2_Mux;
      input                    Load_IR, Load_Add_R, Load_Reg_Y, Load_Reg_Z;
      input                    clk, rst;

      wire                     Load_R0, Load_R1, Load_R2, Load_R3;
      wire [word_size-1: 0]    Bus_2;
      wire [word_size-1: 0]    R0_out, R1_out, R2_out, R3_out;
      wire [word_size-1: 0]    PC_count, Y_value, alu_out;
      wire                     alu_zero_flag;
      wire [op_size-1 : 0]     opcode = instruction [word_size-1: word_size-op_size];

      Register_Unit            R0      (R0_out, Bus_2, Load_R0, clk, rst);
      Register_Unit            R1      (R1_out, Bus_2, Load_R1, clk, rst);
      Register_Unit            R2      (R2_out, Bus_2, Load_R2, clk, rst);
      Register_Unit            R3      (R3_out, Bus_2, Load_R3, clk, rst);
      Register_Unit            Reg_Y   (Y_value, Bus_2, Load_Reg_Y, clk, rst);
      D_flop                   Reg_Z   (Zflag, alu_zero_flag, Load_Reg_Z, clk, rst);
      Address_Register         Add_R   (address, Bus_2, Load_Add_R, clk, rst);
      Instruction_Register     IR      (instruction, Bus_2, Load_IR, clk, rst);
      Program_Counter          PC      (PC_count, Bus_2, Load_PC, Inc_PC, clk, rst);
      Multiplexer_5ch          Mux_1   (Bus_1, R0_out, R1_out, R2_out, R3_out,
                                        PC_count, Sel_Bus_1_Mux);
      Multiplexer_3ch          Mux_2   (Bus_2, alu_out, Bus_1, mem_word,
                                        Sel_Bus_2_Mux);
      Alu_RISC                 ALU     (alu_zero_flag, alu_out, Y_value, Bus_1,
                                        opcode);
endmodule

module Register_Unit (data_out, data_in, load, clk, rst);
      parameter                word_size = 8;
      output [word_size-1: 0]  data_out;
      input  [word_size-1: 0]  data_in;
      input                    load;
      input                    clk, rst;
      reg                      data_out;

      always @ (posedge clk or negedge rst)
        if (rst == 0) data_out <= 0; else if (load) data_out <= data_in;
endmodule

module D_flop (data_out, data_in, load, clk, rst);
      output        data_out;
      input         data_in;
      input         load;
      input         clk, rst;
      reg           data_out;
```

```
        always @ (posedge clk or negedge rst)
          if (rst == 0) data_out <= 0; else if (load == 1)data_out <= data_in;
      endmodule

      module Address_Register (data_out, data_in, load, clk, rst);
        parameter word_size = 8;
        output  [word_size-1: 0]  data_out;
        input   [word_size-1: 0]  data_in;
        input                     load, clk, rst;
        reg                       data_out;
        always @ (posedge clk or negedge rst)
          if (rst == 0) data_out <= 0; else if (load) data_out <= data_in;
      endmodule

      module Instruction_Register (data_out, data_in, load, clk, rst);
        parameter word_size = 8;
        output  [word_size-1: 0]  data_out;
        input   [word_size-1: 0]  data_in;
        input                     load;
        input                     clk, rst;
        reg                       data_out;
        always @ (posedge clk or negedge rst)
          if (rst == 0) data_out <= 0; else if (load) data_out <= data_in;
      endmodule

      module Program_Counter (count, data_in, Load_PC, Inc_PC, clk, rst);
        parameter word_size = 8;
        output  [word_size-1: 0]  count;
        input   [word_size-1: 0]  data_in;
        input                     Load_PC, Inc_PC;
        input                     clk, rst;
        reg                       count;
        always @ (posedge clk or negedge rst)
          if (rst == 0) count <= 0; else if (Load_PC) count <= data_in; else if  (Inc_PC)
          count <= count +1;
      endmodule

      module Multiplexer_5ch (mux_out, data_a, data_b, data_c, data_d, data_e, sel);
        parameter word_size = 8;
        output  [word_size-1: 0]   mux_out;
        input   [word_size-1: 0]   data_a, data_b, data_c, data_d, data_e;
        input   [2: 0] sel;

        assign  mux_out = (sel == 0)      ? data_a: (sel == 1)
                                          ? data_b : (sel == 2)
                                          ? data_c: (sel == 3)
                                          ? data_d : (sel == 4)
                                          ? data_e : 'bx;

      endmodule
```

```
module Multiplexer_3ch (mux_out, data_a, data_b, data_c, sel);
  parameter      word_size = 8;
  output         [word_size-1: 0]   mux_out;
  input          [word_size-1: 0]   data_a, data_b, data_c;
  input          [1: 0] sel;

  assign  mux_out = (sel == 0) ? data_a: (sel == 1) ? data_b : (sel == 2) ? data_c: 'bx;
endmodule
```

The ALU is modeled as combinational logic described by a level-sensitive cyclic behavior that is activated whenever the datapaths or the select bus change. Parameters are used to make the description more readable and to reduce the likelihood of a coding error.

```
/*ALU Instruction          Action
ADD                        Adds the datapaths to form data_1 + data_2.
SUB                        Subtracts the datapaths to form data_1 - data_2.
AND                        Takes the bitwise-and of the datapaths, data_1 & data_2.
NOT                        Takes the bitwise Boolean complement of data_1.
*/
// Note: the carries are ignored in this model.

module Alu_RISC (alu_zero_flag, alu_out, data_1, data_2, sel);
  parameter word_size = 8;
  parameter op_size = 4;
  // Opcodes
  parameter NOP      = 4'b0000;
  parameter ADD      = 4'b0001;
  parameter SUB      = 4'b0010;
  parameter AND      = 4'b0011;
  parameter NOT      = 4'b0100;
  parameter RD       = 4'b0101;
  parameter WR       = 4'b0110;
  parameter BR       = 4'b0111;
  parameter BRZ      = 4'b1000;
  output                      alu_zero_flag;
  output [word_size-1: 0] alu_out;
  input  [word_size-1: 0] data_1, data_2;
  input  [op_size-1: 0]     sel;
  reg                         alu_out;

  assign alu_zero_flag = ~|alu_out;
  always @ (sel or data_1 or data_2)
    case (sel)
      NOP:        alu_out = 0;
      ADD:        alu_out = data_1 + data_2;  // Reg_Y + Bus_1
      SUB:        alu_out = data_2 - data_1;
      AND:        alu_out = data_1 & data_2;
      NOT:        alu_out = ~ data_2;          // Gets data from Bus_1
      default    alu_out = 0;
    endcase
endmodule
```

The control unit is rather large, but its design has a simple form, and its development follows directly from the ASM charts in Figure 7-12. First, declarations are made for the ports and variables needed to support the description. Then the datapath multiplexers are described with nested continuous assignments using the conditional (*? ...:*) operator. Two cyclic behaviors are used: a level-sensitive behavior describes the combinational logic of the outputs and the next state, and an edge-sensitive behavior synchronizes the clock transitions.

```
module Control_Unit (
  Load_R0, Load_R1,
  Load_R2, Load_R3,
  Load_PC, Inc_PC,
  Sel_Bus_1_Mux, Sel_Bus_2_Mux,
  Load_IR, Load_Add_R, Load_Reg_Y, Load_Reg_Z,
  write, instruction, zero, clk, rst);

  parameter word_size = 8, op_size = 4, state_size = 4;
  parameter src_size = 2, dest_size = 2, Sel1_size = 3, Sel2_size = 2;
  // State Codes
  parameter S_idle = 0, S_fet1 = 1, S_fet2 = 2, S_dec = 3;
  parameter  S_ex1 = 4, S_rd1 = 5, S_rd2 = 6;
  parameter S_wr1 = 7, S_wr2 = 8, S_br1 = 9, S_br2 = 10, S_halt = 11;
  // Opcodes
  parameter NOP = 0, ADD = 1, SUB = 2, AND = 3, NOT = 4;
  parameter RD  = 5, WR =  6,  BR =  7, BRZ = 8;
  // Source and Destination Codes
  parameter R0 = 0, R1 = 1, R2 = 2, R3 = 3;

  output Load_R0, Load_R1, Load_R2, Load_R3;
  output Load_PC, Inc_PC;
  output [Sel1_size-1: 0] Sel_Bus_1_Mux;
  output Load_IR, Load_Add_R;
  output Load_Reg_Y, Load_Reg_Z;
  output [Sel2_size-1: 0] Sel_Bus_2_Mux;
  output write;
  input [word_size-1: 0] instruction;
  input zero;
  input clk, rst;

  reg [state_size-1: 0] state, next_state;
  reg Load_R0, Load_R1, Load_R2, Load_R3, Load_PC, Inc_PC;
  reg Load_IR, Load_Add_R, Load_Reg_Y;
  reg Sel_ALU, Sel_Bus_1, Sel_Mem;
  reg Sel_R0, Sel_R1, Sel_R2, Sel_R3, Sel_PC;
  reg Load_Reg_Z, write;
  reg err_flag;

  wire [op_size-1: 0] opcode = instruction [word_size-1: word_size - op_size];
  wire [src_size-1: 0] src = instruction [src_size + dest_size -1: dest_size];
  wire [dest_size-1: 0] dest = instruction [dest_size -1: 0];
```

```
                // Mux selectors
                assign Sel_Bus_1_Mux[Sel1_size-1: 0] = Sel_R0 ? 0:
                                        Sel_R1 ? 1:
                                        Sel_R2 ? 2:
                                        Sel_R3 ? 3:
                                        Sel_PC ? 4: 3'bx;  // 3-bits, sized number
                assign Sel_Bus_2_Mux[Sel2_size-1: 0] = Sel_ALU ? 0:
                                        Sel_Bus_1 ? 1:
                                        Sel_Mem ? 2: 2'bx;
                always @ (posedge clk or negedge rst) begin: State_transitions
                  if (rst == 0) state <= S_idle; else state <= next_state; end

        /* always @ (state or instruction or zero) begin:  Output_and_next_state
```

Note: The above event control expression leads to incorrect operation. The state transition causes the activity to be evaluated once, then the resulting instruction change causes it to be evaluated again, but with the residual value of *opcode*. On the second pass the value seen is the value *opcode* had before the state change, which results in *Sel_PC* = 0 in state 3, which will cause a return to state 1 at the next clock. Finally, *opcode* is changed, but this does not trigger a re-evaluation because it is not in the event control expression. So, the caution is to be sure to use *opcode* in the event control expression. That way, the final execution of the behavior uses the value of *opcode* that results from the state change, and leads to the correct value of *Sel_PC*.
```
        */
```

```
            always @ (state or opcode or src or dest or zero) begin: Output_and_next_state
                Sel_R0 = 0;   Sel_R1 = 0;      Sel_R2 = 0;     Sel_R3 = 0;      Sel_PC = 0;
                Load_R0 = 0; Load_R1 = 0;    Load_R2 = 0;   Load_R3 = 0;     Load_PC = 0;

                Load_IR = 0;  Load_Add_R = 0; Load_Reg_Y = 0; Load_Reg_Z = 0;
                Inc_PC = 0;
                Sel_Bus_1 = 0;
                Sel_ALU = 0;
                Sel_Mem = 0;
                write = 0;
                err_flag = 0;   // Used for de-bug in simulation
                next_state = state;

                case (state)  S_idle:          next_state = S_fet1;
                              S_fet1:          begin
                                                next_state = S_fet2;
                                                Sel_PC = 1;
                                                Sel_Bus_1 = 1;
                                                Load_Add_R = 1;
                                               end
                              S_fet2:          begin
                                                next_state = S_dec;
                                                Sel_Mem = 1;
                                                Load_IR = 1;
                                                Inc_PC = 1;
                                               end
```

```
S_dec:              case (opcode)
                     NOP: next_state = S_fet1;
                     ADD, SUB, AND: begin
                       next_state = S_ex1;
                       Sel_Bus_1 = 1;
                       Load_Reg_Y = 1;
                       case (src)
                         R0:          Sel_R0 = 1;
                         R1:          Sel_R1 = 1;
                         R2:          Sel_R2 = 1;
                         R3:          Sel_R3 = 1;
                         default      err_flag = 1;
                       endcase
                     end // ADD, SUB, AND

                     NOT: begin
                       next_state = S_fet1;
                       Load_Reg_Z = 1;
                       Sel_Bus_1 = 1;
                       Sel_ALU = 1;
                       case (src)
                         R0:          Sel_R0 = 1;
                         R1:          Sel_R1 = 1;
                         R2:          Sel_R2 = 1;
                         R3:          Sel_R3 = 1;
                         default      err_flag = 1;
                       endcase
                       case (dest)
                         R0:          Load_R0 = 1;
                         R1:          Load_R1 = 1;
                         R2:          Load_R2 = 1;
                         R3:          Load_R3 = 1;
                         default      err_flag = 1;
                       endcase
                     end // NOT

                     RD: begin
                       next_state = S_rd1;
                       Sel_PC = 1; Sel_Bus_1 = 1; Load_Add_R = 1;
                     end // RD

                     WR: begin
                       next_state = S_wr1;
                       Sel_PC = 1; Sel_Bus_1 = 1; Load_Add_R = 1;
                     end  // WR

                     BR: begin
                       next_state = S_br1;
                       Sel_PC = 1; Sel_Bus_1 = 1; Load_Add_R = 1;
                     end  // BR
```

```
                            BRZ: if (zero == 1) begin
                              next_state = S_br1;
                              Sel_PC = 1; Sel_Bus_1 = 1; Load_Add_R = 1;
                            end // BRZ
                            else begin
                              next_state = S_fet1;
                              Inc_PC = 1;
                            end
                            default : next_state = S_halt;
                          endcase  // (opcode)

      S_ex1:              begin
                            next_state = S_fet1;
                            Load_Reg_Z = 1;
                            Sel_ALU = 1;
                            case (dest)
                              R0: begin Sel_R0 = 1; Load_R0 = 1; end
                              R1: begin Sel_R1 = 1; Load_R1 = 1; end
                              R2: begin Sel_R2 = 1; Load_R2 = 1; end
                              R3: begin Sel_R3 = 1; Load_R3 = 1; end
                              default : err_flag = 1;
                            endcase
                          end

      S_rd1:              begin
                            next_state = S_rd2;
                            Sel_Mem = 1;
                            Load_Add_R = 1;
                            Inc_PC = 1;
                          end

      S_wr1:              begin
                            next_state = S_wr2;
                            Sel_Mem = 1;
                            Load_Add_R = 1;
                            Inc_PC = 1;
                          end

      S_rd2:              begin
                            next_state = S_fet1;
                            Sel_Mem = 1;
                            case (dest)
                              R0:           Load_R0 = 1;
                              R1:           Load_R1 = 1;
                              R2:           Load_R2 = 1;
                              R3:           Load_R3 = 1;
                              default       err_flag = 1;
                            endcase
                          end
```

```
S_wr2:          begin
                   next_state = S_fet1;
                   write = 1;
                   case (src)
                      R0:          Sel_R0 = 1;
                      R1:          Sel_R1 = 1;
                      R2:          Sel_R2 = 1;
                      R3:          Sel_R3 = 1;
                      default      err_flag = 1;
                   endcase
                end

S_br1:          begin next_state = S_br2; Sel_Mem = 1;
                Load_Add_R = 1; end
S_br2:          begin next_state = S_fet1; Sel_Mem = 1;
                Load_PC = 1; end
S_halt:         next_state = S_halt;
default:        next_state = S_idle;

endcase
end
endmodule
```

For simplicity, the memory unit of the machine is modeled as an array of D-type flip-flops.

```
module Memory_Unit (data_out, data_in, address, clk, write);
 parameter word_size = 8;
 parameter memory_size = 256;

 output [word_size-1: 0] data_out;
 input [word_size-1: 0] data_in;
 input [word_size-1: 0] address;
 input clk, write;
 reg [word_size-1: 0] memory [memory_size-1: 0];

 assign data_out = memory[address];

 always @ (posedge clk)
   if (write) memory[address] <= data_in;
endmodule
```

7.3.6   RISC SPM: Program Execution

A testbench for verifying that *RISC_SPM* executes a stored program[8] is given below. *test_RISC_SPM* defines probes to display individual words in memory, uses a one-shot

---

[8]An assembler for the machine is located at the website for this book, and can be used to generate programs for use in embedded applications of the processor.

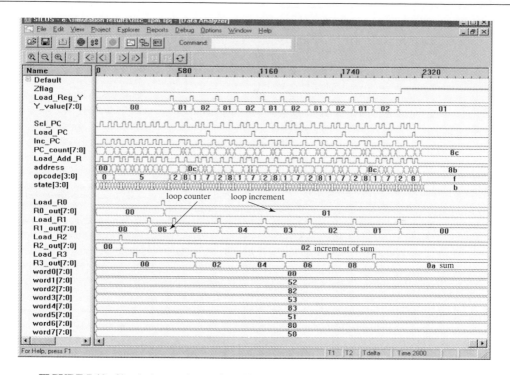

FIGURE 7-13 Simulation results produced by executing a stored program with *RISC_SPM*.

(*initial*) behavior to flush memory, and loads a small program and data into separate areas of memory. The program (1) reads memory and loads the data into the registers of the processor, (2) executes subtraction to decrement a loop counter, (3) adds register contents while executing the loop, and (4) branches to a halt when the loop index is 0. The results of executing the program are displayed in Figure 7-13.

```
module test_RISC_SPM ();
  reg rst;
  wire clk;
  parameter word_size = 8;
  reg [8: 0] k;

  Clock_Unit M1 (clk);
  RISC_SPM M2 (clk, rst);

// define probes
  wire [word_size-1: 0] word0, word1, word2, word3, word4, word5, word6;
  wire [word_size-1: 0] word7, word8, word9, word10, word11, word12, word13;
  wire [word_size-1: 0] word14;

  wire [word_size-1: 0] word128, word129, word130, word131, word132, word255;
  wire [word_size-1: 0] word133, word134, word135, word136, word137;
```

```
wire [word_size-1: 0] word138, word139, word140;
assign word0 = M2.M2_SRAM.memory[0];
assign word1 = M2.M2_SRAM.memory[1];
assign word2 = M2.M2_SRAM.memory[2];
assign word3 = M2.M2_SRAM.memory[3];
assign word4 = M2.M2_SRAM.memory[4];
assign word5 = M2.M2_SRAM.memory[5];
assign word6 = M2.M2_SRAM.memory[6];
assign word7 = M2.M2_SRAM.memory[7];
assign word8 = M2.M2_SRAM.memory[8];
assign word9 = M2.M2_SRAM.memory[9];
assign word10 = M2.M2_SRAM.memory[10];
assign word11 = M2.M2_SRAM.memory[11];
assign word12 = M2.M2_SRAM.memory[12];
assign word13 = M2.M2_SRAM.memory[13];
assign word14 = M2.M2_SRAM.memory[14];

assign word128 = M2.M2_SRAM.memory[128];
assign word129 = M2.M2_SRAM.memory[129];
assign word130 = M2.M2_SRAM.memory[130];
assign word131 = M2.M2_SRAM.memory[131];
assign word132 = M2.M2_SRAM.memory[132];
assign word133 = M2.M2_SRAM.memory[133];
assign word134 = M2.M2_SRAM.memory[134];
assign word135 = M2.M2_SRAM.memory[135];
assign word136 = M2.M2_SRAM.memory[136];
assign word137 = M2.M2_SRAM.memory[137];
assign word138 = M2.M2_SRAM.memory[138];
assign word139 = M2.M2_SRAM.memory[139];
assign word140 = M2.M2_SRAM.memory[140];

assign word255 = M2.M2_SRAM.memory[255];

initial #2800 $finish;

Flush Memory

initial begin: Flush_Memory
 #2 rst = 0; for (k=0; k<=255; k=k+1)M2.M2_SRAM.memory[k] = 0; #10 rst = 1;
end

initial begin: Load_program
 #5

                            // opcode_src_dest
 M2.M2_SRAM.memory[0] = 8'b0000_00_00;          // NOP
 M2.M2_SRAM.memory[1] = 8'b0101_00_10;          // Read 130 to R2
 M2.M2_SRAM.memory[2] = 130;
```

```
                M2.M2_SRAM.memory[3] = 8'b0101_00_11;        // Read 131 to R3
                M2.M2_SRAM.memory[4] = 131;
                M2.M2_SRAM.memory[5] = 8'b0101_00_01;        // Read 128 to R1
                M2.M2_SRAM.memory[6] = 128;
                M2.M2_SRAM.memory[7] = 8'b0101_00_00;        // Read 129 to R0
                M2.M2_SRAM.memory[8] = 129;

                M2.M2_SRAM.memory[9] = 8'b0010_00_01;        // Sub R1-R0 to R1

                M2.M2_SRAM.memory[10] = 8'b1000_00_00;       // BRZ
                M2.M2_SRAM.memory[11] = 134;                 // Holds address for BRZ

                M2.M2_SRAM.memory[12] = 8'b0001_10_11;       // Add R2+R3 to R3
                M2.M2_SRAM.memory[13] = 8'b0111_00_11;       // BR
                M2.M2_SRAM.memory[14] = 140;
                // Load data
                M2.M2_SRAM.memory[128] = 6;
                M2.M2_SRAM.memory[129] = 1;
                M2.M2_SRAM.memory[130] = 2;
                M2.M2_SRAM.memory[131] = 0;
                M2.M2_SRAM.memory[134] = 139;
                //M2.M2_SRAM.memory[135] = 0;
                M2.M2_SRAM.memory[139] = 8'b1111_00_00;      // HALT
                M2.M2_SRAM.memory[140] = 9;                  //  Recycle
            end
        endmodule
```

7.4   Design Example: UART

Systems that exchange information and interact via serial data channels use modems as interfaces between the host machines/devices and the channel, as shown in Figure 7-14. For example, a modem allows a computer to connect to a telephone line and communicate with a receiving computer through its modem [2, 5]. The host machine stores information in a parallel word format, but transmits and receives data in a serial, single-bit, format. A modem is also called a UART, or *universal asynchronous receiver and*

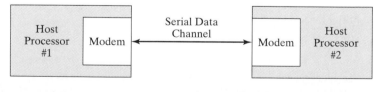

FIGURE 7-14   Processor/modem communication over a serial channel.

Stop Bit	Parity Bit	Data Bit 7	Data Bit 6	Data Bit 5	Data Bit 4	Data Bit 3	Data Bit 2	Data Bit 1	Data Bit 0	Start Bit

FIGURE 7-15 Data format for ASCII text transmitted by a UART.

transmitter, indicating that the device has the capability to both receive and transmit serial data. This design example will address the modeling and synthesis of a UART's transmitter and receiver.

For this discussion, a UART exchanges text data in an American Standard Code for Information Interchange (ASCII) format in which each alphabetical character is encoded by 7 bits and augmented by a parity bit that can be used for error detection. For transmission, the modem wraps this 8-bit subword with a *start-bit* in the least significant bit (LSB), and a *stop-bit* in the most significant bit (MSB), resulting in the 10-bit word format shown in Figure 7-15. The first 9 data bits are transmitted in sequence, beginning with the start-bit, with each bit being asserted at the serial line for one cycle of the modem clock. The stop-bit may assert for more than one clock.

7.4.1 UART Operation

The UART transmitter is always part of larger environment in which a host processor controls transmission by fetching a data word in parallel format and directing the UART to transmit it in a serial format. Likewise, the receiver must detect transmission, receive the data in serial format, strip off the start- and stop-bits, and store the data word in a parallel format. The receiver's job is more complex, because the clock used to send the inbound data is not available at the remote receiver. The receiver must regenerate the clock locally, using the receiving machine's clock rather than the clock of the transmitting machine.

The simplified architecture of a UART presented in Figure 7-16 shows the signals used by a host processor to control the UART and to move data to and from a data bus in the host machine. Details of the host machine are not shown.

7.4.2 UART Transmitter

The input–output signals of the transmitter are shown in the high-level block diagram in Figure 7-17. The input signals are provided by the host processor, and the output signals control the movement of data in the UART. The architecture of the transmitter will consist of a controller, a data register (*XMT_datareg*), a data shift register (*XMT_shftreg*), and a status register (*bit_count*) to count the bits that are transmitted. The status register will be included with the datapath unit.

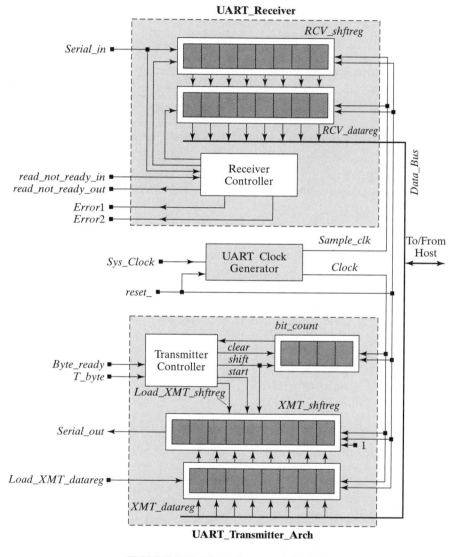

FIGURE 7-16   Block diagram of a UART.

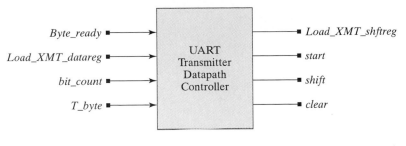

FIGURE 7-17 Interface signals of a state machine controller for a UART transmitter.

The controller has the following inputs. For simplicity, *Load_XMT_datareg* is shown connected directly to *XMT_datareg*:

Byte_ready	asserted by host machine to indicate that *Data_Bus* has valid data
Load_XMT_datareg	assertion transfers *Data_Bus* to the transmitter data storage register, *XMT_datareg*
T_byte	assertion initiates transmission of a byte of data, including the stop, start, and parity bits
bit_count	counts bits in the word during transmission

The state machine of the controller forms the following output signals that control the datapath of the transmitter:

Load_XMT_shftreg	assertion loads the contents of *XMT_data_reg* into *XMT_shftreg*
start	signals the start of transmission
shift	directs *XMT_shftreg* to shift by one bit towards the LSB and to backfill with a stop bit (1).
clear	clears *bit_count*

The ASM chart of the state machine controlling the transmitter is shown in Figure 7.18. The machine has three states: *idle, waiting*, and *sending*. When *reset_* is asserted, the machine asynchronously enters *idle, bit_count* is flushed, *XMT_shftreg* is loaded with 1s, and the control signals *clear, Load_XMT_shftreg, shift*, and *start* are driven to 0. In *idle*, if an active edge of *Clock* occurs while *Load_XMT_data_reg* is asserted by the external host, the contents of *Data_Bus* will transfer to *XMT_data_reg*. (This action is not part of the ASM chart because it occurs independently of the state of the machine.) The machine remains in *idle* until *start* is asserted.

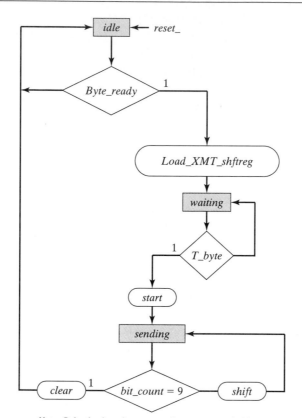

Note: Only the branch corresponding to a true decision is annotated at a decision diamond; signals that are not shown explicitly asserted are de-asserted. Conditional assertions are indicated by the name of the asserted signal

FIGURE 7-18   ASM chart for the state machine controller for the UART transmitter.

When *Byte_ready* is asserted, *Load_XMT_shftreg* is asserted and *next_state* is driven to *waiting*. The assertion of *Load_XMT_shftreg* indicates that *XMT_datareg* now contains data that can be transferred to the internal shift register. At the next active edge of *Clock*, with *Load_XMT_shftreg* asserted, three activities occur: (1) *state* transfers from *idle* to *waiting*, (2) the contents of *XMT_datareg* are loaded into the leftmost bits of *XMT_shftreg*, a (*word_size* + 1)-bit shift register whose LSB signals the start and stop of transmission, and (3) the LSB of *XMT_shftreg* is reloaded with 1, the stop-bit. The machine remains in *waiting* until the external processor asserts *T_byte*.

At the next active edge of *Clock*, with *T_byte* asserted, *state* enters *sending*, and the LSB of *XMT_shftreg* is set to 0 to signal the start of transmission. At the same time, *shift* is driven to 1, and *next_state* retains the state code corresponding to *sending*. At subsequent active edges of *Clock*, with *shift* asserted, *state* remains in *sending* and the contents of *XMT_shftreg* are shifted toward the LSB, which drives the external serial channel. As the data shifts occur, 1s are back-filled in *XMT_shftreg*, and *bit_count* is incremented. With *state* in *sending, shift* asserts while *bit_count* is less than 9. The machine increments *bit_count* after each movement of data, and when *bit_count* reaches 9 *clear* asserts, indicating that all of the bits of the augmented word have been shifted to the serial output. At the next active edge of *Clock*, the machine returns to *idle*.

The control signals produced by the state machine induce state-dependent register transfers in the data path. The activity of the primary inputs (*Byte_ready, Load_XMT_datareg*, and *T_byte*), and the signals from the controller (*Load_XMT_shftreg, start, shift, clear*) are shown in Figure 7-19, along with the movement of data in the datapath registers. The contents of the registers are shown at successive edges of clock, with a time axis going from the top of the figure toward the bottom. Transitions of the active edge of *clock* occur between the successive rows displaying contents of *XMT_datareg*. The bits of the transmitted signal are shown in the sequence in which they are transmitted, with the rightmost cell of *XMT_shftreg* holding the bit that is transmitted at the serial interface at each step. The state of the machine is shown, and the state transitions and datapath register transitions that occur on the rising edges of *clock* are shown in the register boxes. The values of the control signals that cause the register transitions are also shown. The displayed values of the control signals are those they held immediately before the active edge of *Clock*; and which cause the register transfers that are shown. The sequence of output bits is also shown, with 1s being pushed into the MSB of *XMT_shftreg* under the action of *shift*. The sequence of output bits of the transmitted signal are shown as a word at each time step, with the understanding that *the LSB of the word is the first bit that was transmitted*, and the MSB of the word is the most recent bit that was transmitted at the serial interface.

The Verilog description, *UART_Transmitter_Arch*, of the architecture for the transmitter has three cyclic behaviors—a level-sensitive behavior describing the combinational logic for next state and outputs of the controller, an edge-sensitive behavior to synchronize the state transitions of the controller, and another edge-sensitive behavior to synchronize the register transfers of the datapath registers. For simplicity, we include the entire description in a single Verilog module, rather than impose architectural boundaries around the datapath and the controller (see Figure 7-20). The module can be partitioned and synthesized into the individual functional units.

Some simulation results are shown in Figure 7-21 and Figure 7-22 for an 8-bit data word. The waveforms produced by the simulator have been annotated to indicate significant features of the transmitter's behavior. First, observe the values of the signals immediately after *reset_* is asserted. The state is *idle*. Note that *Data_Bus* initially contains the value $a7_h$ (1010_0111_2), a value specified by the testbench used for simulation. With *Byte_ready* not yet asserted, and with *Load_XMT_datareg* asserted, the *Data_Bus* is loaded into *XMT_datareg*. The machine remains in *idle* until *Byte_ready* is asserted. When *Byte_ready* asserts, then *Load_XMT_shftreg* asserts. This causes the

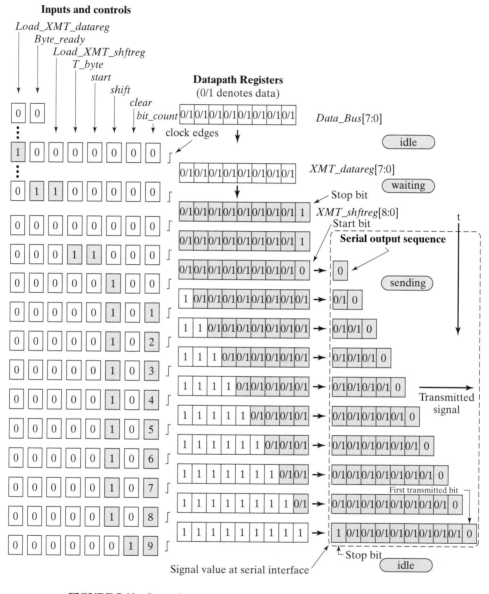

FIGURE 7-19 Control signals and dataflow in an 8-bit UART transmitter.

```
module UART_Transmitter_Arch
  (Serial_out, Data_Bus, Byte_ready, Load_XMT_datareg, T_byte, Clock, reset_);
  parameter        word_size = 8;              // Size of data word, e.g., 8 bits
  parameter        one_hot_count = 3;          // Number of one-hot states
  parameter        state_count = one_hot_count; // Number of bits in state register
  parameter        size_bit_count = 3;         // Size of the bit counter, e.g., 4
                                               // Must count to word_size + 1
  parameter        idle = 3'b001;              // one-hot state encoding
  parameter        waiting = 3'b010;
  parameter        sending = 3'b100;
  parameter        all_ones = 9'b1_1111_1111;  // Word + 1 extra bit

  output           Serial_out                  // Serial output to data channel
  input [word_size − 1:0]  Data_Bus;           // Host data bus containing data word
  input            Byte_ready;                 // Used by host to signal ready
  input            Load_XMT_datareg;           // Used by host to load the data register
  input            T_byte;                     // Used by host to signal the start of transmission
  input            Clock;                      // Bit clock of the transmitter
  input            reset_;                     // Resets internal registers, loads the
                                               // XMT_shftreg with ones

  reg [word_size − 1:0]  XMT_datareg;          // Transmit Data Register
  reg [word_size:0]      XMT_shftreg;          // Transmit Shift Register: {data, start bit}
  reg                    Load_XMT_shftreg;     // Flag to load the XMT_shftreg
  reg [state_count − 1:0] state, next_state;   // State machine controller
  reg [size_bit_count:0]  bit_count;           // Counts the bits that are transmitted
  reg                    clear;                // Clears bit_count after last bit is sent
  reg                    shift;                // Causes shift of data in XMT_shftreg
  reg                    start;                // Signals start of transmission

  assign Serial_out = XMT_shftreg[0];          // LSB of shift register

  always @ (state or Byte_ready or bit_count or T_byte) begin: Output_and_next_state
    Load_XMT_shftreg = 0;
    clear = 0;
    shift = 0;
    start = 0;
    next_state = state;
    case (state)
      idle:      if (Byte_ready == 1) begin
                   Load_XMT_shftreg = 1;
                   next_state = waiting;
                 end

      waiting:   if (T_byte == 1) begin
                   start = 1;
                   next_state = sending;
                 end

      sending:   if (bit_count != word_size + 1)
                   shift = 1;
                 else begin
                   clear = 1;
                   next_state = idle;
                 end

      default:   next_state = idle;
    endcase
  end
end
```

FIGURE 7-20 Verilog Description of the UART transmitter.

```
always @ (posedge Clock or negedge reset_) begin: State_Transitions
  if (reset_ == 0) state <= idle; else state <= next_state; end

always @ (posedge Clock or negedge reset_) begin: Register_Transfers
  if (reset_== 0) begin
    XMT_shftreg <= all_ones;
    bit_count <= 0;
  end
  else begin
    if (Load_XMT_datareg == 1)
      XMT_datareg <= Data_Bus;                    // Get the data bus

    if (Load_XMT_shftreg == 1)
      XMT_shftreg <= {XMT_datareg, 1'b1};         // Load shift reg,
                                                  // insert stop bit
    if (start == 1)
      XMT_shftreg[0] <= 0;                        // Signal start of transmission

    if (clear == 1) bit_count <= 0;
    else if (shift == 1) bit_count <= bit_count + 1;

    if (shift == 1)
      XMT_shftreg <= {1'b1, XMT_shftreg[word_size:1]};  // Shift right, fill with 1's
    end
  end
endmodule
```

FIGURE 7-20 Continued

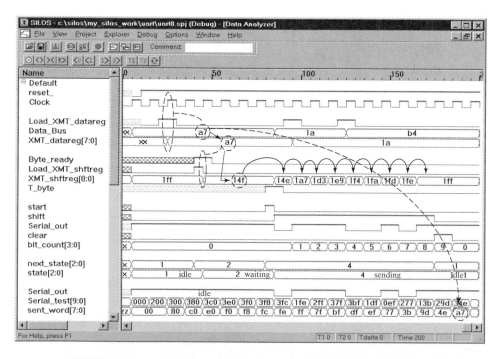

FIGURE 7-21 Annotated simulation results for the 8-bit UART transmitter.

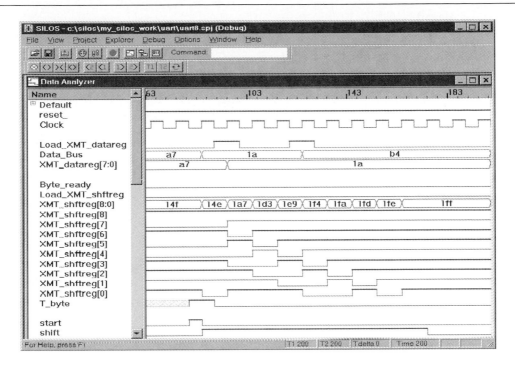

FIGURE 7-22 Data movement through *XMT_shftreg*.

state to change to *waiting* at the next active edge of *Clock*. The 9-bit *XMT_shftreg* is now loaded with the value $\{a7_h, 1\} = 1\_0100\_1111_2 = 14f_h$. Note that the LSB of *XMT_shftreg* is loaded with a 1. The machine remains in *waiting* until *T_byte* is asserted. The assertion of *T_byte* asserts *start*. The machine enters *sending* at the active edge of *Clock* immediately after the host processor's assertion of *T_byte*, and the LSB of *XMT_shftreg* is loaded with a 0. The 9-bit word in *XMT_shftreg* becomes $1\_0100\_1110_2 = 14e_h$. The 0 in the LSB signals the start of transmission. Figure 7-21 has been annotated to show the movement of data through *XMT_shftreg*. Note that 1s are filled behind, as the word shifts to the right. At the active edge of the clock after *bit_count* reaches 9 (for an 8-bit word), *clear* asserts, *bit_count* is flushed, and the machine returns to *idle*.

For diagnostic purposes, the testbench includes a 10-bit shift register that receives *Serial_out* (by hierarchical dereferencing). The eight innermost bits of this register are displayed in Figure 7-21 as *sent_word[7: 0]* (skipping the start-bit and the stop-bit) to reveal the correct transmission of data, $a7_h$. The bit sequence of *Serial_out* likewise has this value. This is evident at the active edge of the clock after the assertion of *clear*. The movement of data through *XMT_shftreg* is shown in Figure 7-22.

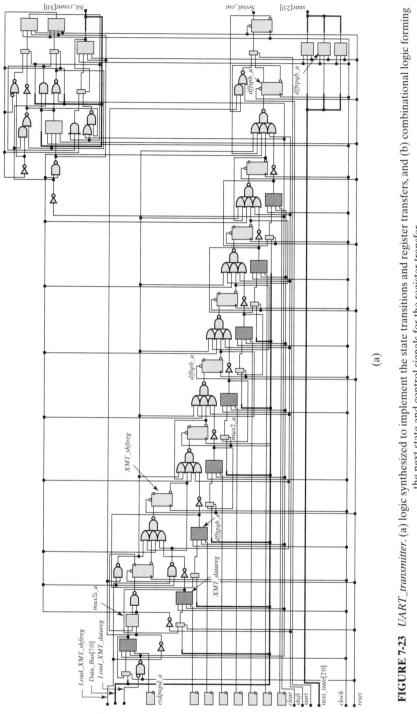

FIGURE 7-23 *UART_transmitter.* (a) logic synthesized to implement the state transitions and register transfers, and (b) combinational logic forming the next state and control signals for the register transfer.

(a)

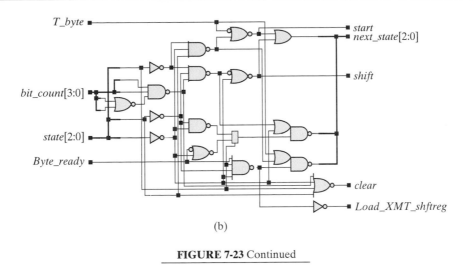

(b)

FIGURE 7-23 Continued

The circuit synthesized from *UART_transmitter* is shown in Figure 7-23. The Verilog model synthesizes as a unit, but for illustration and discussion the description was partitioned and synthesized in two parts, one for the state transitions and register transfers, and another for the combinational logic forming the next state and the control signals for the register transfers. The part governing the state transitions and register transfers consists of an 8-bit register holding *XMT_datareg*, a 9-bit shift register holding *XMT_shftreg*, and a bit counter. The circuit uses *dffrgpqb_a*, a D-type flip-flop with a rising-edge clock, asynchronous active-low reset, and internal gated data of the external data or the output, and *dffspb_a*, a D-type flip-flop with rising-edge clock, and asynchronous active-low set. The shift registers have been highlighted in Figure 7-23.

7.4.3 UART Receiver

The UART receiver has the task of receiving the serial bit stream of data, removing the start-bit, and transferring the data in a parallel format to a storage register connected to the host data bus. The data arrives at a standard bit rate, but it is not necessarily synchronized with the internal clock at the host of the receiver, and the transmitter's clock is not available to the receiver. This issue of synchronization is resolved by generating a *local* clock at a higher frequency and using it to sample the received data in a manner that preserves the integrity of the data. In the scheme used here, the data, assumed to be in a 10-bit format, will be sampled at a rate determined by *Sample_clock*, which is

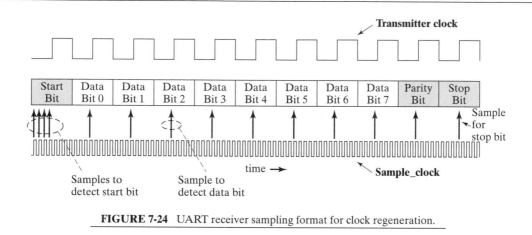

FIGURE 7-24 UART receiver sampling format for clock regeneration.

generated at the receiver's host. The cycles of *Sample_clock* will be counted to ensure that the data are sampled in the middle of a bit time, as shown in Figure 7-24. The sampling algorithm must (1) verify that a start bit has been received, (2) generate samples from 8 bits of the data, and (3) load the data onto the local bus.

Although a higher sampling frequency could be used, the frequency of *Sample_clock* in this example is 8 times the (known) frequency of the bit clock that transmitted the data. This ensures that a slight misalignment between the leading edge of a cycle of *Sample_clock* and the arrival of the start-bit will not compromise the sampling scheme, because the sample will still be taken within the interval of time corresponding to a transmitted bit. The arrival of a start-bit will be determined by successive samples of value 0 after the input data goes low. Then three additional samples will be taken to confirm that a valid start-bit has arrived. Thereafter, 8 successive bits will be sampled at approximately the center of their bit times. Under worst-case conditions of misalignment, the sample is taken a full cycle of *Sample_clock* ahead of the actual center of the bit time, which is a tolerable skew.

The high-level block diagram in Figure 7-25 shows the input–output signals of a state-machine controller that will interface with the host processor and direct the receiver's sampling scheme.

The state machine has the following inputs:

read_not_ready_in	signals that the host is not ready to receive data
Serial_in	serial bit stream received by the unit
*reset_*	active low reset
Sample_counter	counts the samples of a bit
Bit_counter	counts the bits that have been sampled

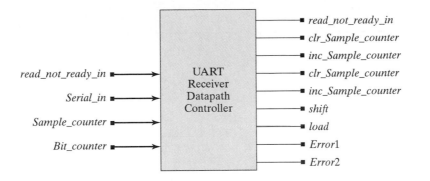

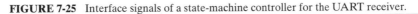

FIGURE 7-25 Interface signals of a state-machine controller for the UART receiver.

The state machine produces the following outputs:

read_not_ready_out	signals that the receiver has received 8 bits
inc_Sample_counter	increments *Sample_counter*
clr_Sample_counter	clears *Sample_counter*
inc_Bit_counter	increments *Bit_counter*
clr_Bit_counter	clears *Bit_counter*
shift	causes *RCV_shftreg* to shift towards the LSB
load	causes *RCV_shftreg* to transfer data to *RCV_datareg*
Error1	asserts if host is not ready to receive data after last bit has been sampled
Error2	asserts if the stop-bit is missing

The ASM chart of a state machine controller for the receiver is shown in Figure 7-26. The machine has three states: *idle, starting*, and *receiving*. Transitions between states are synchronized by *Sample_clk*. Assertion of an asynchronous active-low reset puts the machine in the *idle* state. It remains there until *Serial_in* is low, then makes a transition to *starting*. In *starting*, the machine samples *Serial_in* to determine whether the first bit is a valid start-bit (it must be 0). Depending on the sampled values, *inc_Sample_counter* and *clr_Sample_counter* may be asserted to increment or clear the counter at the next active edge of *Sample_clock*. If the next three samples of *Serial_in* are 0, the machine concludes that the start-bit is valid and goes to the state *receiving*. *Sample_counter* is cleared on the transition to *receiving*. In this state, eight successive samples are taken (one for each bit of the byte, at each active edge of *Sample_clk*), with *inc_Sample_counter* asserted. Then *Bit_counter* is incremented. If the sampled bit is not the last (parity) bit, *inc_Bit_counter* and *shift* are asserted. The assertion of *shift* will cause the sample value to be loaded into the MSB of *RCV_shftreg*, the receiver shift register, and will shift the 7 leftmost bits of the register toward the LSB.

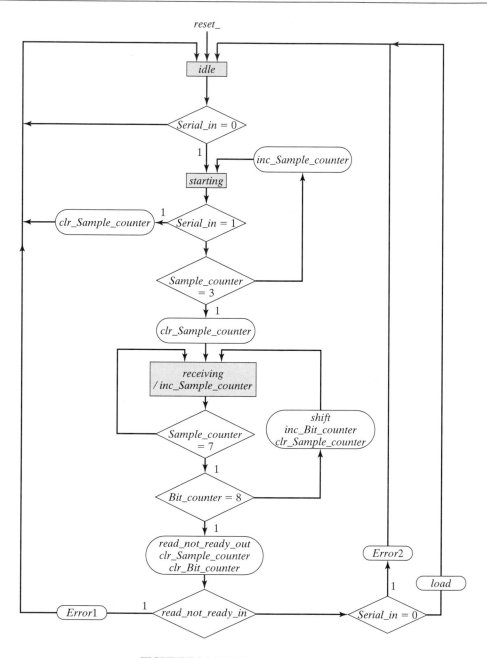

FIGURE 7-26   UART receiver ASM chart.

After the last bit has been sampled, the machine will assert *read_not_ready_out*, a handshake output signal to the processor, and clear the bit counter. At this time, the machine also checks the integrity of the data and the status of the host processor. If *read_not_ready_in* is asserted, the host processor is not ready to receive the data (*Error1*). If a stop-bit is not the next bit (detected by *Serial_in* = 0), there is an error in the format of the received data (*Error2*). Otherwise, *load* is asserted to cause the contents of the shift register to be transferred as a parallel word to *RCV_datareg*, a data register in the host machine, with a direct connection to *data_bus*.

The Verilog description of the 8-bit UART receiver is given in Figure 7-27. The description follows directly from the ASM chart in Figure 7-26.

The simulation results in Figure 7-28 are annotated to show functional features of the waveforms. The received data word is $b5_h = 1011\_0101_2$. The reception sequence is from LSB to MSB, and the data move through the inbound shift register from MSB to LSB. The data word is preceded by a start-bit and followed by a stop-bit. With *reset_* having a value of 0, the state is *idle* and the counters are cleared. At the first active edge of *Sample_clock* after the reset condition is de-asserted, with *Serial_in* having a value of 0, the controller's state enters *starting* to determine whether a start-bit is being received. Three more samples of *serial_in* are taken, and after a total of four samples have been found to be 0, the *Sample_counter* is cleared and the state enters *receiving*. After the eighth sample, *shift* is asserted. The sample at the next active edge of the clock is shifted into the MSB of *RCV_shftreg*. The value of *RCV_shftreg* becomes $80_h = 1000\_0000_2$. The sampling cycle repeats again, and a value of 0 is sampled and loaded into *RCV_shftreg*, changing the contents of the register to $0100\_0000_2 = 40_h$.

The end of the sampling cycle of the received word is shown in Figure 7-29. After the last data bit is sampled, the machine samples once more to detect the stop bit. In the absence of an error, the contents of *RCV_shftreg* will be loaded into *RCV_datareg*. In this example, the value $b5_h$ is finally loaded from *RCV_shftreg* into *RCV_datareg*. Other tests can be conducted to completely verify the functionality of the receiver.

The Verilog description of the partitioned receiver in Figure 7-30 synthesizes into the circuit shown in Figure 7-31. Although the entire description synthesizes as a single unit, the structure of the synthesized result is revealed more easily by partitioning the description into the asynchronous (combinational) and synchronous (state transition) parts shown below. Note that the ports of the parent modules of a partition must be sized properly to accommodate vector ports in the child modules. Otherwise, the ports will be treated as default scalars in the scope of the parent module.

The state-transition part of the synthesized circuit includes *RCV_shftreg*, the shift register receiving *Serial_in*, and *RCV_datareg* (the 8-bit register holding the received word), *Sample_counter*, *Bit_counter* (the 4-bit counter that determines when all of the bits have been received), and the state register for the machine. Two types of flip-flops are used: the four-input *dffrgpqb_a* and the three-input *dffrpqb_a*. The former is a D-type flip-flop with internal gated data between the external datapath and the output, a rising clock, and an asynchronous active-low reset; the latter is a D-type flip-flop with data, rising clock, and asynchronous active-low reset. Only the state register uses the *dffrpqb_a* flip-flop.

```
module UART8_Receiver
(RCV_datareg, read_not_ready_out, Error1 ,Error2, Serial_in, read_not_ready_in, Sample_clk, reset_);
// Sample_clk is 8x Bit_clk

parameter    word_size           = 8;
parameter    half_word           = word_size/2;
parameter    Num_counter_bits    = 4;            // Must hold count of word_size
parameter    Num_state_bits      = 2;            // Number of bits in state
parameter    idle                = 2'b00;
parameter    starting            = 2'b01;
parameter    receiving           = 2'b10;

output       [word_size − 1:0]              RCV_datareg;
output                                      read_not_ready_out,
                                            Error1, Error2;
input        Serial_in,
             Sample_clk,
             reset_,
             read_not_ready_in,


reg                                         RCV_datareg;
reg          [word_size − 1:0]              RCV_shftreg;
reg          [Num_counter_bits − 1:0]       Sample_counter;
reg          [Num_counter_bits:0]           Bit_counter;
reg          [Num_state_bits − 1:0]         state, next_state;
reg                                         inc_Bit_counter, clr_Bit_counter;
reg                                         inc_Sample_counter, clr_Sample_counter;
reg                                         shift, load, read_not_ready_out;
reg                                         Error1, Error2;
```

//Combinational logic for next state and conditional outputs

```
always @ (state or Serial_in or read_not_ready_in or Sample_counter or Bit_counter) begin
  read_not_ready_out = 0;
  clr_Sample_counter = 0;
  clr_Bit_counter = 0;
  inc_Sample_counter = 0;
  inc_Bit_counter = 0;
  shift = 0;
  Error1 = 0;
  Error2 = 0;
  load = 0;
  next_state = state;

  case (state)
    idle:        if (Serial_in == 0) next_state = starting;

    starting:    if (Serial_in == 1) begin
                   next_state = idle;
                   clr_Sample_counter = 1;
                 end else

                 if (Sample_counter == half_word − 1) begin
                   next_state = receiving;
                   clr_Sample_counter = 1;
                 end else inc_Sample_counter = 1;
```

FIGURE 7-27   Verilog description of UART8_Receiver, an 8-bit UART receiver.

```
receiving:        if (Sample_counter < word_ size − 1) inc_Sample_counter = 1;
                  else begin
                    clr_Sample_counter = 1;
                    if (Bit_counter != word_size) begin
                      shift = 1;
                      inc_Bit_counter = 1;
                    end
                    else begin
                      next_state = idle;
                      read_not_ready_out = 1;
                      clr_Bit_counter = 1;
                      if (read_not_ready_in == 1) Error1 = 1;
                      else if (Serial_in == 0) Error2 = 1;
                      else load = 1;
                    end
                  end
         default: next_state = idle;

   endcase
end

// state_transitions_and_register_transfers

   always @ (posedge Sample_clk) begin
     if (reset_ == 0) begin                              // synchronous reset_
       state <= idle;
       Sample_counter <= 0;
       Bit_counter <= 0;
       RCV_datareg <= 0;
       RCV_shftreg <= 0;
     end
     else begin
       state <= next_state;

       if (clr_Sample_counter == 1) Sample_counter <= 0;
       else if (inc_Sample_counter == 1) Sample_counter <= Sample_counter + 1;

       if (clr_Bit_counter == 1) Bit_counter <= 0;
       else if (inc_Bit_counter == 1) Bit_counter <= Bit_counter + 1;
       if (shift == 1 ) RCV_shftreg <= {Serial_in, RCV_shftreg[word_size − 1:1]};
       if (load == 1) RCV_datareg <= RCV_shftreg;
     end
   end
endmodule
```

FIGURE 7-27 Continued

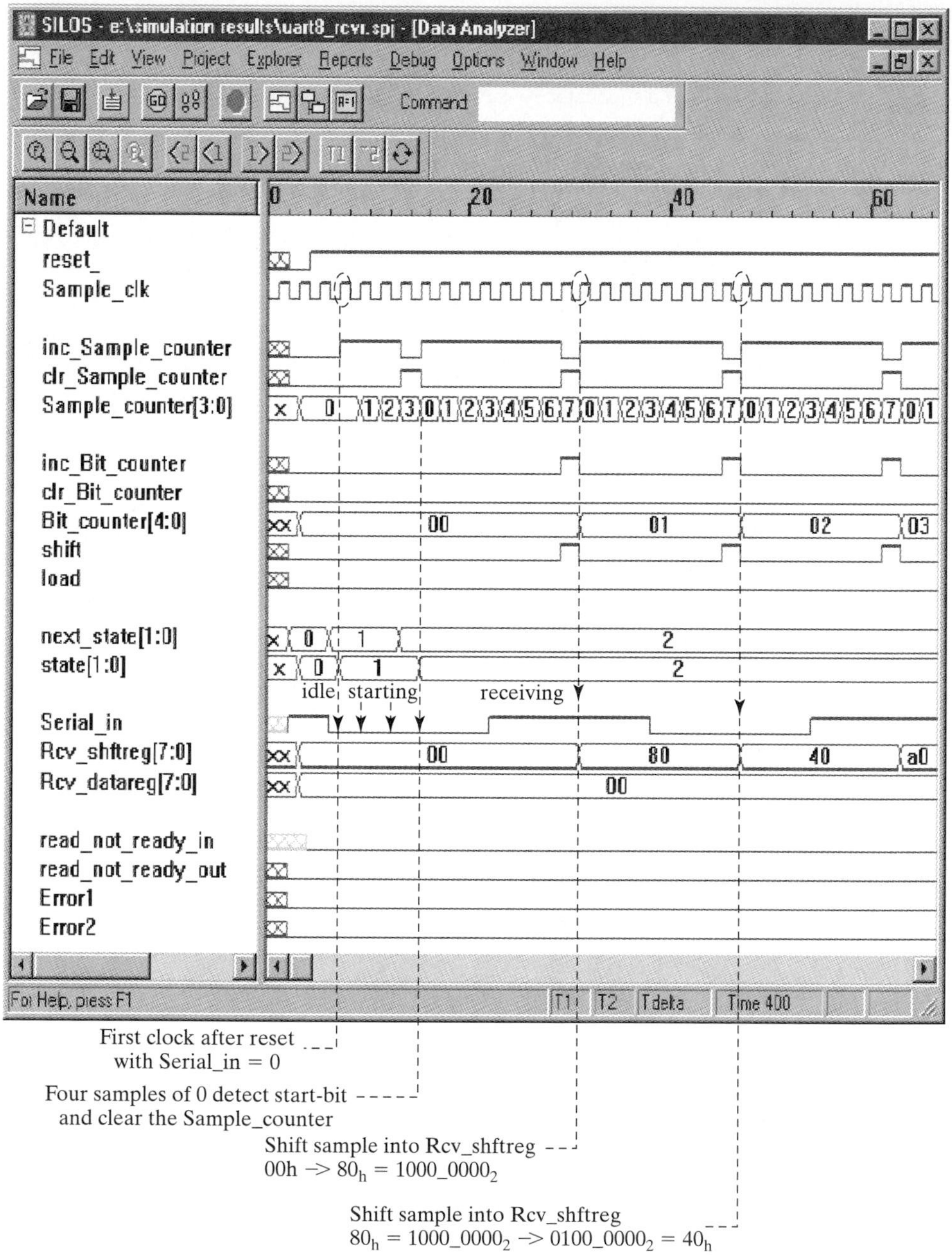

FIGURE 7-28 Annotated simulation results for the UART receiver.

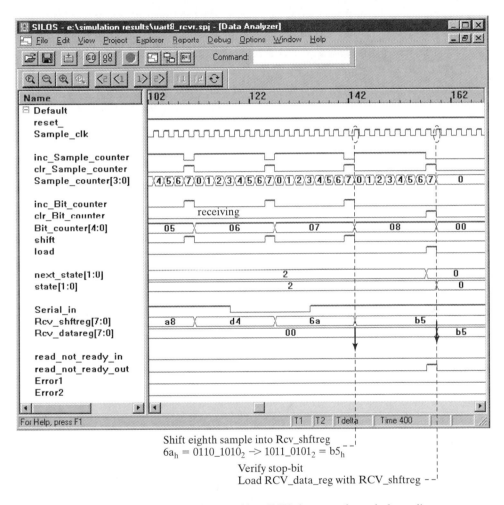

FIGURE 7-29 Transfer of data word into *RCV_datareg* at the end of sampling.

```verilog
module UART8_rcvr_partition (RCV_datareg, read_not_ready_out, Error1, Error2, Serial_in,
  read_not_ready_in, Sample_clk, reset_);

  // partitioned UART receiver                              // Sample_clk is 8x Bit_clk

  parameter           word_size            = 8;
  parameter           half_word            = word_size / 2;
  parameter           Num_counter_bits     = 4;       // Must hold count of word_size
  parameter           Num_state_bits       = 2;       // Number of bits in state
  parameter           idle                 = 2'b00;
  parameter           starting             = 2'b01;
  parameter           receiving            = 2'b10;

  output [word_size − 1:0]  RCV_datareg;
  output              read_not_ready_out,   // Handshake to host processor
                      Error1,               // Host not ready error
                      Error2;               // Data_in missing stop bit

  input               Serial_in,            // Serial data input
                      Sample_clk,           // Clock to sample serial data
                      reset_,               // Active-low reset
                      read_not_ready_in;    // Status bit from host processor

  wire [Num_counter_bits − 1:0]    Sample_counter;
  wire [Num_counter_bits: 0]       Bit_counter;
  wire [Num_state_bits − 1:0]      state, next_state;

controller_part M2
  (next_state, shift, load, read_not_ready_out, Error1, Error2, inc_Sample_counter,
  inc_Bit_counter, clr_Bit_counter, clr_Sample_counter, state, Sample_counter, Bit_counter,
  Serial_in, read_not_ready_in);

state_transition_part M1
  (RCV_datareg, Sample_counter, Bit_counter, state, next_state, clr_Sample_counter,
    inc_Sample_counter, clr_Bit_counter, inc_Bit_counter, shift, load, Serial_in, Sample_clk, reset_);

endmodule

module controller_part (next_state, shift, load, read_not_ready_out, Error1, Error2, inc_Sample_counter,
  inc_Bit_counter, clr_Bit_counter, clr_Sample_counter, state, Sample_counter, Bit_counter,
  Serial_in, read_not_ready_in);

  parameter           word_size            = 8;
  parameter           half_word            = word_size / 2;
  parameter           Num_counter_bits     = 4;       // Must hold count of word_size
  parameter           Num_state_bits       = 2;       // Number of bits in state
  parameter           idle                 = 2'b00;
  parameter           starting             = 2'b01;
  parameter           receiving            = 2'b10;
```

FIGURE 7-30 Verilog description of *UART8_rcvr_partition*, an 8-bit UART receiver partitioned into a controller and a datapath.

```verilog
    output [Num_state_bits − 1:0]   next_state;
    output                          shift, load, inc_Sample_counter;
    output                          inc_Bit_counter, clr_Bit_counter, clr_Sample_counter;
    output                          read_not_ready_out, Error1, Error2;

    input [Num_state_bits − 1:0]    state;
    input [Num_counter_bits − 1:0]  Sample_counter;
    input [Num_counter_bits: 0]     Bit_counter;
    input                           Serial_in, read_not_ready_in;

    reg next_state;
    reg inc_Sample_counter, inc_Bit_counter, clr_Bit_counter, clr_Sample_counter;
    reg shift, load,  read_not_ready_out, Error1, Error2;

    always @ (state or Serial_in or read_not_ready_in or Sample_counter or Bit_counter) begin
        read_not_ready_out = 0;          //Combinational logic for next state and conditional outputs
        clr_Sample_counter = 0;
        clr_Bit_counter = 0;
        inc_Sample_counter = 0;
        inc_Bit_counter = 0;
        shift = 0;
        Error1 = 0;
        Error2 = 0;
        load = 0;
        next_state = state;

        case (state)
          idle:         if (Serial_in == 0) next_state = starting;

          starting:     if (Serial_in == 1) begin
                          next_state = idle;
                          clr_Sample_counter = 1;
                        end else

                        if (Sample_counter == half_word − 1) begin
                          next_state = receiving;
                          clr_Sample_counter = 1;
                        end else inc_Sample_counter = 1;

          receiving:    if (Sample_counter < word_size − 1) inc_Sample_counter = 1;
                        else begin
                          clr_Sample_counter = 1;
                          if (Bit_counter != word_size) begin
                            shift = 1;
                            inc_Bit_counter = 1;
                          end
                          else begin
                            next_state = idle;
                            read_not_ready_out = 1;
                            clr_Bit_counter = 1;
                            if (read_not_ready_in == 1) Error1 = 1;
                            else if (Serial_in == 0) Error2 = 1;
                            else load = 1;
                          end
                        end
          default:      next_state = idle;

        endcase
    end
endmodule
```

FIGURE 7-30 Continued

```
module state_transition_part (RCV_datareg, Sample_counter, Bit_counter, state, next_state,
clr_Sample_counter, inc_Sample_counter, clr_Bit_counter, inc_Bit_counter, shift, load, Serial_in,
Sample_clk, reset_);
  parameter              word_size        = 8;
  parameter              half_word        = word_size / 2;
  parameter              Num_counter_bits = 4;        // Must hold count of word_size
  parameter              Num_state_bits   = 2;        // Number of bits in state
  parameter              idle             = 2'b00;
  parameter              starting = 2'b01;
  parameter              receiving        = 2'b10;

  output [word_size − 1:0]           RCV_datareg;
  output [Num_counter_bits − 1:0]    Sample_counter;
  output [Num_counter_bits: 0]       Bit_counter;
  output [Num_state_bits − 1:0]      state;

  input [Num_state_bits − 1:0]       next_stage;
  input                              Serial_in;
  input                              inc_Sample_counter, inc_Bit_counter;
  input                              clr_Bit_counter, clr_Sample_counter, shift, load;
  input                              Sample_clk, reset_;

  reg                                Sample_counter, Bit_counter;
  reg [word_size − 1:0]              RCV_shftreg, RCV_datareg;
  reg                                state;

// state_transitions_and_datapath_register_transfers

  always @ (posedge Sample_clk) begin
    if (reset_ == 0) begin                     // synchronous reset_
      state <= idle;
      Sample_counter <= 0;
      Bit_counter <= 0;
      RCV_datareg <= 0;
      RCV_shftreg <= 0;
    end
    else begin
      state <= next_state;

      if (clr_Sample_counter == 1) Sample_counter <= 0;
      else if (inc_Sample_counter == 1) Sample_counter <= Sample_counter + 1;

      if (clr_Bit_counter == 1) Bit_counter <= 0;
      else if (inc_Bit_counter == 1) Bit_counter <= Bit_counter + 1;
      if (shift == 1) RCV_shftreg <= (Serial_in, RCV_shftreg[word_size − 1:1]};
      if (load == 1) RCV_datareg <= RCV_shftreg;
    end
  end
endmodule
```

FIGURE 7-30 Continued

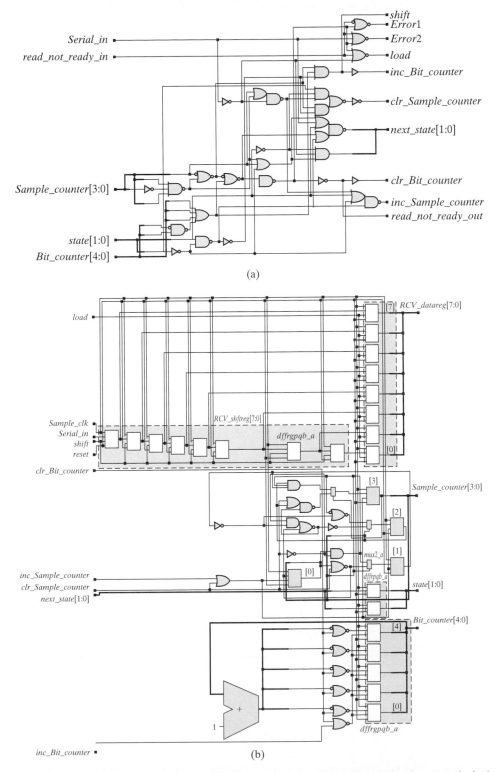

(a)

(b)

FIGURE 7-31 Circuits synthesized from *UART8_receiver*. (a) state transition and register transfer logic, and (b) combinational logic (forming the next state), output register, and control signals for register transfers.

REFERENCES

1. Ernst R. "Target Architectures." *Hardware/Software Co-Design: Principles and Practice.* Boston: Kluwer, 1997.
2. Gajski D, et al. "Essential Issues in Design," In: Staunstrup J, Wolf W, eds. *Hardware/Software Co-Design: Principles and Practice.* Boston: Kluwer, 1997.
3. Hennessy JL, Patterson DA. *Computer Architecture—A Quantitative Approach.* 2nd ed. San Francisco: Morgan Kaufman, 1996.
4. Heuring VP, Jordan HF. *Computer Systems Design and Architecture.* Menlo Park, CA: Addison-Wesley Longman, 1997.
5. Roth CW, Jr. *Digital Systems Design Using VHDL.* Boston: PWS, 1998.

PROBLEMS

1. Develop, verify, and synthesize *Johnson_Counter_ASM*, a Verilog behavioral module of a 4-bit Johnson counter based on a direct implementation of the ASM chart in Figure 7.5. *Hint*: Use a Verilog function to described *next_count*.
2. The functional unit *UART_Clock_Generator* in Figure P7-2 can be used to create a set of baud rate signal pairs for use in the UART in Figure 7-16. Table P7.2

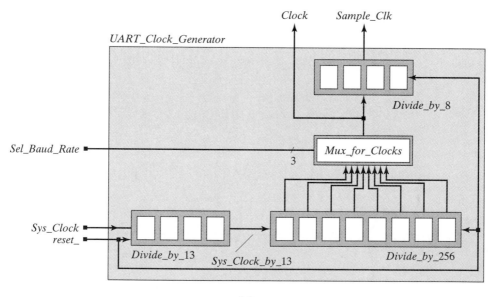

(a)

FIGURE P7-2

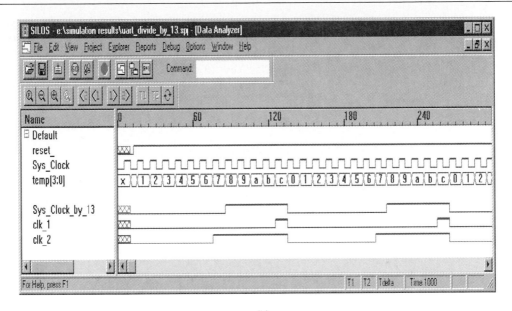

(b)

FIGURE P7-2 Continued

shows the pairs that are generated if *Sys_Clock* is an 8-MHz signal. The code for an experimental version of *Divide_by_13* is also given, with *temp* being a 4-bit counter that counts from 0 to 12 before recycling, at $\frac{1}{13}$ the frequency of *Sys_Clock*. The signal *Sys_Clock_by_13* is generated by the condition that register *temp* has a 1 in its MSB, which is true for the last five contiguous cycles. The signal *clk_1* is generated from the condition that *temp* has the value 12; *clk_2* is generated from the condition that *temp* is greater than 6, and produces a more symmetric waveform than the two others.

(a) Synthesize three versions of *Divide_by_13*, one for each of the three possibilities illustrated by the waveforms in Figure P7-2, and compare their circuits.

TABLE P7.2

Sel_Baud_Rate	Clock	Sample_Clock
000	307,696	38462
001	153,838	19231
010	76,920	9615
011	38,464	4808
100	18,232	2404
101	9,616	1202
110	4,808	601
111	2,404	300.5

(b) Choose one of the methods for forming *Sys_Clock_by_13*, then develop, verify, and synthesize the complete description of *UART_Clock_Generator*.

```verilog
module Divide_by_13 (Sys_Clock_by_13, Sys_Clock, reset_);
  output          Sys_Clock_by_13;
  input Sys_Clock, reset_;
  reg    [3: 0]    temp;

  assign Sys_Clock_by_13 = temp[3];
  // wire clk_1 = (temp == 12);
  // wire clk_2 = (temp > 6);
  always @ (posedge Sys_Clock or negedge reset_ )
    if (reset_ == 0) temp <= 4'b0000;
    else if  (temp == 4'd12) temp <= 4'd0;
    else temp <= temp +1;
endmodule
```

3. A counter is said to enter an abnormal state if its enters a state that is not explicitly decoded in its next-state function. A self-correcting counter has the ability to recover from an abnormal state. The key is to choose default assignments that ensure recovery. Develop and verify a Verilog model of a self-correcting 4-bit Johnson counter using the next state 0001_2 if the current state is 0--0, with "-" denoting a don't-care. Demonstrate that the counter is self-correcting, and synthesize the circuit.
4. The text below describes a fragment of code taken from a Verilog model of a counter. When *mode* is *ring2* the counter is to act like a ring counter, but move two adjacent bits at a time. The simulation results in Figure P7-4 shows that the counter enters an unknown count for one clock, and then fails to count correctly. Find the cause of the error.

```verilog
module Counter8_Prog (count, enable, mode, direction, enable,  clk, reset);
  output [7: 0]    count;
  input   [1: 0]    mode;
  input            enable, direction;
  input            clk, reset;
  reg              count;
  parameter        start_count      = 1;

// Mode of count
  parameter                  binary   = 0;
  parameter                  ring1    = 1;
  parameter                  ring2    = 2;
  //parameter                spiral   = 2;
  parameter                  jump2    = 3;
```

```
// Direction of count
parameter          left     = 0;
parameter          right    = 1;
parameter          up       = 0;
parameter          down     = 1;

always @ (posedge clk or posedge reset)
  if (reset ==1) count <= start_count;
  else if (enable ==1)
    case (mode)
      ring1:       count <= ring1_count      (count, direction);
      ring2:       count <= ring2_count      (count, direction);
      jump2:       count <= jump2_count      (count, direction);
      default:     count <= binary_count     (count, direction);
    endcase
```

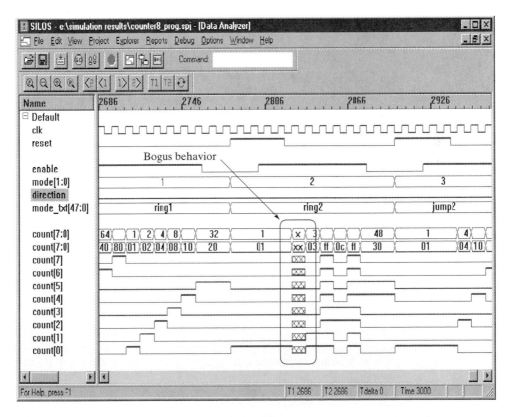

FIGURE P7-4

```
function [7: 0]   ring2_count;
 input [7: 0]    count;
 input           direction;
 begin
  ring2_count = start_count; //8'b0000_0011;
  if (direction == left)
  case (count)
    start_count:          ring2_count = 8'b1100_0000;
    8'b1100_0000:         ring2_count = 8'b0000_0011;
    default:              ring2_count = 8'b1111_1111;
  endcase
  else
  case (count)
    start_count:          ring2_count = 8'b0000_0011;
    8'b1100_0000:         ring2_count = 8'b0011_0000;
    8'b0011_0000:         ring2_count = 8'b0000_1100;
    8'b0000_1100:         ring2_count = 8'b0000_0011;
    8'b0000_0011:         ring2_count = 8'b1100_0000;
       default:           ring2_count = 8'b1111_1111;
  endcase
 end
endfunction
```

5. The register transfers of the UART transmitter in Figure 7-20 decode the controlling signals in separate *if* statements. Discuss whether the signals controlling the datapath can be in conflict, and whether the controlling signals should have been decoded by a priority decoder. Modify the datapath unit to use a priority decoder, and determine whether the synthesized circuit differs from the circuit that was synthesized from *UART_Transmitter_Arch*.

6. The entire Verilog description of *UART_Transmitter_Arch* in Figure 7-20 is written as a single Verilog module. Develop, verify, and synthesize *UART_Transmitter_Part*, a description with hierarchical boundaries around the datapath and the controller.

7. The Verilog description of *UART_Transmitter_Arch* in Figure 7-20 was based on the ASM chart in Figure 7-18, and uses a counter to record the status of the datapath transitions. Develop, verify, and synthesize an alternative description that eliminates the status register and is based on a state-transition graph of the controller.

8. Develop and use a testbench to demonstrate that *RISC_SPM* (see Section 7.3) correctly executes its complete instruction set. After verifying the functionality of the machine, synthesize a gate-level realization. Conduct postsynthesis verification of the gate-level circuit.

9. Modify *ALU_RISC* in *RISC_SPM* (see Section 7.3) to handle carries. Synthesize the new machine.

10. (a) Develop and verify *Binary_Counter_Arch*, a Verilog model with the hierarchical partition, module names, and port structure (*count, enable, clock, rst*) shown in Figure 7-3, with *count* having a parameterized width. Also, develop and verify the nested modules *Control_Unit* and *Datapath_Unit_Arch*, as well as

the child modules, *Mux* and *Count_Register*. (b) Synthesize *Binary_Counter_Arch* and *Binary_Counter_Behav_imp* (an implicit-state machine behavioral model) for a 4-bit wide datapath. (c) Compare the synthesized circuits. (d) Compare the results of simulating the counters, using a common testbench.

11. Develop, verify, and synthesize *Binary_Counter_STG*, using the STG in Figure 7-4 as a guide. Compare to the results of Problem 10.

12. The implicit Moore machine with a modified control unit in *Binary_Counter_Part_RTL_by_3* (see Example 7.1) increments the counter every third clock, but takes one extra cycle to recover from a reset condition (i.e., the first increment of the counter occurs at the fourth clock edge after reset is de-asserted). Using a testbench, verify this claim. Explain why this behavior occurs, then develop and verify an alternative (partitioned) machine that will recover from a reset condition after three clock edges and increment thereafter on every third edge.

13. Modify the datapath and control units in *Binary_Counter_Part_RTL* (see Example 7.1) to implement *Johnson_Counter_RTL_by_3*, a Johnson counter that increments every third edge of the external clock. Synthesize the model and verify the functionality of the synthesized circuit.

14. If the function *next_count* in *Binary_Counter_Part_RTL* (see Example 7.1) is modified to have the nonblocking assignment *next_count* $<=$ *count* $+$ *1*, the machine behaves incorrectly, as shown by the simulation results in Figure P7-14. Under what conditions will a simulator assign *x* to a variable? Explain why *count* is driven to *x*.

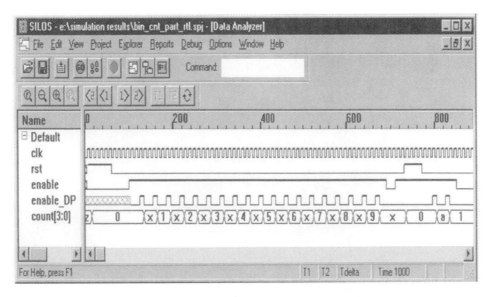

FIGURE P7-14

15. Design a sequential machine that finds the size of the largest gap between two successive 1s in a 16-bit word. Partition the design into a state machine controller and a datapath. The datapath accepts the 16-bit word and produces an output

word whose value is the binary equivalent of the gap size. The interface between the controller and the datapath is shown in Figure P7-15(a).

The ASMD chart in Figure P7-15(b) partially describes the controller. Using the register operations that are annotated on the chart, develop the complete ASMD chart showing the assertions of the output signals of the controller. The data path has a bit counter (k), a storage register (tmp), and a gap register (Gap). Note that the machines are interacting, because the bit count register of the datapath is fed back to the controller.

The machine sequences through the bits of the data word, beginning at the LSB. The register tmp holds the current value of the gap. The register Gap holds the largest gap that has been found as the search evolves. The datapath control signals have the following functionality:

flush_tmp	empties the *tmp* register
incr_tmp	increments the *tmp* register
store_tmp	loads *Gap* with *tmp*
incr_k	increments the bit counter

(a) Using the attached shell file and the completed ASM chart, write a Verilog model of the machine. The shell includes all signal declarations that will be needed, and includes partially declared cyclic behaviors that (1) implement the state transitions, (2) generate the next state and outputs, and (3) control the datapath registers. The internal architecture of the datapath is not shown, but is implicit in the model. (b) Using the testbench below (also available at the web site), verify the design. Organize your graphical output to list the signals in the order shown in Figure P7-15(c). (*Note*: Data are shown in hex and binary format; *Gap* is shown in decimal format.)

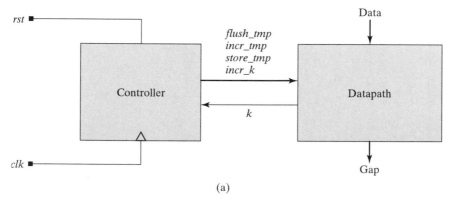

(a)

FIGURE P7-15a

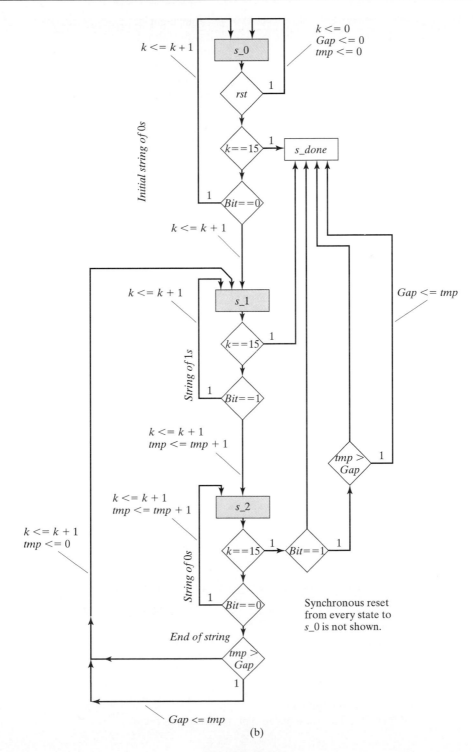

(b)

FIGURE P7-15b

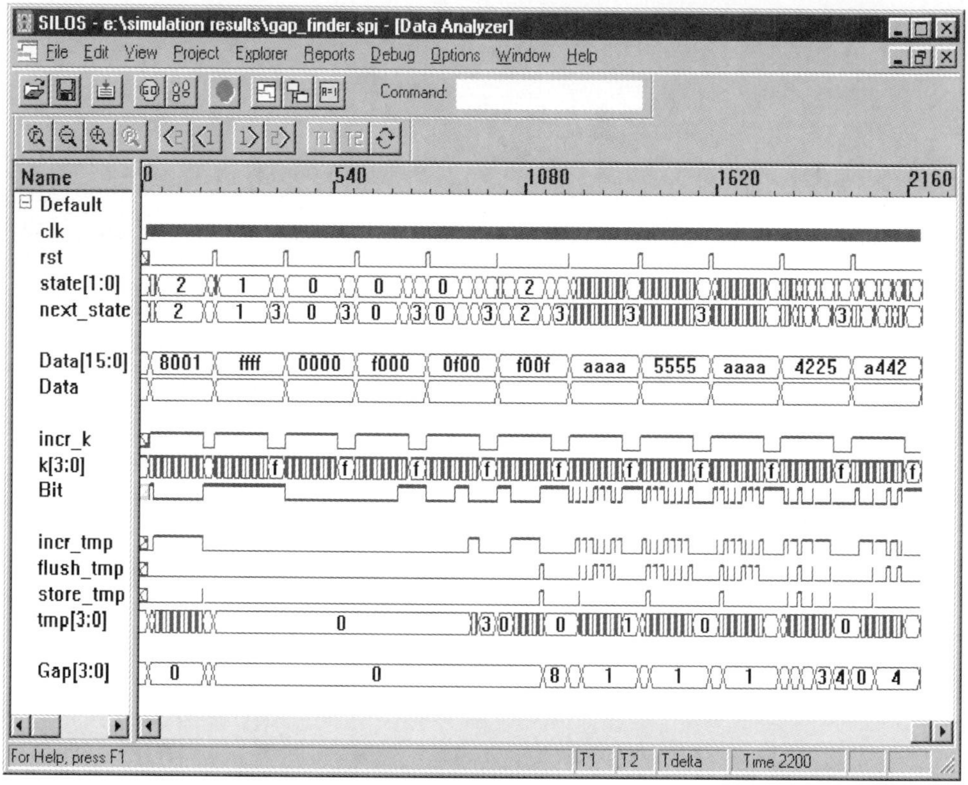

FIGURE P7-15c

```verilog
module Gap_finder(Gap, Data, clk, rst);
    output  [3: 0]      Gap;
    input               [15: 0]    Data;
    input               clk, rst;

    reg     [3: 0]      k, tmp, Gap;                          // datapath registers
    reg     [1: 0]      state, next_state;
    wire                Bit = Data[k];
    reg                 flush_tmp, incr_tmp, store_tmp, incr_k;   // datapath controls

    parameter           s_0         = 2'b00;                  // states
    parameter           s_1         = 2'b01;
    parameter           s_2         = 2'b10;
    parameter           s_done = 2'b11;

    // State transitions

    always @ (posedge clk)
        // YOUR CODE (use active-high synchronous reset)
```

```
// Combinational logic for next state and outputs

always @ (state or Bit or k) begin   // Must have k to see 16'hffff
                                     // Try simulating without it too
   next_state = state;               // Remain until conditions are met
   incr_k = 0;                       // Set all variables to 0 on entry
   incr_tmp = 0;                     // to avoid bogus latches
   store_tmp = 0;
   flush_tmp = 0;

   case (state)
        // YOUR CODE for next state and outputs to control the datapath
   endcase
end

        // Edge-sensitive behavior (i.e., synchronized) for datapath operations
   always @ (posedge clk)
     if (rst == 1) begin k <= 0; Gap <= 0; tmp <= 0; end
     else begin
          // YOUR CODE for register operations controlled by the state machine

end
endmodule

module annotate_Gap_finder ();              // Annotate the clock parameters
   defparam t_Gap_finder.M2.Latency = 10;
   defparam t_Gap_finder.M2.Offset = 5;
   defparam t_Gap_finder.M2.Pulse_Width = 5;
endmodule

   module t_Gap_finder ();
     reg    [15: 0]   Data;
     reg              rst;
     wire   [3: 0]    Gap;

     Gap_finder    M1 (Gap, Data, clk, rst);
     Clock_Prog    M2 (clk);

     initial #2200 $finish;

     initial begin      // expect Gap = 14
       #20 rst = 1;
       #5 Data = 16'b1000_0000_0000_0001;
       #5 rst = 0;
     end

     initial begin     // expect Gap = 0
       #200 rst = 1;
       #5 Data = 16'hffff;
       #5 rst = 0;
     end
```

```
initial begin     // expect gap = 0
  #400 rst = 1;
  #5 Data = 16'h0000;
  #5 rst = 0;
end

initial begin     // expect gap = 0
  #600 rst = 1;
  #5 Data = 16'hf000;
  #5 rst = 0;
end
initial begin     // expect gap = 0
  #800 rst = 1;
  #5 Data = 16'h0f00;
  #5 rst = 0;
end
initial begin     // expect gap = 8
  #1000 rst = 1;
  #5 Data = 16'hf00f;
  #5 rst = 0;
end
initial begin     // expect Gap = 0
  #1200 rst = 1;
  #5 Data = 16'haaaa;
  #5 rst = 0;
end
initial begin     // expect gap = 1
  #1400 rst = 1;
  #5 Data = 16'h5555;
  #5 rst = 0;
end
initial begin     // expect Gap = 4 (decreasing gap size)
  #1600 rst = 1;
  #5 Data = 16'b0100_0010_0010_0101;
  #5 rst = 0;
end
initial begin     // expect Gap = 4 (increasing gap size)
  #1800 rst = 1;
  #5 Data = 16'b1010_0100_0100_0010;
  #5 rst = 0;
end
endmodule;
```

20. Develop, verify, and synthesize a frequency divider with a programmable divisor for the base frequency, and a programmable duty cycle.
21. Develop, verify, and synthesize a Verilog model of a decoder that will decode a 16-bit address to determine in which of eight 8-k segments of a 64-k memory the word resides.

22. Describe the differences between the circuits that will be synthesized from the following Verilog cyclic behaviors:

    always @ (a **or** b **or** c **or** d) y = a + b + c + d;
    always @ (a **or** b **or** c **or** d) y = (a + b) + (c + d);

23. The instruction set of *RISC_SPM* is limited, and might not serve a particular application very well. Develop *RISC_SPM_e* and an enhanced version of *RISC_SPM* with additional instructions that would be useful if it is to be an embedded processor within a vending machine that is to accept currency, make change, and dispense coffee in response to selections made by the customer. The allowed selections are identified in Figure P7-23. The machine is to assert signals that (1) control dispensing units that blend the coffee according to the customer's choices, (2) accept currency and dispense change, and (3) send messages to a display panel.

Size	Venti	Grande	Tall
Coffee	Normal	Decaff	Espresso
Flavor	Hazelnut	Vanilla	Raspberry
Creamer Type	Half-n-Half	Whole	Skim
Creamer Amount	Heavy	Medium	Tall
Sweetener Type	Sugar	Artificial	Light
Sweetener Amount	Heavy	Medium	Equal

FIGURE P7-23

CHAPTER 8

Programmable Logic and Storage Devices

As technology advances, the density, complexity and size of field-programmable gate arrays (FPGAs) provide an attractive, cost-efficient, and increasingly important alternative to semicustom application-specific integrated circuits (ASICs). Mask charges for cell-based ASICs can cost from $250K to $500K, eliminating these devices from the low-volume end of the market. The opportunity to realize large circuits in FPGAs has created pressure for a change in the method by which circuits are designed for FPGA-based applications. Designers who use schematic-entry tools can be productive and efficient when designs are small, but the trend is toward larger and larger designs targeted for FPGAs. The language-based design methodology that has served the ASIC design flow has become essential to FPGA-based design flows, because it is the key to meeting ever-shrinking windows of opportunity for new products. As a result, FPGA vendors have been led to improve their support of language-entry tools for FPGAs, and recognize that more and more designers are shifting from schematic entry tools to language-entry tools. This chapter will emphasize a design flow for FPGAs that is entirely Verilog-based.

The technologies available for implementing digital circuits range from the discrete gates and standard integrated circuits (ICs) used in low-density/low-performance applications, to cell-based and full-custom ICs for high-density/high-performance circuits. Standard integrated circuits can be manufactured cheaply, but they implement very limited, basic functionality at low levels of integration. Production of customized logic, having a small market, creates an inventory risk because the quantities that could be sold do not warrant the expense of their development and production, and IC manufacturers cannot afford to stock multiple variants of specialized functional units, as they do with standard parts. Inevitable progress in technology alone would render their inventory worthless before their investment could be recovered. Our focus in this chapter will be on programmable logic devices, which lie between the two extremes of density and performance that characterize standard parts and full-custom circuits.

Programmable logic devices (PLDs) were born out of a necessity created by two conflicting realities: large, dense, high-performance circuits cannot be built practically or economically from discrete devices, and dedicated ICs cannot be produced economically to satisfy a diversity of low-volume applications. The resolution of these forces lies in programmable logic devices.

Although read-only memories (ROMs), programmable logic arrays (PLAs), programmable array logic (PALs), complex PLDs (CPLDs), FPGAs, and mask-programmable gate arrays (MPGAs) are all programmable, we will use the term *PLD* to indicate the low-density structures that were introduced to implement two-level combinational logic: PLAs, PALs, and similar vendor-named devices. PLDs are distinguished by their having a regular structure of identical basic functional units with fixed architecture. Certainly, MPGAs and standard-cell-based designs can be considered to be programmable because an application-specific program determines the metalization layers interconnecting the underlying transistors to form functional units in the case of an MPGA, and the placement and routing of standard cells can be considered to "program" a cell-based wafer. We consider MPGAs to be in the broader family of programmable devices, but we do not include standard-cell-based designs. Why?

MPGAs are formed from a regular array of transistors. Unprogrammed MPGAs have an identical structure. They are programmed by adding layers of metal interconnects to compose and connect macros with a desired functionality. On a local basis, the interconnect might establish, for example, the connectivity that forms a NAND gate, and on a global basis establish the functionality of an adder. The architecture of the basic functional units remains unchanged, but the interconnection fabric is unique to the application. In contrast, standard cell layouts do not have a fixed, basic architecture of functional units. A standard-cell-based design has regularity in the structure of its layout channels, but the functional units themselves are not uniform and do not have an architecturally determined placement. One cell may implement an inverter, another a flip-flop. Nor is a cell itself programmed. The overall architecture of a cell-based design is completely flexible within the constraints imposed by layout routing channels and a library of cells. No cell pattern need be replicated in a cell-based layout. For these reasons, we make a distinction between programming that overlays an interconnection fabric on a given fixed architecture, and programming that establishes an architecture of functional units, as is the case with standard cells. We mean the former case when we use the term *PLD*.

Storage devices, such as ROMs, are considered to be PLDs because they can implement combinational logic by storing the values of a function at memory locations that are addressed by the inputs of the function. These implementations, of necessity, implement the full truth table of the function. Memory-implemented combinational logic may be inefficient, because minimization techniques are not used to implement a full truth table of a function, and device resources might not be fully utilized.

8.1   Programmable Logic Devices

Programmable logic devices[1] have a fixed architecture but their functionality is programmed for a specific application, either by the manufacturer or by the end user. PLDs whose architecture is programmed by the manufacturer are referred to as *mask-programmable logic devices* (MPLDs); those that are programmed by the end-user are referred to as *field-programmable logic devices* (FPLDs). The architecture of the basic functional unit of a PLD is fixed, and is not customized by the user. Consequently, the development and production costs of PLDs can be amortized over a larger base of customers, and the range of applications for the devices can be very broad. This reduces production and inventory risks for the manufacturer and unit costs for the consumer, while at the same time allowing advances in processing technology to be incorporated into an evolving product line. The design cycle of a system that uses a PLD can be very short because PLDs can be manufactured, tested, and placed in inventory in advance of their being chosen as a technology for an application. Because the devices are premanufactured, they are suitable for rapid prototyping of a design.

Three basic characteristics distinguish PLDs from each other: (1) an architecture of identical basic functional units, (2) a programmable interconnection fabric, and (3) a programming technology. The first type of PLD that we will consider has the *AND-OR* plane structure shown in Figure 8-1. This type of architecture is used to implement ROMs, PLAs, and PALs. It implements Boolean expressions in SOP form: The *AND* plane forms product terms selectively from the inputs, and the *OR* plane forms outputs from sums of selected product terms. A programmable interconnect fabric joins the two planes, so that the outputs implement sum-of-product expressions of the inputs. Whether and how a plane can be programmed determines the particular type of PLD that is implemented by the overall structure.

8.2   Storage Devices

The architecture used to implement PLDs lends itself to implementation of storage devices. Storage devices can be read-only or random-access, depending on whether the contents of a memory cell can be written during normal operation of the device. Read-only memory (ROM) is a device programmed to hold certain contents, which remain unchanged during operation and after power is removed from the device. In contrast, the contents of a random-access memory (RAM) can be changed during operation, and they vanish when power is removed. There is another major distinction between ROMs and RAMs: The circuit of a ROM is structurally modified to program the device prior to its use. In contrast, the circuit of a RAM is not programmed—it is fixed. Only the contents of a RAM are programmed. This occurs dynamically, during normal read and write operations of the circuit.

---

[1]There are many more vendors and families of devices than we can cover in a book of this type. See www.e-insite.net/ednmag for *EDN* Access's annual directory of PLDs, CPLDs, and FPGAs.

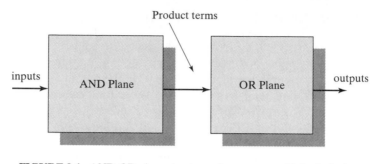

FIGURE 8-1 AND-OR plane structure of a programmable logic device.

8.2.1   Read-Only Memory (ROM)

A $2^n \times m$ ROM consists of an addressable array of semiconductor memory cells organized as 2^n words of m bits each. A read-only memory has n inputs, called "address lines," and m outputs, called "bit lines." The *AND*-plane of the structure shown in Figure 8-2 serves as an address decoder and is nonprogrammable. The address decoder implements a full decode of the n inputs, and each pattern of input bits addresses a unique decoded output, called a "word line." Each input address word selects one of the 2^n memory words to assert a word line, and each cell of a word stores 1 bit of information. Consequently, each word line corresponds to a minterm of a Boolean expression.

ROMs can be manufactured in a variety of technologies: bipolar, complementary metal-oxide semiconductor (CMOS), n-channel MOS (nMOS), and p-channel MOS (pMOS). A mask-programmed ROM implemented in nMOS technology has the circuit structure shown in Figure 8-3. The bit lines form the output word, and n-channel

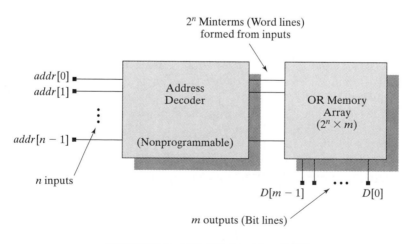

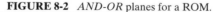

FIGURE 8-2 *AND-OR* planes for a ROM.

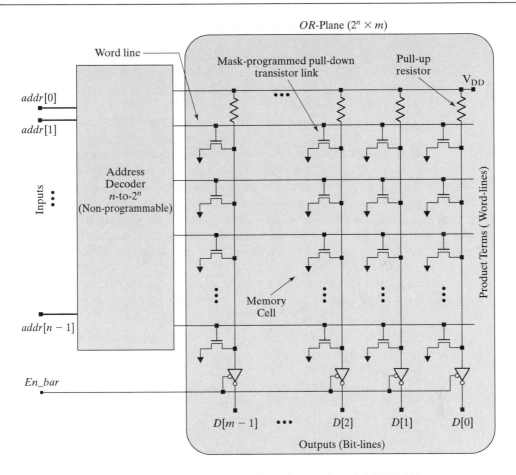

FIGURE 8-3 Circuit structure of a mask-programmed nMOS ROM.

link transistors connect the word lines to the bit lines. A bit line is normally pulled up to V_{dd}, but when a word line is pulled high by the address decoder, the n-channel transistors that are attached to it will be turned on. This action pulls the corresponding bit lines down. The pattern of link transistors attached to a given word line determines the pattern of 1s and 0s that appear on the bit lines for the applied input address word. This pattern is determined by the customized mask set for the device. Given the three-state output inverters, the presence of a link transistor corresponds to a stored 1 at the location of the memory cell. The mask set ensures that transistors will be fabricated only at locations that require a link.

The information stored in a ROM can be read under normal operation of a host circuit, but not written. The outputs of a ROM are normally three-stated, so that the device can be connected to a shared bus serving multiple devices. In commercial ROMs, an additional chip-select input allows multiple devices to be connected to a common bus, each selectable by its unique address. If a ROM has been selected, a pattern of 0s and 1s at its address inputs causes one and only one word line to be asserted.

A $2^n \times m$ ROM can store m different functions of n variables (i.e., truth table storage). Figure 8-4 illustrates a 16×8 ROM having a 4-bit address word and a total of 16 memory words of 8 bits each. Commercial ROMs are available in a range of organization and densities, as shown in Table 8-1.

A ROM is a nonvolatile memory because the stored information remains when power is removed from the device. Mask-programmable ROMs are manufactured with a fixed, nonerasable memory pattern, usually for high-volume applications. Their non-recurring engineering (NRE) cost is relatively high compared to a field-programmable ROM because the mask set that programs the chip is customized to a particular end user's application. The mask set can be produced in about a 4-week cycle. Mask-programmed ROMs are used in applications in which a system needs stored data and has no need to alter the data in ordinary use. For example, they are used as data tables that hold the codes for characters that are to be displayed on a cathode ray tube (CRT) monitor screen of a computer system, and hold the bootstrap program that executes immediately when a personal computer is powered on. They are widely used in electronic point-of-sale terminals in retail stores, instrumentation, domestic appliances, industrial equipment, video games, and security systems.

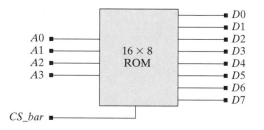

FIGURE 8-4 Schematic symbol for a 16×8 ROM.

TABLE 8-1 Organization and density of commercial ROMs.

Organization	Density
32K × 8	256K bit
64K × 8	512K bit
128K × 8	1M bit
256K × 8	2M bit
512K × 8	4M bit
1024K × 8	8M bit
64K × 16	1M bit
. . .	
256K × 16	4M bit
512K × 16	8M bit

8.2.2   Programmable ROM (PROM)

A field-programmable ROM (PROM) is one that can be programmed (once) by an end user with a special apparatus called a PROM programmer. PROMs are said to be one-time programmable (OTP) or write-once memory (WOM). PROMs are nonvolatile and nonerasable. Usually manufactured in a bipolar technology, a PROM initially has a pull-up device at every crosspoint between a word line and the internal bit line. The pull-up device (diode or transistor) is also connected to a metal fusible link, as shown in Figure 8-5. A PROM programmer selectively applies a voltage (10–30 V) to cause current sufficient to vaporize the link, thereby disconnecting the pull-down device from the word line, and permanently causing a 1 to appear in that cell when it is decoded by a word line. The output of the bit line is the inverted content of the memory cell.

The bit line outputs of a PROM are driven by three-state inverters, and each inverter input is connected to ground by a pull-down resistor and is also connected to the internal bit line. In the absence of a signal on a word line, the bit line will be at ground potential, and the output will be high. The enable line in the circuit of Figure 8-5 is active-low, and a high minterm line pulls a bit line output low. Cells having a blown link are not affected by their minterm line, and their output remains at 1, due to the action of the pull-down resistor. Note that a bit-line can be pulled down by one or more

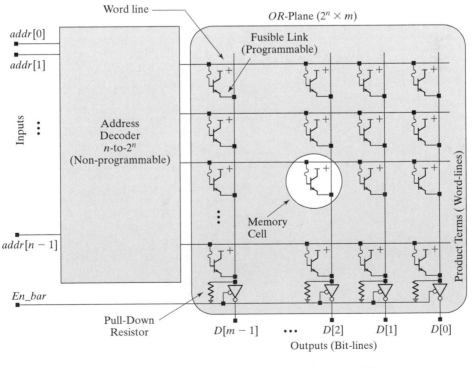

FIGURE 8-5 Circuit structure for a fusible-link bipolar PROM.

cells. In this scheme, the presence of a link transistor implies a 0 in the output word when the wordline is decoded. The programming is permanent (i.e., there is no way to restore the blown links and create a different program). A program can be modified however, if the modification affects only links that have not yet been blown.

8.2.3 Erasable ROMs

The architecture of an erasable PROM is similar to that of a PROM, but it uses a floating-gate nMOS transistor as the link device between a word line and the bit lines (Figure 8-6). A floating-gate transistor has an additional gate inserted between the operational gate and the channel. This gate is surrounded by a high-impedance dielectric material, and is insulated from the operational gate. When special circuitry (not shown) applies a sufficiently high voltage (e.g., 21 V) to the operational gate the insulator breaks down and a negative charge is pulled from the channel and becomes trapped on the floating gate when the programming voltage is removed. The effect of the trapped charge is to turn off the transistor by depleting the channel of carriers, which effectively raises the threshold voltage of the transistor and breaks the link between the word line and the bit line, allowing the bit line to float high and remain high independently of the

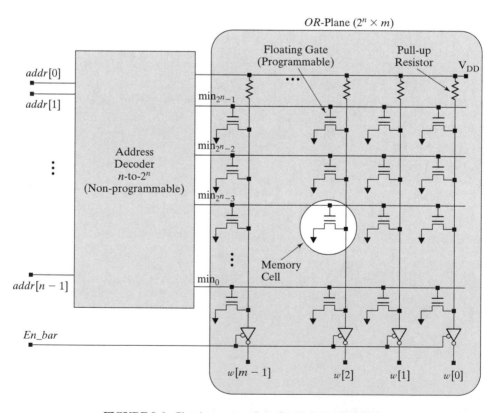

FIGURE 8-6 Circuit structure for a floating-gate EEPROM.

affected word line. Subsequent reads of the cell are 1. In this scheme, the presence of a programmed link transistor (i.e., one that has trapped charge) implies a 0 in the output word when the word line is decoded.

There are two types of erasable ROMs: those that are erasable by ultraviolet (UV) light and those that are erasable by electricity. The former, called an EPROM or UVEPROM, have a quartz opening and rely on a mechanism by which the UV light at a specific wavelength causes a temporary breakdown of the insulation of the floating gate and allows photocurrents to remove the trapped charge and effectively erase the stored information. The latter type, called an EEPROM (electrically erasable PROM), use electrical pulses to break down the insulated floating gate and erase the stored pattern. Application of a high negative voltage to a minterm line will remove the trapped charge from the floating gate. The UV-erase mechanism of an EPROM is nonselective (also referred to as "bulk erase"); all of the memory contents are reprogrammed to 1. A EEPROM, however, has additional circuitry providing a selective erase capability, allowing individual words to be selectively erased and reprogrammed.

EPROMs are commonly used in the debug phase of firmware development for microprocessor-based systems. They require 5 to 20 minutes of exposure to UV light to accomplish an erase. ROMs are substituted for production after the program is correct because their packages omit the quartz window and are cheaper. EEPROMs are attractive because they can be programmed in-circuit, can be erased with low current, and do not require the additional hardware and expense of a PROM programmer or a UV source.

Volatility and fatigue are two important considerations in applications of ROM technology. In the absence of UV light, an EPROM is guaranteed to hold 70% of its charge for at least 10 years [1]. The insulating material in an EEPROM is thinner than that for an EPROM, and can deteriorate, so EEPROMs have a limited number of write/erase cycles, typically 10^2 to 10^5. EEPROMs that have exceeded their fatigue limit may fail to hold a charge on the floating gate or may trap charge on the gate. Because they can be erased electrically they erase much faster than an ordinary EPROM, making them suitable for prototype code development. They are also used in system applications that do not require a high number of write/erase cycles over the useful life of a product, such as storage of default configuration data in a personal computer [1]. EEPROMs are also available with low in-circuit programming voltages (e.g., Atmel AT49LV1024).[2]

8.2.4   ROM-Based Implementation of Combinational Logic

ROMs are commonly used in applications that require a truth table for combinational logic. They are an attractive technology because a ROM can be programmed to implement any of 2^{2^n} different functions of n inputs, and a single ROM can implement any of those functions at any of its bit lines (standard logic would require a new circuit structure for each different function). A ROM-based design can be modified by simply replacing the ROM, without altering the external circuitry. The complexity of the logic

---

[2]See www.atmel.com.

being implemented does not have an impact on the effort to program the device, as it would in the case of discrete or building-block logic. ROMs are usually faster than multiple large and medium scale integrated (LSI/MSI) devices and other PLDs in moderately sized circuit applications, and often they are faster than an FPGA or custom LSI chip in a comparable technology. On the other hand, for moderately complex functions, a ROM-based circuit is usually more expensive, consumes more power, and may run more slowly than a circuit that uses multiple LSI/MSI devices and PLDs or a small FPGA [1]. Their full address decoding circuitry ultimately limits ROMS to applications that have no more than 20 inputs. Like other semiconductor devices, ROMs benefit from advances in technology that are leading to cheaper and denser devices.

8.2.5   Verilog System Tasks for ROMs

Verilog has two file input–output (I/O) system tasks that can be used to load memory data from a text file, reducing the effort required to initialize a large memory, as an alternative to writing the individual words within the ROM model. A single ROM model can serve a variety of applications by substituting text files. The tasks *$readmemb* and *$readmemh* load to specified locations in a memory the contents of a text file formatted as binary or hexadecimal words, respectively (see "Selected System Tasks and Functions" at the companion web site).

Example 8.1

The truth table of the 2-bit comparator presented in Example 4.4 is shown in Figure 8-7 with a symbolic diagram of the fuse links required to program a PROM-based implementation of the circuit with active-low enabled, three-stated outputs. Note that the output column *D0* is unused, and that the pattern of links accounts for the inverted outputs of the device.

The Verilog model *ROM_16_x_4* illustrates how to declare a memory of 16 words, each having a width of 4 bits, and how to load the memory from a text file of data.

```
module ROM_16_x_4   (ROM_data, ROM_addr);
   output [3:0] ROM_data;
   input   [3:0] ROM_addr;
   reg     [3:0] ROM [15:0];

   assign ROM_data = ROM [ROM_addr];

   initial $readmemb ("ROM_Data_2bit_Comparator.txt",  ROM, 0, 15);
endmodule
```

The contents of a binary-formatted text file for the 2-bit comparator would be listed from address 0 to address 15 as:

001x
010x
010x
010x
100x
001x
010x

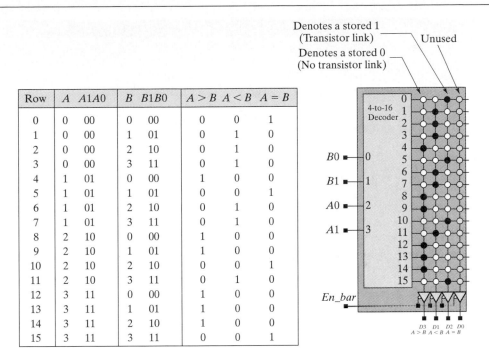

FIGURE 8-7 Truth table and PROM fuse map for a 2-bit comparator.

```
010x
010x
100x
001x
010x
010x
100x
100x
001x
010x
100x
100x
100x
001x
```

A useful tip: in the Silos III simulation environment for Verilog (see www.simucad.com), the text file to be read by *$readmemb* or *$readmemh* is expected to be located in the same directory (folder) in which the project is located, unless a pathname is specified to a different location. The simulation results in Figure 8-8 display the contents of the ROM as a word and as individual bits, illustrating that the unused bit is displayed in Verilog's 4-valued logic as an unknown logic value (denoted by x). The actual fuse map would have to specify a 1 or 0 for the bit.

End of Example 8.1

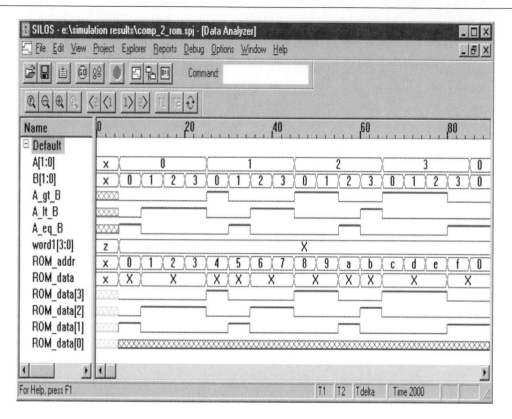

FIGURE 8-8   Simulation results for a ROM-based 2-bit comparator.

8.2.6   Comparison of ROMs

A variety of ROMS are manufactured by several commercial vendors. Table 8-2 compares representative devices and lists some typical performance attributes. The indicated trend of the complexity and cost is qualified by the fact that the unit cost of mask-programmed ROMs can be quite low, depending on the volume of parts that are produced. The performance characteristics are a moving target, linked to advances in process technology.

8.2.7   ROM-Based State Machines

ROMs provide a convenient implementation of a state machine, and can be an economical implementation if the attributes of the device match the application. The ROM-based state machine shown in Figure 8-9 uses a $2^n \times m$ ROM to store the next-state and output functions of a state machine. The state of the machine is stored in a set of D-type flip-flops, because they typically require fewer outputs from the ROM than would a J-K flip-flop.

TABLE 8-2 Comparison of (a) ROM types and (b) performance attributes.

Device	Programming Mode	Erase Mode	Complexity and Cost	Example	Access Time
EEPROM	In-circuit Byte-by-byte	In-circuit Byte-by-byte		Intel 2864 8K $\times$ 8 nMOS	
FLASH	In-circuit	In-circuit Bulk or sector		AT49LV1024 64K $\times$ 16 nMOS	70 ns**
EPROM	Out-of-circuit	Out-of-circuit Bulk, UV Light		Intel 2732 4K $\times$ 8 nMOS	45 ns
PROM	Custom by user (OTP***)	None		TMS47C256 32K $\times$ 8 CMOS AT27BV400 256K $\times$ 16 or 512K $\times$ 8	150 ns
ROM*	Mask	None			

* Requires high volume to offset NRE
** Programming time: 500 ms
*** One-time programmable

(a)

Type	Technology	Read cycle	Write cycle
ROM	NMOS, CMOS	10–200 ns	4 weeks
ROM	Bipolar	< 100 ns	4 weeks
PROM	Bipolar	< 100 ns	10–50 μs/byte
EPROM	NMOS, CMOS	25–200 ns	10–50 ms/byte
EEPROM	NMOS	50–200 ns	10–50 ms/byte

Adapted from Wakerly JF. *Digital Design—Principles and Practice*, Upper Saddle River, NJ: Prentice-Hall, 2000.

(b)

The method for designing a ROM-based state machine is simplified because the truth table is implemented directly, without minimization. The size of the array depends on the number of inputs, not on the complexity of the implemented logic. We form a ROM table in which the row address represents the present state of the machine, and the contents associated with that address hold the output and the next state.

Example 8.2

A Mealy-type state machine describing a binary coded decimal (BCD)-to-Excess_3 code converter was developed by manual methods in Example 3.2. A Verilog model of a ROM memory and of the machine are listed below. The ROM model is external to the state machine model. Its contents are written immediately by the ***initial*** (single-pass) behavior that executes when a simulation begins. The listing identifies the contents

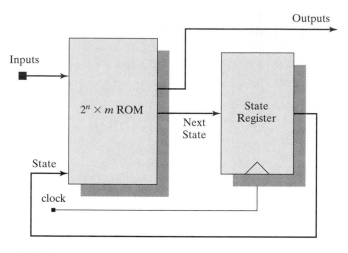

FIGURE 8-9   Block diagram for a ROM-based finite-state machine.

stored at each ROM address. The comments identify the state associated with each address, and hold the output and next state. A continuous assignment updates the address of the ROM (*ROM_addr*) whenever the state or the input of the machine change, ensuring that the machine is of the Mealy type. The testbench specifies a simple input sequence for the purpose of illustrating the machine's behavior, shown in Figure 8-10,

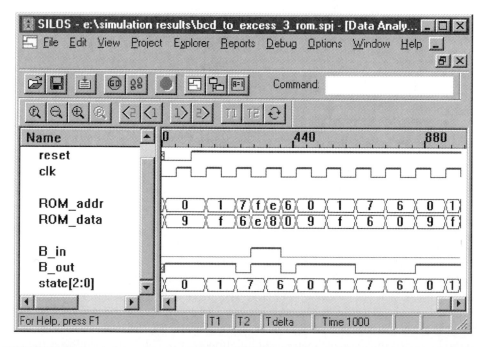

FIGURE 8-10   Simulation results for a ROM-based Verilog model of a BCD-to-Excess_3 code converter.

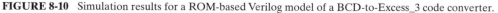

which matches the behavior of the manually designed gate-level machine shown in Figure 3-23. In this example the contents of the ROM are listed within the ROM, rather than in an external file.

```
module ROM_BCD_to_Excess_3 (ROM_data, ROM_addr);
  output [3:0] ROM_data;
  input  [3:0] ROM_addr;
  reg    [3:0] ROM [15:0];
  assign ROM_data = ROM[ROM_addr];
  //                          input state output next_state
  initial begin
    ROM[0] = 4'b1001;    // S_0   0 000 1 001
    ROM[1] = 4'b1111;    // S_1   0 001 1 111
    ROM[2] = 4'b1000;    // S_6   0 010 1 000
    ROM[3] = 4'b1110;    // S_4   0 011 1 110
    ROM[4] = 4'bxxxx;    // not used
    ROM[5] = 4'b0011;    // S_2   0 101 0 011
    ROM[6] = 4'b0000;    // S_5   0 110 0 000
    ROM[7] = 4'b0110;    // S_3   0 111 0 110

    ROM[8] = 4'b0101;    // S_0   1 000 0 101
    ROM[9] = 4'b0011;    // S_1   1 001 0 011
    ROM[10] = 4'b0000;   // S_6   1 010 0 000
    ROM[11] = 4'b0010;   // S_4   1 011 0 010
    ROM[12] = 4'bxxxx;   // not used
    ROM[13] = 4'b1011;   // S_2   1 101 1 011
    ROM[14] = 4'b1000;   // S_5   1 110 1 000
    ROM[15] = 4'b1110;   // S_3   1 111 1 110

  end
endmodule

module BCD_to_Excess_3_ROM (ROM_addr, B_out, ROM_data, B_in, clk, reset);
  output [3:0]  ROM_addr;
  output        B_out;
  input  [3:0]  ROM_data;
  input         B_in, clk, reset;

  reg    [2:0]  state;
  wire   [2:0]  next_state;
  wire          B_out;

  assign next_state = ROM_data [2:0];
  assign B_out = ROM_data[3];
  assign ROM_addr = {B_in, state};

  always @ (posedge clk or negedge reset)
    if (reset == 0) state <= 0; else state <= next_state;
endmodule
```

```
module test_BCD_to_Excess_3b_Converter ();
  wire   B_out, clk;
  wire   [3:0] ROM_addr, ROM_data;
  reg    B_in, reset;

  BCD_to_Excess_3_ROM M1 (ROM_addr, B_out, ROM_data, B_in, clk, reset);
  ROM_BCD_to_Excess_3 M2 (ROM_data, ROM_addr);
  clock_gen M3 (clk);

  initial begin #1000 $finish; end

  initial begin
    #10 reset = 0;  #90 reset = 1;
  end

  initial begin
    #0 B_in = 0;
    #100 B_in = 0;
    #100 B_in = 0;
    #100 B_in = 1;
    #100 B_in = 0;
  end

endmodule
```

End of Example 8.2

8.2.8   Flash Memory

Flash memory devices are similar to EEPROMs, but have additional built-in circuitry to selectively program and erase the device in-circuit, without the need for a special programmer. They have widespread application in modern technology for cell phones, digital cameras, set-top boxes, digital TV, telecommunications, nonvolatile data storage, and microcontrollers. Flash memory is cost-competitive with a magnetic disk for capacities under 5 MB. Its low consumption of power makes it an attractive storage medium for laptop and notebook computers. Flash memories incorporate additional circuitry too, allowing simultaneous erasing of blocks of memory, for example 16 Kbytes to 64 Kbytes. Intel's StrataFlash memory[3] (3-Volt technology), shown in Figure 8-11, encodes the threshold voltage levels of transistors to achieve storage of multiple bits per cell (1 to 3), achieving reduced cell area and die size for a given density. An internal state machine controls the charge placement to achieve a target threshold voltage. Like EEPROMs, flash memories are subject to fatigue, typically having about 10^5 block erase cycles.

---

[3]See www.intel.com.

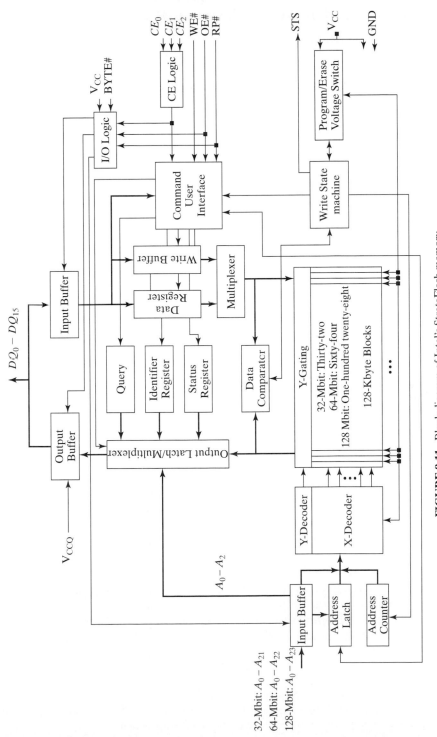

FIGURE 8-11 Block diagram of Intel's StrataFlash memory.

431

8.2.9   Static Random Access Memory (SRAM)

Read-only memories are limited to applications that require retrieval, but not storage, of information during ordinary operation. Computers and other digital systems perform many operations that retrieve, manipulate, transform, and store data, and therefore need read/write memories. For example, an application program must be retrieved from a relatively slow storage medium, such as a floppy disk or a CD-ROM, and moved on demand to a location where it can be accessed quickly by the processor. ROMs are not used to store large application programs, and they cannot dynamically store the data generated by a program's execution. Storage registers and register files support fast, random storage, but cannot be used for mass storage because they are implemented with flip-flops and occupy too much physical area in silicon to support applications that generate and store vast amounts of data. Small register files may be integrated in an ASIC or an FPGA to avoid having to access an external (slower) memory device.

RAM is faster and occupies less area than a register file, and it serves the function of providing fast storage and retrieval of large amounts of data during the operation of a computer (e.g., a video frame buffer). The name *random* indicates that RAMs allow data to be written to or read from any storage location in any order.[4] Most RAMs are volatile—the information they contain vanishes after power is removed from the device. A newer and emerging technology, nonvolatile RAM, will be discussed later in this chapter.

There are two basic types of RAMs: static and dynamic. Static RAMs (SRAMs) are implemented with a transistor-capacitor storage cell structure that does not require refresh; dynamic RAMs (DRAMs) are slower, use fewer transistors, and occupy less physical area, but they require refresh circuitry to retain stored data. They provide the densest storage devices, but their contents must be refreshed every few milliseconds; therefore DRAMs require additional supporting circuitry. SRAMs are used as fast-cache memory in a computer.

The circuit in Figure 8-12 shows the basic structure of an SRAM cell. A pair of inverters are connected in a closed loop and their outputs are tied to pass transistors attached to *Bit_line* and its complement, *Bit_line_bar*. SRAMs commonly use the 6-transistor circuit[5] shown in Figure 8-13. The gate of each pass transistor is connected to the word line of the circuit. Suppose that *Word_enable* is de-asserted, that the stored content of the cell has *cell* = 1 and *cell_bar* = 0, and that the inputs are changed to *Bit_line* = 0 and *Bit_line_bar* = 1. When *Word_enable* is asserted, *cell* is driven to 0 and *cell_bar* is driven to 1. The feedback structure forces the output of one inverter to be the complement of the output of the other inverter.

The values of *Bit_line* and *Bit_line_bar* control the read and write operations. An array of such storage cells is configured with sense amplifiers that are used to read the contents of a cell. Data are written to the cell by precharging *Bit_line* and *Bit_line_bar*

---

[4]In contrast, note that data are read serially from a tape storage media.
[5]Other schemes use as few as four transistors by replacing the *p*-channel pull-up transistors with depletion-load devices that function as resistors and compensate for leakage current.

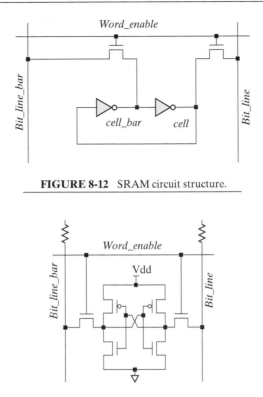

FIGURE 8-12 SRAM circuit structure.

FIGURE 8-13 Transistor-level SRAM cell.

to complementary values, and then strobing *Word_enable*. This forces the inverters to have the values imposed by the bit lines. To understand how reading is done, suppose that *Bit_line* and *Bit_line_bar* are both precharged to a 1. When *Word_enable* is strobed, the internal node at which a 0 is held by an inverter will provide an *n*-channel pull-down path for the bitline to which it is attached by a pass transistor. The differential voltage between *Bit_line* and *Bit_line_bar* can be detected by a sense amplifier and used to determine the configuration of the stored data [2]. The read operation is non-destructive, because the internal state of the stored data is not affected by the circuit activity during a read cycle.

In the following examples we will progressively develop a series of Verilog functional models of SRAMs, beginning with a model of a simple SRAM cell, and proceeding to larger memory blocks with unidirectional and bidirectional data ports.

The basic RAM cell represented by the block diagram symbol in Figure 8-14 has active-low inputs for chip select (*CS_b*), and write enable (*WE_b*). The chip select signal is generated by a decoder that selects among multiple chips in the same system. Note the absence of a clock signal. Storage registers and register files are implemented by flip-flops, but the storage devices of RAMs are implemented as transparent latches, which support asynchronous storage and retrieval of data and minimize the time that a RAM requires service from a shared bus.

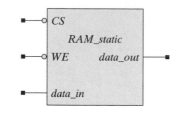

FIGURE 8-14   SRAM cell: block diagram symbol.

Example 8.3

The level-sensitive Verilog description, *RAM_static* models a simple RAM cell, without accounting for propagation delays.[6] With active-low signals denoted by the suffix *_b*, level-sensitive behavior is modeled here by a single continuous assignment declaration with nested conditional operators decoding the status of *CS_b* and *WE_b*. If *CS_b* is not asserted the output is in the three-state mode (has the value Verilog logic value *z*). If *CS_b* and *WE_b* are asserted (low), the cell is in transparent mode, and *data_out* follows *data_in*; if *CS_b* is asserted and *WE_b* is de-asserted, the cell is latched. The contents of the cell can always be read, but a host processor would access *data_out* only when *WE_b* is de-asserted. The functional schematic in Figure 8-15(a) forms *RAM-static*; the simulation results presented in Figure 8-15(b), demonstrates the cell's behavior.

```
module RAM_static (data_out, data_in, CS_b, WE_b);
    output  data_out;
    input   data_in;
    input   CS_b;            Active-low chip select control
    input   WE_b;            Active-low write control

    wire data_out = (CS_b == 0) ? (WE_b == 0) ? data_in : data_out : 1'bz;

endmodule
```

The Verilog description is synthesizable, and is implemented by a single four-input lookup table (LUT) in a Xilinx chip. The maximum combinational path delay after the logic has been placed and routed in a Xilinx XCS10XL chip is 6.756 ns (the delay from the input pad for *WE_b* to the pad for *data_out*).[7]

End of Example 8.3

---

[6]A *wire* declaration with an assignment to an expression implements the implicit combinational logic of the expression. The declaration is equivalent to a separate declaration of a *wire* and a continuous-assignment statement.

[7]This information is contained in the Post-Route Timing Report generated by the implementation process within the Xilinx design flow.

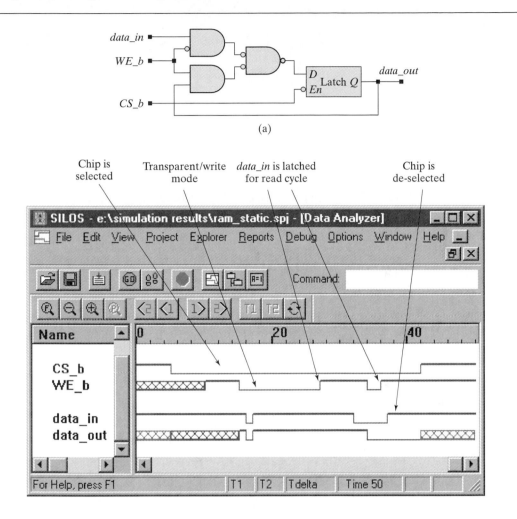

FIGURE 8-15 SRAM cell: (a) Xilinx-generated functional schematic and (b) simulation results illustrating chip select, write, and read behavior of *RAM_static*.

Example 8.4

The Verilog model of an SRAM cell can be modified to incorporate a single bidirectional port for use in a bus-based architecture. An additional active-low signal, *OE_b* (output enable) is added to the block diagram symbol (see Figure 8-16) and controls the datapaths through the three-state I/O buffers. The datapath is reduced from two signal ports to one port, which renders a great savings of package pins and total area if the data port is a wide vector. The structure of the model is shown in Figure 8-17, where the latch is implemented by a mux with feedback. The data paths for a write operation

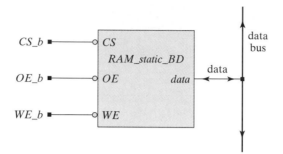

FIGURE 8-16   SRAM cell: block diagram symbol with a bidirectional data port interface to a shared bus.

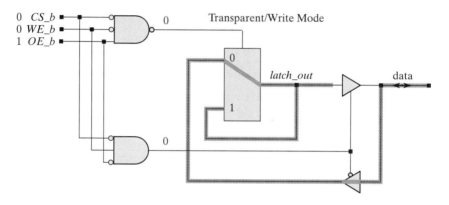

FIGURE 8-17   SRAM cell with bidirectional data port: configured to write external data through a bidirectional data port to the internal cell ($WE\_b = 0, OE\_b = 1$).

are shown. Output enable (OE_b) is asserted (low) during a read operation, and write enable is asserted (low) during a write operation. If WE_b is asserted and OE_b is not, the value at *data* is transparent through *latch_out*, and is held when WE_b is de-asserted (i.e., written to the cell). Conversely, when WE_b is not asserted and OE_b is asserted, the content of the cell can be read through *data*, as shown by the data paths in Figure 8-18.

The Verilog model[8] of the RAM cell with bidirectional data port, *RAM_static_BD*, is given below.

```
module RAM_static_BD (data, CS_b, OE_b, WE_b);
  inout data;          // Bi-directional data port
  input CS_b;          // Active-low chip select
  input OE_b;          // Active-low output enable
  input WE_b;          // Active-low write enable
```

_____

[8]Continuous assignments are used here to illustrate another style for modeling level-sensitive behavior. The default type of the target of the assignment is a **wire**. (Some tools might require an explicit declaration of type.)

```
assign latch_out = ((CS_b == 0) && (WE_b == 0) && (OE_b == 1))
                  ? data: latch_out;

assign data = ((CS_b == 0) && (WE_b == 1) && (OE_b == 0))
            ? latch_out : 1'bz;
endmodule
```

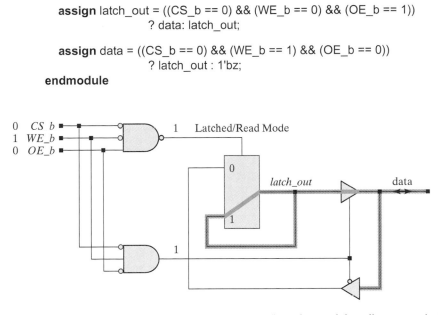

FIGURE 8-18 SRAM cell with bidirectional data port: configured to read the cell contents via the bidirectional data port ($WE\_b = 1, OE\_b = 0$).

Two additional modes are possible. With CS_b asserted (low), the control lines could be $WE\_b = 0, OE\_b = 0$, and $WE\_b = 1, OE\_b = 1$. The configurations that result are shown in Figure 8-19. The cell is latched in both cases; its contents are not affected by the external data path, and *latch_out* does not affect *data*. The contents of the cell are not available at *data*.

The functional schematic of *RAM_static_BD*, created by the Xilinx ISE synthesis tool[9], is shown in Figure 8-20. The schematic consists of a latch with additional logic to steer the I/O datapaths through a bidirectional data port. The synthesized and implemented circuit has *data* mapped to an I/O block (IOB) configured for bidirectional operation in a Xilinx XCS10XL chip.[10] The maximum combination path delay is 8.178 ns (from CS_b to *data*). The slight increase in delay reflects the presence of the additional logic for the bidirectional data port (compared to the delay of the stand-alone cell).

The interface between *RAM_static_BD* and a bidirectional shared bus is illustrated in Figure 8-21, and the structure of the testbench for verifying *RAM_static_BD* is shown in Figure 8-22. A separately declared register variable, *bus_driver*, drives the bidirectional bus and sends data to *RAM_static_BD*. The Verilog testbench, *test_RAM_static_BD*, uses a continuous assignment to assign the value of *bus_driver* to *data_bus* if OE_b is asserted during a write operation, and to disconnect *bus_driver*

---

[9]ISE is the Xilinx "Integrated Synthesis Environment," a tool for HDL-driven design entry and synthesis.
[10]This information is contained in the Pad Report generated by the Place and Route step of the synthesis process.

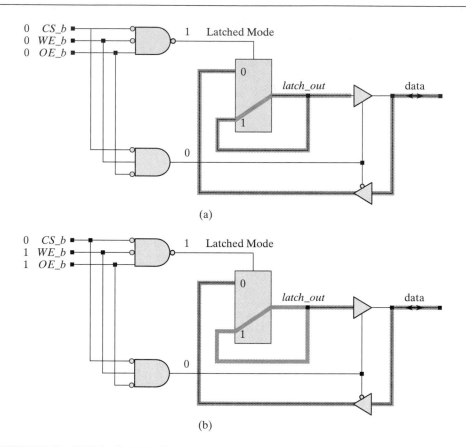

(a)

(b)

FIGURE 8-19 SRAM cell with bidirectional data port: configured for latched data and not reading or writing, (a) with ($WE\_b = 0, OE\_b = 0$) and (b) with ($WE\_b = 1, OE\_b = 1$).

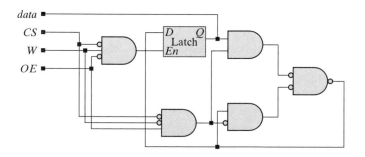

FIGURE 8-20 SRAM cell with bidirectional data port: preoptimization functional schematic created by Xilinx ISE tools.

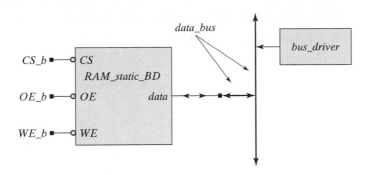

FIGURE 8-21 SRAM interface to a bidirectional data port.

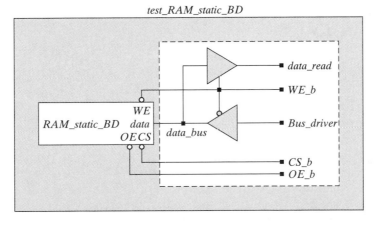

FIGURE 8-22 Testbench structure for SRAM cell with bidirectional data port.

otherwise. Note that *data_bus* has two drivers, *bus_driver* from the testbench, and *data*, the value driven through the bidirectional port of *RAM_static_BD*. The assignments from *bus_driver* to *data_bus* must be synchronized by *WE_b* and *OE_b* to avoid bus contention (i.e., so that the bus has only one driver at a time). The bidirectional nature of the testbench is illustrated in Figure 8-22, which shows *RAM_static_BD* instantiated within the testbench, *test_RAM_static_BD*. The signals *OE_b, WE_b,* and *CS_b* are declared as register variables in the testbench.

```
module test_RAM_static_BD ();
// Demonstrate write / read capability.
  reg  bus_driver;
  reg  CS_b, WE_b, OE_b;

  wire data_bus = ((WE_b == 0) && (OE_b == 1)) ? bus_driver : 1'bz;
```

```
RAM_static_BD M1 (data_bus, CS_b, OE_b, WE_b);

initial #4500 $finish;
initial begin
  CS_b = 1; bus_driver = 1; OE_b = 1;
  #500 CS_b = 0;
  #500 WE_b = 0;
    #100 bus_driver = 0;
  #100 bus_driver = 1;
  #300 WE_b = 1; #200 bus_driver = 0;
  #300 OE_b = 0; #200 OE_b = 1;
  #200 OE_b = 0; #300 OE_b = 1; WE_b = 0;
  #200 WE_b = 1; #200 OE_b = 0; #200 OE_b = 1;
  #500 CS_b = 1;
  #500 bus_driver = 0;
end

initial begin
  #3600 WE_b = 1; OE_b = 1;
  #200 WE_b = 0; OE_b = 0;
end
endmodule
```

The simulation results in Figure 8-23 show a sequence of values for CS_b, WE_b, and OE_b to demonstrate the modes of operation of RAM_static_BD. In the transparent mode, with $WE\_b = 0$ and $OE\_b = 1$, the value of data is determined by *bus_driver*, and *latch_out* is the same as *data*; when WE_b de-asserts, the value of data is latched (i.e., data are written to the cell).[11] When OE_b is asserted, with WE_b de-asserted, the value of *latch_out* appears at *data* and at *data_bus*. When OE_b and WE_b are simultaneously asserted or de-asserted, the bus is not driven. The bus could be used by another client.

End of Example 8.4

Large SRAMs cannot be implemented practically as a simple array structure for two important reasons. Large SRAMs require wide input decoders, and the footprints of long rectangular arrays might not be as convenient as square arrays for physical layout in silicon. As an alternative, large SRAMs arrays are reorganized into nearly rectangular block structures using two levels of decoding.

---

[11]The output of the cell can also be latched by de-asserting CS, but this is not the ordinary way to end a write cycle.

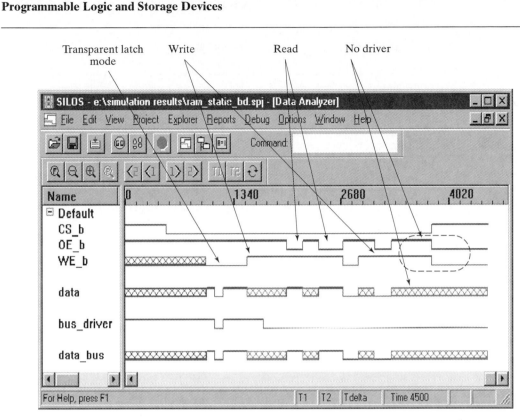

FIGURE 8-23 SRAM cell with bidirectional data port: simulation results.

Example 8.5

A 32K × 8 SRAM can be organized in the structure shown in Figure 8-24, where the array has been partitioned into 8 blocks of size 512 × 64. A 32K memory requires a 15-bit address. The lower 6 bits of the address are passed to a bank of 8 muxes, each having a 64-bit wide datapath. The 6-bit address selects 1 bit from each datapath to form an 8-bit output word, *Data_Out*. These same 6-bits steer *Data_In* to 1 of 64 input lines connected to each of the 8 memory blocks. The upper 9 bits of the address are decoded by combinational logic to select one 64-bit word in each of the 8 blocks. In this reorganized structure, the address decoders have a practical size because they decode fewer outcomes, and the overall structure is nearly square, having a height of 512 cells and a width of 512 cells.

End of Example 8.5

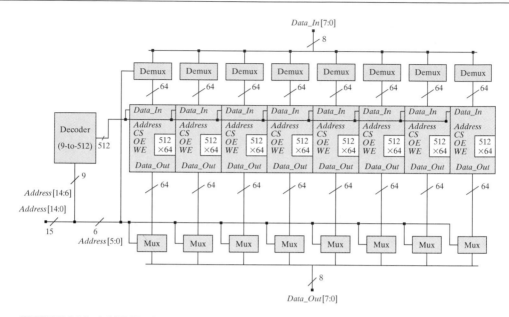

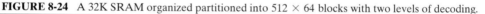

FIGURE 8-24 A 32K SRAM organized partitioned into 512 × 64 blocks with two levels of decoding.

Example 8.6

The alternative architecture shown in the block diagram in Figure 8-25 for a large SRAM has a bidirectional data port, where the column decoder, row decoder, and column I/O circuitry are represented by functional blocks adjacent to a 128 × 128 array of memory cells holding 2048 8-bit words. The upper 7 bits of the address word decode the 128 rows of the array, and the lower 4 bits of the address decode the 16 columns of words. The three-state devices that gate the bidirectional datapaths are not shown, but are contained in the column I/O circuitry. The Verilog model of the SRAM, *RAM_2048_8*, will be based on the organization of the data cells shown in Figure 8-26, where the address organization leads naturally to a row-by-row sequential access, beginning at the upper rightmost cell and proceeding ultimately to the lower leftmost cell. The model will include timing parameters, describing the propagation delays of the device, and timing checks to detect violations of operational constraints during simulation.

The Verilog model *RAM_2048_8* implements the structure illustrated in Figure 8-26. The companion testbench, *t_RAM_static_2048_8*, includes a behavior that writes a pattern of walking 1s through each column of the memory, successively, and another behavior that reads back the patterns stored in memory. Patterns are included in the testbench for simulating with and without delay (the ***specify . . . endspecify*** block containing timing parameters and path delays can be commented out from the code).[12]

---

[12]The delay values used are for illustration and do not represent the fastest devices that are available with the most advanced technology.

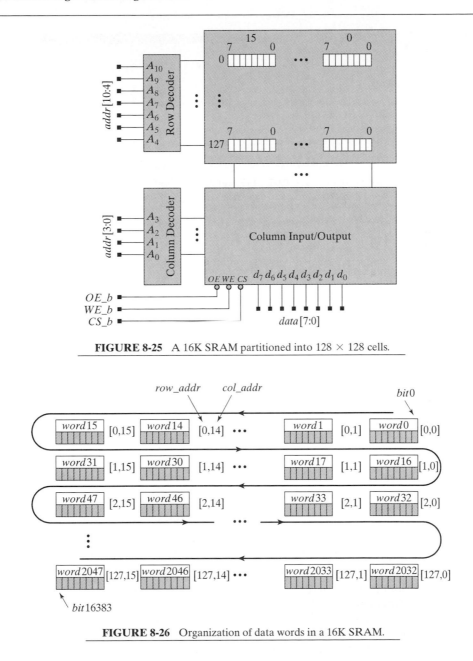

FIGURE 8-25 A 16K SRAM partitioned into 128×128 cells.

FIGURE 8-26 Organization of data words in a 16K SRAM.

The simulation results shown in Figure 8-27 show the patterns written in column 9, starting at row 104, for zero-delay simulation. The three-state action of the bus and the bidirectional datapath causes *data_bus* to have the value zz_H in the displayed waveforms. The results in Figure 8-28 show the patterns read back from the same locations. When nonzero propagation delays are included in the model, the simulation results in

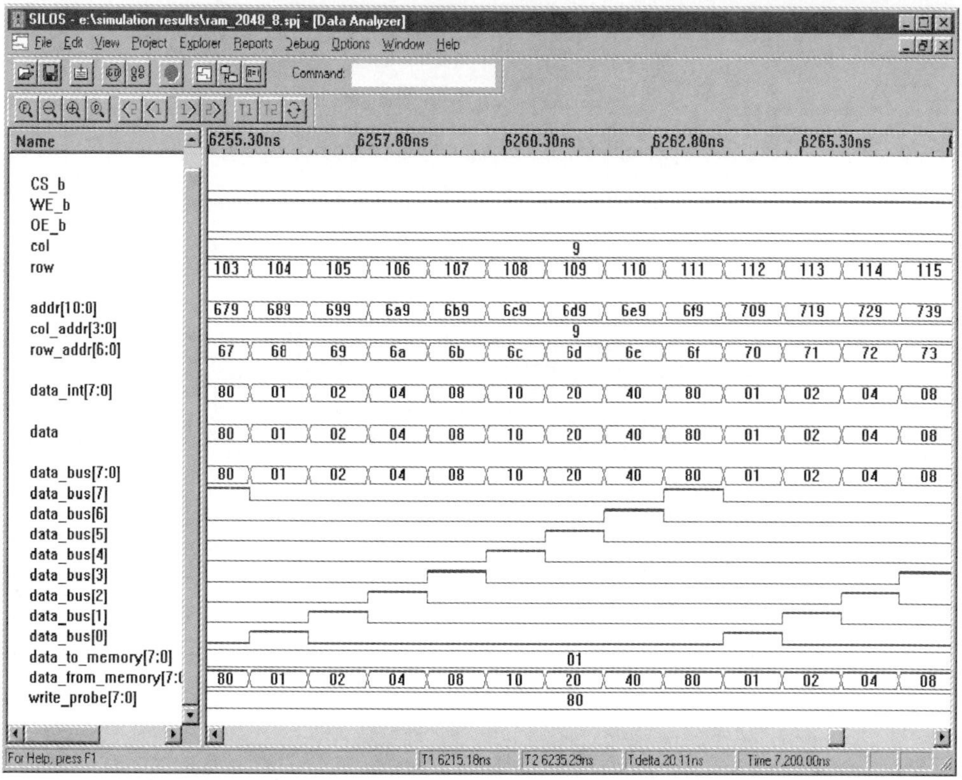

FIGURE 8-27 *RAM_2048_8*: simulation results for writing a walking 1s pattern to memory with zero delay.

Figure 8-29 are obtained for writing and reading data from memory. The testbench includes *write_probe*, which reports the value that is stored in memory at the rising edge of *WE_b* and provides a check on the latching activity of the model.

```
'timescale  1ns / 10ps
    module RAM_2048_8   (data, addr, CS_b, OE_b, WE_b);
      parameter               word_size = 8;
      parameter               addr_size = 11;
      parameter               mem_depth = 128;
      parameter               col_addr_size = 4;
      parameter               row_addr_size = 7;
      parameter               Hi_Z_pattern = 8'bzzzz_zzzz;
      inout [word_size -1: 0]  data;
      input [addr_size -1: 0]  addr;
      input                    CS_b, OE_b, WE_b;

      reg  [word_size -1 : 0] data_int;
```

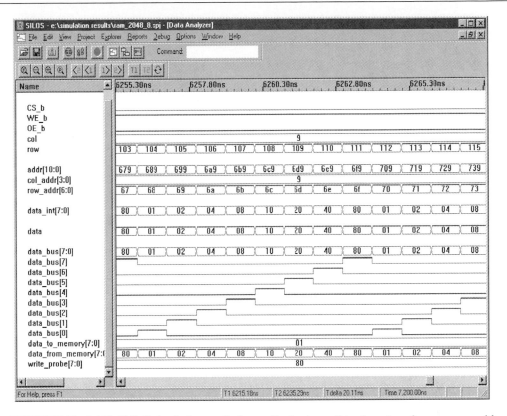

FIGURE 8-28 *RAM_2048_8*: simulation results for reading back a walking 1s pattern from memory with zero delay.

```
reg [word_size -1 : 0] RAM_col0 [mem_depth-1 : 0];
reg [word_size -1 : 0] RAM_col1 [mem_depth-1 : 0];
reg [word_size -1 : 0] RAM_col2 [mem_depth-1 : 0];
reg [word_size -1 : 0] RAM_col3 [mem_depth-1 : 0];
reg [word_size -1 : 0] RAM_col4 [mem_depth-1 : 0];
reg [word_size -1 : 0] RAM_col5 [mem_depth-1 : 0];
reg [word_size -1 : 0] RAM_col6 [mem_depth-1 : 0];
reg [word_size -1 : 0] RAM_col7 [mem_depth-1 : 0];
reg [word_size -1 : 0] RAM_col8 [mem_depth-1 : 0];
reg [word_size -1 : 0] RAM_col9 [mem_depth-1 : 0];
reg [word_size -1 : 0] RAM_col10 [mem_depth-1 : 0];
reg [word_size -1 : 0] RAM_col11 [mem_depth-1 : 0];
reg [word_size -1 : 0] RAM_col12 [mem_depth-1 : 0];
reg [word_size -1 : 0] RAM_col13 [mem_depth-1 : 0];
reg [word_size -1 : 0] RAM_col14 [mem_depth-1 : 0];
reg [word_size -1 : 0] RAM_col15 [mem_depth-1 : 0];

wire [col_addr_size -1: 0]    col_addr = addr[col_addr_size -1: 0];
wire [row_addr_size -1: 0]    row_addr = addr[addr_size -1: col_addr_size];
```

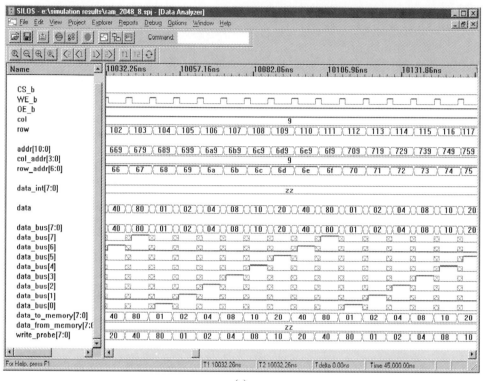

(a)

FIGURE 8-29 *RAM_2048_8*: simulation results for (a) writing to, and (b) reading from memory a walking
1s pattern with nonzero propagation delay.

```
assign data = ((CS_b == 0) && (WE_b == 1) && (OE_b == 0))
    ? data_int: Hi_Z_pattern;
always @ (data or col_addr or row_addr or CS_b or OE_b or WE_b)

  begin
   data_int = Hi_Z_pattern;
   if ((CS_b == 0) && (WE_b == 0))          // Priority write to memory
     case (col_addr)                        // column address
      0: RAM_col0[row_addr] = data;
      1: RAM_col1[row_addr] = data;
      2: RAM_col2[row_addr] = data;
      3: RAM_col3[row_addr] = data;
      4: RAM_col4[row_addr] = data;
      5: RAM_col5[row_addr] = data;
      6: RAM_col6[row_addr] = data;
      7: RAM_col7[row_addr] = data;
      8: RAM_col8[row_addr] = data;
      9: RAM_col9[row_addr] = data;
```

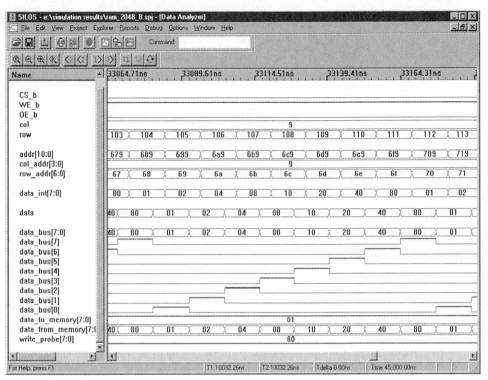

(b)

FIGURE 8-29 Continued

```
     10: RAM_col10[row_addr] = data;
     11: RAM_col11[row_addr] = data;
     12: RAM_col12[row_addr] = data;
     13: RAM_col13[row_addr] = data;
     14: RAM_col14[row_addr] = data;
     15: RAM_col15[row_addr] = data;
   endcase

else if ((CS_b == 0) && (WE_b == 1) && (OE_b == 0))  // Read from memory
   case (col_addr)
     0: data_int = RAM_col0[row_addr];
     1: data_int = RAM_col1[row_addr];
     2: data_int = RAM_col2[row_addr];
     3: data_int = RAM_col3[row_addr];
     4: data_int = RAM_col4[row_addr];
     5: data_int = RAM_col5[row_addr];
     6: data_int = RAM_col6[row_addr];
     7: data_int = RAM_col7[row_addr];
     8: data_int = RAM_col8[row_addr];
```

```
            9: data_int = RAM_col9[row_addr];
            10:data_int = RAM_col10[row_addr];
            11:data_int = RAM_col11[row_addr];
            12:data_int = RAM_col12[row_addr];
            13:data_int = RAM_col13[row_addr];
            14:data_int = RAM_col14[row_addr];
            15:data_int = RAM_col15[row_addr];
         endcase
      end
///*  Comment out of the model for a zero delay functional test.
specify
   // Parameters for the read cycle
   specparam t_RC = 10;        // Read cycle time
   specparam t_AA = 8;         // Address access time
   specparam t_ACS = 8;        // Chip select access time
   specparam t_CLZ = 2;        // Chip select to output in low-z
   specparam t_OE = 4;         // Output enable to output valid
   specparam t_OLZ = 0;        // Output enable to output in low-z
   specparam t_CHZ = 4;        // Chip de-select to output  in hi-z
   specparam t_OHZ = 3.5;      // Output disable to output in hi-z
   specparam t_OH = 2;         // Output hold from address change

   // Parameters for the write cycle
   specparam t_WC = 7;         // Write cycle time
   specparam t_CW = 5;         // Chip select to end of write
   specparam t_AW = 5;         // Address valid to end of write
   specparam t_AS = 0;         // Address setup time
   specparam t_WP = 5;         // Write pulse width
   specparam t_WR = 0;         // Write recovery time
   specparam t_WHZ = 3;        // Write enable to output in hi-z
   specparam t_DW = 3.5;       // Data set up time
   specparam t_DH = 0;         // Data hold time
   specparam t_OW = 10;        // Output active from end of write

//Module path timing specifications
   (addr *> data) = t_AA;                      // Verified in simulation
   (CS_b *> data) = (t_ACS, t_ACS, t_CHZ);
   (OE_b *> data) = (t_OE, t_OE, t_OHZ);       // Verified in simulation

//Timing checks (Note use of conditioned events for the address setup,
//depending on whether the write is controlled by the WE_b or  by CS_b.

//Width of write/read cycle
   $width (negedge addr, t_WC);

//Address valid to end of write

   $setup (addr, posedge WE_b &&& CS_b == 0, t_AW);
   $setup (addr, posedge CS_b &&& WE_b == 0, t_AW);
```

```
//Address setup before write enabled

    $setup (addr, negedge WE_b &&& CS_b == 0, t_AS);
    $setup (addr, negedge CS_b &&& WE_b == 0, t_AS);

//Width of write pulse
    $width (negedge WE_b, t_WP);

//Data valid to end of write
    $setup (data, posedge WE_b &&& CS_b == 0, t_DW);
    $setup (data, posedge CS_b &&& WE_b == 0, t_DW);

//Data hold from end of write
    $hold (data, posedge WE_b &&& CS_b == 0, t_DH);
    $hold (data, posedge CS_b &&& WE_b == 0, t_DH);

//Chip sel to end of write
    $setup (CS_b, posedge WE_b &&& CS_b == 0, t_CW);
    $width (negedge CS_b &&& WE_b == 0, t_CW);

  endspecify
//*/
endmodule
///////////////////////////// testbench /////////////////////////////////
module test_RAM_2048_8 ();
    parameter           word_size = 8;
    parameter           addr_size = 11;
    parameter           mem_depth = 128;
    parameter           num_col = 16;
    parameter           col_addr_size = 4;
    parameter           row_addr_size = 7;
    parameter           initial_pattern = 8'b0000_0001;
    parameter           Hi_Z_pattern = 8'bzzzz_zzzz;

    reg [word_size -1 : 0]  data_to_memory;
    reg                     CS_b, WE_b, OE_b;

    integer                 col, row;
    wire [col_addr_size -1:0] col_addr = col;
    wire [row_addr_size -1:0]     row_addr = row;
    wire [addr_size -1:0]   addr = {row_addr, col_addr};

    parameter           t_WPC = 8;              // Write pattern cycle time
                                                // (Exceeds min)
    parameter           t_RPC = 12;             // Read pattern cycle time
                                                // (Exceeds min)
    parameter           latency_Zero_Delay = 5000;
    parameter           latency_Non_Zero_Delay = 18000;
//parameter             stop_time = 7200;       // For zero-delay simulation
    parameter           stop_time = 45000;      // For non-zero delay
                                                // simulation
```

```verilog
// Three-state, bi-directional I/O bus

  wire [word_size -1 : 0]
    data_bus = ((CS_b == 0) && (WE_b == 0) && (OE_b == 1))
     ? data_to_memory : Hi_Z_pattern;

  wire [word_size -1 : 0]
    data_from_memory = ((CS_b == 0) && (WE_b == 1) && (OE_b == 0))
     ? data_bus : Hi_Z_pattern;

  RAM_2048_8 M1 (data_bus, addr, CS_b, OE_b, WE_b);     // UUT

  initial #stop_time $finish;
/*
// Zero delay test: Write walking ones to memory
  initial begin
   CS_b = 0;
   OE_b = 1;
   WE_b = 1;
   for (col= 0; col <= num_col-1; col = col +1) begin
   data_to_memory = initial_pattern;

   for (row = 0; row <= mem_depth-1; row = row + 1) begin
    #1 WE_b = 0;
    #1 WE_b = 1;
     data_to_memory =
      {data_to_memory[word_size-2:0],data_to_memory[word_size -1]};
    end
   end
  end

// Zero delay test: Read back walking ones from memory
  initial begin
   #latency_Zero_Delay;
   CS_b = 0;
   OE_b = 0;
   WE_b = 1;
   for (col= 0; col <= num_col-1; col = col +1) begin
    for (row = 0; row <= mem_depth-1; row = row + 1) begin
     #1;
    end
   end
  end
*/
///*
// Non-Zero delay test: Write walking ones to memory
// Writing controlled by WE_b
```

```
    initial begin
     CS_b = 0;
     OE_b = 1;
     WE_b = 1;

     for (col= 0; col <= num_col-1; col = col +1) begin
      data_to_memory = initial_pattern;
      for (row = 0; row <= mem_depth-1; row = row + 1) begin
       #(t_WPC/8) WE_b = 0;
       #(t_WPC/4);
       #(t_WPC/2) WE_b = 1;
       data_to_memory =
         {data_to_memory[word_size-2:0], data_to_memory[word_size -1]};
       #(t_WPC/8);
      end
     end
    end

// Non-Zero delay test: Read back walking ones from memory
    initial begin
     #latency_Non_Zero_Delay;
     CS_b = 0;
     OE_b = 0;
     WE_b = 1;
     for (col= 0; col <= num_col-1; col = col +1) begin
      for (row = 0; row <= mem_depth-1; row = row + 1) begin
       #t_RPC;
      end
     end
    end
   //*/

// Testbench probe to monitor write activity
   reg  [word_size -1:0] write_probe;
   always @ (posedge M1.WE_b)
     case (M1.col_addr)
     0: write_probe = M1.RAM_col0[M1.row_addr];
     1: write_probe = M1.RAM_col1[M1.row_addr];
     2: write_probe = M1.RAM_col2[M1.row_addr];
     3: write_probe = M1.RAM_col3[M1.row_addr];
     4: write_probe = M1.RAM_col4[M1.row_addr];
     5: write_probe = M1.RAM_col5[M1.row_addr];
     6: write_probe = M1.RAM_col6[M1.row_addr];
     7: write_probe = M1.RAM_col7[M1.row_addr];
     8: write_probe = M1.RAM_col8[M1.row_addr];
     9: write_probe = M1.RAM_col9[M1.row_addr];
     10:write_probe = M1.RAM_col10[M1.row_addr];
     11:write_probe = M1.RAM_col11[M1.row_addr];
     12:write_probe = M1.RAM_col12[M1.row_addr];
```

```
      13:write_probe = M1.RAM_col13[M1.row_addr];
      14:write_probe = M1.RAM_col14[M1.row_addr];
      15:write_probe = M1.RAM_col15[M1.row_addr];
   endcase
 endmodule
```

End of Example 8.6

The structure of *test_RAM_2048_8* is shown in Figure 8-30. The unit under test, *RAM_2048_8*, and the testbench both include bidirectional three-state I/O.[13] The active-low write enable signal, *WE_b*, has priority over the active-low output-enable signal, *OE_b* (i.e., if $WE\_b = 0$, the output is in the high-impedance condition independently of *OE_b*). This precludes bus contention by not allowing simultaneous reading and writing.

The timing parameters incorporated in the model for *RAM_2048_8* govern the transitions of the output waveforms in response to changes of the input waveforms and establish operational constraints that must be satisfied for correct operation of the device. For example, if the address is not stable when *CS_b* and *WE_b* are low, multiple memory cells can be affected while the device is in the transparent/write mode. The address access time is a key parameter that dictates the rate at which the memory can be read. Table 8-3 lists parameters describing the write cycle of a static RAM, and Table 8-4 describes the read cycle.

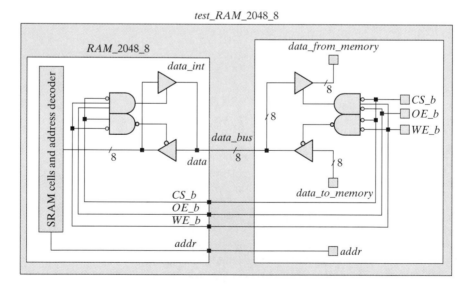

FIGURE 8-30 *test_RAM_2048_8*: structure of the testbench for writing and reading patterns of walking 1s through a shared bi-directional bus.

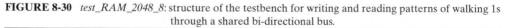

[13]Note that we have simplified the schematic by showing a single three-state buffer instead of an actual configuration having a buffer on each bit line of each bus.

TABLE 8-3 Parameters for the write cycle of a static RAM.

	SRAM Write-Cycle Parameters	
t_WC	**Write cycle time**: Specifies the minimum period for successive writing of data to memory.	
t_CW	**Chip select to end of write**: Specifies the minimum interval between the falling edge of CS_b and the rising edge of WE_b.	
t_AW	**Address valid to end of write**: Specifies the minimum interval between a change in the address and the end of write (the rising edge of WE_b).	
t_AS	**Address setup time before write**: Specifies the width of the interval over which the address must be stable prior to the falling edge of WE_b.	
t_WP	**Write pulse width**: Specifies the minimum width of the write pulse.	
t_WR	**Write recovery time**: Specifies the minimum interval between the rising edge of WE_b and the end of the write cycle.	
t_WHZ	**Write enable to output in high-z**: Specifies the minimum interval between the falling edge of WE_b and the output entering the high-impedance state.	
t_DW	**Data setup time**: Specifies the minimum width of the interval over which the data must be stable prior to the rising edge of WE_b.	
t_DH	**Data hold time after end of write**: Specifies the minimum interval that the data must be stable after the rising edge of WE_b.	
t_OW	**Output active from end of write**: Specifies the earliest time that the output is available after the rising edge of WE_b.	

The timing parameters of a write cycle are illustrated in Figure 8-31. Two cases must be considered: (1) the operation controlled by WE_b with $CS\_b = 0$ (the device is selected) and $OE\_b = 1$ (the read cycle is not active), and (2) the operation controlled by CS_b with $WE\_b = 0$ (write is enabled), and $CS\_b = 0$.

In the former case (shown in Figure 8-31(a)), the address must be stable and the chip must be selected before the falling edge of WE_b. The write cycle occurs over an interval of width t_{WC}, which includes the times at which the address lines may be changed. The address setup time, t_{AS}, establishes the minimum time between the stable address and the falling edge of WE_b. This constraint ensures that the address-decoding circuitry is stable before the write is attempted. The enable input of a transparent latch must satisfy a minimum pulsewidth constraint (t_{WP}); similarly, the time from chip select to the end of the write cycle (t_{CW}) must also satisfy a pulsewidth constraint (t_{CW}). While WE_b is low the device is in the transparent mode and the three-state device driving *data_int* is in the high-impedance state. The device enters this state, with a delay specified by t_{WHZ}, when WE_b is asserted. The data to be written to the SRAM must satisfy a setup time constraint[14] (t_{DW}) and a hold time constraint (t_{DH}) relative to the rising edge of WE_b. *Note*: Figure 8-31a is drawn to illustrate the rising edge of CS_b occurring after the rising edge of WE_b.[15] The address must be stable for an interval

---

[14]We will consider timing constraints in more detail in Chapter 11.
[15]If the rising edge of CS occurs before the rising edge of WE the timing constraints must be applied relative to the rising edge of CS.

TABLE 8-4 Parameters for the read cycle of a static RAM.

	SRAM Read-Cycle Parameters
t_RC	**Read-cycle time**: Specifies the minimum period for successive reading of data from memory.
t_AA	**Address access time**: A key performance parameter specifying the minimum interval between a change in the address and the availability of valid data retrieved from memory.
t_ACS	**Chip select access time**: Specifies the minimum interval between assertion of chip select and the availability of valid data from memory, assuming that $OE\_b = 0$ and $WE\_b = 1$ before $CS\_b = 0$.
t_CLZ	**Chip select low z**: Specifies the minimum interval between assertion of chip select and the output leaving the high-impedance state.
t_OE	**Output enable to output valid**: Specifies the minimum interval between the falling edge of OE_b and the availability of valid data from memory.
t_OLZ	**Output enable to output in low Z**: Specifies the minimum interval between the falling edge of OE_b and the output leaving the high-impedance state.
t_CHZ	**Chip deselect to output in high Z**: Specifies the minimum interval between the rising edge of OE_b and the output entering the high-impedance state.
t_OHZ	**Output disable to output in high Z**: Specifies the minimum interval between the rising edge of OE_b and the output entering the high-impedance state.
t_OH	**Output hold from address change**: Specifies the minimum interval that the output remains valid after a change in the address.

(t_{WR}) the write recovery time, after the rising edge of WE_b, and the bus becomes available after an interval (t_{OW}) expires from the rising edge of WE_b. The interval from the onset of a stable address to the end of the write cycle is represented by the parameter t_{AW}.

When WE_b is low before the falling edge of CS_b, and rises after the rising edge of CS_b, the SRAM is controlled by CS_b and is characterized by the waveforms in Figure 8-31b. In this case, the setup and hold time constraints for the data on the bus are relative to the rising (latching) edge of CS_b.

The two modes of the read cycle are illustrated in Figure 8-32. In Figure 8-32(a) the data is determined by the address (with $CS\_b = 0$ and $WE\_b = 1$, and is valid t_{AA} time units after the address is stable. In Figure 8-32(b), the data becomes valid after t_{ACS} time units from the falling edge of CS_b.

8.2.10 Ferroelectric Nonvolatile Memory

Ferroelectric materials are so named because their electrical characteristics resemble those of ferromagnetic materials. Despite the suggestion implied by their name, ferroelectric materials have nothing to do with ferromagnetics. Their similarity is primarily in the fact that certain ferroelectric materials can exhibit a significant hysteresis effect, but it is not associated with magnetic properties. Instead, the hysteresis effect in a

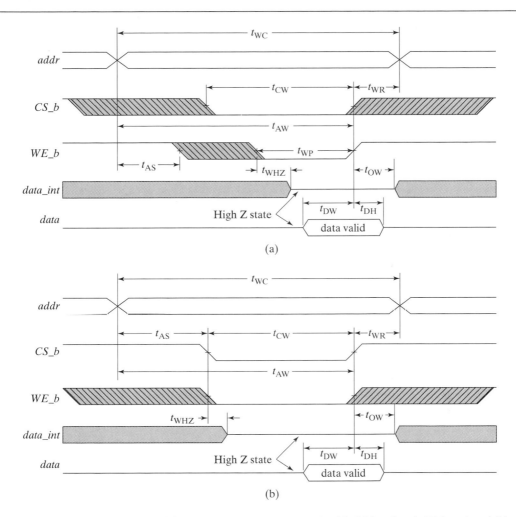

FIGURE 8-31 SRAM timing: (a) write cycle controlled by WE_b, with $CS\_b = 0$ and $OE\_b = 1$, and (b) write cycle controlled by CS_b, with $WE\_b = 0$ and $OE\_b = 1$.

ferrorelectric is due to the so-called spontaneous electrical polarization of a ferroelectric material under the influence of an applied voltage. When power is removed the residual polarization behaves like a bistable memory device. Ferroelectric memories hold the promise of replacing other nonvolatile memories, such as EEPROMs in applications that require short programming time and low power consumption. Contactless smart cards, digital cameras, and utility meters are considered to be appropriate applications for this technology. EEPROMS and flash memories are also nonvolatile, and have lower power to read data than ferroelectrics. Ferroelectric memories can also be embedded with other devices. This technology is expected to mature to have

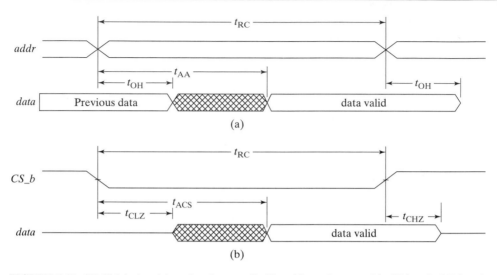

FIGURE 8-32   SRAM timing: (a) read cycle controlled by address changes, with $CS\_b = 0$, $OE\_b = 0$, and $WE\_b = 1$, and (b) read cycle controlled by changes in CS_b, with $OE\_b = 0$, and $WE\_b = 1$.

competitive circuit densities compared to other alternatives. See Sheilholeslami and Gulak [3] for a survey of circuits exploiting ferroelectric technology.

8.3   Programmable Logic Array (PLA)

PLAs were developed for integrating large two-level combinational logic circuits. Like ROMs, their architecture consists of two arrays, shown in Figure 8-33. One array implements the *AND* operation that forms a product term (i.e., a Boolean cube, possibly

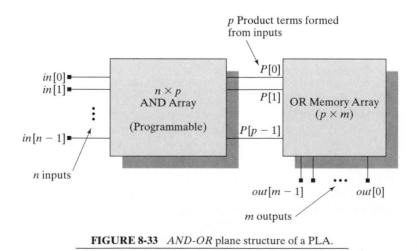

FIGURE 8-33   *AND-OR* plane structure of a PLA.

a minterm), and another array implements the *OR* operation that forms a sum of the product (SOP) terms. A PLA implements a two-level Boolean function in SOP form.

Unlike ROMs, *both arrays of a PLA are programmable* (mask-programmable or one-time field-programmable). However, *the AND plane does not implement a full decoder*, but instead forms a limited number of product terms. The programmable *OR*-plane forms expressions by *OR*-ing together product terms (cubes). An $n \times p \times m$ PLA has n inputs, p product terms (outputs of the *AND* plane), and m output expressions (from the *OR* plane). A $16 \times 48 \times 8$ PLA has 48 product terms. A 16-input ROM would have $2^{16} = 65,536$ input patterns decoded as minterms and available to form the outputs. A PLA would have 8 outputs formed from the 48 product terms (not necessarily minterms).

A PLA implements general product terms, not just minterms or maxterms. Because it has limited *AND*-plane resources, minimal SOP forms must be found so that device resources might accommodate an application's requirements for product terms. PLA minimization algorithms led to development of widely used synthesis algorithms having general application to ASICs [4].

The circuit structure of a PLA implemented in CMOS technology is shown in Figure 8-34.

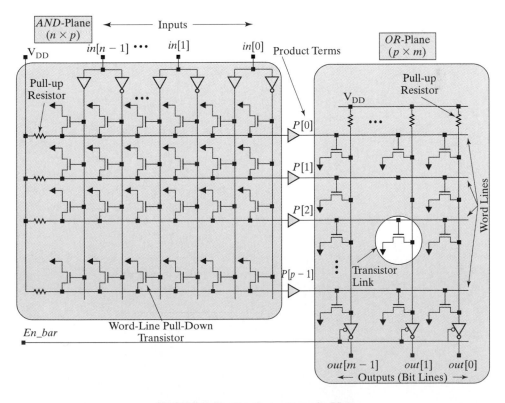

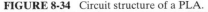

FIGURE 8-34 Circuit structure of a PLA.

The *AND-OR* plane structure shown implements *NOR-NOR* logic, which reverts to equivalent *AND-OR* logic with inverted inputs and three-stated inverters at the outputs. Each input is available as a literal in complemented and uncomplemented form. A programmable link in the *AND* plane determines whether the associated input literal (or its complement) is connected to a buffered word line.

Programming determines whether inputs have a link to word lines and whether word lines connect to the output lines. A word line may be linked to an input or its complement, but not both. Each word line is connected to a pull-up resistor (active device). The aggregate of linked input literals and complements of input literals forms a Boolean cube at the word line to which the links are attached. Unconnected inputs have no effect on a word line. In the absence of an asserted and connected input literal (or its complement) a word line is pulled up. In the absence of an asserted (high) level on its linked word lines, a column line is pulled high. An asserted input turns on a connected link transistor in the *AND* plane and pulls the word line to ground by overriding its pull-up resistor. A column line is asserted (high) if all of its connected word lines are de-asserted (low). An asserted word line turns on a connected link transistor in the *OR* plane, causing its connected word line to be pulled down. A column line is low if any of its connected word lines is asserted (high). If any word line is asserted (high), a connected column line is pulled down. A column line is asserted (high) only if all of its connected word lines are de-asserted (low).

To see that the circuit shown in Figure 8-35 exhibits wired-*AND* logic at its word lines, note that

$$W1 = A'B'$$
$$W1' = (A + B)'$$

$$W2 = C'D'$$
$$W2' = (C + D)'$$

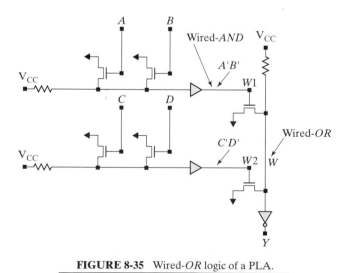

FIGURE 8-35   Wired-*OR* logic of a PLA.

Similarly, the column lines exhibit wired-*OR* behavior, where W is low if $W1$ or $W2$ is high; otherwise W is high (pull-up):

$$W' = W1 + W2$$
$$W = (W1 + W2)'$$
$$Y = W' = W1 + W2 = A'B' + C'D'$$
$$Y = (A + B)' + (C + D)'$$

The overall structure is that of *NOR-NOR* logic, with

$$Y' = [(A + B)' + (C + D)']'$$

The equivalent circuit is shown in Figure 8-36(a), and an equivalent *OR-AND* structure is shown in Figure 8-36(b).

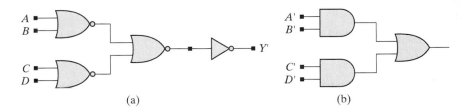

(a)   (b)

FIGURE 8-36   Equivalent circuit structures for PLA logic: (a) *NOR-NOR* logic, and (b) *OR-AND* logic with inverted inputs.

8.3.1   PLA Minimization

The area of a PLA depends primarily on the number of word lines (i.e., distinct product terms), so it is advantageous to find ways to reduce the number of product terms by sharing logic as much as possible. One approach would be to use Karnaugh maps or other minimization methods to reduce each Boolean expression. However, minimization of individual Boolean functions does not necessarily produce an optimal PLA implementation. Minimization of a set of Boolean functions, as an aggregate, can exploit don't-cares and opportunities to share logic because a product term that is generated to form one output expression can be used in another output expression that uses the same term. Alternatively, a common factor in a product of sums form can be shared by multiple functions that have the same factor.

Example 8.7

Consider the three Boolean functions shown below, with their K-maps shown in Figure 8-37. Before minimization, the implementation would require 13 product terms (word lines) to support the cubes of the three functions.

$$f_1(a, b, c, d) = \Sigma\, m(1, 6, 7, 9, 13, 14, 15)$$
$$f_2(a, b, c, d) = \Sigma\, m(6, 7, 8, 9, 13, 14, 15)$$
$$f_3(a, b, c, d) = \Sigma\, m(1, 2, 3, 9, 10, 11, 12, 13, 14, 15)$$

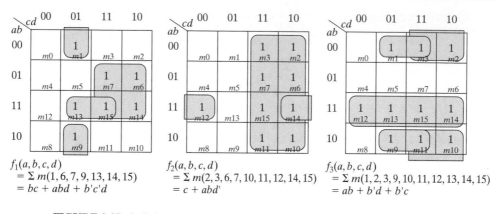

$f_1(a, b, c, d)$
$= \Sigma\, m(1, 6, 7, 9, 13, 14, 15)$
$= bc + abd + b'c'd$

$f_2(a, b, c, d)$
$= \Sigma\, m(2, 3, 6, 7, 10, 11, 12, 14, 15)$
$= c + abd'$

$f_3(a, b, c, d)$
$= \Sigma\, m(1, 2, 3, 9, 10, 11, 12, 13, 14, 15)$
$= ab + b'd + b'c$

FIGURE 8-37   Individual Karnaugh map minimization of three Boolean functions.

After each function is individually minimized, the total number of cubes is 8, a savings of nearly 40%. To minimize the functions as an aggregate, a task easily done by modern synthesis tools, we re-cover the functions and identify common cubes by considering pairwise and threewise intersections, as shown in Figure 8-38. The final result needs only five word lines, having eliminated an additional four word lines.

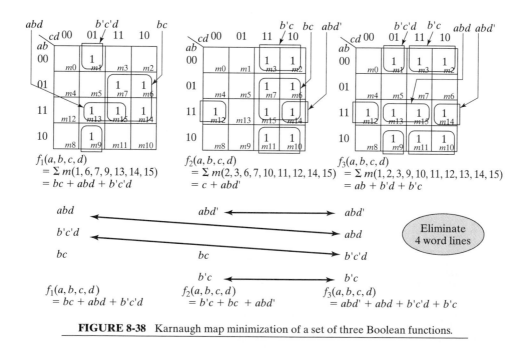

$f_1(a, b, c, d)$
$= \Sigma\, m(1, 6, 7, 9, 13, 14, 15)$
$= bc + abd + b'c'd$

$f_2(a, b, c, d)$
$= \Sigma\, m(2, 3, 6, 7, 10, 11, 12, 14, 15)$
$= c + abd'$

$f_3(a, b, c, d)$
$= \Sigma\, m(1, 2, 3, 9, 10, 11, 12, 13, 14, 15)$
$= ab + b'd + b'c$

$f_1(a, b, c, d)$
$= bc + abd + b'c'd$

$f_2(a, b, c, d)$
$= b'c + bc + abd'$

$f_3(a, b, c, d)$
$= abd' + abd + b'c'd + b'c$

FIGURE 8-38   Karnaugh map minimization of a set of three Boolean functions.

End of Example 8.7

Systematic manual minimization of multiple output functions is feasible for up to three functions, with a maximum of four inputs [5]. Otherwise, a computer-based approach is needed (e.g., espresso [4] and MIS-II [6–8]).

The tabular format shown in Figure 8-39 can be used to specify the functionality of a PLA. Table rows correspond to PLA rows (wordlines). Table columns list inputs and functions indicating whether an input is in a cube, and whether a cube is in a function. Inputs are coded as 1 (care-on), 0 (care-off), and—(don't-care). Outputs are coded as 1 (contains the wordline), or 0 (does not contain the wordline).

	a	b	c	d	f_1	f_2	f_3
abd	1	1	—	1	1	0	1
$b'c'd$	—	0	0	1	1	0	1
$b'c$	—	0	1	—	0	1	1
bc	—	1	1	—	1	1	0
abd'	1	1	—	0	0	1	1

FIGURE 8-39   Tabular format for specifying the structure of a PLA.

ROMs require canonical data (i.e., a complete truth table), but PLAs require only minimum SOP Boolean forms. The cubes in a minimized PLA table may cover multiple minterms, and a given input vector may assert multiple output functions. For a given input vector, the cubes are formed by *AND*-ing the complemented and uncomplemented literals in a row; the outputs are determined by column-wise *OR*-ing the cube (word line) entries having a 1. A simplified representation of a PLA is shown in Figure 8-40. The filled circles indicate whether a literal or its complement is used in a cube and whether a cube is used in an expression.

8.3.2   PLA Modeling

An application for a PLA must be compatible with the limited number of product terms (word lines) that can fit within the device. PLAs are used to implement the next-state and output-forming logic of large state machines that control more complex sequential machines, such as computers. PLAs are a more attractive implementation than ROMs for large state machines because the area of a PLA can be minimized and tailored to an application.

Verilog includes a set of system tasks for modeling multiinput, multioutput PLAs. PLAs implement two-level combinational logic by a array structure of *AND, NAND, OR*, and *NOR* logic array planes. The "personality" file, or matrix, of the PLA specifies the physical connections of transistors forming the product of input terms (cubes) and the sums of those products to form the Boolean expressions of the outputs. See "Selected System Tasks and Functions" at the companion web site.

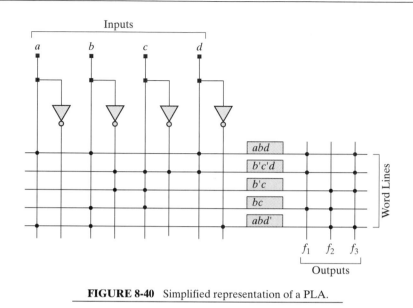

FIGURE 8-40   Simplified representation of a PLA.

Example 8.8

The statements below illustrate calls to Verilog's built-in PLA system tasks describing synchronous and asynchronous arrays and planes:

$async$and$array	(PLA_mem, {in0, in1, in2, in3, in4, in5, in6, in7}, {out0, out1, out2});
$sync$or$plane	(PLA_mem, {in0, in1, in2, in3, in4, in5, in6, in7}, {out0, out1, out2});
$async$and$array	(PLA_mem, {in0, in1, in2, in3, in4, in5, in6, in7}, {out0, out1, out2});
$async$and$array	(PLA_mem, {in0, in1, in2, in3, in4, in5, in6, in7}, {out0, out1, out2});

The outputs of the asynchronous arrays are updated whenever an input signal changes value or whenever the personality matrix of the PLA changes during simulation. The synchronous types are updated when evaluated in a synchronous behavior. Both forms update their outputs with zero delay.

The personality matrix of a PLA specifies the cubes that form the inputs to the PLA and the expressions forming the outputs of the PLA. The data describing the personality is stored in a memory whose width accommodates the inputs and outputs of the PLA and whose depth accommodates the number of outputs.

There are two ways to load data into the personality matrix: (1) using the *$ readmemb* task, read the data from a file, and (2) load the data directly with procedural assignment statements. Both methods can be used at any time during a simulation to reconfigure the PLA dynamically.

Two formats may be used to describe the contents of an array: *array* and *plane*. The array format stores either a 1 or 0 in memory to indicate whether a given input is in a cube, and whether a given cube is in an output. For example, the array format shown below indicates that the cube *in1* & *in2* & *in3* is formed and used in *out1*, but not in *out2*. The cube *in1* & *in3* is used in *out2*.

in1	in2	in3	out1	out2
1	1	1	1	0
1	0	1	0	1

End of Example 8.8

Example 8.9

Suppose we want to implement the logic of the following Boolean equations with a PLA:

$out0 = in0$ & $in1$ & $in2$ & $in3$ + $in4$ & $in5$ & $in6$ & $in7$ + $in1$ & $in3$ & $in5$ & $in7$

$out1 = in1$ & $in3$ & $in5$ & $in7$ + $in4$ & $in5$ & $in6$ & $in7$

$out2 = in0$ & $in2$ & $in4$ & $in6$ + $in4$ & $in5$ & $in6$ & $in7$ + $in1$ & $in3$ & $in5$ & $in7$

The expressions use four distinct cubes:

in0	in1	in2	in3
in0	in2	in4	in6
in4	in5	in6	in7
in1	in3	in5	in7

The personality data of the PLA is shown below, and is placed in a text file, *PLA_data.txt*. The data indicate the presence of a literal by a 1, and the absence of a literal by 0, listed in ascending order of the inputs. There is one row for each cube, a column for each input, and the last three columns indicate whether a row cube is present in each of the three output functions.

11110000 100
10101010 011
00001111 111
01010101 101

The Verilog model *PLA_array* describes a PLA that forms three output functions of eight Boolean input variables. The personality of the array is stored in the array of words *PLA_mem*, whose width corresponds to the width of the personality matrix and whose depth is determined by the number of Boolean expressions that will be formed as outputs. Hence, the array of words has a width of 11 bits and a depth of three words.

```
module PLA_array (in0, in1, in2, in3, in4, in5, in6, in7, out0, out1, out2);
  input                       in0, in1, in2, in3, in4, in5, in6, in7;
  output                      out0, out1, out2;
  reg                         out0, out1, out2;
  reg         [0: 10]   PLA_mem [0: 2];                    // 3 functions of 8 variables

initial begin
  $readmemb ("PLA_data.txt", PLA_mem);
  $async$and$array
    (PLA_mem, {in0, in1, in2, in3, in4, in5, in6, in7}, {out0, out1, out2});
  end
endmodule
```

End of Example 8.9

The PLA in Example 8.4 is configured by the ***initial*** behavior at the beginning of a simulation. The simulator reads the file *PLA_data.txt* and loads the data into the declared memory, *PLA_mem*. Note that the inputs and outputs are declared in ascending order. When an input to the module changes value the array is evaluated to form updated values of *out0, out1*, and *out2*.

The array format requires that the complement of a literal be provided separately as an input if it is needed to form a cube. On the other hand, the plane format encodes the personality matrix, according to the format in Table 8-5, which was adopted from the Espresso format developed at the University of California at Berkeley [4].

TABLE 8-5 Personality matrix symbols for PLA plane format.

Table Entry	Interpretation
0	The complemented literal is used in the cube.
1	The literal is used in the cube.
x	The worst case of the input is used.
z	Don't care; the input has no significance.
?	Same as z.

Example 8.10

Suppose the logic to be implemented in a PLA is described by the following statements:

$$out0 = in0 \ \& \ \sim in2;$$
$$out1 = in0 \ \& \ in1 \ \& \ \sim in3;$$
$$out2 = \ \sim in0 \ \& \ \sim in3;$$

In the plane (Espresso) format, the personality of the PLA is described by:

$$4'b1?0?$$
$$4'b11?0$$
$$4'b0??0$$

The rows correspond to the outputs and are listed in descending order. A row defines the conditions of the inputs that assert that output. For example, the inputs 1000 and 1101 will both assert the first output. A Verilog description of the PLA is given below.

```
module PLA_plane (in0, in1, in2, in3, in4, in5, in6, in7, out0, out1, out2);
    input        in0, in1, in2, in3, in4, in5, in6, in7;
    output       out0, out1, out2;
    reg          out0, out1, out2;
    reg    [0: 3] PLA_mem [0:2];            // 3 functions of 4 variables
    reg    [0: 4] a;
    reg    [0: 3] b;

    initial begin

    $async$and$array
      (PLA_mem, {in0, in1, in2, in3, in4, in5, in6, in7}, {out0, out1, out2});

    PLA_mem [0] = 4'b1?0?;                  // Load the personality matrix
    PLA_mem [1] = 4'b11?0;
    PLA_mem [2] = 4'b0??0;
    end
endmodule
```

End of Example 8.10

8.4   Programmable Array Logic (PAL)

PALs[16] emerged after PLAs, and simplified the dual-array structure by fixing the *OR* plane and allowing only the *AND* plane to be programmed. Each output is formed from a specified number of word lines, and each word line is formed from a small number of product terms. One of the more popular devices, the PAL16L8, has the structure shown in Figure 8-41. The device has 16 inputs and 8 outputs; its package has 20 pins,

---

[16]*Note*: PAL is a trademark of Applied Micro Devices (AMD).

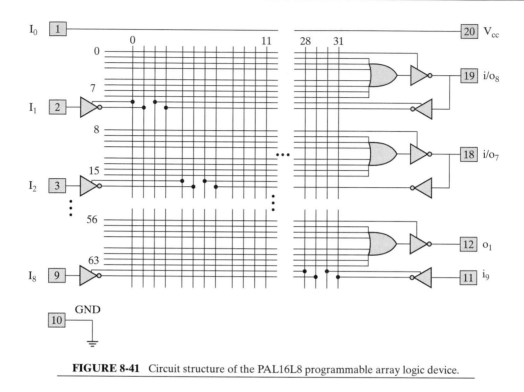

FIGURE 8-41 Circuit structure of the PAL16L8 programmable array logic device.

including power and ground. Each input is available in true or complemented form. There are eight 7-input *OR* gates connected to word lines from the *AND* plane. Each word line can be connected to any input or its complement. An eighth word line in each group controls a three-state inverter that is driven by the group's *OR* gate. Each output implements a sum of products expression from at most seven terms. The device has only 20 pins, so six of the pins are bidirectional. The *AND* gate (not shown) that is associated with each word line is permanently connected to an *OR* gate and cannot be shared with any other *OR* gate, but six of the outputs are connected to three-state inverters and can be fed back to the *AND* array to be shared with other *AND* gates, which accommodates expressions having more than seven product terms. A bidirectional pin also makes it possible for the device to implement a transparent latch by combinational feedback. PLD-based latches have application as address decoder/latches in microprocessor systems [1]. Modern PAL devices are manufactured with registered outputs and selectable output polarity.

Early PAL devices were implemented in bipolar technology; like ROMs, they were programmed by vaporizing metal links. Contemporary devices are implemented in CMOS technology with floating-gate link transistors.

8.5 Programmability of PLDs

ROMs, PLAs, and PLDs are implemented in similar array structures. Table 8-6 compares the options that are presented for programming the devices. PLAs provide the greatest flexibility and are used for large, complex, combinational logic circuits.

8.6 Complex PLDs (CPLDs)

As technology has evolved, more dense and complex devices have been developed to implement large structures (e.g., up to 1024 functions) of field-programmable combinational and sequential logic, and are referred to as complex PLDs, or CPLDs. The high-level architecture of a typical CPLD (shown in Figure 8-42) is formed as a structured array of PLD blocks that have a programmable on-chip interconnection fabric. Aside from increased performance, these architectures overcome the limitation of

TABLE 8-6 Programmability options for various PLDs.

	Programmable Block	
	AND Plane	*OR* Plane
ROM	NA	P
PLA	P	P
PAL	P	NP

NA = not applicable, P = programmable,
NP = not programmable.

Programmable interconnect

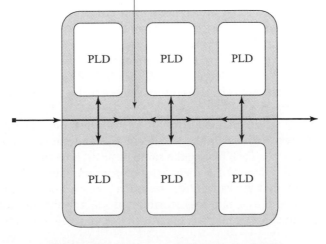

FIGURE 8-42 High-level architecture of a CPLD.

conventional PLDs, which have a relatively small number of inputs. CPLDs have wide inputs, but not at the expense of a dramatic (i.e., exponential) increase in area. The device area of a conventional PLD will be scaled by a factor of 2^n if its input dimension is scaled by a factor of n. An array of identical interconnected PLDs will accommodate the increased dimension of the input space too, but the cell area increases by a factor of only n, in addition to the area required by the interconnection fabric. Thus, CPLDs are distinguished by having wide fan-in AND gates. Large CPLDs do not connect the output of every macrocell to an output pin, but they typically have 100% connectivity between macrocells.

Each PLD block of a CPLD has a PAL-like internal structure that forms combinational logic functions of its inputs. The outputs of the macrocells in the PLDs can be programmed to route to the inputs of other logic blocks to form more complex, multilevel logic beyond the limitations of a single logic block. Some CPLDs are electrically erasable and reprogrammable (EPLD). CPLDs are suited for wide fan-in And-Or logic structures, and exploit a variety of programming technologies: SRAM/transmission gates, EPROM (floating-gate transistors), and antifuses.[17]

8.7   Altera MAX 7000 CPLD

Altera PLDs use CMOS technology with floating-gate EEPROM configuration memory cells to establish routing. Their CPLD families include the MAX 5000, 7000, and 9000 series. The architecture of Altera's 7000[18] series devices (EPM7032, EPM7064, and EPM7096 Devices) is shown in Figure 8-43.

The structure consists of an array of logic array blocks (LABs), a programmable interconnect array (PIA), and an array of programmable I/O blocks. Each LAB has 36 inputs and 16 outputs and contains 16 macrocells, each containing combinational logic and a flip-flop for either combinational or sequential operation. The PIA is a global bus that establishes connectivity between multiple LABs, dedicated inputs, and I/O pins. The PIA provides full connectivity with predictable timing between the logic-cell outputs and the inputs to the LABs. The level of connectivity reduces the density of the structure compared to the denser channel-based architectures used in FPGAs.

The I/O control blocks establish connectivity between the I/O pins and the PIA and LABs. The (dedicated) global inputs clock ($GCLK$) and active-low clear ($GCLRn$) connect to all macrocells. The outputs are enabled by active-low signals $OE1n$ and $OE2n$, which connect to all I/O control blocks. From 8 to 16 LAB outputs can be programmed to route to I/O pins, and from 8 to 16 I/O pins can be programmed to route through the I/O control block to the PIA. Each LAB contains a finer-grain array of programmable macrocells, each having the same basic architecture, shown in Figure 8-44.

A macrocell consists of a logic array (programmable AND plane), a product-term select matrix that drives an OR gate, and a programmable flip-flop. The programmable array functions as a mini-PAL, forming product terms (not minterms), which are OR-ed to form an expression. Each macrocell has up to 36 inputs from the PIA and up

---

[17]Antifuses are programmable low-resistance electrical links.

[18]See www.altera.com for device data sheets and additional resources about PLDs.

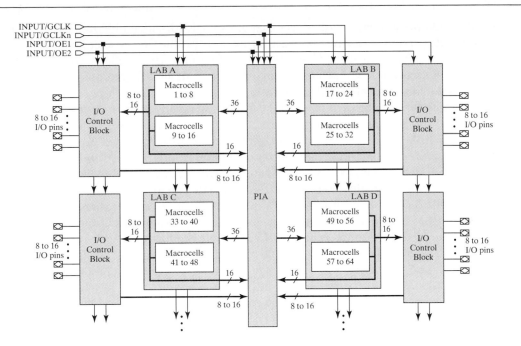

FIGURE 8-43 Altera 7000 Series: architecture for the EPM7032, EPM7064, and EPM7096 devices.

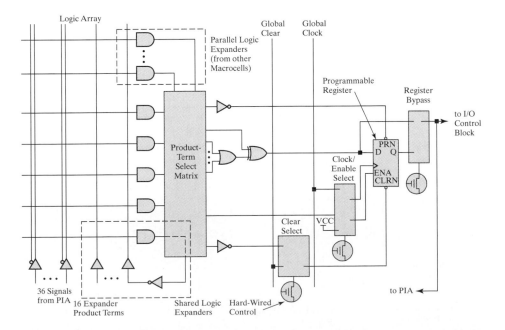

FIGURE 8-44 Altera 7000 Series: macrocell architecture for the EPM7032, EPM7064, and EPM7096 devices.

to 16 additional inputs formed as expander signals (discussed below). Each macrocell generates and provides five product terms to its product-term select matrix (i.e., each macrocell alone can form an expression having up to five cubes). The product-term select matrix can steer a product term to the input of an *OR* gate, an *XOR* gate, a logic expander, or to the preset, clear clock or enable input of a flip-flop.

The flip-flop of each macrocell can be individually programmed to implement a D, T, or J-K flip-flop, or be programmed to implement an set–reset (S-R) latch for use in sequential machines. For example, a macrocell can be converted to a T-type flip flop to give more efficient implementations of counters and adders. (Connect the output of the flip-flop to one input of an *XOR* gate, and drive the other input by the toggle signal (recall that a T-type flip-flop has the characteristic equation: $Q^+ = T \oplus Q$).)

The flip-flop of each macrocell can be synchronized in three ways: (1) by a global clock, (2) by a product term, or (3) by either of the previous modes but gated by a term from the product-term select matrix. Note that the product-term signal could be generated by the logic of the macrocell or could come directly from an I/O pin. Likewise, the clear input can be programmed to be either the global clear signal or a product term from the product-term select matrix. The preset input is a product term from the product-term select matrix. The signal formed by the combinational logic is routed by a programmable register bypass switch to either the I/O control block (combinational output) or to the *D* input of the flip-flop (registered output). The actions of clear and preset are asynchronous.

The macrocells of the Altera EPM7000E and EPM7000S devices have the architecture shown in Figure 8-45. The circuit is nearly identical to that of the EPM7032,

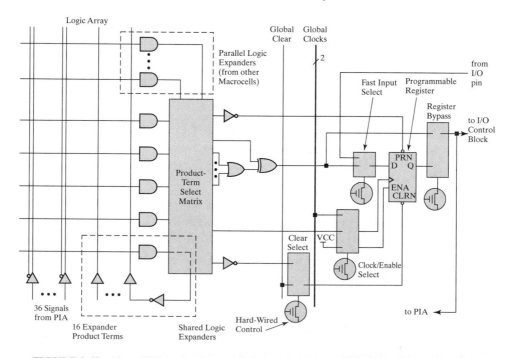

FIGURE 8-45   Altera 7000 Series: macrocell architecture for the EPM7000E and 7000S devices.

EPM7064, and EPM7096 devices, but it has two global clocks and a programmable fast select mux/switch in the datapath to the *D* input of the programmable flip-flop. The fast select switch bypasses the PIA and the macrocell's combinational logic to connect an I/O pin directly to the *D* input of the flip-flop, for fast (2.5 ns) setup times. The true or complemented value of either *GCLK1* and *GCLK2* can be the two clocks.

8.7.1 Shareable Expander

There are two types of logic expanders that let a macrocell exploit the unused product terms of other macrocells in the same LAB to efficiently synthesize fast, complex, logic with more than five product terms.

Shareable expanders increase the number of literals that can form an expression. One unused product term of each macrocell in a LAB can be inverted and fed back to the logic array as a shareable expander, where it can be used by any other macrocell *AND* gate in the same LAB. This creates a more complex three-level *NAND-AND-OR* structure by replacing a literal with a entire product term from another macrocell. Sharing product terms in a LAB can reduce the overall resources required to implement logic. Figure 8-46 shows how a shared expander is formed in a LAB.

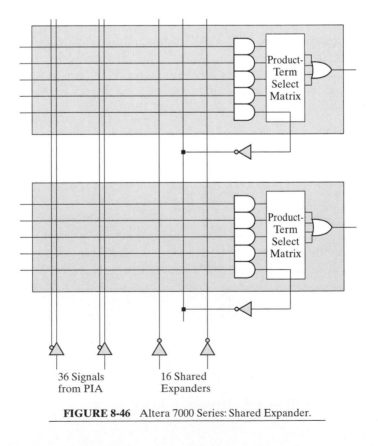

FIGURE 8-46 Altera 7000 Series: Shared Expander.

8.7.2   Parallel Expander

The output of each macrocell ordinarily consists of the sum of only its five product terms. This accommodates most, but not all logic. Parallel expanders (Figure 8-47) increase the number of cubes that can form an expression by chaining (successively *OR*-ing) the output of a macrocell with the output of a neighboring macrocell. Each macrocell's five terms can be summed with up to 15 more product terms provided by parallel expanders from neighboring macrocells in the same LAB. The parallel expanders are allocated in groups of up to five additional product terms. The 16 macro-cells of a LAB are organized into two groups of 8 cells each for the purpose of lending or borrowing parallel expanders (e.g., macrocells 8 down to 1, and macrocells 16 down to 9). A macrocell borrows parallel expanders from lower-number macrocells in the same group of 8. For example, macrocell 1 in LAB 1 can only lend a parallel expander; and macrocell 8 can only borrow a parallel expander from macrocell 7, macrocells 7 and 6, or from macrocells 7, 6, and 5. A macrocell needing 20 product terms can obtain 5 more terms from each of three neighboring macrocells, plus its own 5 terms, for a total of 20 terms.

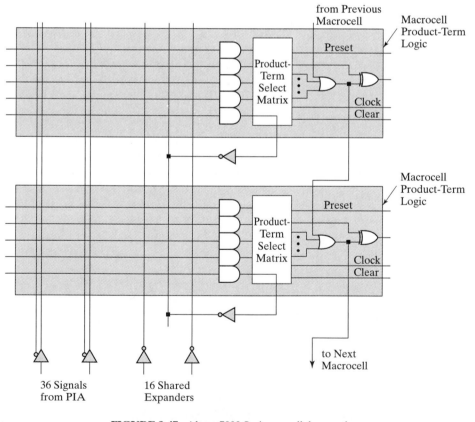

FIGURE 8-47   Altera 7000 Series: parallel expander.

To summarize the shared resources in a LAB: (1) shareable expanders make one unused product term from a macrocell available to any or all macrocells in the same LAB, and (2) parallel expanders chain the outputs of neighboring macrocells in a LAB to form an expression summing up to 20 product terms.

8.7.3   I/O Control Block

Each I/O pin can be programmed as a dedicated input or output pin, or as a bidirectional pin under the control of global signals *OE1n* and *OE2n*. If the enable signal of the three-state buffer in Figure 8-48 is programmed to VCC by the block's logic, the output of the macrocell is hard wired to drive the I/O pin. If the enable signal is programmed to ground the buffer is three-stated (disconnected) and the I/O pin functions as an input pin, providing an external signal to the PIA. Otherwise, either *OE1n* or *OE2n* can dynamically control the bidirectional operation of the circuit, providing connection to a shared bus.

8.7.4   Timing Considerations

Altera's architecture uses a PIA to connect LABs and I/O pins. All signals are available throughout the device. As a result, timing is predictable, an advantage over channel-based architectures, which have routing-dependent delays. The tradeoff is that a channel-routed architecture can achieve a denser implementation by having a sparser interconnect fabric.

8.7.5   Device Resources

A CPLD vendor's synthesis tool optimizes the Boolean logic describing a design, partitions the results to fit into LABs, allocates the functional units to LABs, and establishes the programming needed to configure the device. The tool attempts to optimize the design's utilization of resources, subject to speed constraints imposed by the application and the available devices.

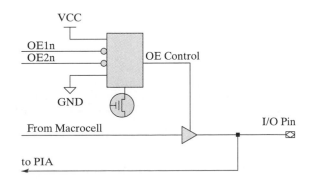

FIGURE 8-48   Altera 7000 Series: the I/O control block for the EPM7032, EPM7064, and EPM7096 devices.

8.7.6   Other Altera Device Families

Altera manufactures an in-system programmable (ISP) version of the 7000 series devices, called the 7000S series. It eliminates the need for a special programmer box. The MAX 9000 series devices have more routing resources and greater density than the 7000 series. Their Flex 6000, Flex 8000, Flex 10, and Flex 20 families of devices use SRAM-based technology (i.e., RAM-programmed transmission gates) to form a channel-based interconnection fabric, rather than EEPROM technology, and use lookup tables (LUTs) to implement logic. We will consider these devices with FPGAs.

8.8   XILINX XC9500 CPLDs

The Xilinx XC9500 family of CPLDs are flash-based (EEPROM) and in-system programmable. The devices are organized as an array of functional blocks in a PAL-like structure with wide *AND* gates and fast flip-flops. Each function block contains up to 18 independent macrocells (see Figure 8-49), and can accommodate 54 inputs and drive 18 outputs (depending on the packaging). A FastCONNECT switch-matrix technology ensures that an application can be fully routed even when the device utilization is high. I/O blocks (IOBs) buffer the inputs and outputs to the device and also receive the global clock and S-R signals. The output buffers have a programmable slew rate.

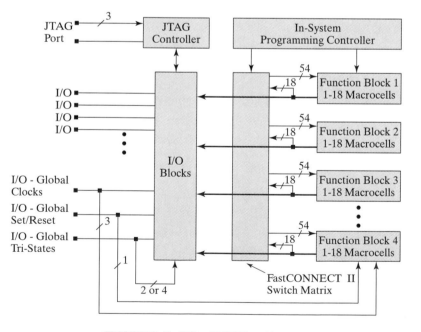

FIGURE 8-49   Xilinx XC9500 architecture.

The PAL-like architecture of a function block receives 54 complemented and uncomplemented inputs and can form up to 90 product terms from the inputs. A "product-term allocator" allocates up to 90 product terms to each macrocell in the function block to form an SOP expression. Each macrocell can receive five direct product terms from the *AND* array, and up to 10 more product terms can be made available from other uncommitted product terms in other macrocells in the same functional block, with a negligible increase in delay. Moreover, partial sums of products can be combined over several macrocells to produce expressions with more than 18 product terms.

Each macrocell can be independently configured for combinational or registered functionality, and receives global clock, output enable, and S-R signals. The pin-to-pin delays of the XC9500 device family are short, and support high system clock rates up to 150 MHz. They should be used in high fan-in state machines, in which speed is a dominant constraint.

The architecture of a macrocell is shown in Figure 8-50. The flip-flop can be configured as a D- or T-type flip-flop, with synchronous or asynchronous S-R operation. The register's clock can be any of three global clocks, or a product term. The register can also be bypassed to provide direct output. Each macrocell has five direct inputs form the *AND*-array, which can be used to implement combinational functions or control inputs (clock, clock enable, set–reset, and output enable).

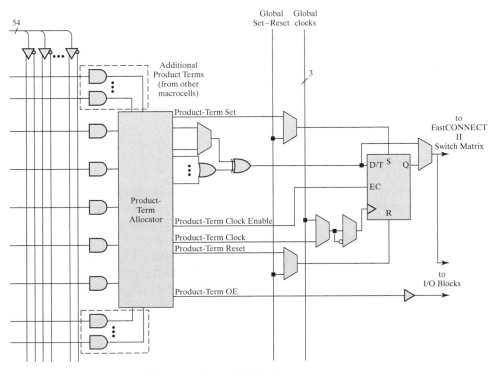

FIGURE 8-50   Xilinx XC9500: Macrocell architecture.

TABLE 8-7 Xilinx XC9500: significant device characteristics.

XC9500 CPLDs	XC9536XV	XC9572XV	XC95144XV	XC95288XV
Macrocells	36	72	144	288
Usable Gates	800	1,600	3,200	6,400
Registers	36	72	144	288
T_{PD} (ns)	3.5	4	4	5
TSU (ns)	2.8	3.1	3.1	3.7
TCO (ns)	1.8	2.0	2.0	2.5
f_{SYSTEM} (MHz)	278	250	250	222

Characteristics of the XC9500 device family are shown in Table 8-7. Note that the propagation delay parameters are fixed for a particular device, allowing predictable performance independently of the internal placement and routing of the design.

8.9 Field-Programmable Gate Arrays

CPLDs are characterized by an array of PAL-like blocks of combinational logic implementing wide-input SOP expressions. They have predictable timing and a crossbar type of interconnection fabric, and are suited for low- and medium-density applications. FPGAs have a more complex and register-rich tiled architecture of functional units and a flexible channel-based interconnection fabric. Featuring flash-based reconfigurability, CPLDs can be reprogrammed a limited number of times, but FPGAs have no practical limit on their reconfigurability. FPGAs are suitable for medium- and high-density applications. They differ in two significant way from CPLDs because (1) their performance is dependent on the routing that is implemented in the device for a particular application, and (2) their functionality is implemented by lookup tables rather than by PAL-like wide-input *AND* gates.

Mask-programmable gate arrays are fabricated in a foundry, where final layers of metal customize the wafer to the specifications of the end user. FPGAs are sold as fully fabricated and tested generic products. Their functionality is determined by programming done in the field by the customer and/or end user. FPGAs allow designers to turn a design into working silicon in a matter of minutes, making rapid prototyping a reality.

FPGAs are distinguished on the basis of several features: architecture, number of gates, mechanism for programming, program volatility, the granularity and robustness of a functional/logical unit, physical size (footprint), pinout, time-to-prototype, speed, power, and the availability of internal resources for connectivity [9, 10]. We will focus on the dominant technology: SRAM-based FPGAs, which lose their programming when power is removed from the part.

SRAM-based FPGAs have a fixed architecture that is programmed in the field for a particular application. A typical, basic architecture, as shown in Figure 8-51, consists of (1) an array of programmable functional units (FUs) for implementing combinational

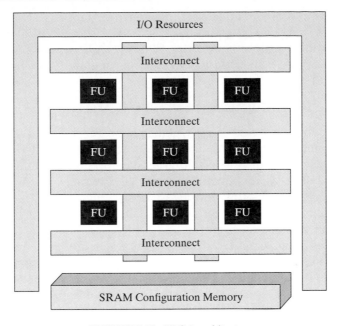

FIGURE 8-51 FPGA architecture.

and sequential logic, (2) a fixed, but programmable, interconnection fabric, which establishes the routing of signals, (3) a configuration memory, which programs the functionality of the device, and (4) I/O resources, which provide an interface between the device and its environment. The performance and density of FPGAs have advanced with improvements in process technology. Today's leading-edge devices include block memory as well as distributed memory, robust interconnection fabrics, global signals for high-speed synchronous operation, and programmable I/O resources matching a variety of interface standards.

Volatile FPGAs are configured by a program that can be downloaded and stored in static CMOS memory, called a configuration memory. The contents of the static RAM are applied to the control lines of static CMOS transmission gates and other devices to: (1) program the functionality of the functional units, (2) customize configurable features, such as slew rate, (3) establish connectivity between functional units, and (4) configure I/O/bidirectional ports of the device. The configuration program is downloaded to the FPGA from either a host machine or from an on-board PROM. When power is removed from the device the program stored in memory is lost, and the device must be reprogrammed before it can be used again.

The volatility of a stored-program FPGA is a double-edged sword—the FPGA must be reprogrammed in the event that power is disrupted, but the same generic part can serve a boundless variety of applications, and it can be reconfigured on the same circuit board under the control of a processor. One of the programs that can be executed by an FPGA can even test the host system in which it is embedded. The ease of reprogramming a stored-program FPGA supports rapid-prototyping, enabling design teams to compete effectively in an environment characterized by narrow and ever-shrinking windows of opportunity. Time-to-market is critical in many designs, and FPGAs provide a path to early entry. FPGAs can be reconfigured remotely, via the Internet, allowing designers to repair, enhance, upgrade, or completely reconfigure a device in the field.

8.9.1   The Role of FPGAs in the ASIC Market

The architectural resources of FPGAs match the general need for computational engines with memory, datapaths, and processors. The flexibility of an FPGA adds a dimension beyond what is available in masked-programmed devices, because mask-programmed devices cannot be reprogrammed. The same FPGA can be programmed to implement a variety of processors. Expensive, high-risk mask sets for ASICs have made flexibility an important consideration. On the other hand, FPGAs cost more per unit gate, and consume more power.

Table 8-8 summarizes key distinctions between FPGA technology, cell-based and mask-programmed ASIC technology, and standard parts. The myriad applications for ASICs and FPGAs require customization to address the needs of the market. The diversity, rewards, evolving technology, and short lifespan of the market preclude producing and stockpiling standard parts to meet these needs without unacceptably high risk. The lower volumes demanded by individual, specialized applications provides a smaller base over which to amortize the costs of development and production, so units costs for FPGAs are higher than for standard parts, and for high-volume, mask-programmed ASICs, but their NRE costs are significantly lower.

The early technology of MPGAs used a fixed array of transistors and routing channels. Routing was a major issue in early devices, and frequently led to incomplete utilization of the available transistors. Today, multilevel metal routing (five and six layers) in a sea-of-gates technology is commonplace, with high utilization of resources. MPGAs are preprocessed to the point of customizing the final metal layers to a particular application. The customization/metalization steps connect individual transistors to form gates, and interconnect gates to implement logic. This technology provides a much quicker turnaround than cell-based and full-custom technologies, because only the final metalization step is customized, but not as fast as that for an FPGA. Depending on the foundry, an MPGA can be turned around in a few days to several weeks. On the other hand, designs can be implemented, programmed, and reprogrammed virtually instantaneously in an SRAM-based FPGA while the part is mounted in an emulator or in its target host application. But FPGAs will always be slower and less dense than a comparable MPGA because of the additional circuitry and delays introduced by their programmable interconnect.

TABLE 8-8   Comparisons of standard parts, ASICs, and FPGAs.

Technology	Functionality	Relative Cost
Standard Part	Supplier-defined	Low
FPGA	User-defined	Higher
ASIC	User-defined	Low

FPGAs are fully tested by the manufacturer before they are shipped, so the designer's attention is focused on the creativity of the design, not on testing for manufacturing defects. Designers can quickly correct design flaws and reconfigure the part to a different functionality in the field [10]. FPGAs address a market than cannot be met by mask-programmed technologies, which are one-time write. Mask-programmed technologies do not support reconfiguration, and corrections are costly. The risk of an MPGA-based design is significantly higher than for an FPGA because a design flaw requires retooling of the final masks, with attendant costs, and lost time to reenter the fabrication-process queue.

MPGAs have a broad customer base for amortizing the NRE of most of the processing steps compared to cell-based and full-custom solutions. Gate arrays are widely used to implement designs that have a high content of random logic, such as state-machine controllers.

MPGAs require the direct support of a foundry, and the completion of a design can depend on the schedule of the foundry's other customers. FPGAs are fully manufactured and tested in anticipation of being shipped immediately to a buyer.

The software interface between the designer and FPGA technology is simple, and it is now readily and cheaply available on PCs and workstations that support schematic and HDL entry. Programmable logic technology continues to grow in density at exponential rates as compared with other technologies, such as dynamic random-access memories (DRAMs) [10]. The speed of parts is growing at a linear rate and is now at a level that supports system-level integration.

Standard cell-based technology uses a library of predesigned and precharacterized cells that implement gates. The design of the individual cells in a library is labor-intensive, as efforts are made to achieve a dense, area-efficient layout. Consequently, a cell library has a high NRE cost, which a foundry must amortize over a large customer base during the lifetime of the underlying process technology. The mask set of a standard cell library is fully characterized and verified to be correct. Place and route tools select, place and interconnect cells in rows on a chip to implement functionality. The structure is semiregular because the cell heights are fixed, while the width of cells may vary, depending on the functionality being implemented. Placement and routing are customized for each application. Place and route are done automatically to achieve dense configurations that meet speed and area constraints. Cell-based technology requires a fully customized mask set for each application. Consequently, volume must be sufficient to offset high production and development costs and ultimately drive an economically low unit cost.

8.9.2 FPGA Technologies

State-of-the-art FPGAs can now implement the functionality of over a million (two-input equivalent) gates on a single chip. Three basic types of FPGAs are available: antifuse, EPROM, and SRAM-based. The capacity and speed of these parts continues to evolve with process improvements that shrink minimum feature sizes of the underlying transistors.

Antifuse devices[19] are programmed by applying a relatively high voltage between two nodes to break down a dielectric material. This eliminates the need for a memory

---

[19]See www.actel.com for more information about antifuse devices.

to hold a program, but the one-time write configuration is permanent. When an antifuse is formed a low-resistance path is irreversibly created between the terminals of
the device. The antifuse itself is relative small, about the size of a via, and over a million
devices can be distributed over a single FPGA [11]. The significant advantage of this
technology is that the on-resistance and parasitic capacitance of an antifuse are much
smaller than for transmission gates and pass transistors. This supports higher switching
speeds and predictable timing delays along routed paths.

EPROM and EEPROM-based technology uses a charged floating gate, programmed by a high voltage. Devices based on these technologies are reprogrammable
and nonvolatile and can be programmed offline while imbedded in the target system.

SRAM-based FPGA technology uses CMOS transmission gates to establish
interconnect. The status of the gates is determined by the contents of the SRAM configuration memory.

There are multiple vendors of SRAM-based FPGA products (e.g., Xilinx, Altera,
ATMEL, and Lucent). The architecture of these FPGAs is similar to that of an MPGA,
with block structures of logic and routing channels. Bidirectional and multiply driven
wires are included. Devices are advertised on the basis of gate counts, but the actual
use of the gates on a device depends on the router's ability to exploit the resources to
support a given design.

The complexity of logic cells in an FPGA functional unit is based on competing
factors. If the complexity of a cell is low (fine-grained, such as the Actel Act-1 part), the
time and resources required for routing may be high. On the other hand, if the complexity is high, there will be wasted cell area and logic. An example of a fine-grained
architecture would be one that is based on two-input NAND gates or muxes, as
opposed to a large-grain architecture using four-input NAND gates or muxes. The
former uses considerably more routing resources.

8.10   Altera Flex 8000 FPGAs

Altera's architecture for its FPGAs is illustrated by the now-mature technology of its
flexible logic element matrix (Flex) 8000 family of devices, and its closely related, and
more advanced, Flex 10 and Flex 20 devices. The basic architecture shown in Figure
8-52 has a grid of LABs, each consisting of eight independently programmable logic
elements (LEs) providing an efficient fine-grained logic structure. A logic element has
a four-input LUT, a programmable register, and dedicated carry and cascade chains.
(The LEs will be discussed in more detail with the Flex 10 device family). The grid of
LABs has access to continuous channels of FastTrack Interconnect. The ends of the
interconnect channel are connect to input–output elements (IOEs), each having a bidirectional I/O buffer and a flip-flop for registering either the input or the output. Four
signals are common to each LE in a LAB: two of which can be used as clocks and two
of which can be used for clear/preset control; each of these can be driven by a dedicated input, an I/O pin, or an internal signal from the LAB's local interconnect. Each
device can have up to four low-skew global clock, clear, or preset control signals.

The Flex 8000 family is targeted for applications in register-intensive digital signal processing, wide datapath manipulation and data transformations (bus interfaces,

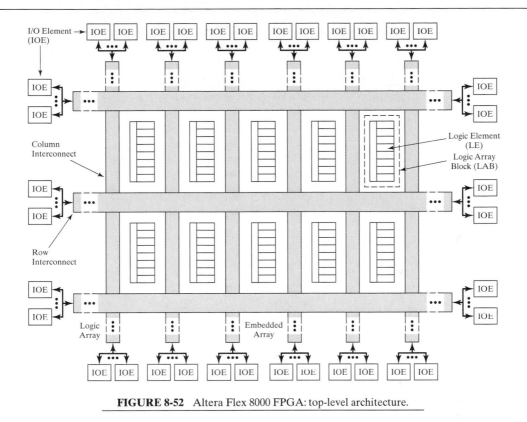

FIGURE 8-52 Altera Flex 8000 FPGA: top-level architecture.

coprocessor functions, transistor-transistor logic (TTL) integration, and high-speed controllers.

8.11 Altera Flex 10 FPGAs

The push for speed in all electronic equipment has driven designers to relentlessly integrate as much of the design as possible. Consequently, advanced FPGAs now have distributed or block memory on the same chip. PLDs, even when accompanied by a register cell, are not efficient implementations of memory. Consequently, advanced FPGAs combine two structures, one for logic and one that implements embedded memory efficiently. Altera's Flex 10 devices are intended to support system-on-programmable chip integration in a single device that has logic implementing the equivalent of up to 200,000 typical gates (i.e., two-input NAND gates). The Flex 10 features dual-port on-chip memory and operates at speeds compatible with 33 MHz and 66 MHz IEEE PCI local bus specifications. Table 8-9 identifies some significant physical features of the device family[20] over a range of low- and high-density of gates.

The Flex 10 architecture shown in Figure 8-53 is similar to that of the Flex 8000 devices shown in Figure 8-52, but the Flex 10 family includes block RAM. The architecture

---

[20]The Flex 10 family is available in 5, 3.3., and 2.5 V technology. Additional features (e.g., performance and packaging) of the Flex 10 device family are described in data sheets available at www.altera.com.

TABLE 8-9   Altera Flex 10K FPGA: significant architectural features.

Feature	EPF10K10	•••	EPF10K250A
Typical gates[1]	10,000		250,000
Max # system gates	31,000		310,000
# Logic elements (LEs)	576		12,160
Logic Array Blocks LAB)	72		1520
Embedded Array Blocks (EAB)	3		20
Total RAM bits	6,144		40,960
Max # user I/O pins	150		470

[1] Logic and RAM, excluding JTAG circuitry.

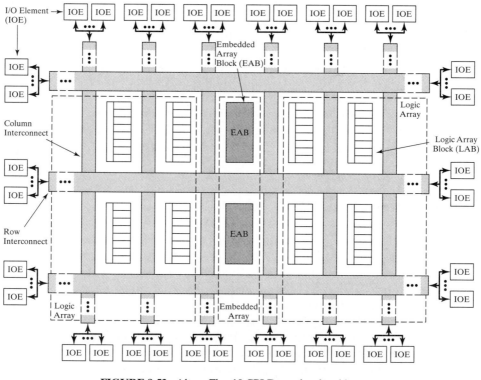

FIGURE 8-53   Altera Flex 10 CPLD: top-level architecture.

consists of an array of embedded array blocks (EABs), an array of LABs, and an interconnection fabric. Several LABs and a EAB occupy a row; each LAB contains eight logic elements and local interconnect, as in the architecture of the Flex 8000. Each EAB has 2K of RAM memory. FastTrack Interconnect channels connect the LABs and EABs and provide global interconnect without making use of switch matrixes, with the result that delays do not depend on the routing. The IOEs can be programmed to serve as input, output or bidirectional signal paths, with programmable slew rate to reduce switching noise (only time-critical paths need a high skew rate—the others can be programmed for low-slew operation to reduce noise).

The internal architecture of a LAB is shown in Figure 8-54. Each LE has four inputs from the local interconnect (which may include bits that have been fed back from the outputs of the logic elements in the LAB (one bit can be fed back from each of the eight logic elements). Each logic element also has four control signals that are programmed to be formed locally or obtained globally. The outputs of a LAB can be connected to the row and column interconnect.

The LE of the Flex 10 family is shown in Figure 8-55. Each element contains an LUT that can implement any function of four variables. The output of the element can be direct or registered. With four bits of data input each, two LEs can implement a full

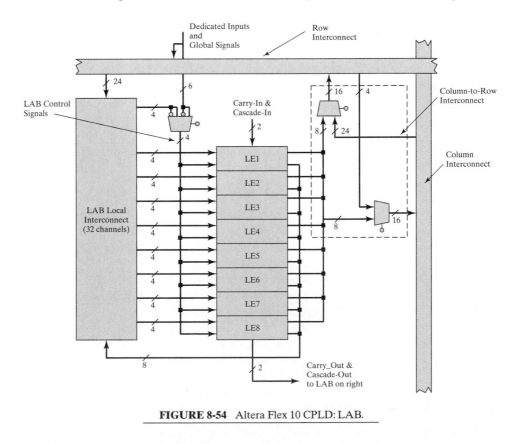

FIGURE 8-54 Altera Flex 10 CPLD: LAB.

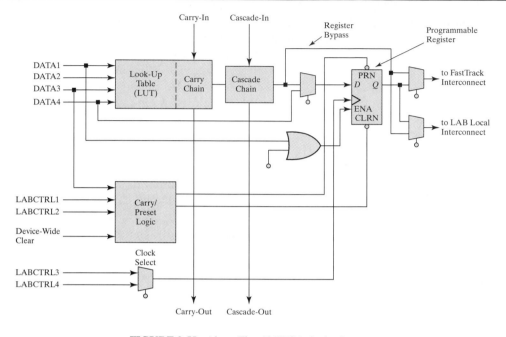

FIGURE 8-55   Altera Flex 10 FPGA: logic element.

adder (one for the sum bit and one for the carry out bit), with the carry-in and carry-out bits linked to those of adjacent LEs.

Dedicated pins support carry-in/out and cascade-in/out connectivity for wide datapaths and functions of more than four variables. The cascade chain can be programmed to *AND* the outputs of logic elements to form a function of more than four variables, or to *OR* the outputs, which forms a sum of cubes, each of which can be a function of up to four variables (Figure 8-56).

Each row of the Flex 10 device has an embedded memory array block, referred to as an EAB (Figure 8-57). The memory (2,048 bits) within an EAB can be programmed to various configurations as RAM or ROM. The EAB local interconnect serves as a local datapath for both the address and input data of the memory. The output of the memory is passed directly to either a row or column interconnect, directly or through a register. Complex functions can be implemented in an EAB in one logic level, avoiding the additional routing delays that would result from implementing the function in a series of chained LEs or by aggregating several units of distributed RAM in an FPGA.

The configuration of a logic array and an embedded memory array in the Flex 10 device family combines the flexibility of a PLD with the density and efficiency of an embedded gate array. The more recent and advanced Flex 10E family of devices shown in Table 8-10 has significantly greater memory capacity. Its EAB blocks have independently configurable dual-port capability, enabling them to support dual-clock applications, such as dual-clock FIFOs.

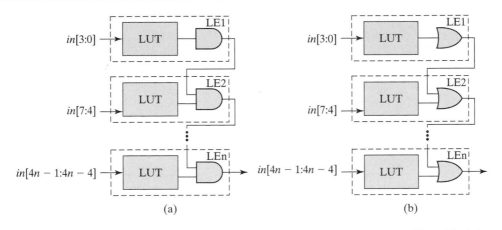

FIGURE 8-56 Cascaded chains in the Altera Flex 10 CPLD: (a) an AND chain, and (b) an OR chain.

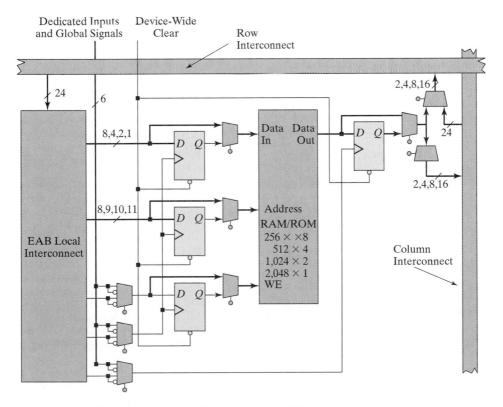

FIGURE 8-57 Altera Flex 10 FPGA with embedded RAM/ROM block.

TABLE 8-10   Altera Flex 10KE FPGA: significant architectural features.

Feature	EPF10K30E	EPF10K50E EPF10K50S	EPF10K100B
Typical gates (1)	30,000	50,000	100,000
Maximum no. of system gates	119,000	199,000	158,000
No. of logic elements	1,728	2,880	4,992
EABs	6	10	12
Total RAM bits	24,576	40,960	24,576
Maximum no. of user I/O pins	220	254	191

Feature	EPF10K100E	EPF10K130E	EPF10K200S
Typical gates (1)	100,000	130,000	200,000
Maximum no. of system gates	257,000	342,000	513,000
No. of logic elements	4,992	6,656	9,984
EABs	12	16	12
Total RAM bits	49,152	65,536	98,304
Maximum no. of user I/O pins	338	413	470

8.12   Altera Apex FPGAs

Advances in process technology have led to denser and faster field-programmable devices. The complexity of applications and the demand for higher levels of integration has produced leading edge devices with on-chip block memory, support for high-speed synchronous operation (phase lock loops [PLLs]), and support for a wide assortment of interfaces to the external environment. The high-density Altera APEX family of Altera FPGAs combines four architectural elements to support complete system-level integration on a single chip. The arrangement partitions an implementation into control and datapath functionality and associates functionality with different architectural elements. The structure in Figure 8-58 has LUTs to support register-intensive datapath and digital signal processing (DSP) functions, and product-term integrators to support high-speed complex (multilevel) combinational logic in control logic and state machines. Embedded memory blocks, referred to as ESBs (embedded system blocks) can support a variety of memory functions (e.g., FIFOs, dual-port RAMs, and content-addressable memory).

The architecture is organized as a series of MegaLABs, shown in Figure 8-59, each consisting of from 16 to 24 LABs, an ESB, and the Flex 8000 architecture's channel-based FastTrack Interconnect fabric.

Significant architectural features of the APEX family are shown in Table 8-11. The I/O elements of the device are located at the end of each row and column of the interconnect. Each element contains a bidirectional I/O buffer, and a register. Global clocks support high-speed registered data transfers. Four dedicated input pins and two

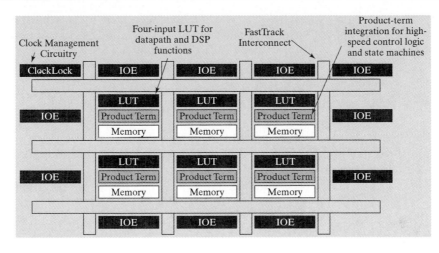

FIGURE 8-58 Altera architecture: Apex 20KE FPGA.

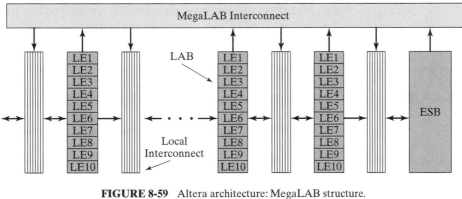

FIGURE 8-59 Altera architecture: MegaLAB structure.

TABLE 8-11 Altera APEX 10KE CPLD: significant architectural features.

Feature	EP20K30E	•••	EP20K1500E
Typical gates	30,000		1,500,000
Maximum no. of system gates	113,000		2,392,000
No. of logic elements (LEs)	1,200		51,840
Embedded system blocks	12		216
Maximum RAM bits	53,248		442,368
Maximum macrocells	192		3,456
Maximum no. of user I/O pins	246		808

dedicated clock pins on each device provide high-speed, low-skew control signals. The four dedicated inputs drive four global signals, which can be used in synchronous control operations.

8.13 Altera Chip Programmability

Altera's Max+PLUS II development system provides a turnkey, desktop environment supporting rapid prototyping of FPGA-based designs The system supports schematic and HDL-based (Verilog, VHDL, AHDL) design entry, logic synthesis, simulation, timing analysis, and device programming.

8.14 XILINX XC4000 Series FPGA

Xilinx launched the world's first commercial FPGA in 1985, with the vintage XC2000 device family. The XC3000 and XC4000 families soon followed, setting the stage for today's Spartan and Virtex device families. Each evolution of devices brought improvements in density, performance, voltage levels, pin counts, and functionality. The XC4000, Spartan, and Spartan/XL devices have the same basic architecture; Table 8-12 displays the evolution of density and operating voltage among three product families.

8.14.1 Basic Architecture

The basic architecture of the XC3000, XC4000, Spartan, and Spartan X/L device family consists of an array of configurable logic blocks (CLBs), a variety of local and global routing resources, and IOBs, programmable I/O buffers, and a SRAM-based configuration memory, as shown in Figure 8-60.

8.14.2 XC4000 Configurable Logic Block

Each CLB consists of an LUT, multiplexers, registers, and paths for control signals. Each CLB of the Xilinx 3000 series had a single programmable LUT implementing two combinational logic functions of five inputs (5-ns CLB delay). The outputs of the functions could be programmed to pass through a register or pass directly to the output of the CLB. A registered[21] output of a function could also be routed internally to the input of the LUT, allowing a group of CLBs to implement register-rich logic for a state machine or a data pipeline.

The CLBs of the successor to the XC3000, the XC4000 device family, are more versatile than those of the XC3000. In the architecture shown in Figure 8-61, note that each CLB contains three function generators (F, G, and H). Each is based on an LUT with 5-ns delay independent of the function begin implemented. Two of the function

---

[21]The flip-flops can be set individually or globally.

TABLE 8-12 Migration of density and operating voltage for Xilinx part families.

Part	Maximum No. of System Gates	System Performance	Operating Voltage
Spartan	40K		5V
Spartan/XL	40K	>80 MHz	3.3
Spartan-II	200K	<200 MHz	2.5

generators (F and G) can generate any arbitrary function of four inputs, and the third (H) can generate any Boolean function of three inputs. The H-function block can get its inputs from the F and G LUTs, or from external inputs. The three function generators can be programmed to generate: (1) three different functions of three independent sets of variables (two with four inputs and one with three inputs—one function must be registered within the CLB), (2) an arbitrary function of five variables, (3) an arbitrary function of four variables together with some functions of six variables, and (4) some functions of nine variables.

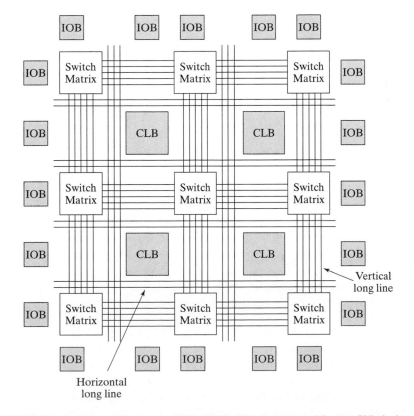

FIGURE 8-60 Basic architecture of Xilinx 3000, 4000, Spartan, and Spartan X/L devices.

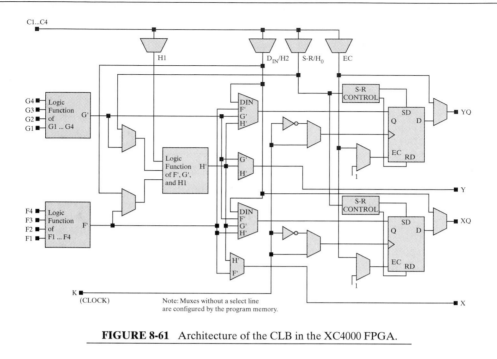

FIGURE 8-61   Architecture of the CLB in the XC4000 FPGA.

Each CLB in the XC4000 series has two storage devices that can be configured as edge-triggered flip-flops with a common clock, or in the XC4000X they can be configured as flip-flops or as transparent latches with a common clock (programmed for either edge and separately invertible) and enable. The storage elements can get their inputs from the function generators, or from the D_{in} input. The other element can get an external input from the *H1* input. The function generators can also drive two outputs directly (X and Y) and independently of the outputs of the storage elements. All of these outputs can be connected to the interconnect network. The storage elements are driven by a global S-R during power-up; the global S-R is programmed to match the programming of the local S-R control for a given storage element.

8.14.3 Dedicated Fast Carry and Borrow Logic

The F and G function generators of the XC4000 family have separate dedicated logic for fast carry and borrow generation, with dedicated routing to link the extra signal to the function generator in the adjacent CLB. The prebuilt carry chain within a CLB can be used to add a pair of 2-bit words in one CLB. One function generator (F) can be used to generate *a0 + b0*, and a second function generator (G) can generate *a1 + b1*. The fast carry will forward the carry to the next CLB above or below. This feature is implemented with a hard macro and the graphical editor. Fast carry and borrow logic increases the efficiency performance of adders, subtractors, accumulators, comparators, and counters.

8.14.4 Distributed RAM

The three function generators within a CLB can be used as RAM, either a 16×2 dual port RAM or a 32×1 single-port RAM. The XC4000 devices do not have block RAM, but a group of their CLBs can form an array of memory.

8.14.5 XC4000 Interconnect Resources

The XC4000 series was designed with interconnect resources to minimize the resistance and capacitance of an average routed path. A grid of switch matrixes overlays the architecture of CLBs to provide general-purpose interconnect for branching and routing throughout the device. The interconnect has three types of general-purpose interconnect: single-length lines, double-length lines, and long lines. A grid of horizontal and vertical single-length lines connect an array of switch boxes. The boxes provide a reduced number of connections between signal paths within each box, not a full crossbar switch. In the XC4000 there is a rich set of connections between single-length lines and the CLB inputs and outputs. These provide capability for nearest-neighbor and across-the-chip connection between CLBs. Each CLB has a pair of three-state buffers that can drive signals onto the nearest horizontal lines above or below the CLB.

Direct (dedicated) interconnect lines provide routing between adjacent vertical and horizontal CLBs in the same column or row. These are relatively high-speed local connections through metal, but are not as fast as a hard-wired metal connection because of the delay incurred by routing the signal paths through the transmission gates that configure the path. Direct interconnect lines do not use the switch matrixes, which eliminates the delay incurred on paths going through a matrix.[22]

Double-length lines traverse the distance of two CLBs before entering a switch matrix, skipping every other CLB. These lines provide a more efficient implementation of intermediate-length connections by eliminating a switch matrix from the path, thereby reducing the delay of the path.

Long lines span the entire array vertically and horizontally. They drive low-skew, high-fanout control signals. Long vertical lines have a programmable splitter that segments the line and allows two independent routing channels spanning half of the array, but located in the same column. The routing resources are exploited automatically by the routing software. There are eight low-skew global buffers for clock distribution, and the skew on a global net is less than 2 ns. The XC4000 device also has an internally generated clock.

The signals that drive long lines are buffered. Long lines can be driven by adjacent CLBs or IOBs and may connect to three-state buffers that are available to CLBs. Long lines provide three-state busses within the architecture, and implement wired-AND logic. Each horizontal long line is driven by a three-state buffer, and can be programmed to connect to a pull-up resistor, which pulls the line to a logical 1 if no driver is asserted on the line.

---

[22]See Xilinx documentation for the pinout conventions to establish local interconnect between CLBs.

Figure 8-62 illustrates the single- and double-length interconnect lines of the XC4000 device.

The connectivity of a CLB with neighboring switch matrixes is shown in Figure 8-63.

The programmable interconnect resources of the device connect CLBs and IOBs, either directly or through switch boxes. These resources consist of a grid of two layers of metal segments and programmable interconnect points (PIP) within switch boxes. A PIP is a CMOS transmission gate whose state (on or off) is determined by the

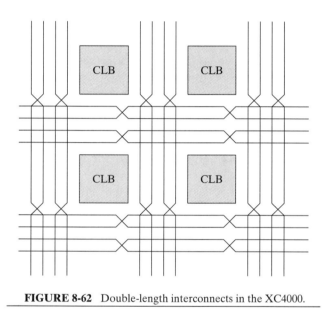

FIGURE 8-62   Double-length interconnects in the XC4000.

content of a static RAM cell in the programmable memory, as shown in Figure 8-64. The connection is established when the transmission gate is on (i.e., a 1 is applied at the gate of the *n*-channel transistor and a 0 is applied at the gate of the *p*-channel transistor). The interconnect path is established without altering the physical medium, as happens in a fuse-type interconnect technology. Thus, the device can be reprogrammed by simply changing the content of the controlling memory cell.

The architecture of a PIP-based interconnection in a switch box is shown in Figure 8-65. The configuration of CMOS transmission gates determines the connection between a horizontal line and the opposite horizontal line, and the vertical lines at the connection. Each switch matrix PIP requires six pass transistors to establish full connectivity.

8.14.6 XC4000 I/O Block (IOB)

Each programmable I/O pin/pin of an XC4000 device has a programmable IOB with buffers for compatibility with TTL and CMOS signal levels. Figure 8-66 shows a simplified

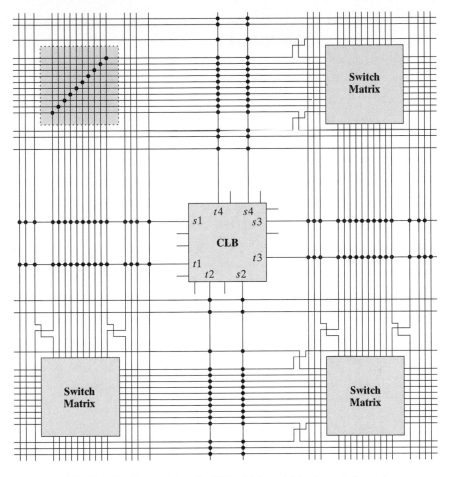

FIGURE 8-63 Connectivity of a CLB with its neighboring switch matrixes.

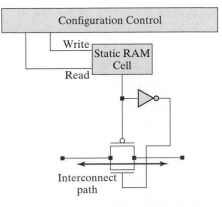

FIGURE 8-64 RAM cell controlling a PIP transmission gate.

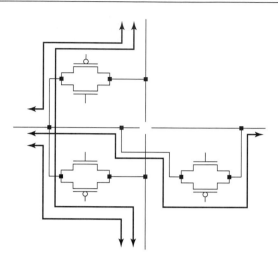

FIGURE 8-65   Circuit-level architecture of a PIP within a switch box.

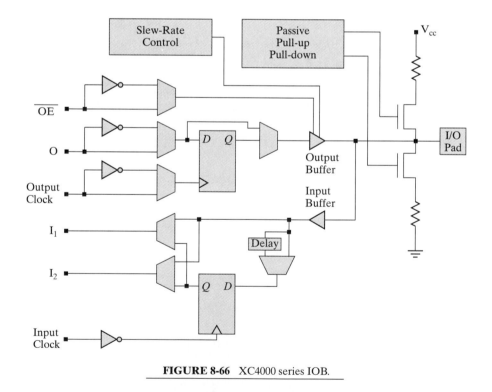

FIGURE 8-66   XC4000 series IOB.

schematic for a XC4000 programmable IOB. It can be used as an input, output or bidirectional port. An IOB that is configured as an input can have direct, latched, or registered input. In an output configuration, the IOB has direct or registered output. The output buffer of an IOB has skew and slew control. The registers available to the input and output path of an IOB are driven by separate, invertible clocks. There is a global set and reset.

The XC4000 architecture has delay elements that compensate for the delay induced when a clock signal passes through a global buffer before reaching an IOB. This eliminates the hold condition on the data at an external pin. The three-state output of a IOB puts the output buffer in a high-impedance state. The output and the enable for the output can be inverted. The slew rate of the output buffer can be controlled to minimize transients on the power bus when noncritical signals are switched. The IOB pin can be programmed for pull-up or pull-down to prevent needless power consumption and noise.

The XC4000 has four edge decoders on each side of the chip. An edge decoder can accept up to 40 inputs from adjacent IOBs and 20 inputs from on-chip. These decoders provide fast decoding of wide address paths. Multiple CLBs might be needed to support decoding when the fan-in exceeds the width of a single CLB.

The XC4000 devices have imbedded logic to support the IEEE 1149.1 (JTAG) boundary scan standard. There is an on-chip test access port (TAP) controller, and the I/O cells can be configured as a shift register. Under test, the device can be checked to verify that all the pins on a PC board are properly connected by creating a serial chain of all of the I/O pins of the chips on the board. A master three-state control signal puts all of the IOBs in high-impedance mode for board testing.

The XC4000EX series, introduced in 1996, provides a twofold increase in routing resources over the earlier members of the XC4000 family. An additional 22 vertical lines are available in each column of CLBs. Another 12 quad lines have been added to each row and column to support fast global routing, illustrated in Figure 8-67. The IOBs have a dedicated early clock and fast-capture latch. The registers have a 4-ns setup time and a 6-ns clock-to-output propagation delay.

8.14.7 Enhancements in the XC4000E and XC4000X Series

The XC4000E and XC4000X devices offer significant increases in speed and capacity, as well as architectural improvements over the basic XC4000 family. Devices in these families can run at system clock rates up to 80 MHz, and internal frequencies can reach 150 MHz, as a result of submicron multilayered metal processes. The XC4000XL (0.35 micron feature size) operates at 3.3 V with a system frequency of 80 MHz.

8.14.8 Enhancements in the Spartan Series

The Spartan series of Xilinx FPGAs has the same architecture of CLBs and IOBs as its XC4000 predecessors, but has higher performance and is targeted for high-volume, low-cost applications requiring on-chip memory and high performance (over 100 MHz system speeds).

Spartan chips can accommodate embedded soft cores, and their on-chip distributed, dual-port, synchronous RAM (SelectRAM) can be used to implement FIFOs,

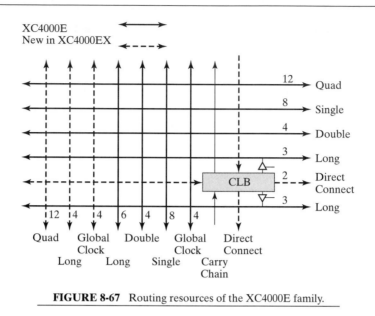

FIGURE 8-67   Routing resources of the XC4000E family.

shift registers, and scratchpad memories. The blocks can be cascaded to any width and depth and located anywhere in the part, but their use reduces the CLBs available for logic. Figure 8-68 displays the structure of the on-chip RAM that is formed by programming a LUT to implement a single-port RAM with synchronous write and asynchronous read. Each CLB can be programmed as a 16×2 or a 32×1 memory.

Dual-port RAMs are emulated in a Spartan device by the structure shown in Figure 8-69, which has a single (common) write port and two asynchronous read ports. A CLB can form a memory with a maximum size of 16×1.

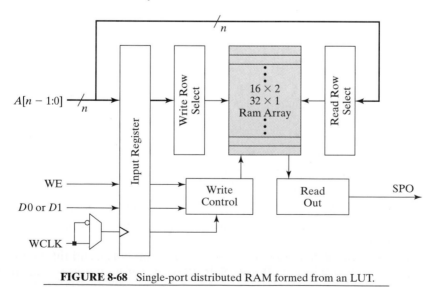

FIGURE 8-68   Single-port distributed RAM formed from an LUT.

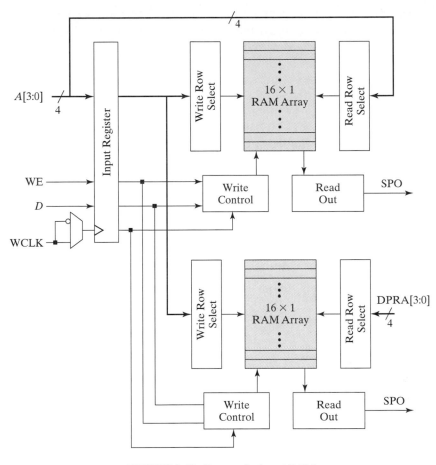

FIGURE 8-69 Spartan dual-port RAM.

8.15 XILINX Spartan XL FPGAs

Spartan XL chips are a further enhancement of the Spartan chips, offering higher speed and density (40,000 system gates per approximately 6000 usable gates), and on-chip, distributed SelectRAM memory.[23] The LUTs of the devices can implement 2^{2^n} different functions of n inputs. The devices are peripheral component interconnect (PCI) bus compliant and have eight flexible global low-skew buffers (BUFGLS) and CLB latches for clock distribution and for secondary control signals; each serves one-fourth of the devices. The input latches support fast capture of data.

---

[23]The maximum number of logic gates for a Xilinx FPGA is an estimate of the maximum number of logic gates that could be realized in a design consisting of only logic functions (no memory). Logic capacity is expressed in terms of the number of two-input NAND gates that would be required to implement the same number and type of logic functions (Xilinx application note).

The XL series is targeted for applications in which low cost, low power, low packaging, and low test cost are important factors constraining the design solution. Spartan XL devices offer up to 80-MHz system performance, depending on the number of cascaded LUTs, which reduces performance by introducing longer paths. Table 8-13 presents significant attributes of devices in the Spartan XL family.

The architecture of the 3000, 4000, Spartan, and Spartan/XL series of Xilinx FPGAs consists of an array of CLB tiles mingled within an array of switch matrixes, surrounded by a perimeter of IOBs. These devices supported only distributed memory, whose use reduces the number of CLBs that could be used for logic. The relatively small amount of on-chip memory limits these devices to applications in which operations with off-chip memory devices do not compromise performance objectives. Beginning with the Spartan II series, Xilinx supported configurable embedded block memory as well as distributed memory in a new architecture.

TABLE 8-13   Significant attributes of the Xilinx Spartan XL device family.

Spartan/XL	XCS05/XL	XCS10/XL	XCS20/XL	XCS30/XL	XCS40/XL
System Gates[1]	2K–5K	3K–10K	7K–20K	10K–30K	13K–40K
Logic Cells[2]	238	466	950	1,368	1,862
Maximum No. of Logic Gates	3,000	5,000	10,000	13,000	20,000
Flip-Flops	360	616	1,120	1,536	2,016
Maximum RAM Bits	3,200	6,272	12,800	18,432	25,088
Maximum Available I/O	77	112	160	192	224

[1] 20–30% of CLBs as RAM
[2] 1 Logic cell = four-input LUT + flip-flop

8.16   XILINX Spartan II FPGAs

Aside from improvements in speed (200 MHz I/O switching frequency), density (up to 200,000 system gates), and operating voltage (2.5 V), four other features distinguish the Spartan II devices from the Spartan devices: (1) on-chip block memory, (2) a novel architecture, (3) support for multiple I/O standards, and (4) delay-locked loops.[24]

The Spartan II device family, manufactured in 0.22/0.18 μm CMOS technology with six layers of metal for interconnect, incorporates configurable block memory in addition to the distributed memory of the previous generations of devices, and the block memory does not reduce the amount of logic and/or distributed memory that is available for the application. The availability of a large on-chip memory can improve system performance by eliminating or reducing the need to access off-chip storage.

Reliable clock distribution is the key to the synchronous operation of high-speed digital circuits. If the clock signal arrives at different times at different parts of a circuit the device may fail to operate correctly. Clock skew reduces the available time budget

---

[24]The Spartan-II devices do not support LVDS (low voltage differential signaling) and LVPECL (low voltage/power emitter-coupled logic) I/O standards.

of a circuit by lengthening the setup time at registers. It can also shorten the effective hold time margin of a flip-flop in a shift register and cause the register to shift incorrectly. At high clock frequencies, the effect of skew is more significant because it represents a larger fraction of the clock cycle time. Buffered clock trees are commonly used to minimize clock skew in FPGAs. Xilinx provides all-digital delay-locked loops (DLLs) for clock synchronization/management in high-speed circuits. DLLs eliminate the clock distribution delay and provide frequency multipliers, frequency dividers, and clock mirrors.

Spartan-II devices are suitable for applications such as implementation of the glue logic of a video capture system, and the glue logic of an ISDN modem. Device attributes are summarized in Table 8-14, and the evolution of technology in the Spartan series is evident in the data in Table 8-15.

The top-level tiled architecture of the Spartan-II device, shown in Figure 8-70, marks a new organization of the Xilinx parts. Each of four quadrants of CLBs is supported by a DLL and is flanked by a 4096-bit block[25] of RAM, and the periphery of the chip is lined with IOBs.

Each CLB contains four logic cells, organized as a pair of slices. Each logic cell, shown in Figure 8-71, has a four-input LUT, logic for carry and control, and a D-type flip-flop. The CLB contains additional logic for configuring functions of five or six inputs.

TABLE 8-14 Spartan II device attributes.

Spartan-II FPGAs	XC2S15	XC2S30	XC2S50	XC2S100	XC2S150	XC2S200
System Gates[1]	6K–15K	13K–30K	23K–50K	37K–100K	52K–150K	71K–200K
Logic Cells[2]	432	972	1,728	2,700	3,888	5,292
Block RAM Bits	16,384	24,576	32,768	40,960	49,152	57,344
Maximum Available I/O	86	132	176	196	260	284

[1] 20–30% of CLBs as RAM
[2] 1 Logic cell = four-input LUT + flip-flop

TABLE 8-15 Comparison of the Spartan device families.

Part	Spartan	Spartan/XL	Spartan-II
Architecture	XC4000 Based	XC4000 Based	Virtex Based
Maximum No. of System Gates	5K–40K	5K–40K	15K–200K
Memory	Distributed RAM	Distributed RAM	Block + Distributed
I/O Performance	80 MHz	100 MHz	200 MHz
I/O Standards	4	4	16
Core Voltage	5V	3.3	2.5
DLLs	No	No	Yes

[25]Parts are available with up to 14 blocks (56K bits).

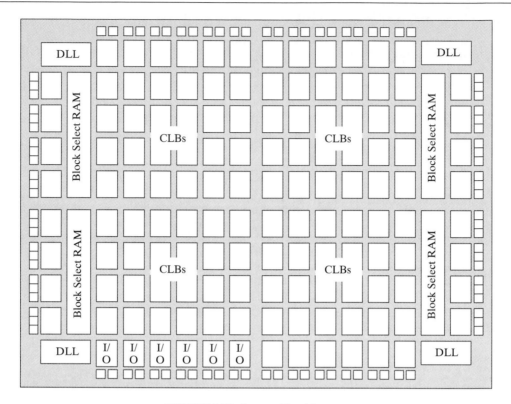

FIGURE 8-70 Spartan II architecture.

The Spartan-II part family provides the flexibility and capacity of an on-chip block RAM; in addition, each LUT can be configured as a 16×1 RAM (distributed), and the pair of LUTs in a logic cell can be configured as a 16×2 bit RAM or a 32×1 bit RAM.

The IOBs of the Spartan-II family are individually programmable to support the reference, output voltage, and termination voltages of a variety of high-speed memory and bus standards (see Figure 8-72). Each IOB has three registers, which can function as D-type flip-flops or as level-sensitive latches. One register (TFF) can be used to register the signal that controls (synchronously) the programmable output buffer. A second (OFF) can be programmed to register a signal from the internal logic (alternatively, a signal from the internal logic can pass directly to the output buffer). The third device can register the signal coming from the I/O pad (alternatively, this signal can pass directly to the internal logic. A common clock drives each register, but each has an independent clock enable. A programmable delay element on the input path can be used to eliminate the pad-to-pad hold time.

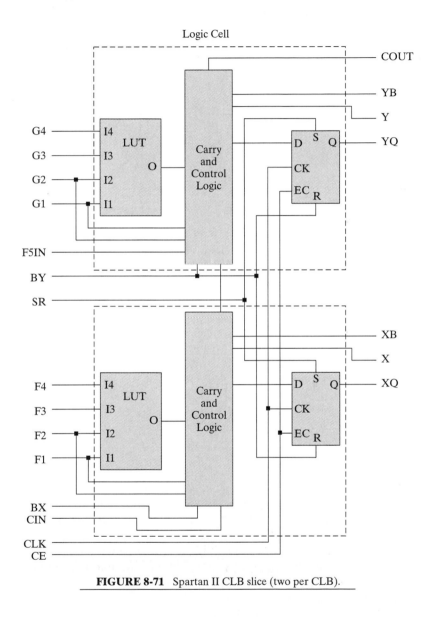

FIGURE 8-71 Spartan II CLB slice (two per CLB).

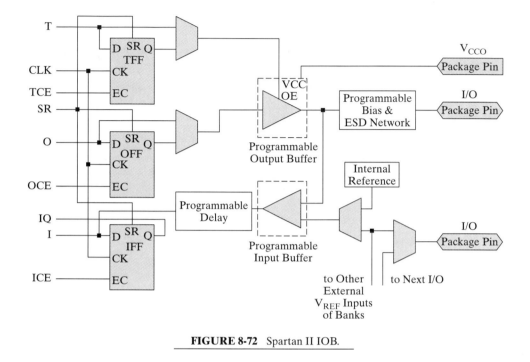

FIGURE 8-72   Spartan II IOB.

8.17   XILINX Virtex FPGAs

The Virtex device series is the leading edge of Xilinx technology. It addresses four key factors that influence the solution to complex system-level and system-on-chip (SoC) designs: (1) the level of integration, (2) the amount of embedded memory, (3) performance (timing), and (4) subsystem interfaces. Process rules for leading-edge Virtex parts stand at 0.18 μm and will continue to shrink. The rules allow up to 138,240 logic cells to be packed into a single die, providing up to 10 million system gates, and the capacity to support memory-intensive (up to 3.5 M bits) system-level applications requiring high density and high performance.

The Virtex family incorporates physical (electrical) and protocol support for 20 different I/O standards, including LVDS and LVPECL, with individually programmable pins. Up to 12 digital clock managers provide support for frequency synthesis and phase shifting in synchronous applications that require multiple clock domains and high frequency I/O. The Virtex architecture is shown in Figure 8-73, and its IOB is shown in Figure 8-74.

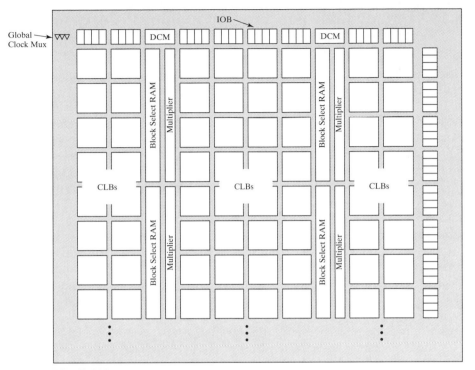

DCM: Clock Manager

FIGURE 8-73 Xilinx Virtex-II overall architecture.

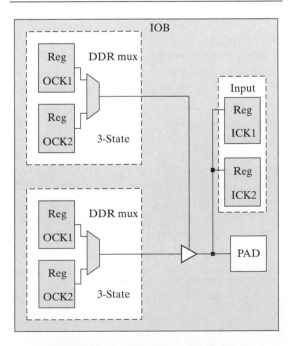

FIGURE 8-74 Xilinx Virtex-II IOB Block.

8.18   Embeddable and Programmable IP Cores for a System on a Chip (SoC)

ASIC cores consist of intellectual property (IP) that has been designed, verified, and marketed by a vendor for re-use by other parties. Cores may be soft (software models) or hard (mask sets). The use of preimplemented and verified embedded cores in an ASIC can shorten the time-to-market of a new product by reducing the amount of circuitry that must be developed. Whether this economy is realized depends on the reliability and documentation of the embedded logic and whether system-level tools exist for integrating and testing the embedded part.

FPGA vendors have expanded their efforts beyond devices and design tools, and also provide an assortment of cores that can be embedded in a device to simplify the designer's task. For example, Xilinx offers, either directly or through partnerships with third parties, cores for basic elements (e.g., accumulators and shift registers), math functions (e.g., multipliers, multiply-and-accumulate [MAC] units, and dividers), memories (e.g., synchronous FIFO), standard bus interfaces, (e.g., PCI), processor peripherals (e.g., interrupt controller), universal asynchronous receiver and transmitters (UARTs), and a variety of networking and communication products (e.g., protocol cores). Special design kits may be required to exploit these available resources. Vendors also provide reference designs illustrating how to exploit embeddable cores.

ASIC designs are characterized by high performance, high NRE cost, and high risk. The risks are high because the cost of a mask set now ranges between $250K and $500K. A mask error and consequent re-spins of a design are prohibitive from the standpoint of cost and lost opportunity to capture market share. Embeddable programmable cores[26] offer flexibility (the design can be modified), lower NRE cost, and lower risk. These hybrid devices are targeted at applications for which standards are evolving or for which NRE costs might have to be amortized over multiple variants of a product. For example, the control logic of a multiprocessor computer in a wireless network for image processing applications could be implemented in a programmable core, allowing the design to be modified to meet a dynamic marketplace. Multiple designs can be produced from a single die. Two aspects are involved here: the compatibility of the processes for manufacturing ASICs and FPGAs, and the IP that is ultimately configuring the FPGA for a specific application. The former will develop and set the stage for proliferation of the latter.[27] There are two variations on the theme: embeddable programmable cores for placement in an ASIC (Actel and Adaptive Silicon) and embeddable complex ASIC cores for placement within an FPGA (Triscend, Xilinx, Lucent, Altera, Atmel, QuickLogic).

Compared to FPGAs, ASICs are relatively expensive to design and manufacture. They gain performance at the expense of flexibility. An emerging technology is that of

---

[26]See the web links for Embeddable Programmable Cores.
[27]The Virtual Socket Interface Alliance (VSIA) is an industry group that promotes technical standards for mixing and matching IP from multiple sources.

embedding an FPGA within an ASIC to gain flexibility, reduce the risk of a design,[28] and extend the life of a design to a wider range of applications.[29] Other programmable architectures are emerging as well. For example, Adaptive Silicon has developed a basic building block, called a Hex block, consisting of sixty-four 4-bit ALUs. Hex blocks can be tiled in rectangular patterns within a fabric of local and global interconnect. A 4 × 4 array of hex blocks supporting arithmetic functions will achieve a density of approximately 25,000 ASIC gates.

8.19   Verilog-Based Design Flows for FPGAs

The design flow for an FPGA-based target technology is shown in Figure 8-75. It relies heavily on bundled software to accomplish the synthesis, implementation, and downloading of the design into a part. The place-and-route step that plays such a dominant role in ASICs is not shown in the design flow because it is transparent to the user. Likewise, the extraction of parasitics is not shown because the fixed architecture of the devices allow their timing to be precharacterized to serve a database within the implementation tool. The simplified flow allows a designer to create design iterations and derivative designs rapidly, ultimately producing a hardware prototype.

The objective of rapid prototyping is to create a working prototype as quickly as possible to meet market conditions and to support broader testing in the host environment. Initially, the tools supporting FPGAs relied on schematic entry, but many vendors are now placing greater emphasis on supporting hardware description languages (HDLs). For example, the recently released Xilinx ISE (Integrated Synthesis Environment) tools are tailored for HDL-based entry, and support floor planning, simulation, automatic block placement and routing of interconnects, timing verification, downloading of configuration data, and readback of the configuration bit stream. The tools ultimately produce the bit-stream file that can be downloaded to the part to configure it on the host board.

8.20   Synthesis with FPGAs

In Chapter 6 we discussed the importance of adopting synthesis-friendly descriptive styles. A model has restricted utility if it cannot be synthesized. In addition, the models must include features that allow them to exploit the unique features of the target architecture. For example, FPGA tools must optimize the partition of memory between distributed and block memory resources. It is especially important in DSP applications that the synthesis tool minimize the use of off-chip memory in order to maximize performance.

---

[28]Portions of the design that are risky and might require future change can be placed in the FPGA.

[29]LSI Logic and Adaptive Silicon have been working to embed an SRAM-based FPGA in an LSI ASIC.

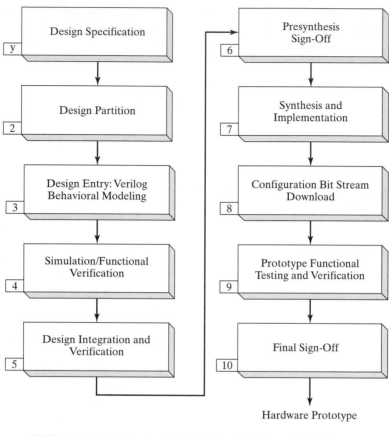

FIGURE 8-75   Design flow for FPGA-based designs with HDL entry.

FPGA vendors provide libraries of macros that implement specific functionality. The designer has a choice between a packaged macro and the circuit that a synthesis engine infers from a behavioral description. If technology-specific cores are used, they should be isolated within the design's hierarchy.

FPGAs are rich in registers, so it is generally advisable to employ one-hot coding of the state of a finite-state machine. There is usually little or no gain in trying to reduce the number of CLBs by employing a sequential binary code, because additional cells will be required to form the more complicated combinational logic that results from such a scheme.

The state decoding of a FSM must cover all possible codes of the state. Otherwise, latches will be introduced into the design.[30] This practice also protects against the machine entering a state from which it cannot recover. It is recommended that the designer assign the default state explicitly rather than use a tool option to do it

---

[30]Synthesis tools will produce reports describing device use, including a report on the number of latches and registers in the implementation. It is a good practice to review these reports to detect unwanted latches.

automatically, if for no other reason than to encourage more thoughtful consideration about the design. (Use the Verilog keyword **default** as the item decoded in a **case** statement.) Also, as discussed in Chapter 6, it is recommended that all of the register variables that are assigned value within a level-sensitive cyclic behavior be initialized at the beginning of the listed code and then assigned value within the behavior by exception—as a way to help prevent synthesis of unwanted latches in the design.

The registers in a CLB do not power up to a specific state, so it is essential that a reset or set signal port be included at the top-level module of the design and be used to drive the machine into a known state. Synchronous resets are used to minimize the possibility of a reset signal causing a metastable condition.

Decoding logic should be implemented with **case** statements rather than **if ... then ... else** statements, unless a priority structure is intended. The former produces parallel logic (faster); the latter tends to produce logic that is nested (e.g., priority decoder) and will be cascaded in a multilevel series structure of LUTs, resulting in a slower circuit.

A net that fans out from a flip-flop to several points in a circuit can be slow and difficult to route; it might ultimately be the source of a timing constraint violation. This problem may be solved by duplicating the flip flop so that the fan-out can be shared. The result will be that the routing step takes less time and is more likely to complete successfully, and the overall performance will be improved. The tradeoff is that the solution occupies a larger area of the chip (more CLBs are required). Candidates for this treatment are the address and control lines of large memory arrays, clock enable lines, output enable lines, and synchronous reset signals. If the driver of a high–fan-out net is asynchronous, synchronize the signal before duplicating it.[31]

The throughput of a design can be improved by partitioning combinational logic systematically and inserting registers at the interface between the partitions. For example, a 16-bit adder can be partitioned into two 8-bit adders and pipelined to reduce the delay of the carry chain by a factor of 2. Pipelining shortens the path that a given signal must travel during a clock cycle. Consequently, the clock can be run faster and timing violations can be eliminated. The tradeoff, which could be unacceptable, is that the pipelined datapath has latency, because the data will take one or more additional clock cycles to propagate through the circuit, depending on the number of pipeline stages that have been added.[32] The second tradeoff is that the pipeline registers occupy CLBs. Thus, the physical part that implements the design must be large enough to supply the additional registers for the pipeline. Reports produced by the software indicate the use of CLBs, so they should be consulted before attempting to increase the clock speed or eliminate a timing violation by pipelining.

If the place-and-route engine within the tool is allowed complete freedom it will generate an optimal assignment of pins. This freedom is curtailed when the part must fit into the socket of a previously configured board. Ideally, the board is not configured

---

[31]Be aware that the Xilinx tools automatically map into the same CLB signals that end with the same numeric suffix, (e.g., sig_1, sig_2). Naming duplicated signals in this manner contradicts the effort to distribute the duplicated signals to different regions of the chip. Instead, use alphabetical labels (e.g., $\_a$, $\_b$) to compose the suffix of duplicated signals.

[32]Pipelining will be considered in more detail in Chapter 9.

until the FPGA has been fully designed. Constraining the pinout constrains the optimization process, and may sacrifice performance. If feasible, careful pin assignment can lead to improved routing of the design. For example, the horizontal long lines in the XC4000 based Xilinx architectures have three-state buffers, which makes them suitable for data busses. On the other hand, vertical long lines for clock enables and vertical carry chains lead naturally to a vertical orientation of the cells of registers and counters. These architectural features suggest that datapaths should be applied to the left and right sides of the part, and control lines should be applied to the top and bottom of the part when manual routing and pin assignment are necessary. The tool has maximum flexibility when no pins are preassigned, but environmental constraints may require that some pins be preassigned, before routing. However, it is recommended that the unconstrained design be routed first, to verify that it can meet timing specifications. If it does not, the constrained design will also be too slow.

Keeping a design synchronous, with a single external clock source, allows timing-driven routing tools to work more efficiently. The parts have clock enables, so there is no need for special measures to gate clocks within a design.

FPGAs are register-rich. Therefore, it is advantageous to employ one-hot encoding in state machines. This leads to simpler next-state and output logic. This form of encoding is sometimes referred to as state-per-bit encoding, because a unique single flip-flop is asserted for each state. Coding style has an impact on the results of targeting a description into an FPGA. One notable example is in the description of a sequencer. If the count sequence does not have to be binary, linear feedback shift registers may be a more attractive alternative because they require less space and route more efficiently than binary counters. Designers should be aware that flip-flops in FPGAs tend to initialize to a cleared output during power-up. A state machine would have to anticipate this condition because it is not one of the explicit one-hot codes.

REFERENCES

1. Wakerly KK. *Digital Design—Principles and Practices*. Upper Saddle River, NJ: Prentice-Hall, 2000.
2. Weste N, Eshraghian K. *Principles of CMOS VLSI Design*. Reading, MA: Addison-Wesley, 1993.
3. Sheilholeslami A, Gulak PG. "A Survey of Circuit Innovations in Ferroelectric Random-Access Memories," *Proceedings of the IEEE, 88*, 667–689.
4. Brayton RK, et al. *Logic Minimization Algorithms for VLSI Synthesis*. Boston: Kluwer, 1984.
5. Tinder RF. *Engineering Digital Design*. 2nd ed. San Diego: Academic Press, 2000.
6. Bartlett K, et al. "Synthesis of Multilevel Logic under Timing Constraints," *IEEE Transactions on Computer Aided Design of Integrated Circuits, CAD-7*, 582–596, 1986.
7. Bartlett K, et al. "Multilevel Logic Minimization using Implicit Don't-Cares," *IEEE Transactions on Computer Aided Design of Integrated Circuits, CAD-5*, 723–740, 1986.

8. Brayton RK, et al. "MIS: A multiple-level interactive logic optimization system." *IEEE Transactions on Computer-Aided Design of Integrated Circuits and Systems, CAD-6*, 1062–1081.

9. Chan PK, Mourad S. *Digital Design Using Field Programmable Gate Arrays*. Upper Saddle River, NJ: Prentice-Hall, 1995.

10. Oldfield JV, Dorf RC. *Field-Programmable Gate Arrays*. New York: Wiley Interscience, 1995.

11. Trimberger SM, ed. *Field-Programmable Gate Array Technology*. Boston: Kluwer.

RELATED WEB SITES

www.accellera.org	Accellera
www.actel.com	Actel Corp.
www.altera.com	Altera, Inc.
www.atmel.com	Atmel Corp.
www.cadence.com	Cadence Design Systems, Inc.
www.mentorg.com	Mentor Graphics corp.
www.opencores.org	Opencores
www.synopsys.com	Synopsys, Inc.
www.synplicity.com	Synplicity, Inc.
www.vsia.com	Virtual Socket Interface Alliance
www.xilinx.com	Xilinx, Inc.

PROBLEMS

1. Using the ROM model given in Example 8.1, develop and verify *comp_2_ROM*, a Verilog model of a 2-bit comparator.

2. The 2-bit comparator presented in Example 8.1 has three outputs. Develop a new model that encodes the outputs in a 2-bit word. Build a testbench that will accept and decode the output of the model and assert one of three outputs corresponding to the outputs of the original model.

3. Estimate the number of memory cells that would be required to implement a 16-bit adder in a ROM.

4. Write a Verilog model of a 256×8 ROM that stores the product of two 4-bit unsigned binary words, as shown in Figure P8-4. Use the multiplier (*mplr*) and multiplicand (*mcnd*) bits to form the address of the ROM.

5. Write a testbench and verify the Verilog model of the static RAM cell, *RAM_static*, given in Example 8.3.

6. Develop an alternative model for *RAM_static* that uses a level-sensitive cyclic (*always*) behavior instead of a continuous assignment (see Example 8.3).

7. **FPGA-Based Design Exercise: A simple ALU**

The top-level module of a sequential machine, *ALU_machine_4_bit*, is depicted in Figure P8-7a, with the input and output ports that interface the module to its

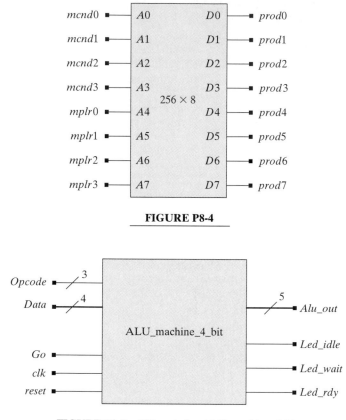

FIGURE P8-4

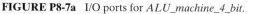

FIGURE P8-7a I/O ports for *ALU_machine_4_bit*.

environment. The machine will be described for implementation on a FPGA prototyping board.

The machine is to operate synchronously as follows: *Led_idle* will indicate that the machine is in its "reset" state. When *Go* is asserted an internal register is to be loaded with the content of *Data[3:0]* and is to assert *Led_wait* until *Go* is de-asserted. After *Go* is de-asserted, *Led_rdy* is to assert. While *Led_rdy* is asserted the slide switches (on a prototyping board) may be used to set a new value for *Data[3:0]* and/or *Opcode[2:0]*. As the slide switches are changed the effect should be apparent at *Alu_out*. The cycle is to repeat if *Go* is re-asserted while *Led_rdy* is asserted (i.e., the storage register is to be reloaded). The machine is to be synchronized by the rising edge of a clock, and have synchronous active-high reset.

Design—Partition

An architectural partition of *ALU_machine_4_bit* is shown in Figure P8-7b. The architecture has three functional subunits: an ALU, a storage register, and

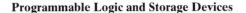

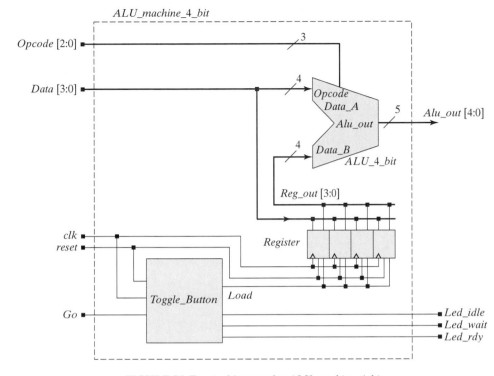

FIGURE P8-7b Architecture for *ALU_machine_4_bit*.

a state-machine controller. The ALU implements an instruction set (described below), and the controller directs the operations of the machine. One input datapath of *ALU_machine_4_bit* is connected to the internal storage register, and the other is connected to one data port of the ALU. The output of the register is connected to the other data port of the ALU. The datapath is to be controlled by *Toggle_Button*, a state machine that will be described below. The opcode and the input datapath of *ALU_machine_4_bit* will be controlled by the manual slide switches on the Digilab[33] prototyping board for Xilinx Spartan-XL parts. Board LEDs will indicate the internal status of the system (i.e., *Led_idle*, *Led_wait*, and *Led_ready*) and are derived from the state within *Toggle_Button*.

　　Our design of *ALU_machine_4_bit* will be progressive. Two functional units, *ALU_4_bit* and *Register*, will be designed and separately verified first. The state machine controller, *Toggle_Button*, will be designed later, along with a programmable clock generator. Then we will integrate the individually verified functional units, verify that the integrated design has the correct functionality, and achieve final presynthesis sign-off. The last step will be to synthesize the design into a working prototype on the Digilab board.

---

[33]See www.digilent.cc

TABLE P8-7a   Functional specification for a 4-bit ALU.

Code	Opcode	ALU Operation
000	*Add*	*Data_A + Data_B*
001	*Sub*	*Data_A − Data_B*
010	*Not_A*	~*Data_A*
011	*Not_B*	~*Data_B*
100	*A_and_B*	*Data_A & Data_B*
101	*A_or_B*	*Data_A | Data_B*
110	*Ror_A*	|*Data_A*
111	*Rand_B*	&*Data_B*

Design—ALU

Using the **module . . . endmodule** encapsulation and port declarations given below, write a Verilog model of *ALU_4_bit*, a 4-bit ALU shown in Figure P8-7c and specified in Table P8-7a.

```
module ALU_4_bit (Alu_out, Data_A, Data_B, Opcode);
    output      [4: 0]    Alu_out;
    input       [3: 0]    Data_A, Data_B;
    input       [2: 0]    Opcode;
    ...
endmodule
```

Write a test plan that specifies the functional features that are to be tested and how they will be tested. Using the test plan, write a testbench, *t_ALU_4_bit*, that verifies the functionality of *ALU_4_bit*.

Verify your design for a suitable number of patterns that cover the data and op-codes of the ALU. As an example, complete Table P8-7b by specifying *Alu_out* for the indicated patterns. Complete the second table with patterns that you choose. Include these patterns in your testbench, along with others that you choose.

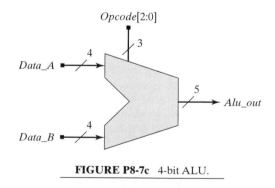

FIGURE P8-7c   4-bit ALU.

TABLE P8-7b Sample data calculations for test patterns to be applied to *ALU_4_bit*.

Data_A	1010	1111	0101	0101
Data_B	0101	0101	1010	1111
Opcode	*Alu_out*	*Alu_out*	*Alu_out*	*Alu_out*
Add Sub	0 1111 0 0101			
Not_A *Not_B*				
A_and_B *A_or_B*				
Ror_A *Rand_B*				
Data_A				
Data_B				
Opcode	*Alu_out*	*Alu_out*	*Alu_out*	*Alu_out*
Add Sub				
Not_A *Not_B*				
A_and_B *A_or_B*				
Ror_A *Rand_B*				

Note: Management appreciates your efforts to arrange the waveforms in the graphical display to enhance the utility of the information, minimizing the amount of interpretation and translation that must be done. (Use the "Set Radix" option in the Silos-III user interface.) Consider annotating the waveforms by hand or by a graphical editor tool to label opcodes etc., or define and display text parameters to indicate mnemonics for opcodes.

Next, write a Verilog model of a 4-bit storage register that has parallel load capability. The register is to be synchronized by the rising edge of a clock, and have active-high synchronous reset. Use the module encapsulation and ports given below.

```
module Register (Reg_out, Data, Load, clk, reset);
   output [3: 0]   Reg_out;
   input  [3: 0]   Data;
   input           Load, clk, reset;

   ...
endmodule
```

Write a test plan that specifies the functional features that are to be tested and how they will be tested. Using the test plan, write a testbench, *t_Register*, that verifies the functionality of *Register*.

Design—Programmable Clock

Using the programmable clock described in Problem 33 in Chapter 5, include the following "annotation module" in your project. The role of this module is to override the default parameters in the clock generator with those that are to be used in a given application. The annotation module uses hierarchical dereferencing, where *M1* is the instance name of the unit under test (UUT) in *t_Clock_Prog*. Your testbench must demonstrate that this works. The example below replaces the default values of *Latency, Offset*, and *Pulsewidth* by 10, 5, and 5, respectively.

```
module annotate_Clock_Prog ();
  defparam t_Clock_Prog.M1.Latency = 10;
  defparam t_Clock_Prog.M1.Offset = 5;
  defparam t_Clock_Prog.M1.Pulse_Width = 5;
endmodule
```

Design—User Interface

The limited switch resources of the prototyping board create a need for a toggle button machine. It will be used to load data into the machine's register.

The state machine described by the ASM chart in Figure P8-7d has synchronous reset and resides in *S_idle* with *Led_idle* asserted, until *Go* is asserted. Then it moves to *S_1*, where it asserts *Load* for one clock cycle and then enters *S_2*, where it asserts *Led_wait* and remains until *Go* is de-asserted. When *Go* is de-asserted the machine enters *S_3* and asserts *Led_rdy*. The machine remains in *S_3* until *Go* is again asserted. Then it returns to *S_1*. This sequence of state transitions lets us load data into a register, wait in *S_2* until the *Go* button is de-asserted, and then pause in *S_3*, where other actions can be taken to operate on a datapath. For example, data can be placed on the input port and the output port can be examined under the action of the opcodes. The outputs *Led_idle*, *Led_wait*, and *Led_rdy* indicate the status of the machine, and can be used to control LEDs on the Digilab prototyping board.

Write a Verilog model of the sequential machine *Toggle_Button*, using the module encapsulation and ports declared below:

```
module Toggle_Button  (Load, Led_idle, Led_wait, Led_rdy, Go, clk, reset);
  output           Load, Led_idle, Led_wait, Led_rdy;
  input            Go, clk, reset;
  reg     [1: 0]   state, next_state;

  ...
endmodule
```

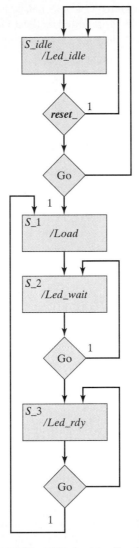

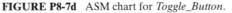

FIGURE P8-7d   ASM chart for *Toggle_Button*.

Write a test plan that specifies the functional features that are to be tested and how they will be tested. Using the test plan, write a testbench, *t_Toggle_Button*, that verifies the functionality of *Toggle_Button*.

Design—Integration and Verification

Now we will integrate the functional units of the architecture for *ALU_machine_4_bit* and verify that the functionality of the integration is correct, leading to presynthesis sign-off. Instantiate *ALU_4_bit, Register*, and

Toggle_Button into *ALU_machine_4_bit* and create a Verilog model of the internal structure represented by the architecture shown in Figure P8-7b. Use the module header and declarations shown below.

```
module ALU_machine_4_bit (
Alu_out,
Led_idle, Led_wait, Led_rdy,
Data,
Opcode,
Go,
clk, reset);

  output [4: 0]    Alu_out;
  output           Led_idle, Led_wait, Led_rdy;
  input   [3: 0]   Data;
  input   [2: 0]   Opcode;
  input            Go, clk, reset;
  wire    [3: 0]   Reg_out;      // Note: Must size the bus connecting
                                 M1 and M2
                                 // because the default size of an
                                 identifier is a scalar.
  ALU_4_bit       M1 (Alu_out, Data, Reg_out, Opcode);
  Register        M2 (Reg_out, Data, Load, clk, reset);
  Toggle_Button   M3 (Load, Led_idle, Led_wait, Led_rdy, Go, clk, reset);

endmodule
```

Write a test plan that specifies the functional features that are to be tested and how they will be tested. Using your test plan, and the headers and instantiations shown below, write a carefully documented testbench, *t_ALU_machine_4_bit*, that implements the machine, and verify the functionality of *ALU_machine_4_bit*. Note that the testbench contains the UUT (*ALU_machine_4_bit*), and the programmable clock generator (*Clock_Prog*), plus the code you write to execute the tests.

```
module annotate_ALU_machine_4_bit ();
  defparam t_ALU_machine_4_bit.M2.Latency = 10;
  defparam t_ALU_machine_4_bit.M2.Offset = 5;
  defparam t_ALU_machine_4_bit.M2.Pulse_Width = 5;
endmodule

module t_ALU_machine_4_bit ();
  reg              Go,  reset;
  reg     [3: 0]   Data;
  reg     [2: 0]   Opcode;
  wire    [4: 0]   Alu_out;
  wire             Led_idle, Led_wait, Led_rdy;
  wire             Load;

  ALU_machine_4_bit      M1 (                        // Instantiate UUT
    Alu_out,
```

```
                        Led_idle, Led_wait, Led_rdy,
                        Data,
                        Opcode,
                        Go, Load,
                        clk, reset);
                        Clock_Prog      M2 (clk);
                        ...             // Your code goes here
                  endmodule
```

Figure P8-7e shows the results of a simple test of *ALU_machine_4_bit*. Note the organization of the display.

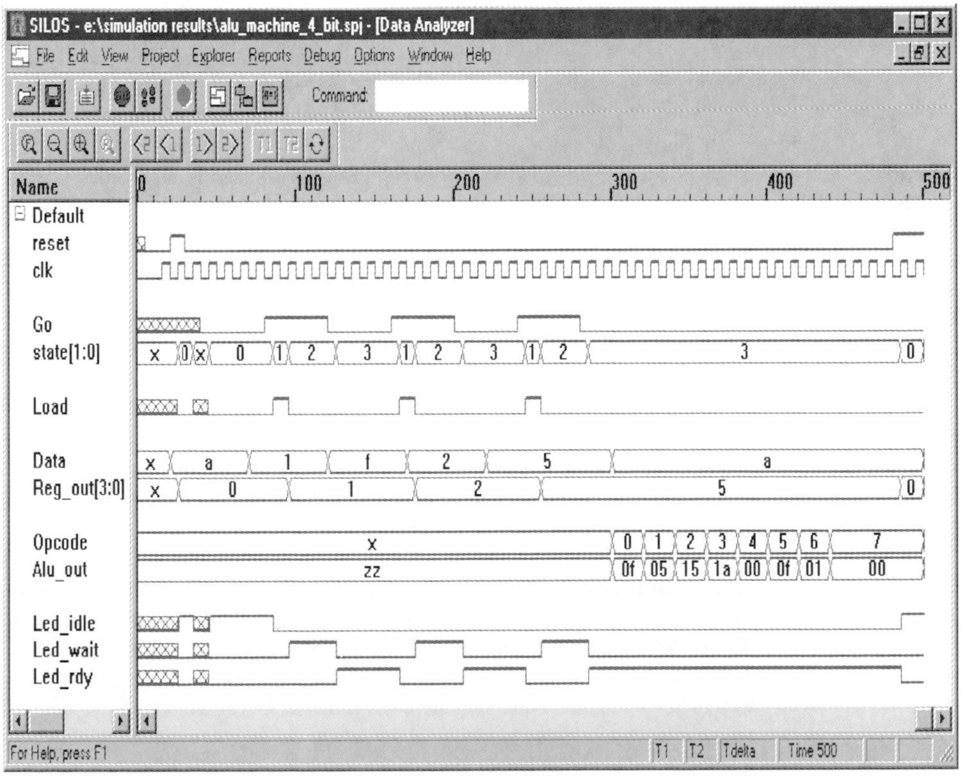

FIGURE P8-7e   Simulation results for *ALU_machine_4_bit*.

Design—Prototype Synthesis and Implementation

The final steps are to synthesize *ALU_machine_4_bit* into a Xilinx FPGA, download the design to the Digilab prototype board, and demonstrate that the prototype functions correctly, concluding with final sign-off. The ports of the module, the pads of the FPGA, and the I/O resources of the board must be integrated. The first step in this process is to decide which board resources will be mapped to the ports of the design. The second step is to map the ports to the IOBs of the FPGA. The pin configuration of the FPGA is described in the manufacturer's data sheets for the Xilinx Spartan-XL10 part. The board's I/O resources are described in the documentation for the Digilab Spartan XL prototyping board. Although a tool (e.g., Xilinx *Foundation Express*) will map ports to pads (pins) automatically, the result might not be compatible with the fixed pad locations on the prototyping board, and may even vary from run to run. It is advisable to constrain the pad mapping. In this application, the pads of the FPGA have been hard wired to certain pins of the prototyping board. *It is critical that the correct signals be mapped to the pins that are being used by the application.*

The datapath of *ALU_machine_4_bit* will be controlled by the finite-state machine *Toggle_Button* (previously designed and verified). The signal *Go* initiates the activity of loading *Data* into *Register*. *Data* and *Opcode* are to be control by the outputs of the manual slide switches. While the state of *Toggle_Button* is *S_3* the machine asserts *Led_rdy*, and the slide switches for *Data* and *Opcode* can be exercised to test the machine by presenting different words to the datapaths of the ALU. A value can be loaded into *Register*, then a different value can be arranged for *Data* by changing the slide switches after *Led_rdy* is asserted.

The input port signals of *ALU_machine_4_bit* must be mapped to the slide switches and push buttons of the Digilab board; the output ports are to be mapped to the LEDs. Figure P8-7f shows the (a) slide-switch configuration, (b) push-button configuration, and (c) the LED configuration that are to be used. These resources are connected to pins at the J2 connector on the board. *Note:* The side of the slide switch that is closest to the nearest edge of the board is logical 1.

Figure P8-7f shows the LED, button and switch pin assignments that have been specified for *ALU_machine_4_bit*, which are listed in Table P8-7c.

Write a test plan that specifies how the FPGA prototype circuit will be tested, making specific reference to the board resources (e.g., slide switch configurations, LED readouts, special instrumentation, etc.). Develop test cases demonstrating that the ALU works correctly. Verify that the *Go* button, the reset button, and the LEDs function correctly. After successfully completing all of the above steps, create the bit-stream file and download it to the prototyping board. Verify the functionality by executing (1) ALU and Opcode tests, (2) *Led_idle, Led_wait*, and *Led_rdy* assertion tests, (3) *reset* assertion test, and (4) any other test that will impress management. Using some of the test cases developed for the test plan, execute the test plan and demonstrate that the prototype functions correctly.

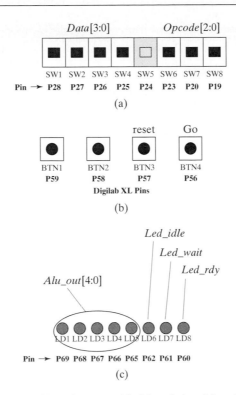

FIGURE P8-7f   Digilab/XLA Pin assignments: (a) slide switches, (b) push buttons, and (c) LEDs.

TABLE P8-7C   Pin assignments for the prototype board.

Port	Pin
Alu_out [4]	P69
Alu_out [3]	P68
Alu_out [2]	P67
Alu_out [1]	P66
Alu_out [0]	P65
Led_idle	P62
Led_wait	P61
Led_rdy	P60
Data [3]	P28
Data [2]	P27
Data [1]	P26
Data [0]	P25
Opcode [2]	P23
Opcode [1]	P20
Opcode [0]	P19
clk	P13
Go	P56
reset	P57

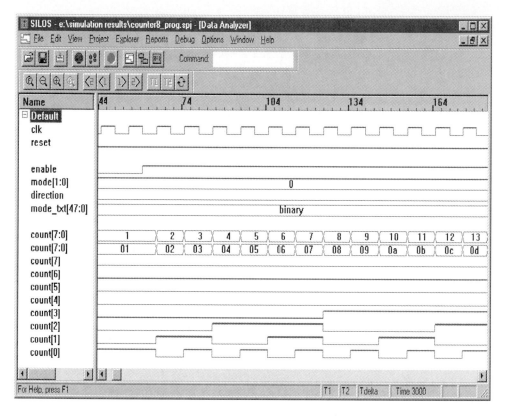

FIGURE P8-8a   Display format for ring counter simulation results.

8. **FPGA-Based Design Exercise: Some Ring Counters**

Using the shell below write and verify a Verilog model of *Counter8_Prog*, a parameterized and programmable 8-bit counter that implements various display patterns to exercise the Digilab prototyping board. The implementation is to use a separate Verilog function for each pattern (e.g., *ring1_count*). Develop a test plan clearly listing each functional feature that is to be tested. Develop a carefully documented testbench and execute the testplan to debug and verify the model and generate final graphical results. (*Note*: Organize the display to have the format shown Figure P8-8a.)

```
module Counter8_Prog (count, mode, direction, enable, clk, reset);
   output [7: 0]   count;      // Hardwired for demo board
   input   [1: 0]   mode;      // Determine pattern sequence displayed by
                                 count
   input            direction; // Determines movement (left/up, right/down)
```

```
input          enable;
input          clk, reset;
reg            count;
parameter      start_count      = 1;      // Sets initial pattern of the
                                                   display to LSB of count

// Mode of count
parameter            binary    = 0;
parameter            ring1     = 1;
parameter            ring2     = 2;
parameter            jump2     = 3;

// Direction of count
parameter            left      = 0;
parameter            right     = 1;
parameter            up        = 0;
parameter            down      = 1;

always @ (posedge clk or posedge reset)
  if (reset ==1) count <= start_count;
  else if (enable ==1)
    case (mode)
      ring1:      count <= ring1_count      (count, direction);
      ring2:      count <= ring2_count      (count, direction);
      jump2:      count <= jump2_count      (count, direction);
      default:    count <= binary_count     (count, direction);
    endcase

function      [7: 0]    binary_count;
  input       [7: 0]    count;
  input                 direction;
  begin
    if (direction == up) binary_count = count +1; else binary_count = count-1;
  end
endfunction

// Other functions are declared here.
endmodule
```

At the active edge of the clock, an 8-bit count will be updated under the control of mode and direction, which selects one of four different functions to form the next value of *count*.

binary: A binary count pattern controlled by *direction* to count up or down.

ring1: A ring counter controlled by *direction* to move left (up) or right (down).

ring2: A ring counter like *ring1*, but that moves two adjacent cells at a time.

jump2   A ring counter that jumps by two cells.

The patterns that are to be implemented for *ring2* and *jump2* are illustrated in Figure P8-8b.

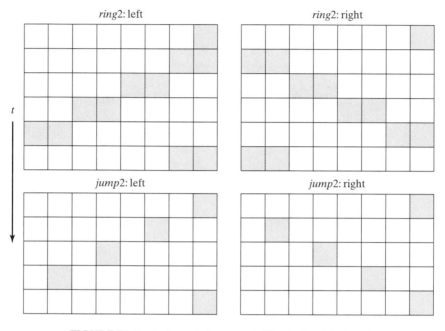

FIGURE P8-8b   Patterns to be generated by *ring2* and by *jump2*.

Design—Prototype Synthesis and Implementation

The frequency of the clock signal at Pin 13 on the prototyping board is 25 + MHz. Model and verify a clock divider that will produce an internal clock signal whose frequency will be low enough to allow changes in the LEDs to be visible.

Your design is to be encapsulated in a module, *TOP*, having the following structure:

```
module TOP (count, mode, direction, enable,  clk, reset);
input ...
output ...

Clock_Divider M0 (clk_internal, clk);
Counter8_Prog UUT (count, mode, direction, enable,  clk_internal, reset);
endmodule
```

Synthesize and implement your design on the Digilab prototyping board. Develop a hardware test plan and use it to test the operation of the prototype.

Using the pin assignments shown in Figure P8-8c, connect the ports of the design to the pins of the FPGA on the Digilab-XLA board. Connect the clock port, *clk*, to *Pin 13*, which is hard wired to the 25+ MHz clock signal from the clock chip.

Develop and verify a Verilog model of *Jumper*, a module that generates the pattern in Figure P8-8d. Verify the hardware prototype.

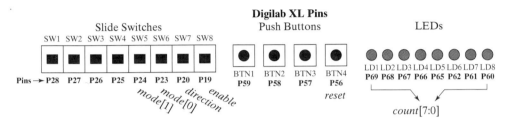

FIGURE P8-8c Pin assignments for *Counter8_Prog*.

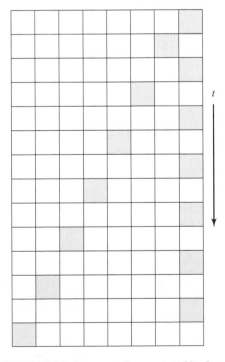

FIGURE P8-8d Patterns to be generated by *Jumper*.

9. **FPGA-Based Design Exercise: SRAM with Controller**

The static RAM modeled by *SRAM_2048_8* (see Example 8.5) is asynchronous. Many applications require a synchronous interface to an SRAM. One such interface (controller) is illustrated in Figure P8-9a(a) below, where a processor provides an address strobe, *ADS*, a read/write signal *R_W*, a clock signal, and a reset to *SRAM_Con*, which forms the signals *OE_b, CS_b,* and

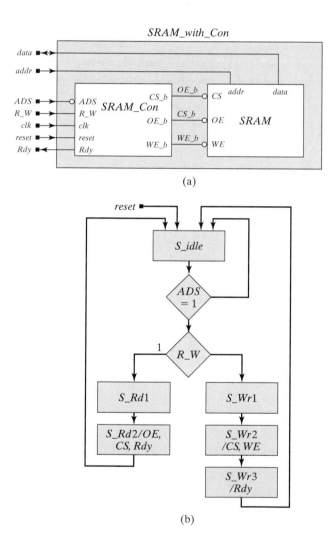

(a)

(b)

FIGURE P8-9a   Synchronous controller for a SRAM: (a) hierarchical block diagram with interface
signals, and (b) ASM chart for the controller.

WE_b, which control the SRAM, and a signal, *Rdy*, which asserts for one clock cycle at the end of a read or write sequence. The ASM chart in Figure P8-9a(b) describes the controller (*Note*: the assertions of the active-low signals of the implementation can be inferred from those on the ASM chart.)

Form a generic module, *SRAM*, by renaming *SRAM_2048_8* and re-using its code. Using the headers and testbench below, develop a generic (parameterized) controller module, *SRAM_Con* that implements the behavior of the ASM chart in Figure P8-9a(b), then instantiate *SRAM* and *SRAM_Con* within *SRAM_with_Con*. Use the results shown in Figure P8-9b to organize the displayed information (including the bus activity) produced by the simulator.

Note that the testbench contains commented statements that would cause the model to fail, because the bus activity presents a high-impedance condition before the SRAM can latch the data. The remedy is shown in the code for the testbench, and consists of conditioning the bus to enter a high-impedance condition after the rising latching edge of *CS_b* or *WE_b* in the SRAM.

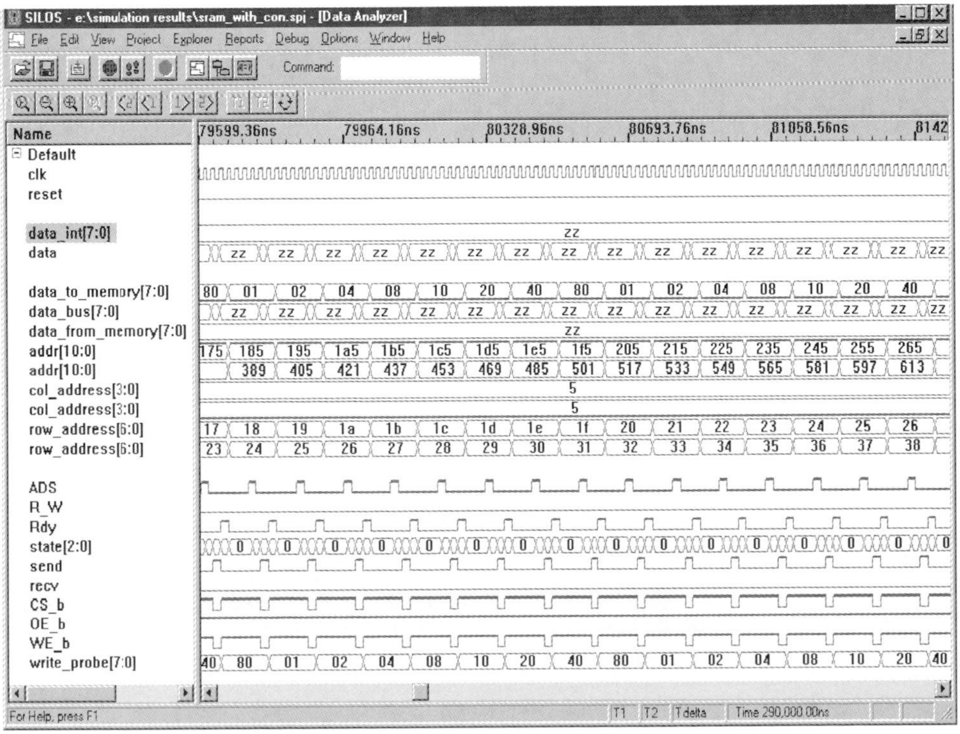

FIGURE P8-9b Simulation results for *SRAM_Con*.

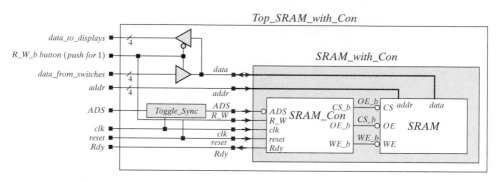

FIGURE P8-9c   FPGA implementation of an SRAM with a synchronous controller.

Design—Prototype Synthesis and Implementation

After verifying the functionality of the model for the parameters of *SRAM_2048_8*, choose parameters that will size the model to fit in the Xilinx Spartan-XL device S10XLPC84 for prototype implementation of the integrated unit on the Digilab-XLA board. Consider the structure shown in Figure P8-9c, where the inbound data and the address are mapped to the pins of the board's slide switches, and *ADS, R_W*, and reset are mapped to push buttons. *Rdy* is to be mapped to a LED. The outbound data is to be mapped to the board's seven segment displays (a decoder will be needed) or to the LEDs. Consider the following issues: (1) bus contention, (2) display of the machine state (consider the LEDs), and (3) clock speed. The module *Toggle_Synch* is to synchronize the asynchronous input *ADS* and to have logic that generates a single pulse on assertion of *ADS* regardless of how long the push button is pressed.

```
'timescale  1ns / 10ps
module SRAM_with_Con (data, addr, Rdy, ADS, R_W, clk, reset);
  parameter      word_size = 8;
  parameter      addr_size = 11;

  inout    [word_size -1: 0] data;
  input    [addr_size -1: 0] addr;
  output Rdy;
  input    ADS, R_W;
  input    clk, reset;

  SRAM_Con M0 (Rdy, CS_b, OE_b, WE_b, ADS, R_W, clk, reset);
  SRAM M1 (data, addr, CS_b, OE_b, WE_b);
endmodule

module SRAM_Con (Rdy, CS_b, OE_b, WE_b, ADS, R_W, clk, reset);

  output Rdy;
```

```
                    output CS_b;
                    output OE_b;
                    output WE_b;
                    input ADS;
                    input R_W;
                    input clk, reset;
                    reg [2:0] state, next_state;
                    reg CS_b, OE_b, WE_b;
                    reg Rdy;

                    ..
                endmodule

                module SRAM   (data, addr, CS_b, OE_b, WE_b);
                    ...
                endmodule
                ///////////////////////////////////////////////////////////////
                module test_SRAM_with_Con ();
                    parameter       word_size = 8;
                    parameter       addr_size = 11;
                    parameter       mem_depth = 128;
                    parameter       col_addr_size = 4;
                    parameter       row_addr_size = 7;
                    parameter       num_col = 16;
                    parameter       initial_pattern = 8'b000_0001;
                    parameter       Hi_Z_pattern = 8'bzzzz_zzzz;
                    parameter       stop_time = 290000;
                    parameter       latency = 248000;
                    reg             [word_size -1: 0]          data_to_memory;
                    reg             ADS, R_W, clk, reset;
                    reg             send, recv;
                    integer         col, row;
                    wire            [col_addr_size -1:0]       col_address = col;
                    wire            [row_addr_size -1:0]       row_address = row;
                    wire            [addr_size -1:0]           addr = {row_address,
                                                                       col_address};

                // Three-state, bi-directional bus

                   wire [word_size -1: 0] data_bus = send? data_to_memory: Hi_Z_pattern;

                   wire [word_size -1: 0] data_from_memory = recv? data_bus:
                   Hi_Z_pattern;

                   SRAM_with_Con M1 (data_bus, addr, Rdy, ADS, R_W, clk, reset);   // UUT

                   initial #stop_time $finish;
                   initial begin reset = 1; #1 reset = 0; end
```

```verilog
  initial begin
    #0 clk = 0;
  forever #10 clk = ~clk;
  end

// Non-Zero delay test: Write walking ones to memory
  initial begin
    ADS = 0;
    R_W = 0;
    send = 0;
    recv = 0;
    for (col= 0; col <= num_col-1; col = col +1) begin
    data_to_memory = initial_pattern;

    for (row = 0; row <= mem_depth-1; row = row + 1) begin
      @ (negedge clk);
      @ (negedge clk);
      @ (negedge clk) ADS = 1; R_W = 0;  // writing
      @ (negedge clk) ADS = 0;
      @ (posedge clk) send  = 1;
      // @ (posedge clk) send = 0;    // Does not work
      @ (posedge M1.M1.WE_b or posedge M1.M1.CS_b) send = 0;
      //@ (posedge clk) #1 send = 0;  //Replacing above line with this works too.
      @ (posedge clk) data_to_memory =
      {data_to_memory[word_size-2:0],data_to_memory[word_size -1]};

    end
    end
  end

// Non-Zero delay test: Read back walking ones from memory
  initial begin
    #latency;
    ADS = 0;
    R_W = 1;
    send = 0;
    recv = 1;
    ADS = 1;
    for (col= 0; col <= num_col-1; col = col +1) begin
      for (row = 0; row <= mem_depth-1; row = row + 1) begin
        #60;
      end
    end
  end

// Testbench probe to monitor write activity
  reg [word_size -1:0] write_probe;
```

```
          always @  (posedge M1.M1.WE_b  or  posedge M1.M1.CS_b)
          case (M1.M1.col_addr)

             0: write_probe = M1.M1.RAM_col0[M1.M1.row_addr];
             1: write_probe = M1.M1.RAM_col1[M1.M1.row_addr];
             2: write_probe = M1.M1.RAM_col2[M1.M1.row_addr];
             3: write_probe = M1.M1.RAM_col3[M1.M1.row_addr];
             4: write_probe = M1.M1.RAM_col4[M1.M1.row_addr];
             5: write_probe = M1.M1.RAM_col5[M1.M1.row_addr];
             6: write_probe = M1.M1.RAM_col6[M1.M1.row_addr];
             7: write_probe = M1.M1.RAM_col7[M1.M1.row_addr];
             8: write_probe = M1.M1.RAM_col8[M1.M1.row_addr];
             9: write_probe = M1.M1.RAM_col9[M1.M1.row_addr];
             10:write_probe = M1.M1.RAM_col10[M1.M1.row_addr];
             11:write_probe = M1.M1.RAM_col11[M1.M1.row_addr];
             12:write_probe = M1.M1.RAM_col12[M1.M1.row_addr];
             13:write_probe = M1.M1.RAM_col13[M1.M1.row_addr];
             14:write_probe = M1.M1.RAM_col14[M1.M1.row_addr];
             15:write_probe = M1.M1.RAM_col15[M1.M1.row_addr];
          endcase
       endmodule
```

10. ## FPGA-Based Design Exercise: Programmable Lock

The objective of this exercise is to design and implement a hardware prototype of a programmable digital combination lock, using the Digilab-XLA prototyping board and the Grayhill 072 hexadecimal keypad. The top-level block diagram of the programmable lock is shown in Figure P8-10a.

The programmable lock has two modes: normal and programming. The action of *reset* is to place the machine into the normal mode. Pressing *mode* once will put the machine in the program mode. In the program mode the *Prog_Mode* LED is asserted and the machine accepts a correct sequence of eight entries from a hex keypad. On receipt of a correct entry code the machine asserts the *Enter_code* LED and awaits entry of six hex values from the keypad. If the sequence of eight values is not correct the machine returns to the reset state and de-asserts the *Prog_Mode* LED. The six values entered after the *Enter_code* LED is asserted will be the key for the lock. After they are entered the machine returns to the reset state and the normal mode automatically. The master code required to enter the program mode is hard wired to an assigned value of your choice.

In the normal mode, a user must enter a sequence of six hex characters. If the sequence of characters matches the key, the *Unlock* LED is asserted and blinks at a low frequency that is visible to the human eye. If the sequence does

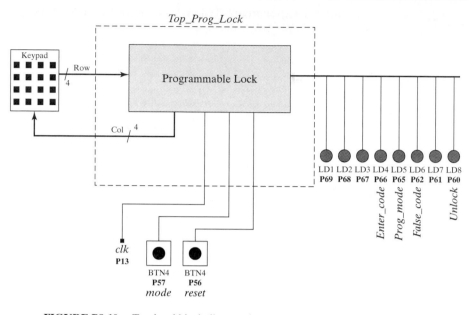

FIGURE P8-10a   Top-level block diagram for a programmable combination lock.

not match the key, the *False_Code* LED is asserted and blinks at a high rate until *reset* is asserted.

    Develop an ASM chart for a programmable lock meeting the specifications given above. Using the ASM chart and the **module ... endmodule** encapsulations and ports given below, develop and verify a Verilog model of *Prog_Lock*.

> **module** Prog_Lock (Col, valid, Enter_mode, Prog_mode, Unlock, Row, mode, clk, reset);

> **endmodule**

> **module** Top_Prog_Lock (Col, Enter_mode, Prog_mode, Unlock, Row, mode, clk, reset);

> **endmodule**

The module *Top_Prog_Lock* is to encapsulate the programmable lock and interface to the LEDs, push buttons, and the keypad.

Design—Issues

Consider debouncing the keypad and/or reducing the clock frequency. Contact bounce is 4 ms at make and 10 ms at break. Specifications for the keypad are available at www.grayhill.com. The action of reset should not erase the stored key. As a worst case, a clear button might have to be added to your design, but the design can be implemented cleverly without it. For security, consider having the displays blank until a key is read.

Design—Prototype Synthesis and Implementation

The Grayhill keypad is to be connected to the Digilab-XLA board as shown below in Figure P8-10b. It is essential that the orientation of the ribbon cable and the connector be exactly as shown. This configuration uses the same pins that connect to the slide switches on the board. Therefore, the slide switches must be place in the 0 position (away from the nearest edge of the board), and may not be used for any other function.

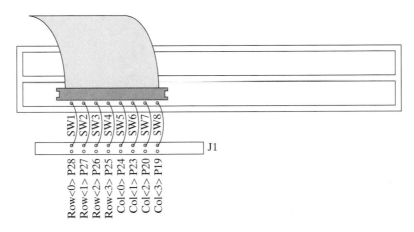

FIGURE P8-10b   Digilab board pin mapping for the Grayhill 072 Hexadecimal keypad.

11. FPGA-Based Design Exercise: Keypad Scanner with FIFO Storage

The objective of this multistage exercise is to systematically design and implement an FPGA-based keypad scanner integrated with a FIFO for data storage and retrieval, a display mux, and the seven-segment displays, slide switches, and LEDs of the Digilab-XLA prototyping board. The top-level block diagram of the system is shown in Figure P8-11a, with the partitioned system. The hardware prototype will be verified to operate with the Grayhill 072 hex Keypad.

The user interface consists of the following inputs: a mode toggle button, a read button, a reset button, and a hexadecimal keypad. The outputs are two seven-segment displays and eight LEDs. The button *mode_toggle* will be used to toggle between display states so that more than eight signals can be presented for view. When a button of the hex keypad is pressed, the system must decode the button and store the data in an internal FIFO. The *read* button will be

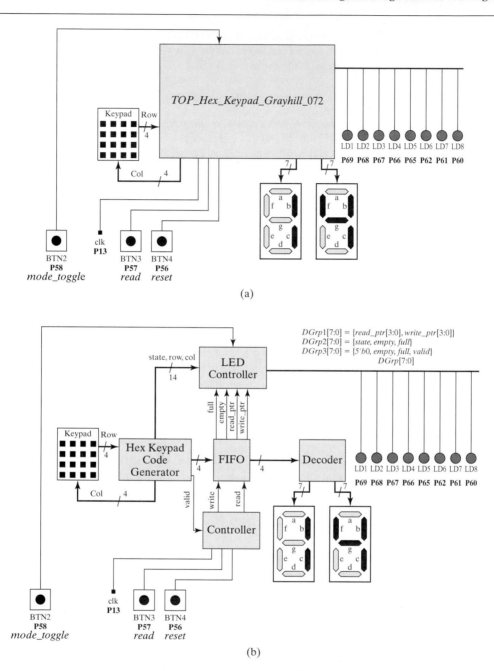

(a)

(b)

FIGURE P8-11a FIFO Keypad Scanner: (a) top-level block diagram, and (b) system partition and architecture.

used to read data from the FIFO, and to display the data on the seven-segment displays. The LEDs will display the status of the FIFO and other information.

The system architecture is subject to future engineering change orders (ECOs) as the customer's specifications evolve to accommodate a rapidly changing marketplace. Note that the specification has not considered the need for a switch debounce circuit and that it does not address the constraints that will be imposed by the Digilab's boards circuitry for the seven-segment displays.

Design: FIFO

The design will use the keypad decoder and synchronizer that were presented in Chapter 5. This part of the exercise will integrate a FIFO with the keypad scanner. A FIFO (first-in, first-out) buffer is a dedicated memory stack consisting of a fixed array of registers. The FIFO that is to be used in this exercise is shown in Figure P8-11b. The registers of the stack operate synchronously (rising edge) with a common clock, subject to reset. The stack has two pointers (addresses), one pointing to the next word to which data will be written and another pointing to the next word that will be read, subject to write and read inputs, respectively. The FIFO has input and output datapaths, and two bit-lines serving as flags to denote the status of the stack (full or empty).

Using the FIFO model and testbench provided below, verify the operation of the FIFO in the Silos-III environment. Configure the FIFO to have a stack of eight registers of 4 bits width.

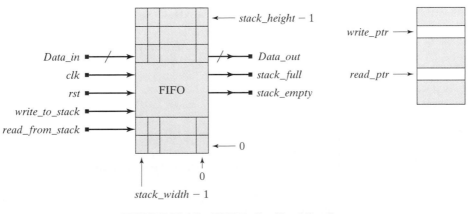

FIGURE P8-11b   FIFO Buffer: Signal Interface.

```
// Note: Adjust stack parameters
// Note:  Model does not support simultaneous read and write.

module FIFO (
  Data_out,                  // Data path from FIFO
  stack_empty,               // Flag asserted high for empty stack
  stack_full,                // Flag asserted high for full stack
  Data_in,                   // Data path into FIFO
  write_to_stack,            // Flag controlling a write to the stack
  read_from_stack,           // Flag controlling a read from the stack
  clk, rst                   // External clock and reset
);
  parameter stack_width = 4;        // Width of stack and data paths
  parameter stack_height = 8;       // Height of stack (in # of words)
  parameter stack_ptr_width = 3;    // Width of pointer to address stack
  output [stack_width -1: 0]        Data_out;
  output                            stack_empty, stack_full;
  input   [stack_width -1: 0]       Data_in;
  input                             clk, rst;
  input                             write_to_stack, read_from_stack;

// Pointers (addresses) for reading and writing
  reg     [ stack_ptr_width -1: 0]  read_ptr, write_ptr;
  reg     [ stack_ptr_width : 0]    ptr_diff;              // Gap between
                                                           ptrs
  reg     [stack_width -1: 0]       Data_out;
  reg     [stack_width -1: 0]       stack [stack_height -1 : 0]; // memory
                                                                  array

  assign stack_empty = (ptr_diff == 0) ? 1'b1 : 1'b0;
  assign stack_full = (ptr_diff == stack_height) ? 1'b1: 1'b0;

  always @ (posedge clk or posedge rst) begin: data_transfer
    if (rst) begin
      Data_out <= 0;
      read_ptr <= 0;
      write_ptr <= 0;
      ptr_diff <= 0;
    end
    else begin
      if ((read_from_stack) && (!stack_empty)) begin
        Data_out <= stack [read_ptr];
        read_ptr <= read_ptr + 1;
        ptr_diff <= ptr_diff -1;
      end
      else if ((write_to_stack) && (!stack_full)) begin
        stack [write_ptr] <= Data_in;
```

```
                    write_ptr <= write_ptr + 1;            // Address for next clock edge
                    ptr_diff <= ptr_diff + 1;
                end
            end
        end    // data_transfer
    endmodule

    module t_FIFO ();
        parameter stack_width = 4;
        parameter stack_height = 8;
        parameter stack_ptr_width = 3;

        wire [stack_width -1 : 0] Data_out;
        wire stack_empty, stack_full;
        reg [stack_width -1 : 0] Data_in;
        reg clk, rst, write_to_stack, read_from_stack;
        wire    [11:0]    stack0, stack1, stack2, stack3, stack4, stack5, stack6,
                          stack7;

        assign stack0 = M1.stack[0];            // Probes of the stack
        assign stack1 = M1.stack[1];
        assign stack2 = M1.stack[2];
        assign stack3 = M1.stack[3];
        assign stack4 = M1.stack[4];
        assign stack5 = M1.stack[5];
        assign stack6 = M1.stack[6];
        assign stack7 = M1.stack[7];

        FIFO M1 (Data_out, stack_empty, stack_full, Data_in, write_to_stack,
        read_from_stack, clk, rst);

        always begin clk = 0;  forever #5 clk = ~clk;  end
        initial #1500 $stop;

        initial  begin
            #10 rst = 1;
            #40 rst = 0;
            #420 rst = 1;
            #460 rst = 0;
        end
        initial fork
            #80 Data_in = 1;
            forever #10 Data_in = Data_in + 1;
        join
        initial fork
            #80 write_to_stack = 1;
            #180 write_to_stack = 0;

            #250 read_from_stack = 1;
            #350 read_from_stack = 0;
```

```
        #420 write_to_stack = 1;
        #480 write_to_stack = 0;
    join
  endmodule
```

Design: Decoder

The decoder must be designed to decode a word read from the FIFO and create driving signals for the active-low, seven-segment displays on the Digilab
board. The decoder must generate display signals for the 16 codes of the
Grayhill 072 keypad. Using the *module ... endmodule* encapsulation and port
given below, write a Verilog module, *Decoder_L*, of a functional unit that forms
the left and right (active-low) codes of two seven-segment displays. The MSB of
Left_out and *Right_out* must be mapped to the "*a*" segment of the display, and
the LSB must be mapped to the "*g*" segment of the string "abcdefg."

```
  module Decoder_L (Left_out, Right_out, Code_in);  // active low displays
    output [6:0]      Left_out, Right_out;
    input   [3:0]      Data_in;

  endmodule
```

Write a test plan specifying how *Decoder_L* is to be tested. Using the test plan,
write a testbench, *t_Decoder_L*, that verifies the functionality of *Decoder_L*,
and execute the test plan.

Design—Display Units

The next objective is to develop functional units supporting the reading and
displaying of the contents of the FIFO on the seven-segment displays of the
Digilab prototyping board. The seven-segment displays on the Digilab prototyping board have a common anode. Each unit has seven cathode pins, corresponding to the segments string "abcdefg." We wish to implement the structure
shown in Figure P8-11c below. The output of the FIFO will be decoded to form
the active-low code of the left and right displays. A mux will select between the
two codes and route them to the appropriate segment simultaneously with an
assertion of the appropriate anode. A clock divider must be used to strobe the
displays at a frequency that is high enough to eliminate the flicker effect (i.e., at
a frequency above the bandwidth of the human eye) and low enough to be displayed by the LED.

　　The FIFO module will read its data on the active edge of the clock while
the read input signal is asserted. The prototyping boards have clocks that are
running at 25 MHz or 50 MHz, depending on the model. In either case, a single
push of the button would cause the entire content of the FIFO to be dumped
before the button could be released. Therefore, a machine must be designed to
accept the button signal and assert a read signal at the FIFO for only one clock
signal. The button must be de-asserted before another read can occur.

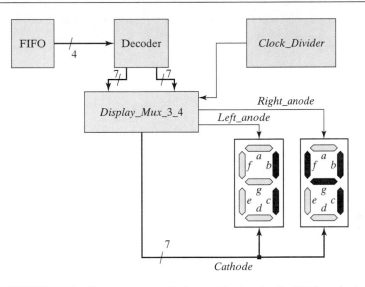

FIGURE P8-11c Seven-segments display architecture for the FIFO readout.

Design—Clock Divider

Now we will design a parameterized clock divider that can be used to strobe the mux controlling the seven-segment displays and to operate the system at a suitable frequency. Using the **module . . . endmodule** encapsulation below, develop a Verilog module of *Clock_Divider*, a parameterized clock divider having a default division by 2^{24}.

    module Clock_Divider (clk_out, clk_in, reset);

     ...

    endmodule

Design—Asynchronous User Interface

Two additional units must be designed: (1) a synchronizer for the "read" signal controlling the FIFO, and (2) a toggle unit that allows only one cell of the FIFO to be read at a time. Using the **module . . . endmodule** encapsulation below, develop a Verilog module of *Synchro_2*, a two-stage synchronizer for the signal that reads the FIFO. The output of *Synchro_2* should be synchronized to the negative edge of *clk*, because the state machine is active on the positive edge.

    module Synchro_2 (synchro_out, synchro_in, clk, reset);

      ...

    endmodule

    Using the **module . . . endmodule** encapsulation below, develop a Verilog module of *Toggle* a state machine that accepts the synchronized signal directing that the FIFO be read, and asserts a signal that reads only one cell of the FIFO, regardless of whether the user holds the "read" button for more than one clock cycle or not. The machine is to ensure that only one cell of the FIFO is read each time the button is pushed.

module Toggle (read_fifo, read_synch, clk, reset);
   ...

endmodule

Write a test plan specifying how *Toggle* is to be tested. Using the test plan, write a testbench, *t_Toggle*, that verifies the functionality of *Toggle*, and that it operates correctly with the FIFO. Execute the test plan.

We will now design a multiplexer to control the common-anode, seven-segment displays of the Digilab prototyping board. Using the **module ... endmodule** encapsulation below, develop a Verilog module of *Display_Mux_3_4*, a functional unit that selects between two cathode codes and asserts the selected code at its output, and also asserts the appropriate anode of the rightmost pair of seven-segment displays on the board.

module Display_Mux_3_4 (Cathode, Left_anode, Right_anode, Display_3, Display_4, sel);
   ...

endmodule

Write and execute a test plan for verifying *Display_Mux_3_4*.

The following objectives remain: (1) integrate the functional units of the FIFO keypad system that were designed above, (2) synthesize the integrated system and implement it in a Xilinx FPGA, and (3) conduct a hardware verification of the working system.

For simplicity, the implementation will omit the *mode_toggle* button controlling the LED display. Instead, the LEDs will be hard wired to display {state[5:0], empty, full}, the state of the controller and the status of the FIFO. An optional version would use the *mode_toggle* button to toggle between the LED data groups.

Design—System Integration

We will integrate the previously designed and verified functional units, and verify that the integrated system functions correctly. Using the top-level encapsulating module *TOP_Keypad_FIFO* given below, form the integrated system. Using the SILOS-III verification environment, eliminate any syntax errors from the integration. Pay careful attention to the mapping of formal and actual port names.

```
module TOP_Keypad_FIFO (Cathode, Col, Left_anode,
  Right_anode, valid, empty, full, Row, read, clk, reset);
  output [6: 0]    Cathode;
  output [3: 0]    Col;
  output           Left_anode, Right_anode;
  output           valid;
  output           empty;
  output           full;
```

```
    input   [3: 0]    Row;
    input             read;
    input             clk, reset;

    wire    [3: 0]    Code, Code_out;
    wire              S_Row;
    wire              valid;
    wire    [6: 0]    Left_out, Right_out;
    wire              clk_slow, clk_display;
    wire              read_fifo, read_synch;

  Synchronizer M0 (
        .S_Row(S_Row),
        .Row(Row),
        .clock(clk_slow),
        .reset(reset));

  Hex_Keypad_Grayhill_072 M1(
        .Code(Code),
        .Col(Col),
        .Valid(valid),
        .Row(Row),
        .S_Row(S_Row),
        .clock(clk_slow/**/),
        .reset(reset));

  FIFO M2 (
        .Data_out(Code_out),
        .stack_empty(empty),
        .stack_full(full),
        .Data_in(Code),
        .write_to_stack(valid),
        .read_from_stack(read_fifo));
        .clk(~clk_slow),
        .rst(reset),
        );

  Decoder_L M3 (
        .Left_out(Left_out),
        .Right_out(Right_out),
        .Code_in(Code_out));

  Display_Mux_3_4  M5 (
        .Cathode(Cathode),
        .Left_anode(Left_anode),
        .Right_anode(Right_anode),
        .Display_3(Left_out),
        .Display_4(Right_out),
        .sel(clk_display/**/));
```

```
            Clock_Divider  #(7) M6 (              // DEFAULT WIDTH  = 24
                  .clk_out(clk_slow),             // Use 20 for slow/visible operation
                  .clk_in(clk),
                  .reset(reset));

            Clock_Divider  #(20) M7 (
                  .clk_out(clk_display),
                  .clk_in(clk),
                  .reset(reset));

            Toggle M8 (
                  .toggle_out (read_fifo),
                  .toggle_in (read_synch),
                  .clk(clk_slow),
                  .reset(reset));

            Synchro_2 M9 (
                  .synchro_out(read_synch),
                  .synchro_in(read),
                  .clk(clk_slow),
                  .reset(reset));
      endmodule


      module Row_Signal (Row, Key, Col);
        output [3: 0]    Row;
        input   [15: 0]   Key;
        input   [3: 0]    Col;
        reg                   Row;

      // Scan for row of the asserted key

        always  @ (Key or Col) begin  //Asynchronous behavior for key assertion
          Row[0] = Key[0]  && Col[0]  || Key[1]  && Col[1] || Key[2] && Col[2]  ||
          Key[3] && Col[3];
          Row[1] = Key[4]  && Col[0]  || Key[5]  && Col[1] || Key[6] && Col[2]  ||
          Key[7] && Col[3];
          Row[2] = Key[8]  && Col[0]  || Key[9]  && Col[1] || Key[10] && Col[2] ||
          Key[11] && Col[3];
          Row[3] = Key[12] && Col[0]  || Key[13] && Col[1] || Key[14] && Col[2] ||
          Key[15] && Col[3];
        end
      endmodule
      */
```

Using the testbench modules given below, verify the functionality of the integrated system. Pay careful attention to the formation of the graphic user interface (GUI) displaying waveforms to display results in a user-friendly format.

```
////////////////////////////  Test Bench
module t_TOP_keypad_FIFO ();
  wire    [5: 0]    state;
  wire    [6: 0]    Cathode;
  wire    [3: 0]    Col;
  wire              Left_anode, Right_anode;
  wire              valid;
  wire              empty;
  wire              full;

  wire    [3: 0]    Row;
  reg               read;
  reg               clock, reset;

  reg     [15: 0]   Key;
  integer           j, k;
  reg     [39: 0]   Pressed;
  parameter                   stack_width = 4;
  parameter         [39: 0] Key_0 = "Key_0";
  parameter         [39: 0] Key_1 = "Key_1";
  parameter         [39: 0] Key_2 = "Key_2";
  parameter         [39: 0] Key_3 = "Key_3";
  parameter         [39: 0] Key_4 = "Key_4";
  parameter         [39: 0] Key_5 = "Key_5";
  parameter         [39: 0] Key_6 = "Key_6";
  parameter         [39: 0] Key_7 = "Key_7";
  parameter         [39: 0] Key_8 = "Key_8";
  parameter         [39: 0] Key_9 = "Key_9";
  parameter         [39: 0] Key_A = "Key_A";
  parameter         [39: 0] Key_B = "Key_B";
  parameter         [39: 0] Key_C = "Key_C";
  parameter         [39: 0] Key_D = "Key_D";
  parameter         [39: 0] Key_E = "Key_E";
  parameter         [39: 0] Key_F = "Key_F";
  parameter         [39: 0] None  = "None";

  wire [stack_width-1:0]  stack0 = UUT.M2.stack[0];    // Probes of the stack
  wire [stack_width-1:0]  stack1 = UUT.M2.stack[1];
  wire [stack_width-1:0]  stack2 = UUT.M2.stack[2];
  wire [stack_width-1:0]  stack3 = UUT.M2.stack[3];
  wire [stack_width-1:0]  stack4 = UUT.M2.stack[4];
  wire [stack_width-1:0]  stack5 = UUT.M2.stack[5];
  wire [stack_width-1:0]  stack6 = UUT.M2.stack[6];
  wire [stack_width-1:0]  stack7 = UUT.M2.stack[7];

  always @ (Key) begin
    case (Key)
```

```verilog
      16'h0000:      Pressed = None;
      16'h0001:      Pressed = Key_0;
      16'h0002:      Pressed = Key_1;
      16'h0004:      Pressed = Key_2;
      16'h0008:      Pressed = Key_3;

      16'h0010:      Pressed = Key_4;
      16'h0020:      Pressed = Key_5;
      16'h0040:      Pressed = Key_6;
      16'h0080:      Pressed = Key_7;

      16'h0100:      Pressed = Key_8;
      16'h0200:      Pressed = Key_9;
      16'h0400:      Pressed = Key_A;
      16'h0800:      Pressed = Key_B;

      16'h1000:      Pressed = Key_C;
      16'h2000:      Pressed = Key_D;
      16'h4000:      Pressed = Key_E;
      16'h8000:      Pressed = Key_F;

    default: Pressed = None;
   endcase
 end
TOP_Keypad_FIFO UUT
(Cathode, Col,  Left_anode, Right_anode, valid, empty, full, Row, read,
clock, reset);

Row_Signal M2(Row, Key, Col);

 initial #42000 $finish;
 initial begin clock = 0; forever #5 clock = ~clock; end
 initial begin reset = 1; #10 reset = 0; end
 initial  begin for (k = 0; k <= 1; k = k+1) begin Key = 0; #25 for (j = 0;
j <= 16; j = j+1) begin
  #67 Key[j] = 1; #160 Key = 0; end end end

 initial begin forever begin
   #307 read = 1;
   #20 read = 0;

  end
 end
endmodule
```

Design—Prototype Synthesis and Implementation

Synthesize the integrated system and target the design into a Xilinx Spartan-10XL FPGA. For the Digilab-XLA prototyping board. Download the bitmap file into the Spartan-10XL FPGA and conduct a demonstration of the functionality of the system.

12. Modify the keypad scanner circuit from the previous exercises to incorporate protection against switch bounce. Consider requiring a switched input to hold its value for a sufficiently long time before the machine accepts the input (e.g., 20 ms). Write and verify a Verilog model of the modified circuit. Synthesize the modified circuit and verify that the debounce circuitry works on the prototyping board.

13. FPGA-Based Design Exercise: Serial Communications Link with Error Correction

The objective of this exercise is to implement a serial communication link between a pair of FPGA prototyping boards and to demonstrate the functionality of an error correction unit. The UARTS that were presented in Chapter 7 are to be used, and include an extended Hamming encoder and an extended Hamming decoder. The block diagram in Figure P8-13a shows the configuration of the transmitter board and the receiver board. At the transmitter board, a sender interacts with the FPGA by pressing a key of the keypad. The keypad decoder, together with a two-stage synchronizer, produces a 4-bit code corresponding to the pressed key.[34] The Hamming encoder accepts a 4-bit input and generates an 8-bit encoded output word; a pair of push buttons can be used to deliberately inject errors into the code, for subsequent decoding and error correction at the receiver board. The UART transmitter will send the output of the Error Injection Unit.

The Error Injection Unit is to be hard wired to corrupt bits 1 and 5, depending on whether the push-button switches are pressed. The data stream bits can be XOR-ed with the logic value presented by the condition of the pushbutton switch. When not pressed, the switch presents a logical 0. The selected bits allow a data bit and a parity bit to be corrupted.

The Hamming decoder at the receiver board accepts an 8-bit word and forms a 4-bit output word. The unit is to be implemented with combinational logic and operate fast enough to form its output in a single cycle of the clock. The decoder is to detect and correct a single-bit error and display the corrected data word on the seven-segment displays. An LED will also be illuminated to indicate the occurrence of such an error. The extended Hamming code includes an additional bit to allow the decoder to detect, but not correct, the occurrence of a double-bit error. Such a condition will be indicated by illumination of another LED.

The Hex keypad interface is to form a unique code for each pressed key. A clock divider forms additional clock signals from the board's nominal 50 MHz clock signal. The information presented to the set of seven-segment displays

---

[34]See Sections 5.16 and 5.17.

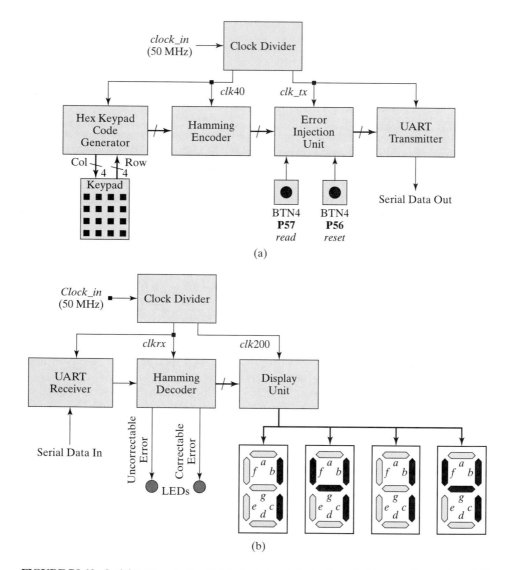

(a)

(b)

FIGURE P8-13 Serial communication link between prototyping boards: (a) transmitter unit, and (b) receiver unit.

will have to be time-multiplexed. The clock signals to be used in the design are shown in Table P8-13a.

Develop, verify, and synthesize the functional units supporting the serial communications link. Synthesize the top modules that are to reside on the transmitter and receiver boards. Discuss the resources required to support the design.

14. Using an FPGA synthesis tool, synthesize RISC_SPM (see Section 7.3). Discuss the machine's performance and its use of FPGA resources.

TABLE P8-13a   Clock signals for the board-to-board communication link.

Clock	Frequency	Description
Input	50 MHz	Input Clock
clk40	3.125 NMz	Keypad interface clock
clk200	3.125 MHz	Seven-segment display clock
clk_rx	25 MHz	UART Receiver clock
clk_tx	3.125 MHz	UART Transmitter clock

CHAPTER 9 — Algorithms and Architectures for Digital Processors

An algorithm is a sequence of processing steps that create and/or transform data objects in memory. A general-purpose machine, or processor, can be programmed (via a high-level language or assembly language) to execute a variety of algorithms, but its architecture might not yield the highest performance for a particular application, might be underused by some applications, and might not have a good balance between the processor's speed and its input–output (I/O) throughput over the domain of application [1]. Compared to dedicated application-specific integrated circuits (ASICs), a general-performance processor might consume more power, require more area, and have a higher cost (depending on volume of sales). Dedicated processors will have simpler instruction sets and simpler microcode.

ASICs are designed to optimize the execution of particular algorithms for a specific application. The hard-wired architecture of an ASIC is customized to achieve a performance/cost tradeoff that weighs in favor of an ASIC chip, rather than a general-purpose processor, in a specific application.

ASIC chips sacrifice flexibility for performance, because their architecture is fixed. Field-programmable gate arrays (FPGAs), at least in principle, can be configured to execute any algorithm. With ASICs and FPGAs, the architecture that implements an algorithm is an important consideration. ASICs are especially suited for applications in digital signal processing, image processing, and data communications, in which parallel datapaths and concurrent processing abound. We saw in Chapter 8 that FPGAs can be configured for a variety of applications, repeatedly. The choice between an FPGA or an ASIC-based implementation is often determined by a bottom-line unit cost, but in some cases the choice depends on the relative performance offered by an ASIC versus that of an FPGA in the application.

A processor can be viewed as an array, or architecture, of elemental function processors or functional units (FUs), which, as a network, realize an algorithm by executing their individual tasks in a coordinated, synchronized manner. For example, an architecture of adders, multipliers, and registers, together with other logic, might implement the algorithm of a lowpass digital filter. High-level design is concerned with implementing an architecture that realizes an algorithm to accomplish in a hard-wired architecture what would be accomplished by executing a program on a general-purpose processor.

High-level design accomplishes two primary tasks: (1) it constructs an algorithm that realizes a behavioral specification (e.g., create a lowpass filter with given performance characteristics), and (2) it maps the algorithm into an architecture (i.e., a structure of FUs) that implements the behavior in hardware. The high-level design space is complex because alternative algorithms may exhibit the same behavior, and because multiple architectures may implement a given algorithm, possibly with different throughput and latency. In this chapter we will consider the overall design process that leads to an application-specific architecture, and we will assume that the high-level design task has been accomplished (i.e., our starting point will be a computational algorithm that must be implemented by a host processor). We will focus on (1) developing an algorithm processor (i.e., a fixed architecture that implements a given algorithm), (2) exploring architectural tradeoffs (both for a network of FUs and for fine-grained implementations of the FUs themselves), (3) developing Verilog descriptions of the architectures, and (4) synthesizing the architectures.[1]

9.1   Algorithms, Nested-Loop Programs, and Data Flow Graphs

Algorithmic processors are composed of functional units, each executing in an environment of coordinated data flow. A sequential algorithm can be described by a nested-loop program (NLP) [1, 2], which consists of a set of nested *for* loops, as depicted in Figure 9-1, and a loop body written in a programming language such as C, in a pseudo language, or in a hardware description language (HDL) such as Verilog (i.e., a cyclic behavior).[2] NLPs are always computable, so such a starting point for a realization of a specification is attractive. Moreover, an NLP provides an unambiguous and executable specification for the machine whose behavior is to be realized.

The sequential ordering of operations and the dependencies of data in an NLP for an algorithm can be represented by a *data flow graph* (DFG) [3, 4]. Language parsers extract a DFG from an NLP or from an HDL-based behavioral description by analyzing the activity flow of the statements and the lifetime of the variables, together with the semantics of the language's constructs. We will see that a DFG is a key tool in

---

[1]Performance issues will also be addressed in Chapter 11.
[2]A generic *for* loop executes repeatedly under the control of lower and upper bound expressions and a loop index mechanism. The Verilog *for* loop is an example.

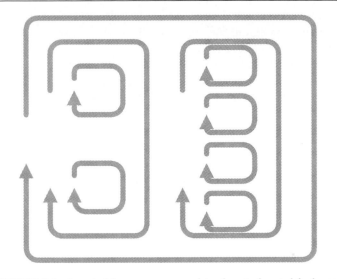

FIGURE 9-1 A nested-loop program consists of a set of nested *for* loops.

developing an architecture for an algorithmic processor and for exploring alternative equivalent architectures.

A DFG is a *directed acyclic graph*, $G(V, E)$, where V is the set of nodes of the graph, and E is the set of edges of the graph [3]. Each node $v_i \in V$ represents a functional unit (FU), which operates on its inputs (data) to produce its outputs. An FU might consist of a single operation or a more complex, ordered composition of operations, in which case the FU itself might be represented by a DFG giving a fine-grained view. Each directed edge $e_{ij} \in E$ originates at a node $v_i \in V$ and terminates at node $v_j \in V$. Given an edge $e_{ij} \in E$, the data produced by node v_i is consumed by node v_j [5].

A data dependency exists between a pair of nodes $(v_i \in V, v_j \in V)$ corresponding to edge e_{ij} if the functional unit v_j uses the results of functional unit v_i, and if the operation of v_j cannot begin until the operation of v_i has finished (i.e., the directed edges of the graph imply a precedence for execution).[3] Thus, the DFG reveals the producers of data, the order in which the data will be generated, and the consumers of data. It also exposes parallelism in the dataflow, which reveals opportunities for concurrent computation and has implications for the lifetime of variables (i.e., memory). In a synchronous dataflow a node consumes its data before new data are presented [5]. The designer's task, in general, is to transform the DFG for an algorithm into a structure of hardware, typically a partitioned structure consisting of a datapath unit and a control unit, as represented by an algorithmic state machine and datapath (ASMD) chart.[4] Given that the control unit must control the datapath, the initial task is to specify an

---

[3]The edges of the graph can be annotated to indicate a time delay (e.g., a register buffer delay) that must elapse before the node receiving data from a provider can commence execution.
[4]See Chapter 7.

architecture for a datapath unit which, if controlled to respect the constraints implied by the DFG, will implement the algorithm. Then the control unit can be designed to coordinate the data flow of the algorithm.[5]

Beginning with the DFG, the high-level synthesis task of *datapath allocation* consists of transforming the DFG of an algorithm into an architecture of processors, datapaths, and registers from which a synthesizable register transfer level (RTL) model can be developed in Verilog or VHDL.[6] A baseline architecture that implements a given DFG can always be formed as a set of FUs connected in a structure that is isomorphic to the DFG.[7] This design is hardware-intensive, and it serves only as a starting point because other architectures may realize the same algorithm with higher performance and less hardware. Datapath allocation binds the FUs of the DFG to a given (selected) set of datapath resources, and schedules their use of the resources.

Many different architectures can implement the same algorithm. They will be distinguished not only by their datapath resources, but also by a temporal schedule for using the resources. The high-level task of *resource scheduling* assigns resources and a time slot to each node of a DFG. Given the parallelism of a DFG, there are many schedules that could map nodes into time slots (control steps), creating several alternatives for realizing the corresponding algorithm in hardware.[8] Scheduling must be conflict-free (i.e., a resource cannot be allocated to multiple FUs in the same time slot). The overall design flow is shown in Figure 9-2.

Three general approaches are used to reorganize the baseline architecture obtained from the DFG: recomposition, pipelining, and replication. *Recomposition* segments the FU into a sequence of functions that execute one after the other to implement the algorithm. The sequence of execution may be further distributed over space (hardware units) and time. In the former case, the nodes of the DFG are mapped isomorphically to the FUs; in the latter case a single FU executes over as many clock cycles as required to complete the operations represented by the DFG. This approach saves hardware by replicating the activity of a single FU over multiple time steps, rather than using multiple processors that execute concurrently in a single step. *Pipelining* inserts registers into a datapath to shorten computational paths and thereby increase the throughput of a system, incurring a penalty in latency and the number of registers. In contrast to recomposition, *replication* uses multiple, identical, concurrently executing processors to improve performance, but at the expense of hardware.

---

[5]We will rely on ASM charts to design a controller; control-flow data graphs and control-data flow graphs can also be used [5].

[6]This retranslation step is potentially a source of error in the design flow, because the behavior of the machine synthesized from the RTL model might not match the intended behavior expressed by the specification. Some electronic design automation (EDA) tool vendors are advocating the System C language (www.systemc.org) as the entry point for system specification, and are developing engines that will translate the executable specification expressed in System C into an RTL description in Verilog or VHDL. Others are building tools to synthesize logic directly from C.

[7]I am grateful to Dr. Hubert Kaeslin, of the Swiss Federal Institute of Technology (ETH), Zurich for correspondence about this approach to design.

[8]See references 3 and 5 for a discussion of various scheduling algorithms that are used in high-level synthesis.

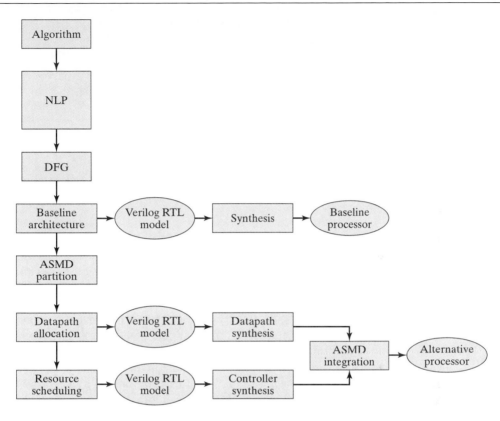

FIGURE 9-2 Design flow for algorithm-based synthesis of a sequential machine.

9.2   Design Example: Halftone Pixel Image Converter

As an example of how an architecture can be synthesized from an algorithm, we will demonstrate the main steps of synthesizing a halftone image converter. The circuit has been used elsewhere to illustrate concepts in automated synthesis of algorithms for digital signal processing (DSP) and will be our platform for developing architectures for an image converter [1]. We will design a base-line machine and then consider an alternative machine that uses less hardware, but executes over multiple clock cycles.

The Floyd–Steinberg algorithm converts an image consisting of an N-row $\times$ M-column array of pixels, each having a resolution of n bits, into an array having only black or white pixels, while incorporating a subjective measure of the quality of the image [1]. The algorithm distributes to a selected subset of each pixel's neighbors the roundoff error induced by converting the pixel from a resolution of n bits to a resolution of 1 bit. The distribution of error to a given pixel is based on a weighted average of the errors at the sites of its selected neighbors. Figure 9-3 identifies the neighboring nodes that are affected by error generated at node (i, j), where i is a column index, and

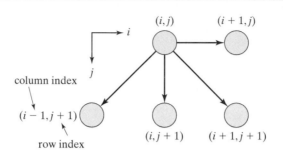

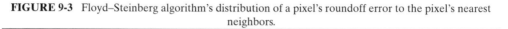

FIGURE 9-3 Floyd–Steinberg algorithm's distribution of a pixel's roundoff error to the pixel's nearest neighbors.

j is a row index into an $N \times M$ array having its origin in the upper left corner of the array (a common reference for images).

A pixel receives a distribution of error from four of its neighbors, as shown in Figure 9-4. The errors received from four neighboring pixels are used to calculate the halftone pixel value at cell (i, j). These relationships hold for each pixel in the array, leading to the DFG shown in Figure 9-5. The data dependencies of the array reveal that the pixels can be converted in a sequential manner, from left to right, from top to bottom, beginning at the top left corner and proceeding to the bottom right corner.

The FUs (nodes) of the DFG execute the pixel conversion, according to the following pseudocode description of the sequence of calculations that will ultimately be described by a cyclic behavior in Verilog. At each pixel location, (i, j), a weighted average of the (previously calculated) error, e, at the selected neighbors is formed according to

$$E\_av = (w1 * e[i - 1, j] + w2 * e[i - 1, j - 1] + w3 * e[i, j - 1] + w4 * e[i + 1, j - 1])/w_T$$

where $w1, \ldots, w4$ are subjective nonnegative weights, and $w_T = w1 + w2 + w3 + w4$. This weighted average is used to calculate a corrected pixel value:

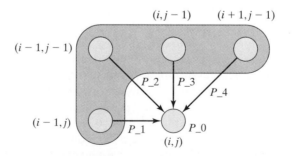

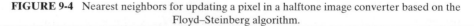

FIGURE 9-4 Nearest neighbors for updating a pixel in a halftone image converter based on the Floyd–Steinberg algorithm.

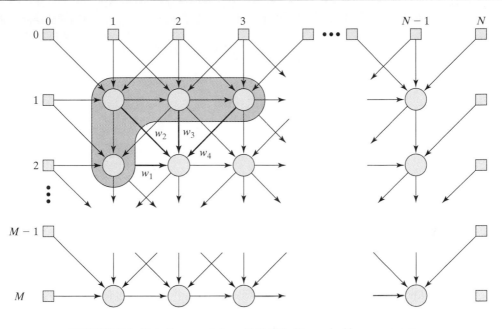

FIGURE 9-5 Data flow graph for a $N \times M$ halftone pixel image converter.

$$CPV = PV[i,j] + E\_av$$

and then we round CPV to 0 or 1, according to a threshold, CPV_thresh. So

$$CPV\_round = 0 \text{ if } CPV < CPV\_thresh$$

and

$$CPV\_round = CPV\_max, \text{otherwise}$$

where $CPV\_max = 255$ for an image with 8-bit resolution, and $CPV\_thresh = 128$. Next, we form a halftone pixel value and save the error:

$$HTPV = 0 \text{ if } CPV\_round = 0, \text{otherwise } HTPV = 1;$$
$$Err[i,j] = CPV - HTPV$$

The main array of nodes in Figure 9-5 is isomorphic to the array of pixels. We have added a column of boundary nodes at the left and right edges of the graph and a row at the top of the graph. These additional nodes have no functionality other than to initialize the computation.

The array of pixels can be updated by sequentially updating the rows or the columns, according to the NLP given below, which includes initialization at the boundaries of the pixel array. The algorithm would execute in $N \times M$ time steps, with each time step including time to retrieve and store data. The algorithm for sequentially updating the pixels can be expressed as the following NLP segment in the C language, where T is the threshold for rounding the corrected pixel value (CPV):

```
for (k = 1; k < N; k++) {Err [k] [0] = 0;}              // Initialization of
                                                           boundary
for (k = 0; k <= M; k++) {Err [0] [k] = 0; Err [N+1] [ 0] = 0;}   // Initialization of
                                                           boundary)
for (j = 1; j < M; ++) {                                // Iterate over rows
  for (i = 1; i < N; i++) {                             // Iterate over
                                                           columns
          E_av = (7 * e[i -1][j] + 1 * Err[i -1] [j-1] + 5 * Err[i] [j-1] + 3 * Err[i +1] [j-1]) / 16;
          CPV = PV[i][j] + E_av;
          CPV_round = (if CPV < T then 0 else 255);  // Threshold = 128;
          HTPV [i][j] = if (
          CPV_round ==0 then 0 else 1);
          Err [i][j] = CPV - CPV_round;
  }
}
```

9.2.1   Baseline Design for a Halftone Pixel Image Converter

There are various options for forming a hardware implementation of an NLP realizing
the image conversion algorithm. Since the computations at each FU are combinational
in nature, the simplest architecture is a structure of FUs that is isomorphic to the DFG,
with input data consisting of the array of pixel values, and output data consisting of the
halftone pixel and error values at each location. The Verilog descriptions of the FU,
PPDU (for pixel processor datapath unit), and *Image_Converter_Baseline*, are given
below. The model is hardware-intensive and structural, consisting of a systolic array[9] of
48 identical processors hard-wired for the dataflow of the DFG. The cycle time of the
host processor providing images to the pixel processor will be limited by the longest
path through the array. The implementation is synthesizable as combinational logic,
and requires no controller.

The testbench produces the sharp contrast image patterns shown in Figure 9-6,
and the graduated images shown in Figure 9-7. The simulation results shown in Figure
9-8 show that the halftone image produced by *Image_Converter_Baseline* matches the
sharp images. A graduated image and its halftone image are shown in Figure 9-8.

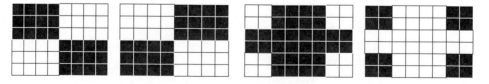

FIGURE 9-6   Sharp test images for the halftone image converter.

---

[9]A systolic array is a set of identical FUs, with high local connectivity and multiple data flows.

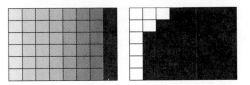

FIGURE 9-7 Graduated test image and halftone image produced by *Image_Converter_Baseline* with weights $(w1, w2, w3, w4) = (2, 8, 4, 2)$.

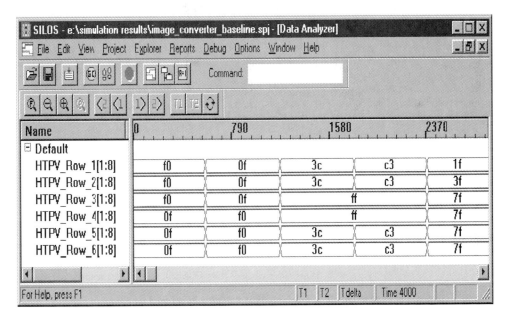

FIGURE 9-8 Simulation results for *Image_Converter_Baseline* with weights $(w1, w2, w3, w4) = (2, 8, 4, 2)$.

```
module Image_Converter_Baseline (
   HTPV_Row_1, HTPV_Row_2, HTPV_Row_3,
   HTPV_Row_4, HTPV_Row_5, HTPV_Row_6,
   pixel_1,  pixel_2,  pixel_3,  pixel_4,  pixel_5,  pixel_6,  pixel_7,  pixel_8,
   pixel_9,  pixel_10, pixel_11, pixel_12, pixel_13, pixel_14, pixel_15, pixel_16,
   pixel_17, pixel_18, pixel_19, pixel_20, pixel_21, pixel_22, pixel_23, pixel_24,
   pixel_25, pixel_26, pixel_27, pixel_28, pixel_29, pixel_30, pixel_31, pixel_32,
   pixel_33, pixel_34, pixel_35, pixel_36, pixel_37, pixel_38, pixel_39, pixel_40,
   pixel_41, pixel_42, pixel_43, pixel_44, pixel_45, pixel_46, pixel_47, pixel_48
   );

   output [1: 8]   HTPV_Row_1, HTPV_Row_2, HTPV_Row_3,
                   HTPV_Row_4, HTPV_Row_5, HTPV_Row_6;
```

```
input [7: 0]
pixel_1,   pixel_2,  pixel_3,  pixel_4,  pixel_5,  pixel_6,  pixel_7,  pixel_8,
pixel_9,   pixel_10, pixel_11, pixel_12, pixel_13, pixel_14, pixel_15, pixel_16,
pixel_17, pixel_18, pixel_19, pixel_20, pixel_21, pixel_22, pixel_23, pixel_24,
pixel_25, pixel_26, pixel_27, pixel_28, pixel_29, pixel_30, pixel_31, pixel_32,
pixel_33, pixel_34, pixel_35, pixel_36, pixel_37, pixel_38, pixel_39, pixel_40,
pixel_41, pixel_42, pixel_43, pixel_44, pixel_45, pixel_46, pixel_47, pixel_48;

wire    HTPV_Row_1, HTPV_Row_2, HTPV_Row_3;
wire    HTPV_Row_4, HTPV_Row_5, HTPV_Row_6;

// Note Left, right, and top borders are initialized to 0
// Errors for the core array:

wire  [7: 0]
Err_1, Err_2, Err_3, Err_4, Err_5, Err_6, Err_7, Err_8,
Err_9, Err_10, Err_11, Err_12, Err_13, Err_14, Err_15, Err_16,
Err_17, Err_18, Err_19, Err_20, Err_21, Err_22, Err_23, Err_24,
Err_25, Err_26, Err_27, Err_28, Err_29, Err_30, Err_31, Err_32,
Err_33, Err_34, Err_35, Err_36, Err_37, Err_38, Err_39, Err_40,
Err_41, Err_42, Err_43, Err_44, Err_45, Err_46, Err_47, Err_48;

PPDU M1 (Err_1, HTPV_Row_1[1], 8'b0, 8'b0, 8'b0, 8'b0, pixel_1);
PPDU M2 (Err_2, HTPV_Row_1[2], Err_1, 8'b0, 8'b0, 8'b0, pixel_2);
PPDU M3 (Err_3, HTPV_Row_1[3], Err_2, 8'b0, 8'b0, 8'b0, pixel_3);
PPDU M4 (Err_4, HTPV_Row_1[4], Err_3, 8'b0, 8'b0, 8'b0, pixel_4);
PPDU M5 (Err_5, HTPV_Row_1[5], Err_4, 8'b0, 8'b0, 8'b0, pixel_5);
PPDU M6 (Err_6, HTPV_Row_1[6], Err_5, 8'b0, 8'b0, 8'b0, pixel_6);
PPDU M7 (Err_7, HTPV_Row_1[7], Err_6, 8'b0, 8'b0, 8'b0, pixel_7);
PPDU M8 (Err_8, HTPV_Row_1[8], Err_7, 8'b0, 8'b0, 8'b0, pixel_8);

PPDU M9 (Err_9, HTPV_Row_2[1], 8'b0, 8'b0, Err_1, Err_2, pixel_9);
PPDU M10 (Err_10, HTPV_Row_2[2], Err_9, Err_1, Err_2, Err_3, pixel_10);
PPDU M11 (Err_11, HTPV_Row_2[3], Err_10, Err_2, Err_3, Err_4, pixel_11);
PPDU M12 (Err_12, HTPV_Row_2[4], Err_11, Err_3, Err_4, Err_5, pixel_12);
PPDU M13 (Err_13, HTPV_Row_2[5], Err_12, Err_4, Err_5, Err_6, pixel_13);
PPDU M14 (Err_14, HTPV_Row_2[6], Err_13, Err_5, Err_6, Err_7, pixel_14);
PPDU M15 (Err_15, HTPV_Row_2[7], Err_14, Err_6, Err_7, Err_8, pixel_15);
PPDU M16 (Err_16, HTPV_Row_2[8], Err_15, Err_7, Err_8, 8'b0, pixel_16);

PPDU M17 (Err_17, HTPV_Row_3[1], 8'b0, 8'b0, Err_9, Err_10, pixel_17);
PPDU M18 (Err_18, HTPV_Row_3[2], Err_17, Err_10, Err_11, Err_12, pixel_18);
PPDU M19 (Err_19, HTPV_Row_3[3], Err_18, Err_11, Err_12, Err_13, pixel_19);
PPDU M20 (Err_20, HTPV_Row_3[4], Err_19, Err_12, Err_13, Err_14, pixel_20);
PPDU M21 (Err_21, HTPV_Row_3[5], Err_20, Err_13, Err_14, Err_15, pixel_21);
PPDU M22 (Err_22, HTPV_Row_3[6], Err_21, Err_14, Err_15, Err_16, pixel_22);
PPDU M23 (Err_23, HTPV_Row_3[7], Err_22, Err_15, Err_16, Err_17, pixel_23);
PPDU M24 (Err_24, HTPV_Row_3[8], Err_23, Err_16, Err_17, 8'b0, pixel_24);
```

```
            PPDU M25 (Err_25, HTPV_Row_4[1], 8'b0, 8'b0, Err_16, Err_17, pixel_25);
            PPDU M26 (Err_26, HTPV_Row_4[2], Err_25, Err_17, Err_18, Err_19, pixel_26);
            PPDU M27 (Err_27, HTPV_Row_4[3], Err_26, Err_18, Err_19, Err_20, pixel_27);
            PPDU M28 (Err_28, HTPV_Row_4[4], Err_27, Err_19, Err_20, Err_21, pixel_28);
            PPDU M29 (Err_29, HTPV_Row_4[5], Err_28, Err_20, Err_21, Err_22, pixel_29);
            PPDU M30 (Err_30, HTPV_Row_4[6], Err_29, Err_21, Err_22, Err_23, pixel_30);
            PPDU M31 (Err_31, HTPV_Row_4[7], Err_30, Err_22, Err_23, Err_24, pixel_31);
            PPDU M32 (Err_32, HTPV_Row_4[8], Err_31, Err_23, Err_24, 8'b0, pixel_32);

            PPDU M33 (Err_33, HTPV_Row_5[1], 8'b0, 8'b0, Err_25, Err_26, pixel_33);
            PPDU M34 (Err_34, HTPV_Row_5[2], Err_33, Err_25, Err_26, Err_27, pixel_34);
            PPDU M35 (Err_35, HTPV_Row_5[3], Err_34, Err_26, Err_27, Err_28, pixel_35);
            PPDU M36 (Err_36, HTPV_Row_5[4], Err_35, Err_27, Err_28, Err_29, pixel_36);
            PPDU M37 (Err_37, HTPV_Row_5[5], Err_36, Err_28, Err_29, Err_30, pixel_37);
            PPDU M38 (Err_38, HTPV_Row_5[6], Err_37, Err_29, Err_30, Err_31, pixel_38);
            PPDU M39 (Err_39, HTPV_Row_5[7], Err_38, Err_30, Err_31, Err_32, pixel_39);
            PPDU M40 (Err_40, HTPV_Row_5[8], Err_39, Err_31, Err_32, 8'b0, pixel_40);

            PPDU M41 (Err_41, HTPV_Row_6[1], 8'b0, 8'b0, Err_33, Err_34, pixel_41);
            PPDU M42 (Err_42, HTPV_Row_6[2], Err_41, Err_33, Err_34, Err_35, pixel_42);
            PPDU M43 (Err_43, HTPV_Row_6[3], Err_42, Err_34, Err_35, Err_36, pixel_43);
            PPDU M44 (Err_44, HTPV_Row_6[4], Err_43, Err_35, Err_36, Err_37, pixel_44);
            PPDU M45 (Err_45, HTPV_Row_6[5], Err_44, Err_36, Err_37, Err_38, pixel_45);
            PPDU M46 (Err_46, HTPV_Row_6[6], Err_45, Err_37, Err_38, Err_39, pixel_46);
            PPDU M47 (Err_47, HTPV_Row_6[7], Err_46, Err_38, Err_39, Err_40, pixel_47);
            PPDU M48 (Err_48, HTPV_Row_6[8], Err_47, Err_39, Err_40, 8'b0, pixel_48);
endmodule


// Pixel Processor Datapath Unit, hard-wired for 8 x 6 array
module PPDU (Err_0, HTPV, Err_1, Err_2, Err_3, Err_4, PV);
    output          [7: 0]    Err_0;
    output                    HTPV;
    input           [7: 0]    Err_1, Err_2, Err_3, Err_4, PV;
    wire            [9: 0]    CPV, CPV_round, E_av;

// Weights for the average error; choose for compatibility with divide-by-16 (>> 4)
    parameter       w1 = 2, w2 = 8, w3 = 4, w4 = 2;
    parameter       Threshold = 128;

    assign  E_av = (w1 * Err_1 + w2 * Err_2 + w3 * Err_3 + w4 * Err_4 ) >> 4;
    assign  CPV = PV + E_av;
    assign  CPV_round = (CPV < Threshold) ? 0: 255;
    assign  HTPV = (CPV_round == 0) ? 0: 1;
    assign  Err_0  = CPV - CPV_round;
endmodule
```

9.2.2 NLP-Based Architectures for the Halftone Pixel Image Converter

The baseline design is at one extreme of the hardware-performance spectrum, for it replicates identical hardware units, each implemented as combinational logic blocks, which execute concurrently in a single long clock cycle and require no memory and no datapath controller. The structure of the array of processors in the base-line design is identical to the structure of the DFG. The NLP itself suggests another alternative: a level-sensitive behavior that executes the NLP. This style is given below as *Image_Converter_0*. The model synthesizes to combinational logic, and converts an image in time $T_{baseline}$, where $T_{baseline}$ is the time associated with the longest datapath through the array. The system clock that presents images to the unit must have a cycle time greater than $T_{baseline}$.

```
module Image_Converter_0 (
    HTPV_Row_1, HTPV_Row_2, HTPV_Row_3,
    HTPV_Row_4, HTPV_Row_5, HTPV_Row_6,

    pixel_1,  pixel_2,  pixel_3,  pixel_4,  pixel_5,  pixel_6,  pixel_7,  pixel_8,
    pixel_9,  pixel_10, pixel_11, pixel_12, pixel_13, pixel_14, pixel_15, pixel_16,
    pixel_17, pixel_18, pixel_19, pixel_20, pixel_21, pixel_22, pixel_23, pixel_24,
    pixel_25, pixel_26, pixel_27, pixel_28, pixel_29, pixel_30, pixel_31, pixel_32,
    pixel_33, pixel_34, pixel_35, pixel_36, pixel_37, pixel_38, pixel_39, pixel_40,
    pixel_41, pixel_42, pixel_43, pixel_44, pixel_45, pixel_46, pixel_47, pixel_48);

    output [1: 8]   HTPV_Row_1, HTPV_Row_2, HTPV_Row_3,
                    HTPV_Row_4, HTPV_Row_5, HTPV_Row_6;
    input   [7: 0]
    pixel_1,  pixel_2,  pixel_3,  pixel_4,  pixel_5,  pixel_6,  pixel_7,  pixel_8,
    pixel_9,  pixel_10, pixel_11, pixel_12, pixel_13, pixel_14, pixel_15, pixel_16,
    pixel_17, pixel_18, pixel_19, pixel_20, pixel_21, pixel_22, pixel_23, pixel_24,
    pixel_25, pixel_26, pixel_27, pixel_28, pixel_29, pixel_30, pixel_31, pixel_32,
    pixel_33, pixel_34, pixel_35, pixel_36, pixel_37, pixel_38, pixel_39, pixel_40,
    pixel_41, pixel_42, pixel_43, pixel_44, pixel_45, pixel_46, pixel_47, pixel_48;

    reg     [7: 0]   PV_Row_1 [1: 8], PV_Row_2 [1: 8], PV_Row_3 [1: 8],
                     PV_Row_4 [1: 8], PV_Row_5 [1: 8], PV_Row_6 [1: 8];

    reg              HTPV_Row_1, HTPV_Row_2, HTPV_Row_3;
    reg              HTPV_Row_4, HTPV_Row_5, HTPV_Row_6;

    reg     [7: 0]
      Err_Row_0 [0: 9], Err_Row_1 [0: 9], Err_Row_2 [0: 9], Err_Row_3 [0: 9],
      Err_Row_4 [0: 9], Err_Row_5 [0: 9], Err_Row_6 [0: 9];
    reg     [9: 0]   CPV, CPV_round, E_av;
    integer i;
```

```
        parameter N = 6;                // rows
        parameter M = 8;                // columns
        parameter Threshold = 128;

   // Weights for the average error; compatibile with divide-by-16 (>> 4)
        parameter    w1 = 2, w2 = 8, w3 = 4, w4 = 2;

        always @
        ( pixel_1 or pixel_2 or pixel_3 or pixel_4 or pixel_5 or pixel_6 or
          pixel_7 or pixel_8 or pixel_9 or pixel_10 or pixel_11 or pixel_12 or
          pixel_13 or pixel_14 or pixel_15 or pixel_16 or pixel_17 or pixel_18 or
          pixel_19 or pixel_20 or pixel_21 or pixel_22 or pixel_23 or pixel_24 or
          pixel_25 or pixel_26 or pixel_27 or pixel_28 or pixel_29 or pixel_30 or
          pixel_31 or pixel_32 or pixel_33 or pixel_34 or pixel_35 or pixel_36 or
          pixel_37 or pixel_38 or pixel_39 or pixel_40 or pixel_41 or pixel_42 or
          pixel_43 or pixel_44 or pixel_45 or pixel_46 or pixel_47 or pixel_48
          )
        begin

   // Note Err_Row_ includes left, right, and top border
   // columns for initialization of the algorithm.

   // Initialize error at left border
     Err_Row_1[0] = 0; Err_Row_2[0] = 0; Err_Row_3[0] = 0;
     Err_Row_4[0] = 0; Err_Row_5[0] = 0; Err_Row_6[0] = 0;

   // Initialize columns in the main array
     Err_Row_1[1] = 0; Err_Row_2[1] = 0; Err_Row_3[1] = 0; Err_Row_4[1] = 0;
     Err_Row_5[1] = 0; Err_Row_6[1] = 0; Err_Row_1[2] = 0; Err_Row_2[2] = 0;
     Err_Row_3[2] = 0; Err_Row_4[2] = 0; Err_Row_5[2] = 0; Err_Row_6[2] = 0;
     Err_Row_1[3] = 0; Err_Row_2[3] = 0; Err_Row_3[3] = 0; Err_Row_4[3] = 0;
     Err_Row_5[3] = 0; Err_Row_6[3] = 0; Err_Row_1[4] = 0; Err_Row_2[4] = 0;
     Err_Row_3[4] = 0; Err_Row_4[4] = 0; Err_Row_5[4] = 0; Err_Row_6[4] = 0;
     Err_Row_1[5] = 0; Err_Row_2[5] = 0; Err_Row_3[5] = 0; Err_Row_4[5] = 0;
     Err_Row_5[5] = 0; Err_Row_6[5] = 0; Err_Row_1[6] = 0; Err_Row_2[6] = 0;
     Err_Row_3[6] = 0; Err_Row_4[6] = 0; Err_Row_5[6] = 0; Err_Row_6[6] = 0;
     Err_Row_1[7] = 0; Err_Row_2[7] = 0; Err_Row_3[7] = 0; Err_Row_4[7] = 0;
     Err_Row_5[7] = 0; Err_Row_6[7] = 0; Err_Row_1[8] = 0; Err_Row_2[8] = 0;
     Err_Row_3[8] = 0; Err_Row_4[8] = 0; Err_Row_5[8] = 0; Err_Row_6[8] = 0;

   // Initialize right border
     Err_Row_1[9] = 0; Err_Row_2[9] = 0; Err_Row_3[9] = 0;
     Err_Row_4[9] = 0; Err_Row_5[9] = 0; Err_Row_6[9] = 0;

   // Initialize top border
     Err_Row_0[0]  = 0; Err_Row_0[1]  = 0; Err_Row_0[2]  = 0;
     Err_Row_0[3]  = 0; Err_Row_0[4]  = 0; Err_Row_0[5]  = 0;
     Err_Row_0[6]  = 0; Err_Row_0[7]  = 0; Err_Row_0[8]  = 0;
     Err_Row_0[9] = 0;
```

```
// Initialize pixels in the main array
  PV_Row_1[1] = pixel_1; PV_Row_1[2] = pixel_2;
  PV_Row_1[3] = pixel_3; PV_Row_1[4] = pixel_4;
  PV_Row_1[5] = pixel_5; PV_Row_1[6] = pixel_6;
  PV_Row_1[7] = pixel_7; PV_Row_1[8] = pixel_8;
  PV_Row_2[1] = pixel_9; PV_Row_2[2] = pixel_10;
  PV_Row_2[3] = pixel_11; PV_Row_2[4] = pixel_12;
  PV_Row_2[5] = pixel_13; PV_Row_2[6] = pixel_14;
  PV_Row_2[7] = pixel_15; PV_Row_2[8] = pixel_16;
  PV_Row_3[1] = pixel_17; PV_Row_3[2] = pixel_18;
  PV_Row_3[3] = pixel_19; PV_Row_3[4] = pixel_20;
  PV_Row_3[5] = pixel_21; PV_Row_3[6] = pixel_22;
  PV_Row_3[7] = pixel_23; PV_Row_3[8] = pixel_24;
  PV_Row_4[1] = pixel_25; PV_Row_4[2] = pixel_26;
  PV_Row_4[3] = pixel_27; PV_Row_4[4] = pixel_28;
  PV_Row_4[5] = pixel_29; PV_Row_4[6] = pixel_30;
  PV_Row_4[7] = pixel_31; PV_Row_4[8] = pixel_32;
  PV_Row_5[1] = pixel_33; PV_Row_5[2] = pixel_34;
  PV_Row_5[3] = pixel_35; PV_Row_5[4] = pixel_36;
  PV_Row_5[5] = pixel_37; PV_Row_5[6] = pixel_38;
  PV_Row_5[7] = pixel_39; PV_Row_5[8] = pixel_40;
  PV_Row_6[1] = pixel_41; PV_Row_6[2] = pixel_42;
  PV_Row_6[3] = pixel_43; PV_Row_6[4] = pixel_44;
  PV_Row_6[5] = pixel_45; PV_Row_6[6] = pixel_46;
  PV_Row_6[7] = pixel_47; PV_Row_6[8] = pixel_48;

// Pixels in Row 1
  for (i = 1; i <= M; i = i +1) begin: row_1_loop
    E_av = (w1 * Err_Row_1[i -1] + w2 * Err_Row_0[i -1]
      + w3 * Err_Row_0[i] + w4 * Err_Row_0 [i +1] ) >> 4;
    CPV = PV_Row_1[i] + E_av;
    CPV_round = (CPV < Threshold) ? 0: 255;
    HTPV_Row_1[i] = (CPV_round == 0) ? 0: 1;
    Err_Row_1[i]  = CPV - CPV_round;
  end // row_1_loop

// Pixels in Row 2
  for (i = 1; i <= M; i = i +1) begin: row_2_loop
    E_av = (w1 * Err_Row_2[i -1] + w2 * Err_Row_1[i -1]
     + w3 * Err_Row_1[i] + w4 * Err_Row_1 [i +1] ) >> 4;
    CPV = PV_Row_2[i] + E_av;
    CPV_round = (CPV < Threshold) ? 0: 255;
    HTPV_Row_2[i] = (CPV_round == 0) ? 0: 1;
    Err_Row_2[i]  = CPV - CPV_round;
  end // row_2_loop

// Pixels in Row 3
  for (i = 1; i <= M; i = i +1) begin: row_3_loop
    E_av = (w1 * Err_Row_3[i -1] + w2 * Err_Row_2[i -1]
      + w3 * Err_Row_2[i]
```

```
            + w4 * Err_Row_2 [i +1] ) >> 4;
            CPV = PV_Row_3[i] + E_av;
            CPV_round = (CPV < Threshold) ? 0: 255;
            HTPV_Row_3[i] = (CPV_round == 0) ? 0: 1;
            Err_Row_3[i]  = CPV - CPV_round;
          end // row_3_loop

      // Pixels in Row 4
        for (i = 1; i <= M; i = i +1) begin: row_4_loop
          E_av = (w1 * Err_Row_4[i -1] + w2 * Err_Row_3[i -1]
            + w3 * Err_Row_3[i] + w4 * Err_Row_3 [i +1] ) >> 4;
          CPV = PV_Row_4[i] + E_av;
          CPV_round = (CPV < Threshold) ? 0: 255;
          HTPV_Row_4[i] = (CPV_round == 0) ? 0: 1;
          Err_Row_4[i]  = CPV - CPV_round;
        end // row_4_loop

      // Pixels in Row 5
        for (i = 1; i <= M; i = i +1) begin: row_5_loop
          E_av = (w1 * Err_Row_5[i -1] + w2 * Err_Row_4[i -1]
            + w3 * Err_Row_4[i] + w4 * Err_Row_4 [i +1] ) >> 4;
          CPV = PV_Row_5[i] + E_av;
          CPV_round = (CPV < Threshold) ? 0: 255;
          HTPV_Row_5[i] = (CPV_round == 0) ? 0: 1;
          Err_Row_5[i]  = CPV - CPV_round;
        end // row_5_loop

      // Pixels in Row 6
        for (i = 1; i <= M; i = i +1) begin: row_6_loop
          E_av = (w1 * Err_Row_6[i -1] + w2 * Err_Row_5[i -1]
            + w3 * Err_Row_5[i] + w4 * Err_Row_5 [i +1] ) >> 4;
          CPV = PV_Row_6[i] + E_av;
          CPV_round = (CPV < Threshold) ? 0: 255;
          HTPV_Row_6[i] = (CPV_round == 0) ? 0: 1;
          Err_Row_6[i]  = CPV - CPV_round;
        end // row_6_loop
      end
    endmodule
```

The NLP for the image converter also suggests a synchronous implementation. This architecture will require memory, but releases the resources of the system bus while the processor is executing. At one extreme, a design would use a single FU, together with memory and a controller, to convert the entire image in 48 clock cycles. The cycle time of the image converter would be limited by the longest path through a single FU. A (naive) alternative is to write a single synchronous cyclic behavior that describes the NLP of the algorithm and let a synthesis tool create an architecture. The expectation is that there is no need to design a controller because a synthesis tool will synthesize the NLP by unrolling the loops and forming a structure that implements the algorithm. The machine would transform the image in one long clock cycle. A Verilog

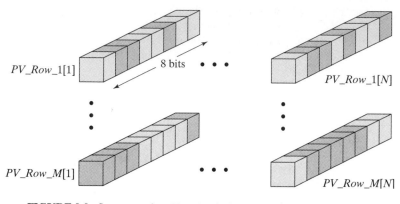

FIGURE 9-9   Structure of an $N \times M$ pixel array with an 8-bit resolution.

model of a machine implementing the algorithm in this style is given by *Image_Converter_1*, below, along with its testbench.

Because Verilog does not have randomly accessible two-dimensional arrays, *Image_Converter_1* declares a separate array of words for each row of pixels, and orders the rows sequentially within multiple *for* loops in a cyclic behavior, from top to bottom, as shown in Figure 9-9. The assignments to variables are made with the procedural assignment operator (=), and the order of the statements reflects the data dependencies of the statements that must be executed by an FU. *The time delays preceding the loops reveal the evolution of the computations, and are removed after debugging and before attempting synthesis.* The sequential ordering of the loops that update the rows of the array implement the NLP and convert the image in a single clock cycle. A signal, *Go*, is asserted by an external agent to launch the image converter, and a signal, *Done*, is asserted when the image has been converted. The signal *reset* flushes the memory of residual values of pixel values and errors to prepare for a new image.

```
module Image_Converter_1 (
  HTPV_Row_1, HTPV_Row_2, HTPV_Row_3,
  HTPV_Row_4, HTPV_Row_5, HTPV_Row_6,
  Done,
  pixel_1,  pixel_2, pixel_3, pixel_4, pixel_5, pixel_6, pixel_7, pixel_8,
  pixel_9,  pixel_10, pixel_11, pixel_12, pixel_13, pixel_14, pixel_15, pixel_16,
  pixel_17, pixel_18, pixel_19, pixel_20, pixel_21, pixel_22, pixel_23, pixel_24,
  pixel_25, pixel_26, pixel_27, pixel_28, pixel_29, pixel_30, pixel_31, pixel_32,
  pixel_33, pixel_34, pixel_35, pixel_36, pixel_37, pixel_38, pixel_39, pixel_40,
  pixel_41, pixel_42, pixel_43, pixel_44, pixel_45, pixel_46, pixel_47, pixel_48,
  Go, clk, reset);

  output          Done;
  output  [1: 8]  HTPV_Row_1, HTPV_Row_2, HTPV_Row_3,
                  HTPV_Row_4, HTPV_Row_5, HTPV_Row_6;
  input   [7: 0]
```

```
      pixel_1,   pixel_2,  pixel_3,  pixel_4,  pixel_5,  pixel_6, pixel_7,  pixel_8,
      pixel_9,   pixel_10, pixel_11, pixel_12, pixel_13, pixel_14, pixel_15, pixel_16,
      pixel_17, pixel_18, pixel_19, pixel_20, pixel_21, pixel_22, pixel_23, pixel_24,
      pixel_25, pixel_26, pixel_27, pixel_28, pixel_29, pixel_30, pixel_31, pixel_32,
      pixel_33, pixel_34, pixel_35, pixel_36, pixel_37, pixel_38, pixel_39, pixel_40,
      pixel_41, pixel_42, pixel_43, pixel_44, pixel_45, pixel_46, pixel_47, pixel_48;
      input        Go, clk, reset;
      reg          Done;
      reg   [7: 0] PV_Row_1 [1: 8], PV_Row_2 [1: 8], PV_Row_3 [1: 8],
                   PV_Row_4 [1: 8], PV_Row_5 [1: 8], PV_Row_6 [1: 8];
      reg          HTPV_Row_1, HTPV_Row_2, HTPV_Row_3;
      reg          HTPV_Row_4, HTPV_Row_5, HTPV_Row_6;

// Weights for the average error; compatible with divide-by-16 (>> 4)
      parameter    w1 = 2, w2 = 8, w3 = 4, w4 = 2;

// Note Err_Row_ includes left, right, and top border
// columns for initialization of the algorithm.
      reg   [7: 0]
       Err_Row_0 [0: 9], Err_Row_1 [0: 9], Err_Row_2 [0: 9], Err_Row_3 [0: 9],
       Err_Row_4 [0: 9], Err_Row_5 [0: 9], Err_Row_6 [0: 9];

      reg   [9: 0] CPV, CPV_round, E_av;
      integer i;

      parameter N = 6;                // rows
      parameter M = 8;                // columns
      parameter Threshold = 128;
      wire PV_1_8 = PV_Row_1[8];

      always begin: wrapper_for_synthesis
      @ (posedge clk) begin: pixel_converter
       if (reset) begin: reset_action
         Done = 0;

// Initialize error at left border
         Err_Row_1[0] = 0; Err_Row_2[0] = 0; Err_Row_3[0] = 0;
         Err_Row_4[0] = 0; Err_Row_5[0] = 0; Err_Row_6[0] = 0;

// Initialize columns in the main array
         Err_Row_1[1] = 0; Err_Row_2[1] = 0; Err_Row_3[1] = 0;
         Err_Row_4[1] = 0; Err_Row_5[1] = 0; Err_Row_6[1] = 0;
         Err_Row_1[2] = 0; Err_Row_2[2] = 0; Err_Row_3[2] = 0;
         Err_Row_4[2] = 0; Err_Row_5[2] = 0; Err_Row_6[2] = 0;
         Err_Row_1[3] = 0; Err_Row_2[3] = 0; Err_Row_3[3] = 0;
         Err_Row_4[3] = 0; Err_Row_5[3] = 0; Err_Row_6[3] = 0;
         Err_Row_1[4] = 0; Err_Row_2[4] = 0; Err_Row_3[4] = 0;
         Err_Row_4[4] = 0; Err_Row_5[4] = 0; Err_Row_6[4] = 0;
         Err_Row_1[5] = 0; Err_Row_2[5] = 0; Err_Row_3[5] = 0;
```

```
        Err_Row_4[5] = 0; Err_Row_5[5] = 0; Err_Row_6[5] = 0;
        Err_Row_1[6] = 0; Err_Row_2[6] = 0; Err_Row_3[6] = 0;
        Err_Row_4[6] = 0; Err_Row_5[6] = 0; Err_Row_6[6] = 0;
        Err_Row_1[7] = 0; Err_Row_2[7] = 0; Err_Row_3[7] = 0;
        Err_Row_4[7] = 0; Err_Row_5[7] = 0; Err_Row_6[7] = 0;
        Err_Row_1[8] = 0; Err_Row_2[8] = 0; Err_Row_3[8] = 0;
        Err_Row_4[8] = 0; Err_Row_5[8] = 0; Err_Row_6[8] = 0;

    // Initialize right border
        Err_Row_1[9] = 0; Err_Row_2[9] = 0; Err_Row_3[9] = 0;
        Err_Row_4[9] = 0; Err_Row_5[9] = 0; Err_Row_6[9] = 0;

    // Initialize top border
        Err_Row_0[0]  = 0; Err_Row_0[1]  = 0; Err_Row_0[2]  = 0;
        Err_Row_0[3]  = 0; Err_Row_0[4]  = 0; Err_Row_0[5]  = 0;
        Err_Row_0[6]  = 0; Err_Row_0[7]  = 0; Err_Row_0[8]  = 0;
        Err_Row_0[9] = 0;

    // Initialize pixels in the main array
        PV_Row_1[1] = pixel_1; PV_Row_1[2] = pixel_2;
        PV_Row_1[3] = pixel_3; PV_Row_1[4] = pixel_4;
        PV_Row_1[5] = pixel_5; PV_Row_1[6] = pixel_6;
        PV_Row_1[7] = pixel_7; PV_Row_1[8] = pixel_8;
        PV_Row_2[1] = pixel_9; PV_Row_2[2] = pixel_10;
        PV_Row_2[3] = pixel_11; PV_Row_2[4] = pixel_12;
        PV_Row_2[5] = pixel_13; PV_Row_2[6] = pixel_14;
        PV_Row_2[7] = pixel_15; PV_Row_2[8] = pixel_16;
        PV_Row_3[1] = pixel_17; PV_Row_3[2] = pixel_18;
        PV_Row_3[3] = pixel_19; PV_Row_3[4] = pixel_20;
        PV_Row_3[5] = pixel_21; PV_Row_3[6] = pixel_22;
        PV_Row_3[7] = pixel_23; PV_Row_3[8] = pixel_24;
        PV_Row_4[1] = pixel_25; PV_Row_4[2] = pixel_26;
        PV_Row_4[3] = pixel_27; PV_Row_4[4] = pixel_28;
        PV_Row_4[5] = pixel_29; PV_Row_4[6] = pixel_30;
        PV_Row_4[7] = pixel_31; PV_Row_4[8] = pixel_32;
        PV_Row_5[1] = pixel_33; PV_Row_5[2] = pixel_34;
        PV_Row_5[3] = pixel_35; PV_Row_5[4] = pixel_36;
        PV_Row_5[5] = pixel_37; PV_Row_5[6] = pixel_38;
        PV_Row_5[7] = pixel_39; PV_Row_5[8] = pixel_40;
        PV_Row_6[1] = pixel_41; PV_Row_6[2] = pixel_42;
        PV_Row_6[3] = pixel_43; PV_Row_6[4] = pixel_44;
        PV_Row_6[5] = pixel_45; PV_Row_6[6] = pixel_46;
        PV_Row_6[7] = pixel_47; PV_Row_6[8] = pixel_48;

    // Initialization complete
      end  // reset_action
      else begin: half_tone_calculations

    // Pixels in Row 1
      if (Go) begin: wrapper
```

```
    #20 for (i = 1; i <= M; i = i +1) begin: row_1_loop
      E_av = (w1 * Err_Row_1[i -1] + w2 * Err_Row_0[i -1]
        + w3 * Err_Row_0[i] + w4 * Err_Row_0 [i +1] ) >> 4;
      CPV = PV_Row_1[i] + E_av;
      CPV_round = (CPV < Threshold) ? 0: 255;
      HTPV_Row_1[i] = (CPV_round == 0) ? 0: 1;
      Err_Row_1[i]  = CPV - CPV_round;
      // Used for Image_Converter_SR
      //@ (posedge clk) if (reset) disable pixel_converter;
    end // row_1_loop

// Pixels in Row 2
    #20 for (i = 1; i <= M; i =  i +1) begin: row_2_loop
      E_av = (w1 * Err_Row_2[i -1] + w2 * Err_Row_1[i -1]
        + w3 * Err_Row_1[i] + w4 * Err_Row_1 [i +1] ) >> 4;
      CPV = PV_Row_2[i] + E_av;
      CPV_round = (CPV < Threshold) ? 0: 255;
      HTPV_Row_2[i] = (CPV_round == 0) ? 0: 1;
      Err_Row_2[i]  = CPV - CPV_round;
      // Used for Image_Converter_SR
      //@ (posedge clk)  if (reset) disable pixel_converter;
    end // row_2_loop

// Pixels in Row 3
    #20 for (i = 1; i <= M; i =  i +1) begin: row_3_loop
      E_av = (w1 * Err_Row_3[i -1] + w2 * Err_Row_2[i -1]
        + w3 * Err_Row_2[i] + w4 * Err_Row_2 [i +1] ) >> 4;
      CPV = PV_Row_3[i] + E_av;
      CPV_round = (CPV < Threshold) ? 0: 255;
      HTPV_Row_3[i] = (CPV_round == 0) ? 0: 1;
      Err_Row_3[i]  = CPV - CPV_round;
      // Used for Image_Converter_SR
      //@ (posedge clk)  if (reset) disable pixel_converter;
    end // row_3_loop

// Pixels in Row 4
    #20 for (i = 1; i <= M; i =  i +1) begin: row_4_loop
      E_av = (w1 * Err_Row_4[i -1] + w2 * Err_Row_3[i -1]
        + w3 * Err_Row_3[i] + w4 * Err_Row_3 [i +1] ) >> 4;
      CPV = PV_Row_4[i] + E_av;
      CPV_round = (CPV < Threshold) ? 0: 255;
      HTPV_Row_4[i] = (CPV_round == 0) ? 0: 1;
      Err_Row_4[i]  = CPV - CPV_round;
      // Used for Image_Converter_SR
      //@ (posedge clk) if (reset) disable pixel_converter;
    end // row_4_loop

// Pixels in Row 5
    #20 for (i = 1; i <= M; i =  i +1) begin: row_5_loop
```

```verilog
          E_av = (w1 * Err_Row_5[i -1] + w2 * Err_Row_4[i -1]
            + w3 * Err_Row_4[i] + w4 * Err_Row_4 [i +1] ) >> 4;
          CPV = PV_Row_5[i] + E_av;
          CPV_round = (CPV < Threshold) ? 0: 255;
          HTPV_Row_5[i] = (CPV_round == 0) ? 0: 1;
          Err_Row_5[i]  = CPV - CPV_round;
          // Used for Image_Converter_SR
          //@ (posedge clk) if (reset) disable pixel_converter;
        end // row_5_loop

   // Pixels in Row 6
      #20 for (i = 1; i <= M; i =  i +1) begin: row_6_loop
        E_av = (w1 * Err_Row_6[i -1] + w2 * Err_Row_5[i -1]
          + w3 * Err_Row_5[i] + w4 * Err_Row_5 [i +1] ) >> 4;
        CPV = PV_Row_6[i] + E_av;
        CPV_round = (CPV < Threshold) ? 0: 255;
        HTPV_Row_6[i] = (CPV_round == 0) ? 0: 1;
        Err_Row_6[i]  = CPV - CPV_round;
        //@ (posedge clk)  if (reset) disable pixel_converter;
      end // row_6_loop

        Done = 1;
      end   // wrapper
      end      // half_tone_calculations
     end      // pixel_converter
    end      // wrapper_for_synthesis
endmodule


  module t_Image_Converter_1 ( );
  // hard-wired for 8 x 6 pixel array with boundary cells.
    wire  [1: 8] HTPV_Row_1;
    wire  [1: 8] HTPV_Row_2;
    wire  [1: 8] HTPV_Row_3;
    wire  [1: 8] HTPV_Row_4;
    wire  [1: 8] HTPV_Row_5;
    wire  [1: 8] HTPV_Row_6;
    reg   Go, clk, reset;
    reg   [7: 0]
    pixel_1,  pixel_2, pixel_3, pixel_4, pixel_5, pixel_6, pixel_7, pixel_8,
    pixel_9,  pixel_10, pixel_11, pixel_12, pixel_13, pixel_14, pixel_15, pixel_16,
    pixel_17, pixel_18, pixel_19, pixel_20, pixel_21, pixel_22, pixel_23, pixel_24,
    pixel_25, pixel_26, pixel_27, pixel_28, pixel_29, pixel_30, pixel_31, pixel_32,
    pixel_33, pixel_34, pixel_35, pixel_36, pixel_37, pixel_38, pixel_39, pixel_40,
    pixel_41, pixel_42, pixel_43, pixel_44, pixel_45, pixel_46, pixel_47, pixel_48;

  Image_Converter_1 M0 (
    HTPV_Row_1, HTPV_Row_2, HTPV_Row_3,
    HTPV_Row_4, HTPV_Row_5, HTPV_Row_6,
    Done,
```

```
        pixel_1,   pixel_2,  pixel_3,  pixel_4,  pixel_5,  pixel_6,  pixel_7,  pixel_8,
        pixel_9,   pixel_10, pixel_11, pixel_12, pixel_13, pixel_14, pixel_15, pixel_16,
        pixel_17, pixel_18, pixel_19, pixel_20, pixel_21, pixel_22, pixel_23, pixel_24,
        pixel_25, pixel_26, pixel_27, pixel_28, pixel_29, pixel_30, pixel_31, pixel_32,
        pixel_33, pixel_34, pixel_35, pixel_36, pixel_37, pixel_38, pixel_39, pixel_40,
        pixel_41, pixel_42, pixel_43, pixel_44, pixel_45, pixel_46, pixel_47, pixel_48,
        Go, clk, reset);

 initial #3200 $finish;
 initial begin clk = 0; forever #5 clk = ~clk; end
 initial begin reset = 1; #50 reset = 0; #100 reset = 1;
   #10 reset = 0; end
 initial begin Go = 0; #100 Go = 1; #100 Go = 0; end
 initial begin: Image_Pattern_1
   pixel_1 =255;pixel_2 =255;pixel_3 =255;pixel_4 =255;pixel_5 =0;pixel_6 =0;
   pixel_7 =0;pixel_8 =0;pixel_9 =255; pixel_10 =255; pixel_11 =255;
   pixel_12 =255;pixel_13 =0;pixel_14 =0;pixel_15 =0;pixel_16 =0;
   pixel_17 =255;pixel_18 =255;pixel_19 =255;pixel_20 =255;pixel_21 =0;
   pixel_22 =0;pixel_23 =0;pixel_24 =0;pixel_25 =0;pixel_26 =0;pixel_27 =0;
   pixel_28 =0;pixel_29 =255;pixel_30 =255;pixel_31 =255;pixel_32 =255;
   pixel_33 =0;pixel_34 =0;pixel_35 =0;pixel_36 =0;pixel_37 =255;
   pixel_38 =255;pixel_39 =255;pixel_40 =255;pixel_41 =0;pixel_42 =0;
   pixel_43 =0;pixel_44 =0;pixel_45 =255;pixel_46 =255;pixel_47 =255;
   pixel_48 =255;
 end
 initial begin: Image_Pattern_2
   #260 pixel_1 =0;pixel_2 =0;pixel_3 =0;pixel_4 =0;pixel_5 =255;pixel_6 =255;
   pixel_7 =255;pixel_8 =255;pixel_9 =0;pixel_10 =0;pixel_11 =0;pixel_12 =0;
   pixel_13 =255;pixel_14 =255;pixel_15 =255;pixel_16 =255;pixel_17 =0;
   pixel_18 =0;pixel_19 =0;pixel_20 =0;pixel_21 =255;pixel_22 =255;
   pixel_23 =255;pixel_24 =255;pixel_25 =255;pixel_26 =255;pixel_27 =255;
   pixel_28 =255;pixel_29 =0;pixel_30 =0;pixel_31 =0;pixel_32 =0;
   pixel_33 =255;pixel_34 =255;pixel_35 =255;pixel_36 =255;pixel_37 =0;
   pixel_38 =0;pixel_39 =0;pixel_40 =0;pixel_41 =255;pixel_42 =255;
   pixel_43 =255;pixel_44 =255;pixel_45 =0;pixel_46 =0;pixel_47 =0;pixel_48 =0;
 end
 initial begin: Image_Pattern_3_Cross
   #520 pixel_1 =0;pixel_2 =0;pixel_3 =255;pixel_4 =255;pixel_5 =255;
   pixel_6 =255;pixel_7 =0;pixel_8 =0; pixel_9 =0; pixel_10 =0; pixel_11 =255;
   pixel_12 =255;pixel_13 =255;pixel_14 =255;pixel_15 =0;pixel_16 =0;
   pixel_17 =255;pixel_18 =255;pixel_19 =255;pixel_20 =255;pixel_21 =255;
   pixel_22 =255;pixel_23 =255;pixel_24 =255;pixel_25 =255;pixel_26 =255;
   pixel_27 =255;pixel_28 =255;pixel_29 =255;pixel_30 =255;pixel_31 =255;
   pixel_32 =255;pixel_33 =0;pixel_34 =0;pixel_35 =255;pixel_36 =255;
   pixel_37 =255;pixel_38 =255;pixel_39 =0;pixel_40 =0;pixel_41 =0;pixel_42 =0;
   pixel_43 =255;pixel_44 =255;pixel_45 =255;pixel_46 =255;pixel_47 =0;
   pixel_48 =0;
 end
 initial begin: Image_Pattern_4_Bar_Cross
```

```
      #780 pixel_1 =255;pixel_2 =255;pixel_3 =0;pixel_4 =0;pixel_5 =0;pixel_6 =0;
      pixel_7 =255;pixel_8 =255;pixel_9 =255; pixel_10 =255; pixel_11 =0;
      pixel_12 =0;pixel_13 =0;pixel_14 =0;pixel_15 =255;pixel_16 =255;
      pixel_17 =255;pixel_18 =255;pixel_19 =255;pixel_20 =255;pixel_21 =255;
      pixel_22 =255;pixel_23 =255;pixel_24 =255;pixel_25 =255;pixel_26 =255;
      pixel_27 =255;pixel_28 =255;pixel_29 =255;pixel_30 =255;pixel_31 =255;
      pixel_32 =255;pixel_33 =255;pixel_34 =255;pixel_35 =0;pixel_36 =0;
      pixel_37 =0;pixel_38 =0;pixel_39 =255;pixel_40 =255;pixel_41 =255;
      pixel_42 =255;pixel_43 =0;pixel_44 =0;pixel_45 =0;pixel_46 =0;pixel_47 =255;
      pixel_48 =255;
  end
  initial begin: Image_Pattern_5_Graduated_Left_to_Right
    #1040    pixel_1 =31;pixel_2 =63;pixel_3 =95;pixel_4 =127;pixel_5 =159;
      pixel_6 =191;pixel_7 =223;pixel_8 =255;pixel_9 =31;pixel_10 =63;
      pixel_11 =95;pixel_12 =127;pixel_13 =159;pixel_14 =191;pixel_15 =223;
      pixel_16 =255;pixel_17 =31;pixel_18 =63;pixel_19 =95;pixel_20 =127;
      pixel_21 =159;pixel_22 =191;pixel_23 =223;pixel_24 =255;
      pixel_25 =31;pixel_26 =63;pixel_27 =95;pixel_28 =127;pixel_29 =159;
      pixel_30 =191;pixel_31 =223;pixel_32 =255;pixel_33 =31;pixel_34 =63;
      pixel_35 =95;pixel_36 =127;pixel_37 =159;pixel_38 =191;pixel_39 =223;
      pixel_40 =255;pixel_41 =31;pixel_42 =63;pixel_43 =95;pixel_44 =127;
      pixel_45 =159;pixel_46 =191;pixel_47 =223;pixel_48 =255;
  end
  initial #1600 $finish;
  initial begin clk =0; forever #5 clk = ~clk; end
  initial fork
    #10 reset = 1;
    #30 reset = 0;
  join
  initial fork
    #50 Go = 1;              // Image #1
    #60 Go = 0;
    #270 reset = 1;          // Image #2
    #290 reset = 0;
    #300 Go = 1;
    #310 Go = 0;
    #600 reset = 1;          // Image #3
    #630 reset = 0;
    #680 Go = 1;
    #690 Go = 0;
    #900 reset = 1;          // Image #4
    #930 reset = 0;
    #980 Go = 1;
    #990 Go = 0;
    #1200 reset = 1;         // Image #5
    #1230 reset = 0;
    #1280 Go = 1;
    #1290 Go = 0;
  join
endmodule
```

The results of simulating *Image_Converter_1* are shown in Figure 9-10. The final halftone image is identical to that produced by *Image_Converter_Baseline*.

The Verilog model *Image_Converter_1* produces the same halftone image as *Image_Converter_Baseline* and *Image_Converter_0*. *Image_Converter_Baseline* and *Image_converter_0* synthesize to equivalent combinational logic, but *Image_Converter_1* cannot be synthesized. The sequential ordering of its procedural assignments in the cyclic behavior updates the values stored in memory as the simulation evolves by immediately overwriting the residual data, ensuring that fresh error data are used at subsequent steps. A physical machine would have to store data in memory and fetch it when needed. These operations cannot execute in a single clock cycle, which explains why the model cannot be synthesized. As a work-around, *Image_Converter_SR* is identical to *Image_Converter_1*, except that the embedded event-control expressions are not removed by the comments. This version updates one row of the array in a single clock cycle and requires eight processors (one for each pixel in the row). The machine shares its resources between the rows, but in a given clock cycle all of the processors are dedicated to a single row of pixels. It will require an image cycle time of $T_c = 6 \times 8 \times T_{FU} = 48\,T_{FU}$ (i.e., the machine saves hardware but not time). It will require a more elaborate control structure to steer data to and from memory and the

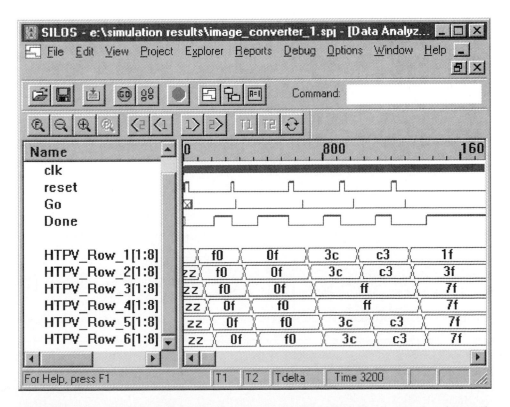

FIGURE 9-10 Simulation results for *Image_Converter_1* with weights $(w1, w2, w3, w4) = (2, 8, 4, 2)$.

shared FUs. Unfortunately, the FPGA synthesis tool[10] did not support embedded event-control expressions within a *for* loop and could not produce an implementation.

9.2.3   Concurrent ASMD-Based Architecture for a Halftone Pixel Image Converter

We will now consider an alternative hardware implementation of the image converter algorithm by partitioning the image converter into an algorithmic state machine with a datapath (i.e., a ASMD).

The data dependencies that are evident in the DFG shown in Figure 9-11 reveal a parallelism that can be exploited to reduce the number of clock cycles and the number of processors required to update the array, at the additional expense of requiring memory. The shaded region defines a *computational wavefront* (i.e., a locus of DFG nodes that can execute concurrently in a given time step). Each node in Figure 9-11 is annotated with a *wavefront index* denoting the time step in which it may execute. The array of

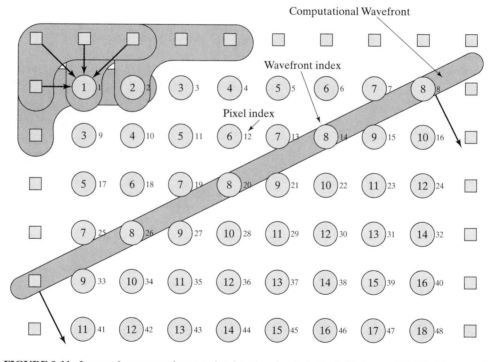

FIGURE 9-11   Locus of concurrently executing functional units in the DFG for an 8 × 6 halftone pixel image converter.

---

[10]Xilinx ISE 3.1i.

nodes can still be processed sequentially, in the order of ascending indexes, but nodes with an identical wavefront index can execute concurrently (i.e., in the same time step).

The wavefront indexes of the DFG partition the temporal domain and identify clock boundaries. Within a clock boundary, the graph reveals the resources required by a synchronous machine that would implement the processor and exploit the parallelism. For example, a machine with the wavefront indexes shown in Figure 9-11 would require a maximum of four identical FUs operating concurrently, with each unit implementing the fine-grained logic required to update a pixel. *The machine would update the entire image in only 18 time steps, while using fewer FU resources and requiring design of a more complex controller than the baseline design.*[11]

The architecture of the machine's datapath will be designed to exploit the concurrency that is evident in the DFG. Then a controller will be designed for the datapath. We will pursue by manual methods what a behavioral synthesis tool should accomplish and then compare the result to the baseline design in more detail.

Our alternative architecture is shown in Figure 9-12. It has four pixel processors (FUs), a memory unit, and a controller. The inputs of each processor provide the value of the pixel at the location indicated by the address passed from the controller and the errors at the locations of the pixel's selected neighbors. The controller is implemented as a sequencer, which creates the addresses of the pixels that are to be updated in a given time step. The entire array of pixels is downloaded to the memory unit in a single clock cycle.

Having identified the need for at most four FUs, we address the resource mapping task by considering Table 9-1, a *reservation table*, which shows one of many possible mappings of the nodes of the DFG of the 8×6 halftone image converter to a set of four processors that exploit the concurrency that was exposed by the DFG. The columns of the table establish a linear execution schedule specifying the DFG nodes that will execute in a given time; the rows of the table establish a binding between the nodes of the DFG and the processors of the architecture. The table establishes a space–time partition in which every node has a unique (processor, time slot) pair.

The architecture represented by the reservation table can process a single image in 18 time slots, but thereafter the pipelining that is evident in the table presents an opportunity to stream the images with a throughput of only 12 time slots. This architecture, however, cannot exploit this opportunity because the memory cannot be selectively loaded to partially update its contents, say in t_{13}, to access the next image cell. In general, the number of clock cycles consumed by a time slot is strongly influenced by bus resources and memory operations.

An ASM chart for the control unit of the machine is shown in Figure 9-13. The state boxes list the values of *index* corresponding to the entries in Table 9-1. This implementation does not exploit the pipelining that is apparent in Table 9-1.

The Verilog description of the alternative image converter is given below. The signal *Ld_Image* is added to the interface to allow an image to be stored on command, thus freeing up the external bus that provides the image. Once an image is loaded into memory, the machine waits until *Go* is asserted and then converts the image.

---

[11]The baseline sequential design updates the array in 48 time steps.

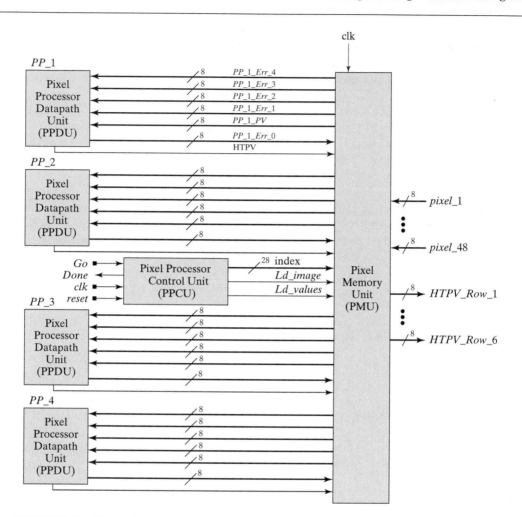

FIGURE 9-12 Alternative architecture for exploiting the concurrency of the DFG of the 8×6 halftone pixel processor shown in Figure 9-10.

TABLE 9-1 A reservation table for mapping DFG nodes to time slots and processors of a 8×6 halftone image converter.

		\multicolumn{18}{c}{Time slots}																
		t_1	t_2	t_3	t_4	t_5	t_6	t_7	t_8	t_9	t_{10}	t_{11}	t_{12}	t_{13}	t_{14}	t_{15}	t_{16}	t_{17}	t_{18}
Processors	P_1	1	2	3	4	5	6	7	8	15	16	23	24						
	P_2			9	10	11	12	13	14	21	22	29	30	31	32				
	P_3					17	18	19	20	27	28	35	36	37	38	39	40		
	P_4							25	26	33	34	41	42	43	44	45	46	47	48

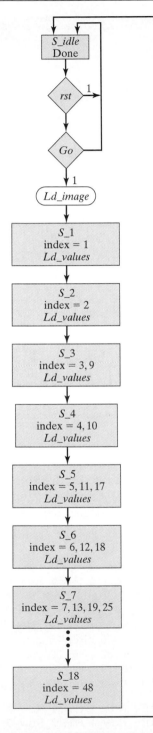

FIGURE 9-13 ASM chart for the control unit of an alternative halftone image converter.

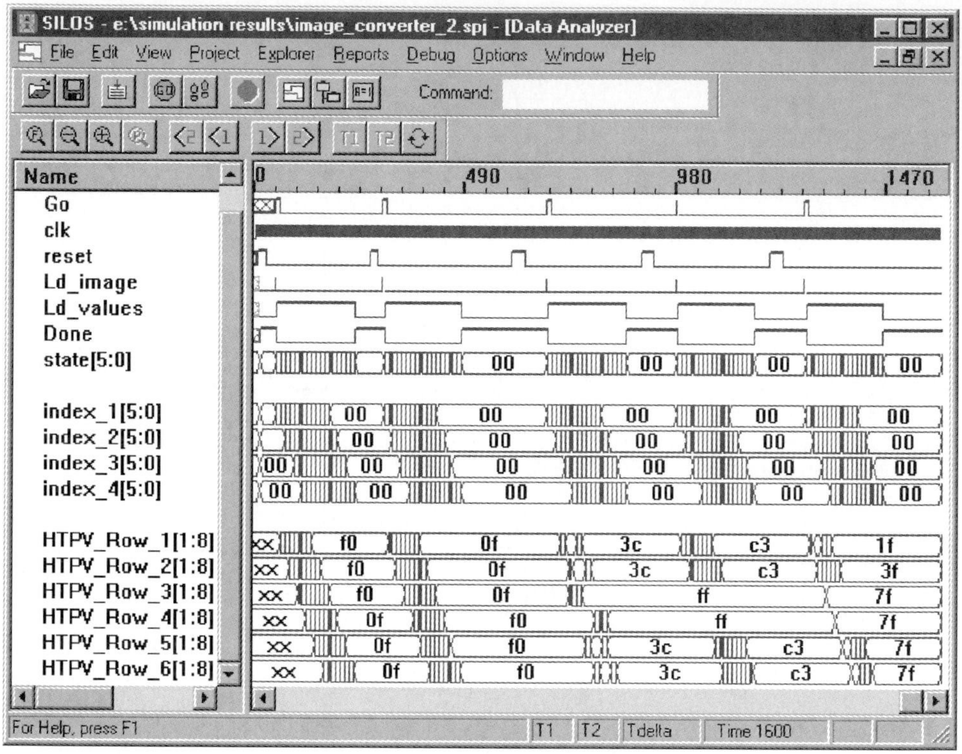

FIGURE 9-14 Simulation results for *Image_Converter_2* operating on the test images given in Figure 9-6.

Simulation results for *Image_Converter_2* are shown in Figure 9-14 for the test images that were given in Figure 9-6. The halftone images match those produced by *Image_Converter_Baseline*, *Image_Converter_0*, and *Image_Converter_1*. Figure 9-15 shows the simulation activity for a single image, and displays the schedule of the four concurrently executing processors.

```
module Image_Converter_2 (
    HTPV_Row_1, HTPV_Row_2, HTPV_Row_3,
    HTPV_Row_4, HTPV_Row_5, HTPV_Row_6, Done,
    pixel_1, pixel_2, pixel_3, pixel_4, pixel_5, pixel_6, pixel_7, pixel_8,
    pixel_9, pixel_10, pixel_11, pixel_12, pixel_13, pixel_14, pixel_15,
    pixel_16, pixel_17, pixel_18, pixel_19, pixel_20, pixel_21, pixel_22,
    pixel_23, pixel_24, pixel_25, pixel_26, pixel_27, pixel_28, pixel_29,
    pixel_30, pixel_31, pixel_32, pixel_33, pixel_34, pixel_35, pixel_36,
    pixel_37, pixel_38, pixel_39, pixel_40, pixel_41, pixel_42, pixel_43,
    pixel_44, pixel_45, pixel_46, pixel_47, pixel_48,Go, clk, reset);

    output [1: 8]   HTPV_Row_1, HTPV_Row_2, HTPV_Row_3,
                    HTPV_Row_4, HTPV_Row_5, HTPV_Row_6;
    output          Done;
```

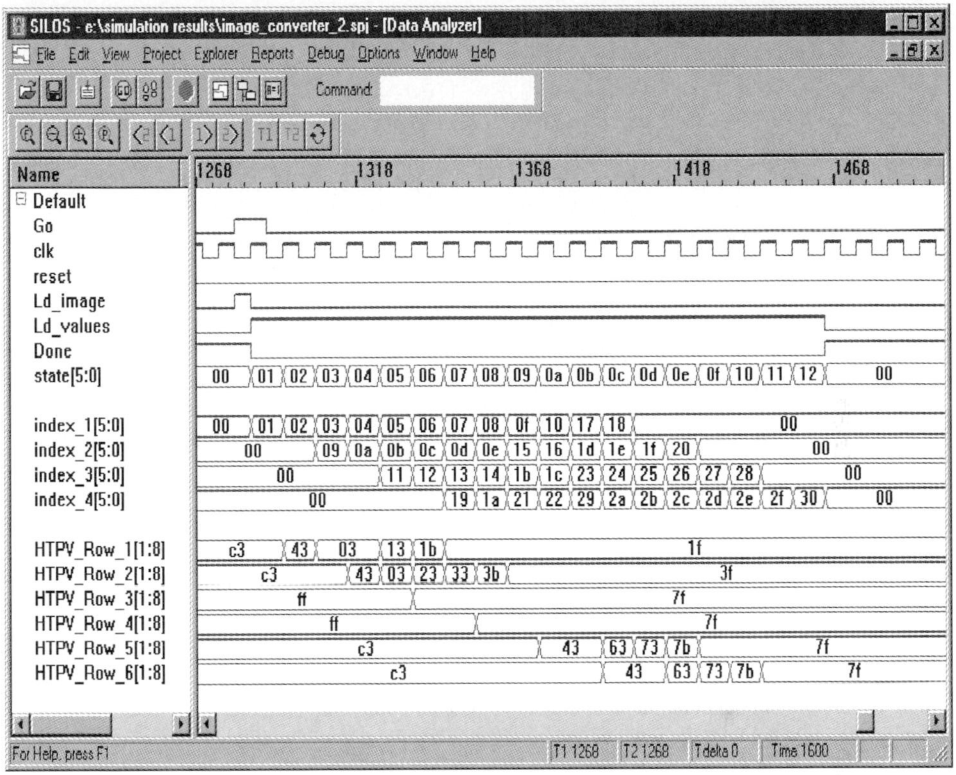

FIGURE 9-15 Simulation results for *Image_Converter_2* operating on the graduated test image given in Figure 9.8, showing the scheduling of four concurrently executing processors.

```
input    [7: 0] pixel_1, pixel_2, pixel_3, pixel_4, pixel_5, pixel_6,
         pixel_7,  pixel_8, pixel_9,  pixel_10, pixel_11, pixel_12, pixel_13, pixel_14,
         pixel_15, pixel_16, pixel_17, pixel_18, pixel_19, pixel_20, pixel_21, pixel_22,
         pixel_23, pixel_24, pixel_25, pixel_26, pixel_27, pixel_28, pixel_29, pixel_30,
         pixel_31, pixel_32, pixel_33, pixel_34, pixel_35, pixel_36, pixel_37, pixel_38,
         pixel_39, pixel_40, pixel_41, pixel_42, pixel_43, pixel_44, pixel_45, pixel_46,
         pixel_47, pixel_48;
input            Go, clk, reset;
wire    [23: 0]  index;
wire             Ld_image, Ld_values;
wire    [7: 0]   PP_1_Err_1, PP_1_Err_2, PP_1_Err_3, PP_1_Err_4,
                 PP_1_PV, PP_2_Err_1, PP_2_Err_2, PP_2_Err_3,
                 PP_2_Err_4, PP_2_PV, PP_3_Err_1, PP_3_Err_2,
                 PP_3_Err_3, PP_3_Err_4, PP_3_PV, PP_4_Err_1,
                 PP_4_Err_2, PP_4_Err_3, PP_4_Err_4, PP_4_PV;
wire    [7: 0]   PP_1_Err_0, PP_2_Err_0, PP_3_Err_0, PP_4_Err_0;
wire             PP_1_HTPV, PP_2_HTPV, PP_3_HTPV, PP_4_HTPV;

wire    [7: 0]   Err_1, Err_2, Err_3, Err_4, PV;
```

```verilog
  PP_Control_Unit M0 (index, Ld_image, Ld_values, Done, Go, clk, reset);

  PP_Datapath_Unit M1 (PP_1_Err_0, PP_1_HTPV, PP_1_Err_1,
  PP_1_Err_2, PP_1_Err_3, PP_1_Err_4, PP_1_PV);
  PP_Datapath_Unit M2 (PP_2_Err_0, PP_2_HTPV, PP_2_Err_1,
  PP_2_Err_2, PP_2_Err_3, PP_2_Err_4, PP_2_PV);
  PP_Datapath_Unit M3 (PP_3_Err_0, PP_3_HTPV, PP_3_Err_1,
  PP_3_Err_2, PP_3_Err_3, PP_3_Err_4, PP_3_PV);
  PP_Datapath_Unit M4 (PP_4_Err_0, PP_4_HTPV, PP_4_Err_1,
  PP_4_Err_2, PP_4_Err_3, PP_4_Err_4, PP_4_PV);

  Pixel_Memory_Unit M5 (
    HTPV_Row_1, HTPV_Row_2, HTPV_Row_3,
    HTPV_Row_4, HTPV_Row_5, HTPV_Row_6,
    PP_1_Err_1, PP_1_Err_2, PP_1_Err_3, PP_1_Err_4, PP_1_PV,
    PP_2_Err_1, PP_2_Err_2, PP_2_Err_3, PP_2_Err_4, PP_2_PV,
    PP_3_Err_1, PP_3_Err_2, PP_3_Err_3, PP_3_Err_4, PP_3_PV,
    PP_4_Err_1, PP_4_Err_2, PP_4_Err_3, PP_4_Err_4, PP_4_PV,
    PP_1_Err_0, PP_2_Err_0, PP_3_Err_0, PP_4_Err_0,
    PP_1_HTPV, PP_2_HTPV, PP_3_HTPV, PP_4_HTPV,
    pixel_1, pixel_2, pixel_3, pixel_4, pixel_5, pixel_6, pixel_7, pixel_8,
    pixel_9, pixel_10, pixel_11, pixel_12, pixel_13, pixel_14, pixel_15, pixel_16,
    pixel_17, pixel_18, pixel_19, pixel_20, pixel_21, pixel_22, pixel_23, pixel_24,
    pixel_25, pixel_26, pixel_27, pixel_28, pixel_29, pixel_30, pixel_31, pixel_32,
    pixel_33, pixel_34, pixel_35, pixel_36, pixel_37, pixel_38, pixel_39, pixel_40,
    pixel_41, pixel_42, pixel_43, pixel_44, pixel_45, pixel_46, pixel_47, pixel_48,
    index, Ld_image, Ld_values, Go, clk);
endmodule

module PP_Control_Unit (index, Ld_image, Ld_values, Done, Go, clk, reset);
  output  [23: 0]          index;
  output                   Ld_image, Ld_values, Done;
  input                    Go, clk, reset;
  reg      [5: 0]          state, next_state;
  reg                      index, Ld_image, Ld_values, Done;
  parameter S_idle = 0, S_1 = 1, S_2 = 2, S_3 = 3, S_4 = 4, S_5 = 5, S_6 = 6,
  S_7 = 7, S_8 = 8, S_9 = 9, S_10 = 10, S_11 = 11, S_12 = 12, S_13 = 13,
  S_14 = 14, S_15 = 15, S_16 = 16, S_17 = 17, S_18 = 18, S_19 = 19,
  S_20 = 20, S_21 = 21, S_22 = 22, S_23 = 23, S_24 = 24, S_25 = 25,
  S_26 = 26, S_27 = 27, S_28 = 28, S_29 = 29, S_30 = 30, S_31 = 31,
  S_32 = 32, S_33 = 33, S_34 = 34, S_35 = 35, S_36 = 36, S_37 = 37,
  S_38 = 38, S_39 = 39, S_40 = 40, S_41 = 41, S_42 = 42, S_43 = 43,
  S_44 = 44, S_45 = 45, S_46 = 46, S_47 = 47, S_48 = 48;

  always @ (posedge clk) if (reset) state <= S_idle; else state <= next_state;

  always @  (state or Go) begin
    Ld_values = 0;  next_state = state;
```

```verilog
      if ((state == S_idle) && Go) next_state = S_1;
      else if ((state > 0) && (state < 18))
        begin next_state = state +1; Ld_values = 1; end
      else if (state == S_18)
        begin next_state = S_idle; Ld_values = 1; end
   end

   always @ (state or Go) begin
   Done = 0; index = 0; Ld_image = 0;
   case (state)
      S_idle:        begin    index = {{6'd0}, {6'd0}, {6'd0}, {6'd0}};
                              Done = 1; if (Go) Ld_image = 1; end
      S_1:                    index = {{6'd1}, {6'd0}, {6'd0}, {6'd0}};
      S_2:                    index = {{6'd2}, {6'd0}, {6'd0}, {6'd0}};
      S_3:                    index = {{6'd3}, {6'd9}, {6'd0}, {6'd0}};
      S_4:                    index = {{6'd4}, {6'd10}, {6'd0}, {6'd0}};
      S_5:                    index = {{6'd5}, {6'd11}, {6'd17}, {6'd0}};
      S_6:                    index = {{6'd6}, {6'd12}, {6'd18}, {6'd0}};
      S_7:                    index = {{6'd7}, {6'd13}, {6'd19}, {6'd25}};
      S_8:                    index = {{6'd8}, {6'd14}, {6'd20}, {6'd26}};
      S_9:                    index = {{6'd15}, {6'd21}, {6'd27}, {6'd33}};
      S_10:                   index = {{6'd16}, {6'd22}, {6'd28}, {6'd34}};
      S_11:                   index = {{6'd23}, {6'd29}, {6'd35}, {6'd41}};
      S_12:                   index = {{6'd24}, {6'd30}, {6'd36}, {6'd42}};
      S_13:                   index = {{6'd0}, {6'd31}, {6'd37}, {6'd43}};
      S_14:                   index = {{6'd0}, {6'd32}, {6'd38}, {6'd44}};
      S_15:                   index = {{6'd0}, {6'd0}, {6'd39}, {6'd45}};
      S_16:                   index = {{6'd0}, {6'd0}, {6'd40}, {6'd46}};
      S_17:                   index = {{6'd0}, {6'd0}, {6'd0}, {6'd47}};
      S_18:                   index = {{6'd0}, {6'd0}, {6'd0}, {6'd48}};
   endcase
   end
endmodule

module PP_Datapath_Unit (Err_0, HTPV, Err_1, Err_2, Err_3, Err_4, PV);
// hard-wired for 8 x 6 array

   output [7: 0]   Err_0;
   output          HTPV;
   input  [7: 0]   Err_1, Err_2, Err_3, Err_4, PV;
   wire   [9: 0]   CPV, CPV_round, E_av;

// Weights for the average error; compatible with divide-by-16 (>> 4)

   parameter       w1 = 2, w2 = 8, w3 = 4, w4 = 2;
   parameter       Threshold = 128;

   assign  E_av = (w1 * Err_1 + w2 * Err_2 + w3 * Err_3 + w4 * Err_4 ) >> 4;
   assign        CPV = PV + E_av;
```

```
    assign          CPV_round = (CPV < Threshold) ? 0: 255;
    assign          HTPV = (CPV_round == 0) ? 0: 1;
    assign          Err_0  = CPV - CPV_round;

endmodule

module Pixel_Memory_Unit (
    HTPV_Row_1, HTPV_Row_2, HTPV_Row_3,
    HTPV_Row_4, HTPV_Row_5, HTPV_Row_6,
    PP_1_Err_1, PP_1_Err_2, PP_1_Err_3, PP_1_Err_4, PP_1_PV,
    PP_2_Err_1, PP_2_Err_2, PP_2_Err_3, PP_2_Err_4, PP_2_PV,
    PP_3_Err_1, PP_3_Err_2, PP_3_Err_3, PP_3_Err_4, PP_3_PV,
    PP_4_Err_1, PP_4_Err_2, PP_4_Err_3, PP_4_Err_4, PP_4_PV,
    PP_1_Err_0, PP_2_Err_0, PP_3_Err_0, PP_4_Err_0,
    PP_1_HTPV, PP_2_HTPV, PP_3_HTPV, PP_4_HTPV,

    pixel_1, pixel_2, pixel_3, pixel_4, pixel_5, pixel_6, pixel_7, pixel_8,
    pixel_9, pixel_10, pixel_11, pixel_12, pixel_13, pixel_14, pixel_15, pixel_16,
    pixel_17, pixel_18, pixel_19, pixel_20, pixel_21, pixel_22, pixel_23, pixel_24,
    pixel_25, pixel_26, pixel_27, pixel_28, pixel_29, pixel_30, pixel_31, pixel_32,
    pixel_33, pixel_34, pixel_35, pixel_36, pixel_37, pixel_38, pixel_39, pixel_40,
    pixel_41, pixel_42, pixel_43, pixel_44, pixel_45, pixel_46, pixel_47, pixel_48,
    index, Ld_image, Ld_values, Go, clk);

    output [1: 8]  HTPV_Row_1, HTPV_Row_2, HTPV_Row_3,
                   HTPV_Row_4, HTPV_Row_5, HTPV_Row_6;

    output [7: 0]
    PP_1_Err_1, PP_1_Err_2, PP_1_Err_3, PP_1_Err_4, PP_1_PV,
    PP_2_Err_1, PP_2_Err_2, PP_2_Err_3, PP_2_Err_4, PP_2_PV,
    PP_3_Err_1, PP_3_Err_2, PP_3_Err_3, PP_3_Err_4, PP_3_PV,
    PP_4_Err_1, PP_4_Err_2, PP_4_Err_3, PP_4_Err_4, PP_4_PV;

    input  [7: 0]  PP_1_Err_0, PP_2_Err_0, PP_3_Err_0, PP_4_Err_0;
    input          PP_1_HTPV, PP_2_HTPV, PP_3_HTPV, PP_4_HTPV;

    input  [7: 0]
    pixel_1, pixel_2, pixel_3, pixel_4, pixel_5, pixel_6, pixel_7, pixel_8,
    pixel_9, pixel_10, pixel_11, pixel_12, pixel_13, pixel_14, pixel_15,
    pixel_16, pixel_17, pixel_18, pixel_19, pixel_20, pixel_21, pixel_22,
    pixel_23, pixel_24, pixel_25, pixel_26, pixel_27, pixel_28, pixel_29,
    pixel_30, pixel_31, pixel_32, pixel_33, pixel_34, pixel_35, pixel_36,
    pixel_37, pixel_38, pixel_39, pixel_40, pixel_41, pixel_42, pixel_43,
    pixel_44, pixel_45, pixel_46, pixel_47, pixel_48;
    input  [23: 0] index;
    input          Ld_image, Ld_values, Go, clk;

    reg [7: 0] PV_Row_1 [1: 8];      reg[7: 0] PV_Row_2 [1: 8];
    reg [7: 0] PV_Row_3 [1: 8];      reg [7: 0] PV_Row_4 [1: 8];
    reg [7: 0] PV_Row_5 [1: 8];      reg [7: 0] PV_Row_6 [1: 8];
```

```
reg    HTPV_Row_1, HTPV_Row_2, HTPV_Row_3,
       HTPV_Row_4, HTPV_Row_5, HTPV_Row_6;

reg    [7: 0] Err_Row_0 [0: 9];    reg    [7: 0] Err_Row_1 [0: 9];
reg    [7: 0] Err_Row_2 [0: 9];    reg    [7: 0] Err_Row_3 [0: 9];
reg    [7: 0] Err_Row_4 [0: 9];    reg    [7: 0] Err_Row_5 [0: 9];
reg    [7: 0] Err_Row_6 [0: 9];

reg    PP_1_Err_4, PP_1_Err_3, PP_1_Err_2, PP_1_Err_1, PP_1_PV,
       PP_2_Err_4, PP_2_Err_3, PP_2_Err_2, PP_2_Err_1, PP_2_PV,
       PP_3_Err_4, PP_3_Err_3, PP_3_Err_2, PP_3_Err_1, PP_3_PV,
       PP_4_Err_4, PP_4_Err_3, PP_4_Err_2, PP_4_Err_1, PP_4_PV;

wire   [5: 0]    index_1 = index [23: 18],
                 index_2 = index [17: 12],
                 index_3 = index [11: 6],
                 index_4 = index [5: 0];

// Note: Incomplete event control expression.
always @ (index_1) begin
case (index_1)
1, 2, 3, 4, 5, 6, 7, 8: begin
PP_1_Err_1 = Err_Row_1[index_1-1]; PP_1_Err_2 = Err_Row_0[index_1-1];
PP_1_Err_3 = Err_Row_0[index_1]; PP_1_Err_4 = Err_Row_0[index_1+1];
PP_1_PV = PV_Row_1[index_1];
end

15, 16: begin
  PP_1_Err_1 = Err_Row_2[index_1-1-8];
  PP_1_Err_2 = Err_Row_1[index_1-1-8];
  PP_1_Err_3 = Err_Row_1[index_1-8];
  PP_1_Err_4 = Err_Row_1[index_1+1-8];
  PP_1_PV = PV_Row_2[index_1-8];
end

23, 24: begin
  PP_1_Err_1 = Err_Row_3[index_1-1-16];
  PP_1_Err_2 = Err_Row_2[index_1-1-16];
  PP_1_Err_3 = Err_Row_2[index_1-16];
  PP_1_Err_4 = Err_Row_2[index_1+1-16];
  PP_1_PV = PV_Row_3[index_1-16];
end

default: begin
  PP_1_Err_1 = 8'bx; PP_1_Err_2 = 8'bx;
  PP_1_Err_3 = 8'bx; PP_1_Err_4 = 8'bx; PP_1_PV = 8'bx;
end
endcase
end
```

```verilog
always @ (index_2) begin
 case (index_2)
  9, 10, 11, 12, 13, 14: begin
  PP_2_Err_1 = Err_Row_2[index_2-1-8];
  PP_2_Err_2 = Err_Row_1[index_2-1-8];
  PP_2_Err_3 = Err_Row_1[index_2-8];
  PP_2_Err_4 = Err_Row_1[index_2+1-8];
  PP_2_PV = PV_Row_2[index_2-8];
 end
  21, 22: begin
  PP_2_Err_1 = Err_Row_3[index_2-1-16];
  PP_2_Err_2 = Err_Row_2[index_2-1-16];
  PP_2_Err_3 = Err_Row_2[index_2-16];
  PP_2_Err_4 = Err_Row_2[index_2+1-16];
  PP_2_PV = PV_Row_3[index_2-16];
 end
  29, 30, 31, 32: begin
  PP_2_Err_1 = Err_Row_4[index_2-1-24];
  PP_2_Err_2 = Err_Row_3[index_2-1-24];
  PP_2_Err_3 = Err_Row_3[index_2-24];
  PP_2_Err_4 = Err_Row_3[index_2+1-24];
  PP_2_PV = PV_Row_4[index_2-24];
 end
 default: begin
  PP_2_Err_1 = 8'bx;    PP_2_Err_2 = 8'bx;
  PP_2_Err_3 = 8'bx;    PP_2_Err_4 = 8'bx;
  PP_2_PV = 8'bx;
 end
 endcase
end

always @ (index_3) begin
case (index_3)
17, 18, 19, 20:             begin
                            PP_3_Err_1 = Err_Row_3[index_3-1-16];
                            PP_3_Err_2 = Err_Row_2[index_3-1-16];
                            PP_3_Err_3 = Err_Row_2[index_3-16];
                            PP_3_Err_4 = Err_Row_2[index_3+1-16];
                            PP_3_PV = PV_Row_3[index_3-16];
                            end
27,  28:                    begin
                            PP_3_Err_1 = Err_Row_4[index_3-1-24];
                            PP_3_Err_2 = Err_Row_3[index_3-1-24];
                            PP_3_Err_3 = Err_Row_3[index_3-24];
                            PP_3_Err_4 = Err_Row_3[index_3+1-24];
                            PP_3_PV = PV_Row_4[index_3-24];
                            end
35, 36, 37, 38, 39, 40:     begin
```

```
                                        PP_3_Err_1 = Err_Row_5[index_3-1-32];
                                        PP_3_Err_2 = Err_Row_4[index_3-1-32];
                                        PP_3_Err_3 = Err_Row_4[index_3-32];
                                        PP_3_Err_4 = Err_Row_4[index_3+1-32];
                                        PP_3_PV = PV_Row_5[index_3-32];
                                        end
           default:                     begin
                                        PP_3_Err_1 = 8'bx; PP_3_Err_2 = 8'bx;
                                        PP_3_Err_3 = 8'bx;
                                        PP_3_Err_4 = 8'bx; PP_3_PV = 8'bx;
                                        end
         endcase
         end

         always @ (index_4) begin
         case (index_4)
           25, 26:                      begin
                                        PP_4_Err_1 = Err_Row_4[index_4-1-24];
                                        PP_4_Err_2 = Err_Row_3[index_4-1-24];
                                        PP_4_Err_3 = Err_Row_3[index_4-24];
                                        PP_4_Err_4 = Err_Row_3[index_4+1-24];
                                        PP_4_PV = PV_Row_4[index_4-24];
                                        end
           33, 34:                      begin
                                        PP_4_Err_1 = Err_Row_5[index_4-1-32];
                                        PP_4_Err_2 = Err_Row_4[index_4-1-32];
                                        PP_4_Err_3 = Err_Row_4[index_4-32];
                                        PP_4_Err_4 = Err_Row_4[index_4+1-32];
                                        PP_4_PV = PV_Row_5[index_4-32];
                                        end

           41, 42, 43, 44,
           45, 46,47, 48:               begin
                                        PP_4_Err_1 = Err_Row_6[index_4-1-40];
                                        PP_4_Err_2 = Err_Row_5[index_4-1-40];
                                        PP_4_Err_3 = Err_Row_5[index_4-40];
                                        PP_4_Err_4 = Err_Row_5[index_4+1-40];
                                        PP_4_PV = PV_Row_6[index_4-40];
                                        end
           default:                     begin
                                        PP_4_Err_1 = 8'bx;        PP_4_Err_2 = 8'bx;
                                        PP_4_Err_3 = 8'bx;        PP_4_Err_4 = 8'bx;
                                        PP_4_PV = 8'bx;
                                        end
         endcase
         end

         always @ (posedge clk )
         if (Ld_image) begin: Array_Initialization
         // Initialize error at left boarder
```

```
       Err_Row_1[0] <= 0;    Err_Row_2[0] <= 0;    Err_Row_3[0] <= 0;
       Err_Row_4[0] <= 0;    Err_Row_5[0] <= 0;    Err_Row_6[0] <= 0;
   // Initialize columns in the main array
       Err_Row_1[1] <= 0;  Err_Row_2[1] <= 0;  Err_Row_3[1] <= 0;
       Err_Row_4[1] <= 0;  Err_Row_5[1] <= 0;  Err_Row_6[1] <= 0;
       Err_Row_1[2] <= 0;  Err_Row_2[2] <= 0;  Err_Row_3[2] <= 0;
       Err_Row_4[2] <= 0;  Err_Row_5[2] <= 0;  Err_Row_6[2] <= 0;
       Err_Row_1[3] <= 0;  Err_Row_2[3] <= 0;  Err_Row_3[3] <= 0;
       Err_Row_4[3] <= 0;  Err_Row_5[3] <= 0;  Err_Row_6[3] <= 0;
       Err_Row_1[4] <= 0;  Err_Row_2[4] <= 0;  Err_Row_3[4] <= 0;
       Err_Row_4[4] <= 0;  Err_Row_5[4] <= 0;  Err_Row_6[4] <= 0;
       Err_Row_1[5] <= 0;  Err_Row_2[5] <= 0;  Err_Row_3[5] <= 0;
       Err_Row_4[5] <= 0;  Err_Row_5[5] <= 0;  Err_Row_6[5] <= 0;
       Err_Row_1[6] <= 0;  Err_Row_2[6] <= 0;  Err_Row_3[6] <= 0;
       Err_Row_4[6] <= 0;  Err_Row_5[6] <= 0;  Err_Row_6[6] <= 0;
       Err_Row_1[7] <= 0;  Err_Row_2[7] <= 0;  Err_Row_3[7] <= 0;
       Err_Row_4[7] <= 0;  Err_Row_5[7] <= 0;  Err_Row_6[7] <= 0;
       Err_Row_1[8] <= 0;  Err_Row_2[8] <= 0;  Err_Row_3[8] <= 0;
       Err_Row_4[8] <= 0;  Err_Row_5[8] <= 0;  Err_Row_6[8] <= 0;
   // Initialize right boarder
       Err_Row_1[9] <= 0; Err_Row_2[9] <= 0; Err_Row_3[9] <= 0;
       Err_Row_4[9] <= 0; Err_Row_5[9] <= 0; Err_Row_6[9] <= 0;
   // Initialize top boarder
       Err_Row_0[0]  <= 0; Err_Row_0[1]  <= 0; Err_Row_0[2]  <= 0;
       Err_Row_0[3]  <= 0; Err_Row_0[4]  <= 0; Err_Row_0[5]  <= 0;
       Err_Row_0[6]  <= 0; Err_Row_0[7]  <= 0; Err_Row_0[8]  <= 0;
       Err_Row_0[9] <= 0;
   // Initialize pixels in the main array
       PV_Row_1[1] <= pixel_1; PV_Row_1[2] <= pixel_2; PV_Row_1[3] <= pixel_3;
       PV_Row_1[4] <= pixel_4; PV_Row_1[5] <= pixel_5; PV_Row_1[6] <= pixel_6;
       PV_Row_1[7] <= pixel_7; PV_Row_1[8] <= pixel_8;
       PV_Row_2[1] <= pixel_9; PV_Row_2[2] <= pixel_10; PV_Row_2[3] <= pixel_11;
       PV_Row_2[4] <= pixel_12; PV_Row_2[5] <= pixel_13; PV_Row_2[6] <= pixel_14;
       PV_Row_2[7] <= pixel_15; PV_Row_2[8] <= pixel_16;
       PV_Row_3[1] <= pixel_17; PV_Row_3[2] <= pixel_18; PV_Row_3[3] <= pixel_19;
       PV_Row_3[4] <= pixel_20; PV_Row_3[5] <= pixel_21; PV_Row_3[6] <= pixel_22;
       PV_Row_3[7] <= pixel_23; PV_Row_3[8] <= pixel_24;
       PV_Row_4[1] <= pixel_25; PV_Row_4[2] <= pixel_26; PV_Row_4[3] <= pixel_27;
       PV_Row_4[4] <= pixel_28; PV_Row_4[5] <= pixel_29; PV_Row_4[6] <= pixel_30;
       PV_Row_4[7] <= pixel_31; PV_Row_4[8] <= pixel_32;
       PV_Row_5[1] <= pixel_33; PV_Row_5[2] <= pixel_34; PV_Row_5[3] <= pixel_35;
       PV_Row_5[4] <= pixel_36; PV_Row_5[5] <= pixel_37; PV_Row_5[6] <= pixel_38;
       PV_Row_5[7] <= pixel_39; PV_Row_5[8] <= pixel_40;
       PV_Row_6[1] <= pixel_41; PV_Row_6[2] <= pixel_42; PV_Row_6[3] <= pixel_43;
       PV_Row_6[4] <= pixel_44; PV_Row_6[5] <= pixel_45; PV_Row_6[6] <= pixel_46;
       PV_Row_6[7] <= pixel_47; PV_Row_6[8] <= pixel_48;
     end   // Array_Initialization
```

```
                    else if (Ld_values) begin: Image_Conversion
                    case (index_1)
                    1, 2, 3, 4, 5, 6, 7, 8:      begin Err_Row_1[index_1] <= PP_1_Err_0;
                                                 HTPV_Row_1[index_1] <= PP_1_HTPV; end

                    15, 16:                      begin Err_Row_2[index_1-8] <= PP_1_Err_0;
                                                 HTPV_Row_2[index_1-8] <= PP_1_HTPV; end

                    23, 24:                      begin Err_Row_3[index_1-16] <= PP_1_Err_0;
                                                 HTPV_Row_3[index_1-16] <= PP_1_HTPV;  end
                    endcase

                    case (index_2)
                    9, 10, 11, 12, 13, 14:       begin Err_Row_2[index_2 -8] <= PP_2_Err_0;
                                                 HTPV_Row_2[index_2 -8] <= PP_2_HTPV; end
                    21, 22:                      begin Err_Row_3[index_2 -16] <= PP_2_Err_0;
                                                 HTPV_Row_3[index_2 -16] <= PP_2_HTPV;  end
                    29, 30, 31, 32:              begin Err_Row_4[index_2 -24] <= PP_2_Err_0;
                                                 HTPV_Row_4[index_2 -24] <= PP_2_HTPV; end
                    endcase

                    case (index_3)
                    17, 18, 19, 20:              begin Err_Row_3[index_3 -16] <= PP_3_Err_0;
                                                 HTPV_Row_3[index_3 -16] <= PP_3_HTPV;  end
                    27, 28:                      begin Err_Row_4[index_3 -24] <= PP_3_Err_0;
                                                 HTPV_Row_4[index_3 -24] <= PP_3_HTPV; end
                    35, 36, 37, 38, 39, 40:      begin Err_Row_5[index_3 -32] <= PP_3_Err_0;
                                                 HTPV_Row_5[index_3 -32] <= PP_3_HTPV; end
                    endcase

                    case (index_4)
                    25, 26:                      begin Err_Row_4[index_4 - 24] <= PP_4_Err_0;
                                                 HTPV_Row_4[index_4 -24] <= PP_4_HTPV; end
                    33, 34:                      begin Err_Row_5[index_4 -32] <= PP_4_Err_0;
                                                 HTPV_Row_5[index_4 -32] <= PP_4_HTPV; end
                    41, 42, 43, 44, 45, 46, 47, 48:
                                                 begin Err_Row_6[index_4 -40] <= PP_4_Err_0;
                                                 HTPV_Row_6[index_4 -40] <= PP_4_HTPV; end
                    endcase
                    end                // Image_Conversion
                    endmodule
```

9.2.4   Halftone Pixel Image Converter: Design Tradeoffs

Key tradeoffs between the alternative designs of the halftone image converter are
summarized in Table 9-2.

TABLE 9-2   A comparison of alternative halftone pixel image converters.

Tradeoffs: 8 × 6 Pixel Halftone Image Converter			
Version	**FU Utilization**	**Memory Utilization**	**Execution Time**
Image_Converter_Baseline[1]	48	None (combinational)	$T_{Baseline}$
Image_Converter_0[2]	48	None (combinational)	$48 \times T_{FU}$
Image_Converter_1[3]	NA*	$6 + 2 \times 48 \times 8$ bytes	$48 \times T_{FU}$
Image_Converter_SR[4]	8	$6 + 2 \times 48 \times 8$ bytes	$48 \times T_{FU}$
Image_Converter_2[5]	4	$6 + 2 \times 48 \times 8$ bytes	$18 \times T_{FU} (12 \times T_{FU})^{**}$

[1] NLP-based Structure of FUs.
[2] NLP-based level-sensitive cyclic behavior.
[3] NLP-based single cycle synchronous.
[4] NLP-based multicycle synchronous.
[5] ASMD-based concurrent processors.
*Not synthesizable.
**Streaming images.

9.2.5   Architectures for Dataflow Graphs with Feedback

The dataflow graph of the pixel processor did not have feedback, so the baseline processor could be realized by feedback-free combinational logic. If the DFG for an algorithm has feedback, the machine will require memory and can be implemented only as a sequential machine.

Example 9.1

An NLP describing the so-called bubble-sort algorithm [6] is given below, in pseudocode. The algorithm sorts a set of N unsigned binary numbers and arranges them in ascending order.

```
begin
 for i=2 to N_key
 begin
  for j = N_key downto i do
   if A[j-1] > A[j] then
   begin
        temp= A[j-1];
        A[j-1]=A[j];
        A[j]=temp
   end
 end
end
```

The DFG of a functional unit for the machine is shown in Figure 9-16. We have included additional structure to represent the memory cells that are associated with the data that are manipulated by the algorithm. The FU compares two adjacent numbers stored in memory and determines whether to swap the contents of the storage

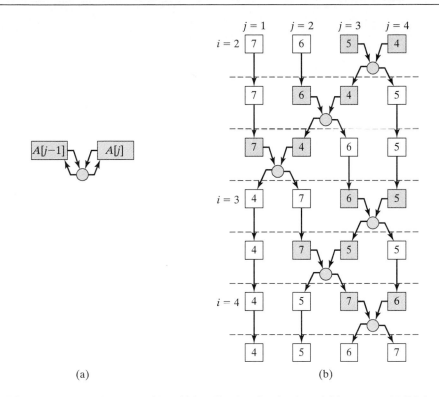

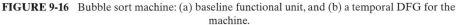

(a)   (b)

FIGURE 9-16   Bubble sort machine: (a) baseline functional unit, and (b) a temporal DFG for the machine.

registers that hold the numbers. The DFG has feedback, because the contents of a memory cell can be written back to the cell. The presence of feedback in the DFG implies the need for data storage. If the loops of the NLP program are unrolled, we get the temporal DFG shown in Figure 9-16(b). Each iteration of the nested loops of the algorithm must occur in a separate clock cycle in order for data to be fetched, transformed, and written back to memory by concurrent operations on registers. The shaded nodes and memory cells in Figure 9-16(b) indicate the datapaths that are exercised during a given time step as the machine executes the algorithm.

The structure of the DFG for the bubble-sort algorithm suggests that a baseline implementation of the machine consists of a single FU that executes repeatedly with different data, until the algorithm expires. The ASMD chart for a machine that implements the baseline architecture is shown in Figure 9-17. The machine has a bank of N registers holding the words of data, plus two counters, which index the inner and outer loops of the NLP.

The Verilog model of the bubble sort machine is listed below. The simulation results in Figure 9-18 and Figure 9-19 illustrate the execution of the algorithm.

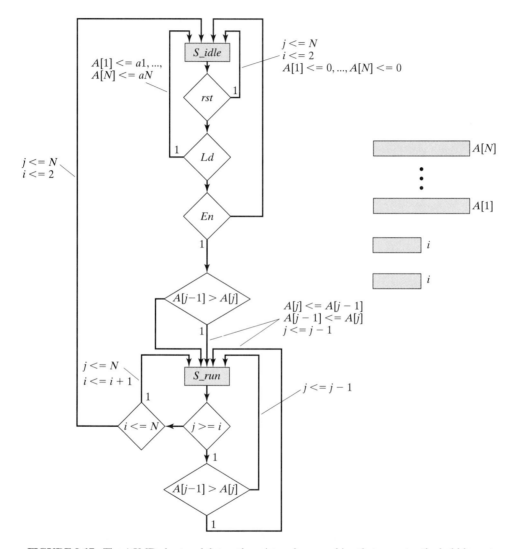

FIGURE 9-17 The ASMD chart and datapath registers for a machine that executes the bubble-sort algorithm.

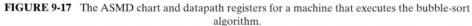

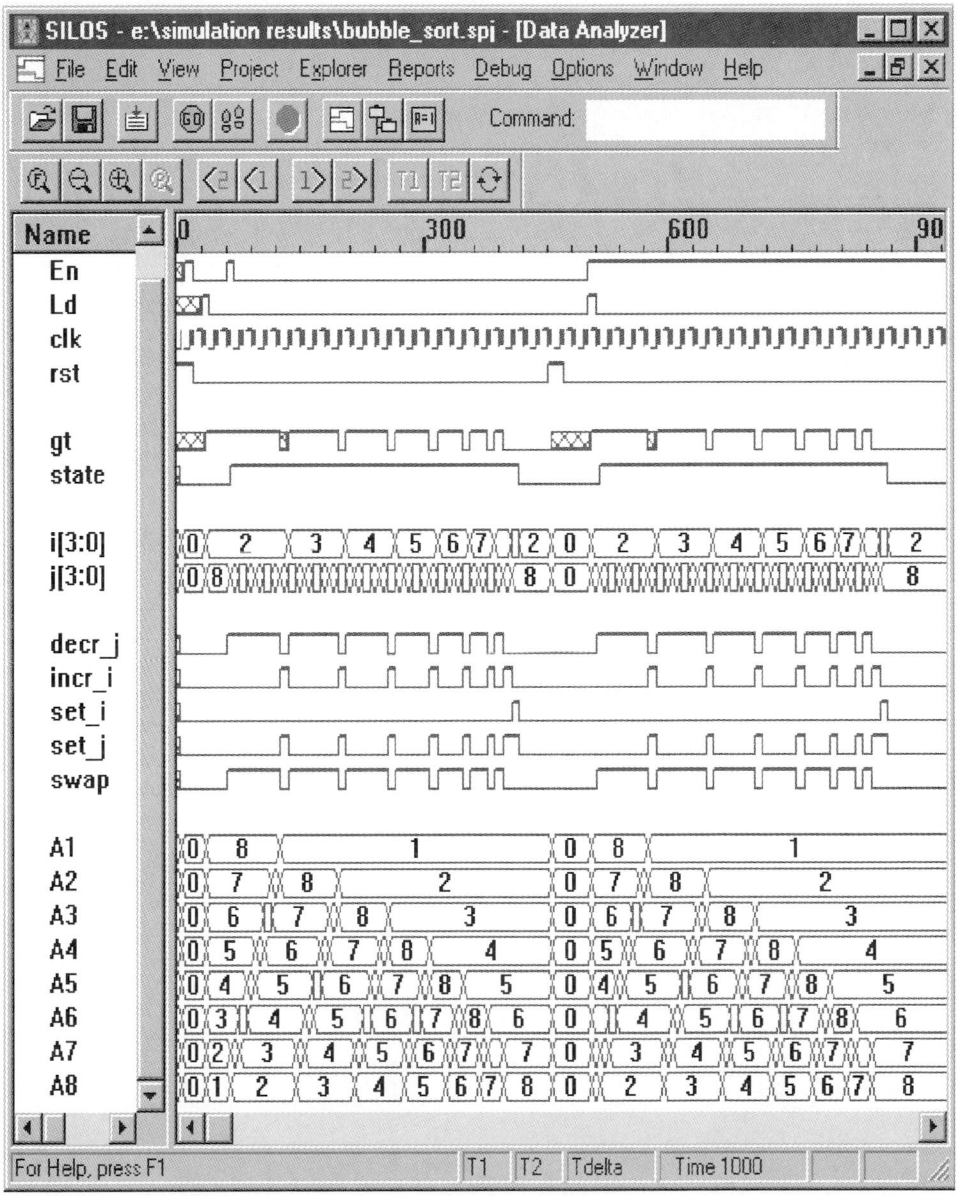

FIGURE 9-18 Simulation results for *Bubble_Sort*, a machine that executes a bubble-sort algorithm, with an initial sort of (8, 7, 6, 5, 4, 3, 2, 1).

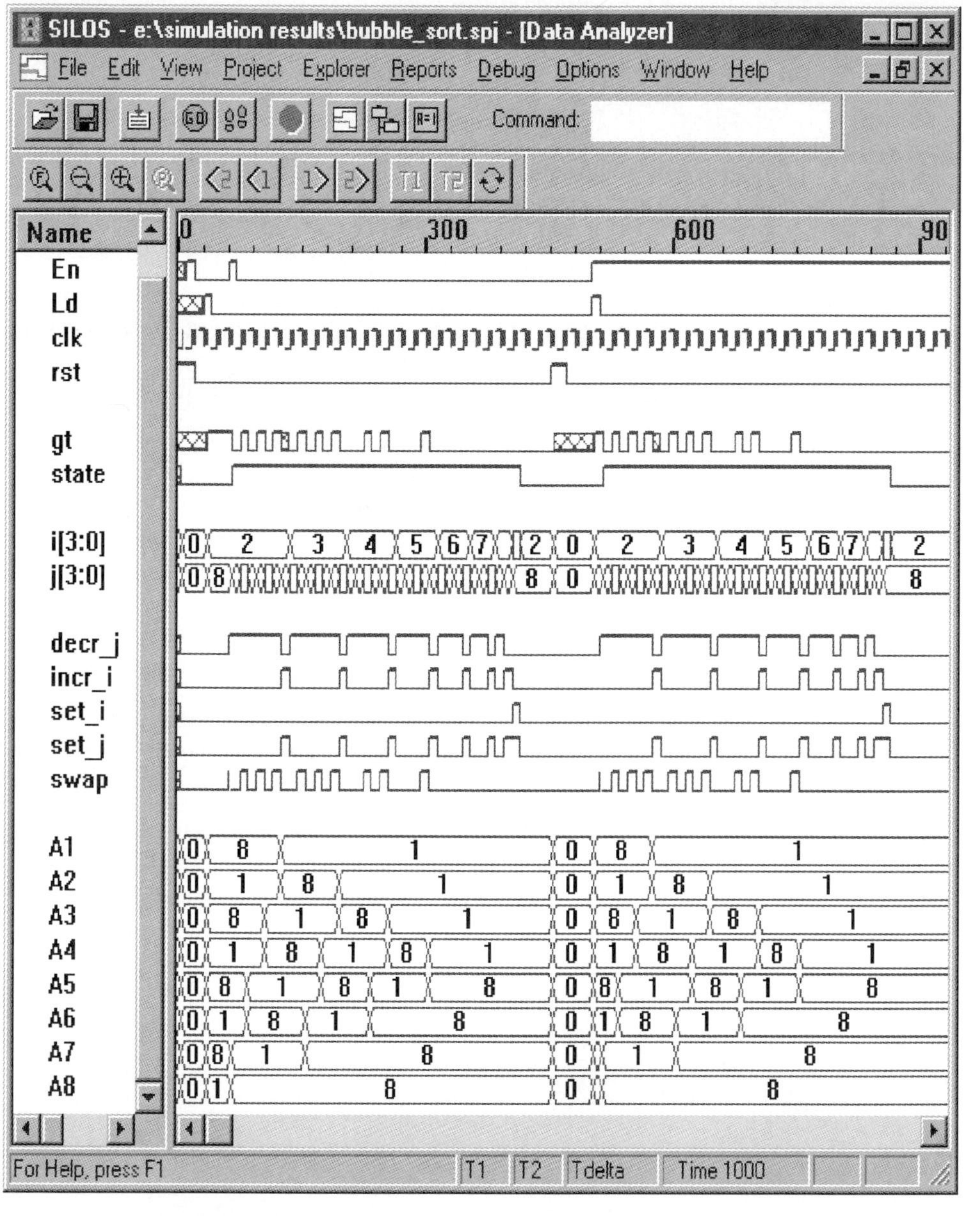

FIGURE 9-19   Simulation results for *Bubble_Sort*, for a machine that executes a bubble-sort algorithm, with an initial sort of (8, 1, 8, 1, 8, 1, 8, 1).

```
module Bubble_Sort (A1, A2, A3, A4, A5, A6, A7, A8,En, Ld, clk, rst);
   output [3:0]  A1, A2, A3, A4, A5, A6, A7, A8;
   input         En, Ld, clk, rst;
   parameter     N = 8;
   parameter     word_size = 4;
   parameter     a1 = 8;
   parameter     a2 = 1;
   parameter     a3 = 8;
   parameter     a4 = 1;
   parameter     a5  = 8;
   parameter     a6 = 1;
   parameter     a7 = 8;
   parameter     a8 = 1;

   reg           [word_size -1: 0]        A [1: N]; // Array of words
   wire [3:0]    A1 = A[1];
   wire [3:0]    A2 = A[2];
   wire [3:0]    A3 = A[3];
   wire [3:0]    A4 = A[4];
   wire [3:0]    A5 = A[5];
   wire [3:0]    A6 = A[6];
   wire [3:0]    A7 = A[7];
   wire [3:0]    A8 = A[8];
   parameter     S_idle = 0;
   parameter     S_run = 1;
   reg    [3: 0]  i, j;

   wire          gt = (A[j-1] > A[j]);        // compares words

   reg           swap, decr_j, incr_i, set_i, set_j;
   reg           state, next_state;

   always @ (posedge clk) if (rst) state <= S_idle;
      else state <= next_state;

   always @ (state or En or Ld or gt or i or j) begin
      swap = 0;
      decr_j = 0;
      incr_i = 0;
      set_j = 0;
      set_i = 0;
      case (state)
        S_idle:      if (Ld) begin next_state = S_idle; end
                     else if (En) begin next_state = S_run;
                     if (gt) begin swap = 1; decr_j = 1; end
                     else next_state = S_idle; end
```

```
        S_run:        if (j >= i) begin next_state = S_run; decr_j = 1;
                      if (gt) swap = 1; end
                      else if (i <= N) begin next_state = S_run; set_j = 1;
                        incr_i = 1; end
                      else begin next_state = S_idle; set_j = 1; set_i = 1; end

   endcase
 end

 always @ (posedge clk)                    // Datapath and status registers
   if (rst) begin i <= 0; j <= 0;
     A[1] <= 0; A[2] <= 0; A[3] <= 0; A[4] <= 0;
     A[5] <= 0; A[6] <= 0; A[7] <= 0; A[8] <= 0;
   end
   else  if (Ld) begin i <= 2; j <= N;
     A[1] <= a1; A[2] <= a2; A[3] <= a3; A[4] <= a4;
     A[5] <= a5; A[6] <= a6; A[7] <= a7; A[8] <= a8;
   end
   else begin /*#1  // for display only; remove for synthesis  */
     if (swap) begin A[j] <= A[j-1]; A[j-1] <= A[j]; end
     if (decr_j) j <= j-1;
     if (incr_i) i <= i+1;
     if (set_j) j <= N;
     if (set_i) i <= 2;
   end
 endmodule

module t_Bubble_Sort ();
   wire [3:0]      A1, A2, A3, A4, A5, A6, A7, A8;
   reg             En, Ld, clk, rst;

   Bubble_Sort M0 (A1, A2, A3, A4, A5, A6, A7, A8,En, Ld, clk, rst);

   initial #1000 $finish;
   initial begin clk = 0; forever #5 clk = ~clk; end
   initial fork
        rst = 1;
     #20 rst = 0;
     #30 Ld = 1;
     #40 Ld = 0;
     #60 En = 1;
     #70 En = 0;
     #450 rst = 1;
     #470 rst = 0;

     #500 Ld = 1;
     #510 Ld = 0;
   join
```

```
            initial fork
              #10 En = 1;
              #20 En = 0;
              #500 En = 1;
            join
          endmodule
```

End of Example 9.1

9.3   Digital Filters and Signal Processors

Digital signal processors (DSPs) are prominent in cellular phones, personal digital assistants, still-image cameras, video cameras, and video recorders, where they provide superior performance, at lower cost and lower power, as compared with analog circuits. In this section we will consider the use of Verilog to model functional units that encode, transmit, and transform digital representations of signals.

DSPs can be categorized according to the sampling frequencies that are required by the spectral content of the signals in an application. Shannon's sampling theorem states that a bandlimited signal[12] can be recovered from its time-domain samples if it is sampled at a frequency, f_s, that is greater than twice the highest frequency in its spectrum. A waveform recovered from samples that were obtained by sampling a signal below its Shannon frequency is called an alias of the signal, because the waveform cannot be distinguished from other signals sampled at the same rate. The sample period, in practice, determines the maximum time available for a processor to operate on a sample of data before the next sample arrives. DSPs, therefore, can be classified according to the portion of the frequency domain in which they are intended to operate, as shown in Table 9-3 [7]. The sampling frequency (1) determines the spectral domain over which a signal can be recovered without aliasing effects and (2) determines the time interval

TABLE 9-3   DSP applications and I/O sample rates.

Application	I/O Sampling Rate
Instrumentation	1 Hz
Control	>0.1 kHz
Voice	8 kHz
Audio	44.1 kHz
Video	1–14 MHz

---

[12]The spectrum of a bandlimited signal is 0 everywhere but within a finite range of frequencies.

available to perform operations on the data. Whether the processor is executing instructions stored in memory or implementing an algorithm in hardware, the sample period constrains the design. A second important constraint arises from the digital nature
of the signals that are being processed.

A digital signal is represented by a binary word with a finite word length in the
machine. Consequently, the information that is processed is subject to truncation,
roundoff, overflow, and underflow errors during processing. For a given dynamic range
of a signal, the word length of its digital format determines the resolution (precision) of
its values. Thus, performance, precision, and functionality characterize a DSP.

DSPs can be implemented in hardware or software or a combination of both. A
software-based approach executes a DSP algorithm on a general-purpose processor.
The focus of the design effort is on the software that programs the processor for the
tasks supporting an application. Various software tools are available to optimize the
program for the machine.[13] A second approach is to implement a DSP algorithm on a
special-purpose, hard-wired, high-performance, customized processor whose architecture has been designed specifically to accomplish a variety of signal-processing tasks
efficiently.[14] In this approach the task of optimizing the design is performed by the synthesize tools that create an architecture and synthesize the logic implementing the
processor.

A dedicated signal processor can be implemented as an ASIC chip to achieve the
most efficient design and the highest performance, but at high unit cost and reduced
flexibility. Field-programmable gate arrays (FPGAs) provide yet another approach—
they can be configured to implement any DSP algorithm, but their performance and
density may lag behind that of a special-purpose DSP or an ASIC chip. FPGAs, however, afford the benefits of flexibility, reduced NRE (nonrecurring engineering) costs,
rapid prototyping, early market entry and reduced risk.

Signal processors are characterized by high throughput and multiple concurrent
operations. DSPs are typically dataflow intensive and have relatively small control
units. DSPs have multiple arithmetic and logic unit (ALU)-like FUs, with high-speed
support for the operations of multiplication and addition, multiple address and data
busses supporting concurrent operations, and multiport registers and random-access
memories (RAMs). Dedicated DSPs are dominated by their datapaths; their control
units are much simpler than those of general-purpose processors.

DSPs operate synchronously on fixed-word-length samples of data that arrive at
regular intervals of time. The instruction sets of a DSP typically includes two fundamental arithmetic operations: multiplication and addition, commonly referred to as
multiply and accumulate (MAC). MAC functional units must be implemented efficiently and must give high performance.

A DSP unit is constrained by the physical technology in which it is implemented,
which fundamentally limits the speed at which its operations can execute and also determines the physical area required to implement devices in hardware. A DSP may be
constrained by the number of channels of data, the rate at which data are exchanged

---

[13]See Texas Instrument's Code Composer Studio.
[14]For example, the Texas Instruments TMS320C6000 processor family.

with the machine's environment, and the size of the input and output words. External channels with a high data rate are multiplexed within the DSP to reduce the internal data rates to levels compatible with the processor's performance.

The data that drive a DSP unit may originate as an analog signal, which is sampled to form a discrete-time signal (i.e., an indexed sequence of numbers). Then the discrete-time signal is converted to a binary fixed-word representation forming a digtal signal (an indexed sequence of numbers with a finite word length). The analog signal itself might have been corrupted by noise, leading to the need for filtering of the received signal. Such filters can be implemented within an ASIC or within an FPGA.

Digital filters transform digital representations of analog signals to remove noise and other unwanted signal components and to shape the spectral characteristics of the resulting signal. Digital filters operate on a finite-precision digital representation of a signal. Consequently, their design must consider finite word length effects that result from the representation of the signal samples, the weighting coefficients of the filter, and the arithmetic operations performed by the filter.

The operations of a DSP unit can be distributed spatially (i.e., over hardware units) or temporally (over a single processor), depending on whether the unit executes its operations in a single cycle of the clock or over multiple cycles. In the former case, in which the unit operates on the entire word of data, the hardware resources must complete the operation within the period of the clock. In the latter case, the machine operates on part of the data word in each clock cycle, so that the capacity and performance of the individual operational units can be relaxed. Distributing operations over the temporal axis allows the machine to operate with higher throughput, but at the expense of a latency between the arrival of data and the availability of the results. Latency can be tolerated in many applications, such as in digital communications.

Digital filters operate in the time-sequence domain, accepting a sequence of discrete, finite-length words to produce an output sequence. The sequence of inputs, $x[n]$ may be the output of an analog-to-digital converter, or the output of some other functional unit. Two common architectures for linear digital filters are represented by the block diagrams in Figure 9-20. A *finite-duration impulse response* (FIR) filter (Figure 9-20a) forms its outputs as a weighted sum of its inputs; an *infinite-duration impulse response* (IIR) filter (Figure 9-20b) forms it output from a weighted sum of its inputs and past values of its output [7, 8, 9, 10, 11]. Consequently, the block diagram symbol of an IIR filter is shown with feedback from the output to the input. Both types of filters have internal storage to hold samples of the inputs, but the IIR filter has additional memory for samples of the output.

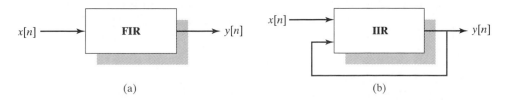

(a)   (b)

FIGURE 9-20 Digital filters: (a) finite impulse response filter, and (b) infinite impulse response filter.

Digital filters are usually designed to have two important characteristics: causality and linear phase. A filter is *causal* if its impulse response is 0 before the impulse is applied. FIR filters are widely used in practical applications because they can be designed to have a linear phase characteristic,[15] which ensures that the filter's output signal is a time-shifted (delayed), but undistorted copy of its input signal [12].[16] On the other hand, an IIR filter having a *linear phase* characteristic cannot be causal. Noncausal filters cannot be realized in hardware. Another distinction between the two types of filters is that FIR filters cannot accumulate roundoff error; IIR filters can accumulate roundoff error as the output is successively passed through the filter.

9.3.1   Finite-Duration Impulse Response (FIR) Filter

A FIR digital filter forms its output as a weighted sum of present and past samples of its input, as described by the *feed-forward difference equation* written below. FIR filters are called moving average filters because their output at any time index depends on a window containing only the most recent M samples of the input, as shown in Figure 9-21. Because its response depends on only a finite record of inputs, a FIR filter will have a finite-length, nonzero response to a discrete-time impulse (i.e., the response of an Mth order FIR filter to an impulse will be 0 after M clock cycles).

$$y_{\text{FIR}}[n] = \sum_{k=0}^{M} b_k x[n - k]$$

A FIR filter can be described by the z-domain functional block diagram[17] shown in Figure 9-22, where (in synchronous operation) each box labeled with z^{-1} denotes a register cell having a delay of one clock cycle. The diagram represents the datapaths and operations that must be performed by the filter. Each stage of the filter holds a delayed sample of the input, and the connection at the input and the connections at the outputs of the stages are referred to as *taps*, and the set of coefficients $\{b_k\}$ are called the *tap coefficients* of the filter. An Mth order FIR will have $M + 1$ taps. The samples of data flow through the shift register, and at each clock edge (i.e., time index), n, the samples are weighted (multiplied) by the tap coefficients and added together to form the output, $y_{\text{FIR}}[n]$. The adders and multipliers of the filter must be fast enough to form $y[n]$ before the next clock, and at each stage they must be sized to accommodate the width of their datapaths. In applications in which numerical accuracy is a driving consideration, lattice architectures may reduce the effects of finite word length, but at the expense of increased computational cost [12].

In most applications, the goal of the implementation is to do the filtering as fast as possible to achieve the highest sampling frequency [7]. The longest signal path through the combinational logic includes M stages of addition and one stage of multiplication.

---

[15]The phase characteristic is required to be linear in the passband of the filter's frequency response.
[16]Techniques exist for designing a FIR filter to have symmetric coefficients, which guarantees that the phase characteristic of the filter is linear.
[17]See Stearns and David [12] for other architectures and a discussion of their merits.

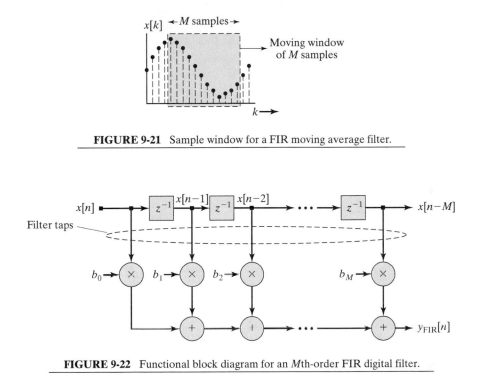

FIGURE 9-21 Sample window for a FIR moving average filter.

FIGURE 9-22 Functional block diagram for an Mth-order FIR digital filter.

The architecture of a FIR must specify a finite word length for each of the machine's arithmetic units, and manage the flow of data during operation. The architecture shown in Figure 9-23 consists of a shift register, multipliers, and adders implementing an Mth-order FIR. The datapaths must be wide enough to accommodate the output of the multipliers and adders. The samples are encoded as finite-length words, then shifted in parallel through a series of M registers. A cascaded chain of MACs form the machine.

9.3.2 Digital Filter Design Process

A process for designing an ASIC- or FPGA-based digital filter has the main steps shown in Figure 9-24. A design begins with development of performance specifications for cutoff frequency, transition band limits, in-band ripple, minimum stop band attenuation, etc. The filter is described by a C-language specification/algorithm, which must be converted into a Verilog RTL model that can be synthesized into hardware that implements the algorithm. The design flow is not ideal, because the algorithm's description in C must be translated into Verilog, creating the possibility for errors to occur. In C, the variables can be represented as floating-point numbers, but in Verilog the parameters

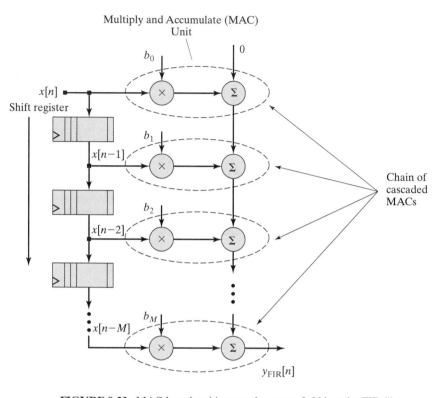

FIGURE 9-23 MAC-based architecture for a type-I, Mth-order FIR filter.

and other data values are expressed in a fixed-point, finite-word-length format. New tools are under development by EDA vendors to support design flows that create an executable and directly synthesizable specification, omitting intermediate translations to an HDL.[18] These tools hold promise, but they are not at a state of development that has gained widespread acceptance of their use.

Various architectures implement FIR and IIR filters [7, 9, 10, 12]. For a given architecture, tools such as MATLAB [7, 13, 14] can be used to determine the filter coefficients that implement a filter that satisfies the specifications of the design. Digital filters operate on finite-word-length representations of physical (analog) values. The finite word length of the data limits the resolution and the dynamic range that can be represented by the filter, leading to quantization errors. Similarly, the representations of the numerical coefficients of the filter have a finite word length, which contributes to additional quantization and truncation error. When data are represented by integers there is an error caused by truncation of the fractional part produced by an arithmetic operation. The arithmetic operations that are performed by the filter can lead to overflow and underflow errors, which must be detected by the machine.

---

[18]See www.synopsys.com

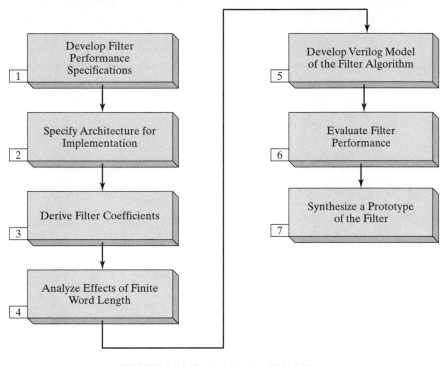

FIGURE 9-24 Design flow for digital filters.

Example 9.2

An eighth-order Gaussian, lowpass FIR filter is modeled by *FIR_Gaussian_Lowpass* on page 599. The design is fully synchronous, with active-high synchronous reset. The filter's tap coefficients are implemented as unsigned 8-bit words (for unsigned integer math), chosen to be even-symmetric to guarantee that the phase characteristic will be linear.

Various algorithms exist for designing lowpass filters to meet specifications on their passband cutoff frequency, stopband frequency, passband gain, stopband attenuation, and sampling rate [7, 15]. The coefficients in *FIR_Gaussian_Lowpass* were chosen to give the impulse response of the filter an approximately Gaussian shape. This choice simplifies the design because the coefficients are positive, and can be scaled to be represented by unsigned binary values.[19] Their magnitudes are determined by a Gaussian distribution over a range (0–9), with an arbitrarily chosen standard deviation of 2. The fractions obtained from the distribution were scaled in proportion to their size relative to the sum of the weights, and then multiplied by 255, the maximum value for an 8-bit word.

---

[19]Other schemes lead to signed fractions, which can be represented in a 2s complement Q-format (see Kehtarnavaz, note 4).

The impulse response of a FIR filter is a sample sequence whose values are the filter's tap coefficients. These can be seen in the waveform for *Data_out* in Figure 9-25. Because *Data_in* is switching at the falling edge of the clock and sampled at the rising edge, the values of *Data_out* are valid immediately before the rising edge of the clock.[20] Note that *Data_out* is formed as a Mealy output of the machine, and that the value of *Data_out* immediately after the first rising edge of clock reflects the value of *Data_in* and the first stored sample of *Data_in* (i.e., the output is not valid). Also note that the output has a finite duration (equal to the eight sample periods).

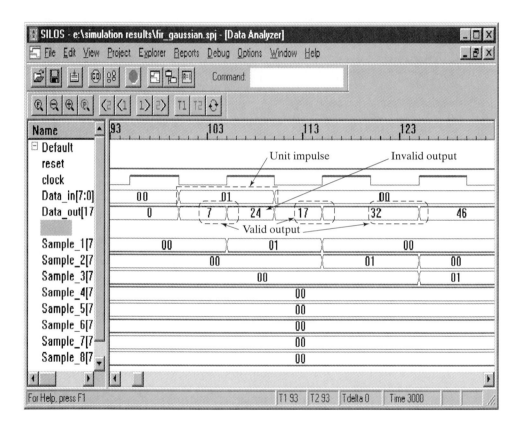

(a)

FIGURE 9-25 Impulse response of *FIR_Gaussian_Lowpass*, an eighth-order Gaussian Lowpass FIR filter: (a) initial nonzero samples of *Data_out*, showing invalid and valid data, and (b) final nonzero samples of *Data_out*.

---

[20]The displayed values reflect the scaling that was done in forming the tap coefficients.

(b)

FIGURE 9-25 Continued

```
module FIR_Gaussian_Lowpass (Data_out, Data_in, clock, reset);
    // Eighth-order, Gaussian Lowpass FIR
    parameter    order = 8;
    parameter    word_size_in = 8;
    parameter    word_size_out = 2*word_size_in + 2;

    parameter    b0 = 8'd7;              // Filter coefficients
    parameter    b1 = 8'd17;
    parameter    b2 = 8'd32;
    parameter    b3 = 8'd46;
    parameter    b4 = 8'd52;
    parameter    b5 = 8'd46;
    parameter    b6 = 8'd32;
    parameter    b7 = 8'd17;
    parameter    b8 = 8'd7;

    output       [word_size_out -1: 0]      Data_out;
    input        [word_size_in-1: 0]        Data_in;
    input                                   clock, reset;
```

```
reg              [word_size_in-1: 0]        Samples [1: order];

integer          k;
assign           Data_out =   b0 * Data_in
                 + b1 * Samples[1]
                 + b2 * Samples[2]
                 + b3 * Samples[3]
                 + b4 * Samples[4]
                 + b5 * Samples[5]
                 + b6 * Samples[6]
                 + b7 * Samples[7]
                 + b8 * Samples[8];

always @ (posedge clock)
  if (reset == 1) begin for (k = 1; k <= order; k = k+1) Samples [k] <= 0; end
  else begin
    Samples [1] <= Data_in;
    for (k = 2; k <= order; k = k+1) Samples [k] <= Samples[k-1];
  end
endmodule
```

End of Example 9.2

9.3.3   Infinite-Duration Impulse Response (IIR) Filter

IIR filters are the most general class of linear digital filters. Their output at a given time step depends on their inputs and on previously computed outputs (i.e., they have memory) [10]. IIR filters are recursive, and FIR filters are nonrecursive.[21] The output of a IIR filter is formed in the data-sequence domain as a weighted sum according to the Nth-order difference equation shown below:

$$y_{\text{IIR}}[n] = \sum_{k=1}^{N} a_k y[n - k] + \sum_{k=0}^{M} b_k x[n - k]$$

The filter is recursive because the difference equation has feedback. Consequently, the filter's response to an impulse may have infinite duration (i.e., it does not become 0 in a finite time).

An IIR filter is modeled in the z domain by its z-domain system function, or transfer function, which is a ratio of polynomials formed as

---

[21]Nonrecursive filters are stable (i.e., their response does not become unbounded); recursive filters may be unstable, depending on the filter's coefficients.

$$H_{\text{IIR}}(z) = \sum_{k=0}^{M} b_k z^{-k} / \sum_{k=1}^{N} a_k z^{-k}$$

The z-domain transforms of the input and output time sequences are related by

$$Y(z) = H_{\text{IIR}}(z)X(z).$$

The tap coefficients of the IIR filter form the sets of the filter's tap coefficients, $\{a_j\}$ and $\{b_k\}$, commonly referred to as the *feedback* and *feedforward* coefficients, respectively. The parameter N is the order of the filter; it specifies the number of prior samples of the output that must be saved to form the current output; it also determines the latency of the output. The value of the parameter M specifies how many prior samples of the input will be used to form the output. The roots of the polynomials of $H_{\text{IIR}}(z)$ determine the location of the filter's poles and zeros in the z domain and shape both the data-sequence of the filter's response to its input, and the frequency domain function that specifies how the filter responds to a periodic input [10, 15].

Various architectures implement an IIR filter, and exhibit different requirements for physical resources, and different sensitivities to numerical errors caused by finite word length for the data and the parameters. The structure shown in Figure 9-26 is known as a Type-1 IIR, and consists of separate feedforward and feedback blocks

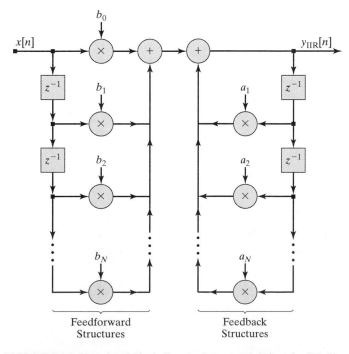

FIGURE 9-26   Functional block diagram for a type-I, Nth-order IIR filter.

implemented as a pair of shift registers—one to hold samples of the input, $x[n]$, and another to hold samples of the output, $y[n]$.

Example 9.3

The Verilog model *IIR_Filter_8* can be used to implement eighth-order IIR filters, depending on the selection of tap coefficients.

```
module IIR_Filter_8 (Data_out, Data_in, clock, reset);
   // Eighth-order, Generic IIR Filter
   parameter    order = 8;
   parameter    word_size_in = 8;
   parameter    word_size_out = 2*word_size_in + 2;

   parameter    b0 = 8'd7;        // Feedforward filter coefficients
   parameter    b1 = 0;
   parameter    b2 = 0;
   parameter    b3 = 0;
   parameter    b4 = 0;
   parameter    b5 = 0;
   parameter    b6 = 0;
   parameter    b7 = 0;
   parameter    b8 = 0;

   parameter    a1 = 8'd46;       // Feedback filter coefficients
   parameter    a2 = 8'd32;
   parameter    a3 = 8'd17;
   parameter    a4 = 8'd0;
   parameter    a5 = 8'd17;
   parameter    a6 = 8'd32;
   parameter    a7 = 8'd46;
   parameter    a8 = 8'd52;

   output    [word_size_out -1: 0]    Data_out;
   input     [word_size_in-1: 0]      Data_in;
   input                              clock, reset;


   reg       [word_size_in-1: 0]      Samples_in [1: order];
   reg       [word_size_in-1: 0]      Samples_out [1: order];
   wire      [word_size_out -1: 0]    Data_feedforward;
   wire      [word_size_out -1: 0]    Data_feedback;

   integer                            k;

   assign Data_feedforward =          b0 * Data_in
                                    + b1 * Samples_in[1]
                                    + b2 * Samples_in[2]
                                    + b3 * Samples_in[3]
```

```
                                        + b4 * Samples_in[4]
                                        + b5 * Samples_in[5]
                                        + b6 * Samples_in[6]
                                        + b7 * Samples_in[7]
                                        + b8 * Samples_in[8];

        assign Data_feedback =            a1 * Samples_out [1]
                                        + a2 * Samples_out [2]
                                        + a3 * Samples_out [3]
                                        + a4 * Samples_out [4]
                                        + a5 * Samples_out [5]
                                        + a6 * Samples_out [6]
                                        + a7 * Samples_out [7]
                                        + a8 * Samples_out [8];

        assign Data_out = Data_feedforward + Data_feedback;

        always @ (posedge clock)
          if (reset == 1) for (k = 1; k <= order; k = k+1) begin
            Samples_in [k] <= 0;
            Samples_out [k] <= 0;
          end
          else begin
            Samples_in [1] <= Data_in;
            Samples_out [1] <= Data_out;
            for (k = 2; k <= order; k = k+1) begin
              Samples_in [k] <= Samples_in [k-1];
              Samples_out [k] <= Samples_out [k-1];
            end
          end
      endmodule
```

End of Example 9.3

Two alternative architectures for an Nth order IIR are shown in Figure 9-27. They are known as Direct Form II (DF-II), and Transposed Direct Form II (TDF-II) [10].

9.4   Building Blocks for Signal Processors

In this section we will consider models of basic operations of integration, differentiation, decimation, and interpolation, which are common to many digital processors.[22]

---

[22]These examples were motivated by the models presented in Chris Hagan's master's thesis [11].

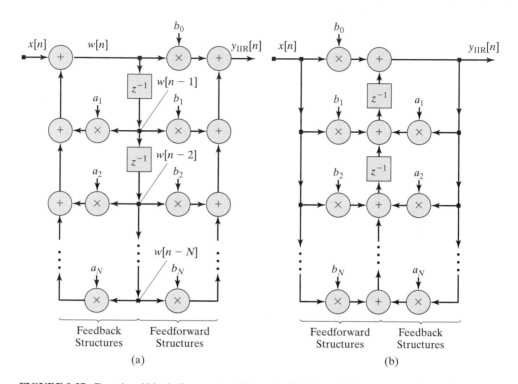

FIGURE 9-27 Functional block diagrams for Nth-order IIR filters: (a) Direct Form II (DF-II), and (b) Transposed Direct Form II (TDF-II).

9.4.1 Integrators (Accumulators)

Digital integrators are used in a popular type of analog-to-digital converter, called a sigma-delta modulator [7]. Digital integrators accumulate a running sum of sample values. Two implementations are common: parallel and sequential.

Example 9.4

The model *Integrator_Par* below describes an integrator for a parallel datapath. At each clock cycle the machine adds *data_in* to the content of the register *data_out*. The signal *hold* pauses the accumulation of samples until it is de-asserted.

```
module Integrator_Par (data_out, data_in, hold, clock, reset);
    parameter    word_length = 8;
    output       [word_length-1: 0]      data_out;
    input        [word_length-1: 0]      data_in;
    input                                hold, clock, reset;
    reg                                  data_out;
```

```
          always @ (posedge clock) begin
            if (reset) data_out <= 0;
            else if (hold) data_out <= data_out;
            else data_out <= data_out + data_in;
          end
        endmodule
```

End of Example 9.4

Example 9.5

The architecture of a byte-sequential integrator is shown in Figure 9-28, and a Verilog model of the machine, *Integrator_Seq*, is given below. It is common for a processor to receive data via a narrower datapath than the datapath within the processor. In this example, the unit is to accumulate 32-bit words, but receives data sequentially, in 8-bit bytes. The signal *hold* pauses the accumulation of samples until it is de-asserted. This architecture performs byte-wide addition, with the current data sample being added to the leftmost byte of the shift register *Shft_Reg*, to form *sum*. At the next clock edge the content of the shift register is shifted toward its MSByte,[23] and the previously formed

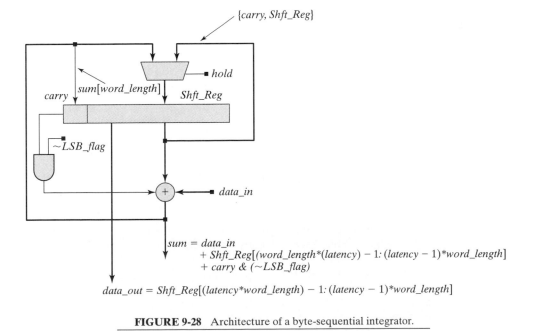

$$sum = data\_in$$
$$+ Shft\_Reg[(word\_length*(latency) - 1 : (latency - 1)*word\_length]$$
$$+ carry \& (\sim LSB\_flag)$$

$$data\_out = Shft\_Reg[(latency*word\_length) - 1 : (latency - 1)*word\_length]$$

FIGURE 9-28 Architecture of a byte-sequential integrator.

---

[23]MSByte and LSByte will denote the most significant and least significant byte, respectively, of a word.

t	*Shft_Reg*			
	Byte_1	*Byte_2*	*Byte_3*	*Byte_4*
	Byte_1 + *Byte_5*	*Byte_2* + *Byte_6*	*Byte_3* + *Byte_7*	*Byte_4* + *Byte_8*
	Byte_1 + *Byte_5* + *Byte_9*	*Byte_2* + *Byte_6* + *Byte_10*	*Byte_3* + *Byte_7* + *Byte_11*	*Byte_4* + *Byte_8* + *Byte_12*
	Byte_1 + *Byte_5* + *Byte_9* + *ByteE_13*	*Byte_2* + *Byte_6* + *Byte_10* + *Byte_14*	*Byte_3* + *Byte_7* + *Byte_11* + *Byte_15*	*Byte_4* + *Byte_8* + *Byte_12* + *Byte_16*

FIGURE 9-29   Accumulation of bytes in *Shft_Reg* over 16 cycles of operation.[24]

sum is loaded into the register's LSByte. These two actions occur concurrently, and the MSByte that the shift register held before the clock is pushed out of the register. The accumulation of bytes in *Shft_reg* is illustrated in Figure 9-25 for a scheme in which four successive bytes compose a word. The input signal *LSB_flag* controls the addition of a carry so that corresponding bytes are added correctly from word to word.

Figure 9-29 demonstrates how successive bytes of *data_in* are aligned within 32-bit words and accumulated in *Shft_Reg*. The simulation results in Figure 9-30 are annotated to show how samples of *data_in* are loaded into *Shft_Reg*, how the leftmost byte of *Shft_Reg* is added to *data_in* to form *sum*, and how *sum* is loaded into the rightmost byte of *Shft_Reg*.

```
module Integrator_Seq (data_out, data_in, hold, LSB_flag, clock, reset);
    parameter      word_length = 8;
    parameter      latency = 4;
    output         [word_length -1: 0]          data_out;
    input          [word_length -1: 0]          data_in;
    input                                        hold, LSB_flag, clock, reset;
    reg            [(word_length * latency) -1: 0]   Shft_Reg;
    reg                                          carry;
    wire           [word_length: 0]             sum;

    always @ (posedge clock) begin
      if (reset) begin Shft_Reg <= 0; carry <= 0; end
      else if (hold) begin
        Shft_Reg <= Shft_Reg;
        carry <= carry;
      end
      else begin
```

---

[24]For simplicity, the carries between bytes are not shown.

```
                    Shft_Reg <= {Shft_Reg[word_length*(latency -1) -1: 0], sum[word_length-1: 0]};
                end
            end

        assign sum = data_in + Shft_Reg[(latency * word_length) -1:
            (latency -1)*word_length] + (carry & (~LSB_flag));

        assign data_out = Shft_Reg[(latency * word_length) -1:
            (latency -1)*word_length];
        endmodule
```

End of Example 9.5

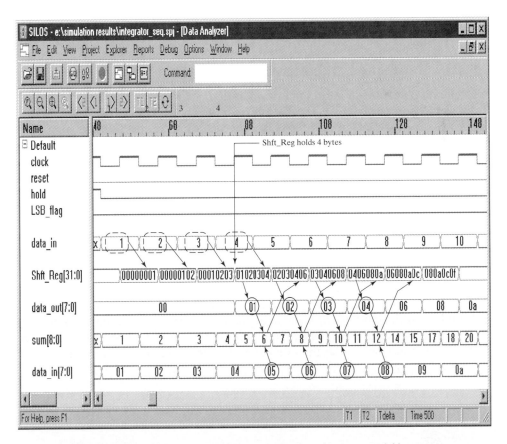

FIGURE 9-30 Simulation results for *Integrator_Seq*, a byte-sequential integrator.

9.4.2 Differentiators

A differentiator provides a measure of the sample-to-sample change in a signal. A bytewide serial differentiator is given below. The backward difference is implemented with a buffer and a subtractor.

```
module differentiator (data_out, data_in, hold, clock, reset);
    parameter word_size = 8;
    output  [word_size -1: 0]  data_out;
    input   [word_size -1: 0]  data_in;
    input                      hold;
    input                      clock, reset;

    reg     [word_size -1: 0]  buffer;
    wire    [word_size -1: 0]  data_out = data_in - buffer;

    always @ (posedge clock) begin
      if (reset) buffer <= 0;
      else if (hold)  buffer <= buffer;
      else buffer <= data_in;
    end
endmodule
```

9.4.3 Decimation and Interpolation Filters

Decimation and interpolation filters are used to achieve sample rate conversion in digital signal processors [15]. Decimation filters decrease the sample rate; interpolation filters increase the sample rate. Such conversions are important, because Shannon's Sampling Theorem [10] states that a bandlimited signal that is sampled at a rate greater than twice its upper spectral limit[25] can be recovered from its samples. Interpolation filters enable a signal to be oversampled, thereby reducing the effects of aliasing. If a signal is not sampled properly, it cannot be recovered with fidelity. Decimation is used to reduce the bandwidth of a signal that has been oversampled. Decimation achieves sample rate reduction.

Example 9.6

The Verilog model *decimator_1* describes the behavior of a parallel-in-parallel-out decimator, which samples its input at a rate determined by *clock* unless *hold* is asserted [11]. Note that samples of *data_in* in Figure 9-31 are dropped because *clock* is running at a rate that is slower than the rate at which *data_in* had been sampled.

```
module decimator_1 (data_out, data_in, hold, clock, reset);
    parameter word_length = 8;
    output [word_length-1: 0]     data_out;
```

---

[25]The upper spectral limit of a bandlimited signal determines the bandwidth of the signal.

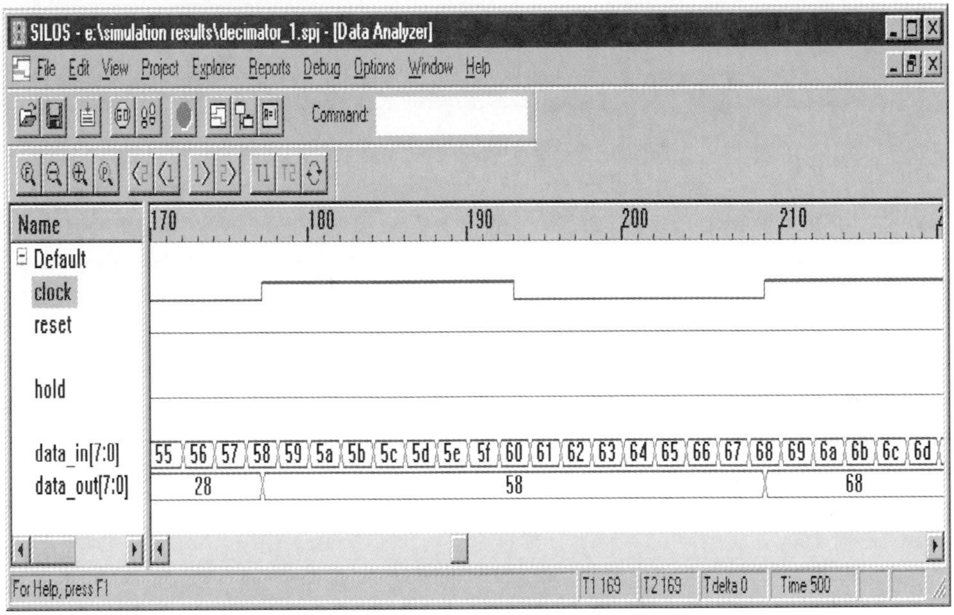

FIGURE 9-31 Simulation results for *decimator_1*.

```
input    [word_length-1: 0]    data_in;
input                          hold;      // Active high
input                          clock;     // Positive edge
input                          reset;     // Active high
reg                            data_out;
always @ (posedge clock)
  if (reset) data_out <= 0;
  else if (hold) data_out <= data_out;
  else data_out <= data_in;
endmodule
```

End of Example 9.6

Example 9.7

The Verilog model *decimator_2* samples a parallel input and produces a parallel output, but includes an option to form a serial output by shifting the output word through the LSB while *hold* is asserted. This action is apparent in the waveforms shown in Figure 9-32.

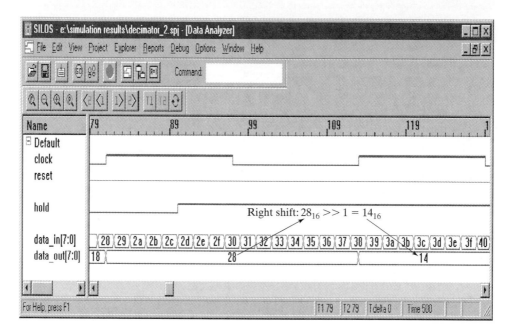

FIGURE 9-32 Simulation results for *decimator_2*, showing that with *hold* asserted the register holding *data_out* is shifted to the right at successive clock edges.

```
module decimator_2 (data_out, data_in, hold, clock, reset);
    parameter                    word_length = 8;
    output [word_length-1: 0]    data_out;
    input   [word_length-1: 0]   data_in;
    input                        hold;      // Active high
    input                        clock;     // Positive edge
    input                        reset;     // Active high
    reg                          data_out;
    always @ (posedge clock)
      if (reset) data_out <= 0;
      else if (hold) data_out <= data_out >> 1;
      else data_out <= data_in;
endmodule
```

End of Example 9.7

Example 9.8

The decimator shown in Figure 9-33 is designed to work in conjunction with a sequential integrator [11]. The decimator's architecture consists of three registers, *Shft_Reg*, *Int_Reg*, and *Decim_Reg*. All three are sized to hold multiple bytes (samples), as determined by a parameter *latency*. Samples from *data_in* are loaded sequentially into the MSByte of *Shft_Reg*, and shifted toward the LSByte on subsequent clocks. When *Shft_Reg* is full, two register transfers occur concurrently: (1) the contents of *Shft_Reg* are loaded into an intermediate holding register, *Int_Reg*, and (2) the LSByte of a new word is loaded into the MSByte of *Shft_Reg* and transfers load *Shft_Reg* until it is full.

The Verilog model, *decimator_3*, includes two edge-sensitive cyclic behaviors—one to describe the byte-buffering activity and the other to describe the functionality of a decimator. The register transfers of *decimator_3* are shown in Figure 9-33. If *load* is asserted (see Figure 9-33a) three register operations occur concurrently: the current sample of *data_in* is loaded into the leftmost bits of *Shft_Reg*, and the contents of *Shft_Reg* are loaded into *Int_Reg*. The decimation register, *Decim_Reg*, holds its contents

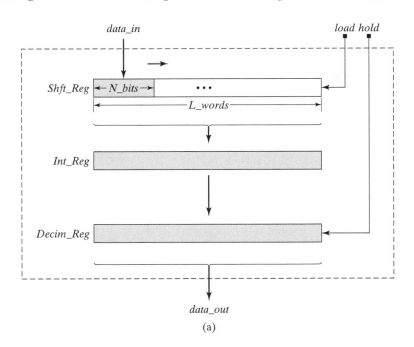

(a)

FIGURE 9-33 Sequential decimator: (a) overall architecture, (b) concurrent register transfers of *data_in* to *Shft_Reg* and from *Shft_Reg* to *Int_Reg* with *load* asserted, (c) shifting contents and loading *data_in* into *Shft_Reg* with *load* de-asserted, and (d) loading contents from *Int_Reg* to *Decim_Reg*, with *hold* de-asserted.

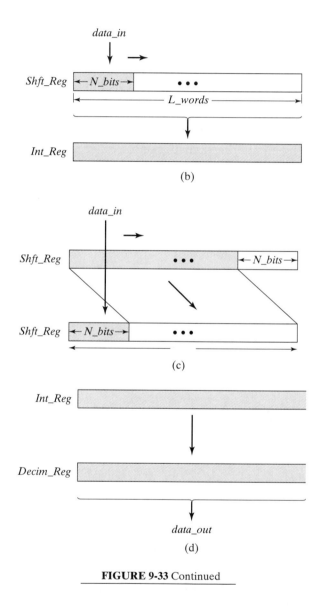

(b)

(c)

(d)

FIGURE 9-33 Continued

while *hold* is asserted; otherwise, it gets the contents of the integration register. The decimation action of the machine is a consequence of the parameter *latency* (i.e., the difference in the rate at which bytes are sequenced at *data_in*) and the rate at which words are formed at *data_out*. Register *Decim_Reg* can be connected directly to the input of a sequential integrator, such as *Integrator_1*.

```verilog
module decimator_3 (data_out, data_in, hold, load, clock, reset);
  parameter                                       word_length = 8;
  parameter                                       latency = 4;

  output      [(word_length*latency) -1: 0]       data_out;
  input       [word_length-1: 0]                  data_in;
  input                                           hold;
  input                                           load;
  input                                           clock;
  input                                           reset;

  reg    [(word_length*latency) -1: 0] Shft_Reg;   // Shift reg
  reg    [(word_length*latency) -1: 0] Int_Reg;    // Intermediate reg
  reg    [(word_length*latency) -1: 0] Decim_Reg;  // Decimation reg

  always @ (posedge clock)
    if (reset) begin
      Shft_Reg <= 0;
      Int_Reg <= 0;
    end
    else if (load) begin
      Shft_Reg[(word_length * latency) -1: (word_length*(latency-1))] <= data_in;
      Int_Reg <= Shft_Reg;
    end
    else begin
      Shft_Reg <= {data_in, Shft_Reg[(word_length*latency)-1: word_length]};
      Int_Reg <= Int_Reg;
    end

  always @ (posedge clock)
    if (reset) Decim_Reg <= 0;
    else if (hold) Decim_Reg <= Decim_Reg;
    else Decim_Reg <= Int_Reg;

  assign data_out = Decim_Reg;
endmodule
```

End of Example 9.8

9.5   Pipelined Architectures

The shortest cycle time of the clock of a synchronous sequential machine is a measure of its performance, and it is bounded by the propagation delay through the combinational logic of the machine. The throughput of a synchronous machine is the rate at which data is supplied to and produced by the machine [3]. Throughput is ultimately limited by the path with the largest propagation delay between (1) a primary input and a register, (2) a path between a pair of registers, (3) a path from a register to a primary output, or (4) a path from a primary input to a primary output. In each case, combinational logic limits the performance of the machine.

Synthesis engines transform a set of two-level, Boolean functions for combinational logic into a set of multilevel Boolean functions with shared logic. The circuit that results is free of redundant logic and exploits don't-care conditions to achieve a minimal description whose input/output logic is equivalent to the original set of two-level equations. The logic that is produced by this process is minimal because its output functions share common internal Boolean subexpressions as much as possible, but it might not be as fast as an equivalent realization that has fewer levels of logic. In general, collapsing levels of logic will produce a faster circuit. This is not always possible, because wide input gates are not practical.

As an alternative approach to gaining performance, pipeline registers can be inserted into the combinational logic datapaths at strategic locations to partition the logic into groups with shorter paths [4, 16]. The placement of the registers is determined by the feedforward cutsets (discussed below) of the DFG of the datapath, to ensure that data remains coherent. Pipelining reduces the number of levels in the blocks of combinational logic, shortens the datapaths between storage elements, and increases the throughput of the circuit, by allowing the clock to run faster.

Pipelining becomes increasingly important in high-speed, wide-word data transmission and processing. For example, a block of combinational logic in Figure 9-34 is partitioned into two blocks and separated by a pipeline register to form an alternative circuit. Suppose the longest path through the original multilevel combinational logic has time-length $T_{\max}$, and a operating frequency of $f_{\text{multilevel}} = 1/T_{\max}$. If the partition creates two blocks of (multilevel) logic, each having a maximum time length of $1/2\, T_{\max}$, the pipelined circuit can operate at a frequency $f_{\text{pipeline}} = 2/T_{\max} = 2 f_{\text{multilevel}}$.[26]

A word of caution: *the partition of a datapath must maintain coherency of the data— a datapath traced from any primary input to any primary output must pass through the same number of pipeline registers.* In general, a cutset of a connected graph is a set of branches that, if removed from the graph, isolates a node of the graph. For our purposes, a *pipeline cutset*, or *feedforward cutset*, is a minimum set of edges that, if removed from the graph, partitions it into two connected subgraphs such that there is no path between an input node and an output node. Cutsets are used to determine alternative placements of pipeline registers. The simple DFG in Figure 9-35 illustrates two cutsets, one of which is a pipeline cutset. The edges through which the dashed arc passes specify a locus for pipeline registers. The feedforward cutset ensures that every path between the input

---

[26]For simplicity, we are neglecting clock skew and the flip-flop's timing constraints.

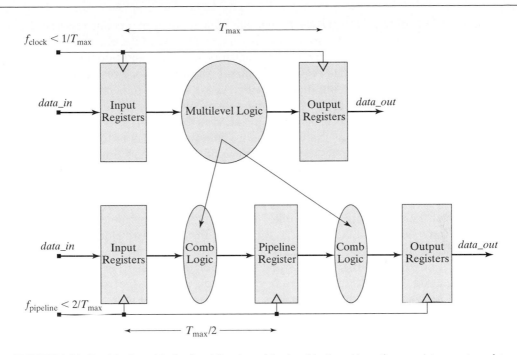

$f_{clock} < 1/T_{max}$

T_{max}

$f_{pipeline} < 2/T_{max}$

$T_{max}/2$

FIGURE 9-34 Partitioning a block of multilevel combinational logic and inserting a register creates a data pipeline.

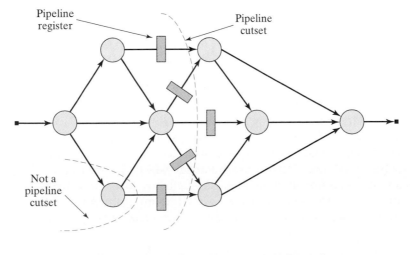

FIGURE 9-35 Cutset-based placement of pipeline registers.

node and the output node passes through the same number of pipeline registers. Removing any of the pipeline registers will destroy the coherency of the data.[27]

Pipelining has a cost. The pipeline registers introduce additional area in the physical layout of an ASIC, and require additional routing of clock resources.[28] This could be an issue for an ASIC, but high-end FPGAs are register-rich, and they readily support pipelined architectures. Partitioning a circuit to create a pipeline must be done carefully, so that balance is achieved in the distribution of path lengths among the groups that are created by the partition. In general, the delay of the slowest combinational logic stage determines the performance of the pipelined circuit and the speed at which the circuit's common clock can run.

Pipelining shortens the clock cycle and increases the throughput, but introduces input–output latency. Each stage of a pipeline adds one cycle of delay before the first output of the circuit will be available. In a two-stage pipeline the effect of a transition of the input signal will not appear at the output until after two clock cycles. The latency accumulates through the pipeline. Latency effectively introduces a time shift between a circuit's input transitions and its output transitions (i.e., the outputs of the combinational logic after time step N are due to the inputs that were applied at time step $N-m$, where m is the number of stages of pipelining). Latency does not alter the function of the circuit. After the pipeline is full an output is formed at every clock cycle, and its maximum throughput is $1/T_{\text{stage}}$, where T_{stage} is the path length of the longest stage of the partitioned DFG.

Pipelining trades spatial (hardware) complexity for temporal complexity (performance) by computing smaller functions in less time. It distributes across multiple, shorter clock cycles the breadth of logic that would be required to implement the complete function in one clock cycle.

There are three major benefits derived from a pipeline of dedicated hardware: (1) dedicated hardware performs the same single task in every clock cycle, without requiring scheduling to coordinate its use among other tasks [17]. The operation begins with the arrival of data at every active edge of the clock, and ends in time to pass the results to the next stage of the pipeline before the arrival of the next clock, (2) the logic to perform a single, dedicated task can be streamlined and optimized as a unit, to meet constraints on performance, area, and power, and (3) the datapaths between adjacent stages of the pipeline are short and direct, reducing the need for shared data buses, control and storage, and having relatively low interconnect capacitance.

The design of a circuit with pipelined datapaths must address the following issues: (1) When should pipelining be considered? (2) Where should the pipeline registers be inserted? and (3) How much latency will be introduced by the pipeline? The design must use a minimum number of pipeline registers to achieve a minimum cycle time. Pipelining should be considered when the timing margins on the critical paths are unsatisfactory, and all other means (e.g., device resizing and alternative architectures)

---

[27]DFGs with no feedback are amenable to pipelining. Those with feedback are difficult to pipeline.
[28]Wave pipelining, a register-free form of pipelining based on coherent signal propagation, will not be considered here.

have been considered. Unsatisfactory timing margins point to a risk of metastability during operation. Various options exist for placing the registers in the circuit's data-paths. These must be evaluated and used to determine the overall latency of the design. Whether the latency is acceptable depends on the specifications for the system's performance.

9.5.1 Design Example: Pipelined Adder

Digital systems that operate on arrays of data typically contain a large number of adders in an array structure. The processing speed in these applications is usually critical, and may warrant pipelining.

The 16-bit adder in Figure 9-36(a) is formed by chaining two 8-bit adders in a serial connection. If each 8-bit adder has a throughput delay of 100 ns the worst-case delay of the configuration will be 200 ns. In a synchronous environment, this structure is organized to have all operations occur in the same clock cycle. An alternative structure can be pipelined to operate at a higher throughput by distributing the processing

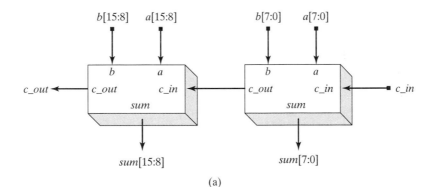

(a)

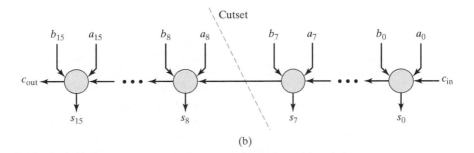

(b)

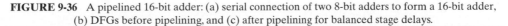

FIGURE 9-36 A pipelined 16-bit adder: (a) serial connection of two 8-bit adders to form a 16-bit adder, (b) DFGs before pipelining, and (c) after pipelining for balanced stage delays.

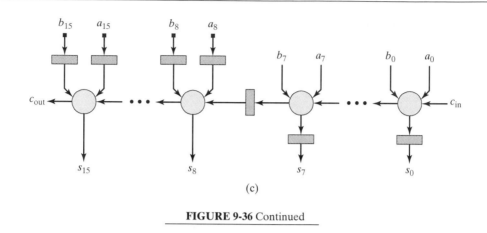

(c)

FIGURE 9-36 Continued

over multiple cycles of the clock. The tradeoff between speed and physical resources (more registers) can warrant this approach. The DFG of the 16-bit adder shown in Figure 9-36(b) reveals a single datapath connection between the functional units of the machine, suggesting a variety of options for pipelining. However, a balanced design will be achieved if the cutset between the 8-bit stages is used, resulting in the register placements shown in Figure 9-36(c).

The pipelined architecture in Figure 9-37 contains an additional register (PR) between the data input register (IR) and the data output registers. The structure sequences the data, so that in a given clock cycle a carry bit must propagate through only half of the datapath. The interface to the input datapath still provides the entire word to the unit in a synchronous manner, but the sum of only the rightmost data byte is formed. That sum, together with the leftmost datapaths, is then stored in a 25-bit internal register. In the next clock cycle the sum of the leftmost data bytes is formed and stored in the pipeline register with the rightmost sum and carried from the previous cycle. With the extra internal register, the pipelined unit can operate at approximately twice the frequency of the original adder, because the longest path supported by the clock interval is through an 8-bit adder instead of a 16-bit adder. After the period of initial latency, a new sum appears at the output of the unit every 100 ns.

The movement of data through the pipelined adder is depicted in Figure 9-38, where $a_L a_R(1)$ denotes the first sample of the left and right bytes of input word a. In the simulation results shown in Figure 9-39, note that the unit has a latency of two clock cycles between the application of the input data and the appearance of valid output. The first data words, 1122_h and 3344_h, are formed at $t_{sim} = 100$ ns, sampled and loaded into register PR at $t_{sim} = 150$ ns, partially added at $t_{sim} = 250$ ns (see $R1\_sum[7: 0]$), and fully added at $t_{sim} = 350$ ns. After the latency period, the data is correctly updated to achieve an overall throughput of approximately twice that of the serially connected 8-bit adders. (The setup times of physical registers will reduce the throughput slightly.)

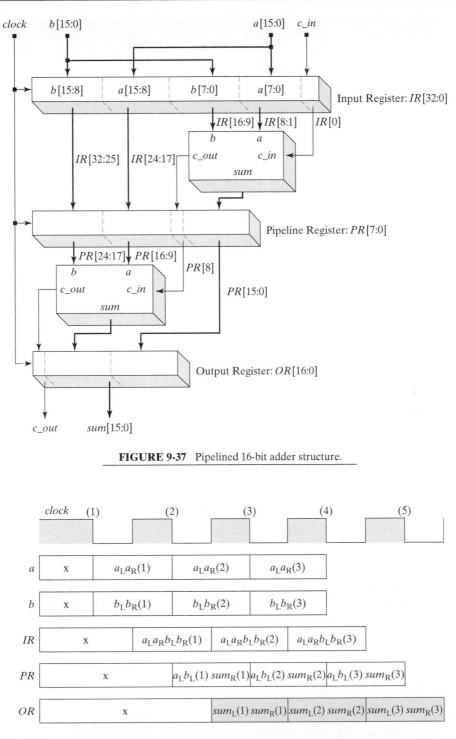

FIGURE 9-37 Pipelined 16-bit adder structure.

FIGURE 9-38 Data movement through a pipelined 16-bit adder structure.

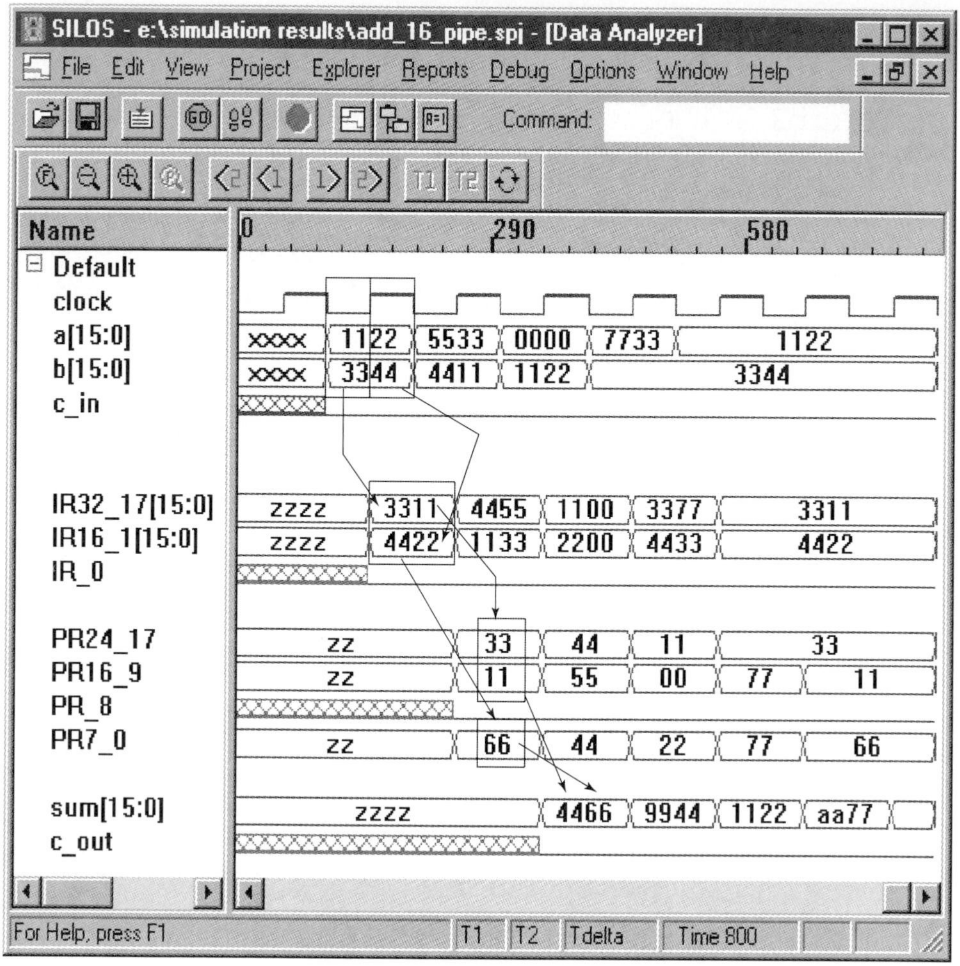

FIGURE 9-39 Simulation results showing movement of data through *add_16_pipe*, a pipelined 16-bit adder.

Nonblocking procedural assignments in Verilog make concurrent assignments to register variables and are the key to modeling concurrent register transfers in architectures that are pipelined to achieve high throughput on datapaths. The Verilog model of the pipelined adder, *add_16_pipe*, uses nonblocking assignments to concurrently sample the datapaths and registers immediately before the active edge of *clock*. These samples are used to form the values that will exist in the registers immediately after the clock event. The model is scaled by the value of *size*, which can be changed to a desired value (must be an even number).

Figure 9-39 shows the result of simulating *add_16_pipe* in a testbench that uses hierarchical dereferencing to display the contents of the internal registers in a format that reveals the dataflow through the pipeline. The displayed outputs *IR32_17, IR16_1,* and *IR_0* show the segments of *IR*. The displayed outputs *PR24_17, PR16_9, PR8,* and *PR7_0* show the segments of *PR*. The waveforms have been annotated to illustrate the register transfers.

```
module add_16_pipe (c_out, sum, a, b, c_in, clock);
    parameter    size    = 16;
    parameter    half    = size / 2;
    parameter    double  = 2 * size;
    parameter    triple  = 3 * half;
    parameter    size1 = half -1;           // 7
    parameter    size2 = size -1;           // 15
    parameter    size3 = half + 1;          // 9
    parameter    R1 = 1;                     // 1
    parameter    L1 = half;
    parameter    R2 = size3;
    parameter    L2 = size;
    parameter    R3 = size + 1;
    parameter    L3 = size + half;
    parameter    R4 = double - half +1;
    parameter    L4 = double;

    input   [size2: 0]       a, b;
    input                    c_in, clock;
    output [size2: 0]        sum;
    output                   c_out;

    reg     [double: 0]      IR;
    reg     [triple: 0]      PR;
    reg     [size: 0]        OR;

    assign {c_out, sum} = OR;

    always @ (posedge clock) begin

    // Load input register

    IR[0] <= c_in;

    IR[L1:R1] <= a[size1: 0];
    IR[L2:R2] <= b[size1: 0];

    IR[L3:R3] <= a[size2: half];
    IR[L4:R4] <= b[size2: half];
```

```
// Load pipeline register

   PR[L3: R3] <=IR[L4: R4];
   PR[L2: R2] <=IR[L3: R3];
   PR[half: 0] <= IR[L2:R2] + IR[L1:R1] + IR[0];
   OR <= {{1'b0,PR[L3: R3]} + {1'b0,PR[L2: R2]} + PR[half], PR[size1: 0]};
 end
endmodule
```

For convenience of illustration, the results of synthesizing a 4-bit pipelined adder with asynchronous reset, *add_4_pipe* are shown in Figure 9-40. The D-type flip-flop used (*dffrpqb_a*) has active-low reset.

9.5.2   Design Example: Pipelined FIR Filter

MACs dominate the performance of digital signal processors. In many applications long chains of MACs must be pipelined to increase the throughput of the unit. For example, the architecture of the FIR filter that was presented in Figure 9-22 consists of a shift register and an array of cascaded MAC units. The longest path through the circuit is proportional to the length of the chain of MACs between the input and the output. The performance specifications of a filter might require that high-speed multipliers and/or adders be used to implement MACs. Another alternative is to pipeline the datapaths to increase the throughput of the filter.

Pipeline registers can be inserted into the structure at locations determined by cutsets, as shown in Figure 9-41. The FIR filter can be pipelined in a variety of ways. In Figure 9-41(a) the cutsets place the pipeline registers at the output of the multipliers. The cutsets of the alternative structure in Figure 9-41(b) have pipeline registers at the inputs of the adders. Other implementations can be formed by omitting some of the pipeline stages to reduce the number of registers, and/or balance the stage delays, at the expense of reduced throughput. *Note:* The structure in Figure 9-41(b) must be modified to achieve coherency of the datapaths.

9.6   Circular Buffers

The algorithms of many digital filters and other signal processors repeatedly shift and store samples of data that are taken over a moving window in the time-sequence domain. For example, a filter might form its output from a weighted sum of samples of the present and most recent $N - 1$ samples of the input. Thus, N samples are used at each time step. If the algorithm is implemented in software on a general-purpose processor, these values would be stored and retrieved repeatedly as the algorithm executes, consuming several clock cycles at each cycle of the filter. Instead of this direct approach, and to realize hardware efficiency and speed, circular buffers are used to create the effect of moving an entire window of samples through memory, without actually moving all of the data [7].

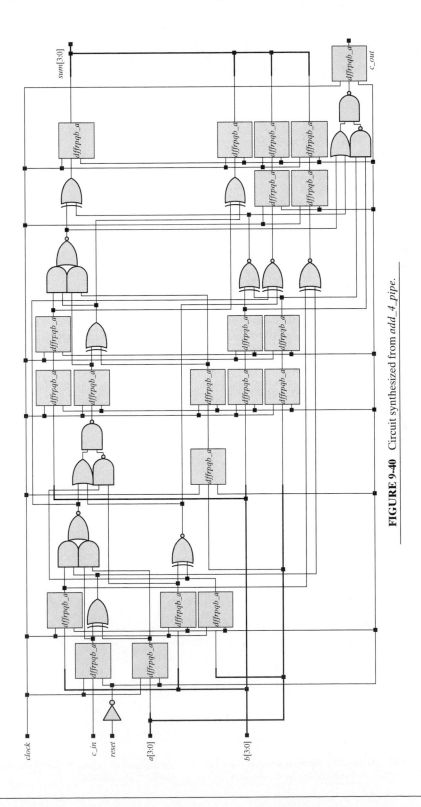

FIGURE 9-40 Circuit synthesized from *add_4_pipe*.

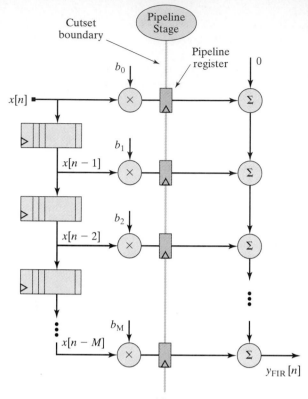

(a)

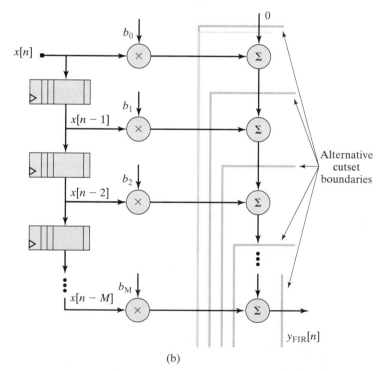

(b)

FIGURE 9-41 Alternative pipeline structures for FIR filter, with pipeline registers placed (a) at the outputs of the multipliers, and (b) at the inputs of the adders.

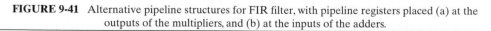

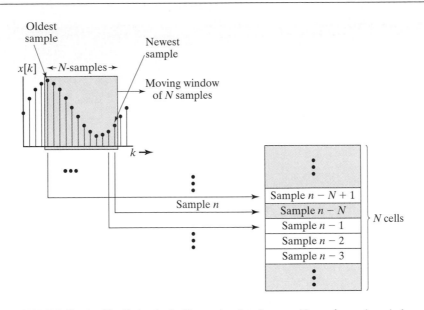

FIGURE 9-42   An N-cell circular buffer storing data from an N-sample moving window.

Circular buffers use an address mechanism that moves pointers to register cells, instead of moving the actual data. Figure 9-42 illustrates the situation in which the nth sample of the sequence $x[k]$ is to be stored in memory. Only the most recent N samples are kept in an N-cell circular buffer, and an address pointer circulates continuously through an ascending sequence of addresses, before wrapping around to the bottom (starting) address. When the nth sample is received, the data held in the entire array of registers is not shifted. Instead, the cell addressed by the pointer receives the nth sample, overwriting the previous contents at the location previously occupied by the n-Nth sample. The net effect is that the storage of data can occur in significantly fewer memory cycles.

A more expensive alternative would load the sample into a fixed location, and simultaneously move the contents of each register to its neighbor, effectively flushing the oldest data. The hardware to manage the address pointer is simpler than the hardware required to move all of the data. However, an ASIC or an FPGA might implement the buffer with a set of dedicated, parallel-load registers, chained together as a wordwide shift register.

Example 9.9

Two Verilog models of an N-sample moving window memory are described below. The first version, *Circular_Buffer_1*, uses an array of parallel-load shift registers to shift the entire contents of the buffer at each time step. The second version, *Circular_Buffer_2*,

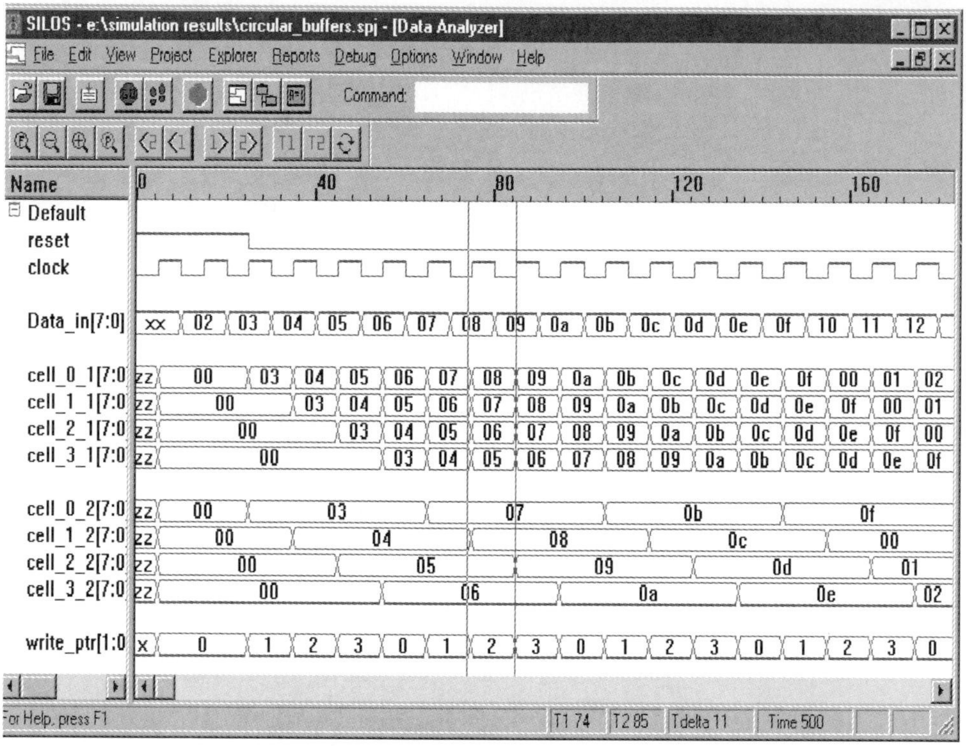

FIGURE 9-43 Results of simulating *Circular_Buffer_1* and *Circular_Buffer_2*, two versions of a data buffer holding an *N*-sample moving window.

includes *write_ptr*, which points to the next cell to be read. The contents of the register do not move. The pointer is incremented at each time step, so the data below the pointer is aged. Note that in the simulation results shown in Figure 9-43 the contents of *Circular_Buffer_1* change at every cycle, while only the one cell addressed by *write_ptr* in *Circular_Buffer_2* changes as new data arrive (at the negative edges of the clock). A digital filter using the data held in *Circular_Buffer_1* would always tap the oldest data at the same location, but the filter using *Circular_Buffer_2* would need logic to track the location of the aged data relative to *write_ptr*.[29]

```
module Circular_Buffer_1 (cell_3, cell_2, cell_1, cell_0, Data_in, clock, reset);
    parameter    buff_size = 4;
    parameter    word_size = 8;
    output       [word_size -1: 0]  cell_3, cell_2, cell_1, cell_0;
    input        [word_size -1: 0] Data_in;
```

---

[29]See Problem 9.15 at the end of the chapter.

```
        input           clock, reset;
        reg             [buff_size -1: 0] Buff_Array [word_size -1: 0];
        wire            cell_3 = Buff_Array[3], cell_2 = Buff_Array[2];
        wire            cell_1 = Buff_Array[1], cell_0 = Buff_Array[0];
        integer         k;

    always @ (posedge clock) begin
      if (reset == 1) for (k = 0; k <= buff_size -1; k = k+1)
       Buff_Array[k] <= 0;
      else for (k = 1; k <= buff_size -1; k = k+1) begin
       Buff_Array[k] <= Buff_Array[k-1];
       Buff_Array[0] <= Data_in;
      end
    end
  end
endmodule


module Circular_Buffer_2 (cell_3, cell_2, cell_1, cell_0, Data_in, clock, reset);
    parameter       buff_size = 4;
    parameter       word_size = 8;
    output          [word_size -1: 0]  cell_3, cell_2, cell_1, cell_0;
    input           [word_size -1: 0] Data_in;
    input           clock, reset;
    reg             [buff_size -1: 0] Buff_Array [word_size -1: 0];
    wire            cell_3 = Buff_Array[3], cell_2 = Buff_Array[2];
    wire            cell_1 = Buff_Array[1], cell_0 = Buff_Array[0];
    integer         k;
    parameter       write_ptr_width = 2;                      // Width of write pointer

    parameter       max_write_ptr = 3;
    reg             [write_ptr_width -1 : 0]     write_ptr;        // Pointer for writing

    always @ (posedge clock) begin
      if (reset == 1 ) begin
       write_ptr <= 0;
       for (k = 0; k <= buff_size -1; k = k+1) Buff_Array[k] <= 0;
      end
      else begin
        Buff_Array[write_ptr] <= Data_in;
         if (write_ptr < max_write_ptr) write_ptr <= write_ptr + 1; else write_ptr <= 0;
       end
    end
  end
endmodule
```

End of Example 9.9

9.7   FIFOs and Synchronization across Clock Domains

When data are passed from one domain into another and the clock of the destination domain is not related to the clock of the source domain, the activity in the domains is incoherent, eliminating the possibility of synchronous operation and guaranteeing that the system will enter a metastable state repeatedly. To circumvent this problem, handshake signals can be used to govern the exchange of data, but the transfer rate is lower than can be achieved by alternatives. In practice, high-performance parallel interfaces between independent clock domains are implemented with a first-in, first-out memory, called a FIFO.

A FIFO consists of a block of memory and a controller that manages the traffic of data to and from the FIFO. The reading and writing operations are similar to those of circular buffers,[30] but a FIFO's architecture provides access to only one register cell at a time, not to the entire array of registers. A FIFO has two address pointers, one for writing to the next available cell, and another one for reading the next unread cell.[31]

FIFOs operate differently from a circular buffer, because a FIFO's output is a single word, which is read on command. The pointers for reading and writing are relocated dynamically as commands to read or write are received, rather than continuously, as in the case of a circular buffer. The FIFO buffer shown in Figure 9-44 can receive data until it is full and can be read until it is empty. A pointer is moved after each operation. A FIFO has separate address busses and datapaths for reading and writing data and status lines indicating the condition of the stack (full, almost full, etc). The FIFO developed below accommodates simultaneous reading and writing of data.

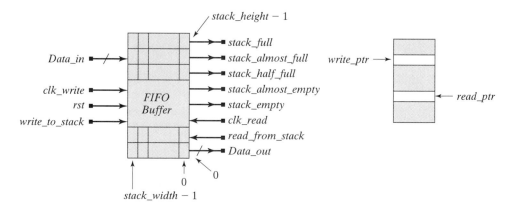

FIGURE 9-44   FIFO Buffer: Input–output ports.

---

[30]See Problem 11 in Chapter 8.
[31]FIFOs are usually implemented with dual-port RAMs with independent read- and write-address pointers and registered data ports (see www.idt.com).

The computational activity in the input and output domains can be synchronized by separate clocks, allowing the FIFO to act as a buffer between two clock domains. If the FIFO is to support simultaneous reading and writing to the same cell, it is necessary to consider synchronization.

Example 9.10

The multichannel circuit in Figure 9-45 has four channels of serial bit streams originating in a 100 MHz clock domain, with each passing though a serial-to-parallel converter that forms a 32-bit word for transfer to a dual-port FIFO [18]. Data are directed to each serial-to-parallel converter at a rate of 100 MHz; a processor operating with a clock of 133 MHz is to read data from the FIFOs and multiplex the four channels of data onto a common datapath. The format of the arriving data has the LSB (least significant bit) of a word arriving first. The commands to write data to a FIFO originate in the domain of *clock_100MHz*; the commands to read data from the FIFOs are issued by the processor (not shown) in the domain of *clock_133MHz*. The processor must

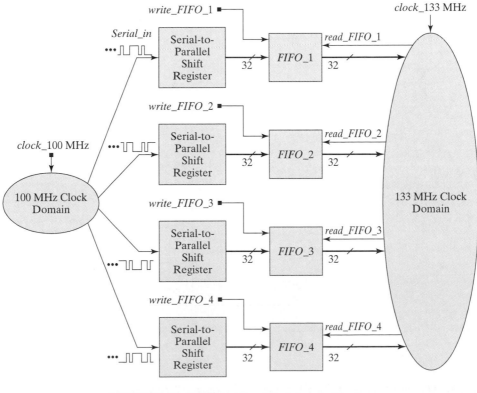

FIGURE 9-45 A FIFO-buffered clock domain interface.

prevent the loss of data, which will occur if a FIFO is full when it receives a write request. We will consider a single channel of data flow.

The serial-to-parallel converter will be implemented as a 32-bit shift register, and will be controlled by the state machine described by the ASMD chart in Figure 9-46. The outputs of the state machine are the signals *shift* and *incr*, which control the register *Data_out* and a counter, *cnt*, respectively. The state of the machine is directed to *S_idle* under the action of *rst*, and remains there until *En* is asserted by an external agent. At the first clock with *En* asserted, the machine makes a transition to *S_1*, where it remains for the next 31 clocks (counted by a 5-bit counter).[32] The signal *write* indicates the status of the datapath, and is to control the writing of a word of data to a FIFO. *write* is asserted during the cycle that loads the 32nd bit into the register. If *write* and *En* are both asserted, *shift* and *incr* are asserted by the state machine, the counter wraps around to 0, and the sequence repeats (with the state remaining in *S_1*); otherwise, *incr* is asserted and the counter wraps to 0 and the machine returns to *S_idle*, with *Data_out* holding the word that was loaded in the previous 32 clock cycles.

The Verilog description of the 32-bit serial-to-parallel converter is given by *Ser_Par_Conv_32* below, and a sample of simulation results is shown in Figure 9-47.

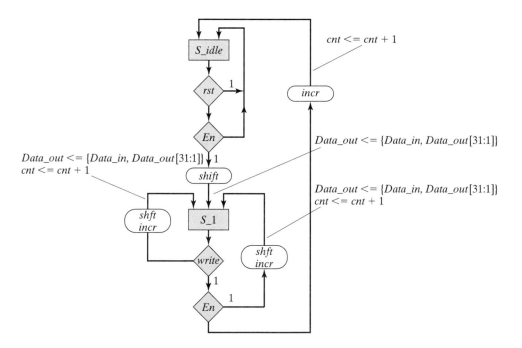

FIGURE 9-46 ASMD chart for a 32-bit serial-to-parallel converter controlled by a state machine.

[32]By not launching the counter until the machine has entered *S_1*, we need only a 5-bit register to count the remaining bits that enter the shift register.

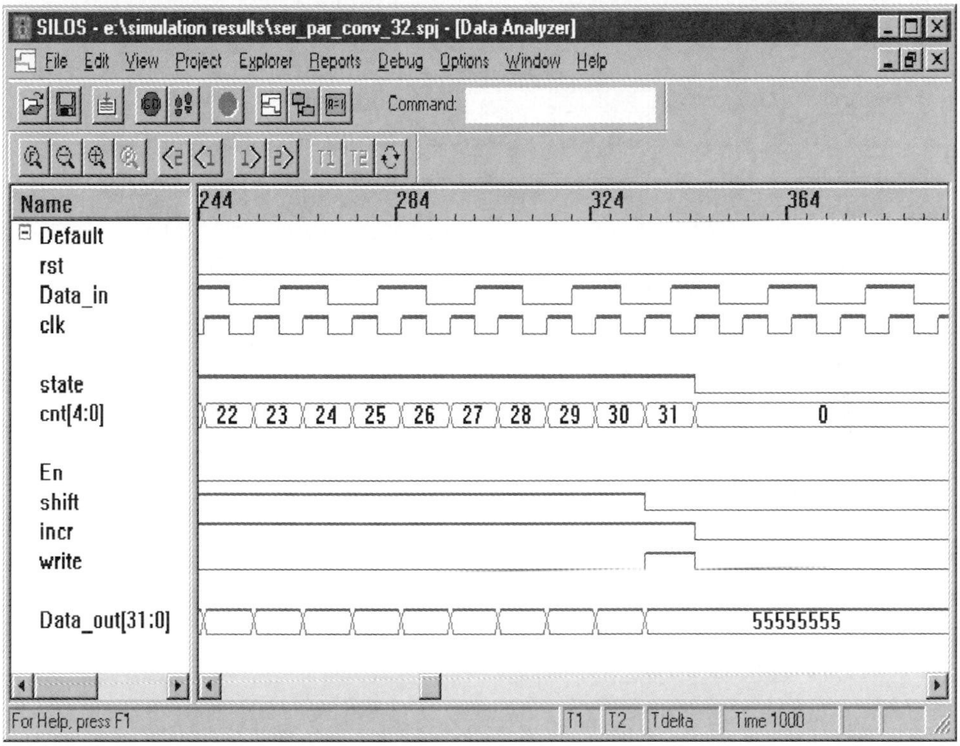

FIGURE 9-47 Simulation results illustrating the behavior of a 32-bit serial-to-parallel converter.

Note that when the 5-bit counter reaches a count of 31, the signal *write* asserts to indicate that the shift register is full; *write* will be used to control the write operation of the FIFO that is to store the data words. Because *En* is de-asserted for the data shown, *shift* is de-asserted. The transitions of the serial bit stream of data at the input to the serial-to-parallel converter occur on the falling edge of *clock_100MHz*, and the data are shifted into the register on the rising edge. *write* is asserted for a duration of one cycle of *clk*.

```
module Ser_Par_Conv_32 (Data_out, write, Data_in, En, clk, rst);
    output [31: 0]  Data_out;
    output          write;
    input           Data_in;
    input           En, clk, rst;

    parameter       S_idle = 0;
    parameter       S_1 = 1;

    reg             state, next_state;
    reg     [4: 0]  cnt;
```

```
reg              Data_out;
reg              shift, incr;

always @ (posedge clk or posedge rst)
  if (rst) begin state <= S_idle; cnt <= 0; end
  else state <= next_state;

always @ (state or En or write) begin
  shift = 0;
  incr = 0;
  next_state = state;
  case (state)
    S_idle:      if (En) begin next_state  = S_1; shift = 1; end
    S_1:         if (!write) begin shift = 1; incr = 1; end
                 else if (En) begin shift = 1; incr = 1; end
                 else begin next_state = S_idle; incr = 1; end
  endcase
end

always @ (posedge clk or posedge rst)
  if (rst) begin cnt <= 0;  end
  else if (incr) cnt <= cnt +1;

always @ (posedge clk or posedge rst)
  if (rst) Data_out <= 0;
  else if (shift) Data_out <= {Data_in, Data_out [31:1]};

assign write = (cnt == 31);
endmodule
```

Each FIFO in Figure 9-45 must synchronize the write operations to *clock_133MHz*, to ensure that the datapath is stable while a cell is being read and that the status of the stack does not change during the setup interval of the storage devices that implement the memory of the FIFO. The signal *write* is generated in the domain of *clock_100MHz*, and will be synchronized to *clock_133MHz*. The synchronizers that were shown in Figure 5-38 are candidates for synchronizing the write signal of the FIFO.

To determine which of the two synchronizer circuits to use, note that the write pulse in Figure 9-45 is to assert for one cycle of *clock_100MHz*, so the pulse has a width of $\Delta_{write} = 1/T_{clock\_100MHz} = 1/(10^8) = 10$ ns. The period of *clock_133MHz* is $T_{clock\_133MHz} = 1/(133 \times 10^6) = 7.5$ ns. Because the width of the asynchronous input pulse is greater than the period of the clock to which it is to be synchronized, we choose the circuit in Figure 5-38(a), a two-stage shift register synchronizer.

Figure 9-48 illustrates a condition that must be anticipated in the design. Given the width of the *write* pulse compared to the period of *clock_133MHz*, it is possible for two active edges of *clock_133MHz* to occur while *write* is asserted. If the machine responded to both pulses the same input bit would be stored twice, in different cells of the FIFO, and the machine would not operate correctly. Consequently, the two-stage

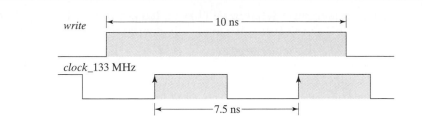

FIGURE 9-48 The duration of the asynchronous write pulse covers two edges of *clock_133_MHz*.

synchronizer shown in Figure 5-38(a) must be modified to include logic to limit the output to a single pulse with a width of one cycle of *clock_133MHz*. The mux that is inserted between the stages of the synchronizer in Figure 9-49 passes the output of the first stage until the output of the second stage is set to 1. Then it passes a 0 to the second stage, which clears the second stage and switches the input to the path from the first stage. The pulse at the output of the synchronizer (i.e., *write-synch*) has a width of one cycle of *clock_133MHz* and has a latency of two cycles.

The cancellation circuit in Figure 9-46 ensures that a write pulse that covers two successive edges of clock_133MHz will not generate two successive write commands (i.e., the synchronized pulse has a duration of one cycle of *clock_133MHz*). During operation, the (unsynchronized) write signal will be the signal *write*, which is generated by the serial-to-parallel converter. *write* asserts only every 32nd cycle of *clock_100MHz*, and has a duration of one cycle of *clock_100MHz*. If the input stage of the synchronizer enters a metastable state and recovers within a clock cycle, the output of the synchronizer will have a latency of three cycles of *clock_133_MHz*; otherwise, the output will have latency of two cycles. The inherent latency of the synchronizer can be compensated for by inserting two stages of buffering at the output of the serial-parallel

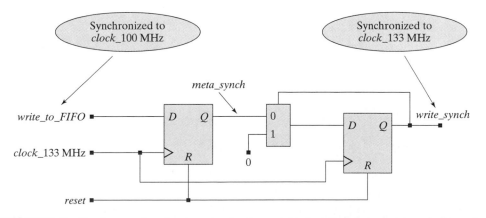

FIGURE 9-49 Circuit to synchronize the write signal across the clock domain boundary and allow only one write pulse.

converters. The compensated dataflow will have a latency of 0 or 1. The pointer structure of a dual port FIFO will correctly handle a late write pulse concurrently with read pulses, so the possible latency does not upset the system.

The waveforms shown in Figure 9-50 demonstrate the operation of the synchronizer. The active edges of *clock_133MHz* slide past those of *clock_100MHz*. The signal *write_to_FIFO* has a duration of one cycle of *clock_100*, and the synchronized write signal, *write_synch*, has a duration of one cycle of *clock_133MHz*. The displayed time has been scaled by a factor of 12.5×10^6, giving Tdelta = 256 time steps between the cursor and the marker of the display.

```
module write_synchronizer (write_synch, write_to_FIFO, clock, reset);
    output          write_synch;
    input           write_to_FIFO;
    input           clock, reset;
    reg             meta_synch, write_synch;

    always @ (negedge clock)
      if (reset == 1) begin
        meta_synch <= 0;
        write_synch <= 0;
      end
      else begin
        meta_synch <= write_to_FIFO;
        write_synch <= write_synch ? 0: meta_synch;
      end
endmodule
```

FIGURE 9-50 Waveforms illustrating synchronization of the FIFO write signal across the clock domain boundary.

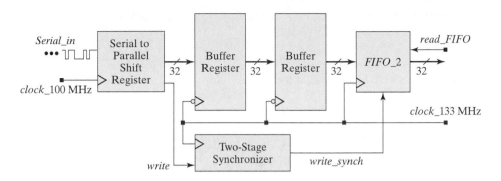

FIGURE 9-51 Circuit with a synchronizer that synchronizes the signal *write*, asserted for one cycle in the domain of *clock_100MHz*, to *clock_133_MHz*, the signal that synchronizes the operation of reading data from the FIFO.

By synchronizing *write* to *clock_133MHz*, we generate a write pulse at the rate of *clock_100MHz*, but its active edge is determined by the active edge of *clock_133MHz*. The width of the synchronized pulse is the width of a cycle of *clock_133MHz*. Note that *write_synch* is synchronized to the falling edge of *clock_133MHz* to avoid a race condition at the FIFO, which operates on the rising edge of *clock_133MHz*.

The latency of the synchronizer introduces a design trap. When *write_synch* finally asserts, the contents of the shift register will have already shifted by two more clocks, destroying coherency between *write_synch* and the contents of the register. To restore coherency, we can lengthen the shift register datapath by adding buffer registers synchronized to the falling edge of *clock_133MHz*. The circuit in Figure 9-51 includes the two-stage synchronizer and the buffer registers.

The Verilog models of the FIFO buffer and two testbenches are shown below. Ordinarily, the FIFO's pointers for writing and reading, *write_ptr* and *read_ptr*, will not be co-located, but if the stack is not full and not empty, the model allows simultaneous reading and writing *at the same location*. Note that in this model the pointers *write_ptr* and *read_ptr* have been sized to wrap at the stack boundary, so the condition of an empty stack cannot be distinguished from the condition of a full stack, as shown in Figure 9-52.

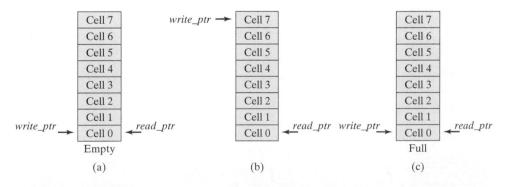

FIGURE 9-52 Pointer configurations (a) for an empty stack, (b) after seven successive writes, and (c) after eight successive writes.

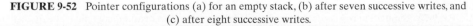

A register, *ptr_gap*, is included in the description to indicate the status of the gap between the pointers, and its contents are incremented, decremented, or held, depending on the condition of the stack and whether a write operation or a read operation is attempted. When the value stored in *ptr_gap* reaches eight, the FIFO is full. If the stack is full, only a read is allowed; if the stack is empty, only a write is allowed.

The first testbench, *t_FIFO_Dual_Port*, is used to test the operation of the FIFO, but without synchronization (*clk_write = clk_read = clk*). The simulation results in

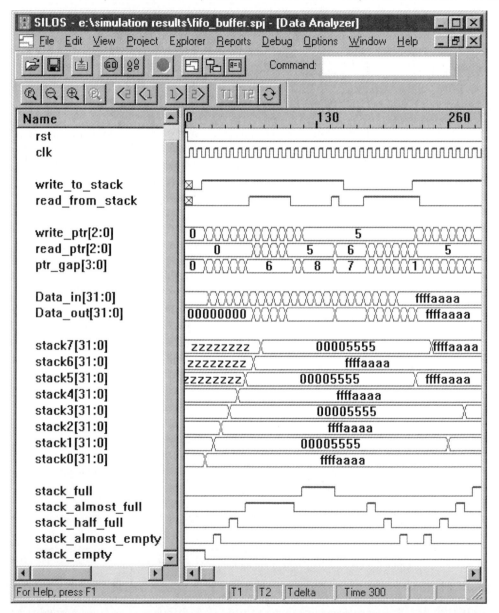

FIGURE 9-53 Simulation results demonstrating FIFO operation with overlapping read and write activity.

Figure 9-53 demonstrate write and read operations of the FIFO and show the activity of the pointers and the status signals, with and without concurrent read and write activity. The second test bench, *t_FIFO_Clock_Domain_Synch*, is used to verify the

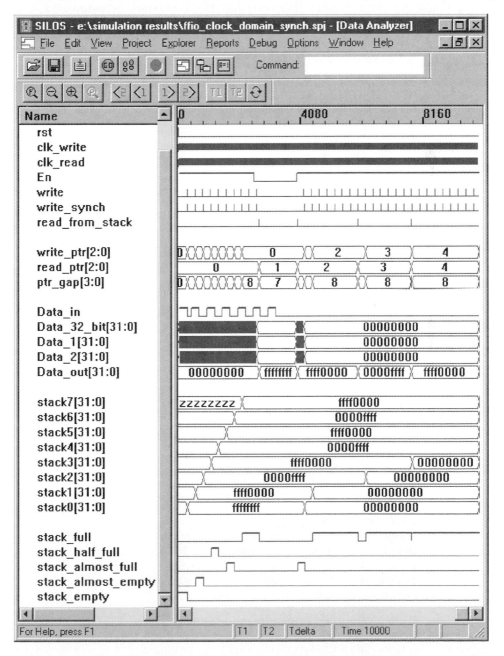

FIGURE 9-54 Simulation results demonstrating clock domain synchronization, with concurrent reading and writing operations.

transfer of data across the boundaries separating the input and output clock domains, with concurrent reading and writing of data. The write signal originates in the domain of the slow clock, and is synchronized to the clock in the fast domain. Simultaneous reading and writing of data is demonstrated with a full stack in Figure 9-54. Additional testing is left to the reader to verify that the model is correct for all conditions of the stack and assertions of read and/or write operations.

```
module FIFO_Buffer (
  Data_out,                          // Data path from FIFO
  stack_full,                        // Status flags
  stack_almost_full,
  stack_half_full,
  stack_almost_empty,
  stack_empty,
  Data_in,                           // Data path into FIFO
  write_to_stack,                    // Flag controlling a write to the stack
  read_from_stack,                   // Flag controlling a read from the stack
  clk, rst
);
  parameter stack_width = 32;        // Width of stack and data paths
  parameter stack_height = 8;        // Height of stack (in # of words)
  parameter stack_ptr_width = 3;     // Width of pointer to address stack
  parameter  AE_level = 2;           // almost empty level
  parameter  AF_level = 6;           // Almost full level
  parameter  HF_level = 4;           // Half full level
  output    [stack_width -1: 0]      Data_out;

  output                             stack_full, stack_almost_full, stack_half_full;
  output                             stack_almost_empty, stack_empty;
  input     [stack_width -1: 0]      Data_in;
  input                              write_to_stack, read_from_stack;
  input                              clk, rst;

  reg       [ stack_ptr_width -1: 0] read_ptr, write_ptr;  // Addresses for
                                                           // reading and writing
  reg       [ stack_ptr_width: 0]    ptr_gap;              // Gap between ptrs
  reg       [stack_width -1: 0]      Data_out;
  reg       [stack_width -1: 0]      stack [stack_height -1 : 0]; // memory array

// Stack status signals
assign stack_full = (ptr_gap == stack_height);
assign stack_almost_full = (ptr_gap == AF_level);
assign stack_half_full = (ptr_gap == HF_level);
assign stack_almost_empty = (ptr_gap == AE_level);
assign stack_empty = (ptr_gap == 0);

always @ (posedge clk or posedge rst)
  if (rst) begin
    Data_out <= 0;
    read_ptr <= 0;
```

```
                 write_ptr <= 0;
                 ptr_gap <= 0;
              end
            else if (write_to_stack && (!stack_full) && (!read_from_stack)) begin
              stack [write_ptr] <= Data_in;
              write_ptr <=  write_ptr + 1;
              ptr_gap   <= ptr_gap + 1;
            end
            else if ((!write_to_stack) && (!stack_empty) && read_from_stack) begin
              Data_out <= stack [read_ptr];
              read_ptr  <= read_ptr + 1;
              ptr_gap   <= ptr_gap - 1;
            end
            else if (write_to_stack && read_from_stack && stack_empty) begin
              stack  [write_ptr] <= Data_in;
              write_ptr <= write_ptr + 1;
              ptr_gap   <= ptr_gap + 1;
            end
            else if (write_to_stack && read_from_stack && stack_full) begin
              Data_out <= stack [read_ptr];
              read_ptr  <= read_ptr + 1;
              ptr_gap   <= ptr_gap - 1;
            end
            else if (write_to_stack && read_from_stack
              && (!stack_full) && (!stack_empty)) begin
              Data_out <= stack [read_ptr];
              stack [write_ptr] <= Data_in;
              read_ptr  <= read_ptr + 1;
              write_ptr  <= write_ptr + 1;
            end
        endmodule

        module t_FIFO_Buffer ();  // Used to test only the FIFO, without synchronization
          parameter                 stack_width = 32;
          parameter                 stack_height = 8;
          parameter                 stack_ptr_width = 4;

          wire    [stack_width -1: 0]    Data_out;
          wire                          write;
          wire                          stack_full, stack_almost_full, stack_half_full;
          wire                          stack_almost_empty, stack_empty;
          reg     [stack_width -1: 0]    Data_in;
          reg                           write_to_stack, read_from_stack;
          reg                           clk, rst;
          wire    [stack_width -1: 0]    stack0, stack1, stack2, stack3, stack4,
                                         stack5, stack6, stack7;

          assign stack0 = M1.stack[0];   // Probes of the stack
          assign stack1 = M1.stack[1];
          assign stack2 = M1.stack[2];
```

```
         assign stack3 = M1.stack[3];
         assign stack4 = M1.stack[4];
         assign stack5 = M1.stack[5];
         assign stack6 = M1.stack[6];
         assign stack7 = M1.stack[7];

         FIFO_Buffer M1 (Data_out, stack_full, stack_almost_full, stack_half_full,
           stack_almost_empty, stack_empty, Data_in, write_to_stack, read_from_stack,
           clk, rst);

         initial #300 $finish;
         initial begin rst = 1; #2 rst = 0; end
         initial begin clk = 0; forever #4 clk = ~clk; end

         // Data transitions
          initial begin Data_in = 32'hFFFF_AAAA;
           @ (posedge write_to_stack);
             repeat (24) @ (negedge clk) Data_in = ~Data_in;
          end

         // Write to FIFO
          initial fork
           begin #8   write_to_stack = 0; end
           begin #16 write_to_stack = 1; #140 write_to_stack = 0; end
           begin #224 write_to_stack = 1; end
          join

         // Read from FIFO
          initial fork
           begin #8 read_from_stack = 0; end
           begin #64 read_from_stack = 1; #40 read_from_stack = 0; end
           begin #144 read_from_stack = 1; #8 read_from_stack = 0; end
           begin #176 read_from_stack = 1; #56 read_from_stack = 0; end
          join
         endmodule

         module t_FIFO_Clock_Domain_Synch ();              // Test for clock domain
          parameter stack_width = 32;                      // synchronization
          parameter stack_height = 8;
          parameter stack_ptr_width = 3;
          defparam M1.stack_width = 32;    // Override any defaults
          defparam M1.stack_height = 8;
          defparam M1.stack_ptr_width = 3;

          wire   [stack_width -1: 0]        Data_out, Data_32_bit;
          wire                              stack_full, stack_almost_full, stack_half_full;
          wire                              stack_almost_empty, stack_empty;
          wire                              write;
          reg                               Data_in;
          reg                               read_from_stack;
          reg                               En;
          reg                               clk_write, clk_read, rst;
```

```
wire    [31: 0]                      stack0, stack1, stack2, stack3;
wire    [31: 0]                      stack4, stack5, stack6, stack7;

// Probes of the stack
assign stack0 = M1.stack[0];      assign stack1 = M1.stack[1];
assign stack2 = M1.stack[2];      assign stack3 = M1.stack[3];
assign stack4 = M1.stack[4];      assign stack5 = M1.stack[5];
assign stack6 = M1.stack[6];      assign stack7 = M1.stack[7];

// 2-stage pipeline to compensate for latency at synchronizer
reg [stack_width-1: 0] Data_1, Data_2;
always @ (negedge clk_read)
  if (rst) begin Data_2 <= 0; Data_1 <= 0; end
  else begin  Data_1 <= Data_32_bit; Data_2 <= Data_1; end

Ser_Par_Conv_32 M00 (Data_32_bit, write, Data_in, En, clk_write, rst);
write_synchronizer M0 (write_synch, write, clk_read, rst);
FIFO_Buffer M1 (Data_out, stack_full, stack_almost_full, stack_half_full,
  stack_almost_empty, stack_empty, Data_2, write_synch, read_from_stack,
  clk_read, rst);

initial #10000 $finish;
initial   fork rst = 1; #8 rst = 0; join
initial begin clk_write = 0; forever #4 clk_write = ~clk_write; end  //  100  MHz
clock
initial    begin clk_read = 0; forever #3 clk_read = ~clk_read; end  //  133  MHz clock
initial   fork #1 En = 0; #48 En = 1; #2534 En = 0; #3944 En = 1;  join
initial   fork
  #6 read_from_stack = 0;
  #2700 read_from_stack = 1; #2706 read_from_stack = 0;
  #3980 read_from_stack = 1; #3986 read_from_stack = 0;
  #6000 read_from_stack = 1; #6006 read_from_stack = 0;
  #7776 read_from_stack = 1; #7782 read_from_stack= 0;  //  Overlaps
  write_synch
join
// Serial data transitions are synchronized to the falling edge of clk_write
initial begin // Generate data and hold
  #1 Data_in = 0;
  @ (posedge En) Data_in = 1;  // wait for enable
  @ (posedge write);
  repeat (6) begin
  repeat (16) @ (negedge clk_write) Data_in = 0;
  repeat (16) @ (negedge clk_write) Data_in = 1;
  repeat (16) @ (negedge clk_write) Data_in = 1;
  repeat (16) @ (negedge clk_write) Data_in = 0;
  end
 end
endmodule
```

End of Example 9.10

REFERENCES

1. van der Hoeven A. *Concepts and Implementation of a Design System for Digital Signal Processor Arrays*. Delft, The Netherlands: Delft University Press, 1990.
2. Bu J. *Systematic Design of Regular VLSI Processors*. Delft, The Netherlands: Delft University Press, 1990.
3. De Micheli G. *Synthesis and Optimization of Digital Circuits*. New York: Mc-Graw-Hill, 1994.
4. Gajski D, et al. "Essential Issues in Codesign." In: Staunstrup J., Wolf W, eds. *Hardware/Software Co-Design: Principles and Practices*. Boston: Kluwer, 1997.
5. Gajski D, et al. *High-Level Synthesis: Introduction to Chip Design*. Boston: Kluwer, 1992.
6. Knuth DE. *The Art of Computer Programming*. Vol. 3. *Sorting and Searching*. Reading, MA: Addison-Wesley, 1973.
7. Kehtarnavaz N, Keramat M. *DSP System Design Using the TMS320C6000*. Upper Saddle River, NJ: Prentice-Hall, 2001.
8. Candy JV. *Signal Processing—The Modern Approach*. New York: McGraw-Hill, 1988.
9. Oppenheim AV, Schafer RW. *Discrete-Time Signal Processing*. Upper Saddle River, NJ: Prentice-Hall, 1989.
10. McClellan JH, Schafer RW, Yoder MA. *DSP First—A Multimedia Approach*. Upper Saddle River, NJ: Prentice-Hall, 1998.
11. Hagan CJ. *Synthesis of Cascade Integrator Comb Digital Decimation Filters*. Technical Report EAS_ECE_1988_05, Dept. of Electrical and Computer Engineering, University of Colorado at Colorado Springs, 1998.
12. Stearns SD, David RA. *Signal Processing Algorithms*. Upper Saddle River, NJ: Prentice-Hall, 1988.
13. McClellan JH, et al. *Computer-Based Exercises for Signal Processing Using MATLAB 5*. Upper Saddle River, NJ: Prentice-Hall, 1998.
14. Stonick VL, Bradley K. *Labs for Signals and Systems Using MATLAB*. Boston: PWS, 1996.
15. Peled A, Liu B. *Digital Signal Processing*. New York: Wiley, 1976.
16. Smith MJ. *Application-Specific Integrated Circuits*. Reading, MA: Addison-Wesley Longman, 1997.
17. Andraka R. "FPGAs cut power with 'pipeline'." *Electronic Engineering Times*, August 7, 2000.
18. Xilinx Foundation ISE v3.1i Workshop Labs, Spring, 2001.
19. Kung SY. *VLSI Array Processors*. Upper Saddle River, NJ: Prentice-Hall, 1988.
20. Haykin S. *Adaptive Filter Theory*. Upper Saddle River, NJ: Prentice-Hall, 1996.

PROBLEMS

1. Determine the size of the largest pixel processor array that can be implemented by synthesizing *Image_Converter_Baseline* (see section 9.2.1) into a Xilinx XCS40/XL FPGA (see Table 8-13).
2. Develop an architecture for implementing a sequential (row-by-row) algorithm realizing the behavior of a halftone image converter having only one processor for an 8 × 6 array. Develop, verify, and synthesize a Verilog model of the machine. (*Optional*: By performing postsynthesis simulation in a given ASIC technology

or an FPGA, determine the maximum rate at which images can be processed by the machine.)

3. The convolution of an N-sample data sequence $\{x[k]\}$ with the impulse response of a digital filter $\{h[k]\}$ with $j = 0,1,\ldots,N - 1$, produces the filter's output and is defined [19, 20] by

$$y[j] = \sum_{k=0}^{j} x[k]h[j - k]$$

for $j = 0, 1, \ldots, 2N - 2$. (a) Using Verilog constructs, write an NLP describing the convolution algorithm. Note that the NLP can be unrolled to give

$$y[0] = x[0]\,h[0]$$

$$y[1] = x[0]\,h[1] + x[1]\,h[0]$$

$$y[2] = x[0]\,h[2] + x[1]\,h[1] + x[2]\,h[0]$$

$$\cdots$$

$$\cdot$$

A fragment of a DFG for the NLP is given in Figure P9-3; (b) complete the DFG for $N = 3$, (c) using the NLP and the DFG, develop, verify, and synthesize *Convolution_Baseline*, a Verilog model that implements the algorithm, (d) using

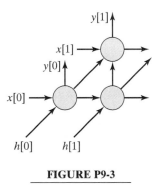

FIGURE P9-3

the DFG, identify a two-stage balanced pipeline architecture for the machine, and (e) develop, verify and synthesize *Convolution_Pipe*, an implementation of the pipeline determined in (d) and compare the two realizations of the algorithm. What is the minimum number of concurrent processors that could implement the algorithm?

4. The bubble sort algorithm (see Example 9.1) sorts the elements of an array of N unsigned binary numbers into ascending order. Develop an NLP in pseudocode to (1) search over N unsigned binary numbers to find the largest, (2) then remove it and repeat the search over the remaining $N - 1$ words, etc., until the list is sorted. Develop a temporal DFG for the algorithm, and specify an architecture for its implementation. Develop, verify, and synthesize a machine that realizes the algorithm. Compare and discuss the relative merits of the two sorters.

5. Using the Floyd–Steinberg algorithm (see Section 9.2), develop a Verilog behavioral model of a machine, *GS_Image_Converter*, that implements a Gray scale conversion of an 8×6 array of pixels having a resolution of 8 bits, into a same-sized array of pixels have a resolution of 4 bits. Implement the following versions of the machine: (1) a baseline level-sensitive realization of the algorithm's NLP and (2) an ASMD-based implementation with maximally concurrent processing.

6. Develop an NLP describing an algorithm that computes the histogram of eight equispaced gray levels in an 8×6 pixel image with 8 bits of resolution. Using a DFG for the algorithm, develop, verify, and synthesize (1) a baseline machine implementing the algorithm and (2) a ASMD-based machine realizing the algorithm with maximally concurrent processors.

7. The reservation table in Figure P9-7 reveals how the throughput of *Image_Converter_2* could be increased by exploiting the idle time of the

Time slots																		Time slots											
t_1	t_2	t_3	t_4	t_5	t_6	t_7	t_8	t_9	t_{10}	t_{11}	t_{12}	t_{13}	t_{14}	t_{15}	t_{16}	t_{17}	t_{18}	t_1	t_2	t_3	t_4	t_5	t_6	t_7	t_8	t_9	t_{10}	t_{11}	t_{12}
1	2	3	4	5	6	7	8	15	16	23	24	1	2	3	4	5	6	7	8	15	16	23	24						
		9	10	11	12	13	14	21	22	29	30	31	32	9	10	11	12	13	14	21	22	29	30	31	32				
				17	18	19	20	27	28	35	36	37	38	39	40	17	18	19	20	27	28	35	36	37	38	39	40		
						25	26	33	34	41	42	43	44	45	46	47	48	25	26	33	34	41	42	43	44	45	46	47	48

FIGURE P9-7

processors and concurrently processing two images. Develop, verify and synthesize a machine that will maximize the throughput that can be obtained in processing an 8×6 array, using four concurrent processors. Include an ASMD chart for the control unit of the machine (see Figure 9-13).

8. Develop, verify, and synthesize a frame processor that converts an 8×6 array of pixels with 8-bit resolution into an image with 4-bit resolution. The processor includes three image buffers, and a controller that directs the processing of one image while a second image is being loaded into memory, and a third (converted) image is being sent through the I/O ports, which accommodate 1 byte each. The pipelining of the operations is illustrated below in Figure P9-8.

t	*Buffer_1*	*Buffer_2*	*Buffer_3*
	Processing	Loading	Reading
	Reading	Processing	Loading
	Loading	Reading	Processing
	Processing	Loading	Reading

$\vdots$

FIGURE P9-8

9. The DFG shown in Figure 9-5 for a halftone pixel image converter can be used to identify alternative cutsets that define a pipeline architecture for the base-line machine. Identify a cutset that will balance a pair of pipeline stages for an 8×6 array of pixels. Specify an architecture, and develop and verify the Verilog model *Image_Converter_1_pipe* that implements the pipeline specified by the cutset. Include an ASMD chart for the control unit of the machine (see Figure 9-13).

10. The Verilog model *FIR_Gaussian_Lowpass* (see Example 9.2) is limited to a seventh-order filter. Develop and verify a re-usable model with (1) parameters and memory for up to 16 filter tap coefficients, and (2) a loop-based algorithm that forms *Data_out* for an allowed order of the filter. When *reset* is asserted, the state of the filter is to return to *S_idle*, where it remains until a signal *Load* is asserted. With the state in *S_idle*, an assertion of *Load* will cause the machine to read a byte of data specifying the order of the filter, and move to state *S_loading*. On subsequent clock edges, the machine remains in *S_loading* and sequentially reads the parameters of the filter. After reading the parameters the state enters *S_running*. While in *S_running*, the machine generates the filtered output (D_out) from the input signal (D_in) until *reset* is asserted again. Synthesize the model and verify the functionality of the gate-level circuit.

11. Develop, verify, and synthesize parameterized Verilog models of the DF-II and TDF-II IIR filters (see Figure 9-27). The filters must import their tap coefficients from the test bench environment.

12. Compare the results of synthesizing *Circular_Buffer_1* and *Circular_Buffer_2* (see Example 9.9).

13. Verify that the 32-bit serial-to-parallel converter *Ser_Par_Conv_32* operates correctly when the data bits are streamed to the machine continuously (i.e., the state does not return to *S_idle* between successive words).

14. Modify the eight-tap Gaussian FIR in Example 9.2 to have a 4-bit representation of the filter's coefficients, but keeping an 8-bit datapath. Compare this modified filter to the filter modeled by *FIR_Gaussian_Lowpass*. Consider their use of physical resources (e.g., configurable logic blocks in a Xilinx FPGA), their accuracy, and their performance.

15. Implement and compare two different architectures for an eight-tap FIR with a 16-bit datapath. The first is to use the architecture shown in Figure 9-23, which has a shift-register structure that stores and shifts the samples of the input sequence. The second is to implement the FIR with a circular buffer controlled by a state machine.

16. Using MATLAB, design a lowpass FIR filter with passband frequency = 1600 Hz, stopband frequency = 2400 Hz, passband gain = 0 dB, stopband attenuation = 20 dB, and sampling rate = 8000 Hz. The datapath at the input is 32 bits wide, and the tap coefficients are to be stored as 16-bit words. Verify that the filter satisfactorily attenuates an input at 2.5 kHz and 3 kHz. Determine the highest sampling rate that can be achieved in the technology that was used to synthesize the filter. (*Optional*: Implement the filter in an FPGA and demonstrate its operation.)

17. Adaptive digital filters are commonly used to (1) filter noise from data whose statistics are either unknown or time-varying and (2) to extract a model of an unknown system from data describing its input–output response. The parameters of an adaptive digital filter are adjusted dynamically, as the statistics of the data evolve during processing. A feedback loop drives an adaptation process

that compares the output of the unknown system with the output of the filter and uses the derived information to adjust the filter's weights. In the structure shown in Figure P9-17 [7], the time-sequence response of the adaptive FIR is to approximate the time sequence of the unknown system, and the tap coefficients of the FIR are to adapt dynamically to reduce the error signal. A least mean square algorithm [20] is used to update the tap coefficients of the FIR by adjusting their values according to

$$b_{k \text{ new}} = b_{k \text{ old}} + \delta * y_{\text{error}}$$

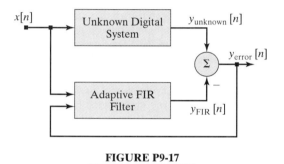

FIGURE P9-17

The stepsize adjustment parameter δ is chosen to cause the error sequence to go to 0. If δ is too big the LMS (least mean square) algorithm might not converge; if it is too small, it might converge very slowly. Consider values of δ between 10^{-2} and 10^{-4}. An adaptive FIR has been designed elsewhere [7] to filter the output of an unknown system modeled by a seventh order bandpass IIR, with a sampling rate of 8 kHz, and having $M = N = 7$. The passband of the IIR filter is from $\pi/3$ to $2\pi/3$ rad. The stopband attenuation of the filter is 20 dB. The filter's coefficients that were presented in reference 7 are given in Table P9-17, normalized to make $a_0 = 1$.

TABLE P9-17   Coefficients for a seventh-order IIR bandpass filter.

j, k	a_j	b_k
0	1.0000	0.1191
1	0.0179	0.0123
2	0.9409	−0.1813
3	0.0104	−0.0251
4	0.6601	−0.1815
5	0.0342	0.0307
6	0.1129	−0.1194
7	0.0058	−0.0178

(a) Develop and compare two implementations of the IIR filter. One is to use a pair of circular buffers, one to hold samples of the output, and one to hold samples of the input. An eight-cell circular buffer will hold the current output and a window of the last seven outputs; a seven-cell circular buffer will hold seven samples of the input. The other implementation is to use shift registers to hold the samples of data. (*Note*: The implementation will require prescaling and postscaling operations to support the arithmetic operations of the datapath with the finite wordlength of the machine.)

(b) Develop a testbench to verify that the IIR acts as a bandpass filter.

18. Find a design error in the circuit in Figure P9-18. Redesign the circuit to implement pipelining correctly.

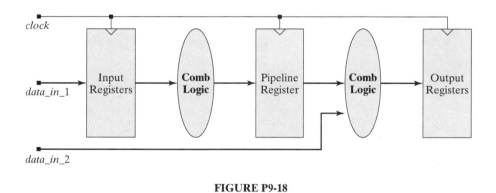

FIGURE P9-18

19. The nodes of the DFG shown in Figure P9-19 have been annotated with propagation delays. Find the optimal placement of pipeline registers in the circuit.

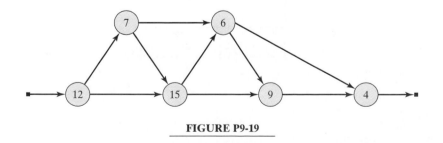

FIGURE P9-19

20. Modify the pipelined FIR filter in Figure 9-41(b) to have coherent datapaths.
21. The DFG in Figure P9-21(a) describes a systolic array processor that implements the matrix product $C = A \times B$, with $c_{ij} = \Sigma a_{ik} b_{kj}$. Each FU of a fully parallel implementation of the processor would require 8 channels of data, 4 multipliers, and 3 adders. Given a shared distribution of data among the cells in a given row and column, the entire array would require 32 channels of data, 64 multipliers, and 48 adders, for only a 4×4 matrix multiplier. As an alternative, consider an array in which the datapaths are pipelined through functional units that are chained together as shift registers. Each FU has the structure shown in Figure 9-21(b), which registers both of its input datapaths and passes the registered values to the adjacent cell in the array. Note that the datapaths between

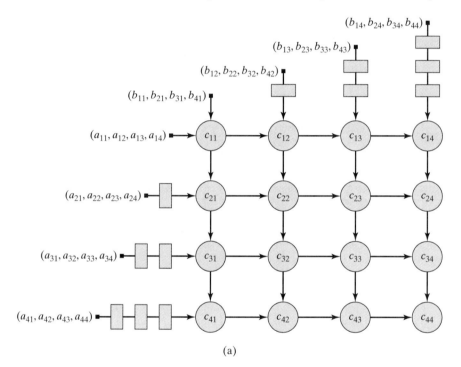

(a)

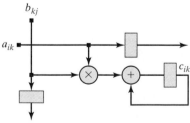

(b)

FIGURE P9-21

cells are short and that the clock must be distributed to each cell. Compare the throughputs (for a complete set of data), latency, and resources of the implementations. Develop, verify, and synthesize Verilog models of each.

22. Develop and verify a Verilog model of a FIR filter whose architecture exploits symmetry in the tap coefficients.

23. Synthesize *Integrator_Par* (see Example 9.4), and then resynthesize the model after replacing the statement clauses

> **else if** (hold) data_out <= data_out;
>   **else** data_out <= data_out + data_in;

by the statement clause:

> **else** if (!hold) data_out <= data_out + data_in;

Compare the two implementations.

24. Develop an ASM chart for a controller that will control *decimator_3* by loading four successive words, then wait two cycles before repeating the sequence, while *Go* is asserted. Otherwise, the machine remains in its reset state.

25. The buffer registers in the FIFO implementation in Example 9.10 maintain coherency of the data path by compensating for the inherent latency of the two-stage synchronizer. Consider whether an alternative design can eliminate the buffer registers by modifying the serial-to-parallel converter to anticipate the signal *write* by two clock cycles.

26. Using *t_FIFO_Buffer*, develop additional tests to verify *FIFO_Buffer* for all conditions of the stack and assertions of read and/or write operations.

27. Synthesize *FIFO_Buffer* and verify a simulation match between the behavioral and gate level models.

CHAPTER 10

Architectures for Arithmetic Processors

This chapter presents alternative architectures and algorithms for the arithmetic operations in a digital machine. Many algorithms in digital signal processing require repeated execution of arithmetic operations, so it is important that they be implemented efficiently. How these operations are implemented depends on how numbers are represented in a machine. So we will briefly examine the commonly used schemes for representing positive and negative numbers and fractions. Then we will examine algorithms and architectures for implementing addition, subtraction, multiplication, and division of fixed-point numbers.

10.1 Number Representation

Numbers are represented by a string of characters in a positional notation system having a given radix, or base. A binary number system has two symbols and a radix of 2. An n-bit unsigned binary number is represented in positional notation as $B = b_{n-1} b_{n-2} \ldots b_1 b_0$, with $b_i \in \{0, 1\}$. All digital machines encode numbers in a word of bits. The word has a fixed length, and the interpretation of the pattern of bits depends on the encoding format used by the machine.

The decimal value, B_{10}, of an n-bit unsigned binary number B is formed as a weighted sum of ascending powers of 2, with the most significant bit (MSB) having the greatest weight (2^{n-1}), and the least significant bit (LSB) having the lowest weight (2^0):

$$B_{10} = b_{n-1} 2^{n-1} + b_{n-2} 2^{n-2} + \cdots + b_1 2^1 + b_0 2^0 = \sum_{i=0}^{i=n-1} b_i 2^i$$

An n-bit word can represent 2^n distinct numbers, but the dynamic range of numbers that are realized is dependent on the encoding format. An n-bit unsigned binary format can represent decimal numbers from 0 to $2^n - 1$. For example, the decimal value of an 8-bit unsigned binary number can range from 0_{10} (0000_0000_2) to 127_{10} (1111_1111_2).

In general, a binary number can be expressed as a weighted sum of ascending and descending powers of 2:

$$B = b_{n-1}b_{n-2}\ldots, b_1\, b_0\, b_{-1}\, b_{-2}\ldots b_{-m+1}\, b_{-m}$$

with decimal value

$$B_{10} = b_{n-1}\, 2^{n-1}\, b_{n-2}\, 2^{n-2} + \cdots + b_1\, 2^1 + b_0\, 2^0 + b_{-1}\, 2^{-1} + b_{-2}\, 2^{-2} + \cdots$$
$$+ b_{-m+1}\, 2^{-m+1}\, b_{-m}\, 2^{-m}$$

and

$$B_{10} = \sum_{i=-m}^{i=n-1} b_i\, 2^i$$

The weights having a negative power of 2 form the fractional part of the number, and the radix point (.) separates the integer part of the number from its fractional part. The radix point for an n-bit integer is located immediately to the right of its LSB; the radix point of an m-bit fraction is located immediately to the left of its MSB. Fixed-point numbers have their radix point in a specific position in a computer word [1]. The radix point is not implemented physically in the machine; the designer must keep track of its location, which may vary as arithmetic operations are performed.

The arithmetic sign of a number in a digital machine must be encoded within the bits of a word. There are three common formats for signed numbers: signed magnitude, 1s complement, and 2s complement. Of these, the 1s complement and 2s complement play a significant role in arithmetic units.

10.1.1 Signed Magnitude Representation of Negative Integers

In signed magnitude representation of positive and negative numbers, the MSB of a word is the encoded sign bit, with 0 representing a positive value, and 1 representing a negative value. The remaining bits of the word represent the magnitude of the number. For example, 0111_2 represents $+7$, and 1111_2 represents -7_{10}. Eight-bit signed-magnitude numbers can be represented by the number wheel shown in Figure 10-1. The dynamic range of numbers in signed magnitude representation from $-2^{n-1} - 1$ to $+2^{n-1} - 1$.

If the signs of two signed-magnitude numbers match, addition is executed directly by adding the magnitudes (not the sign bits) and setting the sign of the result to match the sign of the operands (e.g., $-2_{10} + -3_{10} = 1010_2 + 1011_2 = 1101_2 = -5_{10}$). If the sign bits of the numbers do not match, the signs and relative magnitudes of the words must be used to determine whether to add or subtract the numbers and to determine the sign of the outcome (see Katz [2] for examples). Having two representations for 0

Positive range: 0 to $2^7 - 1 = 127$
Negative range: -0 to -127

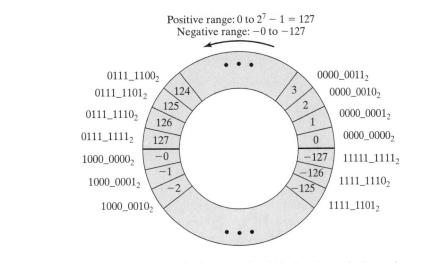

FIGURE 10-1   Number wheel representing 8-bit signed-magnitude numbers.

complicates arithmetic operations on signed binary numbers. Hardware units do not directly implement addition and subtraction of signed magnitude numbers.

10.1.2 Ones Complement Representation of Negative Integers

Positive numbers are represented in a 1s complement system in the same way they are represented in a signed-magnitude system, but negative 1s complement numbers are represented differently, in a 1s complement format. The number wheel for 1s complement numbers is shown in Figure 10-2. Note that 0_{10} has two representations: 0000_0000_2

Positive range: 0 to $2^7 - 1 = 127$
Negative range: -127 to -0

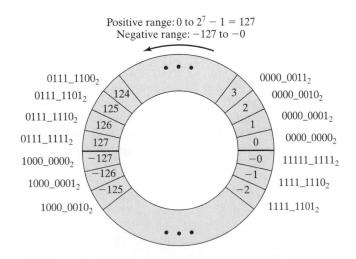

FIGURE 10-2   Number wheel representing 8-bit 1s complement numbers.

and 1111_1111_2. The dynamic range of numbers in an n-bit 1s complement system is from $-2^{n-1} -1$ to $+2^{n-1} -1$, the same as for signed magnitude numbers, but negative numbers are encoded differently in the two formats.

The 1s complement of an n-bit binary number B, denoted by $-B$, is defined by

$$B + (-B) = 2^n - 1$$

Because $2^n - 1$ is an n-bit word consisting of all 1s, the 1s complement of B is the word that must be added to B to form an n-bit word of 1s. That word is the bitwise complement[1] of B, denoted by $\sim B$, and $-B = \sim B$. For example, if $B = 1010_2$, the 1s complement of B is $\sim B = 0101_2$, and $B + (\sim B) = 1010_2 + 0101_2 = 1111_2$. Machines easily implement the bitwise complement operation with inverters.

In a 1s complement format, each positive number has a negative counterpart, including the number 0. The numbers are said to be self-complementing, because the complement of a number is obtained by taking the diminished radix complement of the bits of the word. With a radix of 2, the diminished radix complement (1s complement) of a binary word is the word formed by complementing its bits. The 1s complement format easily implements subtraction, but addition requires treatment of two different representations for 0.[2] Consequently, most digital machines use a 2s complement encoding scheme, which has a unique representation for every number, including 0 [2].

10.1.3 Twos Complement Representation of Positive and Negative Integers

The 2s complement of an n-bit binary integer is defined by $B^* = 2^n -B$, so $B + (B^*) = 2^n = 0$ modulo n. Adding 1 to the 1s complement of a word forms its 2s complement. The range of numbers that can be represented in a 2s complement format is from -2^{n-1} to $+ (2^{n-1} - 1)$. Twos complement numbers can be represented by the number wheel shown in Figure 10-3. For $n = 8$, numbers are incremented counterclockwise by 1, from 0 to $2^7 - 1 = 127_{10}$. In the clockwise direction, the count descends from 0 to $-2^7 = -128_{10}$. Figure 10-4 demonstrates that the 2s complement of the 2s complement of a number is the number itself. One full rotation around the wheel returns the pointer to the same location as the pointer to 0, and increments the count from $2^{n-1} -1$ to 2^n.

The 2s complement system is important in design because subtraction of 2s complement numbers has a very simple hardware realization, and the arithmetic operations of addition, subtraction, multiplication, and division can all be performed in hardware with a unit that can carry out binary operations of addition and bitwise complement. Subtraction involves the operations of bitwise complement and addition. Multiplication involves repeated addition, and division involves repeated subtraction. Arithmetic operations on 2s complement numbers use the same hardware for addition and subtraction but have reduced dynamic range compared to a unsigned binary format.

---

[1] The bitwise complement operator ($\sim$) is included in the Verilog language to support arithmetic operations.
[2] The signed magnitude format also has two representations for 0.

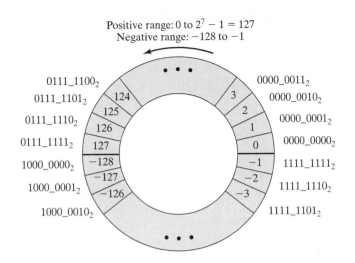

FIGURE 10-3 Number wheel representing 8-bit 2s complement numbers.

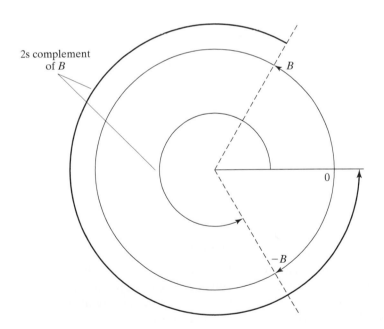

FIGURE 10-4 The 2s complement of the 2s complement of a number B is B.

The signed decimal value of a number in 2s complement format can be obtained directly, as

$$D(B) = -b_{N-1} 2^{N-1} + b_{N-2} 2^{N-2} \cdots + b_1 2^1 + b_0 2^0$$

10.1.4 Representation of Fractions

The 2s complement of a fraction is defined by

$$B^* = 2 - B \quad \text{and} \quad B + B^* = 2.$$

The 2s complement of a fraction can found by starting at the LSB and complementing all the digits to the left of the least significant 1 in the word, which is the same procedure that is used to form the 2s complement of an integer. The 2s complement fraction $1.00000 \ldots$ is a special case. It actually represents the number -1, since the sign bit is negative, and the 2s complement of $1.0000 \ldots$ is $2 - 1 = 1$. The integer $+1$ cannot be represented in the 2s complement fraction system, since $0.111 \ldots$ is the largest positive fraction.

10.2   Functional Units for Addition and Subtraction

Addition and subtraction are implemented in all arithmetic processors. There are several alternatives that provide a tradeoff between hardware cost and performance.

10.2.1 Ripple-Carry Adder

The ripple-carry adder presented in Chapter 4 is limited by the time required to propagate a signal transition from the carry-in bit to the carry-out bit of the unit. If the word length of the processor using the adder is large, it might be necessary to use an alternative architecture to form the outputs quickly enough to satisfy timing constraints. Pipelining the dataflow is one alternative (see Chapter 9), but it introduces latency and requires hardware to implement the pipeline registers. Another alternative is to consider other algorithms for addition. Among those that are used are the carry look-ahead algorithm, the carry select algorithm, and the carry-save algorithm [3]. Of these, we will consider the look-ahead algorithm. In Chapter 11 we will identify important tradeoffs between the ripple-carry adder and the carry look-ahead adder.

10.2.2 Carry Look-Ahead Adder

The algorithm for a carry look-ahead adder is based on the observation that the value of the carry into any stage of a multicell adder depends on only the data bits of the previous stages and the carry into the first stage. This relationship can be exploited to improve the speed of the adder by using additional logic to implement the carry, rather than waiting for the value to propagate through the cells of the adder.

A given cell is said to generate a carry if both of the cell's data bits are 1. A cell is said to propagate a carry if either of the cell's data bits could combine with the carry into the cell to cause a carry out to the next stage of the adder. Let a_i and b_i be the data bits at the ith cell of the adder, let c_i be the carry into the ith cell, let s_i be the sum bit *out* of the ith cell, and let c_{i+1} be the carry out of the cell. We define *generate* and *propagate* bits g_i and p_i using the bitwise-and operator (&) and the exclusive-or operator ($\wedge$) as follows:

$$g_i = a_i \,\&\, b_i$$
$$p_i = a_i \wedge b_I$$

The Venn diagram in Figure 10-5 shows where p_i and g_i are asserted, depending on a_i and b_i. Note that p_i and g_i are mutually exclusive.

The logical expressions forming the sum and carry bits at each stage of the adder can be written in terms of the Verilog bitwise operators as follows:

$$s_i = (a_i \wedge b_i) \wedge c_i = p_i \wedge c_i$$

$$c_{i+1} = ((a_i \wedge b_i) \,\&\, c_i)|(a_i \,\&\, b_i) = (p_i \,\&\, c_i)| \, g_i$$

Note that because p_i and g_i are mutually exclusive the algorithm can also be expressed in arithmetic terms as:

$$s_i = (a_i \wedge b_i) \wedge c_i = p_i \wedge c_i$$
$$c_{i+1} = (a_i \wedge b_i) \,\&\, c_i + a_i \,\&\, b_i = p_i \,\&\, c_i + g_i$$

The carry bit can be formed using either the bitwise OR operator or the arithmetic sum (modulo 2 addition). The schematic for the subcircuit implementing the algorithmic version of the carry is shown in Figure 10-6.

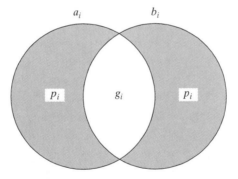

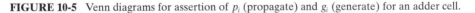

FIGURE 10-5 Venn diagrams for assertion of p_i (propagate) and g_i (generate) for an adder cell.

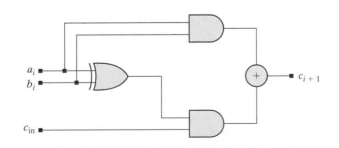

FIGURE 10-6 Schematic for arithmetic implementation of carry bit.

The carry out of the ith cell is formed by adding (OR-ing) the bit propagated by the cell with the bit generated by the cell. Only one of the terms will be 1, because p_i and g_i are mutually exclusive. This second form of the equation for c_{i+1} produces the same result as the first one.

The dependencies of s_i and c_{i+1} on a_i, b_i, and c_i are depicted in the Venn diagram of Figure 10-7, where the three circular regions represent the data inputs to a cell of the adder, a_i, b_i, and c_i; each of the subregions denote where the data outputs s_i and c_{i+1} are asserted. The presence of a variable's label within a subregion of the diagram indicates that the variable is asserted for the combination of the data inputs associated with the subregion. For example, the sum bit is asserted in four subregions of the diagram: where $a_i = 1$, $b_i = 0$, and $c_i = 0$; where $a_i = 0$, $b_i = 1$, and $c_i = 0$; where $a_i = 1$, $b_i = 1$, and $c_i = 1$; or where $a_i = 0$, $b_i = 0$ and $c_i = 1$.

The cells of the adder have:[3]

$$s_0 = p_0 \wedge c_0$$
$$c_1 = (p_0 \,\&\, c_0) + g_0$$
$$s_1 = p_1 \wedge c_1 = p_1 \wedge [(p_0 \,\&\, c_0) + g_0] = p_1 \wedge (p_0 \,\&\, c_0) + p_1 \wedge g_0$$
$$c_2 = (p_1 \,\&\, c_1) + g_1 = p_1 \,\&\, p_0 \,\&\, c_0 + p_1 \,\&\, g_0 + g_1$$

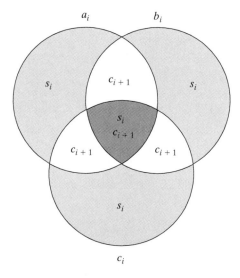

FIGURE 10-7 Data input–output relationships of an adder cell.

[3]The operators $\wedge$ and $\&$ have higher precedence than $+$.

$$s_2 = p_2 \wedge c_2 = p_2 \wedge [p_1 \& p_0 \& c_0 + p_1 \& g_0 + g_1]$$
$$c_3 = p_2 \& c_2 + g_2 = p_2 \& [p_1 \& p_0 \& c_0 + p_1 \& g_0 + g_1] + g_2$$
$$\quad = p_2 \& p_1 \& p_0 \& c_0 + p_2 \& p_1 \& g_0 + p_2 \& g_1 + g_2$$

$$p_3 = p_3 \wedge c_3 = p_3 \wedge [p_2 \& p_1 \& p_0 \& c_0 + p_2 \& p_1 \& g_0 + p_2 \& g_1 + g_2]$$
$$c_4 = p_3 \& c_3 + g_3 = p_3 \& [p_2 \& p_1 \& p_0 \& c_0 + p_2 \& p_1 \& g_0 + p_2 \& g_1 + g_2] + g_3$$

These expressions expose the fact that the sum and the carry-out bits of each cell can be expressed in terms of the data bits of that cell and of the previous cells, and the carry into only the first cell of the adder chain. All of these data are available simultaneously, so there is no need to wait for a carry bit to propagate through the adder to a particular cell. This allows the adder to operate faster, but the cost of this improvement is the extra logic needed to compute the sum and carry-out of each stage. The gate-level implementation of the look-ahead adder requires considerably more silicon area than the ripple-carry adder implemented in the same technology, and requires more testing for process-induced faults. (Look-ahead carry is usually implemented on a bit slice of a word.)

Another important observation is that the sum and carry bits at each cell can be computed recursively. To expose this we write:

$$s_0 = p_0 \wedge c_0$$
$$c_1 = p_0 \& c_0 + g_0$$

$$s_1 = p_1 \wedge c_1$$
$$c_2 = p_1 \& c_1 + g_1$$

$$s_2 = p_2 \wedge c_2$$
$$c_3 = p_2 \& c_2 + g_2$$

$$\cdots$$

The algorithm to implement the addition will take as many steps as there are cells in the adder. The computation at each step of the recursion depends on the data bits of the corresponding cell and on the carry that was calculated at the immediately previous step. If the propagate bits of an n-bit adder are arranged in a vector $p = (p_{n-1}, \ldots, p_2, p_1, p_0)$, the results of the recursion can be used to form an $n + 1$ dimensional vector: $(c_n, \ldots, c_2, c_1, c_0)$ such that the output word and the c_out bit are obtained as:

$$sum = p \wedge (c_{n-1}, \ldots, c_2, c_1, c_0).$$

$$c\_out = c_n$$

The gate-level circuit implementing the 4-bit carry look-ahead logic is shown in Figure 10-8(a), and, for comparison, the schematic of the circuit implementing the recursive algorithm is shown in Figure 10-8(b). The longest path is highlighted in both circuits. Note that the recursive algorithm implements the same circuit as a 4-bit ripple carry adder.

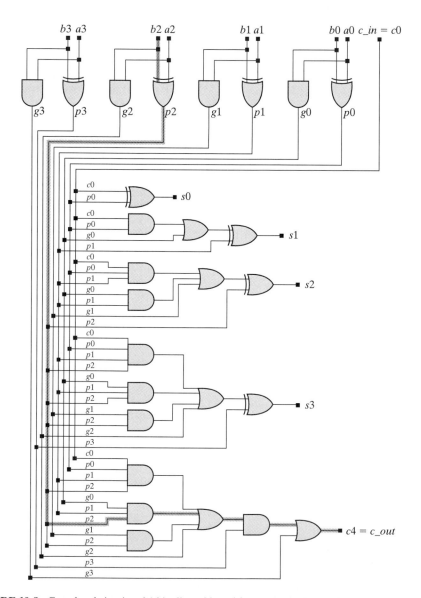

FIGURE 10-8 Gate-level circuits of 4-bit-slice adders: (a) carry look-ahead adder, and (b) ripple-carry adder. The longest paths are highlighted.

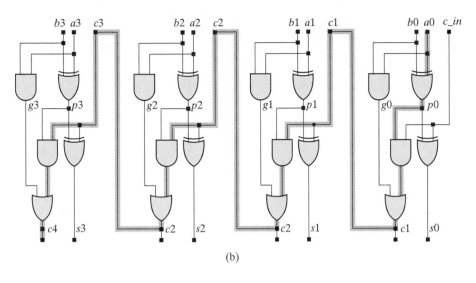

(b)

FIGURE 10-8 Continued

The propagate and generate recursive algorithm [13] for a four-bit adder is given below.

```
module Add_prop_gen (sum, c_out, a, b, c_in);
  output        [3: 0]    sum;
  output                  c_out;
  input         [3: 0]    a, b;
  input                   c_in;

  reg           [3: 0]    carrychain;
  integer                 i;
  wire          [3: 0]    g = a & b; // carry generate, continuous assignment,
                                         bitwise and
  wire          [3: 0]    p = a ^ b; // carry propagate, continuous assignment,
                                         bitwise xor

  always @ (a or b or c_in or p or g)              // event "or"
    begin: carry_generation                        // usage: block name
    integer i;
    carrychain[0] = g[0] + (p[0] & c_in);          // needed for simulation
    for(i = 1; i <= 3; i = i + 1)
      begin
      carrychain[i] = g[i] | (p[i] & carrychain[i-1]);
      end
    end
```

```
    wire [4:0] shiftedcarry = {carrychain, c_in} ;        // concatenation
    wire [3:0] sum = p ^ shiftedcarry;                    // summation
    wire c_out = shiftedcarry[4];                         // carry out, usage: bit select
endmodule
```

Hardware units usually implement subtraction by adding the 1s complement of the subtrahend to the minuend, and then adding 1 to the result. This can be implemented with the architecture shown in Figure 10-9. One adder unit can be used for addition or subtraction, depending on the value of the signal *select*.

10.2.3 Overflow and Underflow

Overflow occurs under two conditions: (1) when adding two positive numbers produces a sum that exceeds the largest positive number that can be represented in the word length of the unit (i.e. the result is negative) and (2) when adding two negative numbers produces a sum that is positive (i.e. the sum exceeds the smallest negative number that can be represented in the word length of the machine). Arithmetic units include logic for underflow and overflow detection.

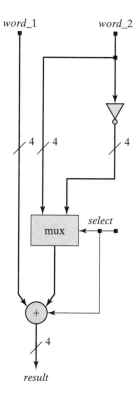

FIGURE 10-9 Architecture for a combined 4-bit datapath adder and subtractor unit.

10.3 Functional Units for Multiplication

Multipliers are important functional elements of arithmetic units, digital signal processors, and other circuits that execute arithmetic operations. Multiplication can be implemented with a combinational circuit or by a sequential circuit. A combinational circuit that multiplies two numbers will require more silicon area, but will operate faster than a sequential multiplier. Sequential multipliers are attractive because they require less area, but a complete multiplication takes several clock cycles to form the product. We will investigate various designs of multipliers for signed and unsigned numbers, beginning with binary multipliers, which form the product of a pair of binary words (i.e., unsigned numbers). We will also design circuits for multiplying fractions. Our approach will be to first present a basic architecture for a multiplier and then present alternative architectures with additional features that enhance the design.

10.3.1 Combinational (Parallel) Binary Multiplier

Consider two binary (unsigned) numbers with the following representation:

$$A = (A_{m-1}, A_{m-2}, \ldots, A_1, A_0)_2 = \sum_{i=0}^{m-1} A_i \, 2^i$$

$$B = (B_{n-1}, B_{n-2}, \ldots, B_1, B_0)_2 = \sum_{j=0}^{n-1} B_j \, 2^j$$

Their product can be written as

$$A \times B = \sum_{i=0}^{m-1} A_i \, 2^i \sum_{j=0}^{n-1} B_j \, 2^j$$

$$A \times B = \sum_{i=0}^{m-1} \sum_{j=0}^{n-1} A_i \, B_j \, 2^{i+j}$$

The product has $m \times n$ terms, and their summation may produce one more term, so the final result can be written as a sum of weighted powers of 2:

$$A \times B = \sum_{k=0}^{m+n-1} P_k \, 2^k$$

where $P_0 = A_0 B_0$, $P_1 = A_1 B_0 + A_0 B_1$, etc. The term $P_{m+n-1} \, 2^{m+n-1}$ accounts for a possible carry.

Figure 10-11 illustrates the basic process for forming the product of two 8-bit binary words for the product $215_{10} \times 23_{10} = 4945_{10}$. Shifted copies of the multiplicand are successively aligned with the locations of the bits of the multiplier word, then the columns are added. If a multiplier bit is 0, the corresponding copy of the multiplicand is skipped, and the location that will be occupied by the next copy is shifted by one position toward the MSB of the multiplier. For example, a double shift is shown in Figure 10-10 at the shaded location where a multiplier bit is 0.

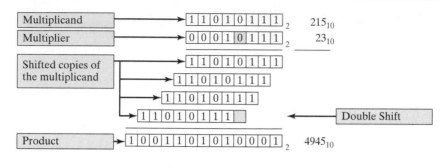

FIGURE 10-10   Formation of a product by columnwise adding shifted copies of the multiplicand.

The process shown in Figure 10-10 works manually, and a combinational circuit can be developed to implement the product, with binary multiplication formed by an AND gate. Note, however, that the process uses columnwise addition of the terms that form the partial products, and would require a hardware scheme that uses multiple adders for each column [2]. An ordinary adder operates on only two words at a time, so a more attractive scheme that forms a sequence of row sums by adding a single copy of the multiplicand to an accumulated product will be presented here.

Figure 10-11 shows how the multiplication evolves. First, a pair of appropriately shifted copies of the multiplicand are added to form a sum, then another shifted copy of the multiplicand is added, and so forth, to accumulate the sums that ultimately form the product. This scheme is attractive, because it adds only two words at a time, has a direct hardware counterpart, and can be described by a sequential behavior in a hardware description language (HDL) model.

The structure of a combinational circuit that multiplies two unsigned binary numbers (words) can be derived from the manual operations for multiplying the numbers in a radix 2 system. For simplicity, we consider two 4-bit words, A and B (the multiplicand and

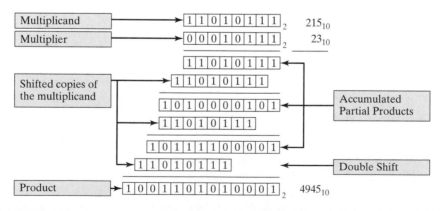

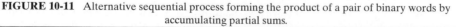

FIGURE 10-11   Alternative sequential process forming the product of a pair of binary words by accumulating partial sums.

2^7	2^6	2^5	2^4	2^3	2^2	2^1	2^0	
				$A3$ $B3$	$A2$ $B2$	$A1$ $B1$	$A0$ $B0$	Multiplicand Multiplier
			A_3B_1 C_{12}	A_3B_0 A_2B_1 C_{11}	A_2B_0 A_1B_1 C_{10}	A_1B_0 A_0B_1	A_0B_0 S_{00}	Partial Product 0 Partial Product 1 1st Row Carries
		C_{13} A_3B_2 C_{22}	S_{13} A_2B_2 C_{21}	S_{12} A_1B_2 C_{20}	S_{11} A_0B_2	S_{10}		1st Row Sums Partial Product 2 2nd Row Carries
	C_{23} A_3B_3 C_{32}	S_{23} A_2B_3 C_{31}	S_{22} A_1B_3 C_{30}	S_{21} A_0B_3	S_{20}			2nd Row Sums Partial Product 3 3rd Row Carries
C_{33}	S_{33}	S_{32}	S_{31}	S_{30}				3rd Row Sums
P_7	P_6	P_5	P_4	P_3	P_2	P_1	P_0	Final Product
2^7	2^6	2^5	2^4	2^3	2^2	2^1	2^0	Weight

FIGURE 10-12 Steps in the multiplication of unsigned 4-bit binary words.

multiplier, respectively), and we form their product, $A \times B$, as shown in Figure 10-12. Beginning with the LSB of the multiplier, each bit is multiplied by the bits of the multiplicand to form a so-called partial product. In a radix 2 system, the operation of multiplication that forms a partial product is equivalent to AND-ing the multiplier bit with each bit of the multiplicand. Each partial product is shifted toward the MSB to the position of the corresponding multiplier bit. Then the partial products are added. As we noted earlier, a manual method would add the terms in the aligned columns of the partial products, which, in general, requires addition of several terms, but hardware adders are designed to add only two words. Consequently, the result of adding the rows of the partial products is accumulated, with attention to any carries that are generated by the addition of the terms in a given column. The structure will require full adders when carries are involved and half adders otherwise. In general, the resulting final product may contain as many as $2*L\_word$ significant bits, where L_word is the length of the multiplier and multiplicand words.

The combinational logic structure consisting of AND-gates, full adders, and half adders shown in Figure 10-13 implements a parallel multiplier for a 4-bit wide data-path. This method of multiplication is called *partial product accumulation* [2], because rows of linked adders generate the accumulated partial products that would evolve from a manual multiplication of the data words, as we showed in Figure 10-11. Most of the array is formed of linked copies of the basic cell shown in Figure 10-13. An alternative structure, composed entirely of basic cells, is shown in Figure 10-14, implementing half adders as full adders with their carry-in line hard wired to 0. The resulting regular array of identical objects is called a systolic array (i.e., copies of a basic cell) and is ideally

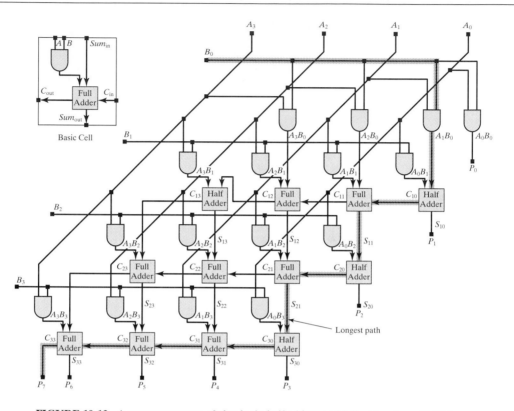

FIGURE 10-13   An array structure of glue logic, half adders, and full adders for a 4-bit binary multiplier.

suited for fabrication as an integrated circuit [5] In practice, the boundary cells can be replaced by their counterparts from Figure 10-13. A 4-bit combinational multiplier has 16 AND gates, 8 full adders, and 4 half adders. An 8-bit multiplier would extend the array structure to accommodate 8-bit input datapaths, producing a 16-bit output data-path for the product. The systolic array in Figure 10-14 is attractive because it has a regular structure of identical cells and easily accommodates expansion to longer word length by direct abutment of cells, which leads to an area-efficient physical layout of the cell on a die, with short interconnect paths between cells. Because the structure is isomorphic to a dataflow graph, it can be used to identify cutsets for pipelining the structure to obtain greater throughput (see Problem 40 at the end of this chapter).

In synchronous operation, the clock cycle that governs the presentation of data to the multiplier must accommodate the longest path through the circuit, which is the path from the LSBs, through the adders, to the MSB of the product. Both the carry and sum paths of the adders affect the longest path, and balanced delays through them are desirable [1]. The area of the device is relatively large, compared to other realizations, such as the sequential multipliers that will be considered below, but the extra area is the price of the superior performance of the combinational multiplier.

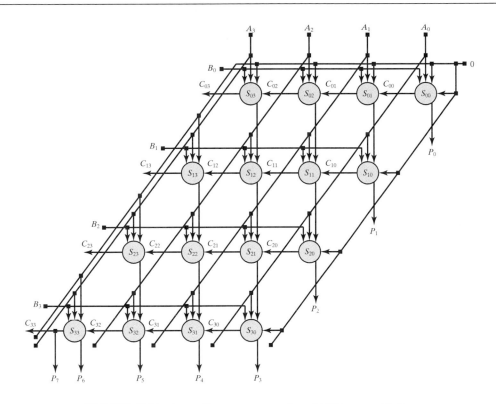

FIGURE 10-14 A systolic array structure for a 4-bit binary multiplier.

10.3.2 Sequential Binary Multiplier

Combinational (array) multipliers operate fast, but require a significant amount of silicon area. If area is an important consideration, it can be reduced at the expense of performance by scheduling the suboperations of the multiplier to execute in successive clock cycles. Sequential multipliers are compact, require fewer adders, and are amenable to pipelining. The area required by combinational multipliers grows geometrically with the word length, but we will see that the area of a sequential multiplier does not grow significantly with word length and that the number of clock cycles required to complete a multiplication also grows in a linear manner, rather than exponentially, with the word length. We will also see that the behavioral description of a sequential multiplier is parameterizable, which makes the model portable, amenable to synthesis, and re-usable.

The sequence of operations forming the product of binary numbers by adding shifted copies of the multiplicand to an accumulated product were implemented by a combinational circuit in Figure 10-13. The schematic of the circuit suggests how to form a sequential behavioral model of the multiplier, one that eliminates the spatial distribution of adders in exchange for a temporal distribution of computation that uses storage registers and a single adder.

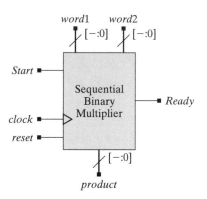

FIGURE 10-15   Interface signals and block diagram symbol for a sequential binary multiplier.

The interface signals and the block diagram symbol of a 4-bit sequential binary multiplier are shown in Figure 10-15, where $[-:0]$ denotes a parameterizable bit range for a datapath (e.g., $[L\_word -1:0]$ for *word1* and *word2*, and $[2*L\_word -1:0]$ for *product*). The datapaths for *word1, word2,* and *product* hold the multiplicand, multiplier, and product words, respectively. The signal *Ready* asserts when the unit is ready to execute a multiplication sequence, and when a valid product has been formed at the end of an execution sequence. *Ready* asserts until *Start* initiates a multiplication sequence, and re-asserts when a valid product is formed. We will now consider architectural alternatives and methodologies for designing a sequential binary multiplier of unsigned words.

10.3.3 Sequential Multiplier Design: Hierarchical Decomposition

The method for designing a sequential binary multiplier has two main steps: (1) choose a datapath architecture and (2) design a state machine to control the datapath. For a given datapath architecture, the state machine must generate the appropriate sequence of control signals to direct the movement of data to produce the desired product.

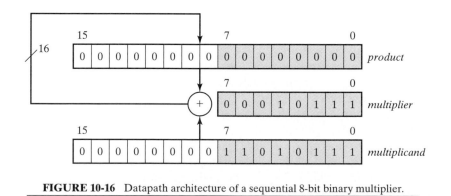

FIGURE 10-16   Datapath architecture of a sequential 8-bit binary multiplier.

The first datapath architecture we will consider is shown in Figure 10-16, for an 8-bit datapath. Along with a single adder, shift registers are allocated for the multiplicand and multiplier, and a fixed register is allocated for the product. The multiplicand register is sized to the length of the *Product* register to accommodate the shifting operations at each cycle. The controller must assert *Ready* and then wait for an external agent to signal *Start*. When *Start* is asserted, the controller must de-assert *Ready*, load the registers, direct the shifting and adding of data to form the product, and finally re-assert the signal *Ready*. The bits of the multiplier register control the shifting and adding operations that form the accumulated product.

The architecture in Figure 10-16 uses only one adder and separate registers to hold *multiplier, multiplicand*, and *product*. The combinational multiplier in Figure 10-13 does not use storage registers, but it requires that the data words be held externally until *product* is formed. This has implications for external storage or bus utilization. In contrast, the sequential adder can release its external datapaths as soon its registers are loaded. Note that long words do not require more area for logic, other than that needed to properly size the registers and the adder.

The controller of the sequential multiplier architecture in Figure 10-16 loads the multiplicand and multiplier words into their registers, then shifts *multiplicand* to the left, relative to the *product* register. At each step, the *multiplier* register is also shifted, but to the right, and the value of *multiplier[0]* determines whether *multiplicand* is added to *product*. The synchronous movement of data is depicted in Figure 10-17 for the product $215_{10} \times 23_{10} = 4945_{10}$, where the addition operation is executed as parallel addition of the entire words. The structure uses a single adder and distributes the operation of addition over multiple clock cycles.

Partitioning the design into a datapath and controller gives the Verilog structural decomposition shown in Figure 10-18, where *Controller* and *Datapath* are separately encapsulated Verilog modules instantiated within a top-level module, *Multiplier_STG_0*. In this structure, *m0* is the LSB of *multiplier*, and is passed to the controller and used to control its state transitions. A fully structural approach to the design would be to write and link structural descriptions of shift registers, adders, and ordinary registers, and then develop (e.g., from timing charts and Boolean expressions) a state machine to control the structure. In contrast, our approach will be to start with the high-level structural partition and then to write behavioral descriptions of the functional units, with reliance on a synthesis tool to determine the actual physical structure. This approach can maximize a designer's productivity, by leveraging the Verilog language and readily available synthesis tools. We will use it to examine alternative datapath architectures, controllers, and design methods and to explore tradeoffs. We will also expose some of the subtle, but serious, traps that await the unwary designer.

10.3.4 STG-Based Controller Design

The first design method will use state-transition graphs (STGs) to specify the state transitions of the controller. Figure 10-19 shows two versions of a STG for the controller, with different behavior in *S_8*. State transitions occur at the active edge of a clock, and are governed by the conditions annotated on the arcs of the graph (i.e., the machine will remain in a given state until the condition is satisfied). An arc without

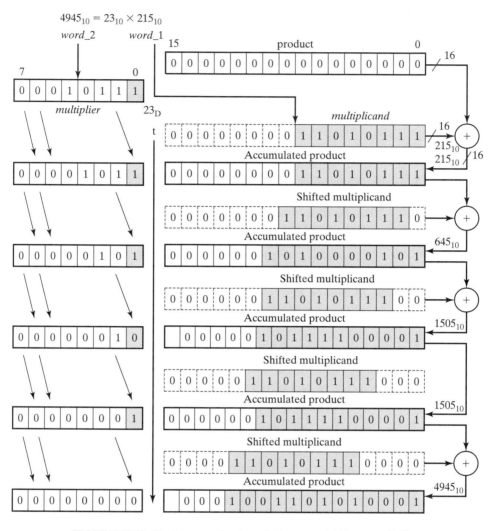

FIGURE 10-17　Register transfers in an 8-bit sequential binary multiplier.

annotation denotes an unconditional transition; – denotes a don't-care condition. Signals that are not explicitly asserted are de-asserted. Under the action of *reset*, the machines enter *S_idle* from any state and remain there, with *Ready* asserted, until *Start* is asserted with *reset* de-asserted. (An alternative design could assert *Ready* as a Mealy output only while *reset* is de-asserted in *S_idle.*) Thereafter, the state transitions depend on the LSB of the shifted multiplier. If the LSB is a 1, the signal *Add* is asserted and a transition is made to a state from which *Shift* will be asserted at the next active edge of *clock*. If the LSB of the multiplier is 0, *Shift* is asserted. When *S_8* is entered, *Ready* is asserted as a Moore-type output. At the next active edge of the clock, the machine in Figure 10-19(a) transitions to *S_idle* to await an assertion of *Start*. The version in Figure 10-19(b) remains in *S_8*, with *Ready* asserted, until *Start* is asserted. Then a transition is made to

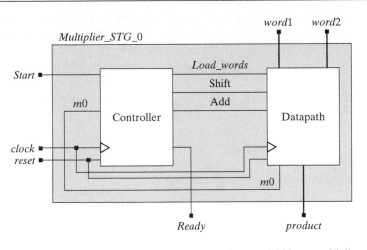

FIGURE 10-18 Structural units of a partitioned sequential binary multiplier.

S_1, rather than to S_idle. Note that once either machine has entered S_1 it ignores activity on the input datapath until the multiplication is complete.

The Verilog description of *Controller*, given below, is based on the STG in Figure 10-19(b). In either machine, a complete multiplication can require passage

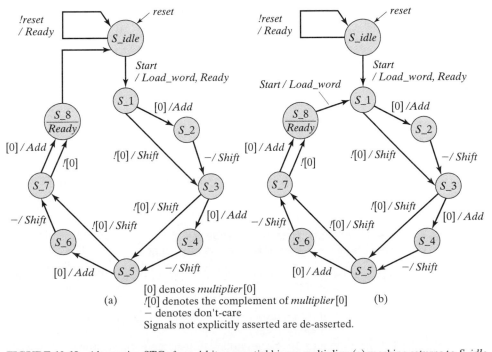

(a)

(b)

[0] denotes *multiplier* [0]
![0] denotes the complement of *multiplier* [0]
– denotes don't-care
Signals not explicitly asserted are de-asserted.

FIGURE 10-19 Alternative STGs for a 4-bit sequential binary multiplier: (a) machine returns to S_idle upon completion of multiplication, and (b) machine resides in S_8 after completion of multiplication.

through as few as five states and as many as eight states after *Start* is asserted. The machine in Figure 10-19(b) can reach a result in one less state after *Ready* is asserted in continuous operation, because the state does not pass through *S_idle*. Note that the size of the STG in Figure 10-19 expands linearly with the size of the datapath. This has implications for the utility of an STG-based method in which the sequence of states depends on the size of the datapath, because the model would have to be edited significantly to accommodate a different word size.

The Verilog behavioral description of *Multiplier_STG_0* is given in Example 10.1, along with a testbench that contains an exhaustive pattern generator for *word1* and *word2*, and a comparator that samples *product* and checks, while *Done* is asserted, whether *product* matches the expected value.

Example 10.1

```
module Multiplier_STG_0 (product, Ready, word1, word2, Start, clock, reset);
  parameter                       L_word = 4;        // Datapath size
  output        [2*L_word -1: 0]  product;
  output                          Ready;
  input         [L_word -1: 0]    word1, word2;
  input                           Start, clock, reset;
  wire                            m0, Load_words, Shift;

  Datapath M1 (product, m0, word1, word2, Load_words, Shift, Add, clock, reset);
  Controller M2 (Load_words, Shift, Add, Ready, m0, Start, clock, reset);
endmodule

module Controller (Load_words, Shift, Add, Ready, m0, Start, clock, reset);
  parameter                       L_word = 4;        // Datapath size
  parameter                       L_state = 4;       // State size
  output                          Load_words, Shift, Add, Ready;
  input                           m0, Start, clock, reset;
  reg           [L_state -1: 0]   state, next_state;
  parameter                       S_idle = 0, S_1 = 1, S_2 = 2;
  parameter                       S_3 = 3, S_4 = 4, S_5 = 5, S_6 = 6;
  parameter                       S_7 = 7, S_8 = 8;
  reg                             Load_words, Shift, Add;
  wire                            Ready = ((state == S_idle) && !reset) ||
                                  (state == S_8);

  always @ (posedge clock or posedge reset)          // State transitions
    if (reset) state <= S_idle; else state <= next_state;

  always @ (state or Start or m0) begin              // Next state and control logic
    Load_words = 0; Shift = 0; Add = 0;              // Default values
    case (state)
      S_idle:     if (Start) begin Load_words = 1; next_state = S_1; end
                  else next_state = S_idle;
      S_1:        if (m0)    begin Add = 1; next_state = S_2; end
                  else       begin Shift = 1; next_state = S_3; end
      S_2:                   begin Shift = 1; next_state = S_3; end
```

```
        S_3:        if (m0)   begin Add = 1; next_state = S_4; end
                    else      begin Shift = 1; next_state = S_5; end
        S_4:                  begin Shift = 1; next_state = S_5; end
        S_5:        if (m0)   begin Add = 1; next_state = S_6; end
                    else      begin Shift = 1; next_state = S_7; end
        S_6:                  begin Shift = 1; next_state = S_7; end
        S_7:        if (m0)   begin Add = 1; next_state = S_8; end
                    else      begin Shift = 1; next_state = S_8; end
        S_8:        if (Start) begin Load_words = 1; next_state = S_1; end
                    else next_state = S_8;
      default:      next_state = S_idle;
    endcase
  end
endmodule

module Datapath (product, m0, word1, word2, Load_words, Shift, Add, clock,
     reset);
  parameter                          L_word = 4;
  output        [2*L_word -1: 0]     product;
  output                             m0;
  input         [L_word -1: 0]       word1, word2;
  input                              Load_words, Shift, Add, clock, reset;
  reg           [2*L_word -1: 0]     product, multiplicand;
  reg           [L_word -1: 0]       multiplier;

  wire                               m0 = multiplier[0];

  // Register/Datapath Operations
  always @ (posedge clock or posedge reset) begin
    if (reset) begin multiplier <= 0; multiplicand <= 0; product <= 0; end
    else if (Load_words) begin
      multiplicand <= word1;
      multiplier <= word2; product <= 0;
    end
    else if (Shift) begin
      multiplier <= multiplier >> 1;
      multiplicand <= multiplicand << 1;
    end
    else if (Add) product <= product + multiplicand;
  end
endmodule

module test_Multiplier_STG_0 ();
  parameter                        L_word = 4;
  wire          [2*L_word -1: 0]   product;
  wire                             Ready;
  integer                          word1, word2;    // multiplicand, multiplier
  reg                              Start, clock, reset;

  Multiplier_STG_0 M1 (product, Ready, word1, word2, Start, clock, reset);
```

```
// Exhaustive Testbench
reg                [2*L_word -1: 0]   expected_value;
reg                                   code_error;

initial #80000 finish;                // Timeout

always @ (posedge clock) // Compare product with expected value
  if (Start) begin
    #5 expected_value = 0;
    expected_value = word2 * word1;
     // expected_value = word2 * word1 + 1; // Use to check error detection
    code_error = 0;
  end
  else begin
    code_error = (M1.M2.state == M1.M2.S_8) ? |(expected_value ^ product) : 0;
end

initial begin clock = 0; forever #10 clock = ~clock; end
initial begin
  #2 reset = 1;
  #15 reset = 0;
end
initial begin #5 Start = 1; #10 Start = 15; end          // Test for reset override
initial begin      // Exhaustive patterns
   for (word1 = 0; word1 <= 15; word1 = word1 +1) begin
   for (word2 = 0; word2 <= 15; word2 = word2 +1) begin
     Start = 0; #40 Start = 1;
     #20 Start = 0;
     #200;
   end // word2
   #140;
   end //word1
 end // initial
endmodule
```

End of Example 10.1

The controller in *Multiplier_STG_0* consists of two behaviors, one is synchronous (edge-sensitive) and the other is combinational (level-sensitive). The synchronous behavior models the state transitions at the active edges of the clock, subject to asynchronous reset action. The combinational behavior forms the next state and the output signals that control the datapath. The Verilog description of its outputs can be written directly from the annotation on the branches of the STG. *Each output variable is initialized to a de-asserted condition at the beginning of the behavior to establish default values*. Only values that are to be asserted are described in the **case** statement branch, for a given state. Note that the combinational behavior uses the procedural assignment operator (=) and includes all of the variables that are read within the behavior. Using a procedural assignment operator in the combinational logic behavior, and using a nonblocking assignment operator (<=) to make all register transfers in an edge-sensitive behavior, will avoid race-induced discrepancies between the behavioral and synthesized models of a sequential machine that has been partitioned into a datapath and a controller; the completeness of the level-sensitive event-control expression of the combinational behavior prevents unwanted latches from being synthesized. Also note that the decoding of S_8 includes an explicit assignment to *next_state* if *Start* is 0, to prevent synthesis of a latch.

The datapath unit is modeled by an edge-sensitive synchronous behavior that handles all of the register operations, under the direction of the signals generated by the controller. The behaviors of the shift registers are compactly described by the shift operator of the Verilog language. The behavior uses the nonblocking assignment operator—the register operations are concurrent, with the values held after the clock edge being determined by the values held before the clock edge.

The testbench, *test_Multiplier_STG_0* combines an exhaustive self-checking pattern generator with a checker to detect whether the model is correct. Figure 10-20 shows simulation results for all possible patterns of *word1* and *word2*. The displayed resolution obscures the actual data, but the signal *code_error* indicates that the model is correct (provided that the testbench itself is indeed correct). Figure 10-21 shows

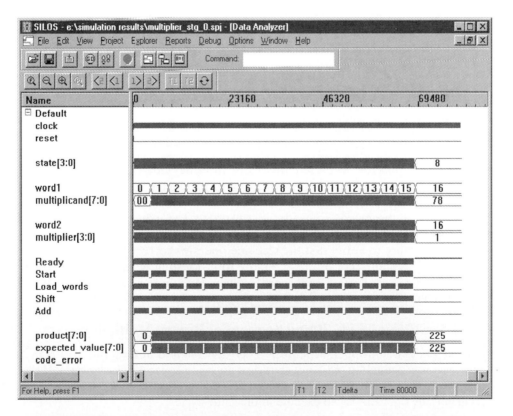

FIGURE 10-20 Results of exhaustive simulation of *Multiplier_STG_0* with a self-checking testbench.

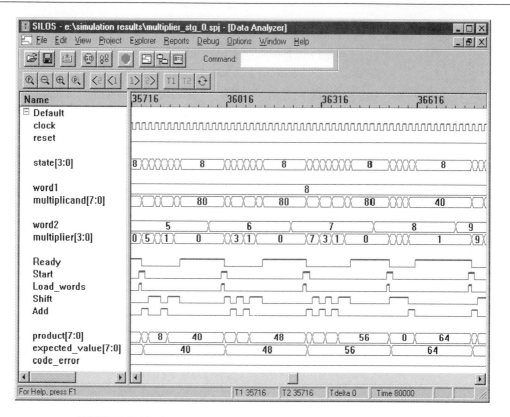

FIGURE 10-21 Sample of a multiplication sequence of *Multiplier_STG_0*.

simulation results for a representative case, with a detailed view of state transitions and control signal assertions leading to the formed product. (*Note: word1, word2, state, expected_value*, and *product* are displayed with a decimal radix; *multiplicand* and *multiplier* are displayed in hexadecimal format.) The exhaustive test results demonstrate that the multiplication operation is correct for the patterns applied. In general, it is also necessary to verify that the machine operates correctly under the action of the asynchronous inputs. For example, a more robust testing scheme would verify that the behavior is correct if *Start, reset, word1*, or *word2* changes randomly. The ability of the testbench to detect an error correctly was also verified.

10.3.5 Efficient STG-Based Sequential Binary Multiplier

The sequential multiplier, *Multiplier_STG_0*, presented in Example 10.1 is inefficient because it executes the add and shift operations in separate clock cycles. If the architecture is modified to direct the output of the adder to the appropriate bits of the product register, the operations can execute in the same cycle.

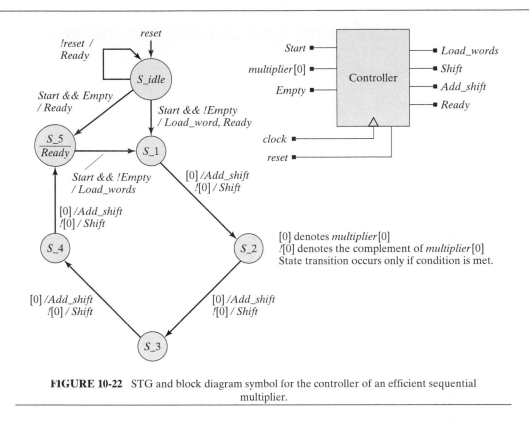

FIGURE 10-22 STG and block diagram symbol for the controller of an efficient sequential multiplier.

The modified controller described by the STG in Figure 10-22 can shorten the execution sequence by almost a factor of 2. The signal *Add* is replaced by the signal *Add_shift* to indicate the combined activity. The datapath modification can be achieved in hardware by moving wires by one position, but we will include this change in our behavioral model and leave the actual details of the wiring to a synthesis tool. Also, this new design has a more intelligent controller—it includes logic to abort the multiplication sequence if either data word presented to the multiplier is 0, in which case there is no need to multiply *word1* by *word2*. The datapath unit is modified to produce a status signal, *Empty*, to indicate the condition that an input datapath is all 0s. The port structure of the binary multiplier remains unchanged, but *Controller* is modified to include another input port, *Empty*. *Datapath* is modified to include an input port for *Start* and an output port for *Empty*.

The Verilog behavioral description of *Multiplier_STG_1* is presented in Example 10-2. Like *Multiplier_STG_0*, this machine's controller ignores *Start* if it is asserted during a multiplication sequence, but additional logic flushes *product* if *Start* is asserted with an empty data word while *Ready* is asserted. This removes the residual value of *product* that resulted from a previous multiplication sequence. Figure 10-23 shows

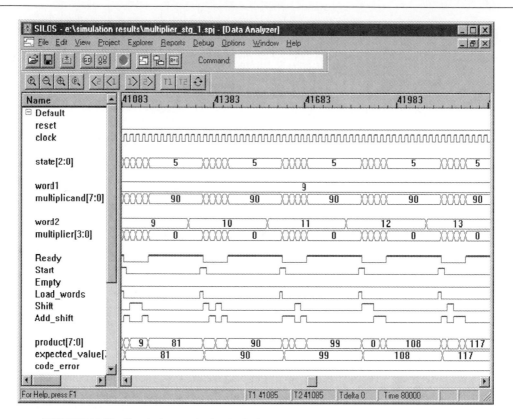

FIGURE 10-23 Simulation results for *Multiplier_STG_1*, an efficient sequential multiplier.

waveforms from an exhaustive simulation; the waveforms in Figure 10-24 demonstrate that the machine transitions from *S_idle* to *S_5* if *Empty* is asserted when *Start* is asserted. Figure 10-25 shows that *Start* is ignored if it is asserted during a multiplication sequence (i.e., here *Start* is asserted while the state is *S_2* during the multiplication of 9 [*word1*] by 15 [*word2*], and *Ready* is not asserted [e.g., 5 × 0]). The simulation results in Figure 10-26 show that *product* is flushed if *Start* is asserted while *Ready* and *Empty* are asserted.

Example 10.2

```
module Multiplier_STG_1 (product, Ready, word1, word2, Start, clock, reset);
  parameter                              L_word = 4;
  output          [2*L_word -1: 0]       product;
  output                                 Ready;
  input           [L_word -1: 0]         word1, word2;
  input                                  Start, clock, reset;
```

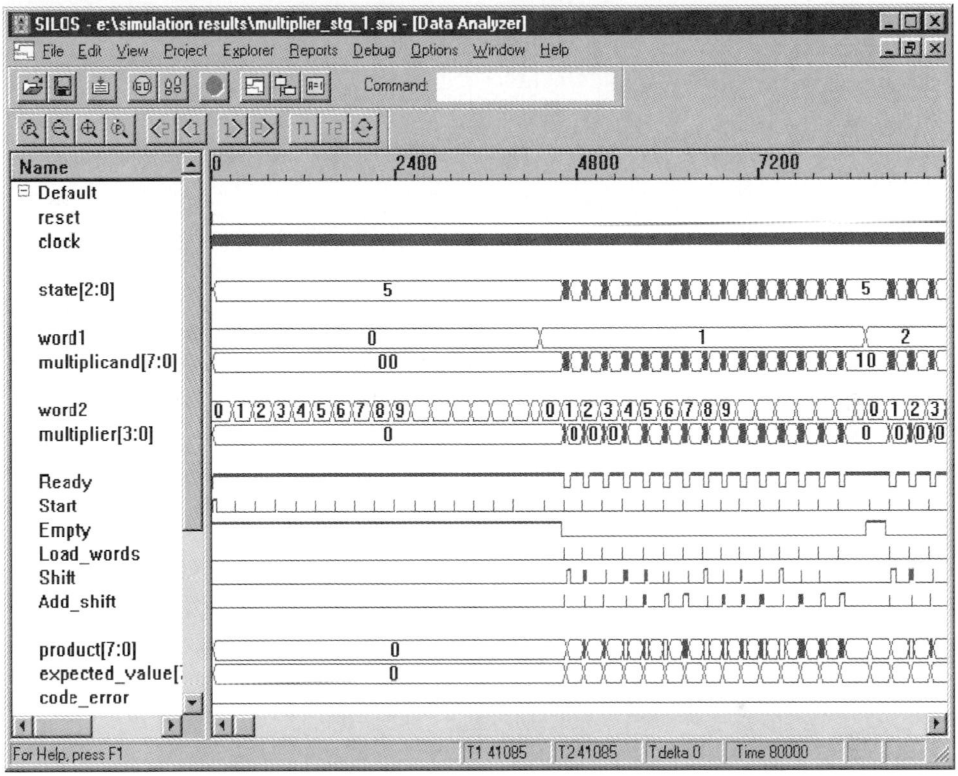

FIGURE 10-24 Simulation results for *Multiplier_STG_1*, demonstrating immediate termination if a data word is 0.

```
wire                          m0, Empty, Load_words, Shift, Add_shift;
wire                          Ready;

Datapath M1
(product, m0, Empty, word1, word2, Ready, Start, Load_words, Shift, Add_shift,
clock, reset);

Controller M2 (Load_words, Shift, Add_shift, Ready, m0, Empty, Start, clock,
reset);
endmodule

module Controller (Load_words, Shift, Add_shift, Ready, m0, Empty, Start, clock,
reset);
   parameter                  L_word = 4;
   parameter                  L_state = 3;
   output                     Load_words, Shift, Add_shift, Ready;
   input                      Empty;
```

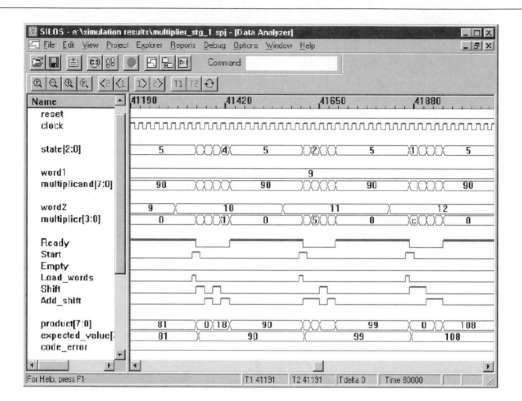

FIGURE 10-25 Simulation results for *Multiplier_STG_1*, demonstrating that the datapath unit ignores
Start while a multiplication is in progress (i.e., *Ready* is not asserted).

```
input                          m0, Start, clock, reset;

reg          [L_state -1: 0]   state, next_state;
parameter                      S_idle = 0, S_1 = 1, S_2 = 2, S_3 = 3, S_4 = 4,
                               S_5 = 5;
reg                            Load_words, Shift, Add_shift;

wire                           Ready = ((state == S_idle) && !reset) || (state
                               == S_5);

always @ (posedge clock or posedge reset)       // State transitions
  if (reset) state <= S_idle; else state <= next_state;
always @ (state or Start or m0 or Empty) begin  // Next state and control logic
  Load_words = 0; Shift = 0; Add_shift = 0;
  case (state)
    S_idle:      if (Start && Empty) next_state = S_5;
                 else if (Start) begin Load_words = 1; next_state = S_1; end
                 else next_state = S_idle;
    S_1:         begin if (m0) Add_shift = 1; else Shift = 1; next_state = S_2; end
    S_2:         begin if (m0) Add_shift = 1; else Shift = 1; next_state = S_3; end
```

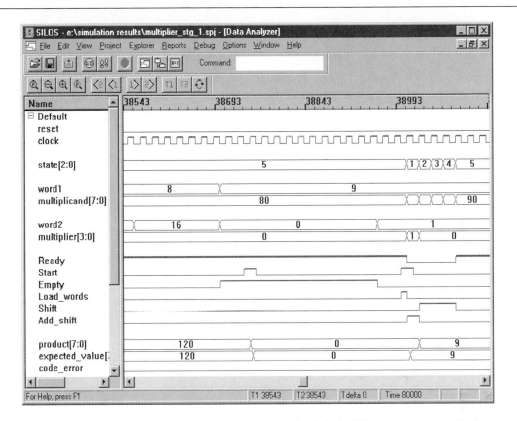

FIGURE 10-26 Simulation results demonstrating that *product* is flushed if *Start* is asserted while *Ready* and *Empty* are asserted.

```
S_3:        begin if (m0) Add_shift = 1; else Shift = 1; next_state = S_4; end
S_4:        begin if (m0) Add_shift = 1; else Shift = 1; next_state = S_5; end
S_5:        if (Empty) next_state = S_5;
            else if (Start) begin Load_words = 1; next_state = S_1; end
            else next_state = S_5;
    default:    next_state = S_idle;
   endcase
  end
endmodule

module Datapath (product, m0, Empty, word1, word2, Ready,
 Start, Load_words, Shift, Add_shift, clock, reset);
  parameter                       L_word = 4;
  output      [2*L_word -1: 0]    product;
  output                          m0, Empty;
  input       [L_word -1: 0]      word1, word2;
  input                           Ready, Start, Load_words, Shift;
  input                           Add_shift, clock, reset;
  reg         [2*L_word -1: 0]    product, multiplicand;
```

```
reg              [L_word -1: 0]        multiplier;
wire                                   m0 = multiplier[0];
wire                                   Empty = (~|word1)|| (~|word2);

// Register/Datapath Operations
always @ (posedge clock or posedge reset) begin
  if (reset) begin multiplier <= 0; multiplicand <= 0; product <= 0; end
  else if (Start && Empty && Ready) product <= 0;
  else if (Load_words) begin
    multiplicand <= word1;
    multiplier <= word2;
    product <= 0;
  end
  else if (Shift) begin
    multiplier <= multiplier >> 1;
    multiplicand <= multiplicand << 1;
  end
  else if (Add_shift) begin
    product <= product + multiplicand;
    multiplier <= multiplier >> 1;
    multiplicand <= multiplicand << 1;
  end
end
endmodule
```

End of Example 10.2

10.3.6 ASMD-Based Sequential Binary Multiplier

STGs are convenient tools for designs that have only a few states, but are cumbersome when the number of states is large. For example, the STG-based designs of the 4-bit binary multiplier in Examples 10.1 and 10.2 do not scale with *L_word*, the size of the datapath. For longer words, additional code must be inserted for each new bit to handle additional state transitions. The growth in code is linear in the size of *L_word*. Portable, re-usable HDL models need to be written in a style that does not require the body of a description to be modified by a third party. Either a scalable STG representation or an alternative method, such as an algorithmic state machine chart and datapath (ASMD), must be found. ASMD charts facilitate scalable, portable, and re-usable designs.

  ASMD charts display the evolution of a digital machine's activity as an algorithm executes under the influence of inputs, and link a datapath to its controller. As an alternative method for the design of a sequential multiplier, we will encapsulate the entire design of the multiplier in a single Verilog module, rather than carry the detail of the hierarchical partition. Note, however, that a formal design method might require that the datapath and the controller be encapsulated in separate modules. Here, placing them in a single module will serve to demonstrate the compactness of the design. The datapath operations will be the same as for *Multipier_STG_0* in Figure 10-19, but the ASMD-based version of the machine will have a slightly enhanced controller. Recall that the machine *Multiplier_STG_1* avoids needless activity if *word_1* or *word_2* are 0 when *Start* is asserted in state *S_idle*. Signal *Empty* will have the same role in this version, but the machine will be further enhanced by terminating activity and asserting *Ready* as soon as the value of the shifted multiplier is 1. (Recall that the versions in *Multiplier_STG_0* and *Multiplier_STG_1* traverse the entire chain of states independently of the content of *multiplier*). If the value of *multiplier* is 0, *product* is completely formed, so the sequence can terminate when *multiplier* is detected to be 0. Thus, termination of the algorithm is data-dependent. On the other hand, the previously considered models based on STGs effectively count all of the bits of the multiplier. The ASMD

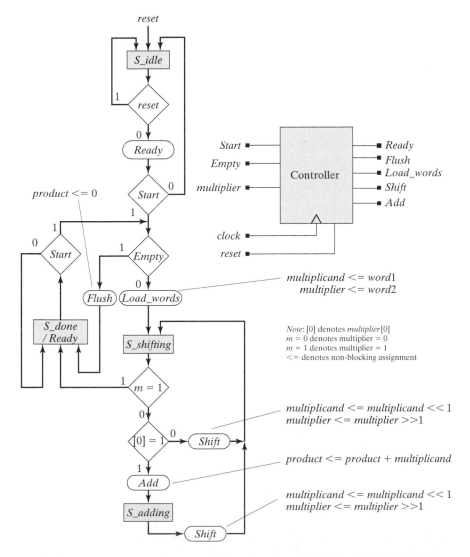

FIGURE 10-27 Block diagram symbol and annotated ASMD chart for the controller of *Multiplier_ASM_0*, a sequential binary multiplier, annotated with datapath operations.

chart of the datapath controller is shown in Figure 10-27. Note that the ordering of the decision diamonds implies that *Start* is decoded with higher priority than *Empty*.

The ASMD chart for the machine's controller specifies its state transitions and the output signals that are to be asserted during its operation. These output signals will control the datapath unit. In this design, the controller's input signals are the primary input *Start* and the internal status signals *Empty* and *multiplier*. Its output signals are *Ready, Flush, Load_words, Shift*, and *Add. Ready* signals that the unit is ready to accept a command to multiply. It must not assert while *reset* is asserted and must not assert after *Start* has been asserted, until the machine has formed *product*. If a data word is empty while *Start* is asserted in *S_idle* or in *S_done*, the machine enters *S_done*, by-passing any further execution. If *Start* is asserted while the state is in *S_done* and a data word is empty, *Flush* is asserted to empty *product* of any residual value from a previous multiplication.

We saw in Chapter 5 that an ASMD chart is an annotated ASM chart linking the controller to the datapath it controls, by specifying the datapath operations that are to occur synchronously with the state transitions, under the control of the indicated asser-tions. This additional information can be an aid in verifying that the functionality is correct, and simplifies the task of designing the overall machine.

The machine here has four states: *S_idle, S_shifting, S_adding*, and *S_done*. It enters *S_idle* when *reset* is asserted (asynchronously), and remains there, with *reset* de-asserted and *Ready* asserted, until *Start* is asserted. If *Start* is asserted and a data word is empty, *Flush* is asserted and the state enters *S_done*. Otherwise, the machine asserts *Load_words* and will enter *S_shifting* at the next active edge of the clock. At the clock edge, with *Load_words* asserted, the *multiplicand* and *multiplier* registers will load *word1* and *word2*, respectively. In *S_shifting*, if *multiplier* has a value of 1 the machine will transition to *S_done*. If not, *multiplier[0]* is tested. If it is 1, *Add* is asserted and the machine will transition to *S_adding* at the next active edge of *clock*. In conjunction with this transition, the contents of register *multiplicand* will be added to the register *product*. In *S_shifting*, if the LSB of *multiplier* is 0, the controller asserts *Shift*, and will transition to *S_shifting*. At the same edge of *clock*, with *Shift* asserted, the contents of *multiplier* will shift by 1 bit toward the register's LSB, and the contents of *multiplicand* will shift by 1 bit toward the register's MSB. In *S_adding, Shift* is asserted, and at the next edge of *clock* the machine will transition back to *S_shifting*. The register opera-tions induced by *Shift* are shown on the ASMD chart as nonblocking assignments with Verilog operator notation.[4] Note that the machine enters *S_done* immediately if a data word is 0, and terminates as soon as the multiplier has a value of 1. This enhanced func-tionality can be implemented at minor cost.

The Verilog code for *Multiplier_ASM_0* is given in Example 10.3. The code re-sides in a single Verilog module, for convenience, but could be partitioned to place the datapath unit in a separate module, because the datapath unit depends only on the signals formed as outputs of the controller, not on its state.

---

[4]The annotation associating datapath operations with the arcs of an ASMD chart is similar to register trans-fer notation used in textbooks on computer architecture.

In general, all of the signals that appear in the decision diamonds of the ASM chart must be included in the event-control expression of the combinational behavior that describes the next-state and output logic of the datapath. All of the variables that are assigned by the behavior are given an initial (default) value of 0, to avoid the synthesis of an inadvertent latch. Here, *reset* is omitted in the next-state function because it is accounted for in the synchronous behavior. The decoding at *S_idle* and *S_4* handles situations in which the machine is directed to multiply with a data word that is 0.

The description of *Multiplier_ASM_0* is parameterizable. A single parameter, *L_word*, can be change to accommodate an arbitrary word length, without other modifications to the model.

Example 10.3

```
module Multiplier_ASM_0 (product, Ready, word1, word2, Start, clock, reset);
  parameter                           L_word = 4;
  output         [2*L_word -1: 0]     product;
  output                              Ready;
  input          [L_word -1: 0]       word1, word2;
  input                               Start, clock, reset;
  reg            [1: 0]               state, next_state;
  reg            [2*L_word -1: 0]     multiplicand;
  reg            [L_word -1: 0]       multiplier;
  reg                                 product;
  reg                                 Flush, Load_words, Shift, Add;
  parameter                           S_idle = 0, S_shifting = 1 S_adding =
                                      2, S_done = 3;

  wire           Empty = ((word1 == 0) || (word2 == 0));
  wire           Ready = ((state == S_idle) && !reset) || (state == S_done);

  always @ (posedge clock or posedge reset)        // State transitions
    if (reset) state <= S_idle; else state <= next_state;

  // Combinational logic for ASM-based controller

  always @ (state or Start or Empty or multiplier) begin
    Flush = 0; Load_words = 0; Shift = 0; Add = 0;
    case (state)
      S_idle:      if (!Start) next_state = S_idle;
                   else if (Start && !Empty)
                   begin Load_words = 1; next_state = S_shifting; end
                   else if (Start && Empty) begin Flush = 1; next_state = S_done; end

      S_shifting:  if (multiplier == 1) begin Add = 1; next_state = S_done; end
                   else if (multiplier[0]) begin Add = 1; next_state = S_adding; end
                   else begin Shift = 1; next_state = S_shifting; end
```

```
      S_adding:        begin Shift = 1; next_state = S_shifting; end

      S_done:          begin if (Start == 0) next_state = S_done; else if (Empty)
                          begin Flush = 1; next_state = S_done; end else
                            begin Load_words = 1; next_state = S_shifting; end
                        end
      default:         next_state = S_idle;
    endcase
  end

  // Register/Datapath Operations

  always @ (posedge clock or posedge reset) begin
    if (reset) begin multiplier <= 0; multiplicand <= 0; product <= 0; end
    else if (Flush)
      product <= 0;
    else if (Load_words == 1) begin
      multiplicand <= word1;
      multiplier <= word2;
      product <= 0;
    end
    else if (Shift) begin
      multiplicand <= multiplicand << 1;
      multiplier <= multiplier >> 1;
    end
    else if (Add) product <= product + multiplicand;
  end
endmodule
```

End of Example 10.3

The simulation results shown in Figures 10-28 to 10-30 show the state transitions specified by the ASMD chart. Figure 10-28 shows the simulation activity following a reset event. With both data words having a value of 0, the state immediately enters *S_done*, where *Ready* is asserted. Figure 10-29 shows simulation activity in which a data word transitions to a 0 from the value that led to the content of product. In this case, if *Start* is asserted *product* is flushed, *state* remains in *S_idle*, and *Ready* is re-asserted. Figure 10-30 shows three cycles of multiplication activity extracted from an exhaustive simulation. The evolution of states leads to assertions of *Ready*, *Load_words, Shift,* and *Add*.

10.3.7 Efficient ASMD-Based Sequential Binary Multiplier

The multiplier described by *Multiplier_STG_1* is inefficient because it performs the shift and add operations in separate clock cycles. Likewise, *Multiplier_ASM_0* executes

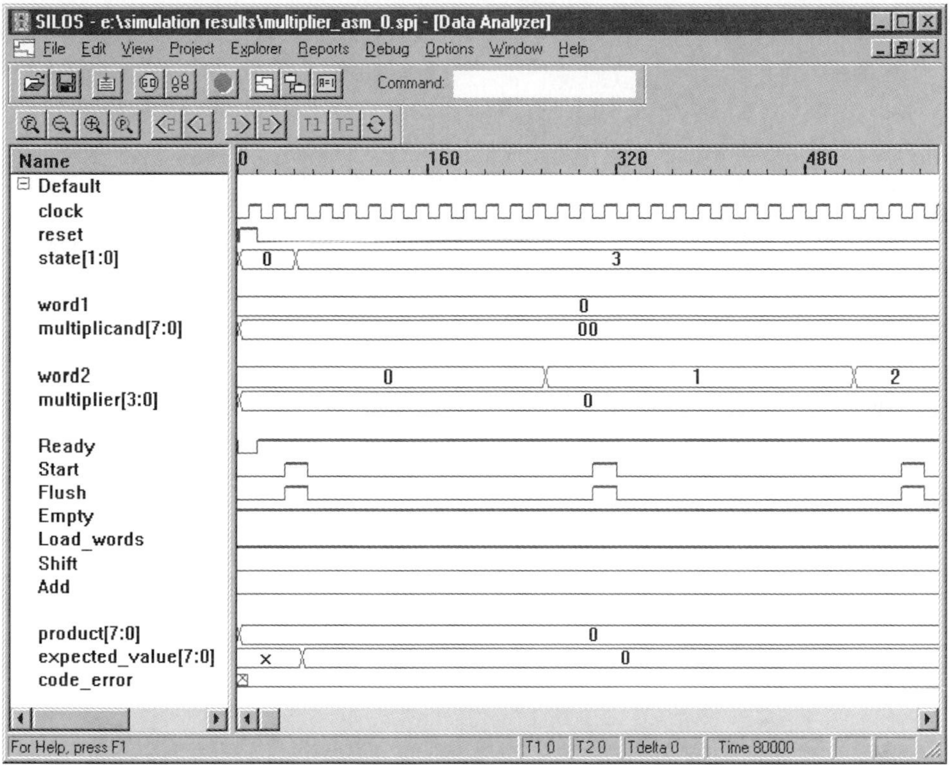

FIGURE 10-28 Simulation results for *Multiplier_ASM_0*, showing activity after initial reset with data words of 0.

more cycles than are needed to form *product*. Figure 10-31 shows the ASMD chart of a more efficient machine, having only two states, whose datapath operations add and shift in the same cycle while *Add_shift* is asserted. The Verilog description of *Multiplier_ASM_1* is presented in Example 10.4, and simulation results are shown in Figures 10-32 to 10-34. Note that *Multiplier_ASM_1* requires fewer clock cycles to compute *product*, and that *Ready* asserts when *multiplier* becomes empty.

Example 10.4

```
module Multiplier_ASM_1 (product, Ready, word1, word2, Start, clock, reset);
  parameter     L_word = 4;
  output        [2*L_word -1: 0]          product;
  output                                  Ready;
  input         [L_word -1: 0]            word1, word2;
  input                                   Start, clock, reset;
```

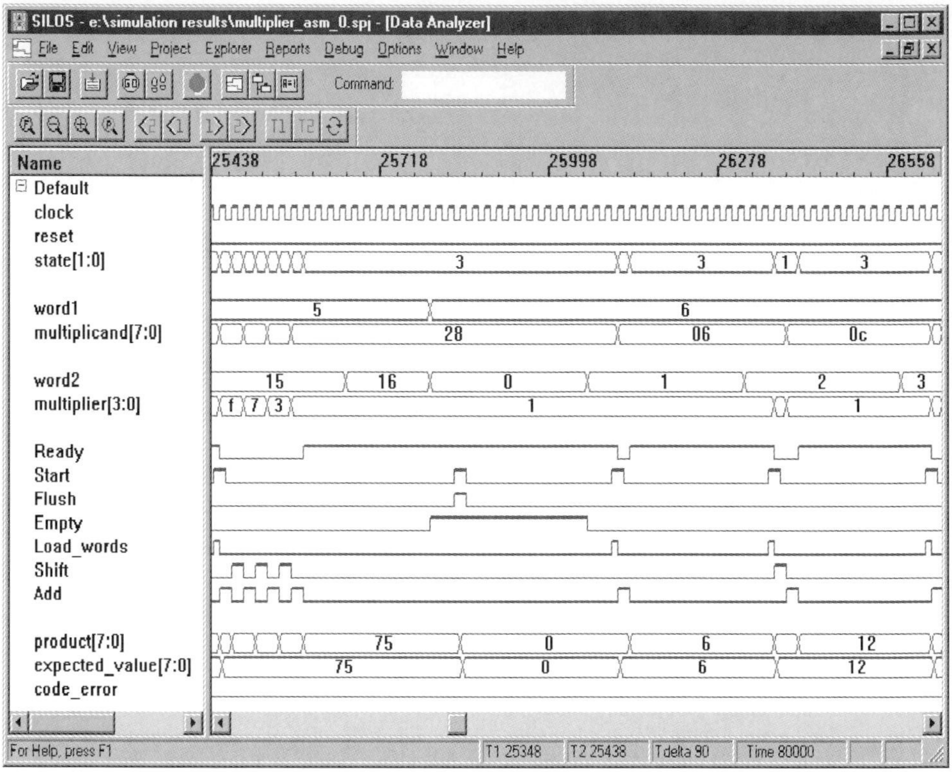

FIGURE 10-29   Simulation results for *Multiplier_ASM_0*, showing proper handling of an empty word.

```
reg                                          state, next_state;
reg          [2*L_word -1: 0]                multiplicand;
reg          [L_word -1: 0]                  multiplier;
reg                                          product, Load_words;
reg                                          Flush, Shift, Add_shift;
parameter                                    S_idle = 0, S_running = 1;

wire                                         Empty = (word1 == 0) || (word2 == 0);
wire                                         Ready = (state == S_idle) && (!reset);

always @ (posedge clock or posedge reset)         // State transitions
   if (reset) state <= S_idle; else state <= next_state;

// Combinational logic for ASM-based controller

always @ (state or Start or Empty or multiplier) begin
   Flush = 0; Load_words = 0; Shift = 0; Add_shift = 0;
   case (state)
```

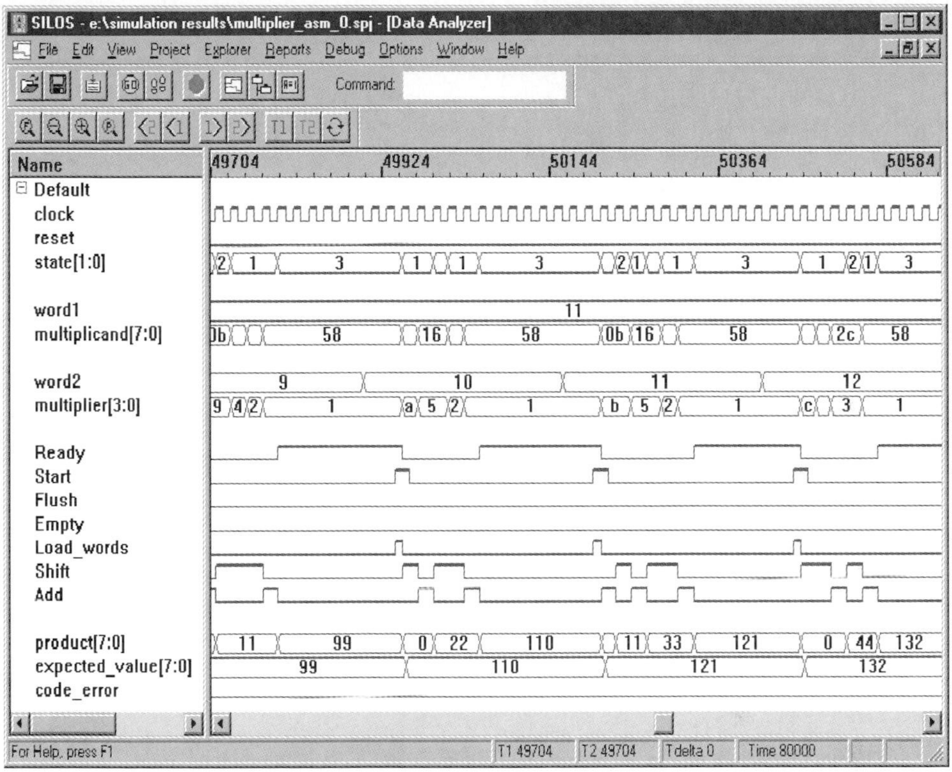

FIGURE 10-30   Simulation results showing three cycles of multiplication for *Multiplier_ASM_0*, an
ASMD-based sequential multiplier.

```
S_idle:        if (!Start) next_state = S_idle;
               else if (Empty) begin next_state = S_idle; Flush = 1; end
               else begin Load_words = 1; next_state = S_running; end

S_running:     if (~|multiplier) next_state = S_idle;
               else if (multiplier == 1)
               begin Add_shift = 1; next_state = S_idle; end
               else if (multiplier[0]) begin Add_shift = 1; next_state = S_running;
               end
               else begin Shift = 1; next_state = S_running; end
    default:   next_state = S_idle;
  endcase
end

// Register/Datapath Operations

always @ (posedge clock or posedge reset)
  if (reset) begin multiplier <= 0; multiplicand <= 0; product <= 0; end
```

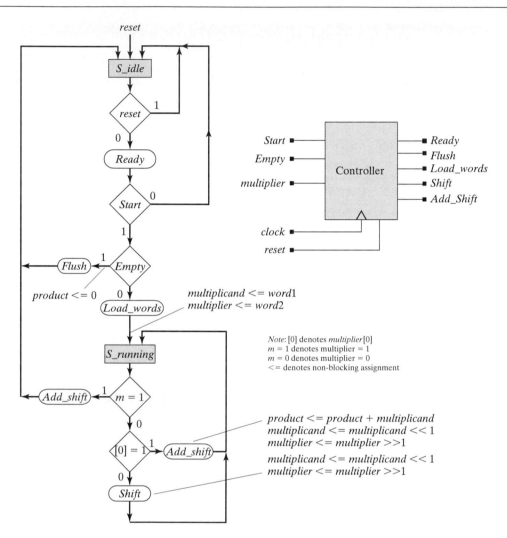

FIGURE 10-31 Block diagram symbol and ASMD chart for the controller of *Multiplier_ASM_1*, an efficient ASMD-based multiplier.

```
else begin
  if (Flush) product <= 0;
  else if (Load_words == 1) begin
    multiplicand <= word1;
    multiplier <= word2;
    product <= 0;
  end
  else if (Shift) begin
    multiplicand <= multiplicand << 1;
```

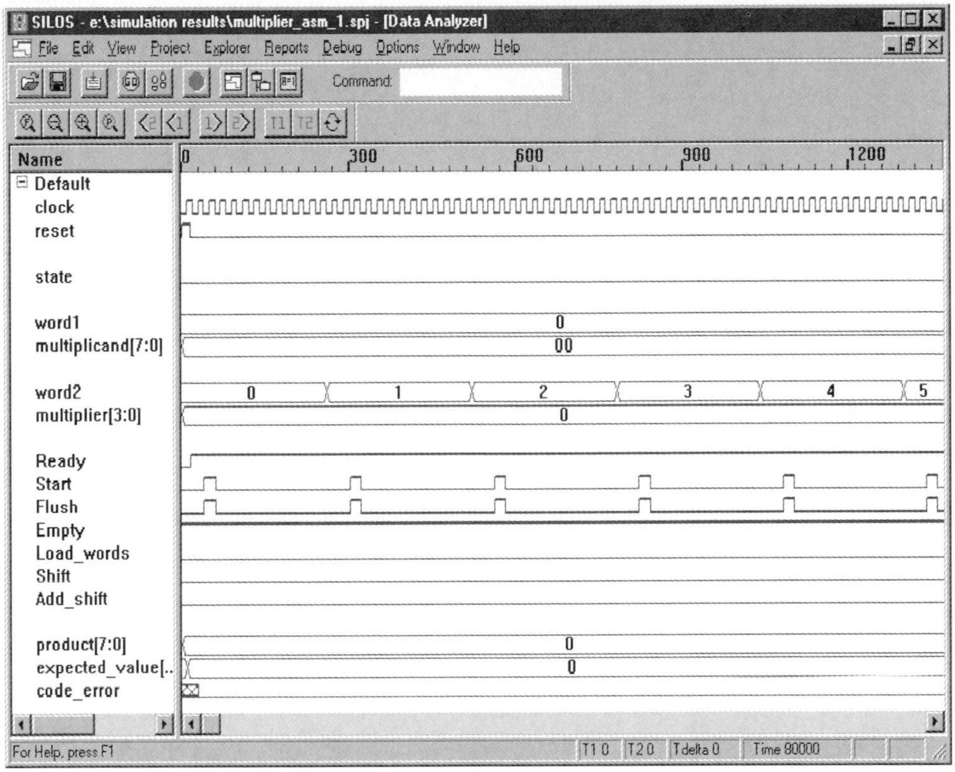

FIGURE 10-32 Simulation results for *Multiplier_ASM_1*, showing activity after assertion of *reset*.

```
            multiplier <= multiplier >> 1; end
        else if (Add_shift) begin product <= product + multiplicand;
            multiplicand <= multiplicand << 1;
            multiplier <= multiplier >> 1;
        end
    end
endmodule
```

End of Example 10.4

10.3.8 Summary of ASMD-Based Datapath and Controller Design

The previous examples have illustrated how STGs and ASMD charts can be used with an HDL to describe and design a state machine controlling a datapath. Because it leads naturally to portable descriptions of behavior, we summarize below the basic elements of an ASMD chart-based design method for datapaths and their controllers:

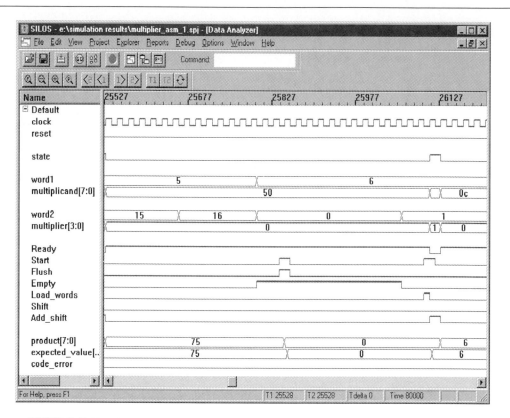

FIGURE 10-33   Simulation results for *Multiplier_ASM_1*, showing correct flushing action when multiplication is attempted with an empty data word.

1. Partition the design effort into (a) a (edge-sensitive) synchronous behavior controlling state transitions, and (b) one or more combinational behaviors and/or continuous assignments specifying the next-state and output logic of the controller, and (c) a (edge-sensitive) synchronous behavior describing the datapath operations that are controlled by the logic developed in (b) above.

2. In a level-sensitive behavioral description of the combinational logic of the controller, initialize all output variables to 0 to ensure that no output will be synthesized as the output of a latch.

3. Use blocking assignments in a level-sensitive behavior describing the combinational logic for next-state and output of the datapath controller.

4. Do not mix datapath operations with the next-state and output logic. Write a separate synchronous (edge-sensitive) behavior describing the datapath operations supporting the architecture.

5. In the behavior describing datapath operations, use nonblocking assignments, and condition the activity flow on the output signals generated by the datapath controller.

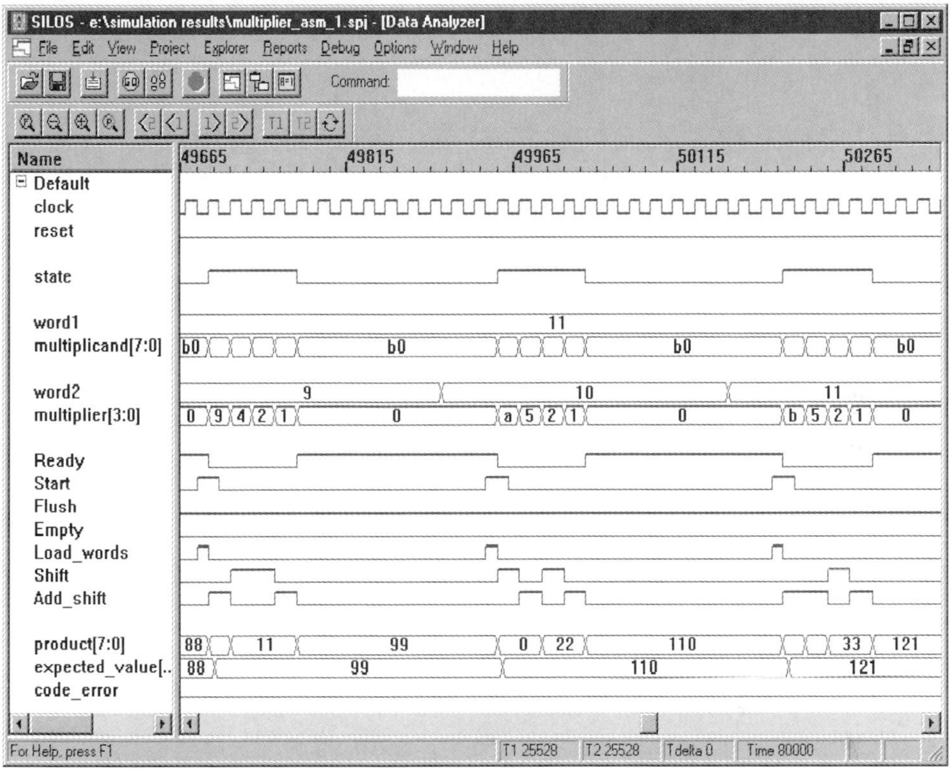

FIGURE 10-34 Simulation results for *Multiplier_ASM_1,* showing a sample of three cycles of multiplication.

10.3.9 Reduced-Register Sequential Multiplier

The previously considered architectures for the binary multiplier use separate registers to hold *multiplicand, multiplier,* and *product.* A shift register sized at $2*(L\_word - 1)$ initially holds *multiplicand* and accommodates the shifting operations that occur at each step of the multiplication sequence. An alternative, and more efficient, architecture is shown in Figure 10-35, where the register for *multiplicand* is hard wired to the adder, as are the leftmost $(L\_word + 1)$ bits of *product.* The value of *multiplier* is initially stored in the leftmost *L_word* bits of *product.* The row sums are placed in the leftmost bits of *product* as they are formed, and the contents of *product* are shifted to the right (i.e., *product* moves relative to a fixed register holding *multiplicand*). At each step, the LSB of *product* determines whether *multiplicand* is to be added to *product.* The sequence of operations continues until all the bits of *multiplier* have been shifted out of *product,* leaving only the result of the multiplication. This scheme eliminates a separate register for multiplier, and reduces the size of the register for multiplicand by a factor of 2. Also, the register for *multiplicand* is fixed (i.e., not a shift register). The adder required to generate the sums is also reduced in size by a factor of 2, saving area

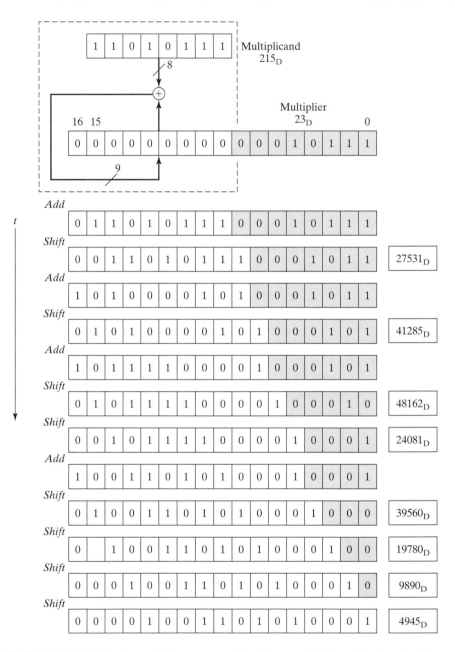

FIGURE 10-35   Architecture and data movement for a binary multiplier with reduced registers.

and improving speed. Figure 10-35 also illustrates the movement of data forming the product $215_{10} \times 23_{10} = 4945_{10}$.

The interface signals and the controller for this datapath architecture are based on the ASMD chart in Figure 10-36. This machine is efficient, like *Multiplier_ASM_1*, having only two states, *S_idle* and *S_running*. Given that the multiplier is stored in the product register and is shifted out of the register as the process evolves, a counter is used to determine when the process is complete. (Problem 6 at the end of the chapter specifies a design that terminates the multiplication process as soon as the shifted multiplier subword is empty of 1s.) The controller generates a signal, *Increment*, which controls the counter.

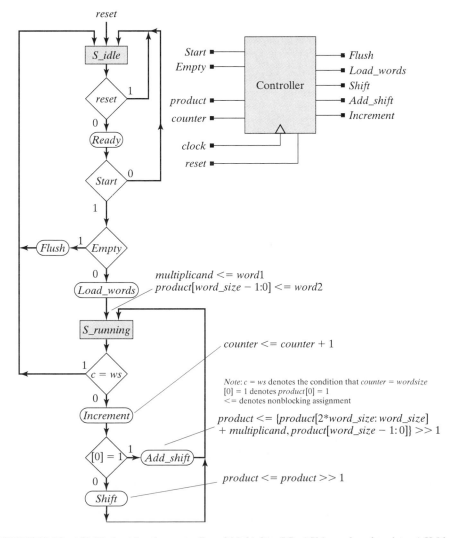

FIGURE 10-36 ASMD chart for the controller of *Multiplier_RR_ASM*, a reduced-register ASM-based sequential multiplier, annotated with datapath operations.

The Verilog code for *Multiplier_RR_ASM* is given in Example 10.5. Note that the register holding *product* is 1 bit larger than in the previous models, because the addition and concatenation operations induced by *Add_shift* occur before the shift operation and might generate a carry that would otherwise be dropped, even though the final result will fit in a register having *L_word* bits. This condition does not arise in the previous designs, because their multiplication process builds sums from right to left, and the product register is large enough to accommodate any intermediate carry. If the product register were sized to only 2*L_word* here, an intermediate carry will overflow and be lost. This condition might not have been detected by a sporadic testing scheme, but it was revealed by the exhaustive testbench's error detection signal.

Example 10.5

```
module Multiplier_RR_ASM (product, Ready, word1, word2, Start, clock, reset);
  parameter                          L_word = 4;
  parameter                          L_cnt = 3;
  output        [2*L_word: 0]        product;
  output                             Ready;
  input         [L_word -1: 0]       word1, word2;
  input                              Start, clock, reset;
  reg                                state, next_state;
  reg           [L_word -1: 0]       multiplicand;
  reg                                product, Load_words;
  reg                                Flush, Shift, Add_shift, Increment;
  reg           [L_cnt -1 : 0]       counter;
  parameter                          S_idle = 0, S_running = 1;
  wire                               Empty = (word1 == 0) || (word2 == 0);
  wire                               Ready = (state == S_idle) && (!reset );

// Controller

  always @ (posedge clock or posedge reset)        // State transitions
    if (reset) state <= S_idle; else state <= next_state;

// Combinational logic for ASM-based controller

  always @ (state or Start or Empty or product or counter) begin
    Flush = 0; Load_words = 0; Shift = 0; Add_shift = 0; Increment = 0;
    case (state)

      S_idle:      if (!Start) next_state = S_idle;
                   else if (Empty) begin next_state = S_idle; Flush = 1; end
                   else begin Load_words = 1; next_state = S_running; end

      S_running:   if (counter == L_word) next_state = S_idle;
                   else begin
                    Increment = 1;
                    if (product[0]) begin Add_shift = 1; next_state = S_running; end
                    else begin Shift = 1; next_state = S_running; end
                   end
```

```
                  default:        next_state = S_idle;
                endcase
              end

      // Register/Datapath Operations

      always @ (posedge clock or posedge reset)
        if (reset) begin multiplicand <= 0; product <= 0; counter <= 0; end
        else begin
          if (Flush) product <= 0;
          if (Load_words == 1)
            begin multiplicand <= word1; product <= word2; counter <= 0; end
          if (Shift) begin product <= product >> 1; end
          if (Add_shift) begin
            product <= {product[2*L_word: L_word] + multiplicand, product[L_word -1: 0]} >> 1;
          end
          if (Increment) counter <= counter +1;
        end
      endmodule
```

End of Example 10.5

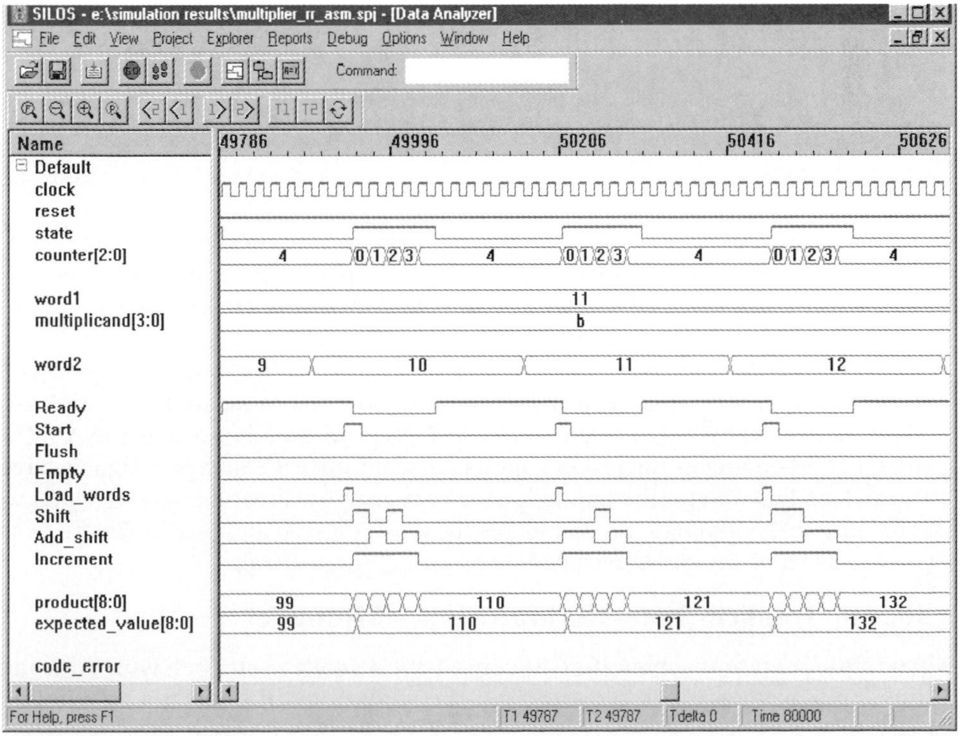

FIGURE 10-37 Register transfers in *Multiplier_RR_ASM* demonstrating multiple cycles of activity.

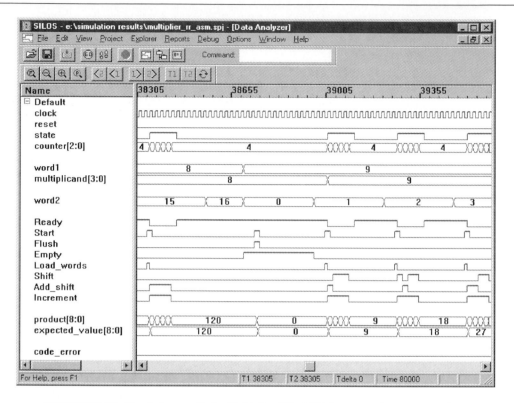

FIGURE 10-38 Simulation results for *Multiplier_RR_ASM*, demonstrating correct register flushing action.

A sample of waveforms produced by *Multiplier_RR_ASM* is shown in Figure 10-37. Note how *counter* evolves while state is *S_running*. The simulation results in Figure 10-38 demonstrate the machine's ability to immediately terminate execution and re-assert *Ready* if a data word is 0 when *start* is asserted in *S_idle*.

The design based on the ASMD chart in Figure 10-36 specifies that *state* return to *S_running* after *product* is finally formed, *before* returning to *S_idle* and asserting *Ready*. Thus, *one cycle of execution is wasted*. As an alternative design, the ASMD chart in Figure 10-39 moves the test of counter to occur *after* the datapath signals have been formed, and completes the multiplication sequence by directing *state* to *S_idle* without returning to *S_running*. This eliminates the wasted cycle, and asserts *Ready* as soon as *product* is valid. The modified code is shown in Figure 10-40.

10.3.10 Implicit-State-Machine Binary Multiplier

An implicit-state machine [6, 7, 8] consists of a single cyclic behavior with multiple embedded, edge-sensitive event-control expressions that specify an evolution of

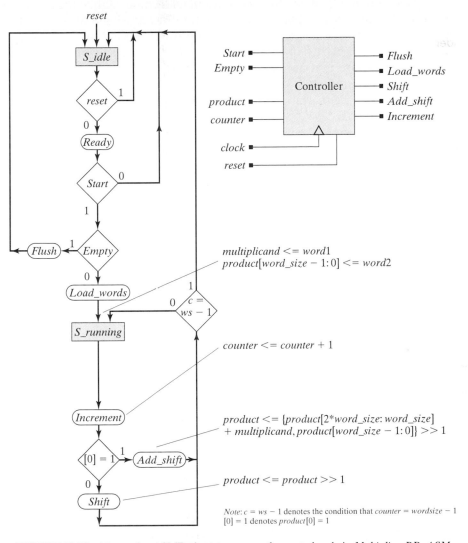

FIGURE 10-39 Alternative ASMD chart to recover the wasted cycle in *Multiplier_RR_ASM*.

```
/* Modified Version to recover final cycle*/
    S_running: begin
                Increment = 1;
                if (product[0]) Add_shift = 1; else Shift = 1;
                if (counter == L_word − 1) next_state = S_idle; else next_state = S_running;
                end
```

FIGURE 10-40 Code to modify *Multiplier_RR_ASM* to recover a wasted cycle of operation.

clock cycles and implicitly define the states of a machine. Unlike the machines in the previous examples, an implicit-state machine does not have an explicitly defined state, nor does it have an explicit state-transition behavior. Some designers prefer the simplicity and clarity of the descriptive style offered by an implicit-state machine, but be aware that an implicit-state machine has the restriction that a given state may be entered from only one other state. Not all systems satisfy this constraint. Implicit-state machines require careful consideration of reset signals, and require that an asserted reset signal must return the machine to the beginning of the sequence of clock cycles, regardless of the clock cycle in which it is asserted. For synthesis, the active edge of the synchronizing signal (clock) of an implicit-state machine must have the same polarity (e.g., rising) at all state transitions.

An implicit-state machine controller for the synchronous binary multiplier whose architecture was shown in Figure 10-35 is presented in Example 10.6. The controller in *Multiplier_IMP_1* was designed to satisfy the following requirements:

1. A signal, *Ready*, is to be asserted at the first active edge of *clock* after *reset* has been cycled through assertion and de-assertion. *Ready* indicates that the multiplier is ready to respond to a *Start* signal, with *reset* de-asserted. *Ready* is to be de-asserted while *reset* is asserted, and, after *Ready* has been asserted, it is to remain asserted until the first active edge occurs with *Start* asserted.

2. The machine must begin a multiplication sequence at the first active edge of *clock* at which *Start* and *Ready* are both asserted.

3. Once the multiplication sequence is initiated, the machine is to assert a signal, *Ready*, at the first active edge of *clock* at which *multiplier* is empty. The assertion of *Ready* is to coincide with the formation of the value of *product*. *Ready* indicates that *product* can be read by an external processor.

4. Once the multiplication sequence is initiated, *Start* is to be ignored until *Ready* is asserted.

5. The datapaths for *word1* and *word2* are to be ignored after the multiplicand and multiplier registers have been loaded (i.e., external datapaths are free during the multiplication sequence).

6. Once asserted, *Ready* is to be held asserted until *Start* (or *reset*) is re-asserted.

7. The machine must recover from a random assertion of *reset* during a multiplication sequence (i.e., must recover from a running reset).

8. The machine must operate correctly if *Start* is asserted randomly, and if *Start* has a duration of more than one cycle of *clock*.

The controller in *Multiplier_IMP_1* is an implicit finite-state machine with the features specified above. The additional code that results from the requirement that an implicit-state machine recover correctly from an assertion of *reset* in any clock cycle is reduced in this example by use of a task, *Clear_Regs*, to reset all registers affected by *reset*. Note that the datapath operations are relatively simple, but the controller is more elaborate and complex.

Controller_IMP_1, has a single cyclic behavior with the following threads of activity: (1) a single-cycle thread in which the machine idles until *Start* is asserted, (2) a two-cycle thread that detects whether a data word is 0 and, if so, terminates immediately, (3) a multicycle thread in which data words are loaded, followed by a sequence of assertions of *Add_shift* or *Shift*, depending on the running content of the bit *multiplier[0]*. Each cycle of activity is initiated by the rising edge of *clock*. If *reset* is asserted, the activity aborts the named block, *Main_Block*, and returns to monitor the first event-control expression.

For convenience and clarity, the three threads of activity within the *Main_Block* of the behavior are labeled as named blocks. In *Idling*, the machine waits for assertion of *Start* (with *Ready* asserted), and asserts *Flush* to remove the residual value of *product*. *Early_Terminate* aborts an attempt to multiply with a zero. In *Load_and_Multiply*, the controller asserts *Load_words* in the first cycle, then issues commands to shift and/or add in successive clock cycles until the multiplication process is complete. In the model shown here, an additional signal, *Done*, is used in the controller to expose a key detail of the operation of the machine. *Done* asserts at the end of the loop that computes *product*. The loop is not data-dependent, and executes through all bits of *multiplier*. If an external agent uses *Ready* to initiate a multiplication sequence after *Ready* is asserted, and while the machine is executing the loop, *Start* will be ignored. This could lead to ambiguity in associating *product* with *word1* and *word2*. For unambiguous operation, *Start* should not be re-asserted until *Done* is asserted.

Example 10.6

```
module Multiplier_IMP_1 (product, Ready, Done, word1, word2, Start, clock, reset);
  parameter                          L_word = 4;
  output        [2*L_word -1: 0]     product;
  output                             Ready, Done;
  input         [L_word -1: 0]       word1, word2;
  input                              Start, clock, reset;
  wire          [L_word -1: 0]       multiplier;
  wire                               Flush, Load_words, Shift, Add_shift;

Datapath_Unit_IMP_1 M1
  (product, multiplier, word1, word2, Flush, Load_words, Shift, Add_shift, clock,
  reset);
```

```verilog
Controller_IMP_1 M2
  (Ready, Flush, Load_words, Shift, Add_shift, Done, word1, word2, multiplier,
  Start, clock, reset);
endmodule

module Controller_IMP_1
  (Ready, Flush, Load_words, Shift, Add_shift, Done, word1, word2, multiplier, Start,
  clock, reset);
  parameter                         L_word = 4;
  output                            Ready, Flush, Load_words, Shift,
                                    Add_shift, Done;
  input      [L_word -1: 0]         word1, word2, multiplier;
  input                             Start, clock, reset;

  reg                               Ready, Flush, Load_words, Shift,
                                    Add_shift, Done;
  integer                           k;
  wire                              Empty = (word1 == 0) || (word2 == 0);

  always
    @ (posedge clock or posedge reset) begin: Main_Block
    if (reset) begin Clear_Regs; disable Main_Block; end

    else if (Start != 1) begin: Idling
      Flush <= 0; Ready <= 1;
    end // Idling

    else if (Start && Empty) begin: Early_Terminate
      Flush <= 1; Ready <= 0; Done <= 0;
      @ (posedge clock or posedge reset)
      if (reset) begin Clear_Regs; disable Main_Block; end
      else begin
        Flush <= 0; Ready <= 1; Done <= 1;
      end
    end // Early_Terminate

    else if (Start) begin: Load_and_Multiply
      Ready <= 0; Flush <= 0; Load_words <= 1; Done <= 0;Shift <= 0;
      Add_shift <= 0;
      @ (posedge clock or posedge reset)
      if (reset) begin Clear_Regs; disable Main_Block; end
      else begin // not reset
        Load_words <= 0;
        if (word2[0]) Add_shift <= 1; else Shift <= 1;
        for (k = 0; k <= L_word -1; k = k +1)

          @ (posedge clock or posedge reset)
          if (reset) begin Clear_Regs; disable Main_Block; end
          else begin // multiple cycles
            Shift <= 0;
```

```
              Add_shift <= 0;
            if (multiplier == 1) Ready <= 1;
            else if (multiplier[1]) Add_shift <= 1;
            else Shift <= 1; // Notice use of multiplier[1]
          end // multiple cycles
          Done <=1;
        end // not reset
      end // Load_and_Multiply
    end // Main_Block

  task Clear_Regs;
    begin
      Ready <= 0; Flush <= 0; Load_words <= 0; Done <= 0; Shift <= 0; Add_shift <= 0;
    end
  endtask
endmodule

module Datapath_Unit_IMP_1
  (product, multiplier, word1, word2, Flush, Load_words, Shift, Add_shift, clock, reset);
  parameter                        L_word = word_size;
  output        [2*L_word -1: 0]   product;
  output        [L_word -1: 0]     multiplier;
  input         [L_word -1: 0]     word1, word2;
  input                            Flush, Load_words, Shift, Add_shift, clock, reset;
  reg           [2*L_word -1: 0]   product;
  reg           [2*L_word -1: 0]   multiplicand;
  reg           [L_word -1: 0]     multiplier;

// Datapath Operations
  always @ (posedge clock or posedge reset)
    if (reset) begin multiplier <= 0; multiplicand <= 0; product <= 0; end
    else begin
      if (Flush) product <= 0;
      else if (Load_words == 1) begin
        multiplicand <= word1;
        multiplier <= word2;
        product <= 0; end
      else if (Shift) begin
        multiplier <= multiplier >> 1;
        multiplicand <= multiplicand << 1; end
      else if (Add_shift) begin
        multiplier <= multiplier >> 1;
        multiplicand <= multiplicand << 1;
        product <= product + multiplicand; end
    end
endmodule
```

End of Example 10.6

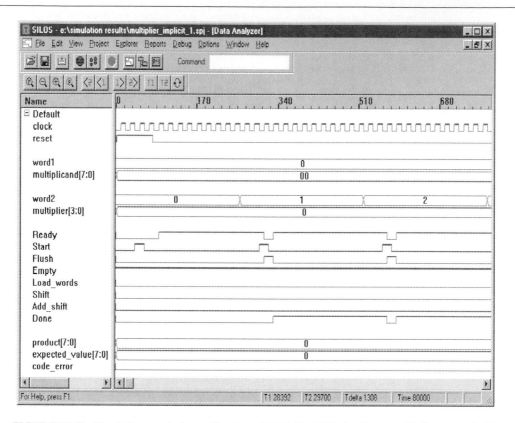

FIGURE 10-41 Simulation results for verification of *Multiplier_IMP_1*, a binary multiplier controlled by an implicit state machine, demonstrating a startup sequence after assertion of power-up *reset*.

The simulation results shown in Figure 10-41 for *Multiplier_IMP_1* demonstrate that the controller (1) "wakes up" correctly after an initial assertion of *reset*, (2) ignores *Start* while *reset* is asserted, and (3) handles a zero multiplicand correctly. The waveforms in Figure 10-42 show that the machine handles a zero multiplier correctly, and those in Figure 10-43 show that the machine recovers correctly when *reset* is asserted during a multiplication sequence (an important consideration for an implicit machine). In Figure 10-44 the waveforms show the machine correctly multiplying a sample of data words. The machine was verified exhaustively for 4- and 8-bit words.

Note in Figure 10-42, that *product* is holding 60, the result of previously multiplying 4 by 15, while *word2* changes to 16, and then to 0. When *Start* is asserted, the machine de-asserts *Ready* and *Done*, and then asserts *Ready* and *Done*. In Figure 10-43, the machine is in the process of multiplying 6 by 11 when *reset* is asserted. The machine flushes *product* and idles to await the next assertion of *Start*, which forms the product of 6 by 12, and then asserts *Ready* and *Done*.

The controller in *Multiplier_IMP_1* wastes cycles of execution by cycling through all of the bits of the multiplier, even after *Ready* has been asserted. A more efficient

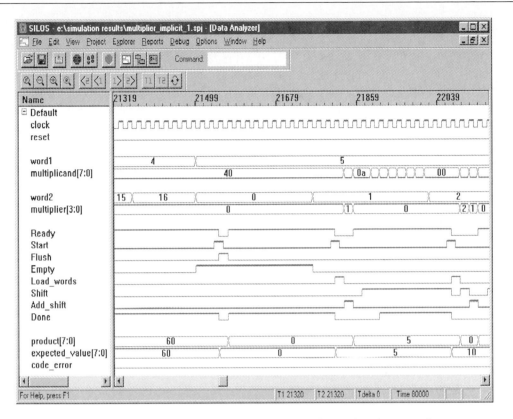

FIGURE 10-42 Simulation results for verification of *Multiplier_IMP_1*, demonstrating correct multiplication with an empty multiplier word (i.e., *word2* = 0).

design is shown in Example 10.7, where *Controller_IMP_2* terminates when *Ready* is asserted. The controller determines when the content of the multiplier has the value 1, and exits the multiplication loop at the next cycle. This machine also has a more robust controller that detects whether the multiplicand or the multiplier have a value of 1 and issues a command to the datapath to load either *word2* or *word1* directly into the register holding *product*.

Example 10.7

```
'define      word_size              4
module Multiplier_IMP_2 (product, Ready, Done, word1, word2, Start, clock, reset);
    parameter                        L_word = 'word_size;
    output      [2*L_word −1: 0]     product;
    output                           Ready, Done;
    input       [L_word −1: 0]       word1, word2;
```

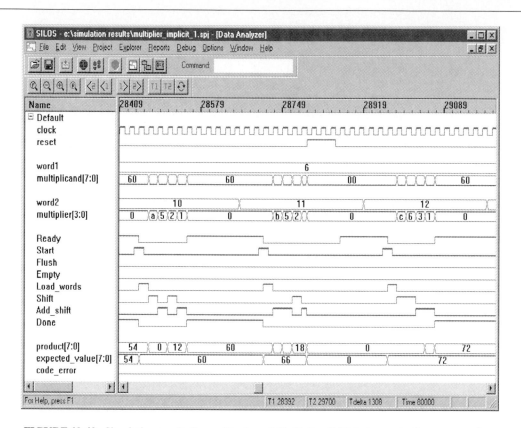

FIGURE 10-43 Simulation results for verification of *Multiplier_IMP_1*, demonstrating recovery from assertion of *reset* before completion of a multiply sequence (i.e., a running reset).[5]

```
input                                          Start, clock, reset;
wire            [L_word -1: 0]                  multiplier;
wire                                           Flush, Load_words, Load_multiplier;
wire                                           Load_multiplicand;
wire                                           Shift, Add_shift;

Datapath_Unit_IMP_2 M1
  (
  product, multiplier, word1, word2, Flush, Load_words, Load_multiplier,
  Load_multiplicand, Shift, Add_shift, clock, reset
  );

Controller_IMP_2 M2
  (
  Ready, Flush, Load_words, Load_multiplier, Load_multiplicand,
  Shift, Add_shift, Done, word1, word2, multiplier, Start, clock, reset
```

---

[5]Note that *Done* does not assert until completion of a subsequent multiplication sequence.

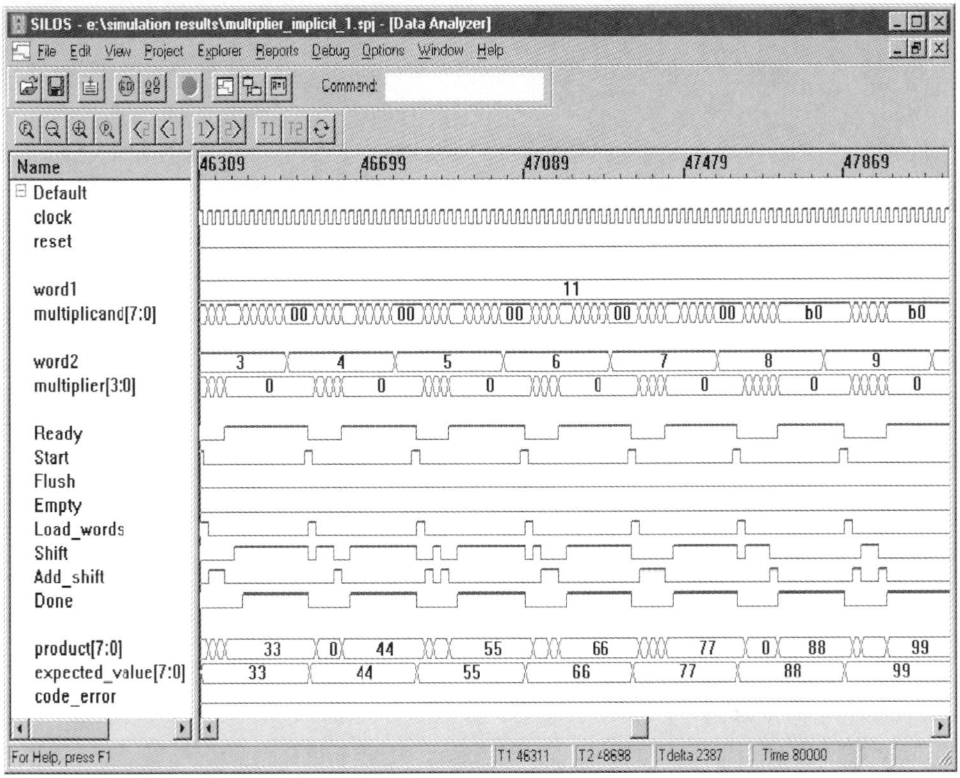

FIGURE 10-44 Simulation results for verification of *Multiplier_IMP_1*, showing multiplication sequences.

```
    );
endmodule

module Controller_IMP_2
    (
    Ready, Flush, Load_words, Load_multiplier, Load_multiplicand,
    Shift, Add_shift, Done, word1, word2, multiplier, Start, clock, reset
    );
    parameter                          L_word = 'word_size;
    output                             Ready, Flush;
    output                             Load_words, Load_multiplier;
    output                             Load_multiplicand;
    output                             Shift, Add_shift, Done;
    input          [L_word -1: 0]      word1, word2, multiplier;
    input                              Start, clock, reset;

    reg                                Ready, Flush;
    reg                                Load_words, Load_multiplier,
                                       Load_multiplicand;
```

```
reg                                                Shift, Add_shift, Done;
integer                                            k;
wire                                               Empty = (word1 == 0) || (word2 == 0);

always @ (posedge clock or posedge reset) begin: Main_Block

  if (reset) begin Clear_Regs; disable Main_Block; end
  else if (Start != 1) begin: Idling
    Flush <= 0; Ready <= 1;
    Load_words <= 0; Load_multiplier <= 0; Load_multiplicand <= 0;
    Shift <= 0; Add_shift <= 0;
  end // Idling

  else if (Start && Empty) begin: Early_Terminate
    Flush <= 1; Ready <= 0; Done <= 0;
    @ (posedge clock or posedge reset)
    if (reset) begin Clear_Regs; disable Main_Block; end
    else begin
      Flush <= 0; Ready <= 1; Done <= 1;
    end
  end // Early_Terminate

else if (Start && (word1== 1)) begin: Load_Multiplier_Direct
  Ready <= 0; Done <= 0;
  Load_multiplier <= 1;
  @ (posedge clock or posedge reset)
  if (reset) begin Clear_Regs; disable Main_Block; end
  else begin Ready <= 1; Done <= 1; end
end

else if (Start && (word2== 1)) begin: Load_Multiplicand_Direct
  Ready <= 0; Done <= 0;
  Load_multiplicand <= 1;
  @ (posedge clock or posedge reset)
    if (reset) begin Clear_Regs; disable Main_Block; end
    else begin Ready <= 1; Done <= 1; end
end

else if (Start ) begin: Load_and_Multiply
  Ready <= 0; Done <= 0; Flush <= 0; Load_words <= 1;
  @ (posedge clock or posedge reset)
  if (reset) begin Clear_Regs; disable Main_Block; end
  else begin: Not_Reset
    Load_words <= 0;
    if (word2[0]) Add_shift <= 1; else Shift <= 1;
    begin: Wrapper
      forever begin: Multiplier_Loop
        @ (posedge clock or posedge reset)
          if (reset) begin Clear_Regs; disable Main_Block; end
          else begin // multiple cycles
```

```
                    Shift <= 0;
                    Add_shift <= 0;
                    if (multiplier == 1) begin Done <= 1;
                      @ (posedge clock or posedge reset)
                      if (reset) begin Clear_Regs; disable Main_Block; end
                      else disable Wrapper;
                    end
                    else if (multiplier[1]) Add_shift <= 1;
                    else Shift <= 1; // Notice use of multiplier[1]
                  end // multiple cycles
                end // Multiplier_Loop
              end // Wrapper
              Ready <= 1;
            end // Not_Reset
          end // Load_and_Multiply
        end // Main_Block

      task Clear_Regs;
        begin
          Flush <= 0; Ready <= 0; Done <= 0;
          Load_words <= 0; Load_multiplier <= 0; Load_multiplicand <= 0;
          Shift <= 0; Add_shift <= 0;
        end
      endtask
    endmodule

    module Datapath_Unit_IMP_2
      (
      product, multiplier, word1, word2, Flush, Load_words, Load_multiplier,
      Load_multiplicand, Shift, Add_shift, clock, reset
      );
      parameter                    L_word = 'word_size;
      output    [2*L_word -1: 0]   product;
      output    [L_word -1: 0]     multiplier;
      input     [L_word -1: 0]     word1, word2;
      input                        Flush, Load_words, Load_multiplier;
      input                        Load_multiplicand, Shift, Add_shift, clock, reset;
      reg       [2*L_word -1: 0]   product;
      reg       [2*L_word -1: 0]   multiplicand;
      reg       [L_word -1: 0]     multiplier;

      // Datapath Operations
      always @ (posedge clock or posedge reset)
        if (reset) begin multiplier <= 0; multiplicand <= 0; product <= 0; end
        else begin
          if (Flush) product <= 0;
          else if (Load_words == 1) begin
            multiplicand <= word1;
            multiplier <= word2;
```

```
                    product <= 0; end
               else if (Load_multiplicand) begin
                 product <= word1; end
               else if (Load_multiplier) begin
                 product <= word2; end
               else if (Shift) begin
                 multiplier <= multiplier >> 1;
                 multiplicand <= multiplicand << 1; end
               else if (Add_shift) begin
                 multiplier <= multiplier >> 1;
                 multiplicand <= multiplicand << 1;
                 product <= product + multiplicand; end
             end
           endmodule
```

End of Example 10.7

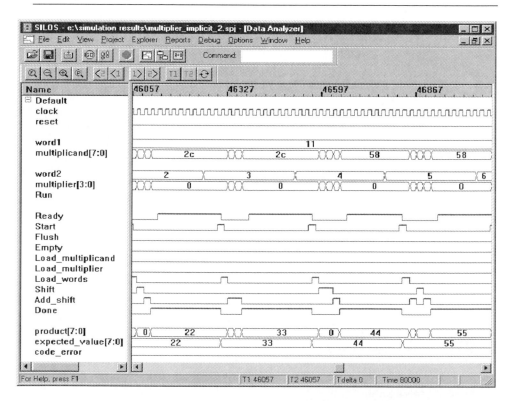

FIGURE 10-45 Simulation results for verification of *Multiplier_IMP_2*, an efficient multiplier controlled by an implicit-state machine, showing multiplication sequences.

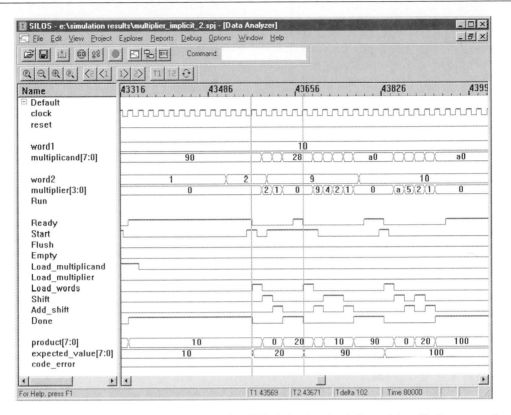

FIGURE 10-46 Simulation results for *Multiplier_IMP_2*, showing simulation activity with *Start* asserted coincidentally with *Done*.

Several cycles of simulation activity are shown in Figure 10-45. Note that *Done* asserts as soon as *product* is formed, and that *Ready* asserts when the machine is prepared to initiate another cycle of multiplication. In Figure 10-46, *Start* launches the multiplication of 10 by 2, and is re-asserted before the computation is complete. The machine ignores the re-assertion of *Start*, completes the multiplication, asserts *Done*, and then asserts *Ready* in the next clock cycle. Then, with *Start* and *Ready* both asserted, the machine multiplies 10 by 9. The simulation results in Figure 10-47 demonstrates the machine's recovery from a running reset. A multiplication sequence begins at the first active edge after the de-assertion of *reset*.

10.3.11 Booth's-Algorithm Sequential Multiplier

Various algorithms have been developed to improve the performance of sequential multipliers and to simplify their circuitry. Booth's recoding algorithm is widely used

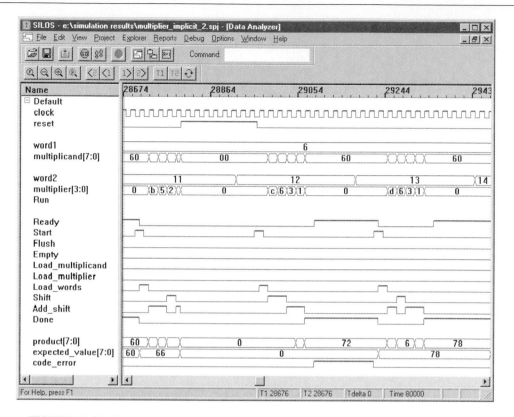

FIGURE 10-47   Simulation results for *Multiplier_IMP_2*, showing recovery from a running reset.

because it has a simple hardware realization, requires less silicon area, and can speed up sequential multiplication significantly [1, 3, 4, 5, 9, 10, 11].

Multipliers that use Booth's algorithm recode the bits of the multiplier to reduce the number of additions required to complete a cycle of multiplication. Only the multiplier is recoded; the multiplicand is left unchanged. A derivative form, called Radix-4 recoding or *bit-pair encoding*, can reduce the number of partial products by a factor of 2 [3] (see Section 10.3.12).

Booth's algorithm is applicable to positive numbers and to negative numbers in 2s complement representation (i.e., signed and unsigned numbers). Thus, a hardware multiplier using Booth recoding does not require modification to accommodate negative numbers. In contrast, multipliers that use signed magnitude representation must extract the magnitudes of the inputs, examine the signs of the data words, and then possibly convert the result to a 2s complement representation (i.e., radix-2 complement form). Multipliers that use Booth's recoding can multiply 2s complement numbers directly. The design given here can multiply two positive numbers, two negative numbers, and a mixture of positive and negative numbers (in 2s complement form).

To gain insight into Booth's algorithm, note that the decimal value of a number in an n-bit 2s complement format can be gotten by (1) multiplying the leftmost bit by -2^{n-1}, (2) multiplying the remaining bits by 2^i, where i is the bit position, and (3) adding the results [12]. For example, the 2s complement representation of -7 is 1001_2. The decimal value, with $n = 4$, is obtained as shown in Figure 10-48.

The negative weight of the leftmost bit is expressed in signed digit notation [4] as an underscore; for example, the signed digit representation of -7 is given as $\underline{1}001$. Ordinarily, the bits of a binary number can have only positive weights, but in Booth's recoding algorithm the bits of a number can have positive or negative weights expressed in signed digit notation.

The key to Booth's algorithm is that it skips over strings of 1s in the multiplier and replaces a series of additions by one addition and one subtraction. For example, the word 1111_0000 is equivalent to $2^8 - 1 - (2^4 - 1) = 2^8 - 2^4 = 256 - 16 = 240$. An arithmetic unit that must multiply negative numbers can exploit this relationship to possibly reduce the number of additions that are required to multiply two numbers. *A Booth recoding scheme recodes the multiplier by detecting strings of 1s and replacing them by signed digits that result in the same decimal value when the indicated addition and subtraction operations are performed.*

Table 10-1 summarizes the recoding rules. The algorithm reads bits from the LSB to the MSB, and the value of two successive bits (m_i, m_{i-1}) determines the Booth recoded multiplier bit, BRC_i. As the algorithm reads two successive bits, the present and

2s complement of $7_{10} = 1\ 0\ 0\ 1\ _2$

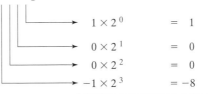

$$1 \times 2^0 \qquad = \quad 1$$
$$0 \times 2^1 \qquad = \quad 0$$
$$0 \times 2^2 \qquad = \quad 0$$
$$-1 \times 2^3 \qquad = -8$$

Decimal value of $1001_2 = -7 = \underline{1}\ 0\ 0\ 1\ _2$

FIGURE 10-48 Extraction of the decimal value from a 2s complement number.

TABLE 10-1 Rules for Booth recoding of a 2s complement number.

m_i	m_{i-1}	BRC_i	Value	Status
0	0	0	0	String of 0s
0	1	1	+1	End of string of 1s
1	0	$\underline{1}$	−1	Begin string of 1s
1	1	0	0	Midstring of 1s

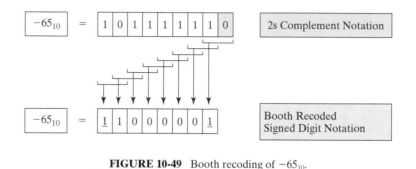

FIGURE 10-49 Booth recoding of -65_{10}.

the immediate past, it forms and uses BRC_i to determine whether to add or subtract before skipping to the next bit. The first step of the algorithm is seeded with a value of 0 to the right of the LSB of the word. If the signed digit $\underline{1}$ is encountered, a subtraction operation is performed (i.e., an appropriately shifted copy of the 2s complement of the multiplicand is added to the product). The process encodes the first encountered 1 as a $\underline{1}$, skips over any successive 1s until a 0 is encountered. That 0 is encoded as a 1 to signify the end of a string of 1s, and then the process continues. The algorithm is valid for the entire range of 2s complement numbers (i.e., those with a 0 in the MSB, and those with a 1 in the MSB [4]).

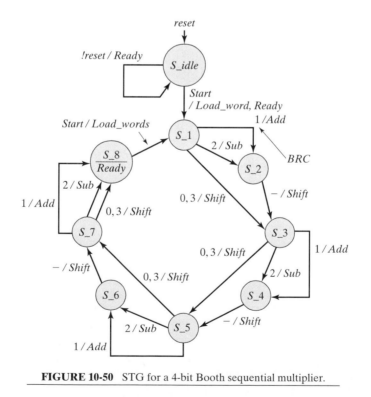

FIGURE 10-50 STG for a 4-bit Booth sequential multiplier.

As an example of Booth recoding, the encoding of $191_{10} = 1011\_1111_2$ is shown in Figure 10-49. Note that ordinary multiplication by this number would require seven additions, but the Booth-recoded multiplier requires only one addition and two subtractions.

An STG for a Booth multiplier is shown in Figure 10-50. The annotation on the branches of the graph indicates that the Booth recoding bits (denoted by *BRC*) control the state transitions. The structural units for *Multiplier_Booth_STG_0* are shown in Figure 10-51, and the Verilog source code of *Multiplier_Booth_STG_0* is given in Example 10.8.

The controller in *Multiplier_Booth_STG_0* generates the signals *Add* and *Sub* to control the addition and subtraction operations implied by the Booth algorithm. An alternative design could use one signal, *Add_sub*, to control these operations, but would need to generate and use a signal *Done* in the datapath unit to ensure that addition and subtraction operations are suspended while *Done* is asserted. Otherwise, the final value of *product* would be overwritten in state *S_8*. Note that the datapath unit uses a priority decoding scheme that decodes *Shift* before *Add* and *Sub*. The controller requires a flip-flop to store the LSB of *multiplier* for use in forming the Booth recoding bits (*BRC[1: 0]*), and the datapath operations require an adder/subtractor unit. Also, note how the 2s complement representation is formed when the multiplicand is negative. The left half of *multiplicand* must be filled with 1s to form the 2s complement correctly (sign extension). Similar actions are taken within the testbench to predict the expected value of *product*. For convenience, a pair of nested *for* loops generate integer values of *word1* and *word2*. Additional code converts the patterns to the wordlength of the parameterized machine. The testbench for *Multiplier_STG_0* is also presented because it contains some noteworthy features that are used when the pattern generator and comparator must form sign extensions of 2s complement negative numbers.

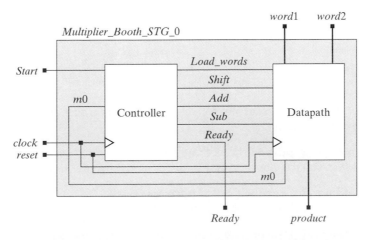

FIGURE 10-51 Structural units of a multiplier with Booth recoding.

Example 10.8

```verilog
module Multiplier_Booth_STG_0 (product, Ready, word1, word2, Start, clock, reset);
  parameter                      L_word = 4;
  parameter                      L_BRC = 2;
  parameter                      All_Ones = 4'b1111;
  parameter                      All_Zeros = 4'b0000;
  output     [2*L_word -1: 0]    product;
  output                         Ready;
  input      [L_word -1: 0]      word1, word2;
  input                          Start, clock, reset;
  wire                           Load_words, Shift, Add, Sub, Ready;
  wire       [L_BRC -1: 0]       BRC;

  Datapath_Booth_STG_0 M1 (product, m0, word1, word2, Load_words,
    Shift, Add, Sub, clock, reset);

  Controller_Booth_STG_0 M2 (Load_words, Shift, Add, Sub,
    Ready, m0, Start, clock, reset);
endmodule

module Controller_Booth_STG_0 (Load_words, Shift, Add, Sub,
  Ready, m0, Start, clock, reset);
  parameter                      L_word = 4;
  parameter                      L_state = 4;
  parameter                      L_BRC = 2;
  output                         Load_words, Shift, Add, Sub, Ready;
  input                          Start, clock, reset;
  input                          m0;
  reg        [L_state -1: 0]     state, next_state;
  parameter                      S_idle = 0, S_1 = 1, S_2 = 2, S_3 = 3;
  parameter                      S_4 = 4, S_5 = 5, S_6 = 6, S_7 = 7, S_8 = 8;
  reg                            Load_words, Shift, Add, Sub;
  wire                           Ready = ((state == S_idle ) && !reset) || (state
                                 == S_8);
  reg                            m0_del;
  wire       [L_BRC -1: 0]       BRC = {m0, m0_del};       // Booth recoding bits

// Necessary to reset m0_del when Load_words is asserted, otherwise it would
start with residual value

  always @ (posedge clock or posedge reset)
    if (reset) m0_del <= 0; else if (Load_words) m0_del <= 0; else m0_del <= m0;

  always @ (posedge clock or posedge reset)
    if (reset) state <= S_idle; else state <= next_state;

  always @ (state or Start or BRC) begin     // Next state and control logic
    Load_words = 0; Shift = 0; Add = 0; Sub = 0;
```

```verilog
       case (state)
         S_idle:        if (Start) begin Load_words = 1; next_state = S_1; end
                        else next_state = S_idle;

         S_1: if ((BRC == 0) || (BRC == 3))      begin Shift = 1; next_state = S_3; end
              else if (BRC == 1)                 begin Add = 1; next_state = S_2; end
              else if (BRC == 2)                 begin Sub = 1; next_state = S_2; end

         S_3: if ((BRC == 0) || (BRC == 3))      begin Shift = 1; next_state = S_5; end
              else if (BRC == 1)                 begin Add = 1; next_state = S_4; end
              else if (BRC == 2)                 begin Sub = 1; next_state = S_4; end

         S_5: if ((BRC == 0) || (BRC == 3))      begin Shift = 1; next_state = S_7; end
              else if (BRC == 1)                 begin Add = 1; next_state = S_6; end
              else if (BRC == 2)                 begin Sub = 1; next_state = S_6; end

         S_7: if ((BRC == 0) || (BRC == 3))      begin Shift = 1; next_state = S_8; end
              else if (BRC == 1)                 begin Add = 1; next_state = S_8; end
              else if (BRC == 2)                 begin Sub = 1; next_state = S_8; end

         S_2:                                    begin Shift = 1; next_state = S_3; end
         S_4:                                    begin Shift = 1; next_state = S_5; end
         S_6:                                    begin Shift = 1; next_state = S_7; end

         S_8: if (Start)                         begin Load_words = 1; next_state =
                                                 S_1; end

              else                               next_state = S_8;

         default:                                next_state = S_idle;
       endcase
     end
endmodule

module Datapath_Booth_STG_0 (product, m0, word1, word2, Load_words,
  Shift, Add, Sub, clock, reset);
  parameter                         L_word = 4;
  output       [2*L_word -1: 0]     product;
  output                            m0;
  input        [L_word -1: 0]       word1, word2;
  input                             Load_words, Shift, Add, Sub, clock,
                                    reset;
  reg          [2*L_word -1: 0]     product, multiplicand;
  reg          [L_word -1: 0]       multiplier;
  wire                              m0 = multiplier[0];
  parameter                         All_Ones = 4'b1111;
  parameter                         All_Zeros = 4'b0000;

// Register/Datapath Operations

  always @ (posedge clock or posedge reset) begin
    if (reset) begin multiplier <= 0; multiplicand <= 0; product <= 0; end
    else if (Load_words) begin
```

```verilog
      if (word1[L_word -1] == 0) multiplicand <= word1;
      else multiplicand <= {All_Ones, word1[L_word -1: 0]};
      multiplier <= word2;
      product <= 0;
     end
    else if (Shift) begin
     multiplier <= multiplier >> 1;
     multiplicand <= multiplicand << 1;
    end
    else if (Add) begin product <= product + multiplicand; end

     else if (Sub) begin product <= product - multiplicand; end
   end
endmodule

module test_Multiplier_STG_0 ();
  parameter                               L_word = 4;
  wire               [2*L_word -1: 0]     product;
  wire                                    Ready;
  integer                                 word1, word2; // multiplicand, multiplier
  reg                                     Start, clock, reset;
  reg                [3: 0]               mag_1, mag_2;

  Multiplier_Booth_STG_0 M1 (product, Ready, word1, word2, Start, clock, reset);

  // Exhaustive Testbench
  reg                [2*L_word -1: 0]     expected_value, expected_mag;
  reg                                     code_error;
  parameter                               All_Ones = 4'b1111;
  parameter                               All_Zeros = 4'b0000;

  initial #80000 $finish;                 // Timeout

// Error detection

  always @ (posedge clock) // Compare product with expected value
   if (Start) begin
    expected_value = 0;
    case({word1[L_word -1], word2[L_word -1]})
     0: begin     expected_value = word1 * word2; expected_mag = expected_value;
               end
     1: begin     expected_value = word1* {All_Ones,word2[L_word -1: 0]};
                  expected_mag = 1+ ~(expected_value); end
     2: begin     expected_value = {All_Ones,word1[L_word -1: 0]} *word2;
                  expected_mag = 1+ ~(expected_value); end
     3: begin     expected_value =
                  ({All_Zeros,1+ ~word2[L_word -1: 0]}) * ({4'b0,1+ ~word1[L_word -1:
                  0]});
                  expected_mag = expected_value; end
    endcase
   code_error = 0;
   end
```

```
      else begin
        code_error = Ready ? |(expected_value ^ product) : 0;
      end

  initial begin clock = 0; forever #10 clock = ~clock; end
  initial begin
    #2 reset = 1;
    #15 reset = 0;
  end

  initial begin       // Exhaustive patterns
    #100
    for (word1 = All_Zeros; word1 <= 15; word1 = word1 +1) begin
      if (word1[L_word -1] == 0) mag_1 = word1;
      else begin mag_1 = word1[L_word -1: 0];
        mag_1 = 1+ ~mag_1; end
      for (word2 = All_Zeros; word2 <= 15; word2 = word2 +1) begin
        if (word2[L_word -1] == 0) mag_2 = word2;
        else begin mag_2 = word2[L_word -1: 0]; mag_2 = 1+ ~mag_2; end
        Start = 0; #40 Start = 1;
```

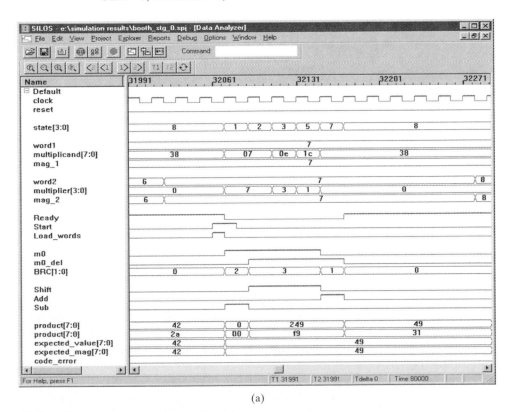

(a)

FIGURE 10-52 Comparison of state transitions in a multiplier: (a) with, and (b) without a Booth recoding scheme.

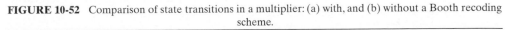

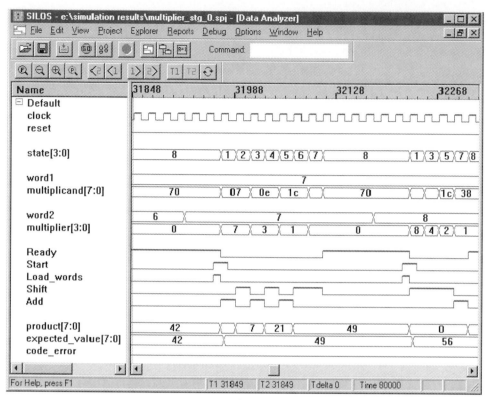

(b)

FIGURE 10-52 Continued

```
    #20 Start = 0;
    #200;
 end // word2
    #140;
 end //word1
end
endmodule
```

End of Example 10.8

Figure 10-52 shows (a) the simulation results produced by *Multiplier_Booth_STG_0* in multiplying 7 × 7, and (b) the results obtained with *Multiplier_STG_0*. For these data, the multiplier forms *product* in five cycles using Booth recoding, and in seven cycles without recoding. The waveforms of *product* are shown in both decimal and hexadecimal format (as 2s complement values), and the waveforms of *expected_value* and *expected_mag* are also in decimal format.

10.3.12 Bit-Pair Encoding

Booth recoding does not always lead to a reduction in the clock cycles required for multiplication in multipliers whose STG has been modified to perform the operation of shifting in the same cycle as *Add_sub*. Depending on the data pattern, Booth recoding may actually increase the number of clock cycles! Thus, the efficiency of the Booth recoding algorithm depends on the data. An alternative scheme, called bit-pair encoding (BPE), overcomes this limitation by encoding the digits as signed radix-4 digits (also called bit-pair encoding) [4, 10]. Bit-pair encoding (recoding) ensures that the number of additions does not increase. In fact, the number of additions is reduced from n to $n/2$.

Bit-pair encoding of a multiplier examines 3 bits at a time, and creates a 2-bit code whose value determines whether to (1) add the multiplicand, (2) shift the multiplicand by 1 bit and then add, (3) subtract the multiplicand (i.e., add the 2s complement of the multiplicand to the product), (4) shift the 2s complement of the multiplicand to the left by 1 bit and then add, or (5) to only shift the multiplicand to the location corresponding to the next bit-pair (i.e., without adding or subtracting at the present location). As in Booth recoding, the first step of the BPE algorithm is seeded with a value of 0 in a register cell to the right of the LSB of the multiplier word. Subsequent actions depend on the value of the recoded bit-pair. The index i increments by 2 until the word is exhausted. If the word contains an odd number of bits, its sign bit must be extended by 1 bit to accommodate the recoding scheme. Recoding divides the multiplier word by 2, so the number of possible additions is reduced by a factor of 2. The rules for bit-pair encoding are summarized in Table 10-2.

Example 10.9

The bit-pair recoding of $-65_{10} = 1011\_1111_2$ is shown in Figure 10-53. Note that multiplication by the recoded multiplier requires only two subtractions to form the product.

End of Example 10.9

TABLE 10-2   Rules for bit-pair (radix-4) recoding of a 2s complement number.

m_{i+1}	m_i	m_{i-1}	Code	BRC_{i+1}	BRC_i	Value	Status	Actions
0	0	0	0	0	0	0	String of 0s	Shift by 2
0	0	1	1	0	1	+1	End of string of 1s	Add
0	1	0	2	0	1	+1	Single 1	Add
0	1	1	3	1	0	+2	End of string of 1s	Shift by 1, Add, Shift by 1
1	0	0	4	1	0	−2	Begin string of 1s	Shift by 1, Subtract, Shift by 1
1	0	1	5	0	1	−1	Single 0	Subtract
1	1	0	6	0	1	−1	Begin string of 1s	Subtract
1	1	1	7	0	0	0	Midstring of 1s	Shift by 2

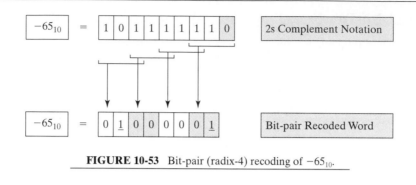

FIGURE 10-53 Bit-pair (radix-4) recoding of -65_{10}.

Example 10.10

The 2s complement product of 5_{10} by the multiplier -65_{10}, with -65_{10} recoded in a bit-pair format, is illustrated in Figure 10-54. The first bit-pair (shaded) indicates subtraction, so the 2s complement of 5_{10} is formed and aligned with the LSB of the multiplicand. Double shifts result from the next two bit-pairs. The final bit-pair (shaded) specifies subtraction, so the 2s complement of 5_{10} is formed at the proper location. Taking the sum of the shifted multiplicands forms the 2s complement of the product. The magnitude of the result is also shown. Note that it is necessary to sign-extend the copies of the multiplicand to fit the word length of the product register.

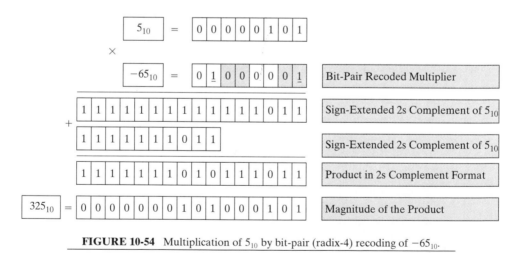

FIGURE 10-54 Multiplication of 5_{10} by bit-pair (radix-4) recoding of -65_{10}.

End of Example 10.10

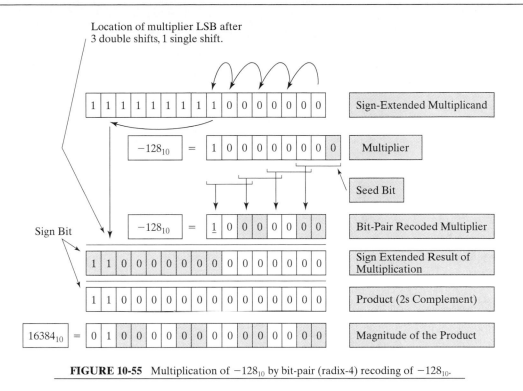

FIGURE 10-55   Multiplication of -128_{10} by bit-pair (radix-4) recoding of -128_{10}.

Example 10.11

Figure 10-55 shows the 2s complement product of -128_{10} multiplied by the multiplier -128_{10}, with -128_{10} recoded in a bit-pair format. The first three bit-pairs of the multiplier (beginning at the LSB) cause a copy of the multiplicand to be shifted by 6 bits toward the MSB; the final bit-pair signifies the beginning and end of a string of 1s, so the multiplicand is shifted by 1 bit and added to the product register. The magnitude of the result is also shown. Note that it is necessary for the product register to have a length (16 bits) of twice the word length of the data (8 bits), and for the multiplicand to be sign-extended to fit the word length of the product register.

End of Example 10.11

Example 10.12

Our final example of a sequential multiplier will present the Verilog description of an 8-bit radix-4 multiplier with the STG shown in Figure 10-56, with separate cycles for

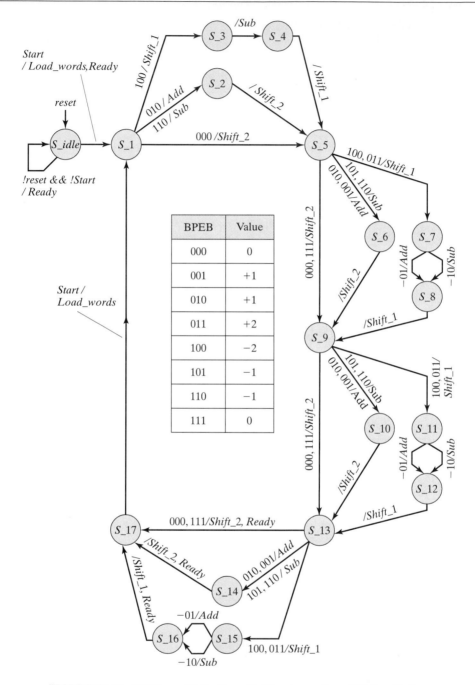

FIGURE 10-56   STG for multiplication with bit-pair recoding of 8-bit multipliers.

shifting, adding, and subtracting. Its STG is also shown along with the recoding rules. The machine forms a 3-bit word, *Bit-Pair Encoding Bits* (BPEB), and uses it to control the state transitions. Note that when BPEB has a value of $+2$ the machine shifts *multiplicand* by one position, adds *multiplicand* to *product*, then shifts by one more bit (e.g., $S\_5 \rightarrow S\_7 \rightarrow S\_8 \rightarrow S\_9$). After these actions, the position index is located at the next BPEB. A BPEB value of -2 is treated in a similar manner. The bits of BPEB are updated at each cycle and effectively remember whether an add or a subtract was decoded at the previous state. Separate signals control the datapath operations of addition and subtraction, so the signal *Ready* is not passed to the datapath unit.

The Verilog description of *Multiplier_Radix_4_STG_0* is given below. The functionality was verified exhaustively for all combinations of positive and negative data words. The controller handles the states necessary for bit-pair encoding, and the datapath unit is expanded to handle single and double bit shifts.

```
'define        All_Ones           8'b1111_1111
'define        All_Zeros          8'b0000_0000

module Multiplier_Radix_4_STG_0 (product, Ready, word1, word2, Start, clock, reset);
parameter                         L_word = 8;

output         [2*L_word -1: 0]   product;
output                            Ready;
input          [L_word -1: 0]     word1, word2;
input                             Start, clock, reset;
wire                              Load_words, Shift, Add_sub, Ready;
wire           [2: 0]             BPEB;

Datapath_Radix_4_STG_0 M1
  (product, BPEB, word1, word2, Load_words, Shift_1, Shift_2, Add, Sub, clock,
  reset);

Controller_Radix_4_STG_0 M2
  (Load_words, Shift_1, Shift_2, Add, Sub, Ready, BPEB, Start, clock, reset);

endmodule

module Controller_Radix_4_STG_0
  (Load_words, Shift_1, Shift_2, Add, Sub, Ready, BPEB, Start, clock, reset);
parameter                         L_word = 8;
output                            Load_words, Shift_1, Shift_2, Add, Sub, Ready;
input                             Start, clock, reset;
input          [2: 0]             BPEB;
reg            [4: 0]             state, next_state;
parameter                         S_idle = 0, S_1 = 1, S_2 = 2, S_3 = 3;
parameter                         S_4 = 4, S_5 = 5, S_6 = 6, S_7 = 7, S_8 = 8;
parameter                         S_9 = 9, S_10 = 10, S_11 = 11, S_12 = 12;
parameter                         S_13 = 13, S_14 = 14, S_15 = 15;
parameter                         S_16 = 16, S_17 = 17;

reg                               Load_words, Shift_1, Shift_2, Add, Sub;
wire                              Ready = ((state == S_idle) && !reset) ||
                                  (next_state == S_17) ;
```

```verilog
always @ (posedge clock or posedge reset)
if (reset) state <= S_idle; else state <= next_state;

always @ (state or Start or BPEB) begin    // Next state and control logic
Load_words = 0; Shift_1 = 0; Shift_2 = 0; Add = 0; Sub = 0;
case (state)
S_idle:          if (Start) begin Load_words = 1; next_state = S_1; end
                 else next_state = S_idle;
S_1:             case (BPEB)
                 0:      begin Shift_2 = 1;        next_state = S_5; end
                 2:      begin Add = 1;            next_state = S_2; end
                 4:      begin Shift_1 = 1;        next_state = S_3; end
                 6:      begin Sub = 1;            next_state = S_2; end
                 default:                          next_state = S_idle;
                 endcase
S_2:             begin    Shift_2 = 1;            next_state = S_5; end
S_3:             begin    Sub = 1;               next_state = S_4; end
S_4:             begin    Shift_1 = 1;           next_state = S_5; end

S_5:             case (BPEB)
                 0, 7:   begin Shift_2 = 1;       next_state = S_9; end
                 1, 2:   begin Add = 1;           next_state = S_6; end
                 3, 4:   begin Shift_1 = 1;       next_state = S_7; end
                 5, 6:   begin Sub = 1;           next_state = S_6; end
                 endcase
S_6:             begin    Shift_2 = 1;            next_state = S_9; end
S_7:             begin    if (BPEB[1: 0] == 2'b01) Add = 1;
                 else     Sub = 1;               next_state = S_8; end
S_8:             begin    Shift_1 = 1; next_state = S_9; end
S_9:             case (BPEB)
                 0, 7:   begin Shift_2 = 1;       next_state = S_13; end
                 1, 2:   begin Add = 1;           next_state = S_10; end
                 3, 4:   begin Shift_1 = 1;       next_state = S_11; end
                 5, 6:   begin Sub = 1;           next_state = S_10; end
                 endcase
S_10:            begin    Shift_2 = 1;            next_state = S_13; end
S_11:            begin    if (BPEB[1: 0] == 2'b01) Add = 1;
                 else     Sub = 1;               next_state = S_12; end
S_12:            begin    Shift_1 = 1;            next_state = S_13; end
S_13:            case (BPEB)
                 0, 7:   begin Shift_2 = 1;       next_state = S_17; end
                 1, 2:   begin Add = 1;           next_state = S_14; end
                 3, 4:   begin Shift_1 = 1;       next_state = S_15; end
                 5, 6:   begin Sub = 1;           next_state = S_14; end
                 endcase
S_14:            begin    Shift_2 = 1; next_state = S_17; end
S_15:            begin    if (BPEB[1: 0] == 2'b01) Add = 1;
                 else     Sub = 1; next_state = S_16; end
S_16:            begin    Shift_1 = 1; next_state = S_17; end
S_17:            if       (Start) begin Load_words = 1; next_state = S_1; end
```

```
                              else      next_state = S_17;
            default:          next_state = S_idle;
            endcase
            end
          endmodule

          module Datapath_Radix_4_STG_0
          (product, BPEB, word1, word2, Load_words, Shift_1, Shift_2, Add, Sub, clock,
          reset);
          parameter                                    L_word = 8;
          output         [2*L_word -1: 0]              product;
          output         [2: 0]                        BPEB;
          input          [L_word -1: 0]                word1, word2;
          input                                        Load_words, Shift_1, Shift_2;
          input                                        Add, Sub, clock, reset;
          reg            [2*L_word -1: 0]              product, multiplicand;
          reg            [L_word -1: 0]                multiplier;
          reg                                          m0_del;
          wire           [2: 0]                        BPEB = {multiplier[1: 0], m0_del};

          // Register/Datapath Operations
          always @ (posedge clock or posedge reset) begin
            if (reset) begin
              multiplier <= 0; m0_del <= 0; multiplicand <= 0; product <= 0;
            end
            else if (Load_words) begin
              m0_del <= 0;
              if (word1[L_word -1] == 0) multiplicand <= word1;
              else multiplicand <= {'All_Ones, word1[L_word -1: 0]};
              multiplier <= word2; m0_del <= 0; product <= 0;
            end
            else if (Shift_1) begin
              {multiplier, m0_del} <= {multiplier, m0_del} > 1;
              multiplicand <= multiplicand << 1;
            end
            else if (Shift_2) begin
              {multiplier, m0_del} <= {multiplier, m0_del} >> 2;
              multiplicand <= multiplicand << 2;
            end
            else if (Add) begin product <= product + multiplicand; end
            else if (Sub) begin product <= product - multiplicand; end
          end
          endmodule
```

Figure 10-57 shows output waveforms in hex and decimal formats verifying the multiplication of -45_{10} by -38_{10}. The values of *word1* and *word2* are also shown in hex and decimal formats (*mag_1* and *mag_2*); the magnitudes of *word1* and *word2* are shown in decimal format. The value of *multiplicand* and the *multiplier* are shown in hex

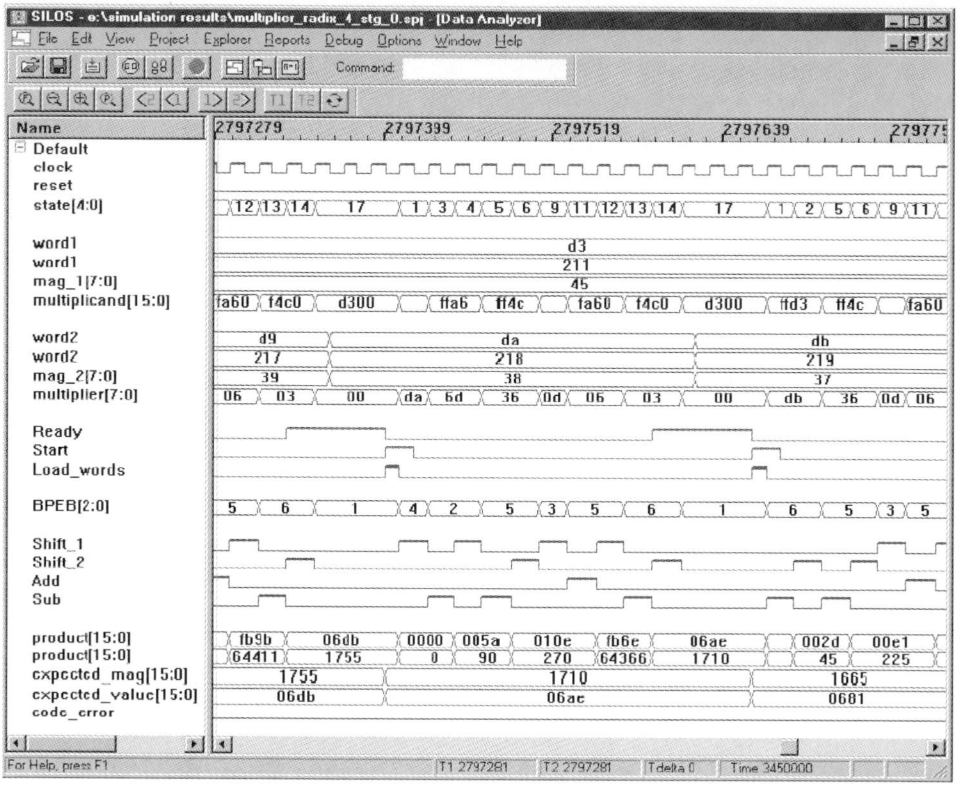

FIGURE 10-57 Simulation of *Multiplier_Radix_4_STG_0* multiplying -45_{10} by -38_{10}.

format. The value of *product* is shown in hex and decimal formats. The expected value (produced by the testbench) is shown in both formats.

End of Example 10.12

10.4   Multiplication of Signed Binary Numbers

Although signed binary numbers in 2s complement format can be multiplied with Booth's algorithm, we will reconsider their multiplication here, to prepare for multiplication of fractions. There are four cases to consider in multiplying signed numbers in 2s complement format, depending on the signs of the multiplicand and multiplier. We have already seen that the product of unsigned binary numbers is formed by adding shifted copies of the multiplicand. We will consider the remaining three cases, in which one or both words are negative.

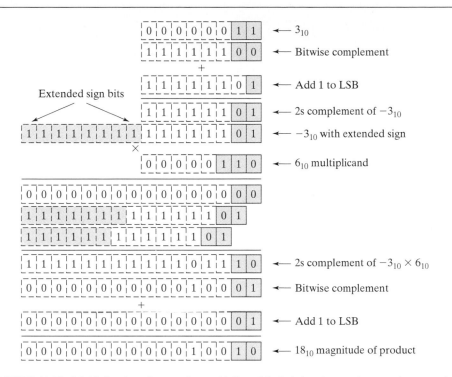

FIGURE 10-58 Multiplication of a negative multiplicand (-3_{10}), in a 2s complement, sign-extended format, by a positive multiplier (6_{10}), forming the product -18_{10}.

10.4.1 Product of Signed Numbers: Negative Multiplicand, Positive Multiplier

The steps to multiply a negative multiplicand by a positive multiplier are the same as the steps taken to multiply unsigned numbers, but the sign bit of the multiplicand must be extended to the word length of the final product before operating on the 2s complement words. The sign-extended multiplicand is used when forming the partial products and accumulating the sums. The result of the multiplication is the 2s complement of the product. Then the magnitude of the product is formed by taking the 2s complement of the result, as illustrated in Figure 10-58 for the product of -3_{10} by 6_{10}. The sign-extended multiplicands are shown, with the carry bits that are generated in each column-wise addition.

10.4.2 Product of Signed Numbers: Positive Multiplicand, Negative Multiplier

To form the product of a positive multiplicand by a negative multiplier, extend the sign of multiplier to the word length of the multiplier. Then add shifted copies of the multiplicand, but instead of adding a copy of the multiplicand at the position corresponding to the

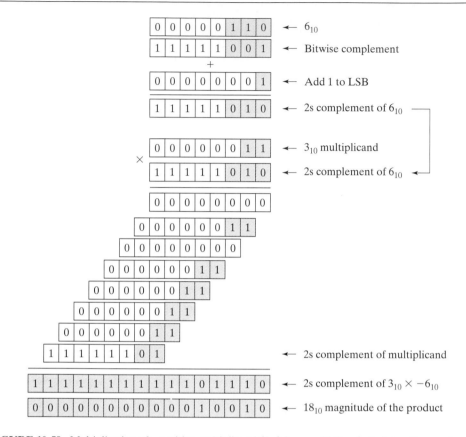

FIGURE 10-59 Multiplication of a positive multiplicand (3_{10}) by a multiplier (-6_{10}), in a 2s complement, sign-extended multiplier, forming the product -18_{10}.

extended sign bit of the multiplier, add the 2s complement of the multiplicand. This last step follows from the observation that the 2s complement of an n-bit multiplier can be written as the sum: $-B_{n-1} \times 2^{n-1} + B_{n-2} \times 2^{n-1} + \ldots + B_1 \times 2^1 + B_0 \times 2^0$. The actions associated with $B_{n-1}, \ldots, B_0$ are the usual ones of adding shifted copies of the multiplicand. The action of the term associated with $-B_{n-1} \times 2^{n-1}$ is equivalent to adding a shifted copy of the 2s complement of the multiplicand to the sum of the previously accumulated partial sums. The results in Figure 10-59 form the product of 3_{10} by -6_{10}.

10.4.3 Product of Signed Numbers: Negative Multiplicand, Negative Multiplier

When the multiplicand and the multiplier are both negative numbers expressed in 2s complement format, the last sum adds the 2s complement of the multiplicand to the accumulated partial sums, but the accumulated sums are formed with shifted copies of the sign-extended multiplicand, instead of the multiplicand. For clarity, Figure 10-60 also shows the columnwise carries that are generated in forming the product of -3_{10} by -6_{10}.

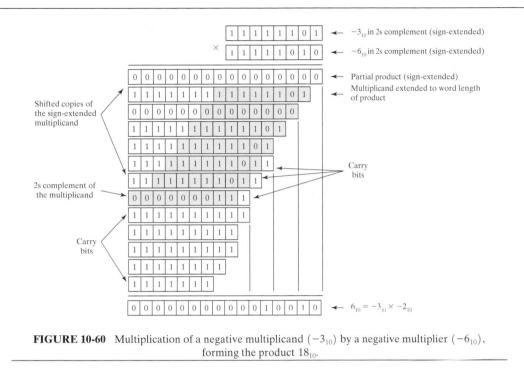

FIGURE 10-60 Multiplication of a negative multiplicand (-3_{10}) by a negative multiplier (-6_{10}), forming the product 18_{10}.

10.5 Multiplication of Fractions

Numbers are normalized in digital signal processors to avoid the overflow that would result when the product of two numbers exceeds the dynamic range provided by the word length of the machine [12]. The dynamic range of the numbers that can be represented by N bits in a 2s complement format is $-2^{N-1} \leq D(B) \leq 2^{N-1} - 1$. For example, 2s complement 4-bit words have $N = 4$, and their dynamic range is from -8 to $+7$. Any product whose value exceeds this range causes overflow, because the product cannot be stored accurately as a 4-bit value. For example, the product of 7 by 3 exceeds the dynamic range provided by a 4-bit word format.

Normalization divides an N-bit 2s complement word by 2^{N-1} to convert a fixed-point integer representation of a value to a fixed-point fractional representation. The dynamic range of the magnitude of the fractional value is bounded between -1 and $+1$. Normalization is equivalent to shifting the word toward its LSB by $N - 1$ positions, and associating its weights with fractions. If a 2s complement word B has the decimal value $D(B) = -b_{N-1} 2^{N-1} + b_{N-2} 2^{N-2} \cdots + b_1 2^1 + b_0 2^0$, its normalized value is given by $F(B) = -b_{N-1} 2^0 + b_{N-2} 2^{-1} + \cdots + b_1 2^{-(N-2)} + b_0 2^{-(N-1)}$, the so-called Q-format representation of the number [12]. For example, a Q-5 number format has 5 bits, including a sign bit. The product of two Q-5 numbers has 10 bits, including an extended sign bit and a sign bit. The radix point of Q-format numbers is to the right of the sign bit.

Normalization of integers prevents overflow in multiplication, because the product of two fractions is always a fraction. It also extends the dynamic range of the numbers that can be multiplied, for a given word length, at the expense of precision. The result of multiplying two normalized numbers may have less precision than if the numbers could be multiplied by a machine that has sufficient word length to avoid overflow. For example, the product of $8_{10} = 1000_2$ by $7_{10} = 0011_2$ is 56_{10}, which cannot be stored as a 4-bit value in 2s complement format. Normalization produces the following fractions in Q-5 format: $F(8_{10}) = 2^{-4} = 0.1000_2$, and $F(7_{10}) = 0.0111_2$. Their product is 00.0011_1000_2, in Q-10 format. Storing the product as a Q-5 value gives $F(8_{10} \times 7_{10}) = 0.0011$. The denormalized decimal value is $F(8_{10} \times 7_{10}) \cong 48_{10}$, obtained by scaling the Q-5 value by 2^8, or left-shifting the word by 8 bits.

The 2s complement of an n-bit number M is a number M^* such that $M + M^* = 2^n$. The 2s complement of a binary fraction M is given by $M^* = 2 - M$, so that $M + M^* = 2$. An m-bit fraction F is represented as $F = b_{-1} 2^{-1} + b_{-2} 2^{-2} + b_{-3} 2^{-3} + \cdots + b_{-m} 2^{-m}$. The 2s complement of a fraction is formed by complementing the bits from the sign bit to the least significant 1, then adding 1 at the position of the least significant 1. This is equivalent to complementing the bits to the left of the rightmost 1 in the word. Both methods are shown in Figure 10-61.

Fractions are multiplied like whole numbers, but overflow is not possible, because the product of two fractions must be a fraction. Care must be taken to adjust the location of the radix point of the result of multiplying fractions. In a fixed-point format, a 4-bit fraction is represented by 5 bits, with the MSB holding the sign of the number in a 2s complement format. The radix point for a fixed-point fraction will be between the sign bit and the MSB of the 4-bit fraction. The product of two 5-bit words will produce a 10-bit result. The MSB will be the extended sign bit, and its neighbor immediately to the right will be the sign bit.

10.5.1 Signed Fractions: Positive Multiplicand, Positive Multiplier

Multiplication of two positive fractions is performed as though the words were unsigned integers. For example, in the product of two 4-bit fractions, the radix point is interpreted to be to the left of the eighth bit. Figure 10-62 shows the product of $3/4_{10}$ by $1/2_{10}$.

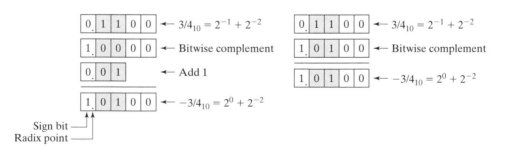

FIGURE 10-61   Equivalent methods of the forming 2s complement of a fraction ($3/4_{10}$).

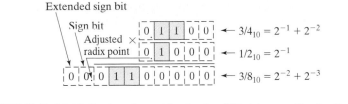

FIGURE 10-62 Multiplication of a positive fraction ($3/4_{10}$) by a positive fraction ($1/2_{10}$).

10.5.2 Signed Fractions: Negative Multiplicand, Positive Multiplier

The product of a negative multiplicand by a positive multiplier is formed by adding shifted copies of the sign-extended multiplicand, and adjusting the radix point in the result. The example in Figure 10-63 forms the product of $-3/4_{10}$ by $3/8_{10}$.

10.5.3 Signed Fractions: Positive Multiplicand, Negative Multiplier

If the multiplicand is positive and the multiplier is negative, we add shifted copies of the sign-extended multiplicand, except at the position of the sign bit of the multiplier. At that position, we add the 2s complement of the multiplier. In this case it is convenient to define the 2s complement of a fraction A as $A^* = 2 - A$, or $10_2 - A_2$. This limits the representation of the multiplier to have only 2 bits to the left of the radix, thereby reducing the number of products that must be added to form the sum, as shown in Figure 10-64.

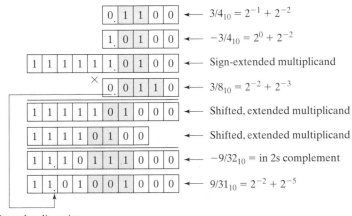

FIGURE 10-63 Multiplication of a negative fraction ($-3/4_{10}$) by a positive fraction ($3/8_{10}$).

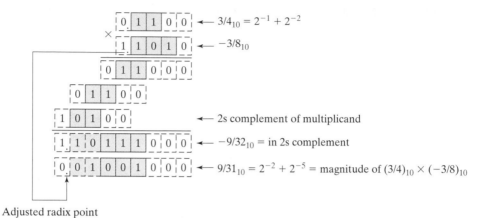

FIGURE 10-64 Multiplication of a positive multiplicand ($3/4_{10}$) by a negative multiplier ($-3/8_{10}$).

10.5.4 Signed Fractions: Negative Multiplicand, Negative Multiplier

To form the product of two negative fractions, add shifted copies of the sign-extended multiplicand, and add the 2s complement of the accumulated sum, as shown in Figure 10.65.

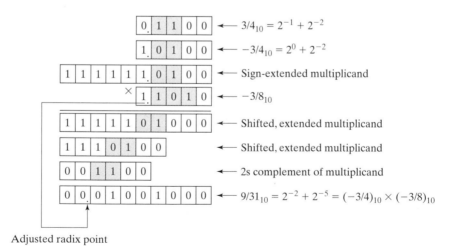

FIGURE 10-65 Multiplication of a negative multiplicand ($-3/4_{10}$) by a negative multiplier ($-3/8_{10}$).

10.6 Functional Units for Division

Sequential multipliers use an *add-and-shift* algorithm to form the product of two words. We will consider various architectures for sequential dividers that use a *subtract-and-shift* algorithm to form a quotient of two numbers.

10.6.1 Division of Unsigned Binary Numbers

A sequential algorithm for dividing two unsigned binary numbers (i.e., positive integers) subtracts the divisor from the dividend repeatedly, until the remainder is detected to be smaller than the divisor. The quotient is formed by incrementing a counter each time a subtraction occurs; the final value of the remainder is formed as the residual content of the dividend when the subtraction sequence ends. Other architectures, such as one implementing a subtract-and-shift algorithm, can be more efficient, but we will examine the basic architecture first.

Example 10.13

Figure 10-66 shows the architecture for *Divider_STG_0*, a machine that forms the quotient of unsigned binary numbers by repeatedly subtracting the content of a divisor register from the content of a dividend register until the remainder is less than the divisor. This architecture is effective, but inefficient. The architecture uses more registers than needed, and it can require a very long execution sequence to form the quotient when the divisor is small compared to the dividend.

   Divider_STG_0 will serve to introduce some features that will be included in a more sophisticated machine. Among the machine's features, we want it to detect an

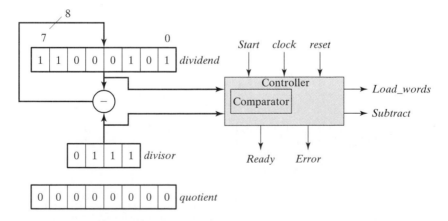

FIGURE 10-66 Architecture of *Divider_STG_0*, a simple, but inefficient, binary divider unit.

attempt to divide by 0, and to terminate without needless computation if the datapath presents a dividend that is 0. The machine should ignore *Start* while a division sequence is in progress. A signal *Ready* should be asserted after a division sequence is completed, and remain asserted until a new sequence begins; *Ready* should also be asserted while the machine is in an idle state, with *reset* not asserted. A signal *Error* should be asserted if a divide-by-zero is attempted, and should remain asserted until *reset* is asserted. An asynchronous reset signal should drive the state to an idle state from any state.

Divider_STG_0 has parameterized word lengths, shown here for a dividend datapath size of 8 bits, and a divisor datapath size of 4 bits. The implementation assumes that the length of the divisor does not exceed the length of the dividend. In Figure 10-66, *dividend* and *quotient* are stored in 8-bit registers, and *divisor* is stored in a 4-bit register. To implement the subtraction of *divisor* from *dividend*, the word for *divisor* will be converted to a 2s complement value, and extended by concatenating the 4-bit 2s complement with four 1s. A comparator determines whether or not to subtract. For an 8-bit dividend, the worst case will take 255 subtraction steps ($dividend = 255$, $divisor = 1$).

The machine's STG is shown in Figure 10-67, with its control logic annotated with symbols for Verilog operators. Unlike an ASMD chart, the STG is not annotated with the register operations of the datapath unit. State transitions along the branches leaving a state are conditioned by the indicated assertions, provided that *reset* is not asserted.

The state of the controller enters *S_idle* on the asynchronous action of *reset* and remains there until *Start* is asserted (with *reset* de-asserted). If *word2* (the datapath value for *divisor*) is 0, the state enters an error state *S_Err* when *Start* is asserted, and

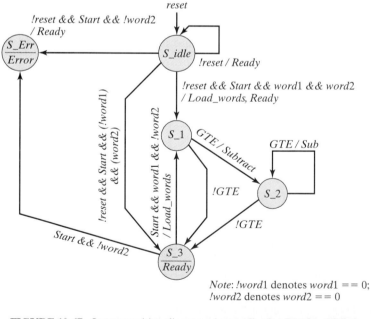

FIGURE 10-67 State-transition diagram of controller for *Divider_STG_0*.

remains there until *reset* is re-asserted. The signal *Error* is asserted as a Moore-type output in the state *S_Err* (the machine also enters *S_Err* as a fail-safe feature if its state is not one of those specified by the STG, but that detail is not shown on the STG). If *word2* is not 0, *word1* (the datapath value of *dividend*) is checked.[6] If *word1* is 0, the state immediately transfers to *S_3*, where *Ready* is asserted as a Moore-type output. If not, *Load_words* is asserted and the machine enters *S_1*, where *Subtract* is asserted while successive subtractions occur. In state *S_idle*, *Load_words* and *Ready* are Mealy-type outputs.

At each step, the algorithm compares *divisor* and *dividend*. When *dividend* is found to be less than *divisor*, the state enters *S_3*, where it remains until the next assertion of *Start* and *Ready* is asserted. It is the responsibility of the external agent controlling the machine to know that *Start* will be ignored until the machine is in *S_idle* or *S_3*. So *Start* should not be asserted until after *Ready* is asserted. Otherwise, the contents of *quotient* and *remainder* could be mistakenly associated with the new values of *word1* and *word2*, rather than the values that were present when the division that led to the assertion of *Ready* was initiated.

End of Example 10.13

Example 10.14

The Verilog description of *Divider_STG_0* is given below for an 8-bit dividend and a 4-bit divisor. We show two ways to implement subtraction. The first corresponds to the actual hardware supporting the machine's datapath operation of subtraction of 2s complement words, and requires sign extension of *divisor*. An alternative uses the statement *dividend* $<=$ *dividend* $-$ *divisor*. This form exploits the built-in 2s complement arithmetic of Verilog, and automatically accommodates the different word lengths of the operands. The testbench (see the website) for *Divider_STG_0* has a triggered stimulus generator. Note that the number of cycles required to form *quotient* is data-dependent, so stimulus patterns are triggered by the completion of a division sequence. The machine is synthesizable, because the data dependency is handled by the controller, not by a data-dependent loop.

```
module Divider_STG_0 (quotient, remainder, Ready, Error, word1, word2, Start,
clock, reset);
/* This version checks for a divide by zero, subtracts the divisor from the dividend
until the dividend is less than the divisor, and counts the number of subtractions per-
formed. The length of divisor must not exceed the length of dividend .*/

parameter                    L_divn = 8;
parameter                    L_divr = 4;
```

---

[6]In Verilog, the Boolean value of *word2* is true if and only if *word2* has the value of a positive integer.

```verilog
parameter                              S_idle = 0, S_1 = 1, S_2 = 2, S_3 = 3, S_Err
                                       = 4;
parameter                              L_state = 3;
output        [L_divn -1: 0]           quotient;
output        [L_divn -1: 0]           remainder;
output                                 Ready, Error;
input         [L_divn -1: 0]           word1;                    // Datapath for
                                                                 dividend
input         [L_divr -1: 0]           word2;                    // Datapath for
                                                                 divisor
input         Start, clock, reset;
reg           [L_state -1: 0]          state, next_state;
reg                                    Load_words, Subtract;
reg           [L_divn -1: 0]           dividend;
reg           [L_divr -1: 0]           divisor;
reg                                    quotient;
wire                                   GTE =   (dividend >= divisor); // Comparator
wire                                   Ready = ((state == S_idle) && !reset) ||
                                       (state == S_3);
wire                                   Error = (state == S_Err);
assign                                 remainder = dividend;

always @ (posedge clock or posedge reset)
  if (reset) state <= S_idle; else state <= next_state;          // State transitions

always @ (state or word1 or word2 or Start or GTE ) begin   // Next state and
                                                            control logic
Load_words = 0; Subtract = 0;                               // Default values
case (state)
  S_idle:     case (Start)
              0:       next_state = S_idle;

              1:       if (word2 == 0) next_state = S_Err;
                       else if (word1) begin next_state = S_1; Load_words = 1;
                       end
                       else next_state = S_3;
              endcase
  S_1:        if (GTE) begin next_state = S_2; Subtract = 1; end
              else next_state = S_3;
  S_2:        if (GTE) begin next_state = S_2; Subtract = 1; end
              else next_state = S_3;
  S_3:        case (Start)
              0:       next_state = S_3;

              1:       if (word2 == 0) next_state = S_Err;
                       else if (word1 == 0) next_state = S_3;
                       else begin next_state = S_1; Load_words = 1; end
              endcase
```

```
        S_Err:                  next_state = S_Err;
        default:                next_state = S_Err;
      endcase
    end

  // Register/Datapath Operations

  always @ (posedge clock or posedge reset) begin
    if (reset) begin divisor <= 0; dividend <= 0; quotient <= 0; end
    else if (Load_words == 1) begin
      dividend <= word1;
      divisor <= word2;
      quotient <= 0; end
    else if (Subtract) begin                    // Note sign extension below
      dividend <= dividend[L_divn -1: 0] + 1'b1 + {{(L_divn -L_divr){1'b1}}, ~divisor[L_divr
      -1: 0]};
      // dividend <= dividend - divisor;        // alternative using built-in 2's
                                                   complement arithmetic

      quotient <= quotient + 1; end
    end
  endmodule
```

End of Example 10.14

Figure 10-68 shows waveforms obtained by simulating *Divider_STG_0* dividing 50_{10} by 6_{10} to produce a quotient of 8_{10} and a remainder of 2_{10}. The values of *word1*, *word2*, *dividend*, *divisor*, *quotient*, and *remainder* are shown in decimal format. The signal *code_error* is generated by the testbench on detecting an error in either *quotient* or *remainder*. These waveforms do not demonstrate all of the features of the design; the testbench provided at the companion website can be used for additional verification.

Although a Verilog behavioral description can be written without concern for the details of implementation, leaving them up to a synthesis tool, it might be wise to consider the fact that *a synthesis tool can fail to produce the most efficient implementation, by not recognizing economies in the architecture*. For example, the architecture of *Divider_STG_0* uses a comparator (see Figure 10-66) to determine whether *divisor* should be subtracted from *dividend*, and uses a subtractor to perform the subtraction. An alternative design would exploit the observation that in 2s complement subtraction the carry bit reveals the relative magnitude of the numbers, eliminating the need for a comparator. The subtractor is implemented by an adder with a carry-in, and an inverted datapath for *divisor* (bitwise-complement). The carry-out of the adder produces the sign bit that controls the datapath. An architecture for the alternative machine, *Divider_STG_0_sub*, is shown in Figure 10-69, and the machine's Verilog description is presented in Example 10.14. Note that a single continuous assignment forms the concatenation {*carry*, *difference*} by adding the 2s complement of *divisor* to *dividend*.

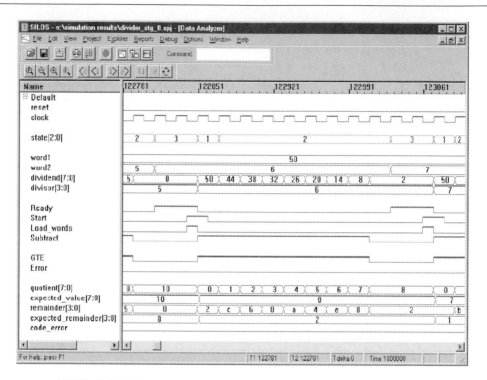

FIGURE 10-68 Simulation results for *Divider_STG_0*, a simple binary divider.

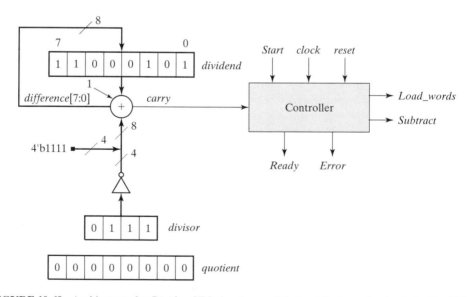

FIGURE 10-69 Architecture for *Divider_STG_0_sub*, a modified architecture of a simple, but inefficient, binary divider unit with an 8-bit dividend. The carry bit formed in 2s complement subtraction replaces the comparator used by *Divider_STG_0*.

Example 10.15

The Verilog description of *Divider_STG_0_sub* uses the carry bit from 2s complement subtraction to replace a comparator and control the datapath of the machine.

```
module Divider_STG_0_sub (quotient, remainder, Ready, Error, word1, word2,
Start, clock, reset);
/* This version checks for a divide by zero, subtracts the divisor from the dividend
until the dividend is less than the divisor, and counts the number of subtractions per-
formed. The length of divisor must not exceed the length of dividend .*/

parameter                          L_divn = 8;
parameter                          L_divr = 4;
parameter                          S_idle = 0, S_1 = 1, S_2 = 2, S_3 = 3,
                                   S_Err = 4;
parameter                          L_state = 3;
output      [L_divn -1: 0]         quotient;
output      [L_divn -1: 0]         remainder;
output                             Ready, Error;
input       [L_divn -1: 0]         word1;               // Datapath
                                                        for dividend
input       [L_divr -1: 0]         word2;               // Datapath
                                                        for divisor
input                              Start, clock, reset;
reg         [L_state -1: 0]        state, next_state;
reg                                Load_words, Subtract;
reg         [L_divn -1: 0]         dividend;
reg         [L_divr -1: 0]         divisor;
reg                                quotient;
wire                               Ready = ((state == S_idle) && !reset) ||
                                   (state == S_3);
wire                               Error = (state == S_Err);
wire        [L_divn -1: 0]         difference;
wire                               carry;
assign                             {carry, difference} = dividend[L_divn-1: 0]
                                   + {{(L_divn -L_divr){1'b1}},
                                   ~divisor[L_divr -1: 0]}
                                   + 1'b1;
assign                             remainder = dividend;

always @ (posedge clock or posedge reset)
  if (reset) state <= S_idle; else state <= next_state;     // State transitions

always @ (state or word1 or word2 or Start or carry) begin   // Next state and
                                                             control logic
```

```
      Load_words = 0; Subtract = 0;
      case (state)
        S_idle:        case (Start)
                         0:      next_state = S_idle;

                         1:      if (word2 == 0) next_state = S_Err;
                                 else if (word1) begin next_state = S_1; Load_words = 1;
                                 end
                                 else next_state = S_3;
                       endcase
        S_1:           if (!carry) next_state = S_3;
                       else begin next_state = S_2; Subtract = 1; end
        S_2:           if (!carry) next_state = S_3;
                       else begin next_state = S_2; Subtract = 1; end
        S_3:           case (Start)
                         0:      next_state = S_3;

                         1:      if (word2 == 0) next_state = S_Err;
                                 else if (word1 == 0) next_state = S_3;
                                 else begin next_state = S_1; Load_words = 1; end
                       endcase
        S_Err:         next_state = S_Err;
        default:       next_state = S_Err;
      endcase
    end

  // Register/Datapath Operations

  always @ (posedge clock or posedge reset) begin
    if (reset) begin divisor <= 0; dividend <= 0; quotient <= 0; end
    else if (Load_words == 1) begin
      dividend <= word1;
      divisor <= word2;
      quotient <= 0; end
    else if (Subtract) begin
      dividend <= difference;
      quotient <= quotient + 1; end
    end
  endmodule
```

End of Example 10.15

10.6.2 Efficient Division of Unsigned Binary Numbers

In the previous section, the machines *Divider_STG_0* and *Divider_STG_0_sub* divide
unsigned binary numbers by repeatedly subtracting the divisor from the dividend.

Both machines are very inefficient in dividing by a relatively small divisor, because they must perform several subtractions. A basic architecture for a more efficient divider is shown in Figure 10-70. Its operations parallel the commonly used manual steps that divide two numbers by aligning the divisor with the most significant bit of the dividend, and then repeatedly subtracting the divisor from the dividend, and shifting the *divisor* toward the LSB of the dividend. However, in the hardware implementation, the contents of the *dividend* register will be shifted repeatedly toward the MSB of the divisor.

Care must be taken in designing the architecture. In *Divider_STG_0* and *Divider_STG_0_sub* the registers holding *divisor* and *dividend* are physically aligned, so their LSBs are aligned too. At any stage of subtraction in the next architecture, it might be necessary to align the divisor and the dividend, depending on their relative size and on the relative location of each word's most significant 1 bit. Also, the dividend register must be extended to the left by 1 bit to accommodate the possibility that the initial value of the aligned content of the divisor register exceeds the value of the associated 4 bits of the dividend register, in which case we must shift a 1 from the MSB of the dividend before subtraction can be performed. For example, to divide 1001_2 by 1010_2 we must first shift the dividend to the left to align the dividend for the next subtraction. Consequently, the controller for the machine will be more complex, and includes signals for shifting both the divisor and the dividend, as shown in Figure 10-70.

The physical architecture of the machine aligns the divisor word with the leftmost four bits of the dividend's 8-bit datapath. In operation, the dividend word is shifted repeatedly from right to left, and the divisor word is subtracted from the dividend bits that it is aligned with at each step, depending on whether the divisor is less than the

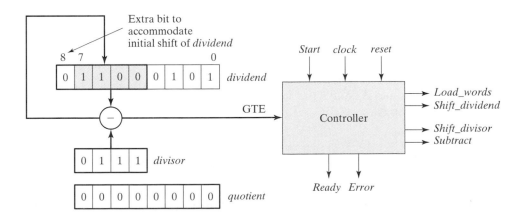

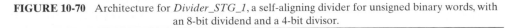

FIGURE 10-70 Architecture for *Divider_STG_1*, a self-aligning divider for unsigned binary words, with an 8-bit dividend and a 4-bit divisor.

corresponding part-select of the dividend. However, instead of subtracting the divisor from the dividend, the machine is aligned to subtract the largest possible product of the divisor and a power of 2, thereby eliminating repeated subtractions when the divisor is relatively small.

The machine is said to be *self-aligning*, because it automatically determines whether *divisor* or *dividend* need to be aligned at the beginning of a division sequence, depending on the relative position of their leftmost nonzero bits. An approach that would always *initially* align both words so that their MSB contains a 1 is inefficient because it can require far more shifts than are needed. The approach we will take is to initially shift the divisor towards the leftmost nonzero bit of the dividend (instead of the LSB of the dividend).

There are two cases that require an initial alignment of the datapath words: (1) the value of the leftmost 4-bit subword of *dividend* is less than the value of *divisor* (e.g., 1100_2 divided by 1110_2) and (2) the LSB of divisor is 0 and the divisor word can be shifted to the left and still divide into *dividend* (e.g., 1100_2 divided by 0101_2). In the first case, *dividend* must be shifted repeatedly to the left by 1 bit until the value of the leftmost 5 bits of the 1-bit-extended *dividend* equals or exceeds that of *divisor*, or until no further shifts are allowed; in the second case, *divisor* must be shifted to the left until the word produced by a further shifting cannot divide into the leftmost 4 bits of the *dividend* word (excluding the extra bit). The physical location of the remainder bits at the end of a division sequence depends on whether the dividend has been shifted for alignment. Therefore, the alignment shifts are counted and used to control the state machine and adjust the value of the remainder at the end of the execution sequence.

The STG for a self-aligning divider, *Divider_STG_1*, is shown in Figure 10-71. At a given state, a control label that is used on a branch leaving a state node will be treated as de-asserted on any other exiting branch where it is not used explicitly. A label that does not appear on any branch leaving a state node will be considered to be a don't-care. The reset signal is shown only at state *S_idle*, but is understood to have asynchronous action at all of the other states too.

The states of the machine are associated with its activity. In state *S_Adivr* the action of *Shift_divisor* aligns the divisor with the most significant 1-bit of the dividend; in state *S_Adivn* the action of *Shift_dividend* aligns the dividend register for subtraction; and in *S_div* the actual subtraction occurs, together with more shift operations. In states *S_Adivn* and *S_Adivr* the variable *Max* detects when the maximum allowed shifts have occurred.

Example 10.16

The testbench and Verilog description of a self-aligning divider, *Divider_STG_1*, corresponding to the STG in Figure 10-71 are given below for an 8-bit dividend and a 4-bit divisor. Note two features of *Divider_STG_1*: (1) the sign bit produced by the subtractor controls the datapath (an alternative design would rely on a synthesis tool to possibly replace the logic of a comparator with the sign bit of the subtractor), and (2) the

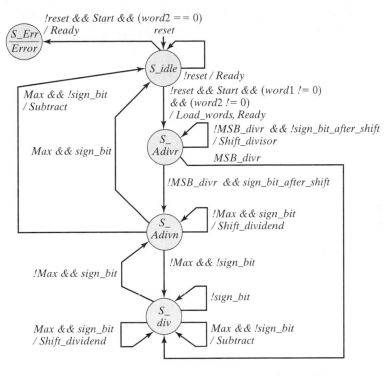

FIGURE 10-71 STG for *Divider_STG_1*, a self-aligning divider.

datapath to the subtractor is multiplexed. This feature eliminates the need for a separate comparator to implement the test to determine whether the result of shifting *divisor* by 1 bit will divide into the 4-bit sub-word of *dividend*. In this case, the value of *comparison* depends on whether *divisor* has been shifted before the difference between *dividend* and *divisor* is formed. The sign bit of *comparison* determines *sign_bit*. Alignment of *dividend* with *divisor* occurs in *S_Adivn*. The datapath operations of subtracting and shifting occur in state *S_div*.[7] The test patterns generated by the testbench are triggered by a de-assertion of *Ready*. This feature was included, because the number of cycles needed to complete a division sequence is data-dependent. A fixed-cycle pattern generator would have to accommodate the worst-case sequence for division, and use that delay for all patterns, making the test much longer than necessary. The testbench developed for this model includes error detection for exhaustive verification.

```
module Divider_STG_1 (quotient, remainder, Ready, Error, word1, word2, Start,
clock, reset);
parameter      L_divn = 8;
parameter      L_divr = 4;                        // Choose L_divr <= L_divn
```

---

[7]See Problem 28 at the end of the chapter for a modification to *Divider_STG_1* that reduces the cycles needed for division.

```verilog
parameter      S_idle = 0, S_Adivr = 1, S_Adivn = 2, S_div = 3, S_Err = 4;
parameter      L_state = 3, L_cnt = 4, Max_cnt = L_divn - L_divr;
output         [L_divn -1: 0]      quotient;
output         [L_divn -1: 0]      remainder;
output                            Ready, Error;
input          [L_divn -1: 0]      word1;                    // Datapath for
                                                             dividend
input          [L_divr -1: 0]      word2;                    // Datapath for
                                                             divisor
input                            Start, clock, reset;
reg            [L_state -1: 0]     state, next_state;
reg                              Load_words, Subtract, Shift_dividend,
                                 Shift_divisor;
reg                              quotient;
reg            [L_divn: 0]        dividend;                  // Extended dividend
reg            [L_divr -1: 0]     divisor;
reg            [L_cnt -1: 0]      num_shift_dividend, num_shift_divisor;
reg            [L_divr: 0]        comparison;
wire           MSB_divr = divisor[L_divr -1];
wire           Ready =((state == S_idle) && !reset) ;
wire           Error = (state == S_Err);
wire           Max = (num_shift_dividend == Max_cnt + num_shift_divisor );
wire           sign_bit = comparison[L_divr];

always @ (state or dividend or divisor or MSB_divr)             // subtract divisor
                                                               from dividend
   case (state)
     S_Adivr:     if (MSB_divr == 0) comparison
                       = dividend[L_divn: L_divn - L_divr] + {1'b1, ~(divisor << 1)}
                       + 1'b1;
                  else comparison = dividend[L_divn: L_divn - L_divr] + {1'b1,
                  ~divisor[L_divr -1: 0]} + 1'b1;
     default:     comparison = dividend[L_divn: L_divn - L_divr] + {1'b1,
                  ~divisor[L_divr -1: 0]} + 1'b1;
   endcase


               // Shift the remainder to compensate for alignment shifts
   assign       remainder = (dividend[L_divn -1: L_divn -L_divr] ) > num_shift_divisor;

always @ (posedge clock or posedge reset)                   // State Transitions
   if (reset) state <= S_idle; else state <= next_state;

// Next state and control logic

always
   @ (state or word1 or word2 or Start or comparison or sign_bit or Max ) begin
   Load_words = 0; Shift_dividend = 0; Shift_divisor = 0; Subtract = 0;
   case (state)
```

```
S_idle:       case (Start)
                  0:    next_state = S_idle;
                  1:    if (word2 == 0) next_state = S_Err;
                        else if (word1) begin next_state = S_Adivr; Load_words
                        = 1; end
                        else next_state = S_idle;
              endcase

S_Adivr:      case (MSB_divr)
                  0:    if (sign_bit == 0) begin
                        next_state = S_Adivr; Shift_divisor = 1;      // can shift
                                                                      divisor

                        end
                        else if (sign_bit == 1) begin
                        next_state = S_Adivn;                         // cannot
                                                                      shift divisor

                        end
                  1:    next_state = S_div;
              endcase

S_Adivn:      case ({Max, sign_bit})
                  2'b00:  next_state = S_div;
                  2'b01:  begin next_state = S_Adivn; Shift_dividend = 1; end

                  2'b10:  begin next_state = S_idle; Subtract = 1; end
                  2'b11:  next_state = S_idle;
              endcase

S_div:        case ({Max, sign_bit})
                  2'b00:  begin next_state = S_div; Subtract = 1; end
                  2'b01:  next_state = S_Adivn;
                  2'b10:  begin next_state = S_div; Subtract = 1; end
                  2'b11:  begin next_state = S_div; Shift_dividend = 1; end
              endcase
default:      next_state = S_Err;
endcase
end

always @ (posedge clock or posedge reset) begin    // Register/Datapath operations
  if (reset) begin
    divisor <= 0; dividend <= 0; quotient <= 0; num_shift_dividend <= 0;
    num_shift_divisor <= 0;
  end

  else if (Load_words == 1)
    begin
      dividend <= word1;
      divisor <= word2;
      quotient <= 0;
      num_shift_dividend <= 0;
```

```verilog
        num_shift_divisor <= 0;
      end

    else if (Shift_divisor) begin
      divisor <= divisor << 1;
      num_shift_divisor <= num_shift_divisor + 1;
    end

    else if (Shift_dividend) begin
      dividend <= dividend << 1;
      quotient <= quotient << 1;
      num_shift_dividend <= num_shift_dividend +1;
    end

    else if (Subtract)
      begin
        dividend [L_divn: L_divn -L_divr] <= comparison;
        quotient[0] <= 1;
      end
    end
  endmodule

module test_Divider_STG_1 ();
  parameter L_divn = 8;
  parameter L_divr = 4;
  parameter word_1_max = 255;
  parameter word_1_min = 1;
  parameter word_2_max = 15;
  parameter word_2_min = 1;
  parameter max_time = 850000;
  parameter half_cycle = 10;
  parameter start_duration = 20;
  parameter start_offset = 30;
  parameter delay_for_exhaustive_patterns = 490;
  parameter reset_offset = 50;
  parameter reset_toggle = 5;
  parameter reset_duration = 20;
  parameter word_2_delay = 20;
  wire [L_divn -1: 0] quotient;
  wire [L_divn-1: 0] remainder;
  wire Ready, Div_zero;
  integer word1;              // dividend
  integer word2;              // divisor
  reg Start, clock, reset;
  reg [L_divn-1: 0] expected_value;
  reg [L_divn-1: 0] expected_remainder;
  wire quotient_error, rem_error;
  integer k, m;

  // probes
  wire    [L_divr-1: 0]        Left_bits = M1.dividend[L_divn-1: L_divn -L_divr];
```

Divider_STG_1 M1 (quotient, remainder, Ready, Error, word1, word2, Start, clock, reset);

initial #max_time **$finish**;
initial begin clock = 0; **forever** #half_cycle clock = ~clock; **end**

initial begin expected_value = 0; expected_remainder = 0;
  forever @ (negedge Ready) **begin**   // Form expected values
    #2 **if** (word2 != 0) **begin** expected_value = word1 / word2; expected_remainder
      = word1 % word2; **end**
    end
  end

assign quotient_error = (!reset && Ready) ? |(expected_value ^ quotient): 0;
assign rem_error = (!reset && Ready) ? |(expected_remainder ^ remainder): 0;

initial begin   // Test for divide by zero detection
  #2 reset = 1;
  #15 reset = 0; Start = 0;
  #10 Start = 1; #5 Start = 0;
end

initial begin   // Test for recovery from error state on reset and running reset
  #reset_offset reset = 1; #reset_toggle Start = 1; #reset_toggle reset = 0;
  word1 = 0;
  word2 = 1;
  while (word2 <= word_2_max) #20 word2 = word2 +1;
  #start_duration Start = 0;
end
initial begin   // Exhaustive patterns
  #delay_for_exhaustive_patterns
  word1 = word_1_min; **while** (word1 <= word_1_max) **begin**
  word2 = 1; **while** (word2 <= 15) **begin**
  #0 Start = 0;
  #start_offset Start = 1;
  #start_duration Start = 0;
  @ (posedge Ready) #0;
  word2 = word2 + 1; **end**   // divisor pattern
  word1 = word1 + 1; **end**   // dividend pattern
  end
endmodule

End of Example 10.16

Figure 10-72 presents simulation results for *Divider_STG_1*, illustrating the initial alignment of *dividend* by the action of *Shift_dividend*. Based on the STG in Figure 10-71, the machine correctly forms the quotient of $28_{10} = 0001\_1011_2$ by $8_{10} = 1000_2$, giving a quotient of 3_{10} with a remainder of 4_{10}. When *Start* is asserted, the machine loads *dividend* and *divisor*, and moves to state *S_Adivr*, where it compares *divisor* and the

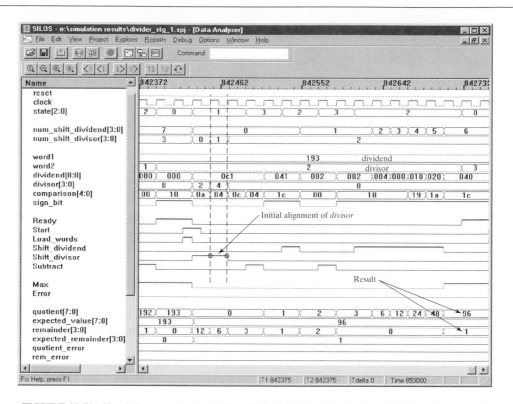

FIGURE 10-72   Simulation results for *Divider_STG_1*, dividing 28_{10} by 8_{10}, with initial alignment of *dividend* by the action of *Shift_dividend*.

leftmost byte of the dividend word to determine whether alignment is necessary. The machine detects a need for alignment of *dividend*, and with *sign_bit* asserted, moves at the next clock to *S_Adivn* to begin aligning *dividend*. With *shift_dividend* asserted for the next three clocks, *dividend* is moved align its MSB with the MSB of *divisor*, giving *dividend* $[8:0] = 0d8_H$. The simulation results reveal a feature of the design: it wastes a clock cycle before and after aligning the dividend. Further modification of the machine is left to a problem at the end of this chapter.

Figure 10-73 shows division of $193_{10} = 1100\_0001_2$ by $2_{10} = 0010_2$ and the action of *Shift_divisor* in aligning *divisor* with *dividend* in two clock cycles. The waveforms have been annotated to show the cycles at which the shifting action occurs.

10.6.3 Reduced-Register Sequential Divider

A more efficient architecture for a divider exploits the fact that the contents of the dividend register are shifted toward its MSB as the division sequence unfolds, leaving room in the register for the bits of the quotient. This architecture is more efficient in its use of physical resources because it eliminates the need for a separate register to hold the quotient, as shown in Figure 10-74. The implementation has the following additional

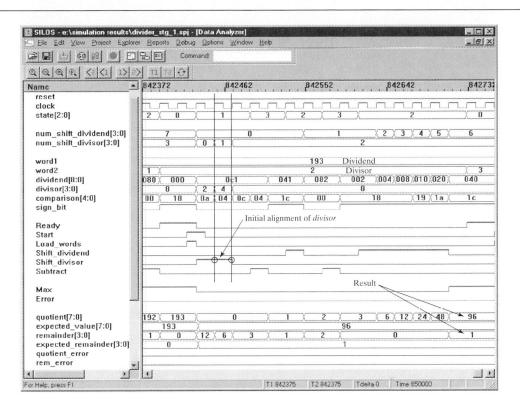

FIGURE 10-73 Simulation results showing division of 193_{10} by 2_{10} and the initial alignment of *divisor* with *dividend* by the action of *Shift_divisor*.

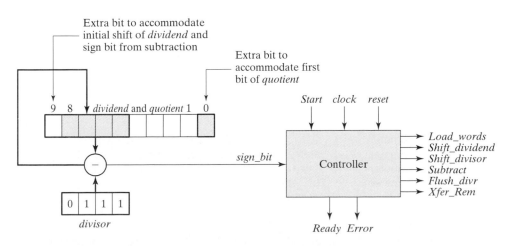

FIGURE 10-74 Architecture for *Divider_STG_RR*, a binary divider with reduced registers.

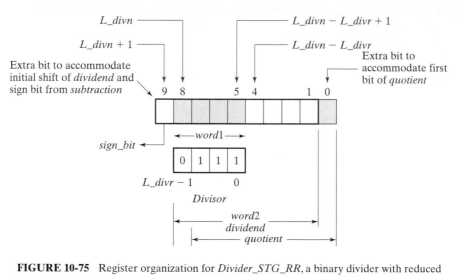

FIGURE 10-75 Register organization for *Divider_STG_RR*, a binary divider with reduced registers.

features: (1) shifting and subtracting occur in the same clock cycle instead of in separate cycles, (2) the remainder is adjusted to correct for its final location in the register, and (3) an overflow bit detects an invalid result.

The organization of the register shared by dividend and quotient is shown in Figure 10-75. The register includes a 1-bit extension to accommodate an initial shift that might be required to align *dividend* and *divisor*, and to hold the sign bit that is formed by subtracting *divisor* from *dividend*. The register is extended on the right by 1 bit to hold the first bit that is formed for *quotient*.

Example 10.17

```
module Divider_RR_STG (quotient, remainder, Ready, Error, word1, word2, Start,
clock, reset);
    parameter       L_divn = 8;
    parameter       L_divr = 4;        // Choose L_divr <= L_divn
    parameter       S_idle = 0, S_Adivr = 1, S_ShSub = 2, S_Rec = 3, S_Err = 4;
    parameter       L_state = 3, L_cnt = 4, Max_cnt = L_divn - L_divr;
    parameter       L_Rec_Ctr = 3;
    output          [L_divn -1: 0]     quotient;
    output          [L_divr -1: 0]     remainder;
    output                             Ready, Error;
    input           [L_divn -1: 0]     word1;          // Datapath for dividend
```

```
input          [L_divr -1: 0]      word2;        // Datapath for divisor
input                              Start, clock, reset;
reg            [L_state -1: 0]     state, next_state;
reg                                Load_words, Subtract_and_Shift,
                                   Subtract, Shift_dividend, Shift_divisor,
                                   Flush_divr, Xfer_Rem;
reg            [L_divn +1: 0]      dividend;     //Doubly extended dividend
reg            [L_divr -1: 0]      divisor;
reg            [L_cnt -1: 0]       num_shift_dividend, num_shift_divisor;
reg            [L_Rec_Ctr -1: 0]   Rec_Ctr;      // Recovery counter
reg            [L_divr: 0]         comparison;   // includes sign_bit
wire           MSB_divr = divisor[L_divr -1];
wire           Ready =((state == S_idle) && !reset) ;
wire           Error = (state == S_Err);
wire           Max = (num_shift_dividend == Max_cnt + num_shift_divisor );

always @ ( state or dividend or divisor or MSB_divr)
  case (state)

    S_ShSub:    comparison = dividend[L_divn +1: L_divn - L_divr +1] + {1'b1,
                ~divisor[L_divr -1: 0]} + 1'b1;

    default:    if (MSB_divr == 0)              // Shifted divisor
                  comparison = dividend[L_divn +1: L_divn - L_divr +1] + {1'b1,
                  ~(divisor << 1)} + 1'b1;
                else comparison = dividend[L_divn +1: L_divn - L_divr +1] +
                {1'b1, ~divisor[L_divr -1: 0]} + 1'b1;
  endcase

wire           sign_bit = comparison[L_divr];
wire           overflow = Subtract_and_Shift && ((dividend[0] == 1) ||
               (num_shift_dividend == 0 ));

assign         quotient = ((divisor == 1) && (num_shift_divisor == 0))?
               dividend[L_divn: 1]:
               (num_shift_divisor == 0) ? dividend[L_divn - L_divr : 0]:
               dividend[L_divn +1: 0];

assign         remainder = (num_shift_divisor == 0) ? (divisor == 1) ? 0:
               (dividend[L_divn: L_divn - L_divr +1] ):
               divisor;

always @ (posedge clock or posedge reset)     // State Transitions
  if (reset) state <= S_idle; else state <= next_state;

// Next state and control logic
```

```
always
  @ (state or word1 or word2 or divisor or Start or comparison or sign_bit or Max
  or Rec_Ctr) begin
  Load_words = 0; Shift_dividend = 0; Shift_divisor = 0;
  Subtract_and_Shift = 0; Subtract = 0; Flush_divr = 0;
  Flush_divr = 0; Xfer_Rem = 0;
  case (state)
    S_idle:        case (Start)
                     0:     next_state = S_idle;
                     1:     if (word2 == 0) next_state = S_Err;
                            else if (word1) begin next_state = S_Adivr; Load_words
                            = 1; end
                            else if (sign_bit == 1) next_state = S_ShSub;
                            else next_state = S_idle;
                   default: next_state = S_Err;
                   endcase

    S_Adivr:       if (divisor == 1)
               begin next_state = S_idle; end else
               case ({MSB_divr, sign_bit})
                 2'b00:  begin next_state = S_Adivr; Shift_divisor = 1; end    // can shift
                                                                               divisor
                 2'b01:  next_state = S_ShSub;                                 // cannot
                                                                               shift divisor
                 2'b10:  next_state = S_ShSub;
                 2'b11:  next_state = S_ShSub;
               endcase

    S_ShSub:   case ({Max, sign_bit})
                 2'b00:  begin next_state = S_ShSub; Subtract_and_Shift = 1; end
                 2'b01:  begin next_state = S_ShSub; Shift_dividend = 1; end
                 2'b10:  if (num_shift_divisor == 0) begin next_state = S_idle; Subtract = 1;
                         end
                         else begin next_state = S_ShSub; Subtract = 1; end
                 2'b11:  if (num_shift_divisor == 0) next_state = S_idle;
                         else if (num_shift_divisor != 0) begin next_state = S_Rec; Flush_divr
                         = 1; end
               endcase

    S_Rec:     if (Rec_Ctr == L_divr - num_shift_divisor) begin next_state =
               S_idle; end
               else begin next_state = S_Rec; Xfer_Rem = 1; end

    default:   next_state = S_Err;
    endcase
  end
```

```verilog
  always @ (posedge clock or posedge reset) begin    // Register/Datapath operations
    if (reset) begin
      divisor <= 0; dividend <= 0;
      num_shift_dividend <= 0; num_shift_divisor <= 0; // use to down-cnt
      Rec_Ctr <= 0;
    end

    else if (Load_words == 1)
      begin
        dividend[L_divn +1: 0] <= {1'b0, word1[L_divn -1: 0], 1'b0};
        divisor <= word2;
        num_shift_dividend <= 0;
        num_shift_divisor <= 0;
        Rec_Ctr <= 0;
    end

    else if (Shift_divisor) begin
      divisor <= divisor << 1;
      num_shift_divisor <= num_shift_divisor + 1;
  end

    else if (Shift_dividend) begin
      dividend <= dividend << 1;
      num_shift_dividend <= num_shift_dividend +1;
    end

    else if (Subtract_and_Shift)
      begin
        dividend[L_divn + 1: 0] <= {comparison[L_divr -1: 0], dividend [L_divn -L_divr: 1],
        2'b10} ;
        num_shift_dividend <= num_shift_dividend +1;
    end

    else if (Subtract)
      begin
        dividend[L_divn +1: 1] <= {comparison[L_divr: 0], dividend [L_divn -L_divr: 1]} ;
        dividend[0] <= 1;
    end

    else if (Flush_divr) begin
      Rec_Ctr <= 0;
      divisor <= 0;
    end

    else if (Xfer_Rem) begin
      divisor[Rec_Ctr] <= dividend[ L_divn - L_divr + num_shift_divisor + 1 + Rec_Ctr];
      dividend[ L_divn - L_divr + num_shift_divisor + 1 + Rec_Ctr] <= 0;
```

```
                        Rec_Ctr <= Rec_Ctr + 1;
                    end

                end
                endmodule
```

End of Example 10.17

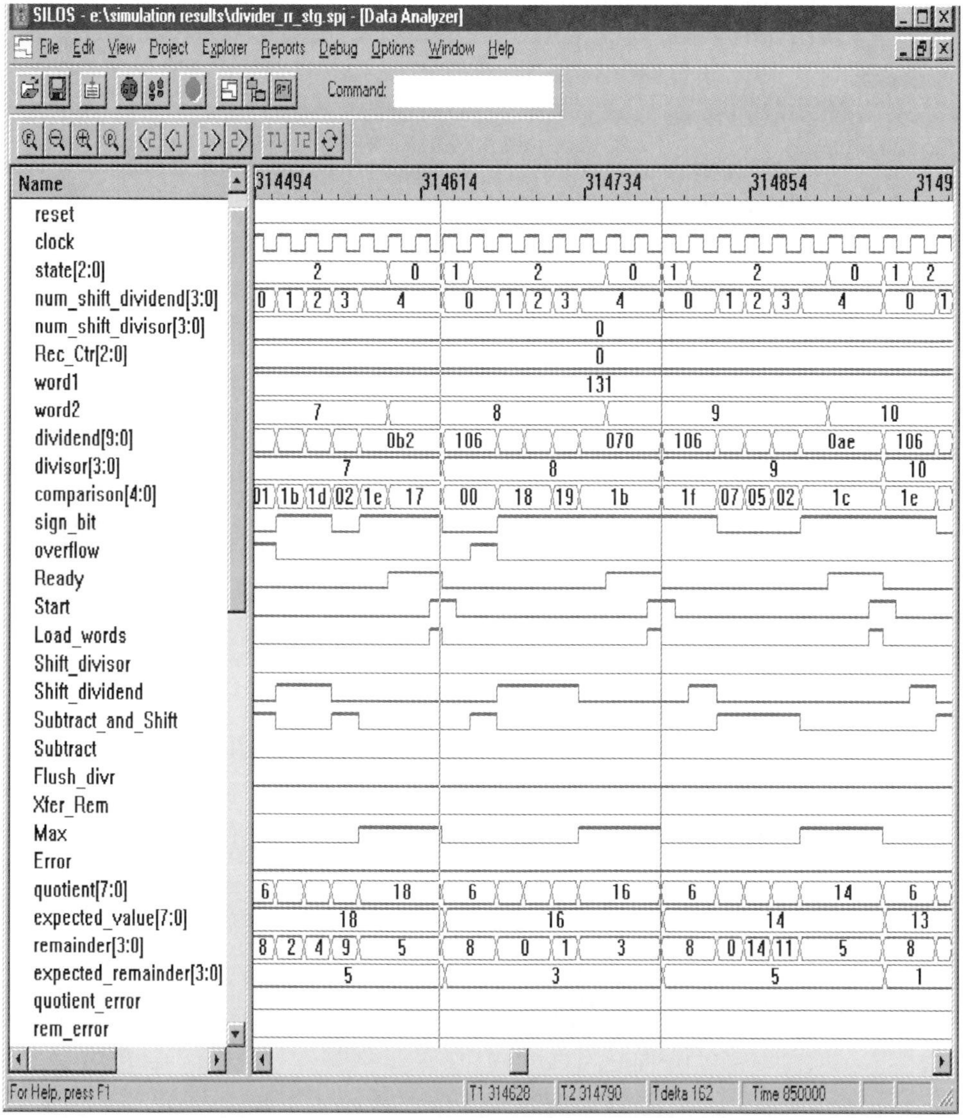

FIGURE 10-76   Simulation results for *Divider_STG_RR*, a binary divider with reduced registers. The model is efficient in its use of hardware and in its execution time.

The simulation results for *Divider_RR_STG* are shown in Figure 10-76, with 131_{10} divided by 8_{10}, giving a quotient of 16_{10} and a remainder of 3_{10}. The displayed waveforms include the control signals and the counters that are used to adjust the remainder.

10.6.4 Division of Signed (2s Complement) Binary Numbers

The simplest method of forming the quotient of two signed numbers is to divide their magnitudes, then adjust the sign of the result, if necessary. Other, more complex, algorithms exist, but will not be considered here [1].

REFERENCES

1. Weste NHE, Eshraghian K. *Principles of CMOS VLSI Design*. Reading, MA: Addison-Wesley, 1993.
2. Katz RH. *Contemporary Logic Design*. Redwood City, CA: Benjamin/Cummings, 1994.
3. Smith MJS. *Application-Specific Integrated Circuits*. Reading, MA: Addison-Wesley, 1997.
4. Heuring V, Jordan H. *Computer Systems Design and Architecture*. Reading, MA: Addison-Wesley 1997.
5. Johnson EL, Karim MA. *Digital Design—A Pragmatic Approach*. Boston: PWS, 1987.
6. Arnold MG. *Verilog Digital Computer Design*. Upper Saddle River, NJ: Prentice-Hall, 1999.
7. Ciletti MD. *Modeling, Synthesis, and Rapid Prototyping with the Verilog HDL*. Upper Saddle River, NJ: Prentice-Hall, 1999.
8. Thomas DE, Moorby PR. *The Verilog Hardware Description Language*, 3rd ed. Boston: Kluwer, 1996.
9. Booth AD. "A Signed Binary Multiplication Technique." *Quarterly Journal of Mechanics and Applied Mathematics, 4*, 1951.
10. Patterson DA, Hennessy JL. *Computer Organization and Design—The Hardware/Software Interface*. San Francisco: Morgan Kaufman, 1994.
11. Cavanaugh JJF. *Digital Computer Arithmetic*. New York: McGraw-Hill, 1984.
12. Kehtarnavaz N, Keramat M. *DSP System Design Using the TMS320C6000*. Upper Saddle River, NJ: Prentice-Hall, 2001.
13. Sternheim, E. et al., *Digital Design and Synthesis with Verilog HDL*, Automaton Pub. Co., San Jose, CA, 1993.

PROBLEMS

1. The 4-bit multiplier shown in Figure 10-13 can be modified to exploit a 4-bit carry look-ahead adder instead of a ripple-carry adder. Compare the performance and area of the two models.
2. Modify the models for *Multiplier_STG_1* and *Multiplier_STG_2* to terminate their activity if the multiplier or the multiplicand is 0 (see Example 10.2).

3. Write gate-level structural models of flip-flops, shift registers, and adders, then build a structural model of the datapath architecture shown in Figure 10-16. Develop and verify an STG-based model for the controller.

4. Modify *Multiplier_STG_0* and *Multiplier_STG_1* (see Example 10.1) to terminate at any intermediate state if the multiplier is empty. Synthesize and compare the area of the new machine to that of the baseline machine.

5. The description in *Multiplier_ASM_0* (see Example 10.3) asserts *Ready* one cycle after asserting *Done*. It might be desirable to assert *Ready* and *Done* simultaneously so that *product* could be read at the same time that new data are being loaded. Modify the ASM chart and the description to form a module *Multiplier_ASM_0a* that has this feature.

6. Develop logic to terminate the activity of multiplication in *Multiplier_RR_ASM* as soon as the subword corresponding to the shifted multiplier is empty of 1s (see Example 10.5).

7. Design and verify *Multiplier_IMP_2*, an alternative sequential multiplier obtained by embedding the datapath operations within the implicit state machine behavior that implements the controller. Use *Multiplier_IMP_1* (see Example 10.6) as a starting point for your design.

8. Develop and verify *Multiplier_Booth_STG_1*, a Booth multiplier that is patterned after *Multiplier_STG_1* (see Example 10.2).

9. Develop and verify *Multiplier_Booth_ASM_0*, a Booth multiplier that is patterned after *Multiplier_ASM_0* (see Example 10.3).

10. Develop and verify *Multiplier_Booth_ASM_1*, a Booth multiplier that is patterned after *Multiplier_ASM_1* (see Example 10.4), with parameterized word length.

11. Develop and verify *Multiplier_Booth_RR_ASM*, a Booth multiplier that is patterned after *Multiplier_RR_ASM* (see Example 10.5). (*Hint*: Consider the role of an arithmetic shift operation for sign extension. Select a set of data that demonstrates a significant reduction in clock cycles to multiply 16-bit words.)

12. Develop and verify *Multiplier_Booth_IMP_1*, a Booth multiplier that is patterned after *Multiplier_IMP_1* (see Example 10.6). Verify that the machine resets correctly from any intermediate state.

13. Write the Booth code for the multiplier word shown in Figure P10-13:

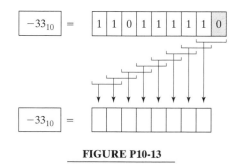

FIGURE P10-13

14. Develop and verify *Multiplier_Radix_4_STG_1*, a modified Booth multiplier using radix-4 encoding that is patterned after *Multiplier_STG_1* (see Example 10.2). Use the testbench given at the companion website to conduct exhaustive verification.

15. Develop and verify *Multiplier_Radix_4_ASM_0*, a modified Booth multiplier using radix-4 encoding that is patterned after *Multiplier_ASM_0* (see Example 10.3). Use the testbench given at the companion website to conduct exhaustive verification.

16. Develop and verify *Multiplier_Radix_4_ASM_1*, a modified Booth multiplier using radix-4 encoding that is patterned after *Multiplier_ASM_1* (see Example 10.4). Use the testbench given at the companion website to conduct exhaustive verification.

17. Develop and verify *Multiplier_Radix_4_RR_ASM*, a modified Booth multiplier using radix-4 encoding that is patterned after *Multiplier_RR_ASM* (see Example 10.5). (*Hint*: Consider the role of an arithmetic shift operation for sign extension. Use the testbench given at the companion website to conduct exhaustive verification.)

18. Develop and verify *Multiplier_Radix_4_IMP_1*, a modified Booth multiplier using radix-4 encoding, that is patterned after *Multiplier_IMP_1* (see Example 10.6). Use the testbench given at the website to conduct exhaustive verification. Verify that the machine resets correctly from any intermediate state.

19. The machine *Multiplier_Radix_4_STG_0* has a default state assignment that directs the state to *S_idle* if an unspecified state is encountered (see Example 10.12). This could cause a mistaken interpretation of *Ready*. Develop and verify a modified machine with a fail-safe behavior.

20. The machine *Multiplier_Radix_4_STG_0* has a simple binary coding of the controller's state (see Example 10.12). Develop and verify a modified machine with a one-hot code, and compare the results of synthesizing both designs.

21. Develop and verify *Multiplier_Radix_4_STG_1*, a more efficient version of the bit-pair encoding multiplier described by *Multiplier_Radix_4_STG_0* that uses additional logic to eliminate needless computation when *word1* or *word2* is 0 or 1 (see Example 10.12).

22. Write a testbench to exercises all of the state transitions of *Divider_STG_0* in Figure 10-56.

23. Using the testbench *test_Divider_STG_0* (given at the companion website), verify the features of *Divider_STG_0* that were specified in Example 10.13.

24. In Example 10.14, *Divider_STG_0* forms *remainder, Error*, and *Done* as combinational outputs by continuous assignment statements. This leads to a higher than necessary signal activity because *remainder* has needless intermediate transitions while *quotient* is being formed. Similarly, *Error* and *Done* require needless simulation activity because their continuous-assignment statements are activated every time *state* changes. Develop and verify a modified machine that reduces signal and simulation activity by forming *quotient, Error,* and *Done* as registered outputs.

25. In Example 10.14 *Divider_STG_0*, an attempt to divide by 0 drives the state to *S_err*, where the machine remains until *reset* is asserted. Develop an STG for an alternative machine that recovers from *S_4* when *Start* is asserted. Develop and verify a Verilog description of the alternative machine.

26. The divider described by *Divider_STG_0* in Example 10.14 asserts *Ready* when the machine enters *S_3*. An external agent that is using *Ready* to determine whether *quotient* is valid must wait until the next clock cycle to read *quotient* (i.e., the agent reads *quotient* at the second clock after it is formed). However, there are conditions in which *product* is formed sooner. Modify the STG shown in Figure 10-67 so that *Ready* asserts as soon as *quotient* is valid, and modify

Divisor_STG_0 to form an alternative machine that implements the STG (i.e., allows an external agent to read *quotient* at the first clock cycle after it is formed).

27. The machine described by *Divider_STG_1* in Example 10.16 cannot be synthesized because *remainder* uses a dynamically determined (i.e., during simulation) shift operation. Develop a synthesizable implementation by (1) modifying the machine's STG to accommodate an additional state in which the reminder is adjusted, (2) developing and verifying the Verilog model of the machine, and (3) verifying that the synthesized machine's behavior matches that of the behavioral description.

28. The self-aligning divider described in Example 10.16 (*Divider_STG_1*) has a more efficient alternative (speedwise) in which the operations of shifting and subtracting execute in the same clock cycle. Develop an STG for an alternative machine that combines the operations in the same state. Develop and verify a Verilog description of the alternative machine. Examine and discuss whether the architecture could have an overflow condition.

29. The self-aligning divider described in Example 10.16 (*Divider_STG_1*) has a more efficient (registerwise) alternative in which the bits of the quotient are loaded into the trailing end of the register holding the dividend as the dividend is shifted. Note that in *Divider_STG_1* the operations of shifting and adding occur in different clock cycles. Be aware that if the shift and subtract operations are not combined in the same clock cycle a data configuration might cause an overflow condition in which the quotient bit formed by subtraction would be targeted to occupy the rightmost cell of the register holding dividend before it is vacated by a shift operation. Develop an STG for this alternative machine, with an additional output to signal the overflow condition, then develop and verify a Verilog description of the machine.

30. Compare the speed and area of a synthesized 4×4 array multiplier to that of a sequential multiplier.

31. The Verilog description of *Divider_STG_0* (see Example 10.14) shows alternative implementations of subtraction. Synthesize each version of the model and compare the results.

32. Synthesize and compare *Divider_STG_0* and *Divider_STG_0_sub* (see Example 10.14) to determine whether eliminating the comparator from the architecture results in a more efficient implementation.

33. Careful examination of the simulation results in Figure 10-72 reveals that *Divider_STG_1* wastes a clock cycle before and after aligning the dividend. Develop and verify an alternative machine that eliminates the wasted clock cycles.

34. The subtraction in *Divider_STG_0* (see Example 10.14) can be written directly as *Dividend - Divisor*, to exploit the built-in 2s complement arithmetic of Verilog. Conduct an experiment to determine whether your synthesis tool synthesizes the alternative implementation more efficiently.

35. The machine *Divider_STG_1* (see Example 10.16) executes shifting and subtraction in separate clock cycles. Explore the possibility of executing the instructions simultaneously, allowing the machine to have higher throughput.

36. Modify the machine *Divider_STG_1* (see Example 10.16) to detect an attempt to divide by 1, allowing the machine to have higher throughput.

37. The machine *Divider_STG_RR* (see Example 10.17) asserts *Ready* when the state returns to *S_idle* after completing a division. Explore the possibility of

asserting *Ready* sooner, allowing the machine to have higher throughput.

38. The machine *Divider_STG_RR* (see Example 10.17) detects an attempt to divide by 1 in state *S_Adivr*. Explore the possibility of detecting a divide by 1 in state *S_idle*, allowing the machine to have higher throughput.

39. Determine whether the descriptions given by *Divider_STG_0, Divider_STG__0_sub, Divider_STG_1,* and *Divider_STG_RR* are synthesizable models. Modify any machine that cannot be synthesized so that it can be synthesized.

40. Using a feedforward cutset for the 4×4 binary multiplier shown in Figure 10-14, develop and verify a balanced one-stage pipelined implementation of the circuit.

41. Using a feedforward cutset for the 8×4 binary multiplier shown in Figure P10-41, develop and verify a balanced one-stage pipelined implementation of the circuit.

42. Identify the computational wavefronts of the systolic array in Figure 10-14, and develop a reservation table (see Table 9-1) that utilizes the maximum number of processors that could execute currently to form the 4×4 product in a synchronous environment with memory.

43. Identify the computational wavefronts of the systolic array in Figure P10-41, and develop a reservation table (see Table 9-1) that utilizes the maximum number of processors that could execute currently to form the 8×4 product in a synchronous environment with memory.

44. A carry-select (conditional-sum) adder can be used to improve the performance of operational units that require an adder circuit. The adder consists of full adders and 2:1 muxes in a configuration in which the datapaths of the muxes are from two full adders with the same input data bits. Certain of the full-adder cells have a hard-wired carry bit. The actual carry bit from a previous stage selects the correct cell for a given data pattern. Each cell generates a carry-out bit for the next stage. The architecture offers a speed improvement because the parallel addition at the inner stages of the adder occurs while the carry to those cells is developed and passed to the mux, rather than after the carry has arrived.

    a. Develop suitable low-level cells (with hypothetical propagation) delays, or use cells from a standard-cell library (with physical delays), and write a Verilog description of the carry-select adder shown in Figure P10-44. Use the *supply0* and *supply1* nets to implement the hard-wired carries. Consider the issue of whether the hard-wired carries should be driven by an internal signal or a signal passed through the port.

    b. Develop and verify the circuit using a nonexhaustive testbench to gain a high level of confidence in the functionality of your design. Carefully select a small but robust set of test vectors. Discuss the testing strategy that you used.

    c. Develop an automated testbench that includes a 6-bit behavioral adder and compares its output to that of the carry-select adder (at suitable times). Use an exclusive-or scheme to generate a *test_results_message* that the circuit operates correctly or not. (*Note*: Do not generate hard copy of the exhaustive simulation.) This scheme should exploit the *$display* system task to

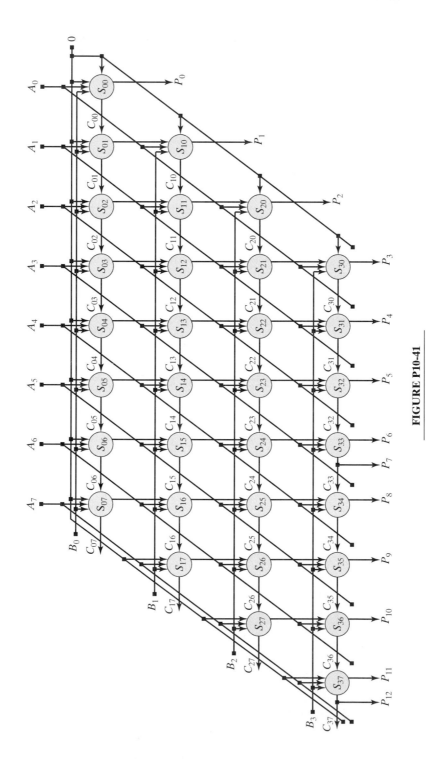

FIGURE P10-41

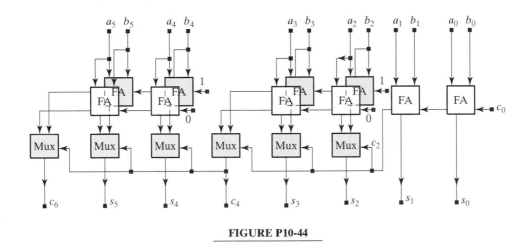

FIGURE P10-44

observe the outputs of the two adders at some time after each pattern is applied to the their inputs. (*Note*: The ***$display*** task provides dynamic control over the display of information. It displays results only when the statement executes within a behavior. The ***$monitor*** tasks executes when its an argument has an event. If the ***$monitor*** task is used instead of ***$display***, the comparison of the two responses will report errors until the output of the gate-level adders stabilizes.)

45. Develop an FPGA-based calculator with the following features: Data entry is through a hex keypad in which 10 keys are used to enter the decimal digits, and the remaining keys are used for the following functions: *Enter_data, addition, subtraction, multiplication, division*, and *decimal point*. The calculator is to have a 10-digit display.

46. Develop, verify, and synthesize an eight-stage finite-duration impulse response (FIR) filter for a 16-bit datapath in 2s complement format (see Figure 9-23). Each multiply and accumulate (MAC) unit is to use a Booth sequential multiplier and a two-stage pipelined adder. Determine the maximum clock frequency at which the data sequence can be supplied to the circuit and the frequency of the internal clock that is to synchronize the operations of the FIR. Compare the area and speed of the synthesized FIR to that of a FIR that uses combinational array multipliers and ripple carry adders to implement the MACs.

47. Develop, verify, and synthesize a balanced one-stage pipelined FIR with eight stages and a 16-bit datapath in 2s complement format (see Figure 9-23). Each MAC unit is to use combinational multipliers and adders. Determine the maximum throughput of the pipelined circuit.

CHAPTER 11

Postsynthesis Design Tasks

The design flow of an application-specific integrated circuit (ASIC) includes tasks for postsynthesis design validation, timing verification, test generation, and fault simulation.[1] Design validation confirms that the functionality of the synthesized gate-level netlist matches that of the behavioral model. Postsynthesis timing analysis checks whether a physical implementation of the design meets timing specifications and determines the maximum frequency at which a synchronous circuit can operate. Test generation develops stimulus patterns that will detect the presence of a process-induced fault in the fabricated circuit. Fault simulation determines how well a set of stimulus patterns exposes the faults that could be in the circuit.

11.1 Postsynthesis Design Validation

Postsynthesis design validation confirms that the netlist of the synthesized circuit has the same functionality as the register transfer level (RTL) model from which it was synthesized. Postsynthesis design validation is not intended to verify that the functionality of a design is correct. Functional verification is achieved earlier in the design flow, before synthesis, with RTL models that simulate more efficiently than gate-level models.

There are two approaches to design validation: formal methods and simulation. Formal methods are beyond the scope of this textbook; we will consider only simulation, which is widely used by the ASIC industry. For our purposes, postsynthesis design validation is to detect a simulation mismatch between the functionality of the RTL model and the gate-level netlist. The consequences of a simulation mismatch can be serious, because the physical circuit may fail to operate correctly. The RTL and gate-level models can be simulated simultaneously, or Verilog system tasks can be used to record the stimulus and response patterns of the RTL circuit and compare them to those of the gate-level circuit when it is exercised by the same stimulus.

---

[1]See Figure 1.1.

There are several potential sources of simulation mismatch between RTL and gate-level models of a circuit. The gate-level model of a circuit uses standard cells, which have built-in descriptions of technology-specific propagation delays of the devices; the RTL model is delay-free. Consequently, at sufficiently high clock frequencies, a timing violation will occur in the gate-level model, but not in the RTL model. Depending on how the models treat timing violations, the values that are propagated in the simulation of the gate-level circuit may differ from those propagated by the RTL model. If the clock speed is fixed, the RTL design must be remodeled and/or resynthesized to achieve timing margins that eliminate the mismatch between the circuits.

Simulation mismatch can occur if the modeling style allows software race conditions in a sequential machine. In general, race conditions exist when multiple cyclic behaviors in a Verilog model make simultaneous assignments to the same variable. The order in which multiple cyclic behaviors are executed by a simulator is indeterminate, and there is no simulator-independent way to determine whether a variable will be assigned value by a procedural assignment in one behavior before or after it is referenced by a procedural assignment in another behavior. So beware of models in which a variable is assigned value in one behavior at the same time that the variable is referenced by another behavior. One way to prevent ambiguous outcomes caused by software race conditions in a latch-free sequential machine is to place all procedural assignments in a single cyclic behavior and order the statements to produce the correct sequence of assignments. However, this approach might not be convenient.

Sequential machines can have race conditions if there is feedback from a datapath to the state machine that controls the datapath. The methodology discussed in Chapter 7 produces race-free logic in such machines by using use nonblocking ($<=$) assignments in edge-sensitive cyclic behaviors, and blocked ($=$) assignments in level-sensitive behaviors, and by having no variable simultaneously referenced and assigned value by multiple blocked assignments.[2]

The style used to model circuits with latches can also lead to race conditions. A circuit will have a race condition if a latch is a reconvergent fanout node.[3] For example, if the enable line and the datapath of a latch have a common variable it is possible for the enable and the data to change simultaneously, and race. This style of design should be avoided because the outcome of the race is indeterminate.

Example 11.1

The output and the level-sensitive event control expression of the latch in Figure 11-1(a) are both conditioned on the datapath [1]. The RTL models *Latch_Race_1* and *Latch_Race_2* below differ only in the order in which the cyclic behaviors appear in the code. The simulation results in Figure 11-1(b) show that *D_out_1* and *D_out_2* differ. The models synthesize to the same structure, a hardware latch with *D_in* connected to the data and enable inputs of the latch. For discussion, we have introduced hypothetical delays in *Latch_Race_3* and *Latch_Race_4*. These models also synthesize to a latch

---

[2]See Howe [1] for further discussion of simulation mismatch in design validation.
[3]A node is a reconvergent fanout node if there are multiple paths to it through combinational logic from some other node.

with *D_in* connected to the data and enable inputs. The physical circuit will have propagation delay, and depending on the relative magnitude of the delays from the inputs to the outputs, will produce waveforms like those shown for *D_out_3* and *D_out_4*. Simulation mismatch is evident in the waveforms.

```
module Latch_Race_1 (D_out, D_in);
 output D_out;
 input   D_in;
 reg     D_out;
 reg     En;

 always @ (D_in) En = D_in;
 always @ (D_in or En) if (En == 0) D_out = D_in;
endmodule

module Latch_Race_2 (D_out, D_in);
 output D_out;
 input   D_in;
 reg     D_out;
 reg     En;

 always @ (D_in or En) if (En == 0) D_out = D_in;
 always @ (D_in) En = D_in;
endmodule

module Latch_Race_3 (D_out, D_in);
 output D_out;
 input   D_in;
 reg     D_out;
 wire    En;

 buf #1 (En, D_in);
 always @ (D_in or En)
  if (En == 0) D_out = D_in;

endmodule

module Latch_Race_4 (D_out, D_in);
 output D_out;
 input   D_in;
 reg     D_out;
 wire    En;

 buf #1 (En, D_in);
 always @ (D_in or En)
 #3  if (En == 0) D_out = D_in;

endmodule
```

End of Example 11.1

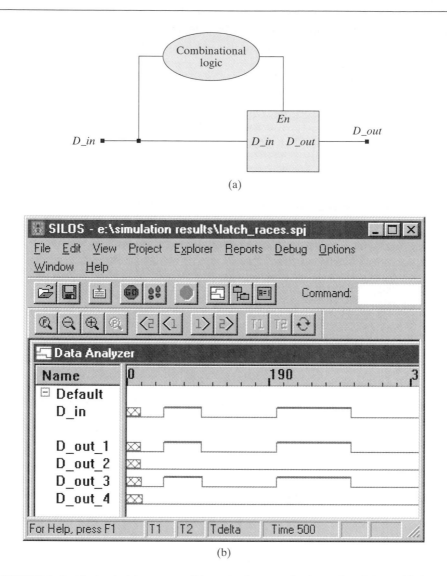

(a)

(b)

FIGURE 11-1 Latch circuit: (a) a race condition exists between the datapath and the enable input, and (b) the value that is latched in the RTL model depends on the order in which the statements are evaluated; the race condition in the physical circuit depends on the relative delay between the enabling input and the datapath through the device. *This style of design should be avoided.*

11.2 Postsynthesis Timing Verification

Postsynthesis timing verification is necessary because functional verification with RTL models does not consider propagation delays, does not verify that hardware timing constraints are met, and does not verify that performance specifications for input–output

(I/O) timing are met. Synthesis tools incorporate timing analysis in the engine that maps a generic design into a physical implementation. The accuracy of the timing analysis performed by synthesis tools is limited, because the tools use prelayout statistical estimates (called wireload models) of the delay caused by the resistance and parasitic capacitance of the interconnect and loading of the nets in the design. The technology-mapping engine does not have access to the actual delays that result from the place-and-route step, so the netlist produced by the synthesis tool does not account for those delays accurately. Estimates of the actual delays are used. The actual values of the resistance and capacitance of the interconnect must be extracted from the layout and used to back-annotate the delay models of the gates to obtain a more accurate analysis of the postlayout timing.

The speed at which a circuit can operate correctly is ultimately limited by the longest logic path, the so-called *critical path*, and by the physical constraints of the storage devices in the chip. The critical path and the paths whose length (in time) are close to that of the critical path must be verified to satisfy the timing constraints. Timing verification must consider the propagation delays of gates, the interconnect between gates, clock skew, I/O timing margins, and device constraints (e.g., setup, hold, and clock pulsewidth of flip-flops) to verify that the circuit will meet the timing specifications of the design without violating device constraints. If the setup or hold condition of an edge-triggered flip-flop is violated, the flip-flop may enter a metastable state (see Chapter 5).

Timing verification uses models of the devices and the interconnect in a circuit to analyze its timing and to determine whether hardware timing constraints and input–output timing specifications are met in the physical design (prelayout and postlayout). Timing verification can be achieved *directly*, by simulating the behavior of a circuit and confirming that hardware constraints and performance specifications are met, or *indirectly*, by analyzing all possible signal paths in the circuit and determining whether timing constraints are satisfied, without actually simulating the behavior of the circuit. These two approaches to timing verification are referred to as dynamic timing analysis (DTA) and static timing analysis (STA), respectively [2, 3]. Table 11-1 compares features of dynamic and static timing analysis.

TABLE 11-1 A comparison of methods for timing verification.

	Approaches to Timing Verification	
	Dynamic	Static
Method	Simulation	Path analysis
Needs Test Vectors	Yes	No
Coverage	Pattern-dependent	Pattern-independent
Risk	Missed alarms	False alarms
Min-Max Analysis	Discontinued	Yes
Couple with Synthesis	Not feasible	Yes
CPU Run time	Days/weeks	Hours
Memory Use	Heavy	Light–moderate

DTA uses stimulus patterns with event-driven simulators to exercise the behavior of the system. DTA uses behavioral, gate-level, and switch-level models of a circuit to simulate and analyze functional paths; transistor-level models are simulated with analog simulators.[4] STA uses the same models as DTA, but creates a directed acyclic graph (DAG) of the circuit by systematically and exhaustively extracting the topology of a gate-level representation of the circuit and calculating the propagation delays on all paths. If all of the possible paths meet their timing constraints and specifications, then the circuit will meet them independently of the stimulus pattern that is applied in operation.

Dynamic and static timing analysis have different risks and cost. DTA may have missed alarms (i.e., fail to detect and report a timing violation). DTA is pattern-dependent, so the stimulus patterns used to exercise the circuit might not exercise the longest path. Developing a robust set of stimulus patterns to verify the timing of complex circuits is difficult. On the other hand, STA, which exhaustively considers all possible topological paths of a circuit, may have false alarms. It may report a timing violation on a nonfunctional path (i.e., one that cannot be exercised in operation). Also, event-driven simulation of large circuits requires significant memory capacity and is relatively slow compared to the time required for a static timing analyzer to analyze the design and detect timing violations.

Timing closure refers to the problem of placing and routing the cells, signal paths, and clock tree of a design to meet the timing specifications. Timing closure is an issue because synthesis tools do not have postlayout information about the delays of the routed interconnect. Failure to achieve timing closure requires resynthesis and/or replacement and routing of cells, and incurs additional cost. Some synthesis tools address the problem of timing closure by having more accurate models of the interconnect loading, and by incorporating physical synthesis within the overall design flow.[5]

DTA is limited to circuits that have about 1 million gates. It has very limited application in an SoC (system on chip) environment, shown in Figure 11-2, where intellectual property (IP) from multiple vendors is integrated on a single chip. It can be very difficult to integrate the test patterns of the various IPs, thus limiting the coverage that can be achieved by simulation. On the other hand, STA does not depend on patterns, achieves full coverage, and is the appropriate method for timing analysis and verification in SoC applications. STA of multicore SoCs with millions of gates is feasible.

11.2.1 Static Timing Analysis

Our focus will be on full-chip, gate-level, static timing analysis for synchronous designs. STA forms a DAG from a netlist of a circuit. The nodes of the DAG represent gates, and the edges of the graph represent signal paths. The topological paths of a DAG include the timing paths of the circuit (i.e., paths that can be exercised by applying stimulus patterns to the primary inputs). The edges are annotated with the propagation delays of the paths. The DAG must be acyclic, (i.e., it must not have a feedback path).[6]

---

[4]Analog simulation is time consuming, and usually done to verify only the critical paths of a high-performance circuit.

[5]See www.magma.com.

[6]The companion website contains a brief tutorial for getting started with the Synopsys PrimeTime static timing analyzer.

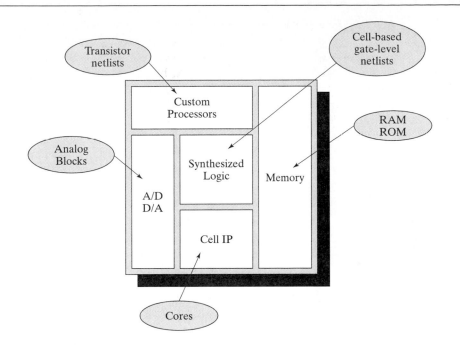

FIGURE 11-2 An SoC approach to design integrates on a single chip intellectual property from several sources.

Example 11.2

The circuit in Figure 11-3(a) has the timing DAG shown in Figure 11-3(b). The edges of the DAG represent the propagation delays of the paths. For simplicity, the gates in this example are shown with symmetric rising and falling delay.

End of Example 11.2

A circuit may have four types of timing paths: (1) paths that originate at a primary input and terminate at the data input of a storage element, (2) paths that connect the output of a storage element to the data input of a storage element, (3) paths that originate at the output of a storage element and terminate at a primary output of the chip, and (4) paths that connect primary inputs to the primary outputs of the circuit. Each type of path passes through combinational logic. STA examines timing paths between their sources and destinations, commonly called *startpoints* and *endpoints*, as shown in Figure 11-4.

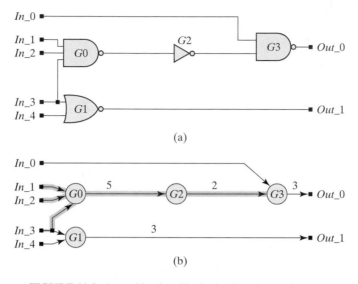

(a)

(b)

FIGURE 11-3 A combinational logic circuit and its timing DAG.

The startpoints of the paths of a circuit are its primary inputs (i.e., package pins) and the clock pins of its storage elements (sequential devices). The physical path originating at a sequential device is attached to the output of the device, but the startpoint of the timing path is the clock pin, because the action of the clock initiates signal propagation along the physical path. The endpoints of the timing paths of a circuit are the primary outputs (package pins) and the data inputs of the storage elements. Not every topological path of the DAG is a timing path, and separate timing paths may exist between a given startpoint and endpoint, depending on whether rising and falling transitions of a signal have symmetric propagation delays.

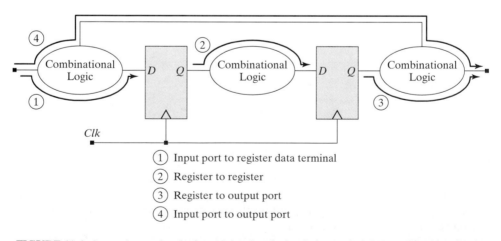

① Input port to register data terminal
② Register to register
③ Register to output port
④ Input port to output port

FIGURE 11-4 Startpoints and endpoints of signal paths for timing analysis in a synchronous circuit.

11.2.2 Timing Specifications

Performance specifications may constrain the offsets of signals at the interfaces of a circuit with its environment, and may constrain the delays of internal paths.[7] An *input delay (offset) constraint* applies to paths from primary inputs (input pads) to a storage element of the design, and specifies the time of arrival of the input signal relative to the active edge of the clock that launched the signal. In a fully synchronous system the signal arriving at the input pad of a circuit would have been launched by a clock edge in the circuit to which the input is connected. A timing analyzer uses the specified input constraint, t_{input_delay}, to determine the timing margin, denoted by t_{input_margin} in Figure 11-5, between the arriving signal and the next active edge of the clock. t_{input_margin} determines the time available for the arriving signal to pass through the internal combinational logic of the circuit and arrive at the path endpoint in time to meet the setup conditions of the flip-flop.

Output delay (offset) constraints apply to paths from storage elements to primary outputs. An output constraint specifies the latest time that a signal propagating from a startpoint may reach its endpoint, relative to the active edge of the clock at the startpoint, as shown in Figure 11-6. The timing analyzer uses t_{output_delay} to calculate

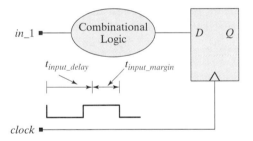

FIGURE 11-5 Input delay constraint.

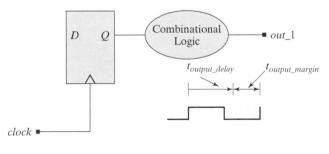

FIGURE 11-6 Output delay constraint.

---

[7]Path constraints influence the time required for a synthesis tool to converge on an implementation that satisfies the constraints; path constraints also influence the placement of logic within an FPGA. A design targeted to an FPGA should be synthesized initially without timing constraints and without pin assignments, to determine whether the design fits within the selected part.

t_{output_margin}, the time available for the output signal to reach its destination before the next active edge of the clock in the circuit to which the output is connected.

An *input–output (pad to pad) constraint* applies to a path whose startpoint is a primary input and whose endpoint is a primary output. The I/O delay constrains the maximum timing length of a combinational path between a primary input and a primary output.

A *cycle time (period) constraint* specifies the maximum period of the clock of a synchronous circuit, and applies to the paths between registers. The constraint specifies a clock (by name) and its period. A timing analyzer might also allow specification of the duty cycle and offset (skew) of the clock waveform. If a circuit has multiple independent clock domains[8] the paths are grouped with the clock that synchronizes the storage elements in the domain, as shown in Figure 11-7. Paths whose endpoints are not the data input of a storage device are placed in a default group.

Timing analysis determines whether the timing constraints of the paths in a circuit are satisfied. No combinational feedback loops are allowed, and all register feedback paths are broken by the clock boundary. All four types of paths pass through combinational logic, and the propagation delay of each path is calculated for both rising and falling edge transitions by back-tracing from the endpoint of the path to its startpoint, accumulating the propagation delays along the path. Paths are sorted by their length and verified to meet I/O timing constraints. If the propagation delays of the devices are given as a range of values (i.e., min:max), the delay of the path is reported as a (min:max) range too.

Verification of an input constraint considers the setup time of the sequential device at the endpoint of the path. Likewise, an output constraint is verified by considering

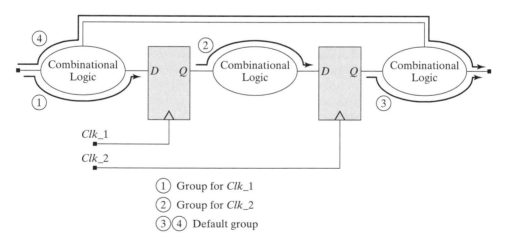

FIGURE 11-7 Path groups for a synchronous circuit with multiple clock domains.

[8]Timing analyzers consider clocks that are derived from the same source (i.e., dependent clocks), but we will not consider them.

the clock-to-output delay of the clock that launches the signal along the path from the sequential device to the primary output. Verification of a cycle time constraint must consider the clock-to-output delay of the device at the path endpoint, the propagation delay through the combinational logic on the path, the setup time of the device at the endpoint of the path, and any skew of the clock. *The longest path through the combinational logic between storage elements will determine the minimum period of the clock.*

The paths in a circuit are assumed to be statically sensitized for timing analysis; that is, the side inputs of any gate along the path are fixed and do not block propagation of a signal through the gate. In Figure 11-8, the side inputs of the NAND gate must be 1.

11.2.3 Factors That Affect Timing

In a typical synchronous circuit signals propagate from a source register to a destination register through combinational logic as shown in Figure 11-9. The delay along the path is a result of the delay between the clock and the output of the register that propagates the signal, and the propagation delay of the gates along the path. The propagation delay of a gate is affected by the capacitance of the input pins of the gates that it drives (i.e., the fanout loads on the path) and the loading associated with the resistance and parasitic capacitance of the path.

The physical gates and nets that are attached to the output of a gate add to its intrinsic capacitance, thereby increasing the propagation delay between a change at the input of the gate and the effect of the change at the output of the gate. The slew rate of the signals driving a gate affect the propagation delay of the gate too, because the capacitor in an RC network will charge slower if the input has a large rise-time than if the

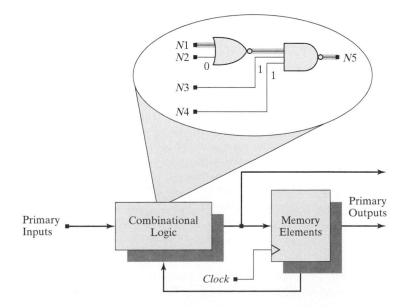

FIGURE 11-8 The side inputs of the gates along a statically sensitized path must not block propagation of a signal through the gate.

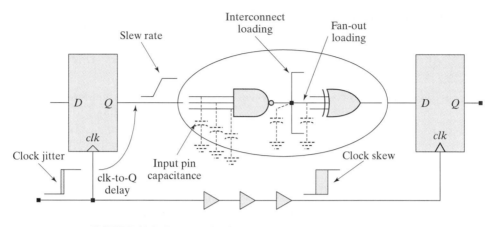

FIGURE 11-9 Factors affecting the timing of a synchronous circuit.

input is a step signal. The registers in a given clock domain of a circuit are synchronized by a common clock, but the edges of the clock cannot be exactly aligned in a physical chip, because the clock signal must propagate along different physical paths. The misalignment of clock edges in a synchronous circuit is referred to as clock skew. Clock skew reduces the timing margin between the data and the clock at the destination register.

The worst-case delay along a path in a circuit is determined by the intrinsic propagation delays[9] of the combinational logic and memory elements along the path, the fanout loading of the gates on the path, the interconnect loading of the signal path, and the signal slew rate. The period of the clock must accommodate the worst-case path delay between registers in the circuit. A path is a worst-case delay path if it satisfies the following conditions: (1) the sum of the functionally appropriate delay values through all of the gates in the worst-case path must not be less than that sum over any other combinational path, and (2) there must exist a vector of primary inputs and memory element values such that the logical value of the output depends on the logical value at every node on the path (i.e., the path from the startpoint to the endpoint can be sensitized), and, if the endpoint is a memory element, a clock transition will be enabled.

The following terms and notation will be used to describe the factors affecting the minimum period of the clock of a synchronous machine:

$t_{clk\_to\_Q}$ Delay between the active edge of the clock and the valid output of a flip-flop synchronized by the clock.

t_{comb_max} Longest path delay through combinational logic.

t_{setup} Setup time of a flip-flop driven by the combinational logic.

t_{skew} Clock skew.

The delay t_{comb_max} depends on the intrinsic gate delay, slew, and loading due to fanout and routing-dependent interconnect. In deep submicron designs (i.e., those with

---

[9] The intrinsic delay is that which the gate exhibits independently of any fanout loads.

physical dimensions $\leq .18$ microns), the interconnect delay is considered to play a dominant role. The clock period must be long enough to allow the signal to satisfy the setup-time margin of the destination register (i.e., the signal must arrive at the data input of the register in time to be stable during the setup time of the register). Put another way, the longest path must satisfy the constraint: $t_{comb\_max} < T_{clock} - t_{clk\_to\_output} - t_{setup} - t_{skew}$, or $t_{setup\_time\_margin} > 0$, where $t_{setup\_time\_margin} = T_{clock} - t_{clk\_to\_Q} - t_{setup} - t_{skew} - t_{comb\_max}$. The circuit has a cycle-time violation if $t_{setup\_time\_margin} \leq 0$.

The hold-time margin of a register imposes a constraint on the shortest path through the logic. The shortest path must satisfy the constraint: $t_{comb\_min} > t_{hold} - t_{clk\_to\_output} + t_{skew}$, or $t_{hold\_time\_margin} > 0$, where $t_{hold\_time\_margin} = t_{comb\_min} - t_{hold} + t_{clk\_to\_output} - t_{skew}$. This constraint prevents a race condition between the output of the register at the startpoint of the path and the data input of a register at the endpoint of the path.[10] A change in the value of the signal at the startpoint of the path must not arrive too soon at the register at the endpoint.

The relationship between the clock period T_{clock} and the delays of a signal path is illustrated in Figure 11-10, for a constraint in which the skew of the clock is neglected. The clock period must be greater than the sum of: the clock-to-output delay, the maximum delay of the combinational path, and the setup time of the endpoint (assumed to be a flip-flop). The slack, t_{slack}, of a path is the difference between the clock period and the delay of the path. If $t_{slack} \leq 0$ for any path the circuit will have a setup timing violation if that path is exercised by the stimulus applied to the circuit.

Synchronous operation of a circuit requires that all memory elements be synchronized by the same clock edge. *Clock skew* is the variation in the arrival of the clock edges at their destinations, relative to the edges of the source of the clock signal. The skew can be due to the intrinsic jitter of the clock itself, or it can be due to routing-induced differential propagation delay of the cells that are driven by the clock. The routing-induced skew of the clock is due to the loading (RC metal interconnect and memory elements), and by buffer chains in the clock distribution path. Long metal runs introduce proportionately more delay. Path-induced skew is unavoidable and must be anticipated [4].

Figure 11-11 illustrates the ambiguity that describes the location of the edge of a clock with skew. The jitter establishes an interval of ambiguity on each side of the nominal

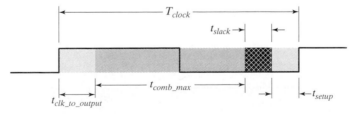
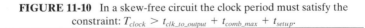

FIGURE 11-10 In a skew-free circuit the clock period must satisfy the constraint: $T_{clock} > t_{clk\_to\_output} + t_{comb\_max} + t_{setup}$.

---

[10]Note that minimum path delays are used in the hold-time constraint.

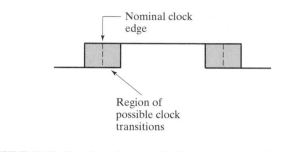

FIGURE 11-11   Skew introduces ambiguity in the location of the clock edge.

location of the clock edge. The actual transition of the clock occurs within the shaded regions. Figure 11-12 illustrates the factors that determine the maximum frequency of a clock that has skew. The effect of the skew is to increase the minimum period of the clock, compared to a skew-free clock. The clock period must satisfy the following constraint in the presence of skew: $T_{clock} > t_{clk\_to\_output} + t_{comb\_max} + t_{setup} + t_{skew}$.

Buffers and any other logic in the path of the clock signal introduce clock skew. In Figure 11-13, the clock edge occurs at a different time at each of the registers. The ambiguity of the location of the edge of the clock increases along the path. For a given clock period, the allowed skew is bounded by: $T_{skew} < T_{clock} - t_{clk\_to\_output} - t_{comb\_max} - t_{setup}$.

Example 11.3

The shift register in Figure 11-14(a) is shown with unbalanced buffer delays on the clock distribution net. The simulation results in Figure 11-14(b) compare the output of a register with equal delays and with unbalanced delays. In the register with unbalanced delays, a sample of D_in passes through the register with clock skew in three cycles instead of in four cycles.

End of Example 11.3

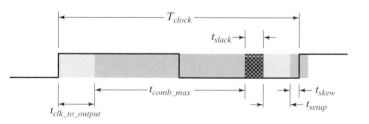

FIGURE 11-12   The period of the clock must be increased to compensate for skew, and must satisfy the constraint: $T_{clock} > t_{clk\_to\_output} + t_{comb\_max} + t_{setup} + t_{skew}$.

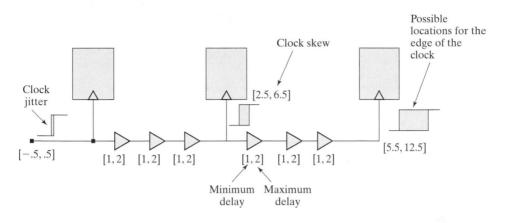

FIGURE 11-13 Buffers in a clock path progressively increase the skew of the clock at its destinations.

Example 11.4

The clock divider in Figure 11-15(a) creates *clock_by_2* from *clock*, but the delay of the buffer skews *clock_by_2* relative to *clock*. The recommended circuit [5] in Figure 11-15(b) delivers a common clock to the flip-flops and eliminates a buffer.

End of Example 11.4

Clock distribution networks for ASICs must be designed carefully to achieve short transition times and to minimize the effect of skew. Equalizing the delays from the clock pads to all memory elements, as shown in Figure 11-16, minimizes skew. Special compilers are used to place clock trees within a layout. Balanced clock trees achieve a balanced physical topology, leading to synchronous clock edges. Otherwise, the edges of the clocks that are farthest from the source of the clock will occur later than the edges that are closest to the source. Clock trees (also called H-trees) are designed to balance the delays through the tree and to reduce the peak currents at the load stage [4]. The trees of multiple clocks should be routed to have similar loads to minimize the skew between the clock phases. Each branch should have a balanced load. A testbench can monitor the signals shown in Figure 11-17 to verify that they do not drift too far apart; static analysis can be performed to determine the arrival times of the clock signals at the endpoints of the clock tree to ensure balance.

11.3 Elimination of ASIC Timing Violations

Table 11-2 summarizes options that can be exercised to eliminate timing violations in an ASIC. The first option is to lengthen the clock cycle. This approach might eliminate

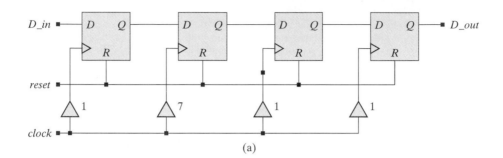

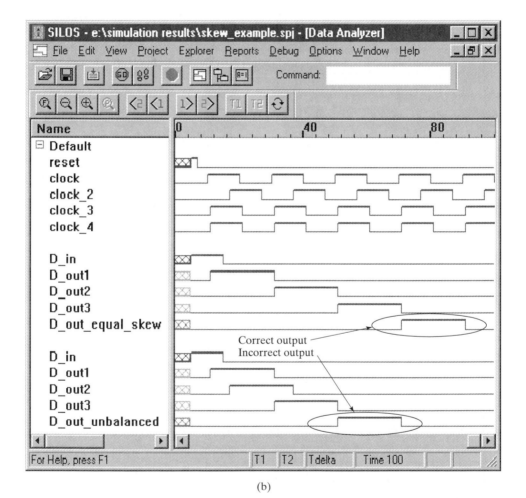

FIGURE 11-14 Effect of clock skew: (a) shift register with unbalanced clock distribution, and (b) waveforms showing incorrect register operation.

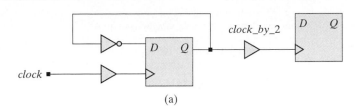

(a)

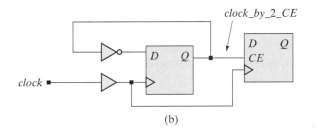

(b)

FIGURE 11-15 Alternative circuits that implement a clock divider: (a) *clock_by_2* will have skew relative to clock, and (b) a common clock synchronizes both flip-flops, without skew.

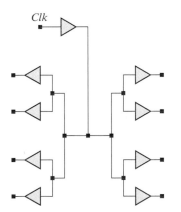

FIGURE 11-16 A clock tree balances the delay from the source of a clock signal to its destinations.

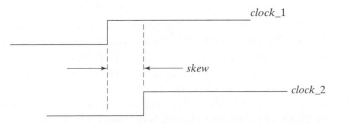

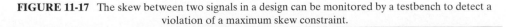

FIGURE 11-17 The skew between two signals in a design can be monitored by a testbench to detect a violation of a maximum skew constraint.

TABLE 11-2   Options for eliminating timing violations in an ASIC.

Elimination of ASIC Timing Violations	
Action	**Effect**
Lengthen the clock cycle	Eliminates the violation, constrained by performance specfications
Reroute critical paths	Reduced net delays
Resize and substitute devices	Reduced device delays and improved setup and hold margins
Redesign the clock tree	Reduced clock skew
Substitute a different algorithm	Reduced path delays (e.g., carry look-ahead vs. ripple carry)
Substitute architectures	Reduced path delays (e.g., pipelining)
Change technologies	Reduced device and path delays

the timing violation if the maximum period of the clock is unconstrained. Otherwise, an alternative approach is to reroute the critical paths of the circuit to reduce their net delays. The devices along a path can be resized to reduce the path delay, and to use flip-flops having shorter setup and hold times, subject to the availability of parts in the cell library. In addition to the options shown in Table 11-2, the behavioral model should be reexamined to see whether it can be modified to synthesize into faster logic. For example, replacing *if...else* statements by *case* statements might lead to parallel logic; state codes might be changed (e.g., use one-hot coding), with the result that the synthesized circuit is faster.

There are fewer structural options for eliminating timing violations in an field-programmable gate array (FPGA), because its architecture is fixed. However, FPGAs include fast carry logic (to improve timing margins) and special clock buffer nets and delay-lock loops that are intended to minimize the clock skew in the design. FPGAs are rich in registers, so pipelining is an attractive option for increasing the throughput of multilevel combinational logic. I/O flip-flops in FPGAs have guaranteed setup, hold and clock-to-output times, with programmable input delay and programmable output slew. For example, the programmable input delay of the Xilinx parts guarantees that the device will have 0 hold time, but an extended setup time. The software tools for synthesizing a design into an FPGA attempt to satisfy I/O and internal timing constraints. If the constraints cannot be met for a given device the only option is to lengthen the clock period or substitute a device that has a higher-rated clock frequency.

Error-free synchronous operation is the initial focus of timing verification. Setup constraints, hold constraints, and pulsewidth constraints are imposed on the design by the physical storage elements. Constraints on the clock skew are imposed by the interplay of the setup constraints, the longest path delay, and the physical layout and distribution of the clock through the circuit. STA can ensure that a circuit has sufficient slack to satisfy setup conditions, for example. Testbenches can monitor other conditions too, such as the relative skew between signals (e.g., the skew between the control lines of a memory). A testbench can also monitor glitches. STA can analyze the distribution of path lengths, which can be used as a guide to determine whether device sizes can be reduced without violating the timing constraints.

We saw in Chapter 6 that nested *if...else* statements imply priority and will synthesize priority encoders. If this structure cannot be avoided, the timing-critical signals should be put in the first clause of the statement, because it will drive the final stage of logic and will have the shortest path to the output. *case* statements synthesize to smaller and faster circuits, but *if...else* statements offer more flexibility, implement priority logic, and might be needed to accommodate late-arriving signals. Nested *if...else* and nested *case* statements synthesize to multiple levels of logic, which might compromise performance.

11.4   False Paths

Static path analysis can produce a false alarm if the tool ignores the true functionality of a path.

Example 11.5

The path *In_1–w0–w1–w2–Out_0* in Figure 11-18 cannot be sensitized, because the conditions that sensitize the path through the gates that drive *w0* and *w1* also drive a 0 onto the side input of the AND gate that drives the inverter. This condition constrains the output of the AND gate to be 1, and constrains the output of the inverter to be 0. The 0 at the output of the inverter desensitizes the AND gate that drives *w2*. Note that *In_2* has two paths to *w2*, a condition known as reconvergent fanout. When a circuit has reconvergent fanout the path through the gate at which the signal reconverges

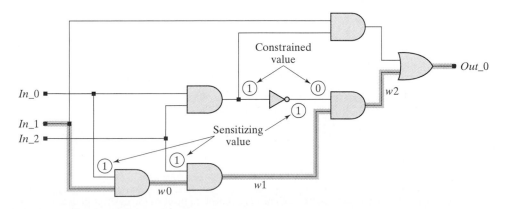

FIGURE 11-18   The path *In_1–w0–w1–w2–Out_0* in cannot be statically sensitized.

might not be statically sensitizable, *because the side inputs of the gate cannot be sensitized independently of the signal that is to propagate through the gate.*

End of Example 11.5

The delay of a topological path may be reported incorrectly if the analysis does not account for the polarity of signal transitions along the path. In general, merely adding the maximum or minimum delay values of the gates along a path may produce a pessimistic calculation of the maximum or minimum path delay and lead to wasted effort to eliminate a timing violation. A timing analyzer must use the appropriate rising or falling delay of the devices when calculating the delay along a path.

Example 11.6

If the analysis of the delays along the path shown in Figure 11-19 simply adds the maximum values of the delays through the gates, the maximum rising delay is $t_{delay\_rising} = 15$. Considering the polarity of the transitions gives $t_{max\_rising} = 5 + 3 + 5 = 13$.

End of Example 11.6

A topological path is said to be a false path if it is not a functional path under the action of the primary inputs. For example, mutually exclusive conditions that steer datapaths might lead to false paths being reported by a timing analyzer. The steering logic eliminates certain paths, which should not be reported by the tool.

Example 11.7

The controlling signals of the multiplexers that steer the datapaths in Figure 11-20 are not independent of each other. Two false paths cannot be exercised. The maximum topological delay through the circuit is $t_{topological} = 30 + 30 = 60$, but the maximum functional delay $t_{max} = 15 + 30 = 45$.

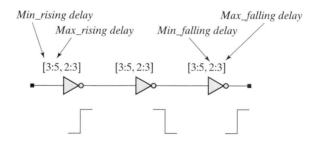

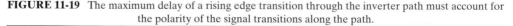

FIGURE 11-19 The maximum delay of a rising edge transition through the inverter path must account for the polarity of the signal transitions along the path.

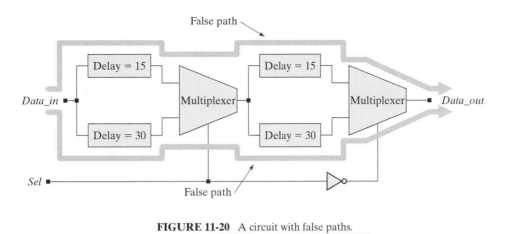

FIGURE 11-20   A circuit with false paths.

End of Example 11.7

A timing analyzer might report false path delays if redundant logic has been added to the design to reduce path delay. If the timing analyzer does not account for the reduced delay attributable to the redundant logic the reported delays will be pessimistic. Also, multicycle paths might trigger a false alarm unless the timing analyzer can identify such paths.

11.5 Dynamically Sensitized Paths

Static timing analysis assumes that the side inputs of the gates along a path are held constant. The analysis considers only paths that are statically sensitizable. A path is said to be dynamically sensitizable if at least one of the side inputs along a path can be switched to a value that allows signal propagation along the path. If a path is dynamically sensitizable, a static timing analyzer will fail to report the delay of the path.

Example 11.8

Static sensitization of the path $In\_2–w1–w2–w3–w4–Out\_0$ in Figure 11-21(a) requires $In\_1 = 1$, which blocks the signal path through the NAND gate driving Out_0. Therefore, the path cannot be sensitized, and it is a false path under static sensitization. However, consider switching In_1 from an initial value that lets In_2 pass through the NAND gate driving $w1$, and then switches in time to pass $w4$ through the NAND gate

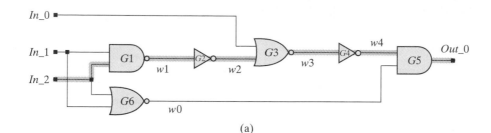

(a)

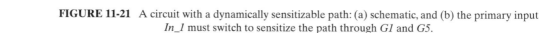

(b)

FIGURE 11-21 A circuit with a dynamically sensitizable path: (a) schematic, and (b) the primary input *In_1* must switch to sensitize the path through *G1* and *G5*.

forming *Out_0*. The simulation results in Figure 11-21(b) show how a transition at *In_2* propagates to *Out_0* along a dynamically sensitized path, with *In_1* switching to sensitize *G1* and *G5*.

End of Example 11.8

11.6 System Tasks for Timing Verification

Verilog has several built-in system tasks for performing timing checks during simulation. Some of these tasks can be included in models to (1) monitor simulation activity automatically, (2) detect a timing violation, and (3) report a timing violation.

11.6.1 Timing Check: Setup Condition

An edge-triggered flip-flop will not operate correctly if the data at its input is not stable for a sufficient time before and after the clock edge. Setup and hold times are logic-level constraints on the operation of storage elements. System failure can result from nondeterministic behavior of storage elements if the setup and hold times of the storage elements are violated. Figure 11-22 shows the setup interval before each active edge of the clock. The system task for detecting violations of the setup time of a device has the following syntax: *$setup (data_event, reference_event, limit).*

A setup time violation occurs when a *data_event* occurs within the specified time *limit* relative to the *reference_event*. In practice, data must be stable before the active edge of the clock of a flip-flop.

Setup-time violations are cause by paths that are long relative to the clock cycle. The delays that contribute to the late arrival of the data must be reduced, or the clock period must be increased to eliminate the timing violation (see Table 11-2).

Example 11.9

Figure 11-23 shows the waveforms of *sys_clk, sig_a,* and *sig_b*. Note that *sig_a* satisfies the setup timing constraint, but *sig_b* does not. The timing check would be activated by the tasks *$setup (sig_a, posedge sys_clk, 5)* and *$setup (sig_b, posedge sys_clk, 5)*; the latter check would report a timing violation.

End of Example 11.9

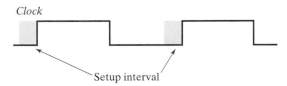

Clock

Setup interval

FIGURE 11-22 The data of a flip-flop must be stable during the setup interval located ahead the active edge of the clock.

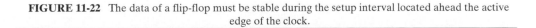

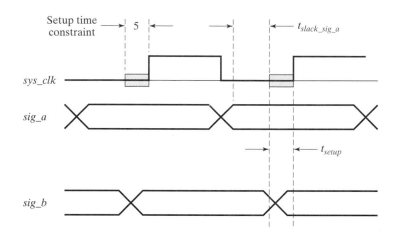

FIGURE 11-23 The setup time constraint on *sys_clk* requires *sig_a* and *sig_b* to be stable during the setup interval located ahead of the active edge of the clock. *sig_b* violates the setup constraint.

11.6.2 Timing Check: Hold Condition

For correct operation of a flip-flop, the data at its input must be stable for a sufficiently long time after the active edge of its clock. A hold violation occurs if the datapath to the flip-flop is so short that a change in the data at the output of the flip-flop at the startpoint of the path propagates too quickly to the input of the flip-flop at the endpoint of the path. Figure 11-24 shows the hold interval in which the data path of a flip-flop must be stable.

Short paths through combinational logic are stretched automatically by synthesis tools to reduce the slack and meet timing constraints. It is desirable to achieve balance in a design, so that paths are not faster or slower than necessary. Needlessly fast paths waste silicon area.

The system task for detecting violations of the *hold time* of a device has the following syntax: **$hold** (*reference_event, data_event, limit*). A hold time violation occurs when a *data_event* occurs within the specified time limit relative to the *reference_event*.

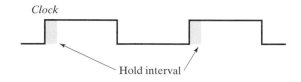

FIGURE 11-24 The data of a flip-flop must be stable during the hold interval located after the active edge of the clock.

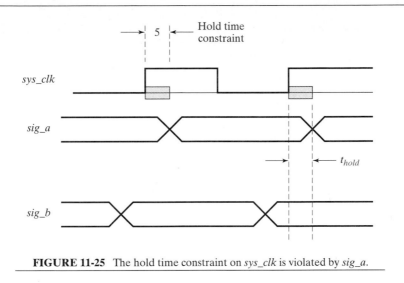

FIGURE 11-25 The hold time constraint on *sys_clk* is violated by *sig_a*.

Example 11.10

For the waveforms in Figure 11-25, *sig_a* is not stable during the hold interval of *sys_clock*. A timing violation would be reported by the task *$hold* (*posedge sys_clk, sig_a*, 5).

End of Example 11.10

11.6.3 Timing Check: Setup and Hold Conditions

The timing task *$setuphold* monitors for violations of both setup and hold constraints between a *reference_event* and a *data_event*, according to the syntax: *$setuphold* (*reference_event, data_event, setup_limit, hold_limit*), where *reference_event* is the edge transition that synchronizes the device.

Example 11.11

In Figure 11-26 *sig_a* and *sig_b* satisfy both the setup and hold constraints.

End of Example 11.11

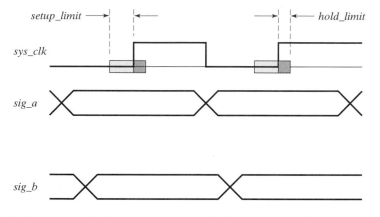

FIGURE 11-26 The setup and hold constraints are satisfied by *sig_a* and *sig_b* relative to the active edges of *sys_clk*.

11.6.4 Timing Check: Pulsewidth Constraint

The minimum width of a clock pulse is constrained for sequential devices. The clock of an edge-triggered device must endure for a time sufficient to charge internal signal nodes. The task *$width* (*reference_event, limit*) will detect a violation of the minimum pulse width. For example, the task *$width (posedge clk, 4)* will detect a violation if the interval between the positive edge of *clk* and the following negative edge is less than 4. This task can also be used in simulation to detect potential glitches and degraded (narrow) clock pulses.

Example 11.12

The minimum pulsewidth constraint in Figure 11-27 is satisfied by *clock_a*, but violated by *clock_b*.

End of Example 11.12

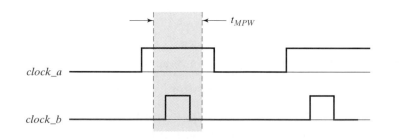

FIGURE 11-27 The task *$pulsewidth* will detect a pulse width violation by *clock_b*.

11.6.5 Timing Check: Signal Skew Constraint

Clock skew is a critical issue in system performance. It results from unbalanced clock trees and degrades setup and hold time margins. The skew between two signals is monitored by the task *$skew (reference_event, data_event, limit)*; it will report a violation if the interval between *reference_event* and *data_event* is greater than *limit*.

Example 11.13

The task invoked by the procedural statement *$skew(**posedge** clk1, **negedge** clk2, 3)* detects a violation if the interval between the rising edge of *clk1* and the falling edge of *clk2* exceeds 3, as shown in Figure 11-28.

End of Example 11.13

11.6.6 Timing Check: Clock Period

The period of a clock is determined by the interval between successive active edges of the waveform. The task *$period (reference_event, limit)* will monitor successive edges of an edge-triggered *reference_event* and detect a timing violation error if the interval between successive events is less than *limit*. In design re-use within an SoC environment, this timing check would verify that a core cell can be used safely at the specified clock period. If the timing check is satisfied, the clock period is at least as long as the minimum period required by the cell.

Example 11.14

The task invoked by the procedural statement *$period (**posedge** clock_a, 25)* does not detect a timing violation in *clock_a*, shown in Figure 11-29.

End of Example 11.14

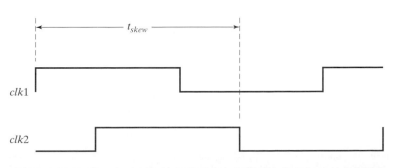

FIGURE 11-28 Skew is checked between signal edges by the system task *$skew*.

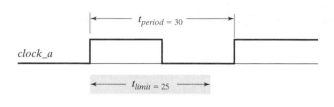

FIGURE 11-29 *clock_a* satisfies the constraint on its minimum period.

11.6.7 Timing Check: Recovery Time

The **$recovery** *(reference_event, data_event, limit)* task checks the time required for synchronous behavior to resume after a reset (asynchronous) or a clear condition has been de-asserted. The recovery check specifies the minimum time that an asynchronous input must be stable before the active edge of the clock. The recovery *limit* specifies the time between the referenced de-assertion event and the next active edge of the *data_event*.

Example 11.15

The task **$recovery** (**negedge** *set*, **posedge** *clock*, 3) checks whether the de-assertion of *set* precedes the active edge of *clock* by three time steps in Figure 11-30.

End of Example 11.15

Timing checks that are specific to a device are usually included in the standard cell library model of the device by including a **specify . . . endspecify** block in a module.

Example 11.16

The Verilog code below describes the cell library flip-flop *dfnf311*. The model includes timing checks for setup and hold constraints and a pulsewidth constraint.

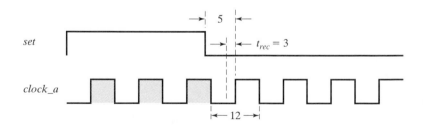

FIGURE 11-30 The interval between the falling (de-asserting) edge of *set* relative to the next rising edge of *clock_a* satisfies the constraint on the recovery time.

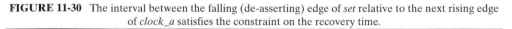

```verilog
                    'timescale  1ns /  1ps
                    'celldefine

module dfnf311 (Q, Q_b, DATA1, CLK2);
// Negative edge-triggered flip-flop with Q and QB

    input                    DATA1, CLK2;
    output        Q, Q_b;
    reg           notify;
    parameter     vlib_flat_module = 1 ;

    U_FD_N_NO inst1 (Q_int, DATA1, CLK2, notify);    // user-defined primitive
    buf inst2 (Q, Q_int);
    not inst3 (Q_b, Q_int);

    specify                                                    // Module paths
      specparam Area_Cost = 4268.16;

      (negedge CLK2 =>  (Q +: DATA1)) = (0.525:1.129:2.889, 0.749:1.525:3.948);

      specparam RiseScale$CLK2$Q = 0.00094:0.00205:0.00540;
      specparam FallScale$CLK2$Q = 0.00086:0.00165:0.00397;
      (negedge CLK2 =>  (Q_b -: DATA1)) = (0.590:1.243:3.207, 0.425:0.914:2.616);
      specparam RiseScale$CLK2$Q_b = 0.00120:0.00248:0.00658;
      specparam FallScale$CLK2$Q_b = 0.00140:0.00289:0.00248;
      specparam    inputCap$CLK2 = 40.18,
                   InputCap$DATA1 = 24.11;
      specparam    InputLoad$CLK2 = 0.009:0.021:0.053,
                   InputLoad$DATA1 = 0.005:0.013:0.032;
      specparam    t_SETUP$DATA1 = 0.76:0.92:1.68,
                        t_HOLD$DATA1  = 0.44:0.74:0.46,
                        t_PW_H$CLK2  = 0.37:0.67:1.99,
                        t_PW_L$CLK2  = 0.37:0.67:1.99;
  // Timing checks

      $setuphold(negedge CLK2, DATA1, t_SETUP$DATA1, t_HOLD$DATA1,
          notify);
      $width (posedge CLK2, t_PW_H$CLK2, 0, notify);
      $width (negedge CLK2, t_PW_L$CLK2, 0, notify);

      'ifdef not_cadence specparam MaxLoad = 'MAX_LOAD_1X;
      'endif
    endspecify
endmodule
'endcelldefine

primitive U_FD_N_NO (Q, D, CP, NOTIFIER_REG);
    output Q;
    input  NOTIFIER_REG, D, CP;
    reg   Q;
```

```
// FUNCTION :  NEGATIVE EDGE TRIGGERED D FLIP-FLOP ( Q OUTPUT UDP ).
     table
  // D   CP        NOTIFIER_REG :  Qt :  Qt+1
     1  (10)        ?        :  ?  :  1;  // clocked data
     0  (10)        ?        :  ?  :  0;

     1  (1x)        ?        :  1  :  1;  // reducing pessimism
     0  (1x)        ?        :  0  :  0;
     1  (x0)        ?        :  1  :  1;
     0  (x0)        ?        :  0  :  0;
     ?  (0x)        ?        :  ?  :  -;  // no change on rising edge
     ?  (?1)        ?        :  ?  :  -;
     *   ?          ?        :  ?  :  -;  // ignore edges on data
     ?   ?          *        :  ?  :  x;  // all transitions
     endtable
  endprimitive
```

End of Example 11.16

11.7   Fault Simulation and Testing

A circuit that has been designed properly may still fail to operate in the field. Failures in the field are expensive to diagnose and repair. The failure of a circuit might be permanent, intermittent, or transient. A *permanent failure* exhibits incorrect operation of the circuit at all times. An *intermittent failure* is exhibited randomly, and for only a finite duration. *Transient failures* occur in the presence of certain environmental conditions that alter the performance characteristics of the devices, such as high temperature or background radiation. The failure might be caused by any of the following: (1) wafer defects, (2) contaminated atmosphere in the clean room (i.e., ambient particulates), (3) impure processing gasses, water, and chemicals, and (4) photomask misalignment.

A process-induced defect might cause a high junction leakage current, a high contact resistance, an open circuit, a short circuit, or an out-of-spec threshold voltage. Our concern will be defects that cause failures that are apparent as functional errors during operation.

Contaminants in the clean-room environment can introduce defects in the manufactured circuit. The atmosphere of the clean room might contain particles of dust, smoke, shaving lotion, or other gaseous products. The wafer might bear residues of materials that were not removed properly by a cleaning process, and the chemicals used in the fabrication process might not be pure enough. Clean rooms are constructed to purify and maintain a clean atmosphere, but the presence of equipment operators and technicians in the room inevitably introduces contaminants. The floor of a clean room is specially constructed to minimize the possibility that vibrations will cause a photomask to be misaligned, which would cause the devices to be exposed improperly and violate spatial constraints on the layout.

The consequences of device failure are so expensive that the semiconductor industry tests devices thoroughly before they are shipped. This testing does not revalidate the functionality of the device or test for design errors, because design errors are assumed to have been found during the verification steps that are executed in the design flow before the mask set is released for fabrication. In fact, it is likely that the set of stimulus patterns that verify the functionality of a device will detect only a few defects within the fabricated circuit. Failure analysis of defects might point to a design error, but that is not its purpose.

Production testing applies test signals and monitors the response of a circuit to confirm that it operates correctly. The set of test patterns (vectors) that are applied to a circuit must be sufficient to test the known failure modes of the chip and detect all defects. The vectors that are used to verify the functionally of the design are usually a starting point for production testing, but they are not sufficient if they fail to exercise some failure modes of the circuit. It is unlikely that the vectors used for design validation will provide sufficient coverage of a complex circuit (e.g., one with an embedded processor, memory controller, and perhaps an embedded DSP, and having about 500,000 gates).

Production testing is concerned with detecting permanent errors that are caused by manufacturing defects. It involves two major steps: test generation and fault simulation. We will examine how tests are generated, but first we must consider *what* we are testing, before we determine *how* to test it. The tests used in production testing are intended to detect failure modes, called *faults*, within the circuit. The set of test vectors is generated in conjunction with a process called *fault simulation*, which determines whether a test detects a fault at a particular site in the circuit.

11.7.1 Circuit Defects and Faults

Models of the failure modes of a circuit consider the *logical effects* that result from the *physical faults* in a circuit. The main physical faults that can occur are summarized in Table 11-3, which is based on a discussion in Smith [4]. Physical defects and the logical faults they cause are assumed to be uniformly distributed over a wafer.

When a circuit fails to operate correctly, the logic realized by the circuit is different from the logic that was specified for the design. There are a variety of modes by which a digital circuit may exhibit failure. One common failure mode occurs when a signal line is shorted to either the power rail or the ground rail. These failure modes are called "stuck" faults, and their location is called a *fault site*. Such defects in the manufactured part can occur when physical vibrations of the mask aligner occur during the photomasking steps, causing the conductor of a signal path to be misaligned and placed too close to the conductor of a neighboring signal path, or when over/under etching of the photoresist alters the physical location of a conductor. The resulting defects can also occur when the fabrication process does not operate within the range of statistical specifications that reduce the possibility of such failures.

Faults due to short-circuits occur in the interconnect between transistors in a logic cell are called *bridging faults*. A CMOS circuit has a *stuck-on* fault if the gate of a transistor is always on. Bridging faults are detected by measuring the quiescent current,

TABLE 11-3   Physical faults and circuit-level effects in an
integrated circuit.

Physical Faults	Degradation Fault*	Open-Circuit Fault	Short-Circuit Fault
Chip-Level Fault			
Leakage or short between package leads	x		x
Broken, misaligned, or poor wire bonding		x	
Surface contamination, moisture	x		
Metal migration, stress, peeling		x	x
Metallization (open or short)		x	x
Gate-Level Fault			
Contact opens		x	
Gate to source/drain junction short	x		x
Field-oxide parasitic device	x		x
Gate-oxide flaw, spiking	x		x
Mask misalignment	x		x

*Parametric fault (threshold voltage shift) or delay fault.

I_{DDQ}, through a circuit. Testing the quiescent current takes more time than tests for stuck faults, but tests for I_{DDQ} are needed for bridging faults, because the tests that reveal stuck faults are not intended to reveal a bridging fault, which are manifested by a high current rather than by an incorrect logic value.

Example 11.17

Figure 11-31(a) illustrates how a bridging fault in a two-input complementary metal-oxide semiconductor (CMOS) NOR gate connects the gates of the pull-up transistors. Ordinarily, a CMOS device will have a short between the power and ground rails of the circuit for only a brief time while the output of the device switches, and the quiescent current through the device, I_{DDQ}, is 0. But the bridging fault in the pull-up logic of this circuit will cause a high current to flow between the rails when $x1 = 0$ and $x2 = 1$, causing thermal damage and early mortality of the device. I_{DDQ} testing, which monitors the quiescent current of the circuit, would detect the presence of the bridging fault [6].

The two-input NOR gate in Figure 11-3(b) has an internal defect in which the pull-down transistor for $x1$ is stuck on, with $x1$ effectively 1. Because the signal applied to the pull-up gate for $x1$ cannot be exercised independently of the signal attached to the pull-down gate for $x1$, this condition cannot be detected externally by testing the

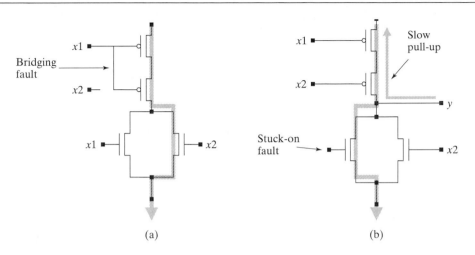

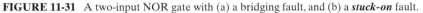

FIGURE 11-31 A two-input NOR gate with (a) a bridging fault, and (b) a **stuck-on** fault.

logic of the circuit. Note that if $x1 = 0$, and $x2 = 0$ then $y = 1$, but the pull-down transistor with the stuck-on gate tends to pull y to 0. The stuck-on fault causes a current that always tends to discharge the output node of the device, making the pull-up action slower than that of a fault-free circuit. In addition, the high current through the pull-down transistor will lead to thermal degradation and early failure of the circuit.

End of Example 11.17

Faults that are caused by shorts of nets to the power or ground rails of the circuit are called *stuck at 1* or *stuck at 0* faults, respectively. A given node in a circuit may be stuck at logical 1, denoted by *s-a-1*, or stuck at 0, denoted by *s-a-0*. The presence of a *s-a-1* or *s-a-0* fault compromises the logic implemented by the circuit.

Example 11.18

The three-input NAND gate in Figure 11-32 is shown in (a) without a fault, in (b) with an *s-a-0* fault on one input, and in (c) with an *s-a-1* fault on one input. The logic of the gate in the fault-free circuit has $y_{good} = (x1\ x2\ x3)'$, but the circuit with the fault $x1$ s-a-0 has $y_{x1\ s-a-0} = 1$; its output is stuck at 1. The circuit with $x1$ stuck at 1 has $y_{x1\ s-a-1} = (x2\ x3)'$ (i.e., the circuit realizes a two-input NAND gate). The logic realized by the circuits having a fault differs from the logic of the fault-free circuit.[11]

End of Example 11.18

---

[11]Note that the internal bridging fault in Figure 11.31(b) affects the gate of the pull-down transistor driven by $x1$ and does not affect the corresponding pull-up transistor. A stuck fault at $x1$ affects both transistors.

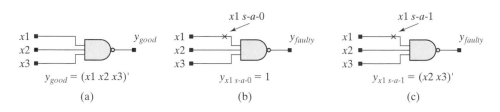

$$y_{good} = (x1\ x2\ x3)' \qquad y_{x1\ s\text{-}a\text{-}0} = 1 \qquad y_{x1\ s\text{-}a\text{-}1} = (x2\ x3)'$$

(a)   (b)   (c)

FIGURE 11-32 A three-input NAND gate with (a) no faults, (b) *x1* with an *s-a-0* fault, and (c) *x1* with an *s-a-1* fault.

All of the nets attached to the inputs and outputs of the gates in a logic circuit are possible sites for *s-a-0* and *s-a-1* faults. Stuck-at faults cause the disappearance of a literal or an implicant in the realized logic, or force the logic function to a fixed value. A test must detect the difference between the logic of the fault-free circuit and the logic of the circuit having an *s-a-0* or *s-a-1* fault. The *s-a-0* fault in Figure 11-32(b) could be detected by a test that applied $x1 = 1$, $x2 = 1$, and $x3 = 1$. If $y_{x1=1, x2=1, x3=1} = 1$ then the test reveals the presence of the fault, because $y_{x1=1, x2=1, x3=1} = 1$ does not match $y_{good} = 0$. Similarly, the *s-a-1* fault in Figure 11-32(c) will be detected by a test that applies $x1 = 0$ and $x2 = 1$, and $x3 = 1$, because $y_{x1=0, x2=1, x3=1} = 0$ does not match $y_{good} = 1$ if the fault is present.

The faults in the isolated three-input NAND gate in Figure 11-32 can be tested easily by applying signals directly to the inputs of the gate, and the output can be monitored directly too. This is not the typical situation, because the fault sites at the inputs and the outputs of most of the gates in a circuit are not directly available at the chip's pins. Moreover, a chip is limited to at most a few hundred pins, but may have millions of fault sites within the device. Nonetheless, the same principles that were used to test the isolated device can be used to develop tests to detect faults at sites embedded within a chip. An embedded fault can be detected if (1) primary inputs can assert a logic value at the site of the fault, and (2) those inputs are compatible with the inputs that propagate the fault to a primary output of the device.

11.7.2 Fault Detection and Testing

If a *s-a-0* or *s-a-1* fault in a combinational logic is testable, it can be detected by applying a set of inputs, known as a *test pattern*, and observing whether the output of the circuit with the fault differs from that of a fault-free circuit. A basic model for testing circuits for stuck faults is shown in Figure 11-33. Identical input signals are applied to a fault-free circuit, and to a circuit in which a fault has been injected at a particular site. The outputs of the circuits are compared to determine whether they differ. If the error signal asserts, the applied pattern is a test for the injected fault. If the outputs of the circuits do not differ, the applied pattern cannot distinguish between the faulty circuit and the fault-free circuit.

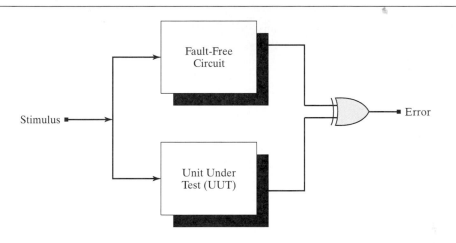

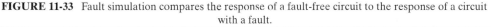

FIGURE 11-33 Fault simulation compares the response of a fault-free circuit to the response of a circuit with a fault.

Example 11.19

A test for the fault $x1$ s-a-0 in the circuit in Figure 11-34(a) is shown in the table in Figure 11-34(b). When the stimulus pattern is applied to the inputs of the circuit, the outputs of the good and faulty circuits differ, so the stimulus pattern $(x1\ x2\ x3) = (1\ \ 1\ \ 1)$ detects the fault and is a test for $x1$ s-a-0.

End of Example 11.19

Note that if $x1$ is stuck at 0 the output of the circuit in Figure 11-34 has a value of 1, independently of the inputs. In general, test for a fault by applying a pattern that asserts the correct (not stuck) value at the fault site, and compare the outputs of the circuit with and without a fault. If the outputs do not match, the stimulus pattern will detect such a fault in the manufactured circuit.

A test for a s-a-0 and a s-a-1 fault in combinational logic requires a single input pattern, and consists of the following four steps:

(a) Inject a fault in the circuit.

(b) Justify the fault (i.e., apply inputs that complement the value of the fault at the fault site).

(c) Sensitize one or more paths to propagate the effect of the fault to an output.

(d) Compare the sensitized output to that of a good circuit.

Circuit with
a fault has
y s-a-1

x1 s-a-0

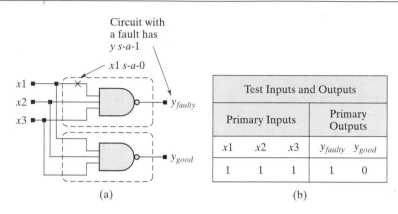

Test Inputs and Outputs				
Primary Inputs			Primary Outputs	
$x1$	$x2$	$x3$	y_{faulty}	y_{good}
1	1	1	1	0

(a)   (b)

FIGURE 11-34 A test for $x1$ s-a-0 detects a difference between the faulty circuit and the fault-free circuit.

To test for a fault in a combinational logic circuit, the primary inputs of the circuit must be chosen to (1) counter-assert (justify) the assumed value of the fault, and (2) propagate the effect of the fault to a primary output (i.e., sensitize the output to the fault). If the test is to detect the condition that a node is stuck at 1, the primary inputs must be those that would assert the value of 0 at the fault site in a fault-free circuit. The primary inputs must also propagate a signal value from the site of the fault to a primary output. If the sensitized output differs from the output of the fault-free circuit the test detects a fault. The test pattern that detects a fault does not necessarily isolate the location of the fault to a particular site, because a test might detect more than one fault.

Example 11.20

Two copies of a circuit are shown in Figure 11-35, with the so-called faulty circuit having the fault $x1$ s-a-0. To detect the fault, the stimulus pattern has the value of $x1$ set to 1 to justify the fault. The side inputs of the NAND gate ($x1$, $x2$, $x3$) are set to 1 to *sensitize* the output of the gate to the value of $x1$. The side input of the OR gate ($x0$) is set to 0 to sensitize the path propagating the value of $x1$ to the primary output of the circuit. The fault-free circuit has $y_{good} = 0$; the circuit with $x1$ s-a-0 has $y_{faulty} = 1$. The stimulus pattern ($x0$ $x1$ $x2$ $x3$) = (0 1 1 1) is a test for the fault $x1$ s-a-0.

End of Example 11.20

11.7.3 D-Notation

A special symbolic notation, called *D-notation* [7], has been developed for use in test generation and fault detection. In *D*-notation the logical values of the signals in a circuit are

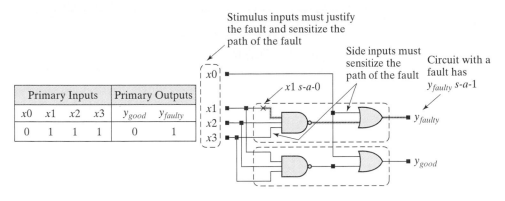

FIGURE 11-35 The stimulus pattern $(x0\ x1\ x2\ x3) = (0\ \ 1\ \ 1\ \ 1)$ detects the fault $x1$ s-a-0 by justifying the fault and sensitizing the output to the fault.

denoted by the symbols D and D'. The symbol D implies that the value of the signal is 1 in a good (fault-free) circuit, and D' implies that the value of the signal is 0 in a good circuit.

Example 11.21

In Figure 11-36, signal $x1$ has the value D when the pattern $(x1\ x2\ x3) = (1\ \ 1\ \ 1)$ is applied, meaning that $x1 = 1$ in the fault-free circuit, and 0 in the circuit having $x1$ s-a-0. Likewise, the output of the circuit has $y_{good} = D'$.

End of Example 11.21

The D notation for faults is useful in identifying faults that can be detected by the same stimulus pattern, and reducing the number of tests that must be applied to the circuit. It is desirable to have a high number of faults detected by a small number of tests.

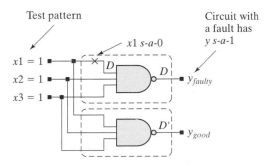

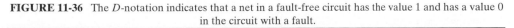

FIGURE 11-36 The D-notation indicates that a net in a fault-free circuit has the value 1 and has a value 0 in the circuit with a fault.

Example 11.22

The two-level circuit in Figure 11-37(a) has faults sites at nine primary inputs, three internal nets, and one primary output. The tables in Figure 11-37(b) and 11.37(c) use D-notation to indicate the values of the internal and primary output nets for the indicated stimulus patterns.[12] The tables are annotated to identify the faults that are detected

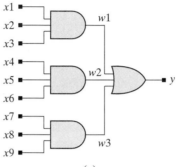

(a)

	Tests for s-a-1 Faults								Tests for s-a-0 Faults					
	#1	site	#2	site	#3	site			#1	site	#2	site	#3	site
$x1$	0	$x1$	1		1			$x1$	1	$x1$	0		x	
$x2$	1		0	$x2$	1			$x2$	1	$x2$	x		x	
$x3$	1		1		0	$x3$		$x3$	1	$x3$	x		0	
$x4$	0	$x4$	1		1			$x4$	0		1	$x4$	x	
$x5$	1		0	$x5$	1			$x5$	x		1	$x5$	x	
$x6$	1		1		0	$x6$		$x6$	x		1	$x6$	0	
$x7$	0	$x7$	1		1			$x7$	0		0		1	$x7$
$x8$	1		0	$x8$	1			$x8$	x		x		1	$x8$
$x9$	1		1		0	$x9$		$x9$	x		x		1	$x9$
$w1$	D'	$w1$	D'	$w1$	D'	$w1$		$w1$	D	$w1$	D'		D'	
$w2$	D'	$w2$	D'	$w2$	D'	$w2$		$w2$	D'		D	$w2$	D'	
$w3$	D'	$w3$	D'	$w3$	D'	$w3$		$w3$	D'		D'		D	$w3$
y	D'	y	D'	y	D'	y		y	D	y	D	y	D	y

(b)   (c)

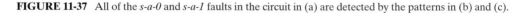

FIGURE 11-37 All of the s-a-0 and s-a-1 faults in the circuit in (a) are detected by the patterns in (b) and (c).

---

[12]In this context, the symbol x denotes a don't-care.

by the pattern. The tests are developed under the assumption that a single fault is present in the circuit (the so-called single stuck fault model [6]), but a given test may actually detect more than one fault. In this circuit, only three tests are needed to detect all of the possible *s-a-1* faults, and another three tests detect all of the *s-a-0* faults.

End of Example 11.22

Example 11.23

When a test is applied to detect the fault *w2 s-a-0* in the multilevel combinational logic circuit in Figure 11-38(a) the signals have the values displayed in *D*-notation in table in Figure 11-38(b). Of the possible tests for the fault, the one with $(x1\ x2\ x3\ x4\ x5) = (1x\ 1\ xx)$ is shown, where x denotes a don't care. The values of the primary inputs that sensitize the fault do not conflict with the values that justify the fault, so the pattern detects the fault. The output of the fault-free circuit is $y = 1$.

End of Example 11.23

Fault sensitization in multilevel networks is accomplished by tracing a path forward from the fault site to a primary output, then setting all of the side inputs of the gates on the path to their sensitizing values (e.g., the side inputs of AND gates are set to 1). Then paths are traced backward from the side inputs through gates to the primary inputs, to establish values of the primary inputs that sensitize the path from the site of the fault to the primary output. This step is called *line-justification*. In general, multipath sensitization must be considered [6], because single-path sensitization might not produce a test for a testable fault.

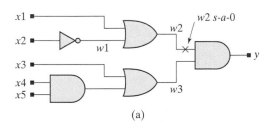

	Justification	Sensitization	Test
x1	1	x	1
x2	x	x	x
x3	x	1	1
x4	x	x	x
x5	x	x	x
w1	x	x	
w2	D	x	
w3	x	1	
y	D	D	

(b)

FIGURE 11-38 Fault detection: (a) a combinational logic circuit, and (b) signal values in *D*-notation for justification and sensitization of the fault *w2 s-a-0*.

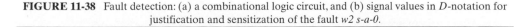

Example 11.24

The fault *G2 s-a-1* in the circuit in Figure 11-39, known as Schneider's circuit [6], can be justified by setting $x2 = 1$, and $x3 = 1$. To sensitize a single path from the fault site to *y_out*, set $x1 = 1$, which drives the output of *G5* to 1. The previous choices for *x1*, *x2*, and *x3* are compatible with setting to 1 the side inputs of the NAND gate forming *y_out*, but setting *G6* to 1 requires setting *x4* to 0, because the value of *G2* is not known. With $x3 = 1$ and $x4 = 0$, the value of *G7* becomes 0, which desensitizes the path of the fault. If the single path through *G6* is chosen to propagate the fault, a similar condition blocks the propagation of the fault to the output. Consequently, the fault *G2 s-a-1* cannot be detected by sensitizing a single path from the fault site to the output of the circuit. The fault can be detected only by simultaneously sensitizing the paths through both *G5* and *G6*, with the stimulus pattern $(x1\ x2\ x3\ x4) = (1\ \ 1\ \ 1\ \ 1)$.

End of Example 11.24

11.7.4 Automatic Test Pattern Generation for Combinational Circuits

It is not feasible to develop tests for faults in large combinational circuits by applying manual methods. The test for the fault *w2 s-a-0* in Figure 11-38 could be developed by manual methods because the circuit was simple. An alternative approach to testing combinational logic is to apply all possible inputs to the circuit and observe whether the outputs of the circuit match those of the fault-free circuit. Although exhaustive test-pattern generation is straightforward, it becomes unwieldy when a circuit has a large number of inputs, and it could be unnecessary. We saw in Example 11.22 that only six patterns were needed to detect the entire set of faults. Exhaustive pattern generation would have produced 512 patterns.

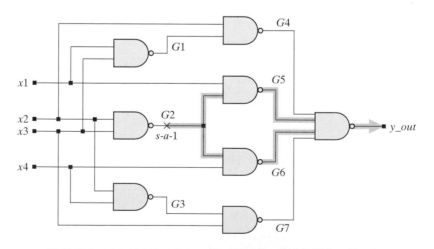

FIGURE 11-39 The fault *G2 s-a-1* cannot be sensitized by a single path.

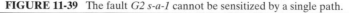

Algorithms exist for automatically and efficiently finding a small set of test patterns that will detect most, if not all, of the faults in a combinational circuit. The D-algorithm [7], which uses the D-notation discussed above, is widely used. It was developed with a calculus that tests for stuck faults at all signal lines embedded in a combinational circuit. The D-algorithm has been incorporated in software for automatic test pattern generation (ATPG). It uses multipath sensitization and is guaranteed to find a test for a logical fault in a combinational circuit if a test exists, but its efficiency degrades if the circuit contains a large number of exclusive-or gates.

Two alternative algorithms, PODEM (path-oriented decision making) [8], and FAN (fanout–oriented test-generation algorithm) [9], are more efficient than the D-algorithm. The PODEM algorithm uses forward implication from a circuit's primary inputs to replace the alternating backtracing and forward propagation steps of the D-algorithm. The FAN algorithm uses additional strategies to reduce backtracing and is even more efficient than PODEM. See Abramovici et al. [6] and Fujiwara and Shimono [10] for details of these algorithms.

ATPG seeks a stimulus pattern that will detect a given fault, but can fail to produce a test for a combinational circuit that has reconvergent fanout. Some faults might not be detectable. Untestable faults are due to (1) redundant logic (see Problem 11.14)[13], (2) uncontrollable nets, and/or (c) unobservable nets. A fault cannot be tested if the net on which the fault is located cannot be controlled or if no output paths can be sensitized to observe the fault.

Example 11.25

The test for the fault $w2$ s-a-0 in Figure 11-40 does not exist. The signal $x5$ has reconvergent fanout at y, so the side inputs affected by $x5$ cannot be set independently. Sensitization of $w3$ requires setting $x5$ to 1, but sensitization of y requires setting $x5$ to 0. Table 11-4 shows the signal values in D-notation and shows the conflicts between the values of the primary inputs required to sensitize $w3$ and y. The reconvergent fanout of $x5$ forces y to 0, independently of the fault $w2$ s-a-0. A test cannot distinguish between the faulty circuit and the good circuit.

End of Example 11.25

11.7.5 Fault Coverage and Defect Levels

Testing for faults ensures the quality of the parts that are shipped to a customer. The probability W of shipping a defective part is related to the test coverage T and the fractional manufacturing yield Y by the expression:

$$W = 1 - Y^{(1-T)}$$

---

[13]Although synthesis tools remove redundant logic, a circuit may be modified after synthesis to add redundant logic to improve the speed of the circuit or to cover hazards. Such modifications are problematic for test generation.

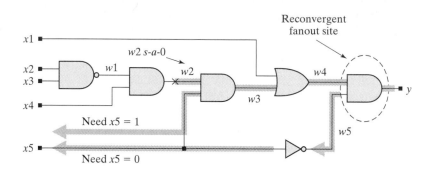

FIGURE 11-40   A test for the fault *w2 s-a-0* does not exist. The reconvergent fanout of *x5* leads to conflicting conditions to test the fault *w2 s-a-0*.

where Y represents the fractional yield of the ASIC manufacturing process that produced the chips (e.g., $Y = 0.75$), and T is the fraction of faults that have been tested (by fault simulation) [10, 11]. Table 11-5 and Figure 11-41 show how the average defect level depends on the coverage of the faults, for a given process yield. It is commonplace for semiconductor vendors to ensure that their fault coverage exceeds 99% of the faults in a circuit.

11.7.6 Test Generation for Sequential Circuits

It can be very difficult to find direct tests for a sequential circuit, because the test might require a long sequence of inputs to drive internal sequential devices to a known state

TABLE 11-4   The reconvergent fanout causes a conflict between the values of *x5* required to sensitize the fault *w2 s-a-0*.

	Justification	Sensitization	Test
x1	x	0	0
x2	0	x	0
x3	x	x	x
x4	1	x	1
x5	x	1 for *w3* 0 for *y*	conflict
w1	1	x	
w2	D	D	
w3	x	D	
w4	x	D	
w5	x	conflict	
y	x	0	

TABLE 11-5 The quantity of undetected defective parts depends on the process maturity and the test vector coverage. Less mature technologies require significantly higher fault coverage to reduce the average defect level.

	Process Yield	Test Vector Coverage		
		70%	90%	99%
		% Defects Undetected		
Advanced Technology	10%	50%	21%	2%
Maturing Technology	50%	19%	7%	0.7%
Mature Technology	90%	3%	1%	0.1%

that justifies a fault site and/or sensitizes a path. It is impractical to verify that two sequential circuits have the same functionality (i.e., are equivalent, by applying input sequences and observing output sequences). An alternative is to treat a sequential circuit as an iterative network.

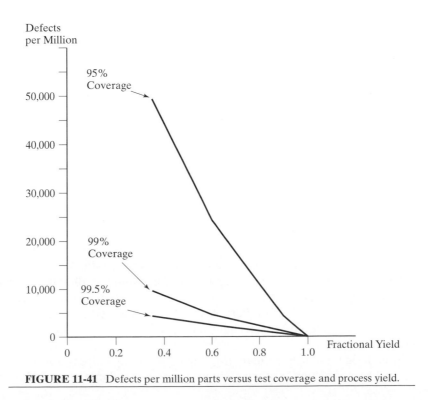

FIGURE 11-41 Defects per million parts versus test coverage and process yield.

Example 11.26

The circuit in Figure 11-42 requires a sequence of three stimulus patterns to detect the fault *w1 s-a-1*. The patterns are developed in reverse order, beginning with the values that the nets must have in cycle 3, the cycle in which the effect of the fault is to be observed at *y*, and working backward in time to determine the values that must exist in

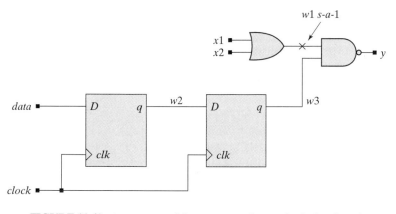

FIGURE 11-42   A sequence of three patterns detects the fault *w1 s-a-1*.

cycle 2, etc. The effect of the fault can be observed in the last cycle. Table 11-6 shows the values of the signals in each cycle of the test.

End of Example 11.26

     Because direct test generation for a sequential circuit can be very difficult, it is usually avoided in favor of scan-path methods. The electronics industry uses scan methods to modify a circuit and make it testable by methods that apply to combinational circuits.

     There are various options to the scan-path method, depending on the extent to which the registers in the machine are linked in a scan chain. An ASIC implements a partial- or a full-scan method depending on whether some or all of its internal flip-flops are replaced by scan cells and connected to form one or more shift registers that are controlled by an external tester. Partial scan is implemented as a tradeoff between the fault coverage provided by a scan chain and the additional logic required to implement the chain.

     Scan design replaces ordinary flip-flops by scan flip-flops to form dual-port registers, called scan registers, making a circuit more controllable and/or observable. Scan

TABLE 11-6 Test-pattern sequence for testing the fault *w1 s-a-1* in Figure 11.42. The test pattern must propagate a 1 through the shift register to sensitize the output *y* to the fault.

		Justification	Sensitization	Test
	w1	D'	D'	
	w2	*x*	*x*	
	w3	*x*	1	
Cycle 3	*x1*	0	*x*	0
	x2	0	*x*	0
	data	*x*	*x*	*x*
	y	D	D	D
	w1	*x*	*x*	
	w2	*x*	1	
	w3	*x*	*x*	
Cycle 2	*x1*	*x*	*x*	*x*
	x2	*x*	*x*	*x*
	data	*x*	*x*	*x*
	y	*x*	*x*	*x*
	w1	*x*	*x*	
	w2	*x*	*x*	
	w3	*x*	*x*	
Cycle 1	*x1*	*x*	*x*	*x*
	x2	*x*	*x*	*x*
	data	*x*	1	1
	y	*x*	*x*	*x*

registers can shift data through a serial port and can load data through a parallel port. In test mode, logic values of a test pattern are shifted into the flip-flops. The values loaded into the flip-flops drive combinational logic paths in one clock cycle, and the logic values at the destinations of those paths can be captured in parallel in the next clock cycle. The captured data can be shifted out of the register and analyzed to detect internal faults in the logic.

Figure 11-43 shows a set of dual-port scan cells connected to form a 4-bit-wide scan register. In normal operation, with $T = 0$, data is loaded in parallel from $D[3:0]$. In test mode ($T = 1$), data is shifted through the register from *x_in3_scan_in* to *y3_scan_out*. Depending on the application and the extent to which scan cells are used, the circuit can be designed to have 100% controllability and observability of its internal nodes via the serial scan path.

Full scan replaces all of the flip-flops in a design by scan cells; boundary scan places scan cells at the I/O ports of an ASIC and links them together to form a boundary scan chain used for board-level testing. The test patterns for an ASIC core with full scan are those used to test the combinational logic of the core, because the restructured circuit has the configuration shown in Figure 11-44. We will revisit scan methods in Section 11.10.6.

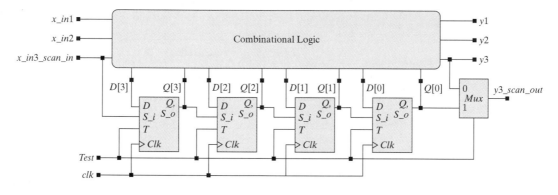

FIGURE 11-43   A scan register is formed with dual-port register cells.

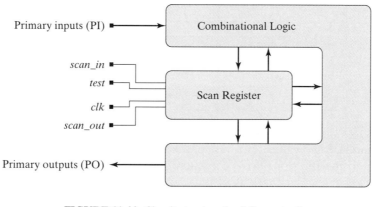

FIGURE 11-44   Circuit structure for full-scan testing.

The test procedure for a circuit with full scan tests the scan circuit, then tests the combinational logic. The scan path is tested by placing the circuit in test mode ($T = 1$) and toggling the clock $n + 1$ times to propagate a test sequence through the chain. Placing the circuit in test mode, shifting a test pattern into the scan register, and applying the primary inputs prepares the combinational logic for testing. With the test pattern in the scan register, the mode is set to normal ($T = 0$), and the response of the circuit is observed at the primary outputs and at the inputs to the scan register. The clock is again toggled to latch the parallel inputs to the scan register. Then the circuit is placed in test mode again, and the captured pattern is shifted out of the register for analysis. At the same time, another pattern is shifted into the register.

11.8   Fault Simulation

Fault simulation compares the behavior of a circuit that has a fault with the behavior of a circuit that is fault-free. If the outputs of the two circuits are different under the application of a stimulus pattern, the pattern is said to be a test for the presence of the fault. Fault simulation produces a measure of how well a given set of stimulus patterns (test patterns) detects faults. We will consider only tests that detect faults in combinational logic with a single stuck fault. *Fault coverage*, the degree to which a set of test patterns detects the possible faults in a circuit is defined below:

Fault coverage = Number of Detected Faults/Total Faults

It is important that a set of tests provide a high level of coverage in order to ensure that bad devices are not shipped to customers. *Fault grading*[14] determines the coverage provided by a set of test patterns by checking whether a fault is detectable and possibly finding a test for the fault. In many applications the coverage must exceed 99.5% of the single stuck faults that a circuit could have.

Fault simulators exercise a circuit by cataloging the fault sites, injecting faults, applying stimulus patterns, and comparing the output of the circuit to a fault-free circuit. Patterns that do not reveal the presence of a fault are not useful in testing. Fault simulation is done in conjunction with test generation to evaluate stimulus patterns. Fault simulation guides construction of a set of test patterns that will achieve a high level of coverage of the faults in a circuit.

11.8.1 Fault Collapsing

The efficiency of testing has an impact on the amortization of expensive testers. Needless testing wastes tester resources and time, so it is important to test only the faults that need to be tested and to find test patterns that reveal as many faults as possible. Efficient fault simulators use *fault collapsing* to form equivalence classes of faults that are detectable by the same test. Faults that are detected by the same test are indistinguishable and are called *equivalent faults*. Fault simulators test only one fault in an equivalence class and eliminate the needless simulation of the other faults in the class. The same test detects any fault in the class, so only one fault in the class needs to be tested.

Example 11.27

The fault *x1 s-a-0* in Figure 11-45 cannot be distinguished from the *fault y s-a-1*, so the test that detects *x1 s-a-0* will also detect *y s-a-1*. The faults are members of the same equivalence class of faults.

---

[14]Also called *fault coverage analysis*.

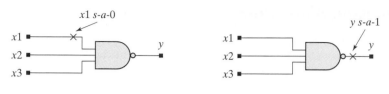

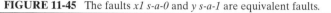

FIGURE 11-45 The faults *x1 s-a-0* and *y s-a-1* are equivalent faults.

End of Example 11.27

There are three major approaches to fault simulation: serial, parallel and concurrent fault simulation [5]. Serial fault simulation is the slowest of the three approaches, but it is the simplest to understand and implement.

11.8.2 Serial Fault Simulation

Serial fault simulation considers a circuit's faults, one fault at a time, and executes the following sequential steps to determine whether an applied stimulus pattern reveals the fault:

(1) Create a list of fault sites;
(2) Inject a fault into the circuit and remove it from the list of faults;
(3) While (the list of fault sites is not empty) {
    Apply a pattern and simulate the good machine
     and the machine with the injected fault;
    Compare the good machine's output to that of the machine with the fault;
    if (there is a difference between the output of the machines)
     the pattern detected the fault;
    else
     the pattern did not detect the fault;
    inject another fault and remove it from the list of faults;
}

11.8.3 Parallel Fault Simulation

Parallel fault simulation simultaneously simulates multiple copies of a machine (circuit), each with its own distinct injected fault and compares their responses to that of the good machine. Although it is faster than serial fault simulation, parallel fault simulation requires more memory and efficient memory management techniques to accommodate the multiple copies of the circuit.

11.8.4 Concurrent Fault Simulation

Concurrent fault simulation is the most widely used algorithm. It limits the scope of fault simulation to the portion of a circuit that is relevant to backward justification (fanin) and forward sensitization (fanout) from a fault site. Concurrent fault simulation requires sophisticated topological analysis of a circuit to limit the scope of the searches that are used to justify and sensitize faults. The payoff is that a concurrent fault simulator is faster than serial or parallel fault simulators.

11.8.5 Probabilistic Fault Simulation

Probabilistic fault simulation identifies test vectors that have high toggle coverage and uses them as the basis for test vectors to detect faults. There is a high degree of correlation between a test pattern that toggles a high number of nodes in a circuit and test patterns that detect a high number of faults [4]. Toggle tests are simpler to perform and faster than other methods of fault detection.

11.9 Fault Simulation with Verifault-XL

Verifault-XL,[15] a concurrent fault simulator widely used by industry, will be discussed here to prepare the reader for the problems at the end of this chapter. Verifault-XL (1) simulates faults at the gate and switch level, (2) propagates the effect of faults through behavioral models, (3) performs fault collapsing to improve performance, (4) detects faults, and (5) generates reports. The tool accepts a Verilog netlist as its input, so the same model is used for fault simulation and design verification. Verifault-XL recognizes models written in the Verilog hardware description language (HDL), and recognizes additional tool-specific tasks that augment the language for use in fault simulation.

With Verifault-XL the user can selectively simulate a given fault, although the default set of faults is the entire set of faults in the circuit described by the netlist on which Verifault-XL operates. The tool automatically forms equivalence classes of faults and collapses to a single *prime fault* the set of faults in a class. Once the set of faults has been partitioned into equivalence classes, the task of fault simulation can be distributed across a farm of processors. At a given processor, prime faults are injected into the circuit and simulated concurrently.

11.9.1 Tasks for Fault Simulation

The Verifault-XL simulator recognizes the tasks listed in Table 11-7. These tasks are not part of the Verilog language standard, and are ignored by the Verilog-XL simulator, but might have to be commented out of the model if it is to be simulated by another vendor's simulator. The task *$fs_add*, without an argument, is the default invocation. It performs fault simulation of the entire circuit. The other options shown in Table 11-7 direct the simulator to examine (a) only the faults in the named module, (b) a specific fault described by its name, and (c) a specific fault in a specific module.

---

[15]See www.cadence.com.

TABLE 11-7 Vendor-dependent tasks for fault simulation with Verifault-XL.

Verifault Task	Simulation Scope
$fs_add;	Entire circuit (default)
$fs_add (module_instance_name);	Group of faults represented by module
$fs_add ("fault_descriptor");	Specific fault
$fs_add (module_instance_name, "fault_descriptor");	Module-specific fault

These options give the user flexibility to direct the effort of the simulator to a narrower scope of activity than the default task invocation.

Example 11.28

A simulation of all of the faults in only module *M1* in *Some_Circuit* in Figure 11-46 is executed by the task *$fs_add (Some_Circuit.M1)*; simulation of only an *s-a-1* fault at input pin 3 of *G4* in module *M1* in *Some_Circuit* is executed by the task *$fs_add ("fault interminal sa1 Some_Circuit.M1.G4.3")*;

End of Example 11.28

11.9.2 Fault Collapsing and Classification with Verifault-XL

Fault simulators eliminate faults that produce redundant results (i.e., faults that cannot be distinguished by a test pattern). The simulator selects a prime fault from each

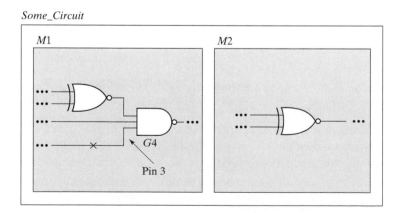

FIGURE 11-46 A fault simulator can simulate a fault at any location in the design hierarchy.

TABLE 11-8 Fault classification is based on a comparison of the output of the good machine and the faulty machine.

		Faulty Machine				
		0	1	Z	L	H	X
	0	U	D	P	P	P	P
	1	D	U	P	P	P	P
Good Machine	Z	U	U	U	U	U	U
	L	U	U	U	U	U	U
	H	U	U	U	U	U	U
	X	U	U	U	U	U	U

D: Detected
P: Potentially detected
U: Undetected
L: 0 or Z
H: 1 or Z
Z: High impedance
X: Unknown

equivalence class of faults. The *prime fault set* is the minimal set of faults that represents the entire set. The test that detects a prime fault also detects every other fault in its equivalence class.

Verifault-XL identifies and removes all redundant faults, and forms three classes of faults: detected, undetected, and potentially detected. Detected faults and potentially detected faults are removed from the fault list and are not tested by subsequent tests.[16] This classification of faults is based on Table 11-8, which classifies faults by comparing the output of the good machine and the faulty machine, with special symbols defined for fault simulation. A fault is detected by a test if the output of the fault-free machine and the machine with an injected fault do not match assertions of 0 and 1.

A fault is *controllable* if the value of the signal at the site of the fault can be toggled by the primary inputs. A fault is *observable* if the fault can be sensitized at a primary output. If a fault is uncontrollable or unobservable it cannot be tested. A fault that is controllable and observable is testable if the primary inputs required for controllability do not conflict with the primary inputs required for observability. Such conflicts may occur if the circuit has reconvergent fanout.

Any net that has been declared to be of type **supply0** or **supply1** cannot be tested, because the net has a fixed logic value and is uncontrollable. Nets that are driven by only **pull-up** or **pull-down** devices cannot be controlled either. Any net that has been declared, but not actually used in the design, is *undriven*, and cannot be controlled by a primary input signal. If a fixed-state net dominates one or more gates in its fanout path, as shown in Figure 11-47, Verifault-XL marks the input and output terminal faults corresponding to the fixed-state net as untestable because they are uncontrollable. The nets at the side inputs to the gate in Figure 11-45 are unobservable.

If a side input to a gate is fixed at a value that dominates the other inputs of the gate, any fault that must propagate through the gate cannot be observed, because the output of the gate is held to a fixed value by the dominating input. The input terminals

---

[16]The simulator option *drop_detected_count* = 0 overrides this default behavior.

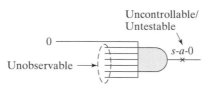

Uncontrollable/
Untestable

FIGURE 11-47 The net at the output of a gate with a dominating side input is untestable because the value of the net cannot be controlled.

of gates that are dominated by fixed-state nets are unobservable. Verifault-XL automatically identifies untestable faults and does not include them in the list of faults that are to be simulated.

Verifault-XL performs back-tracing to determine whether the logic driving a net is observable. There are several conditions under which a net will be found to be unobservable. A net that has no fanout and no strobe point cannot be observed, as shown in Figure 11-48. A net is unobservable if it has no behavioral fanout or bidirectional connections, and if its structural fanouts are unobservable. A fault site is unobservable if it is at the output terminal of a unidirectional gate or a user-defined primitive that drives unobservable nets. If a fault site is an input of a unidirectional gate or a user-defined primitive whose outputs are all unobservable, it is unobservable too. A fault site at a bidirectional net that has no behavioral fanouts, and whose unidirectional structural fanouts are unobservable is also unobservable. Control inputs of bidirectional gates are unobservable when both *inout* terminals connect to unobservable nets.

11.9.3 Structural and Behavioral Fault Propagation

All fault sites in a model are structural. The hardware model of a circuit can be a mix of structural and behavioral models. Gate-level models are structural, but behavioral models are abstract, and have no explicit structural detail. Therefore, a fault simulator does not associate faults with a behavioral model. A fault simulator will propagate the logical effects of a fault through a behavioral model. Structural fault propagation accounts for the fact that fault effects can travel from the site of their origin to other sites.

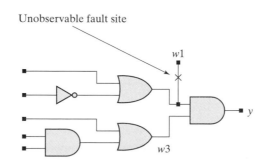

FIGURE 11-48 An internal net with no fanout to a primary output is unobservable.

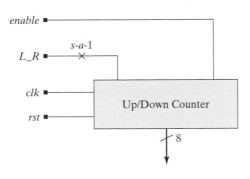

FIGURE 11-49 A fault on the control line of a behavioral model of a counter will cause the counter to count in only one direction.

A fault that is propagated to a behavioral description can affect the value of an expression within the behavior, and/or affect the activity flow within the behavior. For example, if the direction control line *L_R* in Figure 11-49 has the fault *s-a-1*, the counter will count in only one direction, that which results from $L\_R = 1$.

11.9.4 Testbench for Fault Simulation with Verifault-XL.

Fault simulation with Verifault-XL can be implemented with a testbench that includes a single-pass (*initial*) behavior containing Verifault tasks.[17] The tasks must set up the fault model, apply the input pattern, and strobe faults at the outputs. Figure 11-50 shows the general structure of a simple testbench for fault simulation. *$fs_add* specifies the scope of the set of faults that will be simulated.

```
initial                            // Example:
  begin
    $fs_add (all);                 // select the fault set
    $fs_inject;                    // inject faults and perform simulation
    $fs_dictionary;                // create a fault dictionary; defaults to standard output

    #10 // test pattern goes here
    $fs_strobe (list of variables) // compares machines at the named variables

    // more patterns and strobes
    #10 fs_list;                   // generate fault descriptors
  end
```

FIGURE 11-50 A testbench for fault simulation includes a single pass behavior that declares the set of faults to be tested, injects the faults into the circuit, forms a fault dictionary, strobes the outputs of the good and faulty machines, and lists the detected faults.

---

[17]The Cadence Verilog simulator, Verilog-XL, ignores Verifault-XL system tasks.

Example 11.29

If the input pattern $(c\_in\ a\ b) = (1\ 1\ 1)$ is applied to the gate-level model of the binary full adder circuit in Figure 11-51, the effect of the fault *G1 s-a-1* propagates forward to the outputs *sum* and *c_out*. The output at *sum* detects the fault, but *c_out* does not. The circuit is small enough that exhaustive simulation is feasible for the fault *G1 s-a-1*, and Table 11-9 presents the results of applying an exhaustive set of test patterns to the circuit with *G1 s-a-1*.

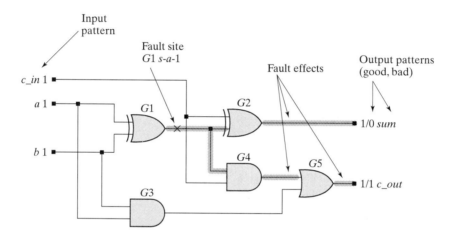

FIGURE 11-51 The test pattern (1 1 1) detects the fault *G1 s-a-1* at the output *sum*, but not at the output *c_out*.

TABLE 11-9 An exhaustive set of test patterns reveals that the pattern (0 0 1) detects the fault *G1 s-a-1* at both *sum* and *c_out*. Shaded entries identify differences between the output of the good machine and the output of the faulty machine.

Inputs			Good Machine		Faulty Machine	
a	*b*	*c_in*	*sum*	*c_out*	*sum*	*c_out*
0	0	0	0	0	1	0
0	0	1	1	0	0	1
0	1	0	1	0	1	0
0	1	1	0	1	0	1
1	0	0	1	0	1	0
1	0	1	0	1	0	1
1	1	0	0	1	1	1
1	1	1	1	1	0	1

```
module Add_full (sum, c_out, a, b, c_in);
output  sum, c_out;
 input  a, b, c_in;
 xor    G1      (w1, a, b);
 xor    G2      (sum, w1, c_in);
 and    G3      (w2, a, b);
 and    G4      (w3, w1, c_in);
 or     G5      (c_out, w3, w2);
endmodule

module test_Add_full (sum, c_out, a, b, c_in);
 reg a, b, c_in;

 Add_full M1 (sum, c_out, a, b, c_in);
 initial begin
  $fs_add (M1);              // Select fault set
  $fs_inject;                // Inject faults
  $fs_dictionary;            // Create dictionary

  #10 {a, b, c_in} = 3'b000;  // Apply pattern #1

  // Compare machines
  #10 $fs_strobe {sum, c_out};
  #10 {a, b, c_in} = 3'b111;  // Apply pattern #2
  #10 $fs_strobe {sum, c_out};
  // Generate fault descriptors
  #10 $fs_list ("list_status");
 end
end module
```

FIGURE 11-52 Full adder and testbench for fault simulation.

A fault simulator will be used to identify tests for the faults in the full-adder circuit. Figure 11-52 shows the gate-level model of the full adder, a testbench that includes tasks for fault simulation with Verifault-XL, and two stimulus patterns.

End of Example 11.29

Table 11-10 summarizes the results of running the fault simulator (Verifault-XL) on the full-adder circuit with a test consisting of the two vectors $(a\,b\,c\_in) = (0\ 0\ 0)$ and $(a\,b\,c\_in) = (1\ 1\ 1)$. Just two vectors detect 73% of the prime faults in the circuit. For small circuits, the fault simulator can be used iteratively to apply a small set of patterns, determine the coverage, examine the undetected faults, and generate more patterns, until the desired coverage is reached. The fault simulator can also be used to confirm the results of using an ATPG.

11.9.5 Fault Descriptors

Fault descriptors can be used to select individual faults to add to, or delete from, the circuit being simulated. The single-pass behavior in Figure 11-53 uses the Verifault task

TABLE 11-10   Results of fault simulation of a binary
full adder using two test patterns.

	Total no.	Total %	Prime no.	Prime %
Untestable	0		0	
Drop_detected	22	73.3	19	73.1
Detected	0	0.0	0	0.0
Potential	0	0.0	0	0.0
Undetected	8	26.7	7	26.9
All	30		26	

```
initial begin
  $fs_add (test_Add_full.M1)'
  $fs_add ("fault interminal sa 1
    test_Add_full.G1.2;");
end

initial begin
  $fs_add; // selects all faults
  $fs_remove (test_Add_full.M1);
  $fs_remove (fault interminal sa1
    test_Add_full.G1.2;");
end

initial begin $fs_read ("fault_list"); end
```

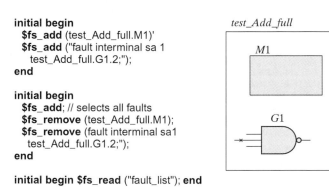

FIGURE 11-53   A hierarchical path name can be used to specify a fault at any location in the design
hierarchy.

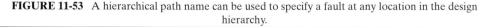

$fs_add to specify that the fault list should include all of the faults in the module instance denoted by *M1* and the specific fault in which input pin 2 is *s-a-1* on gate *G1*.

The fault simulator produces the list of faults shown in Figure 11-54 as output from simulating the binary full adder.

Example 11.30

Figure 11-55 shows the complete list of faults that can be tested in the full adder circuit. *Note:* If a primitive does not have a user-defined name, Verifault-XL will create a name from the primitive type followed by *$pin_number* (e.g., *nor$.2*).

End of Example 11.30

As the simulator runs it composes a *fault dictionary* consisting of all of the tested faults and assigns a unique numerical identifier (*faultid*) to each fault.

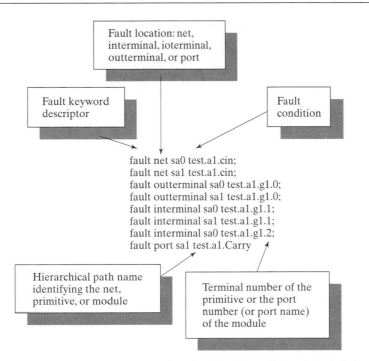

FIGURE 11-54 The fault simulator reports a fault, its location, the condition of the fault (*s-a-0* or *s-a-1*), the hierarchical path name, and the terminal (pin) number of the primitive or port number of the module that drives the net.

Example 11.31

Figure 11-56 lists the fault dictionary for the full adder circuit.

End of Example 11.31

11.10 JTAG[18] Ports and Design for Testability

Design for testability (DFT) ensures that manufactured circuits can be tested for defects. DFT usually requires that a circuit is designed/modified to support testing, because the relatively small number of available I/O pins on a typical chip is inadequate for testing the internal nodes. There are methods for evaluating the testability of a chip by measuring the difficulty of controlling and observing internal nodes [12]. There are also several approaches to improving the testability of a circuit [6], but we will focus on

---

[18]The JTAG port is named after the Joint Test Action Group, a team of industry experts who defined the IEEE 1149.1 and 1149.1a standards.

```
fault net sa0 test_Add_full.M1.a;
fault net sa1 test_Add_full.M1.a;
fault net sa0 test_Add_full.M1.b;
fault net sa1 test_Add_full.M1.b;
fault net sa0 test_Add_full.M1.c_in;
fault net sa1 test_Add_full.M1.c_in;
fault outterminal sa0 test_Add_full.M1.G1.0;
fault outterminal sa1 test_Add_full.M1.G1.0;
fault interminal sa0 test_Add_full.M1.G1.1;
fault interminal sa1 test_Add_full.M1.G1.1;
fault interminal sa0 test_Add_full.M1.G1.2;
fault interminal sa1 test_Add_full.M1.G1.2;
fault outterminal sa0 test_Add_full.M1.G2.0;
fault outterminal sa1 test_Add_full.M1.G2.0;
fault interminal sa0 test_Add_full.M1.G2.1;
fault interminal sa1 test_Add_full.M1.G2.1;
fault interminal sa0 test_Add_full.M1.G2.2;
fault interminal sa1 test_Add_full.M1.G2.2;
fault outterminal sa0 test_Add_full.M1.G3.0;
fault outterminal sa1 test_Add_full.M1.G3.0;
fault interminal sa0 test_Add_full.M1.G3.1;
fault interminal sa1 test_Add_full.M1.G3.2;
fault outterminal sa0 test_Add_full.M1.G4.0;
fault outterminal sa1 test_Add_full.M1.G4.0;
fault interminal sa1 test_Add_full.M1.G4.1;
fault interminal sa1 test_Add_full.M1.G4.2;
fault outterminal sa0 test_Add_full.M1.G5.0;
fault outterminal sa1 test_Add_full.M1.G5.0;
fault interminal sa0 test_Add_full.M1.G5.1;
fault interminal sa0 test_Add_full.M1.G5.2;
```

FIGURE 11-55 Fault list for the binary full adder.

scan-based methods, which extend the concept introduced in Section 11.7.6 for testing sequential circuits. We do so because scan-based approaches have become widespread and important. They not only support testing of circuits for defects, but also support debugging of embedded processors during software development, and support the field programmability of complex programmable logic devices (CPLDs) and FPGAs.

There are several practical problems in chip and board-level testing. (1) Sequential machines are difficult to test because they require a sequence of test patterns. (2) The internal nodes of large circuits cannot be observed at output pins, and may not be controlled easily by the available input pins. (3) The manufacturing process for printed circuit boards forms copper traces for signal paths. The circuit has a defect if traces are shorted or open. (4) An ASIC chip might not have a good bond between the board and a pin of the ASIC or between the pin and the core logic. (5) The core logic of a chip that is mounted on a board might have to be tested in the field, without removing the chip from the unit. (6) It might be necessary to isolate the location of a fault to a particular ASIC or module to reduce the cost of repairing a unit. The electronics industry has circumvented these problems by adopting a standard circuit interface that uses scan chains for board-level and chip-level testing.

```
fault net sa0 test_Add_full.M1.a 'faultid=1';
fault net sa1 test_Add_full.M1.a 'faultid=2';
fault net sa0 test_Add_full.M1.b 'faultid=3';
fault net sa1 test_Add_full.M1.b 'faultid=4';
fault net sa0 test_Add_full.M1.c_in 'faultid=5';
fault net sa1 test_Add_full.M1.c_in 'faultid=6';
fault outterminal sa0 test_Add_full.M1.G1.0 'faultid=7';
fault outterminal sa1 test_Add_full.M1.G1.0 'faultid=8';
fault interminal sa0 test_Add_full.M1.G1.1 'faultid=9';
fault interminal sa1 test_Add_full.M1.G1.1 'faultid=10';
fault interminal sa0 test_Add_full.M1.G1.2 'faultid=11';
fault interminal sa1 test_Add_full.M1.G1.2 'faultid=12';
fault outterminal sa0 test_Add_full.M1.G2.0 'faultid=13';
fault outterminal sa1 test_Add_full.M1.G2.0 'faultid=14';
fault interminal sa0 test_Add_full.M1.G2.1 'faultid=15';
fault interminal sa1 test_Add_full.M1.G2.1 'faultid=16';
fault interminal sa0 test_Add_full.M1.G2.2 'faultid=17';
fault interminal sa1 test_Add_full.M1.G2.2 'faultid=18';
fault outterminal sa0 test_Add_full.M1.G3.0 'faultid=19';      'equiv=29' ;
fault outterminal sa1 test_Add_full.M1.G3.0 'faultid=20';      'equiv=28' ;
fault interminal sa1 test_Add_full.M1.G3.1 'faultid=21';
fault interminal sa1 test_Add_full.M1.G3.2 'faultid=22';
fault outterminal sa0 test_Add_full.M1.G4.0 'faultid=23';      'equiv=30' ;
fault outterminal sa1 test_Add_full.M1.G4.0 'faultid=24';      'equiv=28' ;
fault interminal sa1 test_Add_full.M1.G4.1 'faultid=25';
fault interminal sa1 test_Add_full.M1.G4.2 'faultid=26';
fault outterminal sa0 test_Add_full.M1.G5.0 'faultid=27';
fault outterminal sa1 test_Add_full.M1.G5.0 'faultid=28';
fault interminal sa0 test_Add_full.M1.G5.1 'faultid=29';
fault interminal sa0 test_Add_full.M1.G5.2 'faultid=30';
```

FIGURE 11-56 Fault dictionary.

11.10.1 Boundary Scan and JTAG Ports

Boundary scan is an extension of the scan register concept discussed in Section 11.7.6 for testing sequential circuits. A *boundary scan chain* is added to the netlist of an ASIC[19] by inserting *boundary scan cells* (BSCs) at its I/O pins and linking them to form a shift register around the chip. The same cell can also be used to replace the flips-flops within the core logic and to form internal scan paths consisting of one or more test-data registers linked together in a serial connection. When used internally the cells are referred to as *data register* (DR) cells.

A typical BSC, or data register cell, is shown in Figure 11-57. The cell is designed to allow data to be scanned through the chip without affecting the normal operation of the chip (e.g., in on-line monitoring of the chip's operation). Two muxes control the datapaths of the cell. The input mux determines whether the capture/scan flip-flop is connected to *data_in* or to the serial input, *scan_in*. The output mux determines whether *data_in* or the output flip-flop is connected to *data_out*.

With *mode* = 0 the cell is in normal mode, and *data_in* is passed through the output multiplexer to *data_out* and to the capture/scan flip-flop, where it can be loaded by a pulse of *clockDR*. The capture/scan flip-flops support a boundary scan chain; the output

---

[19]We will discuss scan chains for ASICs, but they are also used in FPGAs and other devices.

register flip-flops hold their data while new data are scanned into the capture/scan flip-flop. *data_in* and *data_out* of a BSC are connected to the inputs and outputs of the core logic of the ASIC. In test mode, a pattern of data can be shifted into the capture/scan flip-flops under the control of *clockDR*. When the scan chain holds a desired pattern, the data in the capture/scan register can be loaded in parallel by toggling *updateDR* to update the *output register* flip-flops.

If a BSC resister cell is connected to an input pin of the chip, *data_in* is connected to the input pad of the chip, and *data_out* is connected to the input pad of the ASIC core's logic. If the cell is used as an output, the core logic of the ASIC is connected to *data_in*, and *data_out* is connected to the output pin of the ASIC. A Verilog model of the BSC cell is given below.

```
module BSC_Cell
    (data_out, scan_out, data_in, mode, scan_in, shiftDR, updateDR, clockDR);
    output       data_out;
    output       scan_out;
    input        data_in;
    input        mode, scan_in, shiftDR, updateDR, clockDR;
    reg          scan_out, update_reg;

    always @ (posedge clockDR) begin
      scan_out <= shiftDR ? scan_in : data_in;
    end

    always @ (posedge updateDR) update_reg <= scan_out;
    assign data_out = mode ? update_reg : data_in;
endmodule
```

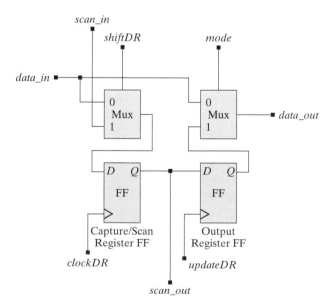

FIGURE 11-57   The DR cells used to implement boundary scan test registers and test-data registers include a capture/scan flip-flop and an output flip-flop.

11.10.2 JTAG Modes of Operation

The modes of operation of a boundary scan cell (data register cell) are summarized in Table 11-11. In *normal mode* (i.e., with *mode* = 0) data pass directly through the cell, from *data_in* to *data_out*. The propagation delay of the multiplexer driving *data_out* adds a slight delay to the signal path. In *test mode*, with *mode* = 1, cell *data_out* is driven by the output register.

In *scan mode* the boundary scan cells are connected as a shift register, with *scan_out* from one cell connected to *scan_in* of the next cell in the chain. A test pattern can be shifted into the register and then loaded into the *output register* to establish a logic value at *data_out* for the purpose of testing the core logic. With *shiftDR* = 1, data are shifted through the capture/scan register flip-flops on the rising edges of *clockDR*, from *scan_in* to *scan_out*.

The *capture mode* captures data from the ASIC without interfering with its operation. The data can be scanned out later, while the chip is operating. This mode is operational with *shiftDR* = 0, which connects the scan path to the capture/scan flip-flop. A subsequent clock pulse of *clockDR* loads *data_in* into the scan register. In this mode, *data_out* can be driven by *data_in* (*mode* = 0) or by the output flip-flop (*mode* = 1).

The *update mode* drives *data_out* by the contents of the output register, with *mode* = 1. The output register is loaded with the content of the scan register by applying a pulse of *updateDR*. If *data_out* is attached to the input pins of the ASIC, the pattern could be a test that is to be applied to the chip. The response of the chip can be captured by a pulse of *clockDR* while *shiftDR* = 0.

Boundary scan methods can test multiple chips on a PC board, the board traces between chips, and the connections between the pins of a chip and its core logic. A tester can isolate and test the core logic of ASICs that are equipped with boundary scan circuitry and a special test access port (TAP), also called a JTAG port. The TAP allows devices on a board to be linked together and tested in-place. A finite-state machine called a *TAP controller* controls a TAP. The JTAG standards IEEE 1149.1 [13] and 1149.1a specify the implementation of a TAP. The JTAG port of a chip can be daisy chained to the JTAG port of another chip, so that a scan chain can connect all of the chips on a board. If the chips on a board are equipped with a JTAG port and a boundary scan chain, a tester can detect shorts and opens in board traces, between the chip's I/O

TABLE 11-11 Modes of operation of a BSC.

Mode	Operation
Normal	With *mode* = 0, *data_in* is connected directly to *data_out*, and the scan chain does not affect the operation of the ASIC.
Scan	With *shiftDR* = 1, data enter through *scan_in* and leave through *scan_out*, at the active edge of *clockDR*.
Capture	With *shiftDR* = 0, *data_in* is loaded into the capture register at the active edge of *clockDR*.
Update	With *mode* = 1, the output of the capture register is shifted to the update register at the active edge of *updateDR*.

pins and the board, and between the ASIC core and its pad frame. An external tester can use the JTAG port to detect internal faults of the ASIC.

The JTAG port has assumed a much larger role than testing ASICs and printed circuit (PC) boards for production defects. The port is used to program configurable PLDs [14] and FPGAs.[20] It is also used to develop and debug software for embedded processors by controlling the processor and providing access to its internal registers.[21]

11.10.3 JTAG Registers

Each chip implementing the JTAG methodology must include a boundary scan register (formed by linking BSCs), a bypass register, and an instruction register. These mandatory registers can be viewed as having the configuration shown in Figure 11-58. The bypass register holds a single bit. The size of the instruction register and the other data registers can be customized to an application. An optional 32-bit wide device identification register can be used to hold data describing the part number, the manufacturer's name, and other information that can be accessed by an external

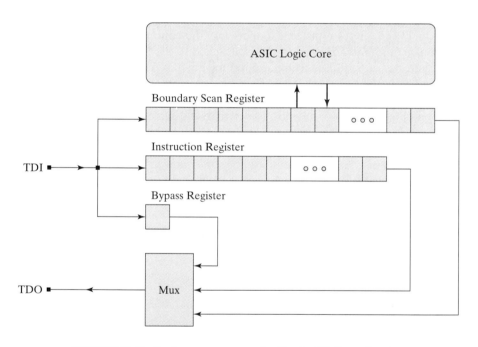

FIGURE 11-58 Register structure required by the JTAG specification.

---

[20]See, for example, www.altera.com and www.xilinx.com.
[21]See www.agilent.com.

tester. The current instruction in the instruction register determines which register is connected between test-data input (TDI) and test-data output (TDO). The actual register can be formed by linking one or more test-data registers (TDRs) of internal scan chains.

The single-bit bypass register has a cell like that shown in Figure 11-59, and the cells of the instruction register have an architecture like that shown in Figure 11-60. The bypass register bypasses an ASIC in the scan chain on a PC board, reducing the length of a test by reducing the number of shifts that must occur before a pattern is located in a given scan register. A Verilog model of a bypass register cell is described below by *BR_Cell*. The signal *shiftDR* gates the scan path, and *clockDR* synchronizes the cell.

The instruction register specifies instructions and controls the internal data-path of the TAP. An instruction defines the serial test-data register path connected between TDI and TDO during scan operations. The cells of the instruction register have asynchronous set/reset, which can be programmed by the parameter *SR_value* to assert either 0 or 1 at the output, depending on the instruction held when the TAP's state machine enters its reset state. The instruction register has serial I/O through *scan_in/scan_out,* and parallel input/output through *data_in/data_out.*

A Verilog model of the cell is given below. Note that a new instruction can be shifted into the scan register while an instruction is held in the output register. The signal *shiftIR* selects the input datapath of the cell, which can be the serial path connected to *scan_in,* or a parallel path connected to *data_in.* The latter provides a means of including data (e.g., status bits) from the ASIC in an instruction. The signal *reset_bar* is generated synchronously by the TAP controller when it enters its reset state; *nTRST* is an optional, asynchronous, active-low fifth input to the TAP.

```
module BR_Cell (scan_out, scan_in, shiftDR, clockDR);
  output scan_out;
  input   scan_in, shiftDR, clockDR;
  reg     scan_out;

  always @ (posedge clockDR) scan_out <= scan_in & shiftDR;

endmodule
```

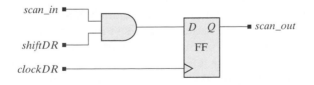

FIGURE 11-59 The bypass register (BR) cell for boundary scan.

```
module IR_Cell
(data_out, scan_out, data_in, scan_in, shiftIR, reset_bar, nTRST, clockIR, updateIR);

output          data_out, scan_out;
input           data_in, scan_in, shiftIR, reset_bar, nTRST, clockIR, updateIR;
reg             data_out, scan_out;
parameter       SR_value = 0;

wire            S_R = reset_bar & nTRST;

always @ (posedge clockIR) scan_out <= shiftIR ? scan_in: data_in;
always @ (posedge updateIR or negedge S_R)
  if (S_R == 0) data_out <= SR_value; else data_out <= scan_out;

endmodule
```

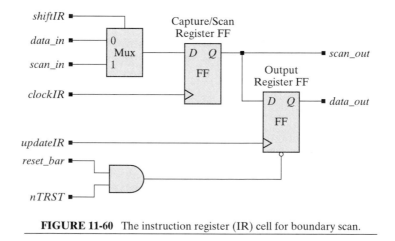

FIGURE 11-60 The instruction register (IR) cell for boundary scan.

11.10.4 JTAG Instructions

The mandatory instructions specified by the JTAG standard are summarized in Table 11-12. The BYPASS instruction scans data from TDI to TDO through a 1-bit bypass register, rather than through the entire boundary scan chain. This bypasses a chip that is not being tested and shortens the scan chain needed to test other components.

The EXTEST (external test) instruction is used to test the interconnect that is external to the chip. A pattern is scanned into the capture/scan register, then the data are loaded (in parallel) into the output register of the boundary scan chain. When the chip is placed in the test mode the pattern appears at the output pins of the chip and drives the interconnect to other chips. Signal values from other chips can be captured and scanned out for analysis of the integrity of the interconnect structure.

The INTEST (internal test) instruction is used to isolate and test the internal circuits of individual components on a board. A pattern is scanned into the capture register, then the data are loaded in parallel into the output register of the boundary scan

TABLE 11-12 Instructions specified by IEEE standard 1149.1.

Instruction	Action
BYPASS	Shifts data through a single-cell bypass register, bypassing the ASIC's boundary scan register, reducing the length of the scan path needed to test other components.
EXTEST	Drives known values onto the output pins of the ASIC for testing board-level interconnect and logic external to the ASIC.
SAMPLE/PRELOAD	SAMPLE captures the data values present at the system pins and loads (in parallel) the data into the capture register flip-flops. PRELOAD places a test data pattern into the output register.
INTEST*	Applies a test pattern to the ASIC logic and captures the response from the logic. Connects only the boundary scan register between TDI and TDO.
RUNBIST*	The host ASIC can execute a self-test while the TAP controller is in the state *S_Run_Idle*.
IDCODE*	Shifts out the data in the IDCODE register (device identification register), providing the tester with the device manufacturer's name, part number, and other data. The instruction defaults to the BYPASS register if there is no IDCODE register in the TAP.
*denotes an optional instruction.	

chain. When the chip is placed in the test mode the cells of the output register that are connected to input ports of the ASIC exercise the logic of the chip, forming outputs that can be captured at the cells of the capture/scan register and then scanned out for analysis. While a pattern is being scanned out another pattern can be scanned into the register.

The SAMPLE/PRELOAD instruction captures data from the I/O pins of the ASIC without interfering with normal operation. The captured data can be scanned out for analysis of the operation of the chip.

The codes for the instructions implemented by the TAP are partially specified by the JTAG standard. The code for the BYPASS instruction is required to be all 1s. The code for the EXTEST instruction is all 0s. The TAP may also include optional test-data registers for implementing internal scan and other tests. Each test-data register corresponds to an internal scan chain that can be exercised by an external tester under the control of the TAP and the instruction register.

11.10.5 TAP Architecture

A TAP has the architecture show in Figure 11-61. The *TDI, TMS,* and *nTRST* inputs have pullups to conform to the JTAG requirement that if, for example, the *TDI* input is disconnected the " ... undriven input produces a response identical to the application of a logic 1." This has implications for the behavior of the TAP controller state machine, which will be discussed later.

An ASIC or other device with a JTAG port requires a test bus of four dedicated input pins (*TDI, TDO, TMS,* and *TCK*) to support boundary scan and internal

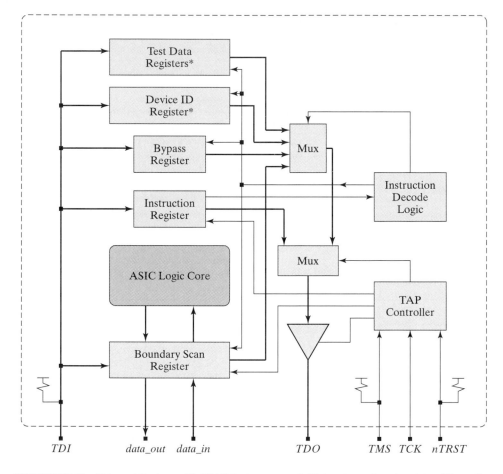

FIGURE 11-61 Chip architecture with JTAG test access port. (* denotes optional registers. The active-
low input *nTRST* is also optional.)

testing.[22] The *TDI* and *TDO* pins of the TAP connect to the first and last cells in the
boundary scan register chain and serve as an interface to the chip. The *test-data input*
(*TDI*) pin serves as input for test patterns that are applied serially to the port; a *test-
data output* (*TDO*) pin serves as a serial output port. The mode of operation of the
TAP is controlled by the *test mode select* (*TMS*) input. A master clock is applied at the
test clock (*TCK*) input pin for testing. A *PC board* implementing the JTAG architecture requires four extra pins to accommodate the *TDI, TDO, TMS*, and *TCK* signals,
and possibly one more pin for *nTRST,* as shown in Figure 11-62(a). Each ASIC acts as
a bus slave; an external agent serves as the bus master, and uses *TMS* and *TCK* to control the slave devices.

---

[22]An optional fifth pin may be used to apply an asynchronous, active-low test–reset input signal (*nTRST*) to
reset the TAP controller. Like *TMS* and *TDI, nTRST* must be attached to a pull-up device.

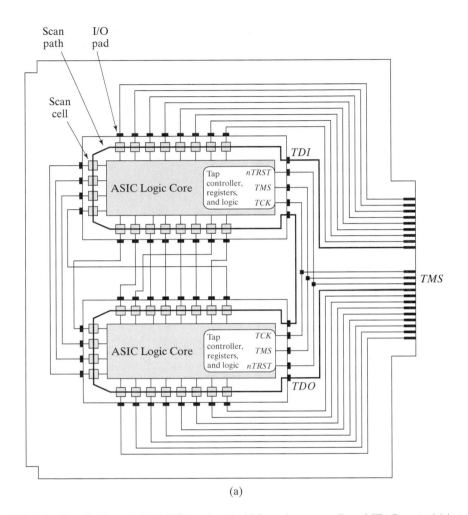

(a)

FIGURE 11-62 A PC board with ASICs equipped with boundary scan cells and JTAG ports: (a) in test mode the chips on the board are daisy-chained in a ring configuration for *TMS*, and (b) in test mode the chips on the board are daisy-chained in a star configuration for *TMS*.

The test access port of each ASIC on the PC board includes a TAP controller, a state machine to which the four pins dedicated to JTAG are attached. The *TMS* input controls the state transitions of the TAP controller, with each transition occurring on the positive edges of *TCK*.[23] Signals generated by the TAP controller drive

___

[23]Required by the JTAG specification.

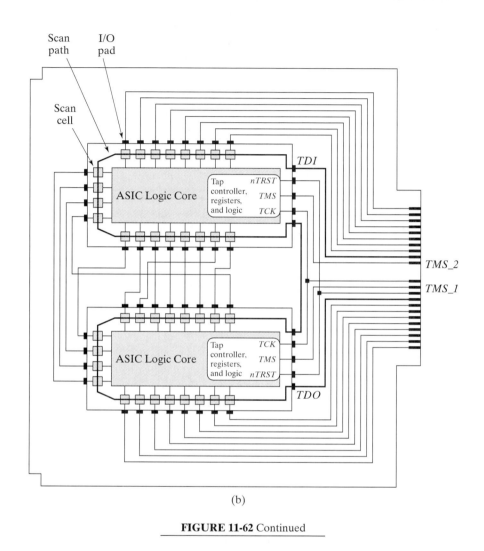

(b)

FIGURE 11-62 Continued

the *shiftDR, mode, clockDR, updateDR, shiftIR, clockIR,* and *updateIR* inputs of the register cells. Figure 11-62(a) shows a PC board with two ASICs with boundary scans chain connected in a ring configuration [6]. For simplicity, the TAP control signals are not shown. In a ring configuration, each chip is driven by the same TAP signals. In a more general configuration known as a star (see Figure 11-62b), the serial ports of the chips are daisy-chained, but each chip in the chain has its own test mode select (*TMS*) signal. A bus master controls the TAPs by controlling the individual TMS signals, thereby allowing the individual TAP controllers to be controlled independently.

11.10.6 TAP Controller State Machine

The instruction register and the TAP controller state machine control the datapath of the test access port. The ASM chart of the machine is shown in Figure 11-63, with decimal annotation showing the state codes.[24] All state transitions occur on the positive (rising) edge of *TCK*; the actions of the test logic in the ASIC are to occur on either the rising or the falling edge of *TCK* in each state of the controller.

The TAP controller's ASM chart is nearly symmetric, with one path controlling the activity of the data register, and the other path controlling the activity of the instruction register of the TAP. If the state of the machine is in *S_Run_Idle* and *TMS* is asserted, the state will return to *S_Reset* if *TMS* is held asserted for two clock cycles, after moving through *S_Select_DR*, and *S_Select_IR*. It will reside in *S_Reset* until *TMS* is de-asserted to cause a transition to *S_Run_Idle*.

Note how alternating values of *TMS* control the activity flow of the TAP controller. That is, the value of *TMS* that causes a transition into *S_Reset, S_Run_Idle, S_Shift_DR, S_Pause_DR, S_Shift_IR,* or *S_Pause_IR* will cause the machine to remain in that state until the value of *TMS* is switched. The states *S_Capture_DR, S_Exit1_DR, S_Exit2_Dr, S_Capture_IR, S_Exit1_IR,* and *S_Exit2_IR* are temporary states. The machine passes through them in one cycle. *S_Capture_Dr* and *S_Capture_IR* are entered and occupied for one cycle when the corresponding capture/scan register is loaded. Note that the so-called exit states (e.g., *S_Exit_2*) enable a single controlling signal, *TMS*, to effectively direct the activity flow of the machine. For example, the flow from *S_Pause_DR* has three ultimate destinations (e.g., *S_Pause_DR, S_Update_DR,* or *S_Shift_DR*). *A single-bit control signal selects between three possibilities by sequencing the decisions over two clock cycles.*

The content of the instruction register determines whether the boundary scan register or one of the (optional) test-data registers is affected by the operation of the controller. Note that the TAP controller input *nTRST* is optional, because the state *S_reset* can be reached from any other state by asserting *TMS* for at most five clock cycles. In post-synthesis simulation *nTRST* might be needed to drive the gate-level model of the TAP controller into a known initial state.

The control states affecting the data register are described in Table 11-13. The control states affecting the instruction register have a similar description.

The output signals generated by the TAP controller state machine to control the operation of the scan registers are shown in Table 11-14.

11.10.7 Design Example: Testing with JTAG

Testing an ASIC with JTAG requires several steps. For example, to test an ASIC core consisting of combinational logic, the state of the machine must be directed to

---

[24]The state codes of the TAP are not specified by the JTAG standard. For clarity, we have added the prefix *S_* to the names of the states of the TAP controller shown in Figure 11.63. For brevity, the states *Test-Logic-Reset, Run-Test-Idle, Select-DR-Scan,* and *Select-IR-Scan* specified in the JTAG standard are named *S_Reset, S_Run_Idle, S_Select_Dr,* and *S_Select_IR,* respectively, in Figure 11.63 and in our model of the TAP controller.

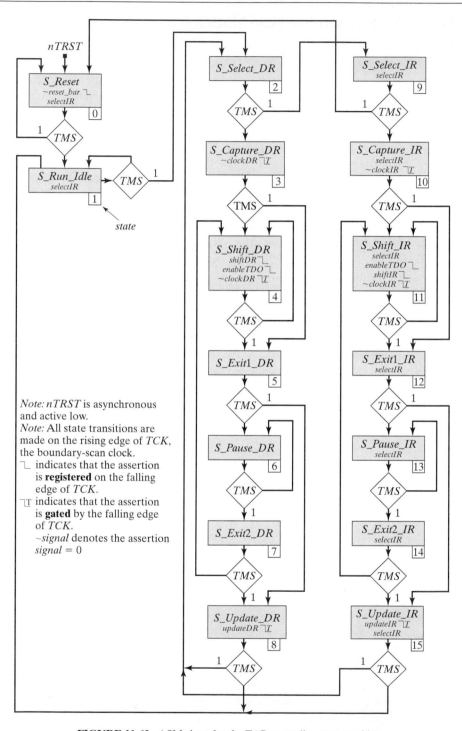

FIGURE 11-63 ASM chart for the TAP controller state machine.

TABLE 11-13 Control states of the TAP controller state machine.

State	Activity
S_Reset	The reset state of the TAP controller. The test logic of the TAP is disabled and the host ASIC operates normally. If the machine has a device identification register, the IDCODE instruction is loaded into the instruction register; otherwise, the BYPASS instruction is loaded.
S_Run_Idle	The TAP controller resides in *S_Run_Idle* while the host ASIC executes an internal test, such as *BIST*. The instruction register must be preloaded with information supporting the test.
S_Select_DR	An assertion of *TMS* while the controller is in *S_Run_Idle* drives the state to *S_Select_DR*, where it resides for one cycle, before passing to *S_Capture_DR* to initiate a scan data sequence, or to *S_Select_IR*, where a sequence can be initiated to update the instruction register or terminate the activity
S_Capture_DR	While the state resides in *S_Capture_DR*, the capture/scan register of the boundary scan register or the test-data register specified by the current instruction can be loaded in parallel (via *data_in*). Capture is initiated by a pulse of *clockDR*, with *shift_DR low*.
S_Shift_DR	A test-data register selected by the instruction register is shifted toward its serial output by one cell at each active edge of *TCK*. A data bit enters the register from the *TDI* port, and leaves from the *TDO* port. The buffer driving *TDO* is active only during shifting.
S_Exit1_DR	A temporary state, entered from *S_Shift_DR* (after a shifting sequence), or from *S_Capture_Dr* (bypassing an initial shifting sequence). After one cycle, the state transitions to *S_Pause_DR* to pause until *TMS* is again asserted, or to *S_Update_DR*, where the captured data are loaded into the output register.
S_Pause_DR	The state resides in *S_Pause_DR* to temporarily halt the scanning process, with *TMS = 0*, until *TMS* is asserted, with the capture/scan register cells holding their state.
S_Exit2_DR	A temporary state. The state resides in *S_Exit2_DR* for one cycle, before a transition to *S_Shift_DR*, where it initiates a scan sequence, or to *S_Update_DR*, where the scan process is terminated and the captured data are loaded into the output register.
S_Update_DR	The state resides in *S_Update_DR* for one cycle after the clock that loads the output register from the capture/scan register, before making a transition to *S_Select_DR*, where it initiates a scan sequence or an instruction sequence, or to *S_Run_Idle*, where it resides while the ASIC executes operations. In test mode the contents of the output register drive the parallel output.

S_Shift_DR and remain there for as many cycles as needed to shift a test pattern into the boundary scan register. At the end of the shifting operations the test inputs should reside in the cells of the capture/scan register that drive the inputs of the chip. Toggling *update_DR* will transfer the content of the capture/scan register into the output register. In test mode, the test patterns in the output register will drive the input pins of the ASIC. The response of the ASIC will appear at the *data_in* pins of the capture/scan cells that connect to the outputs of the ASIC. With *shiftDR* de-asserted, toggling

TABLE 11-14 Moore-type outputs generated by the TAP controller state machine.

Output	Function
reset_bar	Resets the instruction register (IR) to IDCODE or to BYPASS.
shiftIR	Selects the serial input to the capture/scan flip-flop in the instruction register cells.
clockIR	Captures data at the input of the IR or shifts the contents of the IR toward the test data output. Action is gated by the falling edge of *TCK*.
updateIR	Loads the output register flip-flop with the content of the capture flip-flop of the IR. Action is gated by the falling edge of *TCK*.
shiftDR	Selects the serial input to the capture/scan flip-flop in the TDR cells.
clockDR	Captures data at the input of the IR or shifts the contents of the TDR toward the test data output. Action is gated by the falling edge of *TCK*.
updateDR	Loads the output register with the content of the TDR capture/scan flip-flop. Action is gated by the falling edge of *TCK*.
selectIR	Selects either the instruction register or a test-data register to be connected between the TDI and TDO pins of the TAP.
enable TDO	Enables the three-state buffer that drives the test-data output (*TDO*).

clockDR will capture the data at the input pins and load the capture/scan register with the response of the circuit to the test pattern. Then, with *shiftDR* asserted, successive toggling of *clockDR* will scan the data out of the scan chain. Another pattern can be scanned in while the previous pattern is scanned out.

The following example shows how to modify an ASIC with a boundary scan chain and a TAP controller for JTAG. Then the BYPASS and INTEST commands will be demonstrated. Exercises at the end of the chapter will deal with additional instructions.

The overall structure of an ASIC that has been modified to include a test access port for JTAG is shown in Figure 11-64. For simplicity, the ASIC is a 4-bit adder, but the approach applies to more complicated ASICs too. The TAP controller and the control signals for the TAP are omitted in the illustration, but are included in the model of the JTAG-enhanced ASIC. Note that the ASIC retains its port structure, but is wired directly to the boundary scan register through the bus *BSC_Interface [13: 0]*, which accomplishes a *port interface mapping* between the ASIC, the boundary scan register, and the environment. The mode of a port of the ASIC determines whether the associated wires of *BSC_Interface* are connected to an input or output port of the boundary scan register. The outputs of the ASIC are wired to those input pins of the boundary scan register that are connected to the capture/scan register by *data_in [13: 9]*. The corresponding output register cells are connected to the outputs of *ASIC_with_TAP* (i.e., to the primary outputs). Likewise, the output register cells at *data_out [8: 0]* drive the inputs of the ASIC. The corresponding capture/scan register cells are driven by the external (primary) inputs of the chip through *data_in [8: 0]*.

The capture/scan register and output register of the boundary scan register unit are shown separately, with a datapath through them representing the scan path. The scan register (shaded cells) is connected to the outputs of the ASIC and the primary input pins of the chip; the output register is connected to the primary output pins and

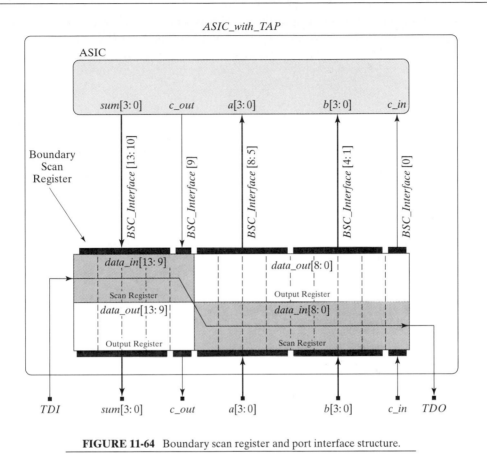

FIGURE 11-64 Boundary scan register and port interface structure.

to the inputs of the ASIC. For example, the *Sum[3: 0]* and *c_out* ports of the ASIC are connected to *data_in[13: 9]*, and *data_out[13: 9]* is connected to *{sum[3: 0], c_out}* at the interface between *ASIC_with_TAP* and its host environment. The structure shown in Figure 11-64 is flexible, and the ordering of the interface signals is arbitrary. The port structure can be modified to accommodate the ports of a different ASIC; the boundary scan register can be resized, and the mapping of *BSC_Interface* can be declared to match the I/O of the ASIC.

As a preliminary step in the overall development and verification of *ASIC_with_TAP*, we show below models of an instruction register and an 8-bit boundary scan register and the results of a brief simulation exercise demonstrating the parallel and serial I/O modes of the boundary scan register. For this exercise, *shiftDr*, *clockDR, updateDR*, and *mode* were controlled by the testbench. The simulation results in Figure 11-65 are annotated to highlight the normal and test modes of operation, and the activity of the registers. The first pulse of *clockDR* captures parallel data at *data_in* (8'haa), which is loaded into *BSC_Scan_Register*. With *shiftDR* de-asserted, the first pulse of *updateDR* demonstrates that the scan register (8'haa) is loaded into the output register, without interfering with normal operation (*data_in* and *data_out* are not affected). With *shiftDR* asserted, the pulses of *clockDR* scan a 1 into the scan register, and data exits through *scan_out*. The second pulse of *updateDR* loads the value of the capture/scan register (8'hff) into the output register. When the test mode is asserted, the value in the output register drives the bus *data_out*. When mode is again de-asserted, *data_out* reverts back to 8'haa, the value of *data_in*. The last pulse of *clockDR* captures *data_in* and loads the value 8'haa into the scan register again.

```verilog
module Boundary_Scan_Register
  (data_out, data_in, scan_out, scan_in, shiftDR, mode, clockDR, updateDR);
  parameter         size = 14;
  output [size -1: 0]   data_out;
  output               scan_out;
  input  [size -1: 0]   data_in;
  input                scan_in;
  input                shiftDR, mode, clockDR, updateDR;
  reg    [size -1: 0]   BSC_Scan_Register, BSC_Output_Register;

  always @ (posedge clockDR)
   BSC_Scan_Register <= shiftDR ? {scan_in, BSC_Scan_Register [ size -1: 1]} :
   data_in;

  always @ (posedge updateDR) BSC_Output_Register <= BSC_Scan_Register;

  assign scan_out = BSC_Scan_Register [0];
  assign data_out = mode ? BSC_Output_Register : data_in;

endmodule


module Instruction_Register
  (data_out, data_in, scan_out, scan_in, shiftIR, clockIR, updateIR, reset_bar);
  parameter           IR_size = 3;
  output [IR_size -1: 0]  data_out;
  output                 scan_out;
  input  [IR_size -1: 0]  data_in;
  input                  scan_in;
  input                  shiftIR, clockIR, updateIR, reset_bar;
  reg    [IR_size -1: 0]  IR_Scan_Register, IR_Output_Register;

  assign                          data_out = IR_Output_Register;
  assign                          scan_out = IR_Scan_Register [0];

  always @ (posedge clockIR)
   IR_Scan_Register <= shiftIR ? {scan_in, IR_Scan_Register [IR_size - 1: 1]} :
   data_in;

  always @ ( posedge updateIR or negedge reset_bar)       // Asynchronous per
                                                           1140.1a.
    if (reset_bar == 0) IR_Output_Register <= ~(0);        // Fills IR with 1s
```

```
                                                          // for BYPASS
                                                          instruction
        else IR_Output_Register <= IR_Scan_Register;

    endmodule
```

A model of the TAP controller state machine is given below. The state of the machine has a binary code. Also, for simplicity, gated clock signals are generated for *clockDr, updateDR, clockIR,* and *updateIR*. For actual implementation, the signals would be connected to the clock inputs of special flip-flops having a multiplexed input (see Section 6.11).

```
        module TAP_Controller
            (reset_bar, selectIR, shiftIR, clockIR, updateIR, shiftDR,
             clockDR, updateDR, enableTDO, TMS, TCK);

        output reset_bar, selectIR, shiftIR, clockIR, updateIR;
```

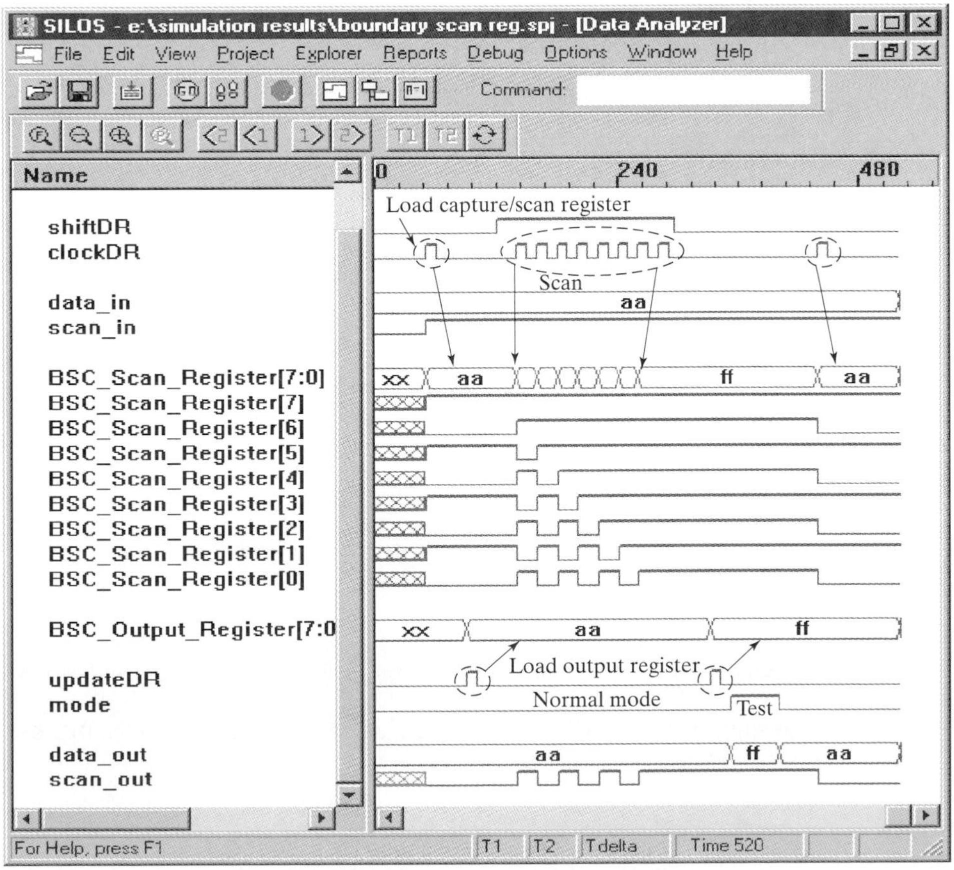

FIGURE 11-65 Simulation results demonstrating correct operation of an 8-bit boundary scan register.

```verilog
output  shiftDR, clockDR, updateDR, enableTDO;
input   TMS, TCK;
reg     reset_bar, selectIR, shiftIR, shiftDR, enableTDO;
wire    clockIR, updateIR, clockDR, updateDR;
parameter   S_Reset      = 0,
            S_Run_Idle   = 1,
            S_Select_DR  = 2,
            S_Capture_DR = 3,
            S_Shift_DR   = 4,
            S_Exit1_DR   = 5,
            S_Pause_DR   = 6,
            S_Exit2_DR   = 7,
            S_Update_DR  = 8,
            S_Select_IR  = 9,
            S_Capture_IR = 10,
            S_Shift_IR   = 11,
            S_Exit1_IR   = 12,
            S_Pause_IR   = 13,
            S_Exit2_IR   = 14,
            S_Update_IR  = 15;

reg [3:0]   state, next_state;

pullup (TMS);  // Required by IEEE 1149.1a; ensures that an undriven input
pullup (TDI);  // produces a response identical to the application of a logic 1."
               // Program for Xilinx implementation

always @ (negedge TCK) reset_bar <= (state == S_Reset) ? 0 : 1; // Registered
                                                            active low


always @ (negedge TCK) begin
  shiftDR <= (state == S_Shift_DR) ? 1 : 0;      // Registered select for scan mode
  shiftIR <= (state == S_Shift_IR) ? 1: 0;
                                                 // Registered output enable
  enableTDO <= ((state == S_Shift_DR) || (state == S_Shift_IR)) ? 1 : 0;
end

// Gated clocks for capture registers
assign clockDR = !(((state == S_Capture_DR) || (state == S_Shift_DR)) &&
(TCK == 0));
assign clockIR =  !(((state == S_Capture_IR) || (state == S_Shift_IR)) &&
(TCK == 0));

// Gated clocks for output registers
assign updateDR = (state == S_Update_DR) && (TCK == 0);
assign updateIR =  (state == S_Update_IR) && (TCK == 0);
```

```
always @ (posedge TCK ) state <= next_state;

always @ (state or TMS) begin
  selectIR = 0;
  next_state = state;

case (state)
  S_Reset:            begin
                        selectIR = 1;
                        if (TMS == 0) next_state = S_Run_Idle;
                      end
  S_Run_Idle:         begin selectIR = 1; if (TMS)  next_state = S_Select
                      _DR; end
  S_Select_DR:        next_state = TMS ? S_Select_IR: S_Capture_DR;
  S_Capture_DR:       begin next_state = TMS ? S_Exit1_DR: S_Shift_DR;
                      end
  S_Shift_DR:         if (TMS) next_state = S_Exit1_DR;
  S_Exit1_DR:         next_state = TMS ? S_Update_DR: S_Pause_DR;
  S_Pause_DR:         if (TMS) next_state = S_Exit2_DR;
  S_Exit2_DR:         next_state = TMS ? S_Update_DR: S_Shift_DR;
  S_Update_DR:        begin next_state = TMS ? S_Select_DR: S_Run_Idle;
                      end
  S_Select_IR:        begin
                        selectIR = 1;
                        next_state = TMS ? S_Reset: S_Capture_IR;
                      end
  S_Capture_IR:       begin
                        selectIR = 1;
                        next_state = TMS ? S_Exit1_IR: S_Shift_IR;
                      end
  S_Shift_IR:         begin selectIR = 1; if (TMS) next_state = S_Exit1_IR;
                      end
  S_Exit1_IR:         begin
                        selectIR = 1;
                        next_state = TMS ? S_Update_IR: S_Pause_IR;
                      end
  S_Pause_IR:         begin selectIR = 1; if (TMS) next_state = S_Exit2_IR;
                      end
  S_Exit2_IR:         begin
                        selectIR = 1;
                        next_state = TMS ? S_Update_IR: S_Shift_IR;
                      end
  S_Update_IR:        begin
                        selectIR = 1;
                        next_state = TMS ? S_Select_IR: S_Run_Idle;
                      end
  default             next_state = S_Reset;
```

```
            endcase
        end
    endmodule
```

The parameterized model listed below, *ASIC_with_TAP*, instantiates the following modules: *ASIC, TAP_Controller, Boundary_Scan_Register, Instruction_Register, Instruction_Decoder*, and *TAP_Controller*. In general, the instruction register of a TAP loads a parallel data path (*data_in*) in the state *S_Capture_IR*, which provides the TAP with design-specific information generated in the host component.[25] In this example, *Dummy_data* = 3'b001 is passed through the port *data_in*.

```
    module ASIC_with_TAP (sum, c_out, a, b, c_in, TDO, TDI, TMS, TCK);
        parameter                           BSR_size = 14;
        parameter                           IR_size = 3;
        parameter                           size = 4;
        output          [size -1: 0]        sum;                // ASIC interface I/O
        output                              c_out;
        input           [size -1: 0]        a, b;
        input                               c_in;

        output                              TDO;                // TAP interface signals
        input                               TDI, TMS, TCK;

        wire            [BSR_size -1: 0]    BSC_Interface;      // Declarations for boundary
                                                                      scan register I/O

        wire                                reset_bar,          // TAP controller outputs
                                            selectIR, enableTDO,
                                            shiftIR, clockIR, updateIR,
                                            shiftDR, clockDR, updateDR;

        wire                                test_mode, select_BR;
        wire                                TDR_out;            // Test data register
                                                                      serial datapath
        wire            [IR_size -1: 0]     Dummy_data = 3'b001;    // Captured in
                                                                      S_Capture_IR
        wire            [IR_size -1: 0]     instruction;
        wire                                IR_scan_out;        // Instruction register
        wire                                BSR_scan_out;       // Boundary scan
                                                                      register
        wire                                BR_scan_out;        // Bypass register

        assign          TDO = enableTDO ? selectIR ? IR_scan_out : TDR_out : 1'bz;
        assign          TDR_out = select_BR ? BR_scan_out : BSR_scan_out;
```

---

[25]The JTAG standard requires that the cells of the two least significant bits of the instruction register shall load the pattern 2'b01 in the state *S_Capture_IR*. The remaining bits have fixed (0 or 1) but application-dependent values.

```
ASIC M0 (
 .sum (BSC_Interface [13: 10]),
 .c_out (BSC_Interface [9]),
 .a (BSC_Interface [8: 5]),
 .b (BSC_Interface [4: 1]),
 .c_in (BSC_Interface [0]));

Bypass_Register M1(
 .scan_out (BR_scan_out),
 .scan_in (TDI),
 .shiftDR (shift_BR),
 .clockDR (clock_BR));

Boundary_Scan_Register M2(
 .data_out ({sum, c_out, BSC_Interface [8: 5], BSC_Interface [4: 1],
 BSC_Interface [0]}),
 .data_in ({BSC_Interface [13: 10], BSC_Interface [9], a, b, c_in}),
 .scan_out (BSR_scan_out),
 .scan_in (TDI),
 .shiftDR (shiftDR),
 .mode (test_mode),
 .clockDR (clock_BSC_Reg),
 .updateDR (update_BSC_Reg));

Instruction_Register M3 (
 .data_out (instruction),
 .data_in (Dummy_data),
 .scan_out (IR_scan_out),
 .scan_in (TDI),
 .shiftIR (shiftIR),
 .clockIR (clockIR),
 .updateIR (updateIR),
 .reset_bar (reset_bar));

Instruction_Decoder M4 (
 .mode (test_mode),
 .select_BR (select_BR),
 .shift_BR (shift_BR),
 .clock_BR (clock_BR),
 .shift_BSC_Reg (shift_BSC_Reg),
 .clock_BSC_Reg (clock_BSC_Reg),
 .update_BSC_Reg (update_BSC_Reg),
 .instruction (instruction),
 .shiftDR (shiftDR),
 .clockDR (clockDR),
 .updateDR (updateDR));

TAP_Controller M5 (
```

```verilog
            .reset_bar(reset_bar),
            .selectIR (selectIR),
            .shiftIR (shiftIR),
            .clockIR (clockIR),
            .updateIR (updateIR),
            .shiftDR (shiftDR),
            .clockDR (clockDR),
            .updateDR (updateDR),
            .enableTDO (enableTDO),
            .TMS (TMS),
            .TCK (TCK));

endmodule

module ASIC (sum, c_out, a, b, c_in);
    parameter               size = 4;
    output [size -1: 0]     sum;
    output                  c_out;
    input   [size -1: 0]    a, b;
    input                   c_in;

    assign {c_out, sum} = a + b + c_in;

endmodule

module Bypass_Register(scan_out, scan_in, shiftDR, clockDR);
    output      scan_out;
    input       scan_in, shiftDR, clockDR;
    reg         scan_out;

    always @ (posedge clockDR) scan_out <= scan_in & shiftDR;

endmodule

module Instruction_Decoder (mode, select_BR, shift_BR, clock_BR, shift_BSC_Reg,
clock_BSC_Reg, update_BSC_Reg, instruction, shiftDR, clockDR, updateDR);
    parameter   IR_size             = 3;
    output                          mode, select_BR, shift_BR, clock_BR;
    output                          shift_BSC_Reg, clock_BSC_Reg,
                                    update_BSC_Reg;
    input                           shiftDR, clockDR, updateDR;
    input       [IR_size −1: 0]     instruction;
    parameter   BYPASS              = 3'b111;     // Required by
                                                  1149.1a
    parameter   EXTEST              = 3'b000;     // Required by
                                                  1149.1a
    parameter   SAMPLE_PRELOAD      = 3'b010;
    parameter   INTEST              = 3'b011;
    parameter   RUNBIST             = 3'b100;
    parameter   IDCODE              = 3'b101;

    reg         mode, select_BR, clock_BR, clock_BSC_Reg, update_BSC_Reg;
```

```
assign          shift_BR = shiftDR;
assign          shift_BSC_Reg = shiftDR;

always @ (instruction or clockDR or updateDR) begin
  mode = 0; select_BR = 0;              // default is test-data register
  clock_BR = 1; clock_BSC_Reg = 1;
  update_BSC_Reg = 0;

  case (instruction)
    EXTEST:                begin mode = 1; clock_BSC_Reg = clockDR;
                                 update_BSC_Reg = updateDR; end
    INTEST:                begin mode = 1; clock_BSC_Reg = clockDR;
                                 update_BSC_Reg = updateDR; end
    SAMPLE_PRELOAD:        begin  clock_BSC_Reg = clockDR;
                                 update_BSC_Reg = updateDR; end
    RUNBIST:               begin  end
    IDCODE:                begin select_BR = 1; clock_BR = clockDR;  end
    BYPASS:                begin select_BR = 1; clock_BR = clockDR; end
    default:               begin select_BR = 1; end

  endcase
  end
endmodule
```

The structure of the testbench (*t_ASIC_with_TAP*) used to test *ASIC_with_TAP* is shown in Figure 11-66. Two arrays, *Array_of_TAP_Instructions* and *Array_of ASIC_Test_Patterns*, hold patterns for scanning instructions and test patterns into the boundary scan register. The test sequence selects a pattern from one of the registers and loads it into *Pattern_Register*. When the test sequence asserts a load signal the pattern held in *Pattern_Register* is loaded into the register *TDI_Reg* within *TDI_Generator*. The TAP controller scans the pattern from *TDI_Reg* into the *TDI* port of the TAP and into *Patttern_Buffer_1*. The pattern from the *TDO* port of the TAP is scanned into *TDO_Reg* within *TDO_Monitor*, and at the same time the pattern in *Pattern_Buffer_1* is scanned into *Pattern_Buffer_2*. When the scan activity is complete the contents of *TDO_Reg* and *Pattern_Buffer_2* are compared to detect an error.

The testbench *t_ASIC_with_TAP* is given below, with comments identifying some of the functional features that need to be verified. Note that the ASM chart of the TAP controller (see Figure 11-63) has the property that the value of *TMS* that causes a transition into a state of the chart is the same for all paths that enter the state. With this observation, we write a set of testbench tasks to specify a sequence of inputs directing the flow along arcs of the ASM chart.

The test patterns in the testbench conform to the port structure shown in Figure 11-66. The patterns demonstrate the flow of data in *ASIC_with_TAP* and demonstrate that the testbench detects an error that has been injected into the circuit. The instruction patterns conform to the instruction code in the model for *Instruction_Decoder*. The test sequences are generated by *TDI_Generator* and monitored by *TDO_Monitor*,

t_ASIC_with_TAP

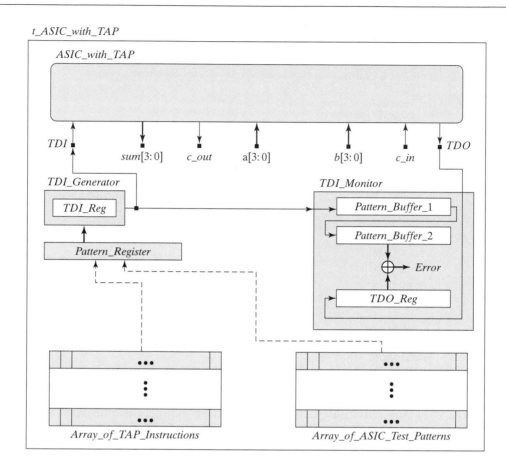

FIGURE 11-66 Structure of a testbench for *ASIC_with_TAP*.

also given below. The task *Load_ASIC_Test_Pattern* executes a test sequence to select and load a scan pattern into the register *TDI_Reg* within *TDI_Generator*. This pattern scans out of *TDI_Generator* while *enableTDO* or *enable_bypass_pattern* is asserted. *TDO_Monitor* includes a two-stage pipeline buffer whose input stage receives the pattern that is shifted into the ASIC. The first stage holds the pattern that is currently in the boundary scan register of *ASIC_with_TAP*; the second stage holds the previous pattern held by *ASIC_with_TAP*, and is used to compare the actual pattern scanned from *ASIC_with_TAP* to the expected pattern. Data from the cells of the boundary scan register that correspond to the outputs of the ASIC are compared with the test pattern data for those cells. A mismatch is detected as an error. The testbench includes an optional segment of code that injects an error into the bits corresponding to the sum generated by the adder, and that checks whether an error is detected by *TDO_Monitor*.

```
module t_ASIC_with_TAP ();                              // Testbench
    parameter                       size = 4;
    parameter                       BSC_Reg_size = 14;
    parameter                       IR_Reg_size = 3;
    parameter                       N_ASIC_Patterns = 8;
    parameter                       N_TAP_Instructions = 8;
    parameter                       Pause_Time = 40;
    parameter                       End_of_Test = 1500;
    parameter                       time_1 = 350, time_2 = 550;

    wire        [size -1: 0]        sum;
    wire        [size -1: 0]        sum_fr_ASIC = M0.BSC_Interface [13: 10];

    wire                            c_out;
    wire                            c_out_fr_ASIC = M0.BSC_Interface [9];
    reg         [size -1: 0]        a, b;
    reg                             c_in;
    wire        [size −1: 0]        a_to_ASIC = M0.BSC_Interface [8: 5];
    wire        [size −1: 0]        b_to_ASIC = M0.BSC_Interface [4: 1];
    wire                            c_in_to_ASIC = M0.BSC_Interface [0];

    reg     TMS, TCK;
    wire    TDI;
    wire    TDO;
    reg     load_TDI_Generator;
    reg     Error, strobe;
    integer pattern_ptr;
    reg     [BSC_Reg_size -1: 0]    Array_of_ASIC_Test_Patterns [0:
                                    N_ASIC_Patterns -1];
    reg     [IR_Reg_size -1: 0]     Array_of_TAP_Instructions [0:
                                    N_TAP_Instructions -1];
    reg     [BSC_Reg_size -1: 0]    Pattern_Register;          // Size to maximum
                                                                        TDR
    reg     enable_bypass_pattern;

ASIC_with_TAP M0 (sum, c_out, a, b, c_in, TDO, TDI, TMS, TCK);

TDI_Generator M1(
  .to_TDI (TDI),
  .scan_pattern (Pattern_Register),
  .load (load_TDI_Generator),
  .enable_bypass_pattern (enable_bypass_pattern),
  .TCK (TCK));

TDO_Monitor M3 (
  .to_TDI (TDI),
  .from_TDO (TDO),
  .strobe (strobe),
  .TCK (TCK));
```

initial #End_of_Test **$finish**;

initial begin TCK = 0; **forever** #5 TCK = ~TCK; **end**

/* **Summary of a basic test plan for ASIC_with TAP**

Verify default to bypass instruction
Verify bypass register action: Scan 10 cycles, with pause before exiting
Verify pull up action on TMS and TDI
Reset to S_Reset after five assertions of TMS
Boundary scan in, pause, update, return to S_Run_Idle
Boundary scan in, pause, resume scan in, pause, update, return to S_Run_Idle
Instruction scan in, pause, update, return to S_Run_Idle
Instruction scan in, pause, resume scan in, pause, update, return to S_Run_Idle
*/

// **TEST PATTERNS**
// **External I/O for normal operation**

initial fork
 // {a, b, c_in} = 9'b0;
 {a, b, c_in} = 9'b_1010_0101_0; // sum = F, c_out = 0, a = A, b = 5, c_in = 0
 join

/* **Option to force error to test fault detection**

initial begin :Force_Error
force M0.BSC_Interface [13: 10] = 4'b0;
end
*/

initial begin   // Test sequence: Scan, pause, return to S_Run_Idle
 strobe = 0;
 Declare_Array_of_TAP_Instructions;
 Declare_Array_of_ASIC_Test_Patterns;
 Wait_to_enter_S_Reset;

// **Test for power-up and default to BYPASS instruction (all 1s in IR), with**
// **default path through the Bypass Register, with BSC register remaining in**
// **wakeup state (all x).**
// **ASIC test pattern is scanned serially, entering at TDI, passing through the**
// **bypass register, and exiting at TDO. The BSC register and the IR are not**
// **changed.**

 pattern_ptr = 0;
 Load_ASIC_Test_Pattern;
 Go_to_S_Run_Idle;
 Go_to_S_Select_DR;
 Go_to_S_Capture_DR;
 Go_to_S_Shift_DR;
 enable_bypass_pattern = 1;

```
        Scan_Ten_Cycles;
        enable_bypass_pattern = 0;
        Go_to_S_Exit1_DR;
        Go_to_S_Pause_DR;
        Pause;
        Go_to_S_Exit2_DR;
        /*
        Go_to_S_Shift_DR;
        Load_ASIC_Test_Pattern;              // option to re-load same pattern and
                                                scan again

        enable_bypass_pattern = 1;
        Scan_Ten_Cycles;
        enable_bypass_pattern = 0;
        Go_to_S_Exit1_DR;
        Go_to_S_Pause_DR;
        Pause;
        Go_to_S_Exit2_DR;
        */
        Go_to_S_Update_DR;
        Go_to_S_Run_Idle;
    end
```

// **Test to load instruction register with INTEST instruction**

```
    initial #time_1 begin
    pattern_ptr = 3;
    strobe = 0;
    Load_TAP_Instruction;
    Go_to_S_Run_Idle;
    Go_to_S_Select_DR;
    Go_to_S_Select_IR;
    Go_to_S_Capture_IR;                      // Capture dummy data (3'b011)
    repeat (IR_Reg_size) Go_to_S_Shift_IR;
    Go_to_S_Exit1_IR;
    Go_to_S_Pause_IR;
    Pause;
    Go_to_S_Exit2_IR;
    Go_to_S_Update_IR;
    Go_to_S_Run_Idle;
    end
```

// **Load ASIC test pattern**

```
    initial #time_2 begin
    pattern_ptr = 0;
    Load_ASIC_Test_Pattern;
    Go_to_S_Run_Idle;
    Go_to_S_Select_DR;
    Go_to_S_Capture_DR;
```

```
      repeat (BSC_Reg_size) Go_to_S_Shift_DR;
      Go_to_S_Exit1_DR;
      Go_to_S_Pause_DR;
      Pause;
      Go_to_S_Exit2_DR;
      Go_to_S_Update_DR;
      Go_to_S_Run_Idle;
```

// **Capture data and scan out while scanning in another pattern**
```
      pattern_ptr = 2;
      Load_ASIC_Test_Pattern;
      Go_to_S_Select_DR;
      Go_to_S_Capture_DR;
      strobe = 1;
      repeat (BSC_Reg_size) Go_to_S_Shift_DR;

      Go_to_S_Exit1_DR;

      Go_to_S_Pause_DR;
      Go_to_S_Exit2_DR;
      Go_to_S_Update_DR;
      strobe = 0;
      Go_to_S_Run_Idle;
    end

/***************************** TAP CONTROLLER TASKS *****************************/
    task  Wait_to_enter_S_Reset;
     begin
     @ (negedge TCK) TMS = 1;
    end
    endtask

    task  Reset_TAP;
     begin
     TMS = 1;
      repeat (5) @ (negedge TCK);
     end
    endtask

    task Pause;                    begin #Pause_Time;             end endtask

    task  Go_to_S_Reset;          begin @ (negedge TCK) TMS = 1; end endtask
    task  Go_to_S_Run_Idle;       begin @ (negedge TCK) TMS = 0; end endtask

    task  Go_to_S_Select_DR;      begin @ (negedge TCK) TMS = 1; end endtask
    task  Go_to_S_Capture_DR;     begin @ (negedge TCK) TMS = 0; end endtask
    task  Go_to_S_Shift_DR;       begin @ (negedge TCK) TMS = 0; end endtask
    task  Go_to_S_Exit1_DR;       begin @ (negedge TCK) TMS = 1; end endtask
    task  Go_to_S_Pause_DR;       begin @ (negedge TCK) TMS = 0; end endtask
```

```
task Go_to_S_Exit2_DR;        begin @ (negedge TCK) TMS = 1; end endtask
task Go_to_S_Update_DR;       begin @ (negedge TCK) TMS = 1; end endtask

task Go_to_S_Select_IR;       begin @ (negedge TCK) TMS = 1; end endtask
task Go_to_S_Capture_IR;      begin @ (negedge TCK) TMS = 0; end endtask
task Go_to_S_Shift_IR;        begin @ (negedge TCK) TMS = 0; end endtask
task Go_to_S_Exit1_IR;        begin @ (negedge TCK) TMS = 1; end endtask
task Go_to_S_Pause_IR;        begin @ (negedge TCK) TMS = 0; end endtask
task Go_to_S_Exit2_IR;        begin @ (negedge TCK) TMS = 1; end endtask
task Go_to_S_Update_IR;       begin @ (negedge TCK) TMS = 1; end endtask
task Scan_Ten_Cycles;         begin repeat (10)   begin @ (negedge TCK)
                              TMS = 0;
                              @ (posedge TCK) TMS = 1;
                              end end endtask

/***************************** ASIC TEST PATTERNS ******************************/
task Load_ASIC_Test_Pattern;
  begin
    Pattern_Register = Array_of_ASIC_Test_Patterns [pattern_ptr];
    @ (negedge TCK ) load_TDI_Generator = 1;
    @ (negedge TCK) load_TDI_Generator = 0;
  end
endtask

task Declare_Array_of_ASIC_Test_Patterns;
  begin
  //s3 s2 s1 s0_ c0_a3 a2 a1 a0_b3 b2 b1 b0_c_in;

  Array_of_ASIC_Test_Patterns [0] = 14'b0100_1_1010_1010_0;
  Array_of_ASIC_Test_Patterns [1] = 14'b0000_0_0000_0000_0;
  Array_of_ASIC_Test_Patterns [2] = 14'b1111_1_1111_1111_1;
  Array_of_ASIC_Test_Patterns [3] = 14'b0100_1_0101_0101_0;
end endtask
/**************************** INSTRUCTION PATTERNS ****************************/
  parameter     BYPASS                  = 3'b111;        // pattern_ptr = 0
  parameter     EXTEST                  = 3'b001;        // pattern_ptr = 1
  parameter     SAMPLE_PRELOAD          = 3'b010;        // pattern_ptr = 2
  parameter     INTEST                  = 3'b011;        // pattern_ptr = 3
  parameter     RUNBIST                 = 4'b100;        // pattern_ptr = 4
  parameter     IDCODE                  = 5'b101;        // pattern_ptr = 5

task Load_TAP_Instruction;
  begin
   Pattern_Register = Array_of_TAP_Instructions [pattern_ptr];
    @ (negedge TCK) load_TDI_Generator = 1;
    @ (negedge TCK) load_TDI_Generator = 0;
  end
endtask

task Declare_Array_of_TAP_Instructions;
  begin
```

```
                    Array_of_TAP_Instructions [0] = BYPASS;
                    Array_of_TAP_Instructions [1] = EXTEST;
                    Array_of_TAP_Instructions [2] = SAMPLE_PRELOAD;
                    Array_of_TAP_Instructions [3] = INTEST;
                    Array_of_TAP_Instructions [4] = RUNBIST;
                    Array_of_TAP_Instructions [5] = IDCODE;
                end
                endtask
        endmodule


        module TDI_Generator (to_TDI, scan_pattern, load, enable_bypass_pattern, TCK);
            parameter                   BSC_Reg_size = 14;
            output                      to_TDI;
            input [BSC_Reg_size –1: 0]   scan_pattern;
            input                       load, enable_bypass_pattern, TCK;
            reg    [BSC_Reg_size –1: 0]  TDI_Reg;
            wire                        enableTDO = t_ASIC_with_TAP.M0.enableTDO;
            assign to_TDI = TDI_Reg [0];

            always @ (posedge TCK) if (load) TDI_Reg <= scan_pattern;
             else if (enableTDO || enable_bypass_pattern)
                TDI_Reg <= TDI_Reg > 1;

        endmodule

        module TDO_Monitor (to_TDI, from_TDO, strobe, TCK);
            parameter                   BSC_Reg_size = 14;
            output                      to_TDI;
             input                      from_TDO, strobe, TCK;
            reg    [BSC_Reg_size –1: 0]  TDI_Reg, Pattern_Buffer_1, Pattern_Buffer_2,
                                        Captured_Pattern, TDO_Reg;
            reg                         Error;
            parameter                   test_width = 5;
            wire                        enableTDO = t_ASIC_with_TAP.M0.enableTDO;
            wire   [test_width –1: 0]    Expected_out =
                                        Pattern_Buffer_2 [BSC_Reg_size –1
                                        : BSC_Reg_size – test_width];

            wire   [test_width –1: 0]    ASIC_out =
                                        TDO_Reg [BSC_Reg_size - 1: BSC_Reg_size –
                                        test_width];
            initial                     Error = 0;

        always @ (negedge enableTDO) if (strobe == 1) Error = |(Expected_out ^
        ASIC_out);

            always @ (posedge TCK) if (enableTDO) begin
                Pattern_Buffer_1 <= {to_TDI, Pattern_Buffer_1 [BSC_Reg_size –1: 1]};
```

```
                    Pattern_Buffer_2 <= {Pattern_Buffer_1 [0], Pattern_Buffer_2
                    [BSC_Reg_size –1: 1]};
                    TDO_Reg <= {from_TDO, TDO_Reg [BSC_Reg_size –1: 1]};
              end
          endmodule
```

The simulation results shown in Figure 11-67 demonstrate that the default instruction is the BYPASS instruction. The signals *c_in*, *b*, *a*, *c_out*, and *sum* are external ports of *ASIC_with_TAP; c_in_to_ASIC, b_to_ASIC*, and *a_to_ASIC* are input ports of *ASIC*, and *c_out_fr_ASIC* and *sum_fr_ASIC* are output ports of ASIC. The system resides in an unknown state when simulation initiates at time $t = 0$. At the first active edge of *TCK* the state of the machine enters *S_Reset* (0),[26] where it remains (Figure 11-67a) until the sequence of inputs of *TMS* scans 10 bits of the *Pattern_Register* (1354_H) through the TAP (Figure 11-67b).[27] In the testbench, *Pattern_Register* holds the pattern selected by *pattern_ptr*. The pulse of *Load_TDI_Generator* loads the pattern into *TDI_Reg* (within *TDI_Generator*). The LSB of *TDI_Reg* drives *TDI*. After the machine enters *S_Shift_DR* (4), 10 cycles of *TCK* with *shiftDR* asserted scan ten bits of the pattern through the bypass register. Note in Figure 11-67(a) that the state transitions occur on the rising edges of *TCK*, and that the waveform of *TDO* is a copy of the waveform of *TDI* delayed by one cycle while *enableTDO* is asserted. Also note that *clock_BSC_Reg* is fixed (i.e., the boundary scan register is idle).

The JTAG specification for the bypass register requires that the output of the register be set to logical 0 on the rising edge of *TCK* following entry into the TAP controller state *S_Capture_DR*. Note in Figure 11-67(a) that this edge occurs at the transition between *S_Capture_DR* and *S_Shift_DR*, and that the output of the bypass register is 0. The output of the register will be the value scanned out of *TDO* and the next rising edge of *TCK* following assertion of *TDO_enable*.

The scanning process does not affect the values of the signals at the ports of *ASIC*. *clockBR* is active for one cycle in state *S_Capture_DR* (3) and for 10 cycles in state *S_Shift_DR* (4). *selectBR* connects the bypass register to *TDI* and *TDO*. *BSC_Scan_Register* and *BSC_output_Register* hold 14'Hx because they have not yet received data. *reset_bar* is asserted (active-low) and resets the 3-bit instruction register to hold all 1s (the BYPASS instruction) in state *S_Reset* (0). The subsequent bits of *TDO* replicate the waveform of *TDI*.

The simulation results in Figure 11-68a show the BYPASS instruction being shifted out of the TAP and the instruction INTEST being loaded into the TAP while the machine is in the state *S_Shift_IR* (11), with *shiftIR* and *enableTDO* asserted. The testbench loads *Pattern_Register* with INTEST, then asserts *load_TDI_Generator* to load *TDI_Reg* with INTEST. In state *S_Shift_IR(11)* the instruction scans into *IR_SCAN_Register* (Figure 11-68b) while BYPASS scans out of the register through *TDO*. When the state of the TAP controller enters *S_Update_IR* the instruction INTEST is loaded into *IR_Output_Register*. The waveforms in Figure 11-68(b) also show the testbench reloading *Pattern_Register* with 1354_H, transferring the pattern to *TDI_Reg*, and

---

[26]See Problem 21 at the end of this chapter.
[27]The data patterns and test sequence intervals have been chosen to illustrate the operation of the TAP.

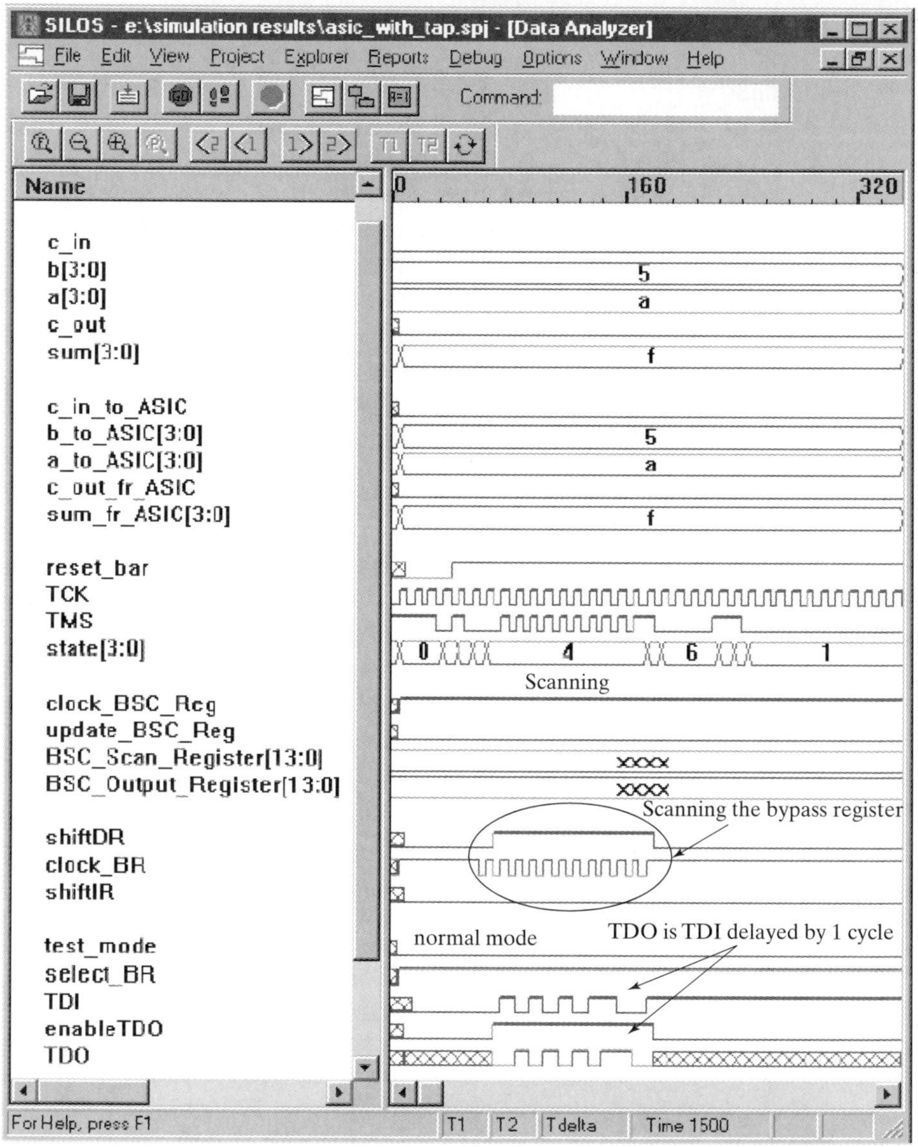

(a)

FIGURE 11-67 Simulation results—scanning a pattern through the bypass register of *ASIC_with_TAP* after power-up: (a) the pattern scans through the chip with a delay of one clock cycle, and (b) control signals, TAP registers, and testbench registers.

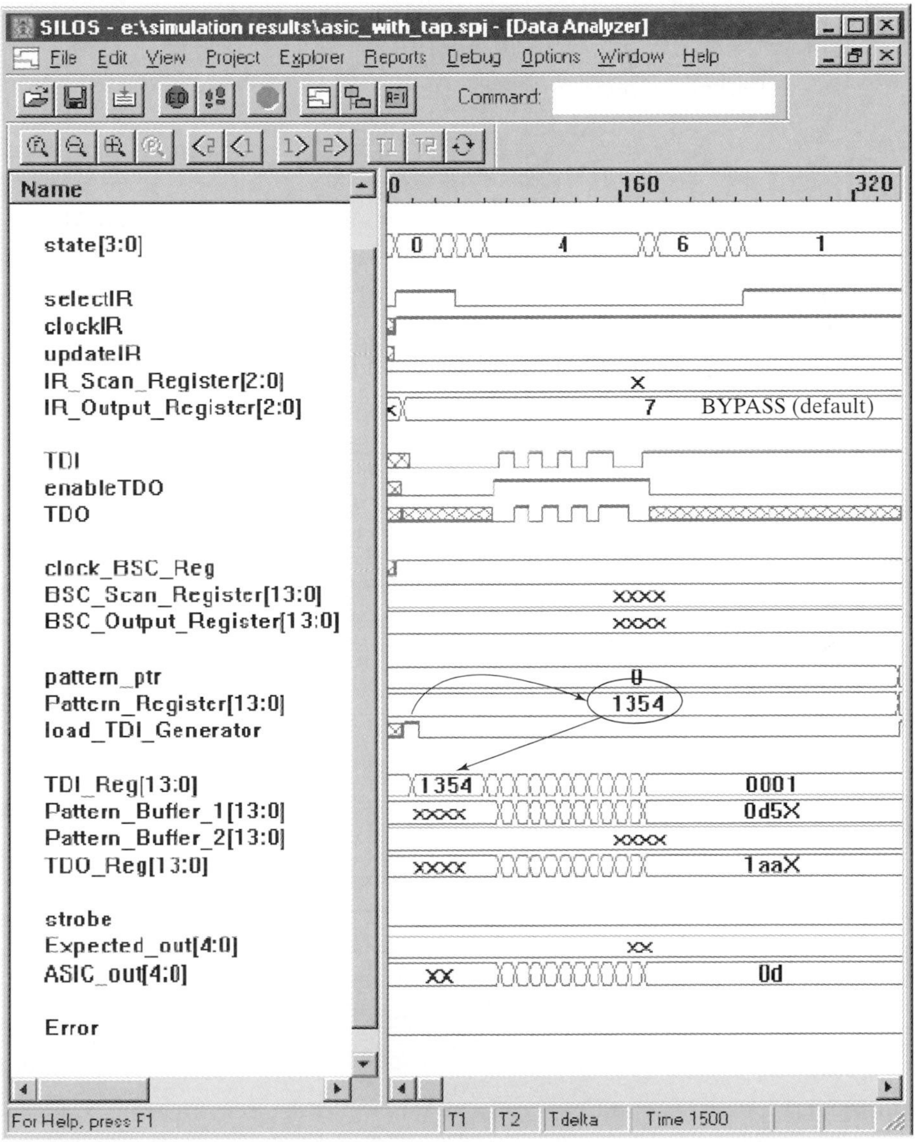

(b)

FIGURE 11-67 Continued

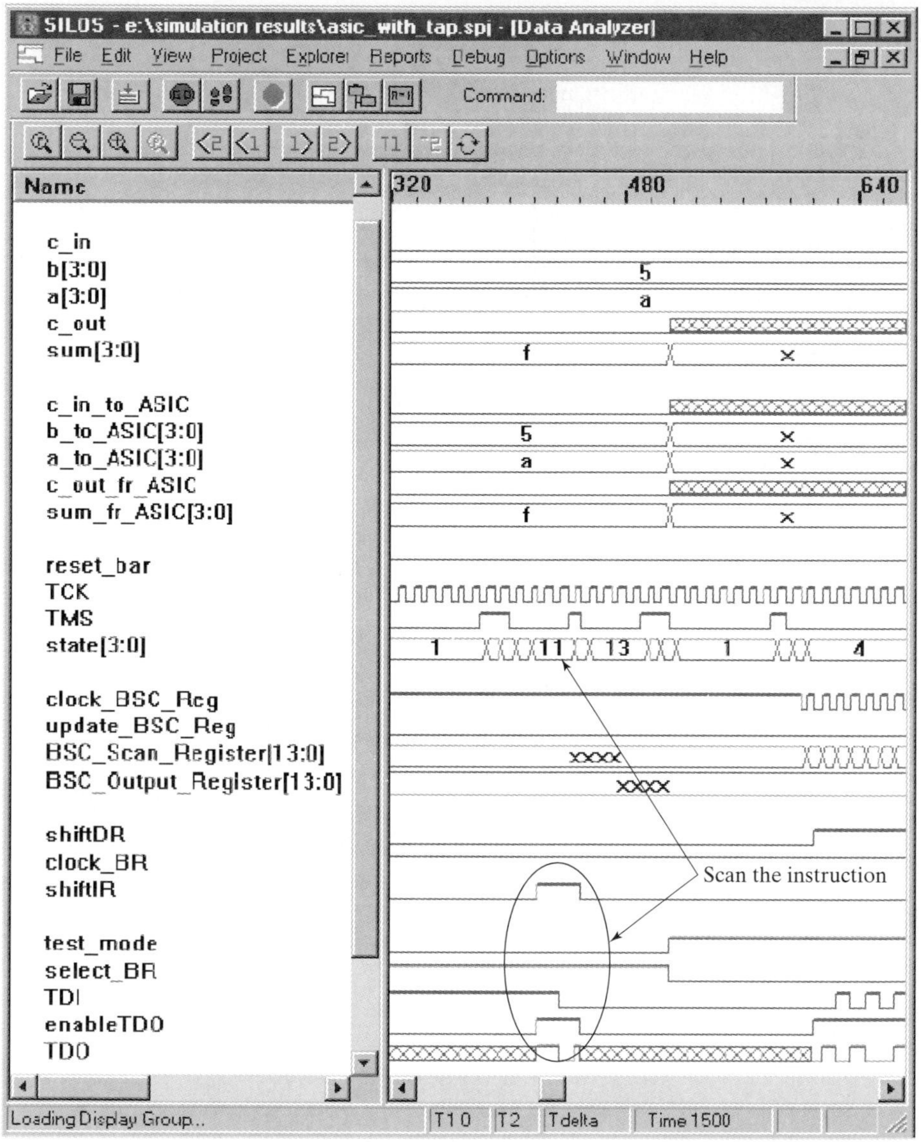

(a)

FIGURE 11-68 Simulation results—loading the instruction INTEST into the instruction register. (a) *enableTDO* is asserted only while scanning (otherwise *TDO* is in the high-impedance state), and (b) the instruction INTEST is loaded, then a data pattern is reloaded and scanned out of the *TDI_Generator* and into *ASIC_with_TAP*, via *TDI*.

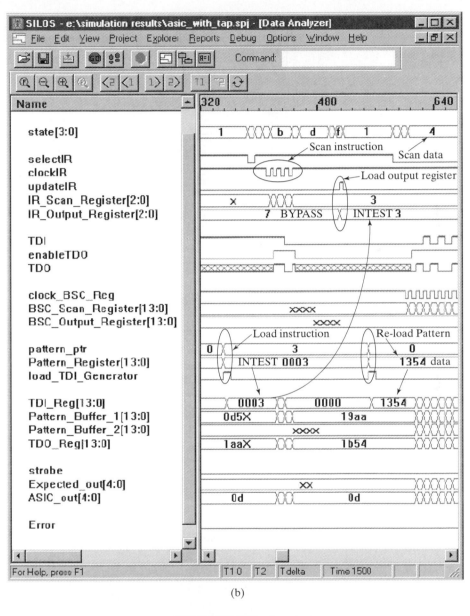

(b)

FIGURE 11-68 Continued

scanning the pattern into the *BSC_Scan_Regsiter* while the state resides in *S_Shift_IR*. Also note that the three-state behavior of *TDO* conforms to the JTAG standard.

With *IR_Output_Register* holding the instruction INTEST, the pattern (1354_H) that was loaded into *BSC_Scan_Register* in Figure 11-68(b) is transferred to *BSC_Update_Register* in Figure 11-69(a) for execution of an internal test of *ASIC*.

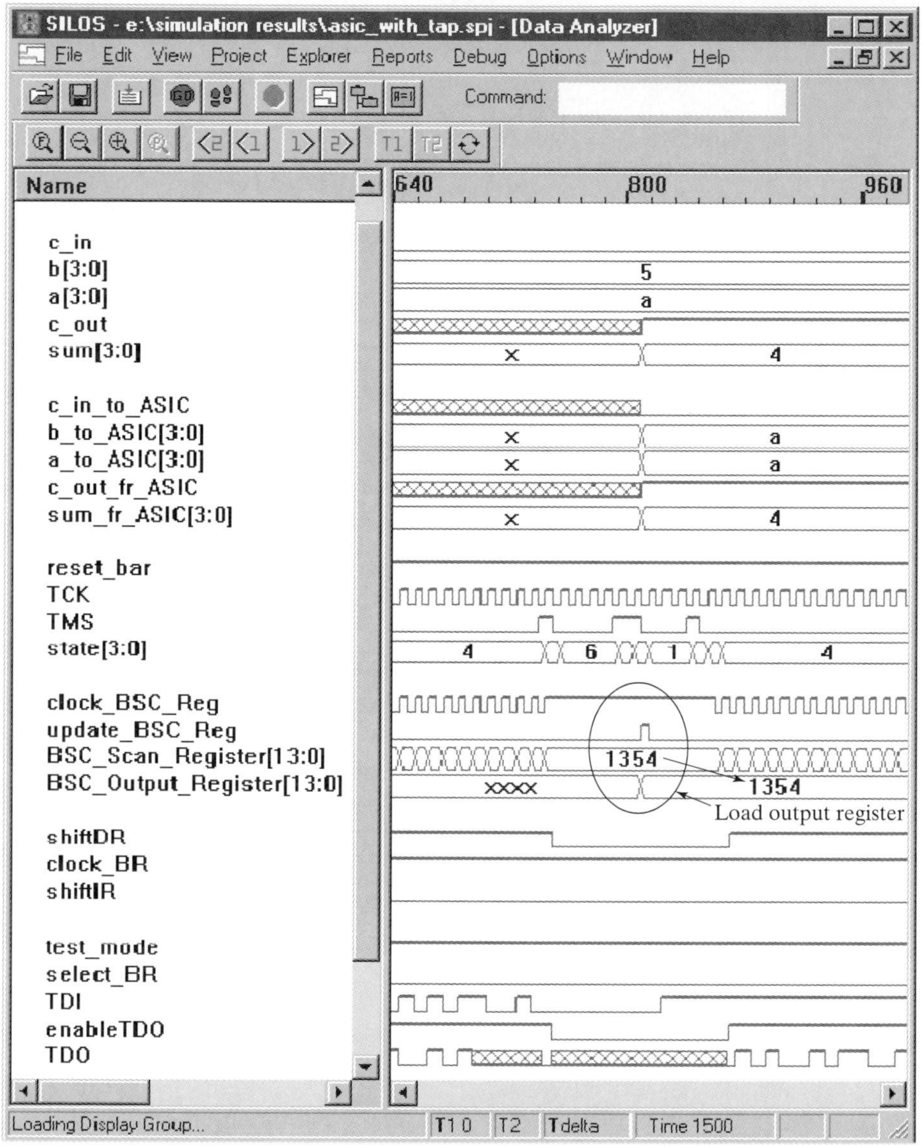

(a)

FIGURE 11-69 Simulation results: (a) after scanning the test pattern 1354_H into the capture/scan register, the boundary scan output register is loaded, and test inputs are applied to *ASIC*, and (b) the outputs of *ASIC* are captured and scanned out through *TDO*, and shifted into *TDO_Reg* for comparison with *Pattern_Buffer_2* (see Figure 11.66).

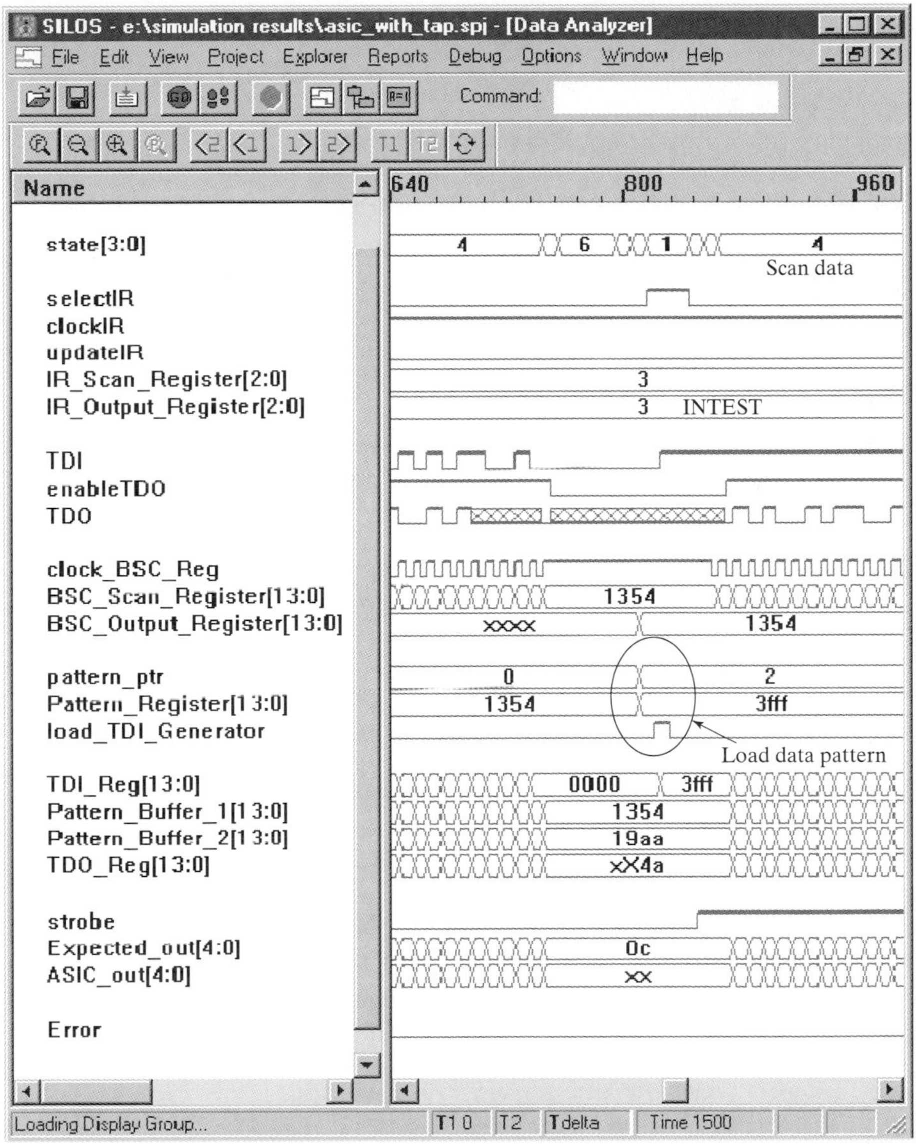

(b)

FIGURE 11-69 Continued

Note that the values at *c_in_to_ASIC*, *b_to_ASIC*, and *a_to_ASIC* change to the values specified by the applied test pattern, and that *c_out_fr_ASIC* and *sum_fr_ASIC* are produced by the adder within *ASIC*.[28] A second test pattern ($3fff_H$) is loaded into

---

[28]The adder implemented in this example has 0 delay.

TDI_reg (see Figure 11-69b) and shifted into *BSC_Scan_Register* while the results of the previous test pattern are scanned out.

The test process is completed in Figure 11-70. The new test pattern is loaded into *BSC_Output_Register* (see Figure 11-70a), and *Expected_out* is compared to *ASIC_out* in Figure 11-70(b). The patterns match, and *Error* remains de-asserted.[29]

11.10.8 Design Example: Built-In Self-Test

Built-in self-test (BIST) logic allows an ASIC to test itself. BIST circuitry is used when it is not practical or possible to test an ASIC with an external tester. Some circuits must be tested in the field each time the host system is restarted; others must be tested as part of a board environment. For example, computers and other sequential machines use BIST to test block RAMs on power-up.

An architecture for a machine with BIST hardware is shown in Figure 11-71. In normal mode the unit under test (UUT) is driven by external (primary) inputs, but in test mode patterns are generated by built-in circuitry and applied to the circuit. The response of the circuit is monitored by additional hardware and compared to an expected response to the input pattern. A difference between the expected and actual response patterns indicates the presence of an internal fault. A controller (state machine) governs the overall process of applying patterns and observing the response of the machine.

The pattern generator for BIST can be implemented by storing stimulus patterns in memory and retrieving them during test mode, but that approach requires a relatively large amount of memory compared to other alternatives. We will consider an alternative using a linear feedback shift register (LFSR)[30] as a pseudo-random pattern generator (PRPG), and a multiple-input signature register (MISR) to monitor patterns. LFSRs are used as PRPGs because they require a small amount of hardware to generate a large set of patterns.

The coefficients of an n-bit autonomous LFSR can be chosen to produce a pseudo-random sequence of n-bit patterns that repeats after $2^n - 1$ steps (i.e., the sequence of patterns is cyclic). This method for generating patterns is attractive because the hardware required to generate the sequence of patterns is significantly less than the hardware that would be required to store the same patterns in memory. LFSRs that have an irreducible and primitive characteristic polynomial produce a sequence of patterns having maximum length [16].

Two types of LFSRs are shown in Figure 11-72. A Type I LFSR augments an ordinary shift register with "external" exclusive-or gates. This type of LFSR can use the same register for ordinary operation. The Type-I shift register in Figure 11-72(a) is tapped to feed cell outputs back to the first (left most) cell in the chain. The Type II structure shown in Figure 11-72(b) has exclusive-or gates in the shift register path at locations where the tap coefficients have the value 1. Both structures generate maximum-length pseudorandom binary sequences, depending on the tap coefficients.

---

[29] The testbench includes an example in which a fault is injected into *ASIC* and detected by an applied test pattern.

[30] See Chapter 5.

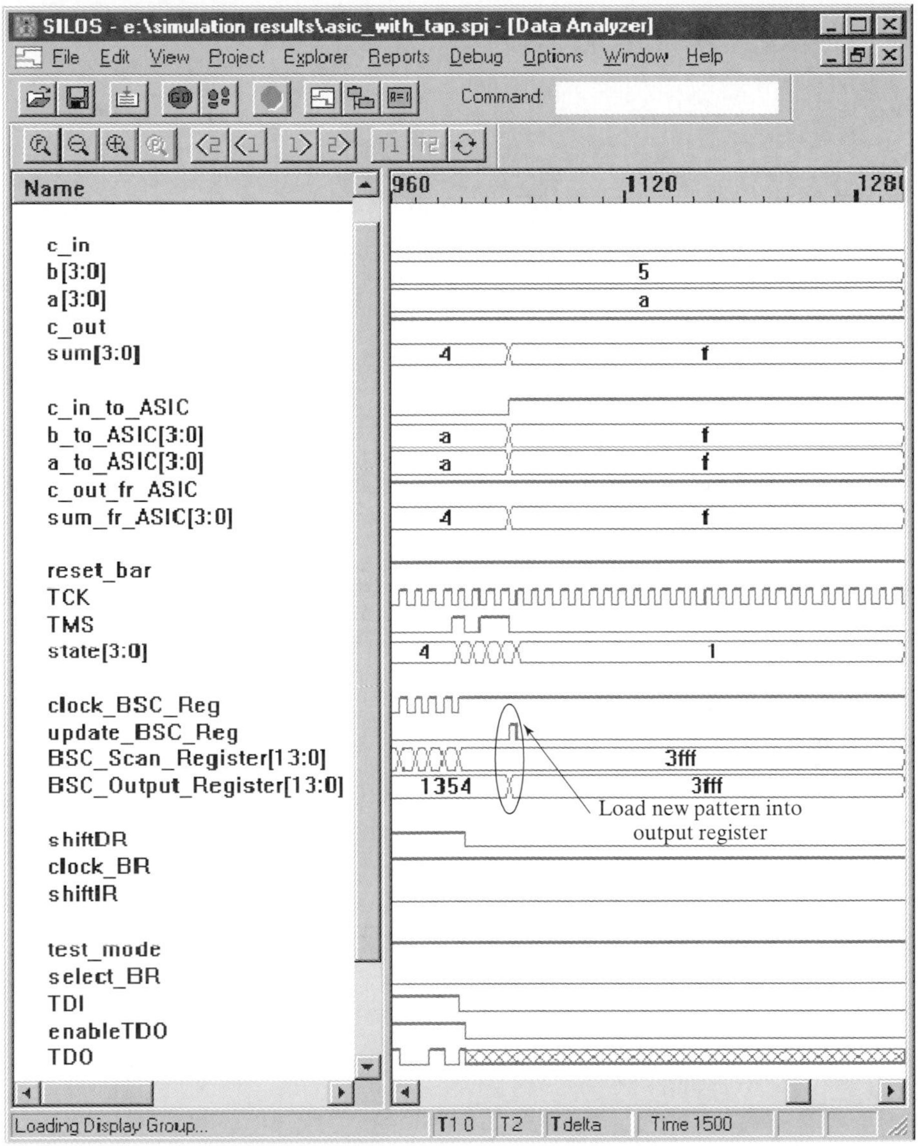

(a)

FIGURE 11-70 Simulation results—after scanning a second data pattern into *ASIC_with_TAP*, (a) the pattern (3fff$_H$ is loaded into *BSC_Output_Register*, and (b) the expected and actual outputs of *ASIC* match.

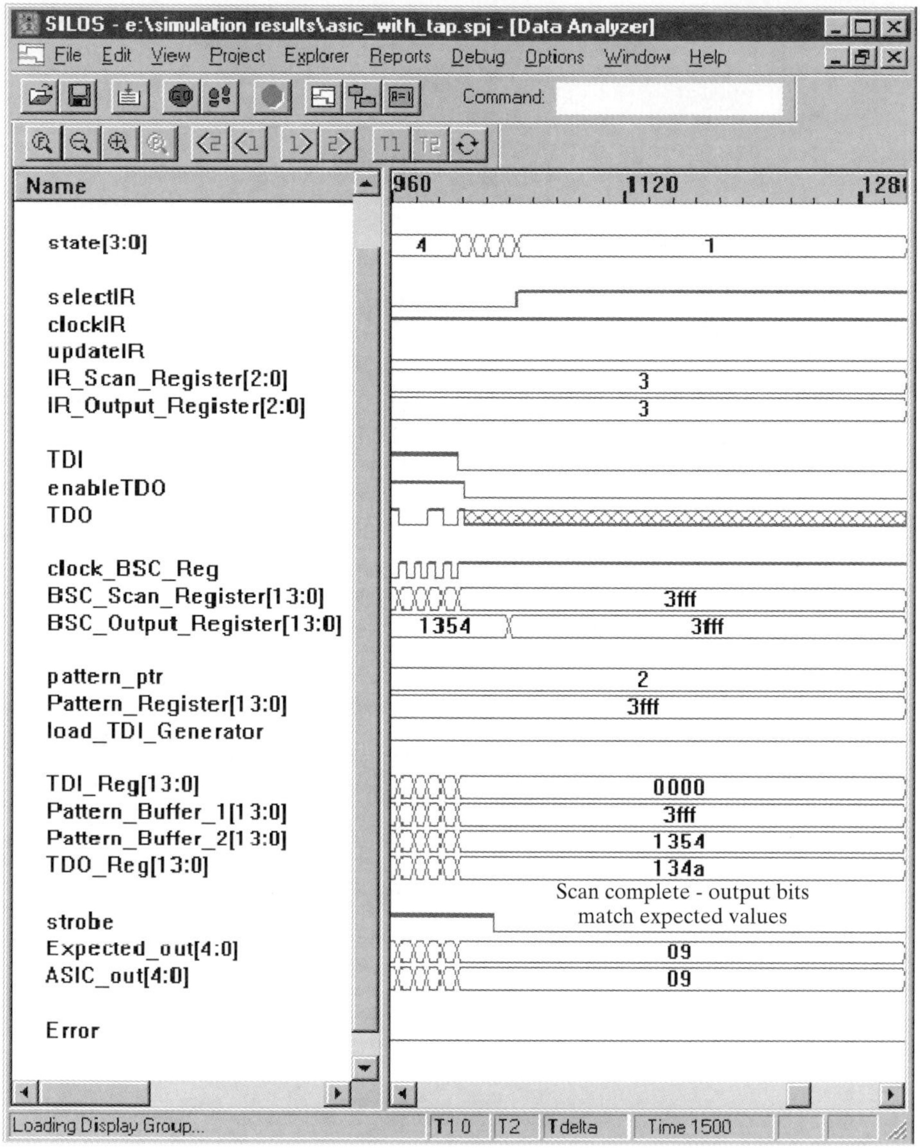

(b)

FIGURE 11-70 Continued

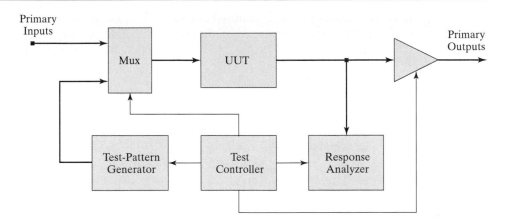

FIGURE 11-71 Architecture for a BIST machine.

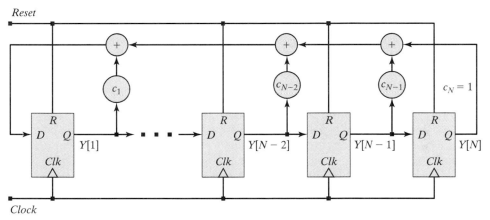

(a)

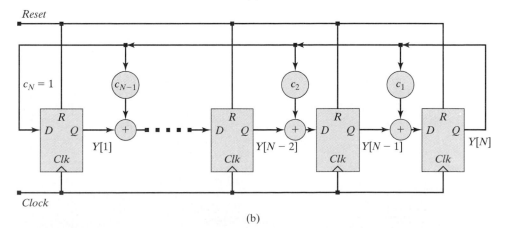

(b)

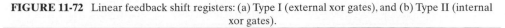

FIGURE 11-72 Linear feedback shift registers: (a) Type I (external xor gates), and (b) Type II (internal xor gates).

Type II LFSRs are preferred in testing because their patterns are more random (i.e., less correlated) than those produced by the Type I machine [4]. Shift register tap coefficients that generate maximal-length pseudorandom binary sequences are given in Table 11-15. Note that the tap coefficients of the two types of LFSRs are labeled in ascending order in opposite directions in Figure 11-72.

The response of a BIST-driven circuit can be compared to its expected response to determine whether the circuit is operating correctly. Instead of storing the patterns of the expected responses, a multiple-input signature register (MISR) compresses the patterns generated by the circuit to form a signature [6]. The signature of a correctly operating circuit is stored for comparison to the actual response. Thus, the MISR circuit and the signature eliminate the need to monitor and compare the responses of the individual test patterns. The MISR in Figure 11-73 is driven by the response vectors of the circuit. The state Y of the circuit after a pattern has been applied is the circuit's signature.

The machine *ASIC_with_BIST* shows how an ASIC can be combined with additional hardware for built-in self-test. For simplicity, the ASIC will be modeled as a 4-bit adder with carries in and out. Figure 11-74 shows the architecture of *ASIC_with_BIST*, including ports for the adder's datapaths, a signal *test_mode*, which controls whether *ASIC_with_BIST* is operating in test mode or in normal mode, and a signal *reset* that drives an internal state machine to a reset state. The signal *done* asserts for one cycle of *clock* to indicate that the BIST test sequence is complete; *error* indicates that the signature produced by *Response_Analyzer* does not match the expected stored signature for the sequence of test vectors generated by the BIST circuit.

The model of *ASIC_with_BIST* includes Verilog modules *ASIC, Pattern_Generator, Response_Analyzer*, and *BIST_Control_Unit*. The BIST implementation does not modify *ASIC*, the circuit that is to be tested by the BIST hardware. *Pattern_Generator*

TABLE 11-15 Tap coefficients for maximum-length pseudo-random binary sequence generators.

n	Coefficient Vectors $(C_n \dots C_2 C_1)$	Coefficents
2	11	$C_2 C_1$
3	101	$C_3 C_1$
4	1001	$C_4 C_1$
5	1_0010	$C_5 C_2$
6	10_0001	$C_6 C_1$
7	100_0100	$C_7 C_3$
8	1000_1110	$C_8 C_4 C_3 C_2$
9	1_0000_1000	$C_9 C_4$
10	10_0000_0100	$C_{10} C_3$
11	100_0000_0010	$C_{11} C_2$
12	1000_0010_1001	$C_{12} C_6 C_4 C_1$
13	1_0000_0000_1101	$C_{13} C_4 C_3 C_1$
14	10_0010_0010_0001	$C_{14} C_{10} C_6 C_1$
15	100_0000_0000_0001	$C_{15} C_1$
16	1000_1000_0000_0101	$C_{16} C_{12} C_3 C_1$
32	1000_0000_0010_0000_0000_0000_0000_0011	$C_{32} C_{22} C_2 C_1$

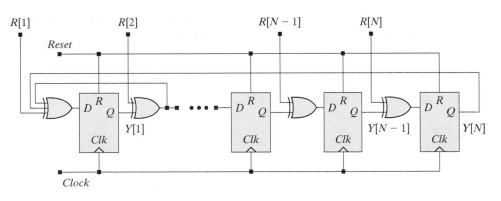

FIGURE 11-73 Multiple-input linear feedback shift register (MISR).

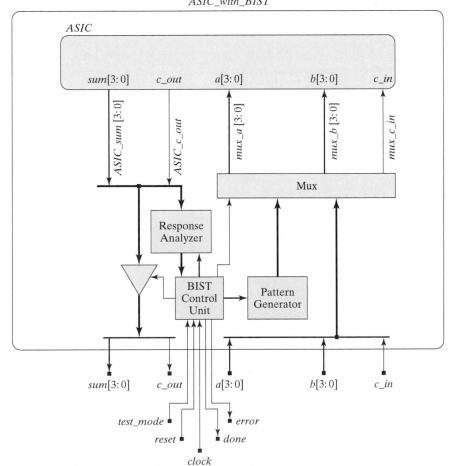

FIGURE 11-74 Architecture for *ASIC_with_BIST*.

is a customized LFSR, with the parameters *size*, specifying the size of the datapaths of the adder in ASIC; *Length*, specifying the length of the shift register; and *initial_state*, specifying the state that results when an external reset is asserted, The maximum-length LFSR in *Pattern_Generator* will generate stimulus patterns, and a MISR in *Response_Analyzer* will generate a signature. At the end of the test sequence, *BIST_Control_Unit* will compare the signature and the stored pattern and assert an error signal if they do not match. The multiplexer and the three-state output buffer in Figure 11-74 are modeled by Verilog continuous-assignment statements in *ASIC_with_BIST*.

The ASM chart in Figure 11-75 describes the state-machine controller for *ASIC_with_BIST*. The signals *clock, reset*, and *test_mode*, are driven by the host environment. The BIST circuit includes a counter, which determines the length of the test sequence. The state remains in *S_test* while patterns are applied and then transitions to *S_compare*, where the signature produced by *Response_Analyzer* is compared to *stored_pattern*. If the patterns match, the state transitions to *S_done* and asserts the Moore-type output *done* for one clock cycle before returning to *S_idle*. If the patterns do not match, the state transitions to *S_error*, where it remains until *reset* is asserted.

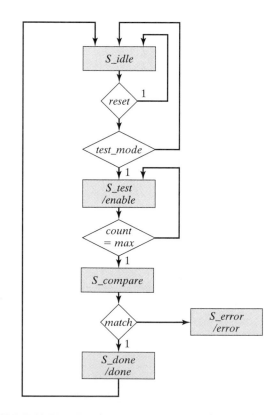

FIGURE 11-75   ASM chart for the controller of *ASIC_with_BIST*.

```
module ASIC_with_BIST (sum, c_out, a, b, c_in, done, error, test_mode, clock, reset);
  parameter                     size = 4;
  output         [size -1: 0]   sum;              //  ASIC interface I/O
  output                        c_out;
  input          [size -1: 0]   a, b;
  input                         c_in;

  output                        done, error;
  input                         test_mode, clock, reset;
  wire           [size -1: 0]   ASIC_sum;
  wire                          ASIC_c_out;
  wire           [size -1: 0]   LFSR_a, LFSR_b;
  wire                          LFSR_c_in;
  wire           [size -1: 0]   mux_a, mux_b;
  wire                          mux_c_in;
  wire                          error, enable;
  wire           [1: size +1]   signature;

  assign         {sum, c_out} = (test_mode) ? 'bz : {ASIC_sum, ASIC_c_out};
  assign         {mux_a, mux_b, mux_c_in} = (enable == 0)
                   ? {a, b, c_in} : {LFSR_a, LFSR_b, LFSR_c_in};

ASIC M0 (
  .sum (ASIC_sum),
  .c_out (ASIC_c_out),
  .a (mux_a),
  .b (mux_b),
  .c_in (mux_c_in));

Pattern_Generator M1 (
  .a (LFSR_a),
  .b (LFSR_b),
  .c_in (LFSR_c_in),
  .enable (enable),
  .clock (clock),
  .reset (reset)
  );

Response_Analyzer M2 (
  .MISR_Y (signature),
  . R_in ({ASIC_sum, ASIC_c_out}),
  .enable (enable),
  .clock (clock),
  .reset (reset));

  BIST_Control_Unit  M3 (done, error, enable, signature, test_mode, clock, reset);

endmodule

module ASIC (sum , c_out, a, b, c_in);
  parameter      size = 4;
  output         [size -1: 0]      sum;
```

```
     output                            c_out;
     input            [size -1: 0]     a, b;
     input                            c_in;

     assign {c_out, sum} = a + b + c_in;
endmodule

module Response_Analyzer (MISR_Y, R_in, enable, clock, reset);
     parameter                        size = 5;
     output           [1: size]       MISR_Y;
     input            [1: size]       R_in;
     input                            enable, clock, reset;
     reg              [1: size ]      MISR_Y;

     always @  (posedge clock)
       if (reset == 0) MISR_Y <= 0;
        else if (enable) begin
          MISR_Y [2: size ] <= MISR_Y [1: size -1] ^ R_in [2: size];
          MISR_Y [1] <= R_in[1] ^ MISR_Y[size] ^MISR_Y[1];
        end
endmodule

module Pattern_Generator (a, b, c_in, enable, clock, reset);
     parameter                        size = 4;
     parameter                        Length = 9;
     parameter                        initial_state = 9'b1_1111_1111;    // initial state
     parameter        [1: Length]     Tap_Coefficient = 9'b1_0000_1000;
     output           [size -1: 0]    a, b;
     output                           c_in;

     input                            enable, clock, reset;
     reg              [1: Length]     LFSR_Y;
     integer                          k;

     assign a = LFSR_Y[2: size+1];
     assign b = LFSR_Y[size + 2: Length];
     assign c_in = LFSR_Y[1];
     always @  (posedge clock)
       if (!reset) LFSR_Y <= initial_state;
         else if (enable) begin
           for (k = 2; k <= Length; k = k + 1)
             LFSR_Y[k] <= Tap_Coefficient[Length-k+1]
               ? LFSR_Y[k-1] ^ LFSR_Y[Length] : LFSR_Y[k-1];
             LFSR_Y[1] <= LFSR_Y[Length];
         end
endmodule

module BIST_Control_Unit  (done, error, enable, signature, test_mode, clock, reset);
     parameter           sig_size = 5;
     parameter           c_size = 10;
     parameter           size = 3;
     parameter           c_max = 510;
```

```
            parameter                    stored_pattern = 5'h1a;  //  signature if fault-free
            parameter                    S_idle = 0,
                                         S_test = 1,
                                         S_compare = 2,
                                         S_done = 3,
                                         S_error = 4;
            output                       done, error, enable;
            input    [1: sig_size]       signature;
            input                        test_mode, clock, reset;
            reg                          done, error, enable;

            reg      [size −1: 0]        state, next_state;
            reg      [c_size −1: 0]      count;
            wire                         match = (signature == stored_pattern);

            always @ (posedge clock) if (reset == 0) count <= 0;
              else if (count == c_max) count <= 0;
              else if (enable) count <= count + 1;
            always @ (posedge clock)
              if (reset == 0) state <= S_idle;
              else state <= next_state;

            always @ (state or test_mode or count or match) begin
              done = 0;
              error = 0;
              enable = 0;
              next_state = S_error;
              case (state)
                S_idle:      if (test_mode) next_state = S_test; else next_state = S_idle;
                S_test:      begin enable = 1; if (count == c_max -1) next_state = S_compare;
                             else next_state = S_test; end
                S_compare: if (match) next_state = S_done;
                             else next_state = S_error;
                S_done:      begin done = 1; next_state = S_idle; end
                S_error:     begin done = 1; error = 1; end
              endcase
            end
        endmodule
```

The testbench for *ASIC_with_BIST* executes the following tests: (1) power-up reset, (2) reset on-the-fly, (3) three-state action of *sum* and *c_out* and selection of the input datapath when *test_mode* is asserted, (4) initiation of activity in the LFSR pattern generator and the MISR when *enable* is asserted by *BIST_Control_Unit*, and (5) detection of an injected fault at an input pin of *ASIC*.

```
        module t_ASIC_with_BIST ();
            parameter                    size = 4;
            parameter                    End_of_Test = 11000;
            wire     [size -1: 0]        sum;              // ASIC interface I/O
            wire                         c_out;
            reg      [size -1: 0]        a, b;
```

```verilog
reg                               c_in;

wire                              done, error;
reg                               test_mode, clock, reset;
reg                               Error_flag = 1;
initial begin Error_flag = 0; forever @ (negedge clock) if ( M0.error) Error_flag =
1; end

ASIC_with_BIST M0 (sum, c_out, a, b, c_in, done, error, test_mode, clock, reset);

initial #End_of_Test $finish;
initial begin clock = 0; forever #5 clock = ~clock; end

// Declare external inputs
initial fork
  a = 4'h5;
  b = 4'hA;
  c_in = 0;
  #500 c_in = 1;
join

// Test power-up reset and launch of test mode
initial fork
  #2 reset = 0;
  #10 reset = 1;

  #30 test_mode = 0;
  #60 test_mode = 1;
join

// Test action of reset on-the-fly

initial fork
  #150 reset = 0;
  #160 test_mode = 0;
join

// Generate signature of fault-free circuit

initial fork
  #180 test_mode = 1;
  #200 reset = 1;
join

// Test for an injected fault

initial fork
  #5350 release M0.mux_b [2] ;
  #5360 force M0.mux_b[0] = 0;
  #5360 begin reset = 0;  test_mode = 1; end
  #5370 reset = 1;
join

endmodule
```

The simulation results in Figure 11-76 demonstrate that power-up reset drives the state of the *BIST_Control_Unit* to *S_idle* (0), and resets the state of the LFSR to $1ff_H$. When *test_mode* is asserted the state transitions to *S_test* (1), where *enable* is asserted. The assertion of *enable* connects the datapath from the LFSR (see *mux_a, mux_b*, and *mux_c_in*) to the ports of *ASIC* and drives the output datapath (see *sum* and *c_out*) of

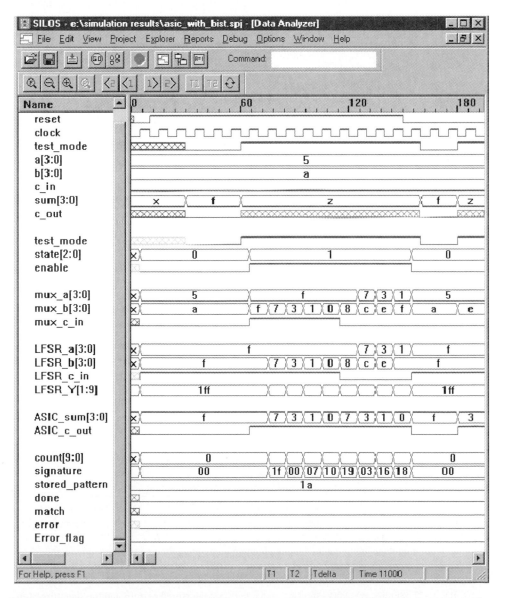

FIGURE 11-76 Simulation results showing (1) power-up reset, (2) reset on-the-fly, (3) three-state action of *sum* and *c_out* and selection of the input datapaths when *test_mode* is asserted, and (4) initiation of activity in the LFSR and the MISR when *test_mode* is asserted.

ASIC_with_BIST into the high-impedance state. With *enable* asserted, the LFSR generates patterns driving the internal datapath for *ASIC_sum* and *ASIC_c_out*, and the MISR within *Response_Analyzer* begins to generate preliminary signatures. A second assertion of *reset* demonstrates that the machine resets on-the-fly.

The simulation results in Figure 11-77 demonstrate that the signature of the fault-free circuit matches the stored pattern. After 510 clock cycles the state enters *S_compare* (2), detects a match, and then enters *S_done* (3) for one cycle, before

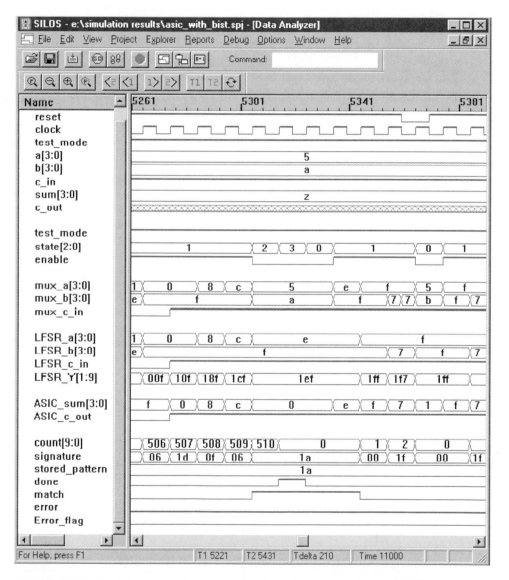

FIGURE 11-77 Simulation results showing match between the stored pattern and the signature of the fault-free circuit.

returning to *S_idle*. The testbench includes *Error_flag* to detect an error when multiple test sequences are applied to detect injected faults. The simulation results in Figure 11-78 were generated after a fault was injected at an input of *ASIC*. They demonstrate that the machine detects the mismatch between the stored pattern of the fault-free machine and the signature produced by the MISR when the machine has the injected fault.

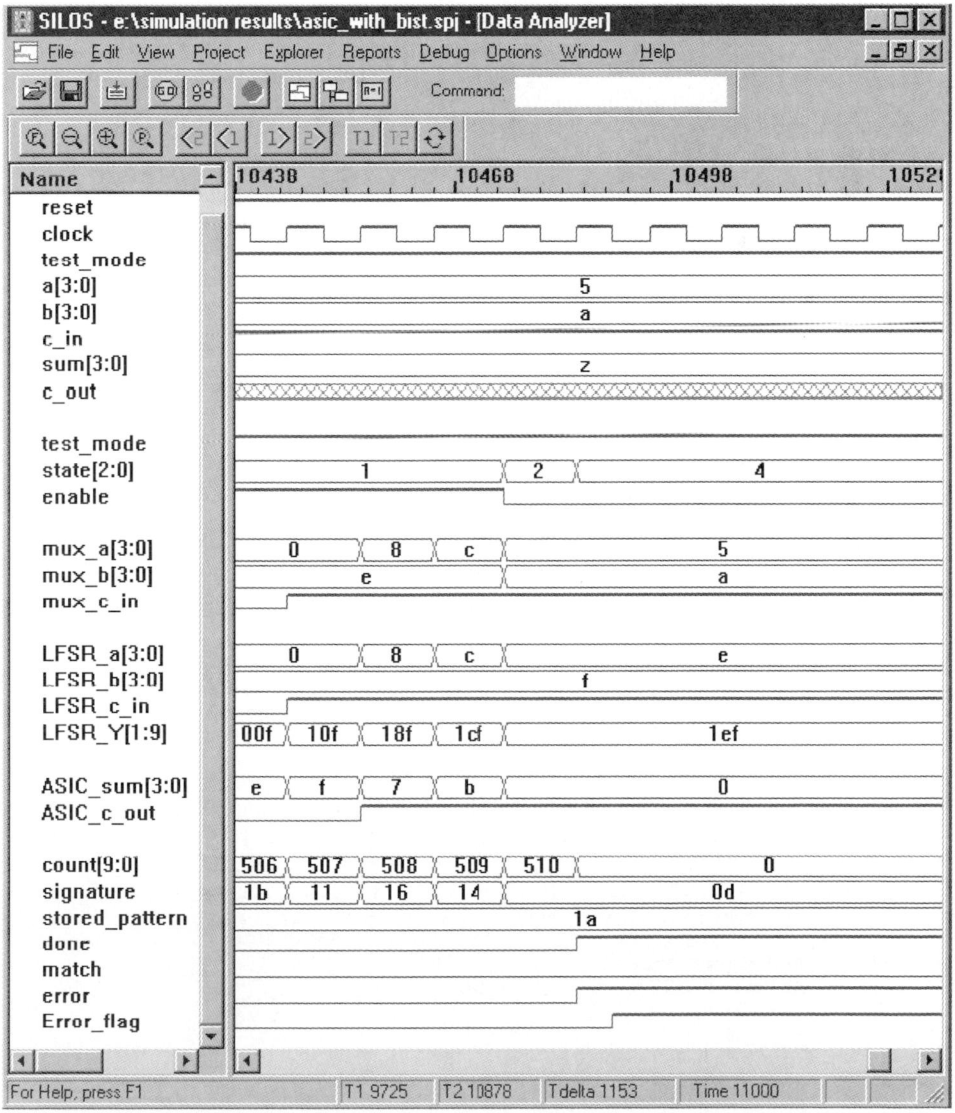

FIGURE 11-78 Simulation results showing detection of an injected fault.

REFERENCES

1. Howe H. "Pre- and Postsynthesis Simulation Mismatches." Proceedings of the Sixth International Verilog HDL Conference, March 31–April 3, 1997, Santa Clara, CA.
2. McWilliams TM. "Verification of Timing Constraints on Large Digital Systems." Ph.D. Thesis, Stanford University, 1980.
3. Osterhout JK. "Crystal: A Timing Analyzer for nMOS VLSI Circuits." In: Bryant R, ed. *Proceedings of the Third Caltech Conference on VLSI*. Rockville, MD: Computer Science Press, 1983, 57–69.
4. Smith MJS. *Application-Specific Integrated Circuits*. Reading, MA: Addison-Wesley, 1997.
5. Xilinx University Program Workshop Notes—ISE 3.1i, Spring 2001.
6. Abramovici M, et al. *Digital Systems Testing and Testable Design*. New York: Computer Science Press, 1990.
7. Roth JP. "Diagnosis of Automatic Failures: A Calculus and a Method." *IBM Journal of Research and Development, 10*, pp. 278–281.
8. Goel P. "An Implicit Enumeration Algorithm to Generate Tests for Combinational Logic Circuits." *IEEE Transactions on Computers, C-30*, 215–222, 1983.
9. Fujiwara H. *Logic Testing and Design for Testability*. Cambridge, MA: MIT Press, 1985.
10. Fujiwara H, Shimono T. "On the Acceleration of Test Generation Algorithms." *IEEE Transactions on Computers, C-32*, 1137–1144.
11. Williams TW, Brown NC. "Defect Level as a Function of Fault Coverage." *IEEE Transactions on Computers, C-30*, 987–988.
12. Goldstein LH. "Controllability/Observability Analysis of Digital Circuits." *IEEE Transactions on Circuits and Systems, CAS-26*, 685–603, 1979.
13. IEEE 1149.1–1990, IEEE Test Access Port and Boundary-Scan Architecture. Piscataway, NJ: Institute of Electrical and Electronics Engineers, 1990.
14. Wakerly JF. *Digital Design Principles and Practices*. 3rd ed. Upper Saddle River, NJ: Prentice-Hall, 2000.
15. Peterson WW. *Error-Correcting Codes*. Cambridge, MA: MIT Press, 1961.

PROBLEMS

1. The gates in Figure P11-1 are annotated with (min:max) delay ranges for rising and falling output transitions. Develop a DAG for the circuit, and enumerate

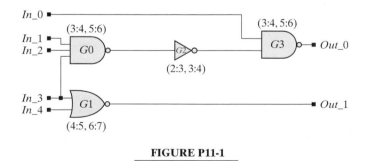

FIGURE P11-1

the delay ranges of the paths through the circuit for rising and falling transitions of the outputs.

2. Using a static timing analyzer, construct test cases demonstrating that the model of the static RAM in Example 8.15, *RAM_2048_8 (data, addr, CS_b, OE_b, WE_b)*, detects timing violations.

3. Using a static timing analyzer, determine the maximum frequency at which a synthesized implementation of the SRAM controller, *SRAM_Con*, can operate (see Problem 9 in Chapter 8).

4. Using a static timing analyzer, determine the maximum clock frequency at which a synthesized implementation of *ALU_machine_4_bit* can operate without a timing violation (see Problem 7 in Chapter 8).

5. Using a static timing analyzer, determine the maximum clock frequency at which a synthesized implementation of *FIFO* can operate without a timing violation (see Problem 10 in Chapter 8).

6. Using a static timing analyzer, determine the maximum clock frequency at which a synthesized implementation of *UART_Transmitter_Arch* can operate without a timing violation (see Chapter 7).

7. Using a static timing analyzer, determine the maximum clock frequency at which a synthesized implementation of *UART_8_Receiver* can operate without a timing violation (see Chapter 7).

8. Using a static timing analyzer, determine the maximum clock frequency at which a synthesized implementation of *RISC_SPM* can operate without a timing violation (see Chapter 7).

9. Using a static timing analyzer, compare the maximum throughput and latency of a gate-level implementation of the 4×4 binary multiplier shown in Figure 10.14, and a balanced pipelined implementation of the same circuit (see Problem 40 in Chapter 10).

10. Gated clocks can be problematic in ASICs and FPGAs. Compare the circuits shown in Figure P11-10, which can be used to generate a clock pulse when a binary counter reaches a specified count.

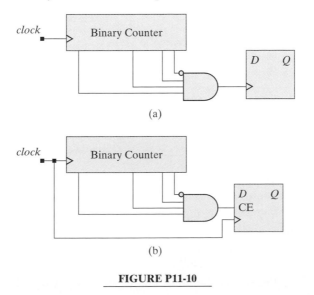

(a)

(b)

FIGURE P11-10

11. Conduct a timing analysis of a 4-bit ripple-carry adder and a 4-bit carry look-ahead adder. The adders are to be implemented using cells from a CMOS standard-cell library (include the source code for the models of the adders). List in the table below the cells that are used in your design, along with their propagation delays (indicate the physical units).

TABLE P11-11

Cell name	Propagation Delay (Rising)	Propagation Delay (Falling)

a. Using a static timing analyzer (e.g., Synopsys PrimeTime)
   i.   Create a timing analysis report and a distribution of the path lengths of the ripple carry adder. Find the longest path, identify it below, and indicate its delay.

Longest path begins at: _____
Longest path ends at: _____
Delay of the longest path   _____ ns

   ii. Find the shortest path, identify it below, and indicate its delay.

Shortest path begins at: _____
Shortest path ends at: _____
Path delay: _____ ns

   iii. Create a timing analysis report and a distribution of the path lengths of the carry look-ahead adder. Find the longest path, identify it below, and indicate its delay.

Longest path begins at: _____
Longest path ends at: _____
Delay of the longest path _____ ns

   iv. Find the shortest path, identify it below, and indicate its delay.

Shortest path begins at: _____
Shortest path ends at: _____
Path delay: _____ ns

b. Using cell area data, compare the areas of the two implementations.
   Area (ripple carry)   _____
   Area (look-ahead)   _____

Provide data showing how the area was calculated.

   c. Develop a testbench for the designs, and identify below input patterns that will exercise the longest and shortest paths, and the delays that are observed *in simulation* with these patterns. Provide waveforms of the results (annotate the results to show the delay).

Pattern for longest path: _____

Delay observed in simulation: _____

Pattern for shortest path: _____

Delay observed in simulation: _____

   d. For 16-bit ripple carry (RCA) and carry look-ahead (CLA) adders implemented in 4-bit slices, find the shortest clock cycle under which data can be fetched from a register, added, and then stored in a register (see Figure 11-7), using the flip-flops of the cell library. (*Note:* The registers are not part of the adder). Provide annotated simulation results showing the above delays.

Shortest clock cycle: RCA: _____

Shortest clock cycle: CLA: _____

   e. Discuss the significant differences between the adders.

12. Find a minimum set of test patterns that will cover all of the *s-a-0* and *s-a-1* faults in the gate-level model of a full adder circuit shown in Figure 11-51.

13. Table 11-10 shows that the binary full adder circuit has 30 *s-a-0* and *s-a-1* faults, and 26 prime faults. Identify the equivalence classes of the adder's faults.

14. The circuit in Figure P11-14 was developed in Example 2.34, where the redundant logic of the AND gate driving *F2* was added to the circuit to cover a hazard. Determine whether the *s-a-0* fault on the input to the AND gate is detectable.

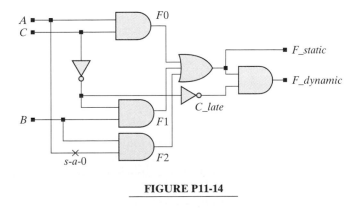

FIGURE P11-14

15. Develop tests for the faults *G4 s-a-0* and *G4 s-a-1* in Figure 11-51.

16. Exhaustive pattern generation/fault simulation is avoided. It is desirable to have a small set of vectors detect as many faults as possible. Conduct a fault simulation of gate-level models of a 4-bit ripple-carry adder and a 4-bit carry look-ahead adder (use a cell library).

a. Provide graphical (waveform) simulation results verifying the functionality of the adders by applying the same set of five patterns to both adders and comparing their outputs to that of a behavioral adder implemented within your testbench by a continuous-assignment statement. The set of vectors is to consist of three patterns of your choice together with the following two test patterns: $a = b = 4'h5$, $c\_in = 0$ and $a = b = 4'hA$, $c\_in = 1$, where a and b are the datapaths to the adders. Your graphical output is to show the values of a, b, c_in, sum, and c_out in decimal format.

b Using a fault simulator, determine the fault coverage provided by the five patterns in part a. Add up to five more patterns to the test suite to increase the coverage, and complete Table P11-16, listing the patterns (hex format) and showing the cumulative and incremental coverage achieved by adding patterns to the suite.

TABLE P11-16

		a	b	c_in	Cumulative and Incremental Coverage			
					RCA_{cum}	RCA_{inc}	CLA_{cum}	CLA_{inc}
1		5	5	0				
2		A	A	1				
3								
4								
5								
6								
7								
8								
9								
10								

17. Develop and verify a Verilog model of the instruction register cell shown in Figure 11.60.

18. Develop a test plan and a testbench to verify the functionality of *Instruction_Register* (see Section 11.10.6).

19. Develop a test plan and a testbench to verify the functionality of *TAP_Controller* (see Section 11.10.6).

20. Verify that *ASIC_with_TAP* (see Section 11.10.6) is synthesizable. Check for latches in the netlist of the synthesized circuit.

21. What feature of the JTAG TAP controller (see Figure 11-63) enables the machine to enter *S_Reset* at the first active edge of the clock following power-up, in the absence of the optional active-low external asynchronous reset *nTRST*?

22. The model for *ASIC_with_TAP* generates gated clocks in *Instruction_Decoder*. Develop *ASIC_with_TAP_NGC*, a model that does not have gated clocks.

23. An optional and dedicated device identification register conforming to the JTAG specification[31] must have the structure shown in Figure P11-23, and have the following features:

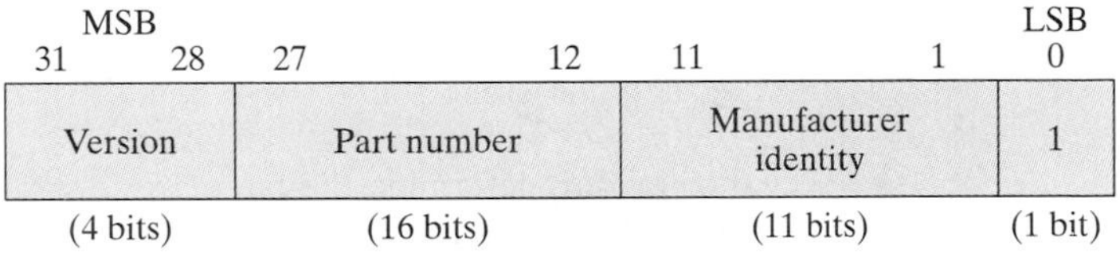

```
       MSB                                                    LSB
    31      28    27              12    11              1      0
   ┌──────────┬─────────────────────┬──────────────────┬─────────┐
   │          │                     │   Manufacturer   │         │
   │ Version  │    Part number      │                  │    1    │
   │          │                     │     identity     │         │
   └──────────┴─────────────────────┴──────────────────┴─────────┘
     (4 bits)        (16 bits)           (11 bits)       (1 bit)
```

FIGURE P11-23

a. A shift-register based path that has serial and parallel inputs, but only a serial output.
b. On the rising edge of *TCK* in the TAP controller state *S_Capture_DR*, the register shall be set such that subsequent shifting causes the content of the register to be scanned through the TDO output of the TAP.
c. The register shall be accessed via the IDCODE instruction.
d. The operation of the register shall have no effect on the operation of the on-chip system logic.

Develop and verify *Device_ID_Register*, a Verilog model conforming to the above specifications.

24. Develop and verify *ASIC_with_TAP_ID* by modifying *ASIC_with_TAP* to include a device identification register.
25. Modify the design of *ASIC_with_TAP* to include the active-low asynchronous reset *nTRST*.
26. The model *ASIC_with_TAP* (see Section 11.10.6) has parameters in several modules. Consequently, each module must be edited to resize the design. Using the **defparam** construct to define all of the parameters of *ASIC_with_TAP*, develop and include an annotation module in *t_ASIC_with_TAP* to parameterize the design.
27. The model *ASIC_with_TAP* (see Section 11.10.6) includes pull-up devices for *TMS*, *TDI*, and *nTRST*. Develop a testbench to verify that the models behave according to the JTAG specification (i.e., the response of the model to an undriven input [at *TMS*, *TDI*, or *nTRST*] shall be identical to the application of a logic 1).
28. The state of the JTAG TAP controller enters *S_Reset* on power-up and remains there until *TMS* is asserted low. Using the ASM chart in Figure 11-63, explain how the host system is protected from a glitch on TMS (i.e., TMS inadvertently changes to 0 for one clock cycle before returning to 1).

[31]Note that by treating the manufacturer identity code 000_0111_1111 as an illegal code, it is possible to detect the end of the identity code sequence from a multichip board, given that the bypass register must load a 1 on the transition from *S_Capture_DR* to *S_Shift_DR*. This enables testers to locate the end of the identification code sequence for a board containing an unknown number of JTAG-enhanced components. The standard includes provisions for a supplementary identification code, and specifications for the manufacturer identification code, which will not be considered here.

29. Develop *ALU_4_bit_with_JTAG*, a JTAG-enhanced version of *ALU_4_bit* (see Table P8.7a in Chapter 8 for a functional specification of *ALU_4_bit*), and a testbench, *t_ALU_4_bit_with_JTAG* having scan patterns that exhaustively verify the functionality of the opcode *A_and_B* and tests for any *s-a-0* or *s-a-1* fault that can be detected at the output. Include in the testbench a test demonstrating that a fault injected at an output pin of *ALU_4_bit* will be detected.

30. Develop *ALU_machine_4_bit_with_JTAG*, a JTAG-enhanced version of *ALU_Machine_4_bit* (see Figure P8-7b in Chapter 8) with a test-data register (TDR) formed by a scan chain of the register in *ALU_machine_4_bit*. Develop a testbench, *t_ALU_machine_4_bit_with_JTAG* with scan patterns for exhaustively verifying the functionality of the opcode *A_and_B* and tests for any *s-a-0* or *s-a-1* fault that can be detected at the output. Include in the testbench a test demonstrating that a fault injected at an output pin of *ALU_4_bit* will be detected.

31. Explain why the state *S_Exit1_DR* is needed in the ASM chart of the TAP controller.

32. Replace the behavioral model of *ASIC* by a gate-level model in *ASIC_with_TAP*. Modify *t_ASIC_with_TAP* to inject structural faults within *ASIC*, and verify that the patterns developed in Problem 16 detect the faults.

33. Using a 4-bit adder ASIC, combine the features of *ASIC_with_TAP* and *ASIC_with_BIST* to form *ASIC_with_JTAG_BIST*, which executes a BIST procedure under the control of the TAP controller.

34. Develop a model for *Board_with_Four_ASICS*, a module that contains four copies of *ASIC_with_TAP* (see Section 11.10.6) connected in a ring configuration (see Figure 11-62). The ASICs are to be connected to form a 16-bit ripple carry adder with the port structure (*sum[15:0], c_out, a[15:0], b[15:0], c_in*). Develop a testbench, *t_Board_with_Four_ASICS*, and test sequences that (1) bypass all four chips, (2) bypass all but the chip producing the most significant bit-slice of the board's output, (3) test the interconnect between the ASICs producing the least significant 8 bits of the machine, (4) tests the ASIC producing *sum[7:4]* for internal faults (use gate-level models for the bit-slice adders). Using the *force . . . release* construct, the testbench is to inject faults as needed to demonstrate that faults are detected.

35. Develop a model for *Board_with_Four_ASICS*, a module that contains four copies of *ASIC_with_TAP* (see Section 11.10.6) connected in a star configuration (see Figure 11-62). The ASICs are to be connected to form a 16-bit ripple carry adder with the port structure (*sum[15:0], c_out, a[15:0], b[15:0], c_in*). Develop a testbench, *t_Board_with_Four_ASICS*, and test sequences that (1) bypass all four chips, (2) bypass all but the chip producing the most significant bit-slice of the board's output, (3) test the interconnect between the ASICs producing the least significant 8 bits of the machine, (4) tests the ASIC producing *sum[7:4]* for internal faults (use gate-level models for the bit-slice adders). Using the *force . . . release* construct, the testbench is to inject faults as needed to demonstrate that faults are detected.

36. Develop and verify a Verilog model for the Type 1 LFSR in Figure 11-71(a).
37. The Type I and Type II LFSRs shown in Figure 11-71 do not enter the state $Y = 0$, because they could not exit from that state. Explain how the modified Type I LFSR shown in Figure P11-37 enters and leaves the state $Y = 0000$.

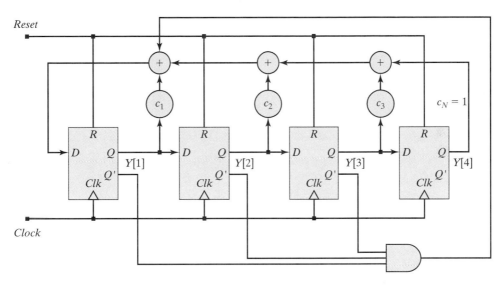

FIGURE P11-37

Verilog Primitives

Verilog has a set of 26 primitives for modeling the functionality of combinational and switch-level logic. The output terminals of an instantiated primitive arc listed first in its primitive terminal list. The input terminals are listed last. The **buf, not, notif0** and **notif1** primitives ordinarily have a single input, but may have multiple scalar outputs. The other primitives may have multiple scalar inputs, but have only one output. In the case of the three-state primitives (**bufif1, bufif0, tranif0, rranif0, tranif1**, and **rtranif1**), the control input is the last input in the terminal list. When a vector of primitives is instantiated the ports may be vectors. If the inputs and outputs of a primitive are vectors, the output vector is formed on a bitwise basis from the input's vector.

Primitives may be instantiated with propagation delay, and may have strength assigned to their output net(s). Their input–output functionality in Verilog's four-valued logic system is defined by the following truth tables, where the symbol L represents 0 or z, and the symbol H represents 1 or z. These additional symbols accommodate simulation results in which a signal can have a value of 0 or z, or a value of 1 or z, respectively.

A.1 Multiinput Combinational Logic Gates

The truth tables of Verilog's combinational logic gates are shown for two inputs, but the gates may be instantiated with an arbitrary number of scalar inputs.

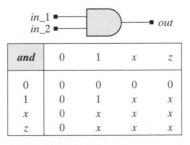

and	0	1	x	z
0	0	0	0	0
1	0	1	x	x
x	0	x	x	x
z	0	x	x	x

FIGURE A-1   Truth table for bitwise-and gate (***and***). Terminal order: *out, in_1, in_2*

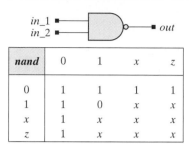

nand	0	1	x	z
0	1	1	1	1
1	1	0	x	x
x	1	x	x	x
z	1	x	x	x

FIGURE A-2   Truth table for bitwise-nand gate (***nand***). Terminal order: *out, in_1, in_2*

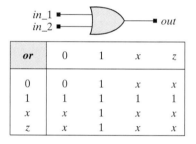

or	0	1	x	z
0	0	1	x	x
1	1	1	1	1
x	x	1	x	x
z	x	1	x	x

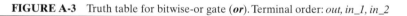

FIGURE A-3   Truth table for bitwise-or gate (***or***). Terminal order: *out, in_1, in_2*

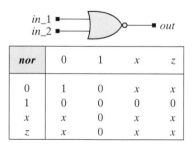

nor	0	1	x	z
0	1	0	x	x
1	0	0	0	0
x	x	0	x	x
z	x	0	x	x

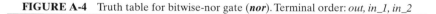

FIGURE A-4   Truth table for bitwise-nor gate (***nor***). Terminal order: *out, in_1, in_2*

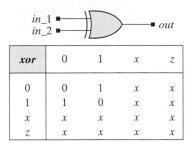

xor	0	1	x	z
0	0	1	x	x
1	1	0	x	x
x	x	x	x	x
z	x	x	x	x

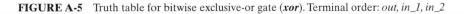

FIGURE A-5   Truth table for bitwise exclusive-or gate (*xor*). Terminal order: *out, in_1, in_2*

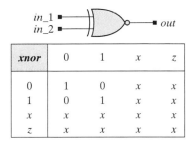

xnor	0	1	x	z
0	1	0	x	x
1	0	1	x	x
x	x	x	x	x
z	x	x	x	x

FIGURE A-6   Truth table for bitwise exclusive-Nor gate (*xnor*). Terminal order: *out, in_1, in_2*

A.2   Multioutput Combinational Gates

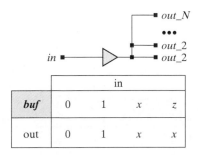

buf	in			
	0	1	x	z
out	0	1	x	x

FIGURE A-7   Truth table for bitwise buffer (*buf*). Terminal order: *out_1, out_2, . . . , out_N, in*

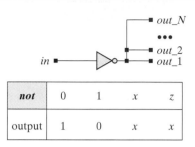

not	0	1	*x*	*z*
output	1	0	*x*	*x*

FIGURE A-8   Truth table for bitwise inverter (***not***). Terminal order: *out_1, out_2, . . . , out_N, in*

A.3   Three-State Gates

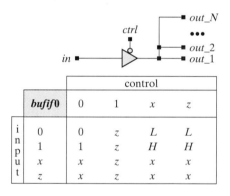

		control			
bufif0		0	1	*x*	*z*
input	0	0	*z*	*L*	*L*
	1	1	*z*	*H*	*H*
	x	*x*	*z*	*x*	*x*
	z	*x*	*z*	*x*	*x*

FIGURE A-9   Truth table for bitwise three-state buffer (***bufif0***) gate with active-low enable. Terminal order: *out_1, out_2, . . . , out_N, in, ctrl*

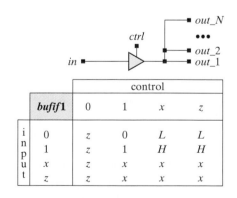

		control			
bufif1		0	1	*x*	*z*
input	0	*z*	0	*L*	*L*
	1	*z*	1	*H*	*H*
	x	*z*	*x*	*x*	*x*
	z	*z*	*x*	*x*	*x*

FIGURE A-10   Truth table for bitwise three-state buffer (***bufif1***). Terminal order: *out_1, out_2, . . . , out_N, in, ctrl*

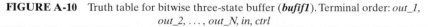

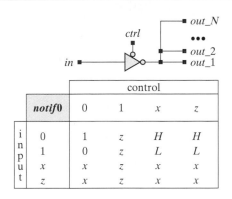

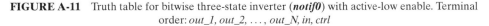

FIGURE A-11 Truth table for bitwise three-state inverter (***notif0***) with active-low enable. Terminal order: *out_1, out_2, . . . , out_N, in, ctrl*

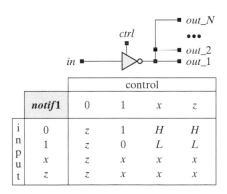

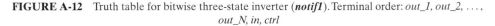

FIGURE A-12 Truth table for bitwise three-state inverter (***notif1***). Terminal order: *out_1, out_2, . . . , out_N, in, ctrl*

A.4   MOS Transistor Switches

The **cmos, rcmos, nmos, rnmos, pmos**, and **rpmos** gates may be accompanied by a delay specification with one, two, or three values. A single value specifies the rising, falling, and turn-off delay (i.e., to the z state) of the output. A pair of values specifies the rising and falling delays, and the smaller of the two values determines the delay of transitions to x and z. A triple of values specifies the rising, falling, and turn-off delay, and the smallest of the three values determines the transition to x. Delays of transitions to L and H are the same as the delay of a transition to x.[1]

---

[1]See Ciletti MD. *Modeling Synthesis and Rapid Prototyping with the Verilog HDL* (Prentice-Hall, Upper Saddle River, NJ: 1999) for a discussion of the rules governing the strength of nets driven by switch-level primitives.

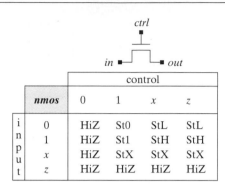

	nmos	control			
		0	1	x	z
input	0	HiZ	St0	StL	StL
	1	HiZ	St1	StH	StH
	x	HiZ	StX	StX	StX
	z	HiZ	HiZ	HiZ	HiZ

FIGURE A-13   nmos pass transistor switch (***nmos***). Terminal order: *out, in, ctrl*

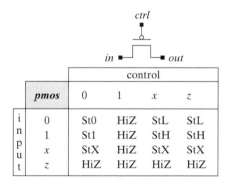

	pmos	control			
		0	1	x	z
input	0	St0	HiZ	StL	StL
	1	St1	HiZ	StH	StH
	x	StX	HiZ	StX	StX
	z	HiZ	HiZ	HiZ	HiZ

FIGURE A-14   pmos pass transistor switch (***pmos***). Terminal order: *out, in, ctrl*

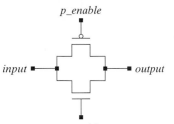

cmos	control	input			
n_enable	*p_enable*	0	1	*x*	*z*
0	0	St0	St1	StX	HiZ
0	1	HiZ	HiZ	HiZ	HiZ
0	*x*	StL	StH	StX	HiZ
0	*z*	StL	StH	StX	HiZ
1	0	St0	St1	StX	HiZ
1	1	St0	St1	StX	HiZ
1	*x*	St0	St1	StX	HiZ
1	*z*	St0	St1	StX	HiZ
x	0	St0	St1	StX	HiZ
x	1	StL	StH	StX	HiZ
x	*x*	StL	StH	StX	HiZ
x	*z*	StL	StH	StX	HiZ
z	0	St0	St1	StX	HiZ
z	1	StL	StH	StX	HiZ
z	*x*	StL	StH	StX	HiZ
z	*z*	StL	StH	StX	HiZ

FIGURE A-15   CMOS transmission gate (***cmos***). Terminal order: *output, input, n_enable, p_enable*

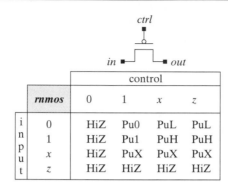

	rnmos	0	1	*x*	*z*
i n p u t	0	HiZ	Pu0	PuL	PuL
	1	HiZ	Pu1	PuH	PuH
	x	HiZ	PuX	PuX	PuX
	z	HiZ	HiZ	HiZ	HiZ

FIGURE A-16   High-resistance nmos pass transistor switch (***rnmos***). Terminal order: *out, in, ctrl*

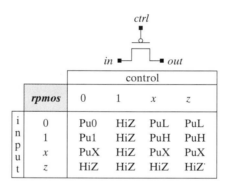

	rpmos	0	1	*x*	*z*
i n p u t	0	Pu0	HiZ	PuL	PuL
	1	Pu1	HiZ	PuH	PuH
	x	PuX	HiZ	PuX	PuX
	z	HiZ	HiZ	HiZ	HiZ˙

FIGURE A-17   High-resistance pmos pass transistor switch (***rpmos***). Terminal order: *out, in, ctrl*

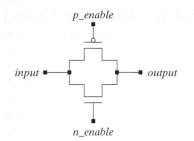

rcmos	control	input			
n_enable	p_enable	0	1	x	z
0	0	Pu0	Pu1	PuX	HiZ
0	1	HiZ	HiZ	HiZ	HiZ
0	x	PuL	PuH	PuX	HiZ
0	z	PuL	PuH	PuX	HiZ
1	0	Pu0	Pu1	PuX	HiZ
1	1	Pu0	Pu1	PuX	HiZ
1	x	Pu0	Pu1	PuX	HiZ
1	z	Pu0	Pu1	PuX	HiZ
x	0	Pu0	Pu1	PuX	HiZ
x	1	PuL	PuH	PuX	HiZ
x	x	PuL	PuH	PuX	HiZ
x	z	PuL	PuH	PuX	HiZ
z	0	Pu0	Pu1	PuX	HiZ
z	1	PuL	PuH	PuX	HiZ
z	x	PuL	PuH	PuX	HiZ
z	z	PuL	PuH	PuX	HiZ

FIGURE A-18   High-resistance cmos transmission gate (**rcmos**). Terminal order: *output, input, n_enable, p_enable*

A.5   MOS Pull-Up/Pull-Down Gates

The pull-up (***pullup***) and pull-down (***pulldown***) gates place a constant value of 1 or 0 with strength *pull*, respectively, on their output. This value is fixed for the duration of simulation, so no delay values may be specified for these gates. The default strength of these gates is *pull. Note*: The ***pulldown*** and ***pullup*** gates are not to be confused with ***tri0*** and ***tri1*** nets. The latter are nets that provide connectivity, and may have multiple drivers; the former are functional elements in the design. The ***tri0*** and ***tri1*** nets may have multiple drivers. The net driven by a ***pullup*** or ***pulldown*** gate may also have multiple drivers. Verilog's ***pullup*** and ***pulldown*** primitives can be used to model pull-up and pull-down devices in electrostatic-discharge circuitry tied to unused inputs on flip-flops.

FIGURE A-19   Pull-up device. Terminal order: *out*

FIGURE A-20   Pull-down device. Terminal order: *out*

A.6   MOS Bidirectional Switches

Verilog includes six predefined bidirectional switch primitives: ***tran, rtran, tranif0, rtranif0, tranif1***, and ***rtranif1***. Bidirectional switches provide a layer of buffering on bidirectional signal paths between circuits. A signal passing through a bidirectional switch is not delayed (i.e., output transitions follow input transitions without delay).

Note: The ***tran*** and ***rtran*** primitives model bidirectional pass gates, and may not have a delay specification. These bidirectional switches pass signals without delay. The ***rtranif0, rtranif1, tranif1***, and ***rtranif1*** switches are accompanied by a delay specification, which specifies the turn-on and turn-off delays of the switch; the signal passing through the switch has no delay. A single value specifies both delays, a pair of values (turn-on, turn-off) specifies both delays, with the turn-on being the first item and turn-off being the second item. The default delay is 0.

FIGURE A-21   Bidirectional switch (***tran***). Terminal order: *in_out1, in_out2*

FIGURE A-22   Resistive bidirectional switch (***rtran***). Terminal order: *in_out1, in_out2*

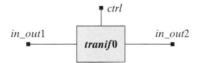

FIGURE A-23   Three-state bidirectional switch (***tranif0***). Terminal order: *in_out1, in_out2, ctrl*

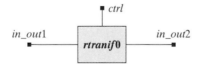

FIGURE A-24   Resistive three-state bidirectional switch (***rtranif0***). Terminal order: *in_out1, in_out2, ctrl*

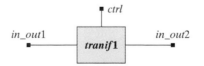

FIGURE A-25   Three-state bidirectional switch (***tranif1***). Terminal order: *in_out1, in_out2, ctrl*

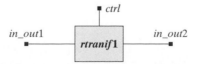

FIGURE A-26   Resistive three-state bidirectional switch (***rtranif1***). Terminal order: *in_out1, in_out2, ctrl*

Verilog Keywords

Verilog keywords are predefined, lower-case, nonescaped identifiers that define the language constructs. An identifier may not be a keyword, and an escaped identifier is not treated as a keyword. In this book, Verilog keywords are printed in boldface.

always	for	parameter	supply0
and	force	pmos	supply1
assign	forever	posedge	table
begin	fork	primitive	task
buf	function	pull0	time
bufif0	highz0	pull1	tran
bufif1	highz1	pulldown	tranif0
case	if	pullup	tranif1
casex	initial	rcmos	tri
casez	inout	real	tri0
cmos	input	realtime	tri1
deassign	integer	reg	triand
default	join	release	trior
defparam	large	repeat	trireg
disable	macromodule	rnmos	vectored
edge	medium	rpmos	wait
else	module	rtran	wand
end	nand	rtranif0	weak0
endcase	negedge	rtranif1	weak1
endfunction	nmos	scalared	while
endmodule	nor	small	wire
endprimitive	not	specify	wor
endspecify	notif0	specparam	xnor
endtable	notif1or	strength	xor
endtask	or	strong0	
event	output	strong1	

Verilog Data Types

Verilog has two families of fixed data types: nets and registers. Nets establish structural connectivity. Registers store information.

C.1   Nets

The family of net data types is described in Table C-1.

TABLE C-1   Data types in Verilog

Net Types	
wire	Establishes connectivity, with no logical behavior or functionality implied.
tri	Establishes connectivity, with no logical behavior or functionality implied. This type of net has the same functionality as **wire**, but is identified distinctively to indicate that it will be three-stated in hardware.
wand	A net that is connected to the output of multiple primitives. It models the hardware implementation of a wired-AND, e.g., open collector technology.
wor	A net that is connected to the output of multiple primitives. It models the hardware implementation of a wired-OR, e.g., emitter coupled logic.
triand	A net that is connected to the output of multiple primitives. It models the hardware implementation of a wired-AND, e.g., open collector technology. The physical net is to be three-stated.
trior	A net that is connected to the output of multiple primitives. It models the hardware implementation of a wired-OR, e.g., emitter coupled logic. The physical net is to be three-stated.
supply0	A global net that is connected to the circuit ground.
supply1	A global net that is connected to the power supply.
tri0	A net that is connected to ground by a resistive pull-down connection.
tri1	A net that is connected to the power supply by a resistive pull-up connection.
trireg	A net that models the charge stored on a physical net.

At time $t_{\text{sim}} = 0$, nets that are driven by a primitive, module, or continuous assignment have a value determined by their drivers, which defaults to the (ambiguous) logic value, x. The simulator assigns the default logic value z (high impedance) to all nets that are not driven. These initial assigned values remain until they are changed by subsequent events during simulation.

C.2 Register Variables

Register variables are assigned values by procedural statements within an *always* or *initial* block. Register variables hold their value until an assignment statement changes them. The following are predefined register types: *reg, integer, real, realtime*, and *time*.

C.2.1 Data Type: reg

The data type **reg** is an abstraction of a hardware storage element, but it does not correspond directly to physical memory. A *reg* variable has a default initial value of x. The default size of a register variable is a single bit. Verilog operators create a **reg** variable as an unsigned value. A register variable may be assigned value only by a procedural statement, a user-defined sequential primitive, a task, or a function. A **reg** variable may never be the output of a predefined primitive gate, an **input** or **inout** port of a module, or the target of a continuous assignment.

C.2.2 Data Type: integer

The data type *integer* supports numeric computation in procedural statements. Integers are represented internally to the word length of the host machine (at least 32 bits). A negative integer is stored in 2s complement format. A *integer* variable has a default initial value of 0.

Verilog operators operate on integers with 2s complement arithmetic, with the (most significant bit) indicating the sign of the value. For example, the negative integer -4_{10} is stored as 1111_1111_1111_1111_1111_1111_1111_1100. When the size of a number assigned to an integer is less than the length of the word used by the machine to store an integer, the number is padded with 0s to the left. The number assigned to an integer variable must have a decimal equivalent (i.e., x and z are not allowed). Because integers have a fixed word size, they may not be declared to have a range specification. Some examples of valid declarations of integers and arrays of integers are shown below:

Example

```
integer A1, K, Size_of_Memory;
integer Array_of_Ints [1:100];
```

End of Example

An integer will be interpreted as a signed value in 2s complement form if it is assigned a value without a base specifier (e.g., $A = -24$). If the value assigned has a specified base, the integer is interpreted as an unsigned value. For example, if A is an integer, the result of $A = -12/3$ is -4; the result of $A = -'d12/3$ is 1431655761. Both words evaluate to the same bit pattern, but the former is interpreted as a negative value in 2s complement.

C.2.3 Data Type: real

Variables having type *real* are stored in double precision, typically a 64-bit value. A *real* variable has a default initial value of 0.0. Real variables can be specified in decimal and exponential notation. An object of type *real* may not be connected to a port of a module or to a terminal of a primitive. Verilog includes two system tasks that convert data types to permit real data transfer across a port boundary in a hierarchical structure: *$realtobits* and *$bitstoreal*. The language reference manual (LRM) describes limitations on the use of operators with real operands.

C.2.4 Data Type: realtime

Variables having type *realtime* are stored in real number format. A *realtime* variable has a default initial value of 0.0.

C.2.5 Data Type: time

The data type *time* supports time-related computations within procedural code in Verilog models. *time* variables are stored as unsigned 64-bit quantities. A variable of type *time* may not be used in a module port, nor may it be an input or output of a primitive. A *time* variable has a default value of 0.

C.2.6 Common Error: Undeclared Register Variables

Verilog has no mechanism for handling undeclared register variables. An identifier that has not been declared is assumed to reference a net of the default type (e.g., *wire*). A procedural assignment to an undeclared variable will cause a compiler error.

C.2.7 Addressing Nets and Register Variables

The most significant bit of a part-select of a net or register is always the leftmost array index; the least significant bit is the rightmost array index. A constant or variable expression can be the index of a part-select. If the index of a part-select is out of bounds the value x is returned by a reference to the variable.

Example

If an 8-bit word *vect_word* has a stored value of decimal 4, then *vect_word[2]* has a value of 1; *vect_word[3:0]* has a value of 4; *vect_word[5:1]* has value 2; i.e., *vect_word[7:0]* = 0000_0100$_2$, and *vect_word[5:1]* = 0_0010$_2$.

End of Example

C.2.8   Common Error: Passing Variables through Ports

Table C-2 summarizes the rules that apply to nets and registers that are port objects in a Verilog module. For example, a register variable may not be declared to be an **inout** port.

A variable that is declared as an **input** port of a module is implicitly a net variable within the scope of the module, but a variable that is declared to be an **output** port may be a net or a register variable. A variable declared to be an **input** port of a module may not be declared to be a register variable. An **inout** port of a module may not be a register type. A register variable may not be placed in an output port of a primitive gate, and may not be the target left-hand side (LHS) of a continuous-assignment statement.

TABLE C-2   Rules for port modes with nets and registers.

Variable Type	Port Mode		
	input	output	inout
Net Variable	Yes	Yes	Yes
Register Variable	No	Yes	No

C.2.9   Two-Dimensional Arrays (Memories)

Verilog does not have a distinct data type for true two-dimensional arrays, with access to each array element. Instead, it has an extension of the declaration of a register variable to provide a *memory* (i.e., multiple addressable cells of the same word size). An example of the syntax for a memory of register variables is shown below. Bit-select and part-select are not valid with a memory. Reference may be made to only a *word* of a memory. The MSB of a part-select is the leftmost array index; the least significant bit (LSB) is the rightmost. If an index is out of bounds the result is the logic value *x*. A *constant* expression may be used for the LSB and the MSB in a declaration of an array.

Example

The code fragment in Figure 4-9 shows how the syntax for declaring a *reg* memory variable simplifies to the form: *reg word_size array_name memory_size* for an array of 1024 32-bit words:

End of Example

C.2.10 Scope of a Variable

The scope of a variable is the module, task, function, or named procedural (***begin . . . end***) block in which it is declared. In Figure C-1 a net at the input port of *child_module* can be driven by a net or register in the enclosing *parent_module*, and a net or a register at the output port of *child_module* can drive a net in *parent_module*.

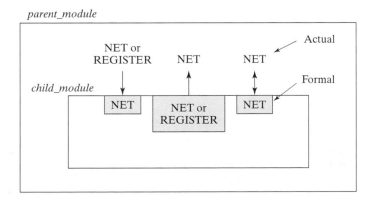

FIGURE C-1   Scope of nets and registers.

C.2.11 Strings

Verilog does not have a distinct data type for strings. Instead, a string must be stored within a properly sized register by a procedural assignment statement. A properly sized **reg** (array) has 8 bits of storage for each character of the string that it is to hold.

Example

A declaration of a **reg**, *string_holder*, that will accommodate a string with *num_char* characters is given below.

reg [8*num_char-1: 0] string_holder;

End of Example

The declaration in the example implies that 8 bits will encode each of the *num_char* characters. If the string "Hello World" is assigned to *string_holder*, it is necessary that *num_char* be at least 11 to ensure that a minimum of 88 bits are reserved. If an assignment to an array consists of less characters than the array will accommodate, 0s are automatically filled in the unused positions, beginning at the position of the MSB (i.e., the leftmost position).

C.3   Constants

A constant in Verilog is declared with the keyword **parameter**, which declares and assigns value to the constant. The value of a constant may not be changed during simulation. Constant expressions may be used in the declaration of the value of a constant.

Example

```
parameter high_index = 31;                              // integer
parameter width = 32, depth = 1024;&]                   // integers
parameter byte_size = 8, byte_max = byte_size-1;        // integer
parameter a_real_value = 6.22;                          // real
parameter av_delay = (min_delay + max_delay)/2;         // real
parameter initial_state = 8'b1001_0110;                 // reg
```

End of Example

C.4   Referencing Arrays of Nets or Regs

A net or a register is referenced by its identifier. A reference to a vector object can include a bit-select (i.e., a single bit, or element) or part-select consisting of a range of contiguous bits enclosed by square brackets (e.g., [7:0]). An expression can be the index of a part-select. If a declared vector identifier has an ascending (descending) order from its LSB to its MSB, a referenced part-select of that identifier must have the same ascending (descending) order from its LSB to its MSB.

Verilog Operators

The built-in operators of the Verilog HDL manipulate the various types of data implemented in the language to produce values on nets and registers. Some of the operators are used within expressions on the righthand side of continuous-assignment statements and procedural statements; others are used in Boolean expressions in conditional statements or with conditional operators. Verilog implements the classes of operators listed in Table D-1. The meaning of Verilog's operators is fixed; there is no provision for "overloading" an operator. The interpretation of the operators and operands is automatic and transparent to the user. The arithmetic supporting these operators is fully implemented in Verilog for scalar and vector **nets** and **regs**, for 2 complement arithmetic modulo 2^n, where n is the length of the operand word.

D.1 Arithmetic Operators

Arithmetic operators create a numeric value by operating on a pair of operands representing numeric values expressed in binary, decimal, octal, or hexadecimal form. The arithmetic operators supported by Verilog are identified in Table D-2.

When arithmetic operations are performed on vectors (net and registers) the result is determined by modulo 2^n arithmetic, where n is the size of the vector. *The bit*

TABLE D-1   Arguments and results produced by Verilog
operators.

Operator	Argument	Result
Arithmetic	Pair of operands	Binary word
Bitwise	Pair of operands	Binary word
Reduction	Single operand	Bit
Logical	Pair of operands	Boolean value
Relational	Pair of operands	Boolean value
Shift	Single operand	Binary word
Conditional	Three operands	Expression

TABLE D-2   Verilog arithmetic operators
and symbols.

Symbol	Operator
+	Addition
−	Subtraction
*	Division
/	Multiplication
%	Modulus

pattern that is stored in a register is interpreted as an unsigned value. A negative value is stored in 2s complement format, but it is interpreted as an unsigned quantity when it is used in an expression. For example, if -1 is stored in a 2-bit register, its value, after its 2s complement is formed, is $11_2 = 3_d$. On the other hand, if -1 is stored in a 3-bit register its value $111_2 = 7_d$.

Example

The simulation results shown below illustrate the modulo 2^n arithmetic operations of addition, subtraction, and negation.

```
module arith1 ();
    reg     [3:0]    A, B;
    wire    [4:0]    sum, diff1, diff2, neg;
    assign sum = A + B;
    assign diff1 = A - B;
    assign diff2 = B - A;
    assign neg = -A;

    initial
    begin
      #5 A = 5; B = 2;
      $display ("            t_sim A    B    A+B   A-B  B-A  -A");
      $monitor ($time,"%d    %d    %d    %d    %d    %d", A, B, sum, diff1, diff2, neg);
      #10 $monitoroff;
      $monitor ($time,,"%b %b %b %b %b %b", A, B, sum, diff1, diff2, neg);
    end
endmodule
```

t_sim	A	B	A + B	A − B	B − A	−A
5	5	2	7	3	29	27
15	0101	0010	00111	00011	11101	11011

Note that $A + B$ and $A - B$ have values that are expected. But $B - A$ does not return the decimal equivalent of $2 - 5 = -3$. The actual value of $B - A$ is obtained by adding the 2s complement of A (11011_2) to the value of B (00010_2), using the word length of the result. Verilog descriptions involving operations on nets and registers must anticipate modulo 2^n arithmetic, or implement an arithmetic scheme, such as signed magnitude arithmetic. Note that the value of $B - A$ is the 2s complement of the correct result, but without the correct sign. The presence of 1 in the MSB indicates that the result is to be interpreted as a negative value expressed in 2s complement format.

End of Example

D.2 Bitwise Operators

The bitwise negation operator negates the individual bits of a word. The other bitwise operators produce a binary word result by operating bitwise on a pair of operands. The operands may be scalar or vector. Table D-3 lists the Verilog bitwise operators.

TABLE D-3   Verilog bitwise operators.

Symbol	Operator
+	Addition
−	Subtraction
*	Division
/	Multiplication
%	Modulus

Example

If $y1$ is the binary word 1011_0001 and $y2$ is the binary word 0010_1001, then the bitwise operation $y1$ & $y2$ produces the result: 0010_0001. A 1 is produced in a given position of the result if a 1 is present at that position in both of the operands.

End of Example

D.3 Reduction Operators

Reduction operators are unary operators. They create a single-bit value by operating on a single data word. Table D-4 lists the Verilog reduction operators.

Table D-4   Verilog reduction operators.

Symbol	Operator
&	Reduction and
~&	Reduction nand
|	Reduction or
~|	Reduction nor
^	Reduction exclusive or
~^, ^~	Reduction xnor

Example

If y is the eight-bit binary word 1011_0001, the "reduction and" operation on y produces: $\& y = 0$. The reduction and operation takes the "and" of the bits in the operand. It returns a 1 if all of bits in the operand are 1.

End of Example

Example

Examples of the reduction operator are given below.

Expression	Result	Operator
&(010101)	0	Reduction And
|(010101)	1	Reduction Or
&(010×10)	0	Reduction And
|(010×10)	1	Reduction Or

End of Example

D.4 Logical Operators

The Verilog logical operators operate on Boolean operands as logical connectives to create a Boolean result. The operand may be a net, a register, or an expression that is evaluated to produce a Boolean result. This family of operators is listed in Table D-5. Logical operators are commonly used with the conditional assignment operator and in conditional (*if*) statements within behaviors, functions, or tasks.

TABLE D-5 Verilog logical operators.

Symbol	Operator
!	Logical negation
&&	Logical and
||	Logical or
==	Logical equality
!=	Logical inequality
===	Case equality
!==	Case inequality

Example

Examples using logical operators are given below:

a. **if** ((a < size − 1) && (b != c) && (index != last_one) ...
b. **if** (! inword) ...
c. **if** (inword == 0) ...

End of Example

The *case* equality operators (e.g., ===) are used to determine whether two words match identically on a bit-by-bit basis, including bits that have values x or z. The logical equality operator (==) is less restrictive—it is used to test whether two words are identical, but it produces an x result when the test is ambiguous. The comparison is made bit by bit, and 0s are filled in as necessary. The result of the test is 1 if the comparison is true, and 0 if the comparison is false. If the operands are nets or registers, their values are treated as unsigned words. If any bit is unknown, the relation is unknown, and the result that is returned is ambiguous (x value). If the operands are integers or reals, they may be signed values, but they are compared as though they are unsigned.

The appropriate use of the logical OR and the logical AND operators is as connectives in a logical expression. Verilog is loosely typed, so the logical operators can be used inappropriately. For example $A \&\& B$ will return a Boolean scalar result. If A and B are scalars, the result will be the same as obtained using $A \& B$. If A and B are vectors, $A \&\& B$ returns Boolean true if both words are positive integers. $A \& B$ returns true if the word formed from the bitwise operation is a positive integer. For example, suppose $A = 3'b110$ and $B = 3'b11x$. Then $A \&\& B = 0$, which has a Boolean value of false, because B is false. On the other hand, $A \& B = 110$, which has a Boolean value of true.

D.5   Relational Operators

The Verilog relational operators compare operands and produce a Boolean (true or false) result. If the operands are nets or registers, their values are treated as unsigned words. If any bit is unknown, the relation is unknown, and the result that is returned is ambiguous (x value). If the operands are integers or reals, they may be signed. Table D-6 lists the Verilog relational operators.

TABLE D-6   Verilog relational operators.

Symbol	Operator
<	Less than
<=	Less than or equal to
>	Greater than
>=	Greater than or equal to

D.6   Shift Operators

Verilog shift operators operate on a single operand and shift (left or right) the bit pattern of the operand by a specified number of positions, filling in with zeros in the positions that are vacated. These operators are listed in Table D-7.

TABLE D-7   Verilog shift operators.

Symbol	Operator
$\ll$	Left shift
$\gg$	Right shift

Example

If word A has the bit pattern 1011_0011, then $A = A \ll 1$ creates the bit pattern $A = 0110\_0110$. Note that the shift operator fills the word with a 0 in the vacant position. The shift operator can also be used to accomplish multiple shifts. For the same data word, $A = A \ll 3$ creates the word $A = 1001\_1000$.

End of Example

D.7   Conditional Operator

The Verilog conditional assignment operator selects an expression for evaluation, based on the value of *conditional_expression*. The conditional assignment operator has the syntax: *conditional_expression ? true_expression : false_expression*. If *conditional_expression* evaluates to Boolean true, then *true_expression* is evaluated; otherwise, *false_expression* is evaluated.

Example

The statement below assigns the value of A to Y if $A = B$; otherwise, it assigns the value of B.

$$Y = (A == B)\,?\,A:B;$$

End of Example

Example

The following Verilog statement uses the conditional operator to assign value to *bus_a*.

wire [15:0] bus_a = drive_bus_a ? data : 16'bz;

The effect of the assignment is summarized below:

> *drive_bus_a* = 1 sets data on *bus_a*
>
> *drive_bus_a* = 0 sets *bus_a* to high impedance
>
> *drive_bus_a* = x sets *x* on *bus_a*

End of Example

The conditional operator can be nested to any depth. The conditional operator can also be used to control the activity flow of the procedural statements within a Verilog behavior. The following rules determine the value that results from an assignment using the conditional operator: (1) logic value z is not allowed in the *conditional_expression*, (2) 0s are automatically filled if the operands have different lengths, (3) if the *conditional_expression* is ambiguous, then both *true_expression* and *false_expression* are evaluated and the result is calculated on a bitwise basis according to Table D-8.

Note that the truth table assigns 0 (1) to the expression when both *true_expression* and *false_expression* have the value 0 (1). In these cases the result of the evaluation does not depend on *conditional_expression*. One restriction that applies to the use of the concatenation operator is that no operand may be an unsized constant.

TABLE D-8   Truth table for the conditional assignment operator.

false_expression	? :	*true_expression* 0	1	*x*
	0	0	*x*	*x*
	1	*x*	1	*x*
	x	*x*	*x*	*x*

D.8   Concatenation Operator

The concatenation operator forms a single word from two or more operands. This operator is particularly useful in forming logical busses. The concatenation result follows the same order in which the words are given. The concatenation operator nests to any depth and repetition.

a. If the operand A is the bit pattern 1011 and the operand B is the bit pattern 0001, then $\{A, B\}$ is the bit pattern 1011_0001.
b. $\{4\{a\}\} = \{a, a, a, a\}$
c. $\{0011, \{\{01\}, \{10\}\}\} = 0011\_0110$.

End of Example

D.9   Expressions and Operands

Verilog expressions combine operands with the language's operators to produce a re-sultant value. A Verilog operand may be composed of nets, registers, constants, num-bers, bit-select of a net, bit-select of a register, part-select of a net, part-select of a register, memory element, a function call, or a concatenation of any of these. The result of an expression may be used to determine an assignment to a net or register variable or to choose among alternatives. The value of an expression is determined by perform-ing the indicated operations on its operands. An expression may consist of a single identifier (operand), or some combination of operands and operators that conforms to the allowed syntax of the language. The evaluation of an expression always produces a value that is represented by one or more bits.

Example

Some examples of expressions are given below:
a. **assign** THIS_SIG = A_SIG ^ B_SIG;
b. **assign** y_out = (select) **?** input_a : input_b;

End of Example

D.10 Operator Precedence

Verilog evaluates expressions from left to right, and the evaluation of a Boolean ex-pression is terminated as soon as the expression is determined to be true or false. The precedence of Verilog operators within an expression is given in Table D-9. The operators in the same row have the same precedence, and the rows are ordered from top to bottom.

The result produced by a compiler may not correspond to the intent expressed in an expression. As a precaution, use parentheses to eliminate ambiguity in expressions.

TABLE D-9   Verilog operators and their precedence.

Operator Symbol	Function	Precedence
$+ - !$ ~ (unary)	Sign, complement	Highest
$* / \%$	Multiplication, Division, Modulus	
$+ -$ (binary)	Addition, Subtraction	
$\ll \gg$	Shift	
$< \; <= \; > \; > =$ $== \; != \; === \; !==$	Relational	
$\& \; \sim\&$ $\wedge \; \wedge\sim \; \sim\wedge$ $\mid \; \sim\mid$	Reduction	
$\& \&$ $\mid\mid$	Logical	
$? :$	Conditional	Lowest

Backus–Naur Formal Syntax Notation

The syntax of the Verilog language conforms to the following Backus–Naur form (BNF) of formal syntax notation.

1. White space may be used to separate lexical tokens.
2. *Name* ::= starts off the definition of a syntax construction item. Sometimes *Name* contains embedded underscores (_). Also, the symbol ::= may be found on the next line.
3. The vertical bar (|) introduces an alternative syntax definition, unless it appears in bold.
4. **Name** in bold text is used to denote reserved keywords, operators, and punctuation marks required in the syntax.
5. [item] is an optional item that may appear once or not at all.
6. {item} is an optional item that may appear once, more than once, or not at all. If the braces are in bold, then they are part of the syntax.
7. *Name1_name2* is equivalent to the syntax construct item *name2*. The *name1* (in italics) imparts some extra semantic information to *name2*. However, the item is defined by the definition of *name2*.

Verilog Language Formal Syntax

This formal syntax specification is provided in Backus–Naur form (BNF). It is reprinted from IEEE Std 1364-1995, by the Institute of Electrical and Electronics Engineers, Inc. The IEEE disclaims any responsibility or liability resulting from the placement and use in this publication. This information is reprinted with the permission of the IEEE.

F.1 Source Text

```
source_text ::= {description}
description ::=
              module_declaration
              | udp_declaration
module_declaration ::=
              module_keyword module_identifier [list_of_ports]; {module_item} endmodule
module_keyword ::= module | macromodule
list_of_ports ::= (port {, port})
port ::=
              [port_expression]
              |.port_identifier ( [port_expression] )
port_expression ::=
              port_reference
              | {port_reference {, port_reference} }
port_reference ::=
              port_identifier
              | port_identifier [constant_expression]
              | port_identifier [msb_constant_expression : lsb_constant_expression]
```

```
module_item ::=

                module_item_declaration
                | parameter_override
                | continuous_assign
                | gate_instantiation
                | udp_instantiation
                | module_instantiation
                | specify_block
                | initial_construct
                | always_construct
module_item_declaration ::=
                parameter_declaration
                | input_declaration
                | output_declaration
                | inout_declaration
                | net_declaration
                | reg_declaration
                | integer_declaration
                | real_declaration
                | time_declaration
                | realtime_declaration
                | event_declaration
                | task_declaration
                | function_declaration
parameter_override ::= defparam list_of_param_assignments;
```

F.2   Declarations

```
parameter_declaration ::= parameter list_of_param_assignments;
list_of_param_assignments ::= param_assignment { , param_assignment}
param_assignment ::= parameter_identifier = constant_expression
input_declaration ::= input [range] list_of_port_identifiers;
output_declaration ::= output [range] list_of_port_identifiers ;
inout_declaration ::= inout [range] list_of_port_identifiers ;
list_of_port_identifiers ::= port_identifier { , port_identifier}
reg declaration ::= reg [range] list_of_register_identifiers ;
time_declaration ::= time list_of_register_identifiers ;
integer_declaration ::= integer list_of_register_identifiers ;
real_declaration ::= real list_of_real_identifiers ;
realtime_declaration ::= realtime list_of_real_identifiers ;
event_declaration ::= event event_identifier {, event_identifier} ;
list_of_real_identifiers ::= real_identifier { , real_identifier}
list_of_register_identifiers ::= register_name {, register_name}
  register_name ::=
                register_identifier
                | memory_identifier [upper_limit_constant_expression :
```

*lower_limit*_constant_expression]

range ::= [*msb*_constant_expression : *lsb*_constant_expression]

net_declaration ::=

      net_type [**vectored** | **scalared**] [range] [delay3] list_of_net_identifiers ;

      | **trireg** [**vectored** | **scalared**] [charge_strength] [range] [delay3]

      list_of_net_identifiers;

      | net_type [**vectored** | **scalared**] [drive_strength] [range] [delay3]

      list_of_net_decl_assignments;

net_type ::= **wire** | **tri** | **tri1** | **supply0** | **wand** | **triand** | **tri0** | **supply1** | **wor** | **trior**

list_of_net_identifiers ::= *net*_identifier { , net_identifier }

drive_strength ::=

      (strength0, strength1)

      | (strength1, strength0)

      | (strength0, **highz1**)

      | (strength1, **highz0**)

      | (**highz1**, strength0)

      | (**highz0**, strength1)

strength0 ::= **supply0** | **strong0** | **pull0** | **weak0**

strength1 ::= **supply1** | **strong1** | **pull1** | **weak1**

charge_strength ::= **(small)** | **(medium)** | **(large)**

delay3 ::= # delay_value | # (delay_value [, delay_value [, delay_value]])

delay2 ::= # delay_value | # (delay_value [, delay_value])

delay_value ::= unsigned_number | parameter_identifier |

      constant_mintypmax_expression

list_of_net_decl_assignments ::= net_decl_assignment { , net_decl_assignment}

net_decl_assignment ::= *net*_identifier = expression

function_declaration ::=

      function [range_or_type] *function*_identifier ;

      function_item_declaration {function_item_declaration}

      statement

      endfunction

range_or_type ::= range | **integer** | **real** | **realtime** | **time**

function_item_declaration ::=

      block_item_declaration

      | input_declaration

task_declaration ::=

      task *task*_identifier ;

      {task_item_declaration}

      statement_or_null

      endtask

task_argument_declaration ::=

      block_item_declaration

      | output_declaration

      | inout_declaration

block_item_declaration ::=

      parameter_declaration

      | reg_declaration

      | integer_declaration
      | real_declaration
      | time_declaration
      | realtime_declaration
      | event_declaration

F.3   Primitive Instances

gate_instantiation ::=

      n_input_gatetype [drive_strength] [delay2] n_input_gate_instance { ,
      n_input_gate_instance} ;
      | n_output_gatetype [drive_strength] [delay2] n_output_gate_instance { ,
      n_output_gate_instance} ;
      | enable_gatetype [drive_strength] [delay3] enable_gate_instance { ,
      enable_gate_instance} ;
      | mos_switchtype [delay3] mos_switch_instance {, mos_switch_instance} ;
      | pass_switchtype pass_switch_instance {, pass_switch_instance} ;
      | pass_en_switchtype [delay3] pass_en_switch_instance
      { , pass_en_switch_instance} ;
      cmos_switchtype [delay3] cmos_switch_instance {, cmos_switch_instance} ;
      | **pullup** [pullup_strength] pull_gate_instance { , pull_gate_instance} ;
      | **pulldown** [pulldown_strength] pull_gate_instance { , pull_gate_instance} ;

n_input_gate_instance ::= [name_of_gate_instance] (output_terminal, input_terminal { ,
      input_terminal })

n_output_gate_instance ::= [name_of_gate_instance] (output_terminal { , output_terminal} ,
      input_terminal)

enable_gate_instance ::= [name_of_gate_instance] (output_terminal, input_terminal,
      enable_terminal)

mos_switch_instance ::= [name_of_gate_instance] (output_terminal, input_terminal ,
      enable_terminal)

pass_switch_instance ::= [name_of_gate_instance] (inout_terminal, inout_terminal)

pass_enable_switch_instance ::= [name_of_gate_instance] (inout_terminal, inout_terminal,
      enable_terminal)

cmos_switch_instance ::= [name_of_gate_instance] (output_terminal, input_terminal,
      ncontrol_terminal, pcontrol_terminal)

pull_gate_instance ::= [name_of_gate_instance] (output_terminal)

name_of_gate_instance ::= *gate_instance*_identifier [range]

pullup_strength ::=

      (strength0, strength1)
      | (strength1, strength0)
      | (strength1)

pulldown-Strength ::=

      (strength0, strength1)
      | (strength1, strength0)
      | (strength0)

input_terminal ::= *scalar*_expression
enable_terminal ::= *scalar*_expression
ncontrol_terminal ::= *scalar*_expression
pcontrol_terminal ::= *scalar*_expression
output_terminal ::= *terminal*_identifier | *terminal*_identifier [constant_expression]
inout_terminal ::= *terminal*_identifier | *terminal*_identifier [constant_expression]
n_input_gatetype ::= **and** | **nand** | **or** | **nor** | **xor** | **xnor**
n_output_gatetype ::= **buf** | **not**
enable_gatetype ::= **bufif0** | **bufif1** | **notif0** | **notif1**
mos_switchtype ::= **nmos** | **pmos** | **rnmos** | **rpmos** |
pass_switchtype ::= **tran** | **rtran**
pass_en_switchtype ::= **tranif0** | **tranif1** | **rtranif1** | **rtranif0**
cmos_switchtype ::= **cmos** | **rcmos**

F.4 Module Instantiation

module_instantiation ::=

    *module*_identifier [parameter_value_assignment] module_instance {,
    module_instance};
parameter_value_assignment ::= # (expression {, expression})
module_instance ::= name_of_instance ([list_of_module_connections])
name_of_instance ::= *module_instance*_identifier [range]
list_of_module_connections ::=

    ordered_port_connection {, ordered_port_connection}
    | named_port_connection {, named_port_connection}
ordered_port_connection ::= [expression]
named_port_connection ::= . *port*_identifier ([expression])

F.5 UDP Declaration and Instantiation

udp_declaration ::=

    primitive *udp*_identifier (udp_port_list);
    udp_port_declaration {udp_port_declaration}
    udp_body
    endprimitive
udp_port_list := *output_port*_identifier, *input_port*_identifier {, *input_port*_identifier}
udp_port_declaration ::=

    output_declaration
    | input_declaration
    | reg_declaration
udp_body ::= combinational_body | sequential_body
combinational_body ::= **table** combinational_entry {combinational_entry} **endtable**
combinational_entry ::= level_input_list : output_symbol;
sequential_body ::= [udp_initial_statement] **table** sequential_entry {sequential_entry} **endtable**
udp_initial_statement ::= **initial** *udp_output_port*_identifier = init_val;

init_val ::= 1'b0 | 1'b1 | 1'bx | 1'bX | 1'B0 | 1'B1 | 1'Bx | 1'BX | 1 | 0
sequential_entry ::= seq_input_list : current_state : next_state;
seq_input_list ::= level_input_list | edge_input_list
level_input_list ::= level_symbol {level_symbol}
edge_input_list ::= {level_symbol} edge_indicator {level_symbol}
edge_indicator ::= (level_symbol level_symbol) | edge_symbol
current_state ::= level_symbol |
next_state ::= output_symbol | -
output_symbol ::= **0** | **1** | **x** | **X**
level_symbol ::= **0** | **1** | **x** | **X** | **?** | **b** | **B**
edge_symbol ::= **r** | **R** | **f** | **F** | **p** | **P** | **n** | **N** |*
udp_instantiation ::= *udp*_identifier [drive_strength] [delay2] udp_instance {, udp_instance};
udp_instance ::= [name_of_udp_instance] (output_port_connection, input_port_connection
                  {, input_port_connection})
name_udp_instance ::= *udp_instance*_identifier [range]

F.6   Behavioral Statements

continuous_assign ::= **assign** [drive_strength] [delay3] list_of_net_assignments;
list_of_net_assignments ::= net_assignment {, net_assignment}
net_assignment ::= net_lvalue = expression

initial_construct ::= **initial** statement
always_construct ::= **always** statement

statement ::=

            blocking_assignment ;
          | non_blocking assignment ;
          | procedural_continuous_assignments ;
          | procedural_timing_control_statement
          | conditional_statement
          | case_statement
          | loop_statement
          | wait_statement
          | disable_statement
          | event_trigger
          | seq_block
          | par_block
          | task_enable
          | system_task_enable

statement_or_null ::= statement | ;

blocking_assignment ::= reg_1 value <= [delay_or_event_control] expression
non_blocking assignment ::= reg_1value <= [delay_or_event_control] expression
procedural_continuous_assignment ::=
         | **assign** reg_assignment ;
         | **deassign** reg_1value ;

```
              | force reg_assignment ;
              | force net_assignment ;
              | release reg_1value ;
              | release net_1value ;
procedural_timing_control_statement ::=
              delay_or_event_control statement_or_null
delay_or_event_control ::=
              delay_control
              | event_control
              | repeat (expression) event_control
delay_control ::=
              # delay_value
              | # (mintypmax_expression)
even_control ::=
              @ event_identifier
              | @ (event_expression)
event_expression ::=
              expression
              | event_identifier
              | posedge_expression
              | negedge_expression
              | event_expression or event_expression
conditional_statement ::=
              | if (expression) statement_or_null [else statement_or_null]
case_statement ::=
              case (expression) case_item {case_item} endcase
              | casez (expression) case_item {case_item} endcase
              | casex (expression) case_item {cast_item) endcase
case_item ::=
              expression {, expression} : statement_or_null
              | default [ : ] statement_or_null
loop_statement ::=
              | forever statement
              | repeat (expression) statement
              | while (expression) statement
              | for (reg_assignment; expression; reg_assignment) statement
reg_assignment ::= reg_lvalue = expression
wait_statement ::= wait (expression) statement or-null
event_trigger ::=
              | -> event identifier;
disable_statement ::=
              | disable task_identifier;
              | disable block_identifier;
seq_block ::= begin [ : block_identifier {block_item_declaration} ] {statement} end
par_block ::= fork [ : block_identifier { block_item_declaration} ] {statement} join
task_enable ::= task_identifier [ (expression { , expression } ) ];
system_task_enable ::= system_task_name [(expression { , expression } ) ];
system_task_name ::= $identifier Note: The $ may not be followed by a space.
```

F.7   Specify Section

specify_block ::= **specify** [specify_item] **endspecify**

specify_item ::=

>> specparam_declaration
>> | path_declaration
>> | system_timing_check

specparam_declaration ::= **specparam** list_of_specparam_assignments ;

list_of_specparam_assignments ::= specparam_assignment { , specparam_assignment}

specparam_assignment

>> *specparam*_identifier = constant_expression
>> | pulse_control_specparam

pulse_control_specparam ::=

>> **PATHPULSE$** = (*reject*_limit_value [, *error*_limit_value]);
>> |
>> **PATHPULSE$**specify_input_terminal_descriptor$specify_output_terminal
>> _descriptor = (*reject*_limit_value[, *error_limit*_value]);

limit_value ::= constant_mintypmax_expression

path_declaration ::= simple_path_declaration ;

>> | edge_sensitive_path_declaration ;
>> | state_dependent_path_declaration ;

simple_path_declaration ::= parallel_path_description = path_delay_value

>> | full_path_description = path_delay_value

parallel_path_description ::= (specify_input_terminal_descriptor [polarity_operator] =>

>> specify_output_terminal_descriptor

full_path_description ::= (list_of_path_inputs [polarity_operator] *> list_of_path_outputs)

list_of_path_inputs ::= specify_input_terminal_descriptor { , specify_input_terminal}

list_of_path_outputs ::= specify_output_terminal_descriptor { ,

>> specify_output_terminal_descriptor}

specify_input_terminal_descriptor ::=

>> input_identifier
>> | input_identifier [constant_expression]
>> | input_identifier [*msb*_constant_expression: *lsb*_constant_expression]

specify_output_terminal_descriptor ::=

>> output_identifier
>> | output_identifier [constant_expression]
>> | output_identifier [*msb*_constant_expression: *lsb*_constant_expression]

input_identifier ::= *input_port*_identifier | *inout_port*_identifier

output_identifier ::= *output_port*_identifier | *inout_port*_identifier

polarity_operator ::= +|-

path_delay_value ::= list_of_path_delay_expression | (list_of_path_delay_expressions)

list_of_path_delay_expressions ::=

>> *t*_path_delay_expression
>> | *trise*_path_delay_expression, *tfall*_path_delay_expression
>> | *trise*_path_delay_expression, *tfall*_path_delay_expression,
>> *tz*_path_delay_expression

       | *t01*_path_expression, *t10*_path_delay_expression,
       *t0z*_path_delay_expression,
       *tzl*_path_delay_expression, *tlz*_path_delay_expression,
       *tz0*_path_delay_expression
       | *t0l*_path_delay_expression, *t10*_path_delay_expression,
       *t0z*_path_delay_expression,
       *tzl*_path_delay_expression, *tlz*_path_delay_expression,
       *tz0*_path_delay_expression,
       *t0x*_path_delay_expression, *txl*_path_delay_expression,
       *tlx*_path_delay_expression,
       *tx0*_path_delay_expression, *txz*_path_delay_expression,
       *tzx*_path_delay_expression

path_delay_expression ::= constant_mintypmax_expression
edge_sensitive_path_declaration ::= parallel_edge_sensitive_path_description =
path_delay_value
       | full_edge_sensitive_path_description = path_delay_value
parallel_edge_sensitive_path_description ::=
       ([edge _identifier] specify_input_terminal_descriptor =>
          specify_output_terminal_descriptor [polarity_operator]:
       data source_expression))
full_edge_sensitive_path_description ::= ([edge_identifier] list_of_path inputs *>
       list_of_path_outputs [polarity_operator] : data_source_expression))
data_source_expression ::= expression
edge_identifier ::= **posedge** | **negedge**
state_dependent_path_declaration ::=
       if (conditional_expression) simple_path_declaration
       | **if** (conditional_expression) edge_sensitive_path_declaration
       | **ifnone** simple_path_declaration
system_timing_check ::=
       $setup (timing_check_event, timing_check_event, timing_check_limit
       [, notify_register]) ;
       | **$hold** (timing_check_event, timing_check_event, timing_check_limit
       [, notify_register]) ;
       | **$period** (controlled_timing_check_event, timing_check_limit
       [, notify_register]) ;
       | **$width** (controlled_timing_check_event, timing_check_limit,
       constant_expression [, notify_register]) ;
       | **$skew** (timing_check_event, timing_check_event, timing_check_limit
       [, notify_register]) ;
       | **$recovery** (controlled_timing_check_event, timing_check_event,
       timing_check_limit [, notify_register]) ;
       |**$setuphold** (timing_check_event, timing_check_event, timing_check_limit,
       timing_check_limit [, notify_register]
timing_check_event ::=
       [timing_check_event_control] specify_terminal_descriptor
       [**&&&** timing_check_condition]
specify_terminal_descriptor ::=
       specify_input_terminal_descriptor

```
                              | specify_output_terminal_descriptor
       controlled_timing_check_event ::=
                        timing_check_event_control specify_terminal_descriptor
                        [&&& timing_check_condition]
       timing_check_event_control ::=
                        posedge
                        | negedge
                        | edge_control_specifier
       edge_control_specifier ::= edge [edge_descriptor [, edge_descriptor]]
       edge_descriptor ::=
                        01
                        | 10
                        | 0x
                        | xl
                        | 1x
                        | x0
       timing_check_condition ::=
                        scalar_timing_check_condition
                        | (scalar_timing_check_condition)
       scalar_timing_check_condition
                        expression
                        | ~expression
                        | expression == scalar_constant
                        | expression === scalar_constant
                        | expression != scalar_constant
                        | expression !== scalar_constant
       timing_check_limit ::= expression
       scalar_constant ::=
                        1'b0 | 1'b1 | 1'B0 | 1'B1 | 'b0 | 'b1 | 'B0 | 'B1 | 1 | 0
       notify_register ::= register_identifier
```

F.8   Expressions

```
net_1 value ::=
                        net_identifier
                        | net_identifier [expression]
                        | net_identifier [msb_constant_expression: lsb_constant_expression]
                        | net_concatenation
reg_1 value ::=
                        reg_identifier
                        | reg_identifier [expression]
                        | reg_identifier [msb_constant_expression: lsb_constant_expression]
                        | reg_concatenation
constant_expression ::=
                        constant_primary
                        | unary_operator_constant_primary
```

            | constant_expression binary_operator constant_expression
            | constant_expression **?** constant_expression: constant_expression
            | string

constant_primary ::=

        number
        | *parameter*_identifier
        | *constant*_concatenation
        | *constant*_multiple_concatenation

constant_mintypmax_expression ::=

        constant_expression
        | constant_expression : constant_expression : constant_expression

mintypmax_expression ::=

        expression
        | expression : expression : expression

expression ::=

        primary
        | unary_operator_primary
        | expression binary_operator expression
        | expression **?** expression : expression
        | string

unary_operator ::=

        + | - | ! | ~ | **&** | ~**&** | | | ~| | ^ | ~^ | ^~

binary_operator ::=

        + | - | * | / | % | == | != | === | !== | **&&** | ||
        | < | <= | > | >= | **&** | | | ^ | ^~ | ~^ | >> | <<

primary ::=

        number
        | identifier
        | identifier [expression]
        | identifier [*msb*_constant_expression : *lsb*_constant_expression]
        | concatenation
        | multiple_concatenation
        | function_call
        | (mintypmax_expression)

number ::=

        decimal_number
        | octal_number
        | binary_number
        | hex_number
        | real_number

real_number ::=

        [sign] unsigned_number . unsigned_number
        | [sign] unsigned_number [. unsigned_number] **e** [sign] unsigned_number
        | [sign] unsigned_number [. unsigned_number] **E** [sign] unsigned_number

decimal_number ::=

        [sign] unsigned_number
        | [size] decimal_base unsigned_number

binary_number ::= [size] binary_base binary_digit { _ | binary_digit}
octal_number ::= [size] octal_base octal_digit { _ | octal_digit}
hex_number ::= [size] hex_base hex_digit { _ | hex_digit}
sign ::= + | -
size ::= unsigned_number
unsigned_number ::= decimal_digit { _ | decimal_digit}
decimal_base ::= 'd | 'D
binary_base ::= 'b | 'B
octal_base ::= 'o | 'O
hex_base ::= 'h | 'H
decimal_digit ::= 0 | 1 | 2 | 3 | 4 | 5 | 6 | 7 | 8 | 9
binary_digit ::= x | X | z | Z | 0 | 1
octal_digit ::= x | X | z | Z | 0 | 1 | 2 | 3 | 4 | 5 | 6 | 7 |
hex_digit ::= x | X | z | Z | 0 | 1 | 2 | 3 | 4 | 5 | 6 | 7 | 8 | 9 | a | b | c | d | e | f | A | B | C | D | E | F |

concatenation ::= {expression {, expression} }
Multiple_concatenation ::= {expression { expression { , expression }}}
function_call ::=

        *function*_identifier (expression { , expression})
        | name_of_system_function [(expression { , expression})]
name_of_system_function ::= $identifier
string ::= "{ Any_ASCII_Characters_except_new line}"
NOTES

    1) —Embedded spaces are illegal
    2) —The $ in name_of_system_function may not be followed by a space.

F.9   General

comment ::=

        short_comment
        | long_comment
short_comment ::= // comment_text \ n
long_comment ::= /* comment_text */
comment_text ::= {Any_ASCII_character }
identifier ::= IDENTIFIER [{ . IDENTIFIER }]
IDENTIFIER ::=

        simple_identifier
        | escaped_identifier
simple_identifier ::= [a-zA-Z][a-zA-Z_$]
escaped_identifier ::= \ {Any_ASCII_character_except_white_space} white_space
White_space ::= space | tab | newline

NOTE—The period in identifier may not be preceded or followed by a space.

_____

Additional Features of Verilog

G.1 Arrays of Primitives

Declaring a range between the key word and ports of a primitive forms an array of instances of the primitive.

Example

The description in *array_of_nor* contains a declaration of 8-bit input and output data-paths. The declared instance of the **nor** primitive with an 8-bit range specification creates a structure of 8 **nor** gates. The individual bits of the datapaths are automatically connected in sequential order to the inputs of the corresponding gate. The array structure is shown in Figure G-1.

```
module array_of_nor (y, a, b,);
    input    [0:7] a, b;
    output   [0:7] y;

    nor [0:7] (y, a, b);

endmodule
```

End of Example

G.2 Arrays of Modules

Declaring a range between the instance name of a module and its ports forms an array of instances of the module. (*Note:* The list of ports in an array of instances must be

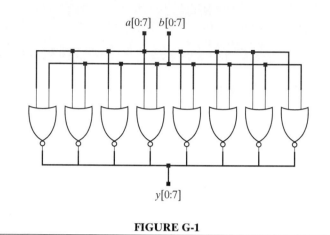

$a[0:7]$ $b[0:7]$

$y[0:7]$

FIGURE G-1

compatible with the structure being instantiated. If the port of an instantiated object is an array, then the size of the port in the instantiated array of instances of the object must be sufficiently large to accommodate all copies of the object.)

Example

An array of full adders connected to form 4-bit slice ripple-carry adders is described in *array_of_adders*.

```
module array_of_adders (sum, c_out, a, b, c_in);
    input          [3:0]    a, b;
    input                   c_in;
    output         [3:0]    sum;
    output                  c_out;
    wire           [3:1]    carry;

    Add_full M[3:0] (sum, {c_out, carry[3:1]}, a, b, {carry[3:1], c_in});

endmodule
```

End of Example

G.3   Hierarchical Dereferencing

An identifier must be associated with only one object within a scope or domain within which the identifier has unique meaning (i.e., within a module, named procedural block, task, or function). Consequently, a variable may be referenced directly by its identifier within the scope in which it is declared. Verilog also supports hierarchical dereferencing by a variable's hierarchical path name. This feature allows testbenches to monitor the activity of variables at any location within the hierarchical decomposition of the unit under test. If a variable is referenced but not declared locally, Verilog will search upward through the boundaries of named blocks, tasks, and functions to resolve the identifier, but it will not search beyond a module boundary. See Section 9.5.1.

G.4   Parameter Substitution

Verilog supports two methods of changing the values of parameters in a module: direct substitution and indirect referencing. Direct substitution overrides the value of the parameter on a module instance basis.

Example

The parameters declared within the *G2* instance of *modXnor* are overridden by including #(4,5) in the instantiation of the module. The values given in the instantiation replace the values of *size* and *delay* that were given in the declaration of *modXnor*. The replacement is made in the order that the parameters were originally declared. This method can be cumbersome if the edited value is near the end of a long list.

```
module modXnor (y_out, a, b);
    parameter                      size = 8, delay = 15;
    output          [size-1:0]     y_out;
    input           [size-1:0]     a, b;
    wire            [size-1:0]     #delay y_out = a ~^b;        //bitwise xnor
endmodule

module Param;

                    [7:0]     y1_out;
    wire            [3:0]     y2_out;
    reg             [7:0]     b1, c1;
    reg             [3:0]     b2, c2;

    modXnor G1 (y1_out, b1, c1);               //Uses default parameters

    modXnor #(4, 5) G2 (y2_out, b2, c2);       //Overrides default parameters

endmodule
```

End of Example

Indirect substitution uses hierarchical dereferencing to override the value of a parameter in a module. Declaring a separate module in which the **defparam** statement is used with the hierarchical path name of the parameters that are to be overridden most conveniently does this. (*Note:* This feature can be misused by annotating from anywhere within a design hierarchy.)

Example

In *hdref_param* the values of *size* and *delay* in instance *G2* of *modXnor* are overridden by the statements in the module *annotate*.

```
module hdref_Param;                    //a top level module
  wire          [7:0] y1_out;
  wire          [3:0] y2_out;
  reg           [7:0] b1, c1;
  reg           [3:0] b2, c2;

  modXnor       G1 (y1_out, b1, c1),
                G2 (y2_out, b2, c2);    //instantiation
endmodule

module annotate:                        //a separate "annotation" module
  defparam
  hdref_Param.G2.size = 4.               //parameter assignment by
  hdref_Param.G2.delay = 5;              //hierarchical reference name
  endmodule

module modXnor (y_out, a, b);
  parameter              size = 8, delay = 15;
  output       [size-1:0]   y_out;
  input        [size-1:0]   a, b;
  wire         [size-1:0]   #delay y_out = a  ~^b;    //bitwise xnor
endmodule
```

End of Example

G.5   Procedural Continuous Assignment

There are two constructs for procedural continuous assignments, which declare dynamic bindings to nets or registers in a model. Ordinarily, a continuous assignment remains in effect for the duration of a simulation. A ***assign . . . deassign*** procedural continuous assignment (PCA) is made by a procedural statement to establish an alternative binding (i.e., dynamically substitute the righthand expression). PCAs using the keyword ***assign*** are used to model the level-sensitive behavior of combinational logic, transparent latches, and asynchronous control of sequential parts. The binding remains in effect until it is removed by the (optional) ***deassign*** key word or until another procedural continuous assignment is made.

Example

The four-channel mux below uses the ***assign . . . deassign*** PCA to bind the output to a selected datapath.

```
module mux4_PCA (a, b, c, d, select, y_out);
    input           a, b, c, d;
    input  [1:0]    select;
    output          y_out;
    reg             y_out;

always @ (select)
    if (select == 0) assign y_out = a; else
    if (select == 1) assign y_out = b; else
    if (select == 2) assign y_out = c; else
    if (select == 3) assign y_out = d; else assign y_out = 1'bx;
endmodule
```

End of Example

The ***force . . . release*** form of procedural continuous assignment applies to register variables as well as nets, and overrides ***assign . . . deassign*** continuous assignments. The ***force . . . release*** construct is used primarily within testbenches to inject logic values or logic into a design. See *t_ASIC_with_JTAG* in Section 11.10.6.

Example

In synchronous operation the *data* input of a D-type flip-flop is transferred to the *q* output at the synchronizing edge of *clock* (e.g., at the rising edge or the falling edge of the synchronizing signal). If either the *preset* or the *clear* signal is asserted this (synchronous) clocking action is ignored and the output is held at a constant value. A Verilog model of this behavior is shown below for active-low *preset* and *clear*. The PCA has immediate effect when the statement executes. While *preset* or *clear* is asserted the ordinary synchronous behavior is ignored. If both *clear* and *reset* are de-assigned, the synchronous activity commences with the next active edge of the clock after the deassignment executes.

```
module FLOP_PCA (q, qbar, data, preset, clear clock);
    output          q, qbar;
    input           data, preset, clear, clock;
    reg             q;

    assign          qbar = ~q;
```

```
      always @ (negedge clock) q <= data;

      always @ (clear or preset) begin
      if (!clear) assign q = 0;
      else if (!preset) assign q = 1;
      else deassign q;
    end
    endmodule
```

End of Example

G.6   Intra-Assignment Delay

When a timing control operator (**#** or **@**) appears in front of a procedural statement in a behavioral model the delay is referred to as a "blocking" delay, and the statement that follows the operator is said to be "blocked." The statement that follows a blocked statement in the sequential flow cannot execute until the preceding statement has completed execution. Verilog supports another form of delay in which a timing control is placed to the righthand side of the assignment operator *within* an assignment statement. This type of delay, called *intra-assignment delay*, evaluates the righthand side expression of the assignment and then schedules the assignment to occur in the future, at a time determined by the timing control. Ordinary delay control postpones the execution of a statement; intra-assignment delay postpones the occurrence of the assignment that results from executing a statement. A statement in a list of blocked procedural assignments (i.e. those using the = operator, must complete before the statement after it can execute).

Example

When the first statement is encountered in the sequential activity flow below, the value of B is sampled and scheduled to be assigned to A 5 time units later. The statement does not complete execution until the assignment occurs. After the assignment to A is made, the next statement can execute $(C = D)$. Thus, C gets D 5 time units after the first statement is encountered in simulation.

$$\cdots$$

$$A = \#5\ B;$$

$$C = D;$$

$$\cdots$$

End of Example

Intra-assignment delay control (**#**) has the effect of causing the righthand-side expression of an assignment to be evaluated immediately when the procedural statement is encountered in the activity flow. However, the assignment that results from the statement is not executed until the specified delay has elapsed. Thus, referencing and evaluation are separated in time from the actual assignment of value to the target register variable.

Intra-assignment delay can also be implemented with the event-control operator and an event control expression. In this case, the execution of the statement is scheduled subject to the occurrence of the event specified in the expression.

Example

In the description below G gets $ACCUM$ when A_BUS changes. As a result of the intra assignment delay, the procedural assignment to G cannot complete execution until A_BUS changes. The statement $C = D$ is blocked until G gets value. The value that G gets is the value of $ACCUM$ when the statement is encountered in the activity flow. This may differ from the actual value of $ACCUM$ when A_BUS finally has activity.

$$\ldots$$
$$G = @ (A\_BUS)\ ACCUM;$$
$$C = D;$$
$$\ldots$$

End of Example

G.7   Indeterminate Assignment and Race Conditions

Multiple concurrent behaviors (i.e., *always* or *initial* blocks) may assign value to the same register variable at the same time step. A simulator must determine the outcome of these multiple assignments and distinguish between blocking ($=$) and nonblocking ($<=$) assignments. The activity of a simultor is triggered by an event (i.e. a change in the value of a net, a register variable, or the triggering of an abstract event). (See Section G.10.) The processing steps of the simulator are organized to establish an event queue to determine the order in which assignments to variables occur in simulation. Consequently, the queue manages the assignments to registers when nonblocking and blocking assignments are made simultaneously to the same target variable (i.e., in the same time step). At a given time step, the simulator will (1) evaluate the expressions on the righthand side (RHS) of all the assignments to register variables in statements that are encountered at that time step, (2) execute the blocking assignments to registers, (3) execute nonblocking assignments that do not have intra-assignment timing controls (i.e., they execute in the current time step), (4) execute past procedural assignments whose timing controls have scheduled an assignment for the current simulator time, and (5) advance the simulator time (t_{sim}). The language reference manual (LRM) for Verilog refers to this organization of the simulation activity as a "stratified-event queue." That is, the queue of pending simulation events is organized into five different regions, as shown in Table G-1.

The first region, the active region, consists of events that are scheduled to occur at the current simulation time, and which have top priority for execution. These events result from (1) evaluating the RHS of nonblocking assignments, (2) evaluating the inputs of a primitive and changing the output, (3) executing a procedural (blocked) assignment to a register variable, (4) evaluating the RHS of a continuous assignment and updating the LHS, (5) evaluating the RHS of a procedural continuous assignment and updating the LHS, and (6) evaluating and executing *$display* and *$write* system tasks. Any procedural assignments blocked by a #0 delay control are placed in the inactive queue, and execute after the active queue is empty, in the next simulation cycle at the current time-step of the simulator. The activity of the active queue is dynamic. When it becomes empty, the contents of the inactive queue are moved to the active queue, and the process continues.

The order of processing events in the active queue is not specified by the LRM and is tool-dependent. For example, if an input to a module at the top level of the design hierarchy has an event at the current simulation time, as prescribed by a testbench, the event would reside in the active area of the queue. Now suppose that the input to the module is connected to a primitive with zero propagation delay and whose output is changed by the event on the input port. This event would also be scheduled to occur at the current simulation time and would also be placed in the active area of the queue. If a behavior is activated by the module input and if the behavior generates an event by means of a nonblocking assignment, that event would be placed in the nonblocking assignment update area of the queue. Events that were scheduled to occur at the current simulation time but that originated in nonblocking assignments at an earlier simulation time would also be placed in the "nonblocking assignment update" area. The monitor area contains events that are to be processed after the active, inactive, and nonblocking assignment update events, such as the *$monitor* task. The last area of the stratified event queue consists of events that are to be executed in the future. Given this organization of the event queue, *the simulator executes all of the active events in a single simulation cycle.* As it executes, it may add events to any of the regions of the queue, but it may delete events only from the active region. After the active region is empty, the events in the inactive region are activated (i.e. they are added to the active region and a new simulation cycle begins). After the active region and the inactive region are both empty, the events in the nonblocking assignment update area of the queue are activated, and a new simulation cycle begins. After the monitor events have executed, the simulator advances to the next time at which an event is scheduled to occur. Whenever an explicit #0 delay control is encountered in a behavior, the associated process is suspended and added as an inactive event for the current simulation time. The process will be resumed in the next simulation cycle at the current time.

TABLE G-1

Class of Event	Time of Occurrence	Order of Processing
Active Event ■ Evaluate RHS of nonblocking assignments ■ Evaluate inputs to primitives and update their outputs ■ Execute blocking assignments ■ Update continuous assignments ■ Update procedural continuous assignments ■ Update procedural continuous assignments	Current t_{sim}	In any order
Inactive	Current t_{sim} with #0blocking assignments	In any order after all active events
Nonblocking Assignment Update	Evaluated during previous or present t_{sim} to be assigned at t_{sim}	In any order after all active and inactive events
Monitor	Current t_{sim}	After all active, inactive, and nonblocking assignment update events
Future Active and Nonblocking Assignment Update	A Future Simulation Time	

Simulation cycle: the processing of all of the events in the active event queue. An explicit #0 delay control requires that the executing process be suspended and added as an inactive event for the current simulation time, so that the process is resumed in the next simulation cycle (IEEE 1364).

In addition to the structure imposed by the stratified event queue, the simulator must also adhere to the rule that the relative ordering of blocking and nonblocking assignments at the same simulation time will be such that the nonblocking assignments will be scheduled after the blocking assignments, with the exception that blocking assignments that are triggered by nonblocking assignments will be scheduled after the nonblocking assignments that are already scheduled. (*Caution:* The **$display** task executes immediately when it is encountered in the sequential activity flow of a behavior.) The **$monitor** task executes at the end of the current simulation cycle (i.e. after the nonblocking assignments have been updated). Thus, in the code below, *execute_display* assigns value to a and b, samples the current RHS of a and b, displays the current values of a and b, then updates a and b. The values of a and b at the end of the behavior are not the values that were displayed (i.e. **$display** executes before the nonblocking assignments). On the other hand, *execute_monitor* assigns value to c and d, samples c and d, updates c and d, and then prints the values of c and d. The values of c and d when the behavior expires are the same as the values that were printed. The standard output is printed below:

$$\text{display: } a = 1 \quad b = 0$$
$$\text{monitor: } c = 0 \quad d = 1$$

```
initial begin: execute_display        initial begin: execute_monitor
  a = 1;                                 c = 1;
  b = 0;                                 d = 0;

  a <= b;                                c <= d;
  b <= a;                                d <= c;
  $display ("display: a = %b b = %b", a, b);   $monitor ("monitor: c = %b d = %b", c, d);
end                                    end
```

G.8   wait STATEMENT

The wait construct suspends a thread of an activity flow within a behavior until an expression evaluates true.

Example

The assignment of *register_b* to *register_a* below is suspended until *enable* is asserted true. After the assignment is made activity is again suspended for 10 time steps before assigning *register_d* to *register_c*.

```
wait (enable) register_a = register_b;
#10 register_c = register_d;
```

End of Example

G.9 fork . . . join Statement

The *fork ..join* construct within a behavior branches an activity flow into multiple parallel threads, each of which may be a *begin . . . end* block statement. The statements within a *begin . . . end* block statements execute in the ordinary way (i.e., sequentially). *fork ... join* statements are helpful in modeling complex waveforms in testbenches and abstract (and nonsynthesizable)models of behavior. The activity of a *fork . . . join* statement is complete when all of the statements within it have completed execution.

Example

The assignment to A below is made at $t_{sim} = 5$, and the assignment to C is made at $t_{sim} = 10$;

```
fork

  #5 A = B;
  #10 C = D;

join.
```

End of Example

G.10 Named (Abstract) Events

The Verilog *named event* provides a high-level mechanism of communication and synchronization within and between modules. A named event, sometimes referred to as an abstract event, is an abstraction that provides interprocess communication without requiring details about physical implementation. In the early stages of design this can free the designer from having to pass signals between modules explicitly through their ports. A named event can be declared only in a module; it can then be referenced within that module directly, or in other modules by hierarchically dereferencing the name of the event. The occurrence of the event itself is determined explicitly by a procedural statement using the *event-trigger* operator, ->.

Example

In the description below, the abstract event *up_edge* is triggered when *clock* has a positive edge transition. A second behavior detects the event of *up_edge* and assigns value to the flip-flop's output, subject to an asynchronous *reset* signal. Hierarchical referencing allows modules to communicate between any locations in a design hierarchy, without requiring structural detail [1].

```
module Flop_event (clock, reset, data, q, q_bar);
   input           clock, reset, data;
   output          q, q_bar;
   reg             q;
   event           up_edge;

   assign q_bar = ~q;

   @ (posedge clock) -> up_edge;

   @ (up_edge or negedge reset)
      begin
      if (reset == 0) q <= 0; else q <= data;
      end
endmodule
```

End of Example

G.11 Constructs Supported by Synthesis Tools

Synthesis tools support a limited subset of the Verilog language. It is essential that models use only supported constructs. Otherwise a synthesis tool will report an error and fail to synthesize a circuit. Table G-2 lists language constructs that are commonly supported by synthesis tools; Table G-3 lists constructs that are to be avoided. Not all of these are inherently unsynthesizable (e.g., *repeat* loops), but they are not supported by vendors because equivalent functionality can be described by other constructs. Verilog has robust delay constructs for modeling inertial delays of primitives, and transport delays of nets, and pin-pin path delays of modules (see Appendix F), but technology-dependent attributes, such as propagation delays, are not to be included in modes for synthesis. The rule is to model only the functionality of the circuit, not it's timing. The synthesis tools will implement a design subject to constraints on area, timing, and availability of parts in a cell library or speed grade in an FPGA. For additional details, see [1].

REFERENCES

1. Ciletti MD. *Modeling, Synthesis, and Rapid Prototyping with the Verilog HDL.* Upper Saddle River, NJ: Prentice-Hall, 1999.

TABLE G-2

Module declaration
Port modes: **input, output, inout**
Port binding by name
Port binding by position
Parameter declaration
Connectivity nets: **wire, tri, wand, wor, supply0, supply1**
Register variables: **reg, integer**
Integer types in binary, decimal, octal, hex formats
Scalar and vector nets
Subrange of vector nets on RHS of assignment
Module and macromodule instantiation
Primitive instantiation
Continuous assignments
Shift operator
Conditional operator
Concatenation operator (including nesting)
Arithmetic, bitwise, reduction, logical and relational operators
Procedural block statements (**begin** ... **end**)
case, casex, casez, default
Branching: **if, if** ... **else, if** ... **else** ... **if**
disable (of procedural block)
for loops
Tasks: **task** ... **endtask** (no timing or event control)
Functions: **function** ... **endfunction**

TABLE G-3

Assignment with variable used as bit select on LHS
global variables
case quality, inequality (===, !==)
defparam
event
fork ... **join**
forever
while
wait
initial
pulldown, pullup
force ... **release**
repeat
cmos, rcmos, nmos, rnmos, pmos, rpmos
tran, tranif0, tranif1, rtran, rtranif0, rtranif1
primitive ... **endprimitive**
table ... **endtable**
intra-assignment timing control
delay specifications
scalared, vectored
small, medium, large
specify, endspecify
$time
weak0, weak1, strong0, strong1, pull0, pull1
$keyword

Flip-Flop and Latch Types

Some of the examples in the text use a variety of flip-flop and latches from a standard-cell library. The main functional features of those flip-flops are summarized in Table H-1.

TABLE H-1

Symbol	Description
CK D G Q RB *dffrgpqb_a*	The *dffrgpqb_a* is a D-type flip-flop with rising-edge *clock* (*CK*), internally gated data input (between the output and the external datapath and the output), (*D*) asynchronous active-low gate control (*G*), reset (*RB*), and *Q* output.
CK D0 D1 Q RB SL *dffrmpqb_a*	The *dffrmpqb_a* is a D-type flip-flop with rising-edge clock (*CK*), dual, internally multiplexed data inputs, (*D0* and *D1*), asynchronous active-low reset (*RB*), data select (*SL*), and *Q* output.
CK D Q SB *dffspqb_a*	The *dffspqb_a* is a D-type flip-flop with rising-edge clock (*CK*), asynchronous active-low set (*SB*), and *Q* output.
CK Q D RB QB *dffrpb_a*	The *dffrpb_a* is a D-type flip-flop with rising-edge clock (*CK*), asynchronous active-low reset (*RB*), *Q* and *QB* outputs.
CK D Q RB *dffrpqb_a*	The *dffrpqb_a* is a D-type flip-flop with rising-edge clock (*CK*), asynchronous active-low reset (*RB*), and *Q* output.
D GB Q RB *latrpqb_a*	The *latrpqb_a* is a D-type transparent latch with active-low latch enable (*GB*), active-low reset (*RB*), and *Q* output.
D Q G RB QB *latrnb_a*	The *latrnb_a* is a D-type transparent latch with active-high latch enable (*G*), active-low reset (*RB*), *Q* and *QB* output.
D G Q RB *latrnqb_a*	The *latrnqb_a* is a D-type transparent latch with active-high latch enable (*G*), active-low reset (*RB*), and *Q* output.

Verilog-2001

The Verilog HDL underwent its first revision in 2000, and emerged as IEEE Std. 1364-2001, known as Verilog-2001, with significant changes aimed at improving the utility and clarity of the language. The Verilog Standards Committee clarified ambiguous syntax and semantics in IEEE Std. 1364-1995[1] and removed errors in the LRM. The language was enhanced to support higher level modeling and abstract modeling, while maintaining backward compatibility with IEEE 1364-1995. We will discuss a selected set of changes here. For a comprehensive treatment of Verilog-2001 see Sutherland and Verilog HDL 2001 [1, 2].

I.1 ANSI C Style Changes

IEEE Std. 1364-2001 introduced ANSI C style syntax for module and UDP declarations.

I.1.1 Module Port Mode and Type Declarations

Verilog-2001 allows the mode and type of a port to be combined in a single declaration, as shown in Figure I-1. Also, the input ports have default type *wire*, so the declaration of type *wire* for the input ports may be omitted to further simplify the description. The optional description in Figure I-2 places the declaration of the mode, type and vector range of the signals in the port.

I.1.2 Module Declarations

See Figure I-2.

I.1.3 Module Port Parameter List

Parameters are declared as module items in Verilog-IEEE 1364 (i.e. within the body of the module's declaration). In Verilog-2001, the declarations of parameters may be included between the module name and the port list, as shown in Figure I-3.

---

[1] Referred to as Verilog-1995.

Verilog-1995	Verilog-2001

module Add_16 (c_out, sum, a, b, c_in);
  output   c_out;
  output [15: 0]   sum;
  input   [15: 0] a, b;
  input   c_out;
 	
  reg   sum, c_out;
  wire   a, b, c_in;
 	
  always @ (a or b or c_in)
    {c_out, sum} = a + b + c_in;
 	
endmodule
	module Add_16 (c_out, sum, a, b, c_in);
	  output reg c_out;
	  output reg [15: 0] sum;
	  input wire [15: 0] a, b;
	  input wire c_in;
	
	  always @ (a or b or c_in)
	    {c_out, sum} = a + b + c_in;
	
	endmodule

FIGURE I-1

Verilog-1995	Verilog-2001

module Add_16 (c_out, sum, a, b, c_in);
  output   c_out;
  output [15: 0]   sum;
  input   [15: 0] a, b;
  input   c_out;
 	
  reg   sum, c_out;
  wire   a, b, c_in;
 	
  always @ (a or b or c_in)
    {c_out, sum} = a + b + c_in;
 	
endmodule
	module Add_16 (output reg [15: 0] sum,
	                output reg c_out,
	                input [15: 0] a, b,
	                input c_in);
	
	  always @ (a or b or c_in)
	    {c_out, sum} = a + b + c_in;
	
	endmodule

FIGURE I-2

Verilog-1995	Verilog-2001

module Add (c_out, sum, a, b, c_in);
  parameter   size = 16;
  output   c_out;
  output [size−1: 0] sum;
  input   [size−1: 0] a, b;
  input   c_out;
 	
  reg   sum, c_out;
  wire   a, b, c_in;
 	
  always @ (a or b or c_in)
    {c_out, sum} = a + b + c_in;
 	
endmodule
	module Add #(parameter size = 16) (c_out, sum, a, b, c_in);
	  output reg c_out;
	  output reg [size−1: 0] sum;
	  input wire [size−1: 0] a, b;
	  input wire c_in;
	
	  always @ (a or b or c_in)
	    {c_out, sum} = a + b + c_in;
	
	endmodule

FIGURE I-3

I.1.4 UDP Declarations

Verilog-2001 allows ANSI-style declarations combining the mode and/or data type of the elements of a port with the port list (see Figure I-4).

I.1.5 Declarations of Functions and Tasks

The syntax of Verilog-1995 for declaring functions and tasks separates the arguments of a function from its name and associates inputs and outputs with their order in separately made declarations. Verilog-2001 adopts an ANSI C style that associates the arguments with the name, using the same syntax as that for modules. Examples of the new syntax for functions and tasks are shown in Figure I-5.

Verilog-1995	Verilog-2001
primitive latch (q_out, enable, data); **output** q_out; **input** enable, data; **table** ... **endtable** **endprimitive**	**primitive** latch **(output reg** q_out, **input** enable, **input** data); **table** ... **endtable** **endprimitive**

FIGURE I-4

Verilog-1995	Verilog-2001
function [16: 0] sum_FA; **input** [15: 0] a, b; **input** c_in; sum_FA = a + b + c_in; **endfunction**	**function** [16: 0] sum_FA **(input** [15:0] a, b, **input** c_in); sum_FA = a + b + c_in; **endfunction**

(a)

Verilog-1995	Verilog-2001
task sum_FA; **output** [16: 0] sum; **input** [15: 0] a, b; **input** c_in; sum = a + b + c_in; **endtask**	**task** sum_FA **(output** [16: 0] sum, **input** [15:0] a, b, **input** c_in); sum = a + b + c_in; **endtask**

(b)

FIGURE I-5

Verilog-1995	Verilog-2001

function real sum_Real;
  input real   a, b;
 	
  sum_Real = a + b;
 	
endfunction
	function real sum_Real (**input real** a, b);
	
	  sum_Real = a + b;
	
	endfunction

(a)

Verilog-1995	Verilog-2001

task sum_Real;
  output real   sum
  input   real   a, b;
 	
  sum = a + b;
 	
endtask
	task sum_Real
	  (**output real** sum, **input real** a, b);
	
	  sum = a + b;
	
	endtask

(b)

FIGURE I-6

The type of the inputs and outputs of a function or task is *reg*, unless specified by a declaration within the function of task. Verilog-2001 allows the type to be declared within the port of the function or task, as shown in Figure I-6.

I.1.6   Initialization of Variables

Variables of type *wire, reg, integer* and *time* are initialized to a default value[2] of *x* in the first cycle of simulation. Variables of type *real* and *realtime* are initialized to the default value 0.0. In Verilog-1995 a separate declaration can declare an initial value for a *reg, integer*, or *time* variable. Verilog-2001 combines the declaration of an initial value with the declaration of the type of a *reg, integer, time, real*, or *realtime* variable that is declared at the module level (i.e., a variable declared elsewhere, such as in a task, may not be declared to have an initial value). The value of a *wire* remains at the value of its default until the *wire* is driven to a different value in simulation. An example of initialization of a variable is given in Figure I-7. A wire may inherit an initial value from a continuous assignment.

An initial value may be assigned to a variable as part of an ANSI C style of a port declaration, as shown in Figure I-8.

I.2   Code Management

Verilog-2001 expands the capabilities of tasks and functions to include re-entrant tasks and recursive functions.

---

[2]The default net type can be overridden by a compiler directive.

Verilog-1995	Verilog-2001

module Clk_gen (clock);
parameter   delay = 5;
output   clock;
reg   clock;
 	
initial begin
  clock = 0;
  forever #delay clock = ~clock;
end
 	
endmodule
	module Clk_Gen **#(parameter** delay = 5) **(output** clock);
	  reg clock = 0;
	
	  initial forever #delay clock = ~clock;
	
	endmodule

FIGURE I-7

Verilog-1995	Verilog-2001

module Clk_gen (clock);
parameter   delay = 5;
output   clock;
reg   clock;
initial begin
  clock = 0;
  forever #delay clock = ~clock;
end
 	
endmodule
	module Clk_Gen **#(parameter** delay = 5) **(output reg** clock = 0);
	
	  initial forever #delay clock = ~clock;
	
	endmodule

FIGURE I-8

I.2.1   Re-Entrant Tasks

Tasks in Verilog-1995 are allocated static memory that persists for the duration of simulation. The memory space of a task is shared by all calls to the task. Task variables retain their value between calls. Tasks may be called from multiple concurrent behaviors, setting up the possibility that data may be overwritten and compromised before a given call to a task is complete. As noted in Sutherland and Verilog HDL 2001 [1, 2], designers work around this problem by placing the same task in multiple modules and isolating their memory space, but this wastes resources and complicates maintenance of the code.

    Verilog-2001 supports *re-entrant tasks* with dynamic allocation and de-allocation of memory during simulation, each time a task is called. The keyword **automatic** designates a task with dynamic memory allocation. Such tasks are not static, and their allocated memory is not shared. Because the memory allocated for an automatic task is released when the task completes execution, models that use such tasks must not

Verilog-1995	Verilog-2001

function [63: 0] Bogus
  input   [31: 0] N;
  if (N == 1) Bogus = 1;
  else Bogus = N * Bogus (N−1);
endfunction
	function automatic [63: 0] factorial
	  input [31: 0] N;
	  if (N == 1) factorial = 1;
	  else factorial = N * factorial (N−1);
	endfunction

FIGURE I-9

reference data generated by the task after the task exits. This imposes restrictions on the style of code that may use automatic tasks [1, 2].

I.2.2 Recursive Functions

Functions in Verilog-1995 are static too, and may not include delay constructs (i.e., #, @ *wait*). Functions effectively implement combinational logic equivalent to an expression. Because a function executes in zero time, there is no possibility for concurrent calls to the same function. However, subsequent calls to a function overwrite its memory space. If a function calls itself recursively, each call will overwrite the memory of the previous call. In Verilog-2001 a function can be declared *automatic*, which causes distinct memory to be allocated each time a function is called. The memory is released when the function exits. The classical example of recursion in Figure I-9 compares a recursive implementation in Verilog-2001 with an illegal description in Verilog-1995.

I.2.3 Constant Functions

Functions may be used in Verilog-1995 only where a nonconstant expression can be used. For example, the widths and depths of arrays can be hard wired by fixed numbers defined by parameters, which are constants. Although parameters can be defined in terms of other parameters, this mechanism for scaling a design can be cumbersome.

Verilog-2001 supports *constant functions*, which can be called wherever a constant is required. Constant functions are evaluated at elaboration time, and do not depend on the values of variables at simulation run-time. Only constant expressions may be passed to a constant function, not the value of a net or register variable. Consequently, a constant function may reference only parameters, localparams, locally declared variables, and other constant functions. The parameters that are used by a function must be declared before the function is called, and the memory used by a function is released after the function has been elaborated.

Avoid using *defparam* statements to redefine the parameters within a function, because the value returned may differ between simulators. The parameters within an instance of a module can be redefined unambiguously by the # construct. Constant functions may not call system functions and tasks, and may not use hierarchical path references.

Verilog-1995	Verilog-2001

module Add (sum, a, b, c_in);
  output   sum;
  input   a, b;
  input   c_in;
  reg   sum;
  always @ (a **or** b **or** c_in) **begin**
    sum = a + b + c_in;
    assign match = a & b; // Error
  end
endmodule
	module Add (sum, a, b, c_in);
	  output sum;
	  input a, b;
	  input c_in;
	  reg sum;
	  always @ (a **or** b **or** c_in) **begin**
	    sum = a + b + c_in;
	    assign match = a & b; // Valid
	  end
	endmodule

FIGURE I-10

I.3   Support for Logic Modeling

I.3.1   Implicit Nets

In Verilog-1995 an undeclared identifier will be an implicit net data type if it (1) appears in the port of an instantiated module, (2) is connected to an instance of a primitive, or (3) appears on the LHS of a continuous-assignment statement and is also declared as a port of the module containing the assignment. If the implicit net is connected to a vector port it will inherit the size of the port; otherwise, it will be a scalar net. The default data type of an implicit net is *wire*, which can be modified by a compiler directive. If the type of the LHS of a continuous assignment is not declared explicitly and is not determined implicitly by the above rules, an error will result. In Verilog-2001 an undeclared identifier that is not a port of a module will be inferred as an implicit scalar net. Figure I-10 shows a module that complies with Verilog-2001, but not with Verilog-1995.

I.3.2   Disabled Implicit Nets

The mechanism for implicitly declaring a net can be disabled by including a new argument, ***none*** with the **'*default_nettype*** compiler directive. This argument requires all nets to be explicitly declared. Disabling the default assignment of type to identifiers will reveal compilation errors that arise from misspelled identifiers but that would be otherwise undetected.

I.3.3   Variable Part Selects

Verilog-1995 allows a part select of contiguous bits from a vector if the range indexes of the part select are constant. Verilog-2001 provides two new part-select operators to support a variable part select of fixed width, +*:* and −*:*, having the syntax:
$[<starting\_bit> +: <width>]$ and $[<starting\_bit> -: <width>]$, respectively. The parameter width specifies the size of the part select, and *start_bit* specifies the leftmost or rightmost bit in the vector from which the part select is taken, depending on

Verilog-1995	**Verilog-2001**
reg [15: 0] sum; **reg** [2: 0] K;	**reg** [15: 0] sum; **reg** [2: 0] K;
// Valid; **wire** a_byte = sum[15: 8];	// Valid; **wire** a_byte = sum[15: 8];
// Error: **wire** b_byte = sum[K + 3: K];	// Valid: **wire** b_byte = sum[K +: 3];

FIGURE I-11

Verilog-2001			
reg	[15: 0]	data	[0: 127] [0: 127];	// 2-dimensional array of words
real	time_array		[0: 15] [0: 15] [0: 15];	// 3-dimensional array
wire	[31: 0]	d_paths	[15: 0];	// 1-dimensional array of words

FIGURE I-12

whether the selection will be made by incrementing or decrementing the index of the bits in the vector. See Figure I-11 for an example.

I.3.4   Arrays

Verilog-1995 supports only one-dimensional arrays of type *reg, integer*, and *time*. Verilog-2001 supports arrays of *real* and *realtime* variables, in addition to arrays of type *reg, integer*, and *time*. Arrays can be of any number of dimensions in Verilog-2001. The range specifications for the indexes of the dimensions of an array follow the declared name of the array. Examples are shown in Figure I-12.

    A bit or a part select of a word in an array cannot be selected directly in Verilog-1995, but Verilog-2001 allows selection of a bit or a part select from an array of any number of dimensions. To select a word, reference the array with an index for each dimension. To select a bit or a part, reference the array with an index for each dimension plus a bit or range specification. See Figure I-13.

I.4   Support for Arithmetic

I.4.1   Signed Data Types

Verilog-1995 is limited to signed arithmetic on 32-bit integers. The *reg, time*, and all net data types are unsigned, and expressions are evaluated as signed arithmetic only if every operand is a signed variable (i.e., has type *integer*). The data types of the variables

Verilog-2001			
reg	[15: 0] data	[0: 127] [0: 127];	// 2-dimensional array of words
real	time_array	[0: 15] [0: 15] [0: 15];	// 3-dimensional array
wire	[31: 0] d_paths	[15: 0];	// 1-dimensional array of words
wire	[15: 0] a_data_word = data [4] [21];		// references a word
wire	a_time_sample = time_array [7] [7] [7];		// references a word
wire	[7: 0] a_byte = data [64] [32] [12: 5];		// references a byte
wire	a_bit = data [31] [8] [3];		// references a bit

FIGURE I-13

in an expression, not the operators, determine whether signed or unsigned arithmetic is performed. Verilog-2001 uses the reserved key word ***signed*** to declared that a ***reg*** or a net type variable is signed, and supports signed arithmetic on vectors of any size, not just 32-bit values. See Figure I-14.

I.4.2   Signed Ports

Ports may be declared to be signed in two ways: by declaration with the mode of the port or by declaration of the type of the associated port variable. Figure I-15 shows examples of declarations of signed variables without and with ANSI C style syntax.

I.4.3   Signed Literal Integers

Verilog-1995 represents literal integers in three ways: a number (e.g., -10), an unsized radix-specified number (e.g., 'hA), and a sized radix-specified number (e.g., 64'hF). If a

Verilog-1995		Verilog-2001	
integer	m, n;	**integer**	m, n;
reg [63: 0] v;		**reg signed** [63: 0] v;	
...	// value stored	...	// value stored
m = 12;	// 0000_..._0000_1100	m = 12;	// 0000_..._0000_1100
n = −4;	// 1111_..._1111_1100	n = −4;	// 1111_..._1111_1100
v = 8;	// 0000_..._0000_1000	v = 8;	// 0000_..._0000_1000
m = m / n;	// result: −3	m = m / n;	// result: −3
v = v / n;	// result: 0	v = v / n;	// result: −2

FIGURE I-14

```
                    ┌─────────────────────────────────────────────┐
                    │                 Verilog-2001                │
                    ├─────────────────────────────────────────────┤
                    │ module Add_Sub (sum_diff, a, b);            │
                    │   output signed [63: 0] sum_diff;    // stored as signed value │
                    │   input  signed [63: 0] a, b;        // stored as signed value │
                    │                                             │
                    │   ...                                       │
                    │ endmodule                                   │
                    └─────────────────────────────────────────────┘
```

(a)

```
                    ┌─────────────────────────────────────────────┐
                    │                 Verilog-2001                │
                    ├─────────────────────────────────────────────┤
                    │ module Add_Sub (output reg signed [63: 0] sum_diff, │
                    │                 input  wire signed [63: 0] a, b);   │
                    │                                             │
                    │   ...                                       │
                    │ endmodule                                   │
                    └─────────────────────────────────────────────┘
```

(b)

FIGURE I-15

```
                    ┌─────────────────────────────────────────────┐
                    │                 Verilog-2001                │
                    ├─────────────────────────────────────────────┤
                    │ reg signed [63: 0] v;       // signed variable │
                    │ ...                                         │
                    │ v = 12;                     // literal integer │
                    │ ...                                         │
                    │ v = v / −64'd2;             // stored as 0    │
                    │ v = v / −64'sd2             // stored as −6   │
                    └─────────────────────────────────────────────┘
```

FIGURE I-16

radix is specified the number is interpreted as an unsigned value; if a radix is omitted, the number is interpreted as a signed value. In Verilog-2001, a sized literal integer can be declared as an integer. The symbol *s* is used to specify that a sized or unsized literal integer is signed, as illustrated in Figure I-16.

I.4.4   Signed Functions

Functions in Verilog-1995 may be called any place that an expression can be used. The value returned by a function is signed if and only if the function is declared to be an integer. With the reserved keyword ***signed***, Verilog-2001 allows signed arithmetic to be performed on returned values of a vector size. Figure I-17 identifies the possible types

Example	Returned Value	Verilog-1995	Verilog-2001
function sum	single bit	x	x
function [31: 0] sum	unsigned vector of 32 bits	x	x
function integer sum	signed vector of 32 bits	x	x
function real sum	64-bit double-precision	x	x
function time sum	unsigned 64-bit vector	x	x
function signed [63: 0] sum	signed 64-bit vector		x

FIGURE I-17

of a function in Verilog-1995 and Verilog-2001. Remember that the data types of the variables in an expression, not the operators, determine whether signed or unsigned arithmetic is performed. Signed arithmetic is performed only when all of the operands are signed variables.

I.4.5   System Functions for Sign Conversion

Verilog-2001 provides two new system functions for converting values to signed or unsigned values. The function **$signed** returns a signed value from the value passed in. The function **$unsigned** returns an unsigned value from the value pass in. The functions are useful because an expression returns a signed value if and only if all of its operands are signed variables. Sign conversion eliminates the need to declare and assign value to additional variables to circumvent the restrictions of Verilog-1995. See Figure I-18.

I.4.6   Arithmetic Shift Operator

Bitwise shift operators ($\ll$, $\gg$) are supported by Verilog-1995. These operators have two operands: an expression (operand) whose value is to be shifted, and an expression

```
                              Verilog-2001

integer          v;

reg [63: 0]      sum_diff;

v = −16;
sum_diff = 48;

sum_diff = sum_diff / v                        // returns 0
signed_sum_diff = $signed (sum_diff) / v;      // returns −3
```

FIGURE I-18

Verilog-2001
integer data_value, data_value_1995, data_value_2001; // signed datatype
...
data_value = −9;   // stored as 1111..._1111_0111
...
data_value_1995 = data_value >> 3;   // stored as 0001..._1111_1110
data_value_2001 = data_value >>> 3;   // stored as 1111..._1111_1110
data_value_1995 = data_value << 3;   // stored as 1111..._1011_1000
data_value_2001 = data_value <<< 3;   // stored as 1111..._1011_1000

FIGURE I-19

(operand) that determines the number of shifts. The bitwise-shift operators insert a 0 into the cell vacated by a shift. Verilog-2001 includes the operators $>>>$ to shift arithmetically to the right and $<<<$ to shift bitwise to the left. The arithmetic right-shift operator ($>>>$) implemented in Verilog 2001 inserts the MSB (sign bit) back into the MSB if the expression that determines the value of the shifted word is signed; otherwise it will insert a 0. The expression forming the second operand of the shift operator may be signed or unsigned; all other expressions are interpreted to be signed if and only if every operand is signed. The arithmetic left-shift operator ($<<<$) is functionally equivalent to the bitwise left-shift operator ($<<$). The examples in Figure I-19 show the distinctions between the bitwise and arithmetic shift operators.[3]

I.4.7   Assignment Width Extension

Verilog-1995 has two rules for extending the bits of a word when the expression on the RHS side of an assignment statement has a smaller width than the expression on the LHS. If the expression on the RHS is signed, the sign-bit determines the extension to fill the LHS. If the expression on the RHS is unsigned (i.e., *reg, time*, and all net types), its extension is formed by filling with 0. This can lead to inappropriate extensions when the LHS exceeds 32 bits.

Verilog-2001 has a more elaborate set of rules for extending the width of a word beyond 32 bits, as summarized in Figure I-20. These rules differ from those for Verilog-1995, so a model that adhered to the rules of Verilog-1995 will not work the same as a model employing the rules of Verilog-2001.

I.4.8   Exponentiation

Exponentiation is not implemented conveniently in Verilog-1995—it requires repeated multiplication within a loop statement. Verilog-2001 includes the operator ******, which

---

[3]The shift-left operator implemented in Verilog-2001 fills with 0, which corresponds to multiplication by a power of 2. An arithmetic shift-left that fills with the LSB produces different results.

	Extended value	
Left-most bit of RHS expression	unsigned RHS expression	signed RHS expression
0	0	0
1	0	1
x	x	x
z	z	z

FIGURE I-20

Verilog-2001		
returned value	base	exponent
double-precision floating point	real, integer, or signed value	real, integer, or signed value
ambiguous	0	not a positive number
ambiguous	negative number	not an integer

FIGURE I-21

Verilog-2001
reg [7: 0] base; **reg** [2: 0] exponent; **reg** [15: 0] value; value = base ** exponent;

FIGURE I-22

implements exponentiation directly. The operator has two operands: a base and an exponent. The type of the returned value depends on the operands, as shown in Figure I-21. The example in Figure I-22 illustrates the syntax for using **. (*Note*: The operator for exponentiation has higher precedence than the operator for multiplication.)

Verilog-1995	Verilog-2001

module Adder (sum, c_out, a, b, c_in);
  output [15: 0] sum;
  output   c_out;
  input [15: 0] a, b;
  input   c_in;
  reg   c_out, sum;
  always @ (a **or** b **or** c_in)
    {c_out, sum} = a + b + c_in;
 	
endmodule
	module Adder (sum, c_out, a, b, c_in);
	  output [15: 0] sum;
	  output c_out;
	  input [15: 0] a, b;
	  input c_in;
	  reg c_out, sum;
	  always @ (a, b, c_in)
	    {c_out, sum} = a + b + c_in;
	
	endmodule

FIGURE I-23

I.5   Sensitivity List for Event Control

The event-control expression in Verilog-1995 uses the *or* operator to compose an expression that is sensitive to multiple variables. Verilog-2001 allows a comma-separated list, as shown by the example in Figure I-23.[4]

I.6   Sensitivity List for Combinational Logic

Level-sensitive cyclic behaviors (*always*) will simulate and synthesize combinational logic if the event-control expression of the behavior is complete (i.e., it contains every signal that is reference implicitly or explicitly in the behavior). If the event-control expression is incomplete, a synthesis tool will infer latched logic, rather than combinational logic. Unintentional omission of a signal from an event-control expression is problematic, so Verilog-2001 uses a wildcard token to indicate a level-sensitive event-control expression that is sensitive to every variable that is reference within the behavior, thereby eliminating the need to explicitly identify them and avoiding the consequences of an incomplete event-control expression. See Figure I-24.

Verilog-1995	Verilog-2001

module Adder (sum, c_out, a, b, c_in);
  output [15: 0] sum;
  output   c_out;
  input [15: 0] a, b;
  input   c_in;
 	
  always @ (a **or** b **or** c_in)
    {c_out, sum} = a + b + c_in;
 	
endmodule
	module Adder (sum, c_out, a, b, c_in);
	  output [15: 0] sum;
	  output c_out;
	  input [15: 0] a, b;
	  input c_in;
	
	  always @ (*) // alternative: **always @** *
	    {c_out, sum} = a + b + c_in;
	
	endmodule

FIGURE I-24

---

[4]The new syntax permits mixing separated by *or* with items separated by commas, but this usage makes the code less readable.

Verilog-1995	Verilog-2001
module Adder (sum, diff, a, b, c_in); 　**output** [16: 0]　sum, diff; 　**input**　 [15: 0]　a, b; 　**input**　　　　c_in; 　**reg**　　　　c_out, sum; 　**always** @ (a **or** b **or** c_in) **begin** 　 sum = a + b + c_in; 　 diff = a − b; 　**end** **endmodule**	**module** Bogus (sum, diff, a, b, c_in); 　**output** [16: 0]　sum, diff; 　**input**　 [15: 0]　a, b; 　**input**　　　　c_in; 　**reg**　　　　c_out, sum; 　**always begin** 　@ *diff = a − b; 　 sum = a + b _ c_in; 　**end** **endmodule**

FIGURE I-25

Caution: The @ operator is associated with the single statement or ***begin . . . end*** block of statements that immediately follow it. Careless use of the operator will lead to models that do not represent the functionality of combinational logic, and that do not synthesize into combinational logic. In Figure I-25 the cyclic behavior within *Bogus* misuses the event control operator @, making the behavior sensitive to only *a* and *b*, but not to *c_in*.

I.7　Parameters

Parameters make models more configurable, readable, extendable, and portable. Declared by the keyword ***parameter***, parameters in Verilog-1995 are run-time constants, and their value can be changed before simulation and during elaboration. There are two mechanisms for redefining the value of a parameter: remotely, using the keyword ***defparam***, and implicitly, by *in-line redefinition*. The declaration redefining a parameter with the ***defparam*** keyword can be placed anywhere in the design hierarchy, and it redefines the value of a parameter at any location in the design hierarchy via hierarchical dereferencing of path names. This poses the risk that parameters can be changed inadvertently from any location in the design, since parameters are not fixed constants. In-line redefinition requires that text adhering to the syntax *#(value_1, value_2, . . . , value_m)* be inserted after the instance name of a module to redefine parameters declared within the module. The order of the sequence of *value_1, value_2 . . .* must correspond to the order of the sequence in which the parameters are declared within the module. This is cumbersome when the modules contains several parameters, not all of which are to be redefined. Because the parameters are not explicitly named in this syntax for redefinition, the practice is prone to error, and renders the model less readable. Verilog-1995 also supports specparams (*specify parameters*) that may be declared and used only within ***specify . . . endspecify*** blocks[5] within a module. A ***specparam*** is local

---

[5]Specify blocks are used to declare input–output paths across a module, assign delay to those paths, and declare time checks to be performed on signals at the module inputs.

Verilog-1995	
Operands	Operation
All operands are signed integers	Signed arithmetic
An operand is unsigned	Unsigned integer arithmetic
At least one operand is a real value	Floating point arithmetic

FIGURE I-26

to the block in which it is declared, and may be used only within the block. A standard delay format (SDF) file can redefine the value of a ***specparam***. The risk, again, is that specparams could be mistakenly redefined.

A parameter inherits its size and type from the final value assigned to it during elaboration, before simulation, which need not be the same type that was assigned to the parameter when it was declared in its parent module. A parameter can be an unsized integer (at least 32 bits), a sized and unsigned integer, a real (floating point) number, or a string. Other parameters can be operands in the expression that declares the value of a parameter. Thus, in Verilog-1995 the size and type of a parameter can be changed when the parameter is redefined, which could produce undesirable side effects, because the operations performed in an expression depend on the size and type of its operands. Figure I-26 displays the rules used by Verilog-1995 to determine the arithmetic performed by the operands in an expression.

I.7.1   Parameter Constants

Verilog-2001 provides explicit definition of the size and data type of a parameter. Figure I-27 shows the rules for determining how the size and type of a parameter are redefined in Verilog-2001 by an expression having arithmetic operators. When the sign, size, or type of a parameter is explicitly declared it cannot be overridden by a subsequent parameter value redefinition.

I.7.2   Parameter Redefinition

Verilog-2001 provides *explicit in-line redefinition* of parameters on a module-instance basis. The syntax for redefining the parameters of an instantiated module is shown below.

Verilog-2001					
Specified by Declaration			Subject to Redefinition	Redefinition Rule	
Sign	Range	Type			
No	No	No	Yes	Same as Verilog-1995[1]	
Yes	Yes	No	No	Parameter is signed	Size is specified by the range
Yes	No	No	Inherits size	Parameter is signed	Size is inherited from last redefinition
No	Yes	No	No	Parameter is unsigned	Size is fixed by range
		Yes	Yes	Parameter retains type	

[1] In Verilog-1995 a parameter inherits the vector size and type of the last parameter redefinition.

FIGURE I-27

```
module_name
  instance_name
    #(.parameter_name (parameter_value), ...) (port_connections);
```

This feature of Verilog-2001 explicitly identifies the redefined parameters. The redefinition does not depend on the order in which the parameters are defined in the associated module. Consequently, the code is self-documenting and more readable than its counterpart in Verilog-1995.

I.7.3 Local Parameters

Verilog-2001 introduces local parameters (keyword: ***localparam***), whose value cannot be directly redefined from outside the module in which they are declared.[6] Figure I-28 compares the parameters, specify parameters, and local parameters. Although a

---

[6] It might be desirable to protect, for example, the state-assignment codes from inadvertent change.

| | | Verilog-2001 | | |
| | | Verilog-1995 | | |
		parameter	specparam	localparam
Location of declaration	Module item	yes	no	yes
	Task item	yes	no	yes
	Function item	yes	no	yes
	Specify block	no	yes	no
Method of direct redefinition	By a defparam	yes	no	no
	Redefined in-line	yes	no	no
	SDF files	no	yes	no
Method of indirect redefinition by assignment of value	From another parameter	yes	yes	yes
	From a localparam	yes	yes	no
	From a specparam	no	yes	no
Allowed reference	Within a module	yes	no	yes
	Within a specify block	no	yes	no

FIGURE I-28

localparam cannot be directly redefined, it can be indirectly redefined by assigning it the value of a *parameter*, which can be changed by the methods described above.

I.8   Instance Generation

Verilog-1995 supports structural modeling with declarations of arrays of instances of primitives and modules. The *generate . . . endgenerate* construct in Verilog-2001 extends this feature to replicate distinct copies of net declarations, register variable declarations, parameter redefinitions, continuous assignments, *always* behaviors, *initial* behaviors, tasks, and functions.[7] Verilog-2001 introduces a new kind of variable, denoted by the keyword *genvar*, which declares a nonnegative integer[8] that is used as an index

---

[7] A *generate . . . endgenerate* block may not include port declarations, constant declarations, and specify blocks.

[8] A variable of type *genvar* may be declared within a module or within the *generate . . . endgenerate* block; it may not be assigned a negative value, an *x* value, or a *z* value.

Verilog-2001

```
module Adder_CLA (sum, c_out, a, b, c_in);
  parameter           size = 32;
  input   [size − 1: 0]  a, b;
  input                 c_in;
  output [size − 1: 0]  sum;
  output                c_out;
  wire    [size/8 − 1: 0]  c_o, c_i;
  assign                c_i[0] = c_in;
  assign                c_out = c_o[size/8 − 1];

  generate
    genvar j;
    for (j = 1; j <= 3; j = j + 1)
      begin: j
        assign c_i[j] = c_o[j − 1];
      end
  endgenerate

  generate
    genvar k;
    for (k = 0; k <= size/8 − 1; k = k + 1)
      begin: M
        Add_cla_8 ADD (sum[((k + 1)*8 − 1) -: 8], c_o[k], a[((k + 1)*8 − 1) -:8], b[((k + 1)*8 − 1) -:8], c_i[k]);
      end
  endgenerate
endmodule
```

FIGURE I-29

in the replicating *for* loop associated with a **generate . . . endgenerate** block. The index of the *for* loop of a **generate . . . endgenerate** block must be a **genvar** variable, and the initializing statement and the loop update statement must both assign value to the same **genvar** variable. The contents of the *for* loop of a **generate . . . endgenerate** statement must be within a named **begin . . . end** block. Note: *The name of the block is used to build a unique name for each generated item.*

The model in Figure I-29 generates a 32-bit adder from copies of an 8-bit adder, with instance names *M[0].ADD, M[1].ADD, M[2].ADD*, and *M[3].ADD*, A separate generate statement connects the internal carry chain of the adder by generating continuous assignments. Note that the entire model is parameterized, so redefining the value of size to 64 will generate and connect 8 copies of the 8-bit slice adder, *Add_cla_8*. This leads to a more compact description than instantiating 8 individual 8-bit adders. Also, manual replication of structural items does not lead to a parameterized model, which ultimately limits the utility of the model.

In Figure I-30 **generate** statements are used to generate a parameterized pipeline of words.

The replication of items by a generate block can be controlled by *if* statements and *case* statements. Figure I-31 uses an *if* statement to determine whether a ripple-carry adder or a look-ahead adder is instantiated, depending on the width of the data-path. Figure I-32 uses a *case* statement to determine the instantiation.

```
                          Verilog-2001

module generated_array_pipeline (data_out, data_in, clk, reset);
  parameter               width = 8;
  parameter               length = 16
  input   [width − 1: 0]  data_in;
  input                   clk, reset;
  output [width − 1: 0]   data_out;
  reg     [width − 1: 0]  pipe    [0: length − 1];
  wire    [width − 1: 0]  d_in    [0: length − 1];

  assign d_in [0] = data_in;
  assign data_out = pipe[size − 1];

  generate
    genvar k;
    for (k = 1; k <= length − 1; k = k + 1) begin: W
      assign d_in[k] = pipe[k − 1]
    end
  endgenerate

  generate
    genvar j;
      for (j = 0; j <= length − 1; j = j + 1)
        begin: stage
          always @ (posedge clk or negedge reset)
            if (reset == 0) pipe[j] <= 0; else pipe[j] <= d_in[j];
        end
  endgenerate
endmodule
```

FIGURE I-30

```
                          Verilog-2001

module Add_RCA_or_CLA (sum, c_out, a, b, c_in);
  parameter size = 8;
  input   [size − 1: 0]  a, b;
  input                  c_in;
  output [size − 1: 0]   sum;
  output                 c_out;

  generate
    if (size < 9) Add_rca #(size) M1 (sum, c_out, a, b, c_in);
        else Add_cla #(size) M1 (sum, c_out, a, b, c_in);
  endgenerate
endmodule
```

FIGURE I-31

Verilog-2001

module Add_RCA_or_CLA (sum, c_out, a, b, c_in);
  parameter size = 8;
  input   [size − 1: 0]   a, b;
  input   c_in;
  output [size − 1: 0]   sum;
  output   c_out;
 	
  generate
   case (1)
     size < 9:   Add_rca **#**(size) M1 (sum, c_out, a, b, c_in);
      default: Add_cla **#**(size) M1 (sum, c_out, a, b, c_in);
   endgenerate
  endmodule

FIGURE I-32

REFERENCES

1. Sutherland S. *Verilog 2001*. Boston: Kluwer, 2002.

2. *IEEE Standard for Verilog Hardware Description Language 2001*, IEEE Std. 1364–2001. Piscataway, NJ: Institute of Electrical and Electronics Engineers, 2001.

Programming Language Interface

The Verilog hardware description language has a built-in programming language interface (PLI) that allows the user to create a "super-Verilog" language with user-defined system tasks that are implemented by the user in the C programming language. These user-defined system tasks are versatile because they are global to the language environment, rather than local to a particular module. This capability greatly expands the utility of the language.

A simulator creates a set of data structures when it compiles a Verilog description. These data structures contain topological and other information about the design. PLI includes a library of C-language functions that can directly access the data structures of a design, giving the user access to a vast amount of information that can support other applications. For example, the data structures that describe the structural connectivity of a description would enable a timing analysis algorithm to enumerate all of the paths from the input ports to the data input of a given flip-flop. The Language Reference Manual lists some of the applications possible with PLI:

- Dynamically scan the data structures of the design and annotate the delays of model instances. (Back-annotation is done after layout to ensure that the models used in timing verification accurately account for the layout-specific parasitic delays induced by the loading of metal interconnect and fanout.)
- Dynamically read test vectors from a file and pass the information to another software tool.
- Create custom graphical environments for user interfaces and displays.
- Create custom debugging environments.
- Decompile the source code to create a Verilog source code description from the data structures of the design.

- Link a C-language simulation model into the design during simulation.
- Interface a hardware unit to the design during simulation.

These are but a few of the uses for PLI. The interested reader is referred to the Language Reference Manual, of which over one-half of the content is dedicated to PLI, and to Sutherland [1], which provides a comprehensive treatment of PLI.

REFERENCES

1. Sutherland S. *The Verilog PLI Handbook*. Boston: Kluwer, 1999.

Websites

Additional resources can be obtained at the various websites listed below. Other sites will be posted on our companion website.

Industry Organization

www.accellera.org	Accellera
www.vsia.com	Virtual Socket Interface Alliance
www.opencores.org	Opencores
www.systemc.org	System C

FPGA and Semiconductor Manufacturers

www.actel.com	Actel Corp.
www.altera.com	Altera, Inc.
www.atmel.com	Atmel Corp.
www.latticesemconductor.com	Lattice Semiconductor Corporation:
www.mcu.motsps.com/hc11/	Motorola site
www.mcu.motsps.com/hc05/	Motorola site

Media and Archives

www.eedesign.com	EE Design
www.ednmag.com	EDN magazine
(Annual PAL, PLD, and FPGA directory)	

www.eetimes.com	EE Times
www.isdmag.com	Integrated System Design magazine
http://xup.msu.edu	Xilinx University Resource Center[1]
www.mrc.uidaho.edu/vlsi/	See this site for additional links

EDA Tools and Resources

www.cadence.com	Cadence Design Systems, Inc.
www.co-design.com	Co-Design Automation, Inc.
www.mentorg.com	Mentor Graphics corp.
www.model.com/verilog	Model Technology
www.montereydesign.com	Monterey Design Systems
www.qualis.com	Qualis, Inc.
www.simucad.com	Simucad, Inc.
www.synopsys.com	Synopsys, Inc.
www.synplicity.com	Synplicity, Inc.
www.xilinx.com[1]	Xilinx, Inc.

Consultants

www.sunburst-design.com	Sunburst Design, Inc.
www.sutherland.com	Sutherland HDL, Inc.
www.whdl.com	Willamette HDL, Inc.

---

[1]The Xilinx University Resource Center website, maintained and hosted by the Department of Electrical and Computer Engineering at Michigan State University, provides a collection of resources already located on the Web, as well as original content. A robust online support system, consisting of a mailing list, discussion board, and e-mail is in place and monitored to answers questions.

Web-Based Tutorials

The companion website for the book, located at www.prenhall.com, contains a synthesis-ready standard cell library and tutorials for the following EDA tools: Silos III simulator (Simucad), Xilinx ISE integrated synthesis environment, Synopsys Design Compiler, and Synopsys Primetime timing analyzer.

Index

Index of Verilog Modules and User-Defined Primitives

List of Tables